To Anita

In memory of our many discussions that helped in the development of this edition.
Greetings from
Joel

NMR PROBEHEADS
FOR BIOPHYSICAL AND BIOMEDICAL EXPERIMENTS

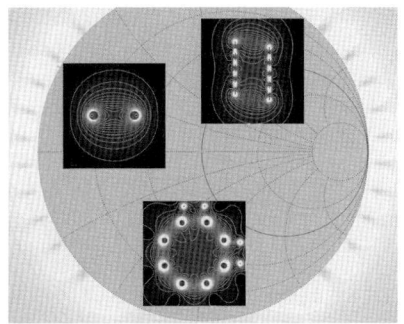

Theoretical Principles
& Practical Guidelines

Second Edition

NMR PROBEHEADS
FOR BIOPHYSICAL AND BIOMEDICAL EXPERIMENTS

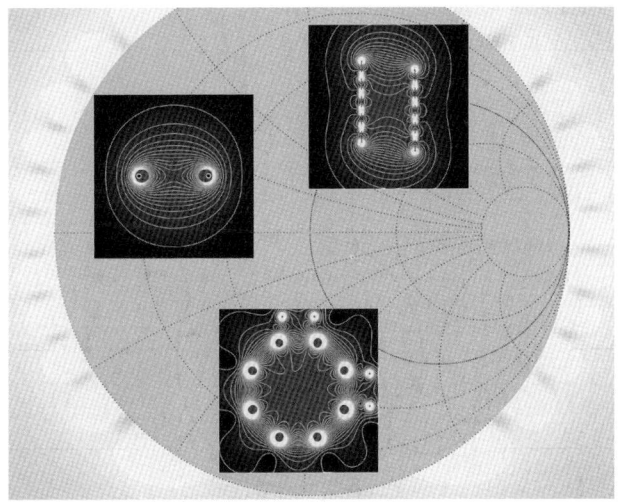

Theoretical Principles
& Practical Guidelines
Second Edition

Joël Mispelter
INSERM, France & Curie Institute, France

Mihaela Lupu
Curie Institute, France

André Briguet
University Claude Bernard Lyon 1, France

Imperial College Press

Published by

Imperial College Press
57 Shelton Street
Covent Garden
London WC2H 9HE

Distributed by

World Scientific Publishing Co. Pte. Ltd.
5 Toh Tuck Link, Singapore 596224
USA office: 27 Warren Street, Suite 401-402, Hackensack, NJ 07601
UK office: 57 Shelton Street, Covent Garden, London WC2H 9HE

Library of Congress Cataloging-in-Publication Data
Mispelter, Joël, author.
 NMR probeheads for biophysical and biomedical experiments : theoretical principles and practical guidelines / Joel Mispelter, Mihaela Lupu, Andre Briguet. -- Second edition.
 p. ; cm.
 Includes bibliographical references and index.
 ISBN 978-1-84816-662-2 (hardcover : alk. paper)
 I. Lupu, Mihaela, author. II. Briguet, André, author. III. Title.
 [DNLM: 1. Biophysical Phenomena. 2. Magnetic Resonance Spectroscopy. 3. Biomedical Technology. 4. Electronics, Medical. 5. Equipment Design. QC 762]
 QC762
 616.07'548--dc23
                                           2015010863

British Library Cataloguing-in-Publication Data
A catalogue record for this book is available from the British Library.

Copyright © 2015 by Imperial College Press

All rights reserved. This book, or parts thereof, may not be reproduced in any form or by any means, electronic or mechanical, including photocopying, recording or any information storage and retrieval system now known or to be invented, without written permission from the Publisher.

For photocopying of material in this volume, please pay a copying fee through the Copyright Clearance Center, Inc., 222 Rosewood Drive, Danvers, MA 01923, USA. In this case permission to photocopy is not required from the publisher.

Typeset by Stallion Press
Email: enquire@stallionpress.com

Printed by FuIsland Offset Printing (S) Pte Ltd Singapore

To Jean-Marc Lhoste, a builder….

Preface to the Second Edition

The subject of Nuclear Magnetic Resonance (NMR) has grown rapidly over the last 50 years, with wide applications in physics, chemistry, and more recently in biophysics and medicine. Since this development has no reason to stop, the increasing interest in designing the NMR head probes that constitute the very heart of NMR equipment is obvious.

The NMR probe simply represents a fine interface along the border separating two different worlds: the macroscopic one and the world of tiny precessing nuclear spins. The probe should be able to accommodate different samples, from very small quantities to the entire human body, while being able to capture the spin's signals. It should produce a specific designed Radio Frequency (RF) magnetic field, as intense as possible, while being homogeneous over the largest possible sample volume. It should simultaneously capture the signal from several types of spins, without any interference, and above all it should be sensitive. These are, only in part, the general demands a user expects from an NMR head probe.

Having all these in mind, the authors have tried to lead the reader through the most basic stages in accomplishing a correct probe design, starting from the very basic oscillating circuit up to more complicated designs.

Chapter 1 is essentially the same as before. Chapter 2 focuses in detail on the RF components often used for the probe designs and realization and their behavior on the workbench. The advantages of the Smith Chart on the designing pathway are shown. Finally, Low Noise Amplifiers are introduced with their specific demands in terms of optimum noise transfer when connected to the probes. Last, but not least, the new trends offered by RF metamaterials are presented.

Chapter 3 is mainly dedicated to the "tools" offered by linear network analysis when designing a probe, as well as hints for calculating current densities along the conductive parts of the probes or mutual inductances between them. Chapter 4 treats the main problems arising when interfacing the probe to the spectrometer: tuning, matching, possible source of losses, etc. Chapter 5 deals with the quadrature driving and the specific hybrid circuits allowing the spectrometer interfacing. Chapter 6 and 7 are essentially the same as the first edition. Chapter 8 is enriched with homogeneous prototype description of linear as well as quadrature probes, single or multiple tuned. Chapter 9, dealing with heterogeneous resonators, remains essentially unchanged.

We did not want to finish the book without presenting a few considerations regarding debugging and evaluating the probe, on both the RF workbench and connected to the spectrometer, in the last chapter.

In the appendix we summarized some useful formulae regarding inductances, capacitances and transmission line characteristic impedances.

The reference list is rather extensive (but not exhaustive), in order to give readers the opportunity to access most of the information on which the book is based.

Generally, the book is intended to be useful to almost anyone involved in NMR or MRI, from students to medical or biological scientists, needing to perform an experiment under certain physical and/or geometrical conditions, not met by conventional or available probes.

Nevertheless, the authors cannot help themselves to recommend a special, favorite application to NMR/MRI probes, as shown in [Weekley *et al.*, 2003].

The authors

Acknowledgements

The authors want to specially acknowledge Michel Decorps and Siew Kan, pioneers in NMR probe design in France, for helpful discussions through the years.

The second edition benefited from comments, corrections and discussions with readers of the first edition. The comments and encouragements provided by Anita Flynn (Berkeley), Tom Barbara (Portland), Marcello Alecci (Rome) and Andrew Webb (Leiden) were particularly appreciated.

One of the authors (M.L.) wants to thank Institut Curie – Research Department for their support.

Resources for Readers

Files supporting the contents of this book have been made available for download from the publisher's website:

http://www.worldscientific.com/worldscibooks/10.1142/p759#t=suppl

The files contain complete installation instructions, a manual, and examples of the software as described in the book. Sources, compatible with Unix, are provided as well as binaries for Linux and Windows operating systems. An archive containing all materials is also available as an iso file prepared for CDROM burning.

Contents

Preface vii
Acknowledgements ix
Resources for Readers x

1. Introduction 1
   1.1 What is an NMR Probe? 1
      1.1.1 The basic pulsed NMR experiment 5
      1.1.2 The head probe from a theoretical point of view 9
         1.1.2.1 The principle of reciprocity and the calculation of the induced emf 9
         1.1.2.2 Losses 13
         1.1.2.3 The sensitivity 16
   1.2 What Probes for a Specific NMR Experiment? 18
      1.2.1 High resolution NMR in solution 19
      1.2.2 Solid state NMR 20
      1.2.3 Biomedical and biophysical applications 21

2. Radio Frequency Components 23
   2.1 The RF Parts of NMR Spectrometers and Scanners 23
   2.2 Characterization of RF Components 26
      2.2.1 Frequency range 26
      2.2.2 Complex impedance and admittance 28
      2.2.3 Impedance measurements and reflection coefficient 30
      2.2.4 S-parameters 36
      2.2.5 The Smith Chart 40
   2.3 Electrical Properties of Materials 44
      2.3.1 Resistance of metal conductors 44
      2.3.2 Losses in dielectric substrate 48
   2.4 Passive Linear Components at RF 51
      2.4.1 Conductors 51
         2.4.1.1 Round wire 52
         2.4.1.2 Ribbon conductor 53
         2.4.1.3 Litz wire 55
      2.4.2 Cables and transmission lines 58
         2.4.2.1 Wave propagation and characteristic impedance 59

2.4.2.2 Low-loss transmission lines	63
2.4.2.3 Attenuation in low-loss lines terminated by Z_c	65
2.4.2.4 Attenuation when the line is not terminated by its characteristic impedance	70
2.4.2.5 Impedance transformation	75
2.4.3 Discrete passive components.	81
2.4.3.1 Resistors	81
2.4.3.2 Inductors (generalities)	82
2.4.3.3 The air-solenoid (high frequency model, Q optimization)	83
2.4.3.4 High frequency model of the air-solenoid	90
2.4.3.5 Toroidal inductors.	91
2.4.3.6 Other inductors (single loop, printed spiral coil)	93
2.4.3.7 Choke inductor	98
2.4.3.8 Adjustable inductor	103
2.4.3.9 Capacitors (generalities)	104
2.4.3.10 Capacitors (high frequency model)	105
2.4.3.11 Fixed capacitor	107
2.4.3.12 Variable capacitor	108
2.4.3.13 Resonators	109
2.4.4 Distributed components made by transmission lines	116
2.5 Non-linear Devices (Diodes)	124
2.5.1 Silicon and Schottky P–N diodes	125
2.5.2 PIN diodes	127
2.5.3 Varactors	127
2.6 Active Devices (Low Noise Amplifier or LNA)	128
2.6.1 Noise factor (F) and noise figure (NF)	128
2.6.2 LNA designs for NMR frequencies	131
2.7 Metamaterials (Flux Guides)	137
2.7.1 Magnetic metamaterials at RF	138
2.7.2 Components of a magnetic metamaterial	143
2.7.2.1 "Swiss-Rolls"	144
2.7.2.2 Ring planar resonators	145
2.7.2.3 Transmission line network	148
2.7.3 RF metamaterial designs	150
2.7.3.1 Flux compressors and flux guides	151
2.7.3.2 Lenses	151
2.7.4 Dielectric resonators	152
3. Introduction to Linear Network Analysis	155
3.1 Introduction to Network Theory	156
3.1.1 Impedance and admittance matrices	160

  3.1.2 The transmission (ABCD) matrix 162
  3.1.3 The scattering matrix 167
 3.2 Linear Network Simulation of Probe Coil Circuits 168
  3.2.1 NMRP: a simple tool for probe evaluation 169
   3.2.1.1 Simprobe: a linear probe network analyzer 169
   3.2.1.2 Building the Z matrix 176
   3.2.1.3 Analysis of simprobe outputs 179
   3.2.1.4 Calculating inductances 185
   3.2.1.5 Calculating current densities 186
   3.2.1.6 Calculating resonant spectrum 188
   3.2.1.7 Mutual inductance and current distribution 190
  3.2.2 Other dedicated linear circuit analyzers 194

4. **Interfacing the NMR Probehead**   **197**
 4.1 Impedance Matching 199
  4.1.1 The Bode–Fano limit 199
  4.1.2 Connected matching circuits 201
   4.1.2.1 Basic L-matching circuits 201
   4.1.2.2 Series tuned/parallel matched 203
   4.1.2.3 Parallel tuned/series matched 207
   4.1.2.4 Efficiency of the capacitive matching circuits 212
   4.1.2.5 Transmission line matching (remote matching) 214
   4.1.2.6 Efficiency of transmission line matching networks 221
  4.1.3 Flux coupled matching circuits 225
   4.1.3.1 Theory of inductive matching 228
   4.1.3.2 Inductive matching with tuned coupling loop 230
   4.1.3.3 Inductive matching with a non-tuned coupling loop 231
   4.1.3.4 Inductive matching with fixed mutual and variable
       capacitor 233
   4.1.3.5 Efficiency of the inductive coupling 235
   4.1.3.6 Coupled resonators 236
   4.1.3.7 Flux concentrators 238
 4.2 Balancing the Probehead 243
  4.2.1 Evidencing the electric losses effect 244
   4.2.1.1 Experimental setup 245
   4.2.1.2 Expected sensitivity 246
   4.2.1.3 Frequency shifts 248
   4.2.1.4 Q factors 250
  4.2.2 Evidencing the electric losses: the antenna effect 251
  4.2.3 Symmetrical capacitive coupling networks 254
   4.2.3.1 Splitting the matching capacitor 254
   4.2.3.2 A versatile capacitive balanced matching network 256

      4.2.3.3 Capacitive bridge 264
4.3 Summary of useful matching circuits 265
4.4 Accessories 268
    4.4.1 Transmit/Receive (TR) switches 268
    4.4.2 Damping circuits 270
    4.4.3 Baluns and cable traps 275
        4.4.3.1 LC-balun 277
        4.4.3.2 The 4:1 $\lambda/2$ balun transformer 278
        4.4.3.3 Broadband balun transformers 279
        4.4.3.4 Tuned cable traps 281
4.5 Interfacing the Probe to a Low Noise Amplifier (LNA) 284
4.6 Ultra-broadband and Ultrafast Recovery Probes 289
    4.6.1 Delay line ultra-broadband NMR probe 290
    4.6.2 Transmission line ultra-broadband NMR probe 292
    4.6.3 Non-resonant probe circuit 293

5. Quadrature Driving   295
5.1 Interfacing the Quadrature Probehead to the Console 297
5.2 Quadrature Hybrids 302
    5.2.1 $\lambda/4$ transmission line hybrid 302
    5.2.2 $\lambda/8$ transmission line hybrid 309
    5.2.3 Lumped element quadrature hybrids 315
        5.2.3.1 Quarter-wave hybrid equivalent 316
        5.2.3.2 $\lambda/8$ hybrid equivalent 319
    5.2.4 Frequency response of the quad hybrids 320
5.3 180° Hybrid 325
    5.3.1 The 180° rat-race hybrid 326
    5.3.2 Using the rat-race hybrid 180° in quadrature NMR 334
    5.3.3 Lumped element equivalent of the 180° hybrid 336
    5.3.4 Frequency response of the 180° ring hybrids 338
5.4 Other 90° Hybrids 341

6. Multiple Frequency Tuning   345
6.1 Shunting Methods 350
    6.1.1 Dual frequency switching circuits 350
    6.1.2 Practical double tuned circuits 355
    6.1.3 Multiple tuning of a single coil 361
    6.1.4 Balancing the shunting configurations 367
        6.1.4.1 Approximate balanced circuit 368
        6.1.4.2 Multiple-frequency full balancing circuit 370
6.2 Multiple-pole Circuits 374
    6.2.1 Approximate double-pole network component values 378

6.2.2 Exact solutions	382
6.2.3 Multiple-pole tuning for more than two frequencies	383
6.2.4 Balancing the multiple-pole circuits	384
6.2.4.1 Inductive coupling	384
6.2.4.2 The null-point method	386
6.3 Coupling Tank Circuits	388
6.3.1 Coupled identical resonators	392
6.3.2 Coupled resonators having the same resonance frequency but different L/C ratio	395
6.3.3 General case (different coupled resonators)	396
6.3.4 Fluxed coupled resonators	398
6.3.5 Special case of a short circuited coil	402
6.3.6 Π-network configuration	403
6.3.7 Summary of coupled resonator properties	407
6.4 Efficiency of Multiple Tuned NMR Probe	411
6.4.1 Efficiency of shunting methods	411
6.4.2 Efficiency of multiple pole circuits	415
6.4.3 Efficiency of coupled resonant circuits	421
6.5 Interfacing the Multiple Frequency Resonator to the Spectrometer	422
6.6 Is the Q Factor Representative of the Sensitivity?	424
7. Magnetic Field Amplitude Estimation	**427**
7.1 The Biot–Savart Approximation	427
7.1.1 Magnetic field produced by straight wires	429
7.1.2 Magnetic field produced by a loop	432
7.2 Effective Field for NMR Experiments	436
7.3 Estimation of the Current Distribution	441
7.3.1 Limits and usefulness of the thin wire approximation	441
7.3.2 Current distribution in the isolated flat strip	443
7.3.3 Proximity effects	452
7.3.3.1 Two coplanar strips	452
7.3.3.2 Three or more strips	455
7.3.3.3 Wire in proximity of a conductive plane	456
7.3.4 Current density in round wires	459
7.3.5 Concluding remarks	461
7.4 Survey of Modern Electromagnetic Simulation Methods	462
8. Homogeneous Resonators	**467**
8.1 Axial Resonators	469
8.1.1 Magnetic field amplitude	469
8.1.2 Approximations of the spherical uniform current density	473
8.1.2.1 Helmholtz coil	473

8.1.2.2 Four coil configuration 475
8.1.2.3 Guidelines for a practical design of Helmholtz probes
 and four coil probes 478
8.1.3 Solenoid types 480
 8.1.3.1 The solenoid coil 480
 8.1.3.2 The loop gap 485
8.1.4 Practical designs of solenoid type coils 490
 8.1.4.1 A microcoil for static magnetic field mapping 490
 8.1.4.2 A 0.4 ml high sensitivity phosphorous coil (162 MHz) 491
 8.1.4.3 A 150 ml double tuned ($^{1}H/^{31}P$)coil operating at 4.7 T 493
8.2 Transverse Resonators... 498
 8.2.1 Magnetic field amplitude... 498
 8.2.2 The saddle-shaped coil 501
  8.2.2.1 The optimum geometry and RF magnetic field............ 501
  8.2.2.2 A practical design... 505
 8.2.3 UHF saddle coil-like resonators 509
  8.2.3.1 The Alderman–Grant coil; a version of the slotted
   cylinder... 511
  8.2.3.2 Coupling the Alderman–Grant resonator to the
   spectrometer; a practical design 515
  8.2.3.3 Shielding the UHF saddle coil-like resonators 520
 8.2.4 The birdcage resonator 527
  8.2.4.1 RF field map of the k = 1 mode................................... 528
  8.2.4.2 Network analysis of the birdcage circuit 534
  8.2.4.3 Estimation of the current in a birdcage coil................. 541
  8.2.4.4 Estimation of the tuning capacitance........................... 548
 8.2.5 Practical use of the birdcage 552
  8.2.5.1 Tuning the birdcage resonator 552
  8.2.5.2 Asymmetry effects.. 556
  8.2.5.3 Interfacing the birdcage to the spectrometer................ 560
  8.2.5.4 A practical design.. 562
 8.2.6 Practical design of a quadrature birdcage 570
  8.2.6.1 Designing and adjusting the birdcage resonator for
   quadrature operation... 572
  8.2.6.2 Interfacing the resonator to the console 575
  8.2.6.3 Design of the shield 577
  8.2.6.4 Evaluation of the probe.. 578
 8.2.7 Double tuning the birdcage resonator............................ 579
  8.2.7.1 Pole insertion methods.. 580
  8.2.7.2 Alternate rung method.. 581
  8.2.7.3 Four ring double resonant birdcage 583
  8.2.7.4 Crossed-coil resonators... 585

8.2.7.5 A practical design	587
8.3 Transmission lines resonator	592
8.3.1 TEM resonators	592
8.3.2 Split transmission line resonators	597
9. Heterogeneous Resonators	**601**
9.1 The Basic Surface Coil	601
9.1.1 The simple loop magnetic field distribution	603
9.1.1.1 "Ideal" case	603
9.1.1.2 Effect of inductive coupling	606
9.1.2 Practical design guidelines	611
9.1.2.1 How many turns?	611
9.1.2.2 Spiral windings	611
9.1.2.3 Non-circular winding shapes	616
9.1.2.4 Wiring shape	618
9.1.2.5 Opened resonators	622
9.1.3 Surface coils for ultra high frequency	623
9.1.3.1 Segmented loop	623
9.1.3.2 The crossover coil	625
9.1.3.3 Split ring resonator (common mode)	626
9.1.3.4 Microstrip coils (differential mode)	628
9.1.4 Superconducting surface coils	630
9.1.5 Interfacing the coil to the spectrometer	632
9.2 Extending the Observed Volume (Multi-Ring Coils)	634
9.2.1 Coaxial rings	634
9.2.1.1 RF field profiling	634
9.2.1.2 X-observed, proton decoupled system	637
9.2.2 Array coils	640
9.3 The Surface Coil as Receive-Only Probe	643
9.3.1 Passive decoupling	644
9.3.2 Active decoupling	651
10. Probe Evaluation and Debugging	**655**
10.1 Instrumentation	656
10.1.1 The pick-up coil	656
10.1.2 Impedance bridge	657
10.1.3 Power divider, hybrid, and directional coupler	660
10.1.4 Sweep generator, crystal detector, and spectrum analyzer	662
10.1.5 Scalar and vector network analyzer	665
10.1.5.1 Transmission/reflection and S-parameters test sets	666
10.1.5.2 Calibration	667
10.1.6 Other impedance measuring instruments	670

10.2 Characterization of a Transmission Line 670
   10.2.1 Velocity coefficient 671
   10.2.2 Evaluation of the characteristic impedance 672
   10.2.3 Estimation of the loss parameters 674
10.3 Noise Figure Measurement 674
10.4 Evaluating the Probe on the RF Workbench 678
   10.4.1 Matching the probe input impedance to 50 Ω 678
   10.4.2 Evaluation of the Q factor 680
   10.4.3 $B_1/\sqrt{P}$ evaluation methods 684
10.5 Evaluating the Probe on the NMR Instrument 686

Appendix A. Physical Constants and Useful Formulae 691

Bibliography 709

Index 733

Chapter 1

Introduction

1.1 What is an NMR Probe?

Any kind of physical investigation usually needs a suitable sensor in order to interface the physical phenomenon to the final display of the results. The task is complicated if the studies are concerning the molecular or atomic level, even the sub-atomic range nowadays. The principal problem arising when dealing with quantum experiments is how to measure the physical properties at sub-atomic level without the investigating system being perturbed by even the measuring process or by the probe we use.

This is precisely the case with a Nuclear Magnetic Resonance (NMR) experiment, which is based on picking up the signal generated by the assemblies of nuclei having a nonzero spin number. The nuclear spins, denoted as s, like hydrogen ($s = 1/2$), phosphorus ($s = 1/2$), carbon 13 ($s = 1/2$), sodium ($s = 3/2$) are involved in the molecular or ionic constitution of a large number of materials (liquids, solids, or living heterogeneous systems).[1] The spin assembly defines the sample which is observed by the NMR approach. The sample contains a very large number of magnetic moments associated with the spin properties of the considered nuclei. When the spins of the protons and neutrons comprising these nuclei are not paired, the overall spin of the charged nucleus generates a magnetic dipole along the spin axis; the intrinsic magnitude of this dipole is the fundamental nuclear property called the

---

[1] A complete list of nuclides able to give NMR is given in Chapter 2.

nuclear magnetic moment. Consequently, the nuclear magnetic moment of a collection of nuclei can align with an externally applied static magnetic field $\vec{B_0}$ in ($2s + 1$) ways, either in the same direction or opposed to $\vec{B_0}$, thus generating a macroscopic magnetization $\vec{M}$ of the whole sample proportional to $\vec{B_0}$. This property is a characteristic of paramagnetic substances for which the magnetization is proportional to the static magnetic field and inversely proportional to the temperature.

With common thermal polarization, this magnetization is governed by the Boltzmann equilibrium law

$$\vec{M} = N \frac{\gamma^2 s(s+1)\hbar^2}{3 k_B T} \vec{B_0}, \qquad (1.1)$$

where N is the number of nuclei present in the sample, γ is the gyromagnetic ratio of the nuclei, $\hbar$ is the Planck constant divided by 2π, k_B is the Boltzmann constant and T the spin temperature, which is equal to the sample temperature at thermal equilibrium between spins and the thermostat made by the sample itself (known also as the lattice). The SI unit to express the amplitude of $\vec{M}$ is labelled as *ampere · meter²* (notation: A m²).

The aim of the NMR technique is to quantify the nuclear magnetization of a sample, which is generally a rather small and very specific physical property that cannot be measured by conventional means. For this purpose, a resonance approach was developed in 1946 by research groups at Stanford and MIT, in the USA [Bloch *et al.*, 1946; Purcell *et al.*, 1946]. The radar technology developed during World War II made many of the electronic aspects of the NMR experiment possible and thus the observation and determination of the predicted nuclear magnetization. The principle of the most popular method is based on the detection of magnetic variable flux provided by the sample, similarly to the light signal given by a bicycle alternator. This is possible once the magnetization is tilted from its equilibrium position along and in the sense of the external applied magnetic field $\vec{B_0}$. After being tilted, the magnetization gets a precession motion around the static magnetic field direction, generally represented by the "vertical" direction on pictures (Fig. 1.1). In this particular case, and in this case only, one may consider that the behavior of the magnetization is comparable to a magnet getting precession around an axis (here the axis is given by the static field direction). Then, if a conducting loop is set in a vertical plane, an Electromotive Force (emf) will appear between the two extremities of the wire as shown in Fig. 1.1. The angular frequency of precession is given by

$$\omega = \gamma \left| \vec{B_0} \right|, \qquad (1.2)$$

where γ is the "gyromagnetic" ratio of the considered nuclei; this parameter must be expressed in *radian/second/tesla* (notation: rad s^{-1} T^{-1}).

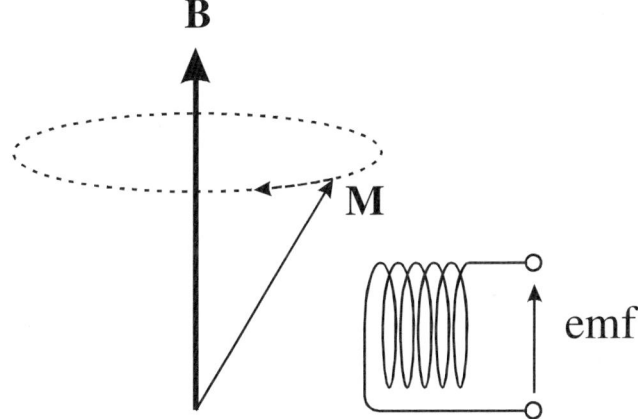

Fig. 1.1 Precession motion of nuclear magnetization around the applied magnetic field $\vec{B_0}$ and generation of an electromotive force in a "radiofrequency coil" due to the time variation of the magnetic flux.

Typical values for static fields in NMR experiments are presently in the 1 to 23 teslas range, giving frequencies from 42 to 1000 MHz for hydrogen. Such frequencies are typically Radio Frequency (RF) and the

NMR devices (probes, amplifiers, electronic detectors, etc.) must operate accordingly.

It is clear from Eq. (1.2) that the larger the static magnetic field, the larger the emf is since the angular frequency is proportional to $|\vec{B_0}|$ for any nuclei observed by NMR. It clearly appears that the sensitivity of the NMR experiment will increase when increasing $|\vec{B_0}|$, and this explains the expensive efforts towards high static field magnets to perform NMR.

Nevertheless, even with the largest magnetic fields presently available (approximately 23 teslas), the NMR signal may still be too poor due to the smallness of the sample volume or to the weakness of the gyromagnetic ratio of the observed nuclei. The signal weakness is also due to the fact that, in parallel, manufacturers, biochemists and biologists try to observe smaller and smaller samples. Consequently, one efficient way to improve the nuclear magnetic signal generated at the sensor output is to pick it up using a resonant device. Practically, as will be demonstrated in the following chapters, matching a resonant circuit to the NMR spectrometer means constructing a resonator by tuning its receiving loop with good quality components such as capacitors to avoid losses. In this case, if Q is the quality coefficient of the receiving coil, the emf induced in the coil will be multiplied by a factor proportional to Q at the resonator output. The voltage thus obtained will be a superposition of signal and noise, both multiplied by the same factor. The advantage brought by this configuration in terms of the Signal-to-Noise Ratio (SNR) could be the reduction of the pass-band of the system (the SNR is inversely proportional to the pass-band square root). In order to take into account thermal noise, this resonator may be considered a time dependent voltage source in series with a noise source. In the example of Fig. 1.2 the capacitor now acts as a filter at the input of the rest of the circuitry, especially for noise transmission.

For more than half a century NMR detection was classically based on the generation of an oscillating magnetic flux by an electrically resonating circuit, which is still widely held [Hill and Richards, 1968]. One of the recent approaches uses the Josephson effect in semiconductors to observe the nuclear magnetization [McDermott et al., 2002]. Conversely, this latter technique is sensitive, not to the flux variation, but directly to the flux itself. This innovative method allows NMR experiments without the need for intense polarizing magnetic fields, but rather uses very low fields, smaller than the Earth's.

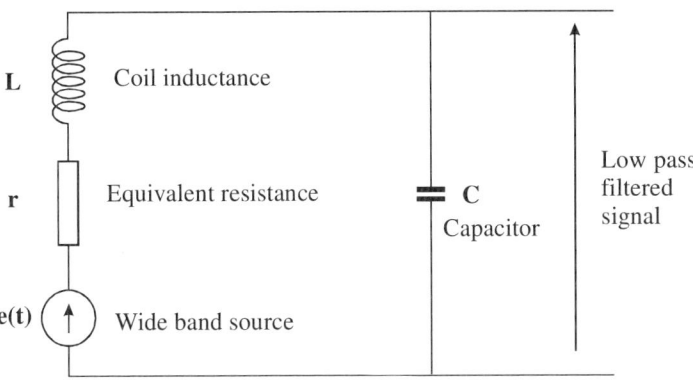

Fig. 1.2 Principle of the NMR resonator using a tuning capacitor.

1.1.1 *The basic pulsed NMR experiment*

Considering that the precession of the magnetization is able to generate an emf, a current is passing through the receiving loop when this loop is closed by an output circuitry. By reciprocity, when a current is passing through the wire loop during a limited period of time, denoted as τ, the sample magnetization can be tilted as well.[2] The magnetization excitation is performed provided the angular frequency of the applied current is very close to the precession angular frequency of the nuclei.

For a given tilt angle θ, the pulsed RF field must be applied along a direction perpendicular to the static magnetic field. A simple model,

---

[2] An extended study of RF pulse generation is developed in the book of Michel Decorps [Decorps, M., 2001].

generally used, assumes that this alternative field is linearly polarized and that its amplitude is constant and equal to $|\vec{B_{RF}}|$. The angle θ, the pulse length τ and the RF amplitude are related by the following formula

$$\theta = \gamma \left(\frac{|\vec{B_{RF}}|}{2} \right) \tau. \qquad (1.3)$$

We do remember that a rectilinear oscillating field orthogonal to $|\vec{B_0}|$ may be decomposed into two opposite circular fields in a plane perpendicular to $|\vec{B_0}|$. The previous result may be derived using the well-known rotating frame representation. In this representation the RF field appears as fixed and its amplitude, half of $|\vec{B_{RF}}|$, denoted as B_1. Usually the B_1 value is of the order of 10^{-4} tesla, which is considerably smaller than usual static magnetic field values. Consequently, the precession frequency around the direction of B_1, the field seen in the rotating frame, is in the kilohertz range for hydrogen ("gyromagnetic" ratio is 2π times 42.57 10^6 rad s^{-1} T^{-1}). A typical duration for the RF pulse necessary to tilt the magnetization by a 90° angle with respect to its initial orientation is about 100 microseconds. When reasoning in the rotating frame, one must assume that the motion of the magnetization has to be fast enough compared to the relaxation effects. This condition implies, first of all, that the magnetization value remains almost constant during the pulse. Secondly, once the preceding condition is fulfilled, during any RF pulse, the angle between $|\vec{B_{RF}}|$ and $\vec{M}$ should remain constant. Some particular tip angle values are represented in Fig. 1.3 to illustrate this remark.

The most traditional representation of an NMR experiment considers a vertical static field and an RF coil with its axis horizontally oriented, as depicted in Fig. 1.1. Starting from equilibrium magnetization, a short RF current in the coil may create a horizontal magnetization component which, in turn, is able to modulate the magnetic flux passing through the wires at almost the same frequency.

Although relaxation effects are weak interactions compared to the nuclei coupling with the RF field, they are nevertheless efficient because they represent the main interaction between the nuclei and the experimental environment.

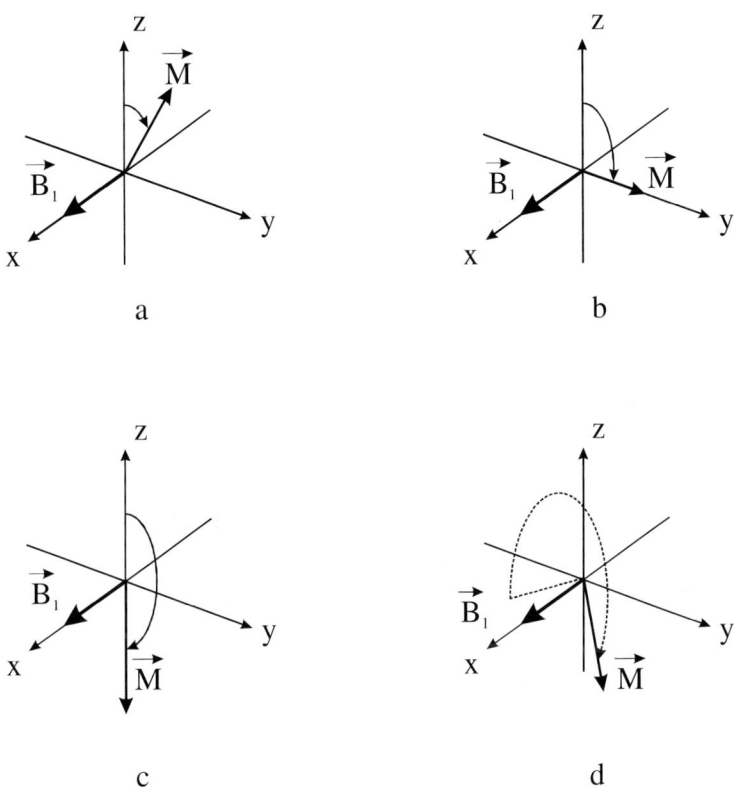

Fig. 1.3 Examples of magnetization tip angles around $\vec{B}_1$, which defines the x orientation in the rotating frame, with magnetization starting from the equilibrium direction: (a) arbitrary pulse; (b) 90° pulse; (c) 180° pulse; (d) 180° pulse with magnetization starting from a direction taken in the Oxy plane and giving a final magnetization in the same plane.

Spin–lattice relaxation corresponds to the recovery of the longitudinal magnetization, i.e. the component measured along the static field, and is a rather slow process. Spin–spin relaxation acts naturally on the transverse magnetization decay, but this phenomenon is generally shortened by an important out-of-phase effect generated by the static field nonuniformities over the whole observed sample, especially in liquids or heterogeneous samples as the living tissues. Several features must be taken into account in order to understand the magnetization behavior during the free evolution period which follows the RF excitation:

- the relaxation time (T_1), which governs the recuperation of the longitudinal magnetization, is extremely long compared to the period of precession (typically one second for water protons and a few tens of a nanosecond for the precession period);
- the apparent transverse relaxation time (T_2^*), shorter than T_1, is also very long respective to the precession period. Notice that this time constant can be defined only when the static field distribution presents through the sample a very particular repartition, said to be "Lorentzian", leading to an exponential decay;
- during a very large number of rotations of the magnetization about the static field direction (more than several thousand) there is generally no noticeable decay of the amplitude of the transverse magnetization and the same observation works for the induced signal itself.

The nuclei RF signal can be further demodulated in order to be recorded on a narrow spectral range (free induction decay). If the sample contains isochronal nuclei, this signal has the appearance shown in Fig. 1.4(a). A Fourier transform of the free induction decay signal [Ernst and Anderson, 1966] gives the complex signal (or phase-amplitude signal) represented in Fig. 1.4(b).

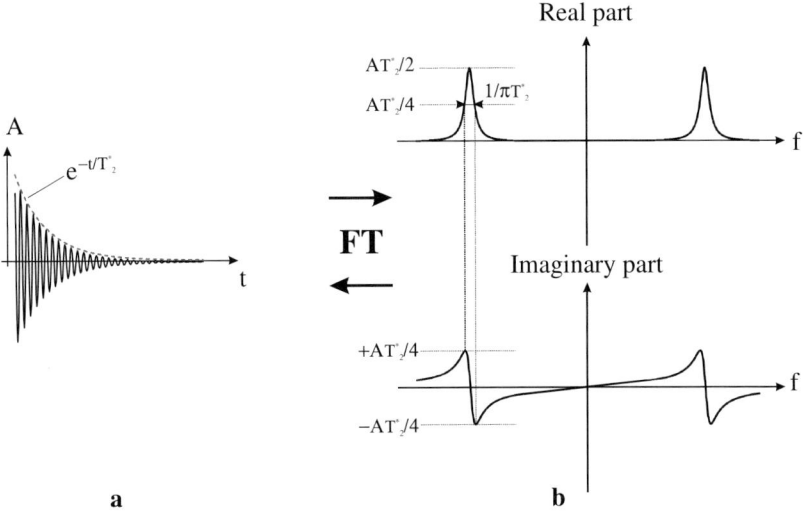

Fig. 1.4 (a) Free induction decay signal in a supposed homogeneous magnetic field leading to an exponential decay. (b) Fourier transform of the signal in phase-amplitude representation.

1.1.2 The head probe from a theoretical point of view

1.1.2.1 The principle of reciprocity and the calculation of the induced emf

A single conducting coil (more generally a set of conductors) is the most frequently used way to pick up the emf generated by the transient motion of the nuclear magnetization. From the theoretical point of view, the emf evaluation can be developed either on the basis of the reciprocity principle [Hoult and Richards, 1976; Hoult, 2000], or from a direct application of the law of electromagnetism [Pimmel, 1990]. The second approach will be presented here because it gives detailed information about signal generation and it leads to a condensed comprehensive treatise. One may start the analysis from the schematic drawn in Fig. 1.5, where both the sample and the RF coil are schematically represented.

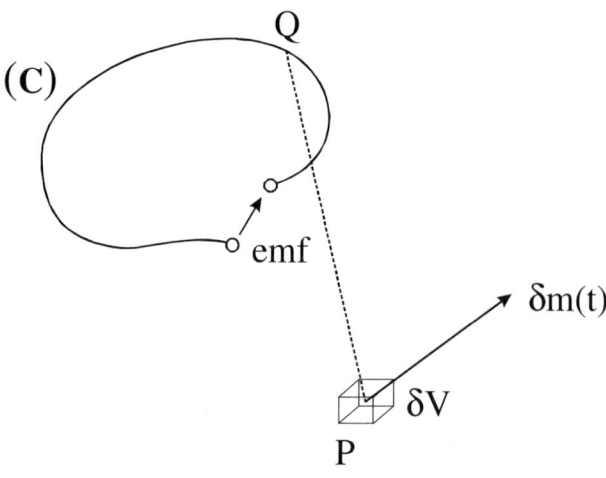

Fig. 1.5 Schematic diagram for the calculation of the electromotive force generated by the magnetization precession.

After excitation due to a RF pulse, the time varying magnetization of a volume element placed in P is denoted as $\overrightarrow{\delta m}(t)$. According to usual electromagnetic rules, such a magnetic moment creates in space, at point Q, a magnetic potential vector expressed as:

$$\overrightarrow{\delta A_Q} = -\frac{\mu_0}{4\pi} \overrightarrow{\delta m}(t) \times \overrightarrow{\nabla} \left(\frac{1}{r_{PQ}} \right)_Q, \tag{1.4}$$

where μ_0 is the vacuum permeability ($4\pi.10^{-7}$ *henry/meter*, notation: H m^{-1}) and where the reversed delta symbol denotes the "nabla" operator which acts at the Q position, and r_{PQ} is the distance between P and Q.

The emf induced in the receiving coil (C), considered here as a thin wire, is given by the Maxwell–Faraday law, or in other words by the circulation along the coil of the magnetic vector potential generated by the magnetic moment

$$\delta e(t) = -\frac{\partial}{\partial t}\int_C \delta \vec{A}_Q \cdot d\vec{\ell}, \qquad (1.5)$$

where (C) is associated with the wire extension.

Using Eq. (1.4) in Eq. (1.5) one gets

$$\delta e(t) = \frac{\partial}{\partial t}\left\{\delta\vec{m}(t) \cdot \frac{\mu_0}{4\pi}\int_C \vec{\nabla}\left(\frac{1}{r_{PQ}}\right)_Q \times d\vec{\ell}\right\}. \qquad (1.6)$$

The integral factor of the preceding equation represents exactly the opposite of the magnetic field existing at point P when a unit current is passing through the wire. Consequently one may express the emf generated by the motion of $\delta\vec{m}(t)$ as

$$\delta e(t) = -\frac{\partial}{\partial t}\left\{\delta\vec{m}(t) \cdot \vec{b}_{RF}(X,Y,Z)\right\}, \qquad (1.7)$$

where $\vec{b}_{RF}(X,Y,Z)$ appears as the RF field at P, the coordinates of P being (X,Y,Z), when a unit current is passing through the coil.

This result has a practical importance in order to evaluate the signal rising from a given volume δV_e since the amplitude of $\delta\vec{m}$ is $\delta m_0 = M_0 \delta V_e$, taking for M_0 the magnetization per unit volume which may be derived from Eq. (1.1) and δV_e the volume element at P. Still using the Lorentzian model, and assuming a positive nuclear "gyromagnetic" ratio leading to right polarized precession, one may write, in complex representation:

$$\delta\vec{m} = \delta m_0 e^{-t/T_2^*}\left[e^{j(\omega t+\varphi)}\vec{I} + e^{j(\omega t+\varphi+\frac{\pi}{2})}\vec{J}\right], \qquad (1.8)$$

where $\vec{I}$ and $\vec{J}$ are the unit vectors associated with the X and Y axis respectively, φ is an initial phase and T_2^* is assigned to the decay of transverse magnetization belonging to δV_e.

The most general expression for the electromagnetic field created at point P when a current $I = I_0 e^{j\omega t}$ is passing through the contour (C) must be written as superposition of two opposed rotating fields. Consequently $\overrightarrow{B_{RF}}(X,Y,Z)$, used above, can be written as[3]

$$\overrightarrow{B_{RF}}(X,Y,Z) = (\Lambda+P)e^{j(\omega t+\varphi')}\vec{I} + (\Lambda-P)e^{j(\omega t+\varphi'-\frac{\pi}{2})}\vec{J}, \quad (1.9)$$

where the amplitudes Λ and P correspond to left and right rotating components respectively, where φ' is an arbitrary current generated phase.

$$\overrightarrow{b_{RF}}(X,Y,Z) = \frac{1}{I_0}(\Lambda+P)e^{j\varphi'}\vec{I} + \frac{1}{I_0}(\Lambda-P)e^{j(\varphi'-\frac{\pi}{2})}\vec{J} \quad (1.10)$$

and

$$\overrightarrow{b_{RF}}(X,Y,Z) \cdot \overrightarrow{\delta m} = 2\Lambda \frac{\delta m_0}{I_0} e^{j(\omega t+\varphi+\varphi')} e^{-t/T_2^*}. \quad (1.11)$$

Neglecting slow time varying terms (terms in $1/T_2^*$), the time derivation finally gives

$$\delta e = 2\omega \frac{\Lambda}{I_0} \delta m_0 e^{j(\omega t+\varphi+\varphi'-\frac{\pi}{2})}. \quad (1.12)$$

In this analysis, Eq. (1.12) plays a particularly important role since the time derivation of the scalar product of $\overrightarrow{\delta m}$ with the RF field component rotating in the precession sense is negligible compared to the term weighted by ω. Consequently, during the receiving period, the only

---

[3] A linearly polarized RF field is defined by $\Lambda = P$. A left-hand polarized circular field is defined by $\Lambda \neq 0$, $P = 0$ while $\Lambda = 0$, $P \neq 0$ defines a right-hand one.

Introduction

efficient RF component to be considered in $\vec{B}_{RF}(X,Y,Z)$ rotates in the opposite sense with respect to the precession motion. For the case of a linearly polarized field, Λ denotes the amplitude of $\vec{B}_1$ used in Fig. 1.3.

Locally speaking, the phase φ depends on the orientation of $\vec{\delta m}$ at the beginning of signal detection. With a single excitation–detection coil, this angle is generally close to $\pi/2$, independent of the nonuniformities of the RF field. With a separated transmitting–receiving coil arrangement, the situation is different since the transmitting and receiving radio fields do not generally present the same spatial distribution in addition to the different orientation. This situation may represent a risk of signal attenuation. The phase difference φ' between the current $I = I_0.e^{j\omega t}$ passing through (C) and the local vector $\vec{B}_{RF}(X,Y,Z)$ are generally weak when propagation effects have not been taken into account, which is the usual situation.

Following a θ pulse, as described above, and from Eq. (1.12) the signal amplitude generated by δV_e is given by:

$$a = \frac{2B_1}{I_0} N_0 \delta V_e \frac{\gamma^3 \hbar^2 s(s+1)}{3k_B T} B_0^2 \sin\theta. \qquad (1.13)$$

Notice that, using SI units for calculations, it is straightforward from Eq. (1.13) that a is expressed in *Volts* (notation V).

Extension of the formula to sample volumes larger than δV_e depends mainly on the spatial uniformity of nuclear density and will not be developed here.

1.1.2.2 *Losses*

During the receiving period, the RF coil can be considered as an electrical voltage source and can be represented through the circuit in Fig. 1.6(a), where L corresponds to the self-conductance of all wires, r is the equivalent resistance of the represented circuitry and e(t) a wide band voltage generator comprising noise superimposed on useful signal. The noise generator in the equivalent circuit of Fig. 1.6(b) accounts for energy losses when the observed sample is surrounded by the RF coil.

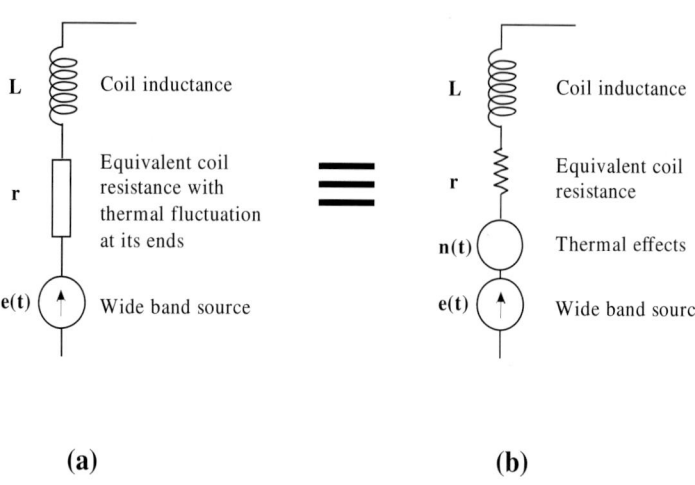

Fig.1.6 (a) Equivalent voltage source for the signal generation. (b) Equivalent circuit for the RF coil used as receiving coil (taking into account the noise equivalent source).

The resistance value includes the "ohmic" values of all wires in the presence of high frequency oscillating currents. Due to a skin effect that reduces the current penetration in conductors, this contribution is larger than the static resistance. A second contribution, designated as *magnetic losses*, is caused by power leakage due to the induced currents in the conducting samples at the operating RF. This effect can be related to the electrical conductivity of the sample medium. For example, in biological samples, the conductivity is approximately equal to the electrical conductivity of a saline solution of 9 kg m^{-3} of sodium chloride. A third contribution to the overall resistance is caused by the electrical losses due to the potential differences between ground and some circuit parts, reaching high potential values during the RF excitation. Potential differences around the coil imply the generation of electric fields *E*. Whenever these fields fringe into the samples they will interact with lossy dielectric materials leading to the corresponding dielectric losses.

Finally, the equivalent resistance of the RF probe may be written as

$$r = r_\Omega + r_M + r_E + r_R, \qquad (1.14)$$

where r_Ω corresponds to conventional Joule effects in the wires, r_M is the resistance due to magnetic losses, r_E is generated by electrical losses and r_R is generated by radiated energy.

A quantitative estimation of each term may be derived on the basis of electromagnetic laws, provided the sample shape and its electrical characteristics are known. For a given coil

$$r_\Omega = R_\Omega \omega^{\frac{1}{2}}, \qquad (1.15)$$

where R_Ω depends on the coil geometry and material characteristics. For a spherical sample of radius b the calculations for the next two terms lead to the following results

$$r_M = R_M \sigma n^2 \omega^2 b^5, \qquad (1.16)$$

where R_M depends on the coil geometry, n the number of conductive turns, σ being the electrical conductivity of the sample.

The next term in Eq. (1.14) has the form

$$r_E = R_E \omega^3 L^2 C_d, \qquad (1.17)$$

where R_E depends on the coil geometry, L and C_d the coil self inductance and capacitance respectively.

The last term in Eq. (1.14) represents the losses generated by the radiation process (Chapter 4). The estimation of a small loop radiation resistance gives

$$r_R = R_R D \omega^4, \qquad (1.18)$$

where R_R represents the propagation medium properties and D is the loop diameter, hence related to coil geometry.

Every term of Eq. (1.14) has a geometrical or material component that will strongly depend on the chosen probe design and a frequency component. Obviously the chosen design should minimize as much as possible the contribution of all these losses. The other important component of all these terms is given by the RF ω that imposes one contribution or another depending on the frequency range.

Generally, one may consider that the resistance due to the presence of the sample, which dominates over r_Ω at high frequency, is mainly proportional to ω^2.

1.1.2.3 The sensitivity

The equivalent noise voltage source n(t) introduced in Fig. 1.6(b) represents the voltage fluctuations due to the presence of the resistance r. This noise is considered as thermal white noise (having a null mean value) and a mean square amplitude given by its variance square root

$$\sigma_n = \sqrt{4 r k_B T \Delta f} \,, \tag{1.19}$$

where k_B is the Boltzmann constant, T the probe temperature and Δf the bandwidth of the NMR receiving system. Using the SI units system, the second member of Eq. (1.19) leads to a voltage value (expressed in Volts).

The sensitivity is defined by the ratio of the useful signal amplitude to the mean square amplitude of noise. According to Eqs. (1.13) and (1.19), it gives the following dimensionless formula:

$$S = \frac{2B_1}{I_0} N_0 \delta V_e \frac{\gamma^3 \hbar^2 s(s+1)}{3k_B T} B_0^2 \sin\theta \frac{1}{\sqrt{4 r k_B T \Delta f}} \,. \tag{1.20}$$

For a given probe characterized by $\theta, \Delta f$ and a given sample characterized by $N_0, \gamma, s, \delta V_e$ the ultimate sensitivity is proportional to $B_1/(I_0 \sqrt{r})$. This implies that losses have to be kept as low as possible, meaning mainly a small value of r. In parallel the probe must exhibit the highest possible efficiency related to B_1 amplitude for a given current passing through the wires. Assuming that the RF field generated by the coil is uniformly distributed through the inner probe volume, $B_1/(I_0 \sqrt{r})$ may be replaced by a term proportional to $\sqrt{\eta Q}$ where η is the filling factor of the resonator coil and Q its quality factor.[4] The first formulation is certainly more rigorous that the second one since it is directly derived from the principle of reciprocity. But it does not explicitly involve the resonance phenomenon induced in the oscillating circuit of the probe. The fact that the resonating frequency of the electrical circuit made by the loop and the capacitor is equal to the precession frequency of the observed magnetization is a very important feature of NMR detection and can justify the use of the word "resonance" on the experimental scope. The emf taken at the terminals of the circuit is almost Q times the emf generated by the magnetization flux variation in the receiving coil, noise included.

The sensitivity dependence of $B_1/(I_0 \sqrt{r})$ results in S being approximately proportional to $\omega^{7/4}$ at low frequencies. Similar behavior can be observed at high frequency when operating on samples showing very low conductivity, especially in some spectroscopic applications. At high frequencies, and particularly with living tissues, S becomes proportional to ω.

Other remarks concern the presence of a temperature factor in the sensitivity formula. Here it is assumed that the coil wires and the NMR sample are both at the same temperature, which is represented by the common T symbol. In fact, these two temperatures may be significantly different, and it is possible to lower the temperature of the receiving wires in order to reduce thermal noise generation and to increase sensitivity.

One may notice that Eq. (1.20) is a signal-to-noise expression given in the time domain, strictly limited to the probe. Moreover it may be interesting to have some idea about the corresponding SNR in the frequency domain. The Discrete Fourier Transform (DFT) is mainly used to pass from time to frequency domain in digital spectroscopy [Bringham, 1974]. Let's consider a single spectral line signal which is detected at zero frequency. The number N_s of acquired data points (0 to N_{s-1}) is sufficiently high to sample the signal up to the noise level. The signal sampled values are obtained from digitization of $a\exp(-t/T_2^*)$, where a is given in Eq. (1.13). The corresponding noise is denoted as n_k,

---

[4] The Q factor is defined as: $Q = L\omega/r$.

($k=0.....N_{s-1}$) and its square root variance given by Eq. (1.19). The DFT applied to the signal samples shows that the maximum amplitude is proportional to the product $N_s aT_2^*$, whereas if applied only to the n_k collection it is proportional to $\sigma_t \sqrt{N_s}$. The ratio of signal amplitude to mean square amplitude of noise in the frequency domain is, in turn, proportional to $\sqrt{N_s}$. This explains that in several circumstances the easiest way to observe the NMR signal from noise is to perform the Fourier Transform on the Free Induction Decay signal, especially when the magnet is well shimmed (long T_2^*). Since signal processing cannot compensate hardware imperfections, a careful design and construction of the probe will represent the ultimate condition to observe the NMR signal with the optimum SNR.

1.2 What Probes for a Specific NMR Experiment?

In the preceding sections we have considered the principle of the NMR probe only, i.e. the technical process which is presently used to observe the magnetization precession motion in order to get a magnetization measurement. This is based on the generation of an induced emf, and it works like in an electric alternator. This is the main principle working since the beginning of the NMR experimental adventure which started almost 70 years ago. Optical or mechanical detections are also possible but not yet significantly developed in biomedical perspectives such as human diagnostic or animal model studies. Other techniques based on Superconducting Quantum Interferences Devices (SQUID) permit one to directly determine the magnetic flux with a very high sensitivity [McDermott *et al.*, 2002; Wong-Foy *et al.*, 2002].

Consequently these systems that are employed for magneto-graphic applications, which may concern extremely low magnetic fields (few femtotesla) generated by living organs (heart, brain,...), may be considered for the future design of NMR probes. The present use of superconducting materials for NMR probes is based on high critical temperature superconducting materials for the coil design, consequently the system is still based on magnetic flux detection. Indeed, the advantage of superconducting loops is that the equivalent resistance can be fixed at a very low value, provided the magnetic and electrical losses through the sample are contained. This perspective was already proved with small size probes and, considering the difficulty of performing high critical temperature superconducting surface deposits, the application domain is very restricted [Ginefri *et al.*, 2003; Hall *et al.*, 1991]. Consequently, to design NMR probes from the inductive principles described above and developed in the rest of the book still represents a wide field of activities and one must, at least, consider three main domains of application.

1.2.1 *High resolution NMR in solution*

Organic chemical and biochemical applications of NMR are generally devoted to elucidating molecular structure and require a very high resolution power, 10^8 to 10^9 or better (the resolution power being the inverse ratio of the spectral line width to the resonance frequency). Fortunately the samples under analysis are often in the liquid state or, more simply, dissolved in a solution of rather mobile liquid molecules. In this case, the weakness of mean interactions leads to extremely short correlation times for molecular translation and reorientation, typically in the range of a few picoseconds. Consequently, in liquids, the relaxation times T_1 (spin–lattice) and T_2 (spin–spin) are rather long: several tens of milliseconds, even of the order of a second or more. Long T_2 values make the resonance line widths very fine if magnetic static field uniformity is sufficient. Line widths are mainly limited by the spatial distribution of the static magnetic field through the sample. This sample fills the bottom part of a calibrated NMR tube and usually the mechanical design of the probe keeps this tube in a vertical position. In order to improve the homogeneity of the static field it is possible to rotate the tube about its vertical axis during the measurements. For variable temperature measurements, the insert can be cooled to rather low temperature values (almost −160 °C) or warmed to temperatures of the order of 200 °C. Such performances require a specific Dewar assembly with particular glass properties in order to prevent undesired effects caused by dilatation. For the vertical orientation of both static magnetic field direction and tube axis, the RF coils must generate a horizontal high frequency field. This is generally achieved using a kind of saddle coil wound around a vertically oriented cylindrical glass form. Nevertheless, there are cases where the orientation of the NMR probe may have other degrees of freedom in the gap of the magnet due to its specifically small dimensions. This is particularly the case with microcoil probes [Webb, 2005], with geometry not imposed by the direction of the external magnetic field but imposed by specific experiments.

For a transmit/receive configuration, a single coil system can be easily built. For a quadrature transmit/receive coil system or a transmit coil associated with a separate receiving one, the design requires two orthogonal coils and the building may be rather complicated considering the tiny space. Particular caution must be taken in order to avoid direct coupling between the two coils. For double or multiple resonance experiments this configuration is still valid. One may also notice that, in order to obtain the best RF field uniformity, coils must respect "canonical" proportions: for a single threaded coil the length value must be two times the diameter and the longitudinal opening angle of the saddle must be equal to 120°.

When using for both transmitting and receiving RF, this coil structure can be advantageously replaced by more efficient ones such as the Alderman and Grant (or slotted tube) designs, which are discussed in this book.

1.2.2 Solid state NMR

The most important setback of solid state high resolution NMR is caused by dipolar interaction between spins which cannot move freely, as in the liquid state, to give zeroed interactions and small line widths. The secular term of Hamiltonian interaction can be nulled out by magic angle spinning (54°) of the tube sample with respect to the static magnetic field. The spinning rate should be quite fast. Simply speaking, this particular design and RF high powered pulse trains render the interactions of the spin system of the solid sample rather similar to interactions occurring in the liquid state. The aim is to reduce spectral dipolar widths considerably and to obtain, subsequently to the Fourier transform, reasonable line widths. Considering fast rotation inertial effects, the sample tube must be rather small. Moreover, the important value of the magic angle permits one to efficiently use a solenoid coil around the sample. In this scope, variable path solenoids can be designed in order to improve the RF field uniformity.

The present practice of solid state NMR necessitates particularly well designed probes equipped with complex mechanics, not only due to rotation capabilities but also to perform measurements under controlled atmosphere in several applications like catalytic material characterization. Consequently it is not highly recommended, except for experts, to develop homemade probes for solid state NMR experiments.

1.2.3 *Biomedical and biophysical applications*

This field of application is rather wide since it includes NMR studies through Magnetic Resonance Spectroscopy (MRS) and Magnetic Resonance Imaging (MRI) [Gadian, 1982; Haase *et al.*, 2000]. Coil design for biochemistry, biology, pharmacology, and eventually medicine will be widely developed in the following chapters. The choice of resonator design for a given application is dictated by several parameters depending on the sample or on the kind of experiment. It also depends on the bore magnet accessibility. In this case the two main different features for the coils are required according to the horizontal or vertical configuration of the magnet. Roughly speaking, the vertical configuration, even with small animal studies and rather high magnetic fields, constrain the users to develop RF probe units very similar to the high resolution system used for liquid spectroscopy, except that the coil itself must be updated to the particular location suiting the observe location in the sample. Generally a surface coil, of one turn of copper or silver wire, is well suited for brain, heart, and liver examination. In this case, the rest of the probe is almost identical to a high resolution one. Notice that the vertical position is not the best physiological position for animals. Nevertheless this configuration works well for perfused organs such as excised heart or excised kidney and, in this case, the common saddle shaped coil can be very efficient. A larger variety of resonator designs can be constructed when using a horizontal magnet because accessibility is easier than in the vertical case. For small samples, one may use a simple solenoid coil which is orthogonal to the field axis direction, in order to take advantage of the potential sensitivity of this design. More generally, the RF coil access is oriented along the direction of the magnet bore axis, the same as the static field one, and resonator structures have been extensively developed on this symmetry basis: the Alderman and Grant resonator, Hayes coils or birdcage configurations are among the most popular designs. They are interesting because they are wound on a cylindrical form that fits the cylindrical symmetry of the magnet and gradient coils geometry well. Such RF coils will be proposed and studied in the following chapters. Simply speaking, there are two conditions to be met for probe construction: the first one to take into account is the geometrical structure that is generally imposed by sample shape or volume requirements and accessibility. The second involves electrical problems to solve: tuning/matching unit adjustment, coupling modes between coils themselves, between coils and the output of the RF power amplifier and types of connection to the input of the electronic chain for signal detection and processing. All these features will be developed in the following chapters.

Chapter 2

Radio Frequency Components

We introduce here some properties characterizing the Radio Frequency (RF) components that are often present in the circuits of probes. First of all, the RF devices in NMR spectrometers or scanners are briefly described and the tools characterizing these components are presented. Then, the RF electrical properties of relevant materials (metals and dielectrics) are discussed as well as the passive components like cables, conductors, transmission lines, resistors, inductors and capacitors. Both discrete and distributed components are considered. Some active devices used in probe circuits, such as diodes and low noise preamplifiers, are briefly introduced in a fifth section. A short introduction on metamaterials is given here, especially dedicated to flux guides. Finally, specific requirements for NMR probe components are addressed in the last section of this chapter.

2.1 The RF Parts of NMR Spectrometers and Scanners

Fig. 2.1 shows a simplified block diagram of a classical NMR spectrometer and MRI scanner. The RF modules of both pieces of equipment are essentially of the same type. There are three main RF parts: the probe(s), the power amplifier(s), also called the transmitter, and the receiver(s). All these parts are connected with transmission lines, mainly coaxial cables.

In addition, the MRI scanner comprises magnetic field gradient coils and associated amplifiers. These are required for frequency coding the sample space. Gradient coils, along one or the three geometric axes, may also be found in NMR spectrometers and are used for coherence pathway selection.

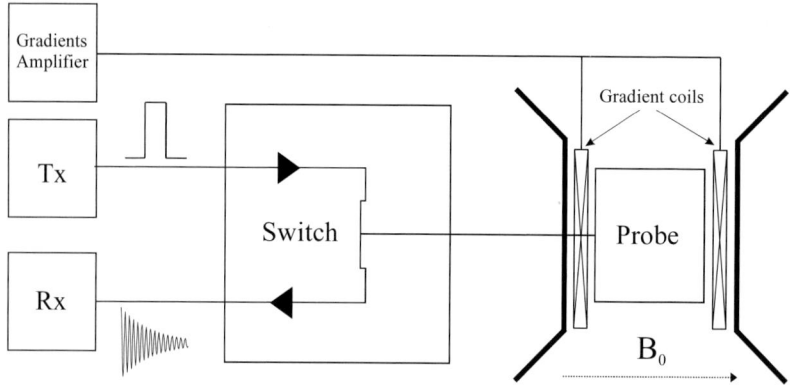

Fig. 2.1 Block representation of the RF parts of an NMR spectrometer.

The probe has a dual role; as an emitter it rotates the macroscopic spin magnetization and as a receiver it detects the magnetic flux created by the precession of the excited nuclear magnetization. The constraints regarding the probe components are very different for these two roles and will be discussed later in detail.

In a spectrometer, excitation (or transmission) and receiving processes are generally realized in the same coil.

In an MRI scanner, the excitation of the spin magnetization and the capture of the resulting tiny signal are frequently realized by two, or more, different coils. The transmitting coil is generally a large whole body coil. It is rarely used for receiving as it is inefficient in terms of sensitivity for small voxels (high resolution imaging). In contrast, specialized receive-only coils are designed for examination of specific organs of interest. Sometimes, a small specific coil is designed to be used as a transmit/receive probe. The small coil ascertains a good compromise between Signal-to-Noise Ratio (SNR) and spatial resolution, but is limited regarding volume exploration.

On a small animal MRI scanner different situations can also be encountered. Small whole body (mouse or rat) coils can be used efficiently as a transmit/receive probe. If a sufficient place is available in the bore magnet, a receive-only surface coil can be inserted in a larger transmit coil with benefits towards the SNR.

On a "microscope MRI" very small transmit/receive microcoils are easy to implement in small available bore magnets.[1] These transmit/receive microcoils have an intrinsic high SNR.

The maximum RF power delivered by the transmitter is below 1 kW in a liquid NMR spectrometer and well above this value in a solid state NMR machine and in human scanners (up to tens of kW). Such high powers should be manipulated with some care, not only regarding the RF components and Radio Frequency Interference (RFI), but also regarding the patient in a scanner or the sample in a spectrometer. Voltages up to several kV should be supported by the probe capacitors or other components. Cables not well terminated in their characteristic impedance will be subjected to standing waves and to common mode currents. These effects may be exacerbated if poor quality cables are used. Part of the applied energy will be radiated in the surroundings, creating RFI on the NMR console and on neighboring sensitive instruments. Common mode currents may create burn injuries if cables are close to patients.

High electric fields created in the sample (or the patient) by a poorly designed probe will result in high power dissipation (dielectric losses), heating of the sample, and lead to possible damage.

The preamplifier, and how it connects to the probe, is also a critical part of an NMR machine (either a spectrometer or an MRI scanner). It should amplify the weak NMR signal to the level required by the digital circuits (ADC), while adding a minimum of noise. The quality of the cable connecting the probe to the preamplifier is of paramount importance and care should be taken regarding the impedances that appear at both ends of the cable. The preamplifier should be a Low Noise Amplifier (LNA) having the best possible Noise Figure (*NF*). Nowadays, LNAs with ultra low *NF* (<0.5 dB) are commercially available at most NMR frequencies.

---

[1] High resolution MRI is best performed at very high field (above 9.4 T). Generally, the bore of a very high field magnet is small, from about 50 mm (narrow bore) to 150 mm (extremely wide bore).

Finally, people working on a commercial medical scanner frequently encounter some difficulties regarding the connection of the probe to the scanner. The probe should have some identification characteristics, such as a specific hardware (the simplest one being a resistor), and the connections to the console are done through proprietary connectors. A strong collaboration with the manufacturer is required in this case. These issues are not encountered with (relatively old) spectrometers and generally not on instruments installed in a (bio)physics laboratory.

2.2 Characterization of RF Components

2.2.1 *Frequency range*

The RF domain is defined somewhat depending on the topics of interest. It is tempting to assign the frequencies used in radio communications as RF. However, nowadays these span an extremely wide range of frequencies, which makes such a definition useless.

The standard communication bands (broadcasting, public services, military, television, phones, wifi) cover almost the entire spectrum up to around 2.5 GHz. As a result, RF measuring instruments are readily available for frequencies up to 3 GHz. This will conveniently cover the frequency spectrum for NMR, for which the 1 GHz frequency has been reached in recent years.

Fig. 2.2 shows the resonant frequencies for all the stable nuclei that may be observed by NMR in a static magnetic field of 2.35 T (100 MHz for the proton). Most of the nuclei resonate at frequencies above 1 MHz, except at much lower magnetic fields.[2]

At frequencies above 1 MHz, the resistive component of wire conductors[3] becomes negligible compared to its reactive component.

---

[2] When NMR experiments are done in the earth's magnetic field [Mohorič and Stepišnik, 2009], the resonant frequency for the proton becomes as low as 2 kHz.

[3] These are the most important components of an NMR probe.

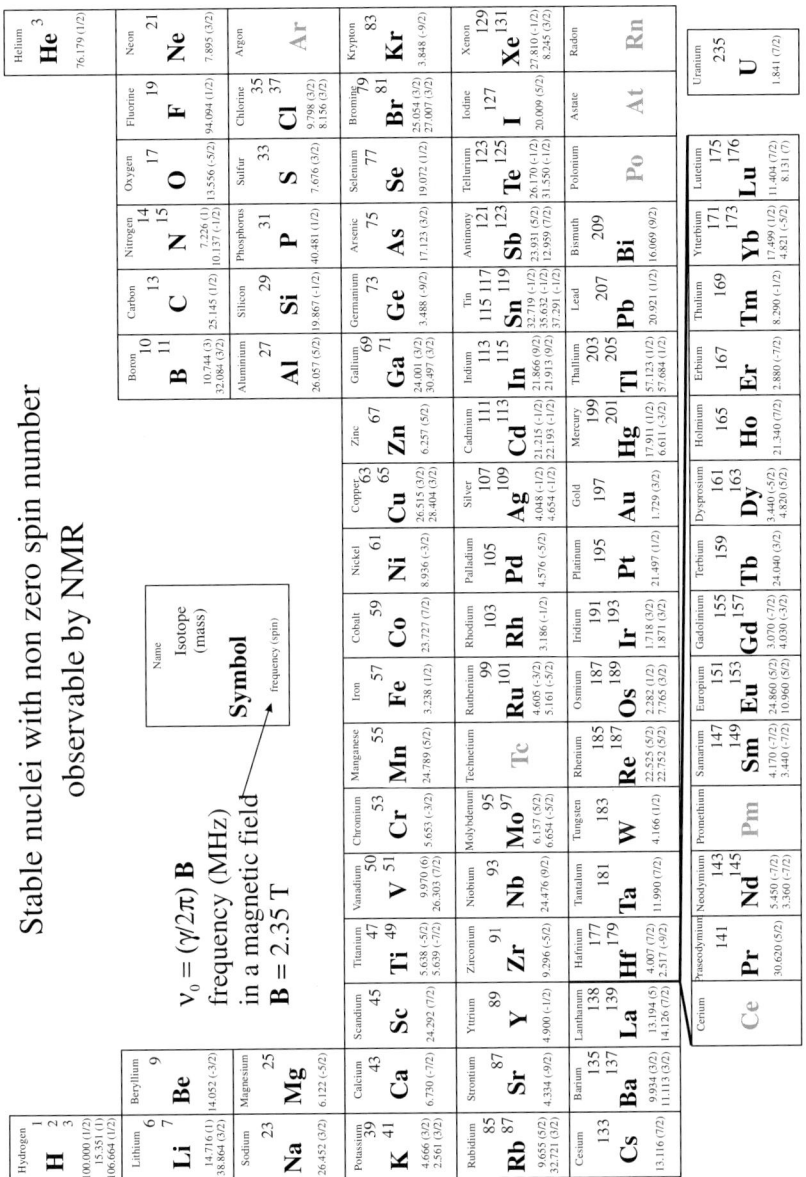

Fig. 2.2 Resonant frequency of all stable NMR nuclei in a static field of 2.35 tesla, corresponding to a proton resonance frequency of 100 MHz.

The result is that some estimations, such as the current distribution on the surface of a conductor or proximity effects, become frequency independent.[4]

Nuclear Quadrupolar Resonance (NQR) and low field Electron Spin Resonance (ESR) span a range of frequencies from a few MHz to GHz. These Magnetic Resonance (MR) methods basically use the same type of detection probes as NMR.

Hence, we will define the RF range extending roughly from 1 to 1000 MHz, covering almost all the applications of NMR, NQR and low field ESR. In this frequency range, the electrical properties of NMR components could be described easily using a single formalism.

2.2.2 Complex impedance and admittance

The impedance Z of a two-terminal network (Fig. 2.3) is defined by Ohm's law[5] as:

$$Z = \frac{V}{I} \in \mathbb{C}, \tag{2.1}$$

where I is the current flowing through the device and V the voltage across the terminals.

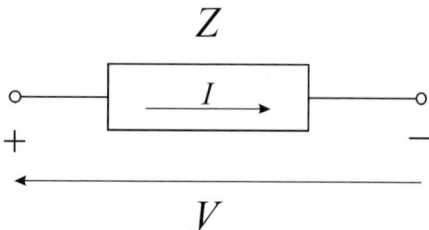

Fig. 2.3 Sign conventions for the current and voltage in a two-terminal network.

---

[4] The interaction with the electric field will somewhat change this simplified view, especially at high frequency (above 200–300 MHz).

[5] Ohm's law applies originally to pure resistances. Eq. (2.1) is a generalization of Ohm's law to the case of complex impedances.

By convention, the current flows from the positive to the negative terminal (Fig. 2.3).

Z is a complex number (Fig. 2.4):

$$Z = r + jX .\qquad(2.2)$$

where r and X are, respectively, the resistance and the reactance.

The power dissipated (active power) in a two-terminal device of impedance Z is the real part of the total power. The stored power in the device is the imaginary part of the power.

$$\begin{aligned}P_{active} &= \text{Re}\{VI^*\}\\ P_{stored} &= \text{Im}\{VI^*\}\end{aligned},\qquad(2.3)$$

where I^* is the complex conjugate of the complex valued current. The imaginary component of VI^* represents the stored energy in the device.

If the impedance is a pure resistance r ($X = 0$), the current and voltage are in phase. The dissipated energy is equal to:

$$P = rI^2 .\qquad(2.4)$$

If the impedance is purely reactive ($r = 0$), the current and voltage are out-of-phase by $\pm 90°$. The dissipated energy is zero.

Any network in the real world irreversibly dissipates the energy in its resistance r as heat (the Joule effect).

The ratio of the stored energy to the energy loss per cycle is given by the Q factor (quality factor), defined as the ratio of X to r:

$$Q = \frac{|X|}{r} .\qquad(2.5)$$

The impedance in Eq. (2.2) is represented by a series combination of real and imaginary components. In certain circumstances, it is easier to represent the device as the parallel combination of its real and imaginary components. In these cases, it is more convenient to work with the admittance defined as the inverse of the impedance. The admittance Y is given by the addition of a conductance (G) and a susceptance (B), in a similar manner as the impedance is the series combination of a resistance and a reactance (Fig. 2.4):

$$Y = G + jB.\qquad(2.6)$$

The quality factor is defined similarly as:

$$Q = \frac{|B|}{G}.\qquad(2.7)$$

Two admittances connected in parallel add together as two impedances in series.

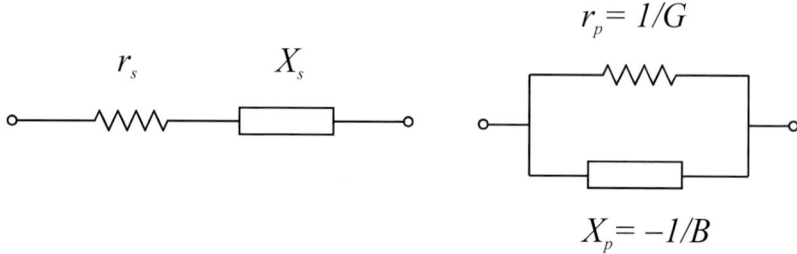

Fig. 2.4 Series and parallel representations of a two-terminal network.

2.2.3 *Impedance measurements and reflection coefficient*

An obvious method to measure the impedance of a two-terminal circuit is a direct application of Eq. (2.1). Direct reading instruments [Gorss, 1967; Alonzo et al., 1967] provide both the amplitude and phase of complex impedances. A known voltage is applied to an unknown impedance and a current transformer converts the current that flows in the circuit into a voltage (Fig. 2.5). Both voltages are measured by synchronous detectors, providing the amplitude and phase. These instruments, based on the so-called I–V method, were initially limited to 100 MHz. Lately they have been improved and are still in use for measurements up to 3 GHz [Agilent application note 5950-3000, Impedance measurement handbook, 4th edition].

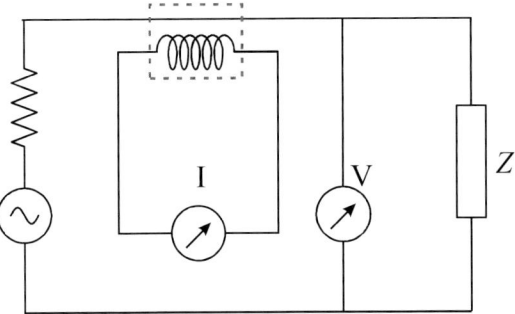

Fig. 2.5 Impedance measurement setup by the I–V method.

Indirect methods to measure impedances subjected to alternating current (AC) have been developed by RF engineers since the beginning of the last century. These methods are based on comparison techniques. They ultimately provide the so-called reflection coefficient (Γ), which can be defined as:

$$\Gamma = \frac{Z - Z_0}{Z + Z_0}, \qquad (2.8)$$

where Z_0 is a known calibrated impedance to which the unknown Z is compared.

In some comparison methods, the reflection coefficient is obtained from the imbalance voltage of a bridge, similar to the well-known Wheatstone bridge (Fig. 2.6). The unknown impedance forms one branch, another one being the reference impedance Z_0. The other branches include variable and calibrated resistances and capacitors[6] [Terman 1943, p. 905; Jordan, 1989, pp. 12–3 to 12–10] which are adjusted to establish a balance condition (a "null", $\Gamma = 0$). The unknown

---

[6] Good variable air capacitors can be easily constructed. Variable inductances are more difficult to build. However, the phase properties of balanced bridges permit a null condition to be obtained using only capacitors and resistors, even if the impedance to be measured has an inductive component.

impedance components (r and X) are directly read on a dial attached to the calibrated components. These impedance bridges are however limited to frequencies lower than 100 MHz due to the difficulty of obtaining variable components having reliable resistive and reactive characteristics over a very broad frequency range.

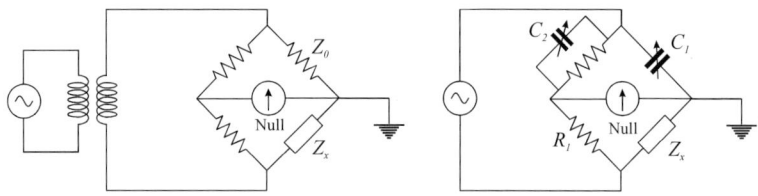

Fig. 2.6 The classical Wheatstone bridge (left) and its modification (right) for complex impedance measurements.

Fixed resistances can be made broadband up to several GHz, but a purely resistive bridge cannot fulfill the null condition, except at DC. Hence, instead of searching a null condition, the ratio of the unbalance voltage to the reference voltage sourcing the bridge is measured with an RF voltmeter. The unknown impedance is deduced from this ratio which is proportional to Γ (Chapter 10). Such resistive bridges (also known as directional bridges) have been used in the past in conjunction with vector voltmeters up to 1 GHz. With modern detectors,[7] directional bridges have been successfully used up to 27 GHz [Dunsmore, 2012]. These can also operate at very low frequencies.

Another comparison technique uses the characteristic impedance Z_0 of a low-loss transmission line, which is purely resistive and which can be determined accurately from its geometric dimensions.

When such a line is terminated by an unknown impedance Z, some part of the power is reflected back, unless Z is equal to Z_0. Standing waves appear on the line (Fig. 2.7).

---

[7] Microwave receivers.

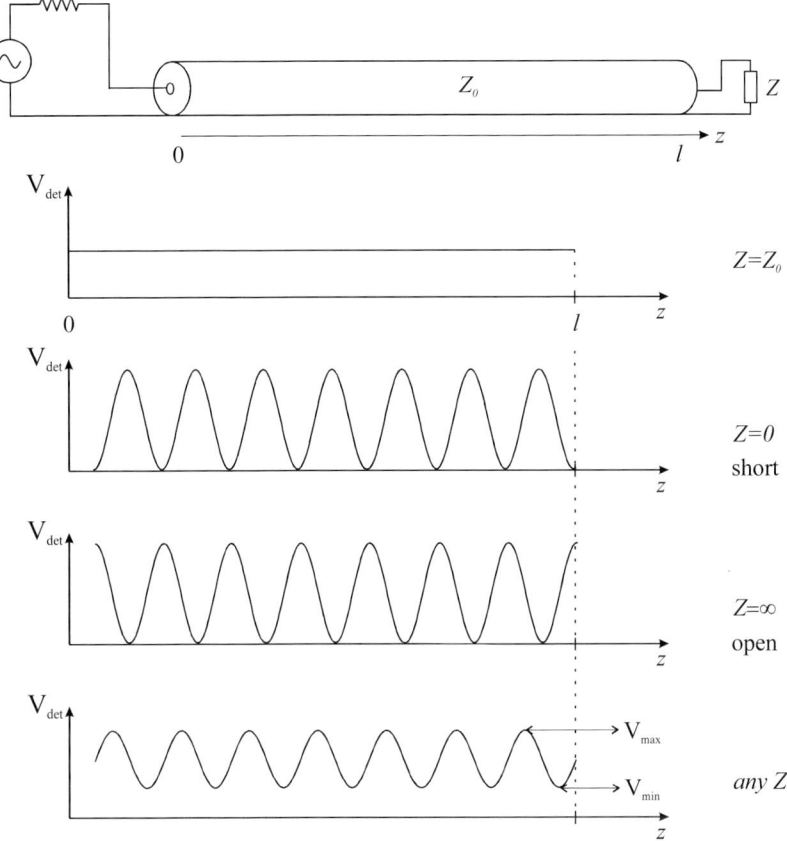

Fig. 2.7 Standing wave pattern on a line terminated by its characteristic impedance (Z_0), a short, an open (infinite impedance) or any impedance Z.

The voltage (and current) along the line is no longer a constant. Its amplitude varies from V_{min} to V_{max}, which defines the so-called Voltage Standing Wave Ratio (VSWR):

$$VSWR = \frac{|V_{max}|}{|V_{min}|}. \tag{2.9}$$

The VSWR is related to the reflection coefficient by:

$$VSWR = \frac{1+|\Gamma|}{1-|\Gamma|}, \qquad (2.10)$$

from which one can obtain the magnitude of the impedance Z.

The complete characterization of the unknown impedance also requires the determination of the phase (φ) of Γ:

$$\Gamma = |\Gamma|e^{j\varphi}. \qquad (2.11)$$

Until the late 1960s,[8] the standing wave pattern was characterized using a slotted line (Fig. 2.8). In this cheap and pedagogical method [Lee, 2004, pp. 246–254] a detector is moved along the line in order to record the amplitudes and positions of the electric field minima and maxima.

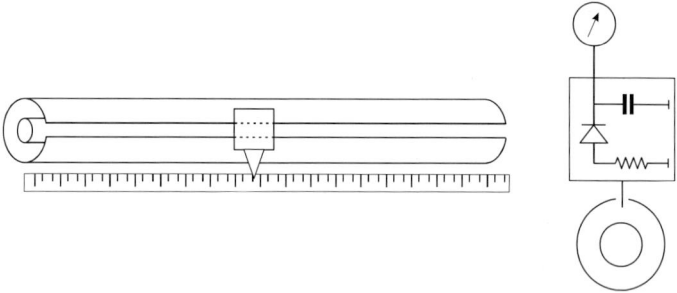

Fig. 2.8 The slotted line. An electric field probe is inserted in the slit and connected to a diode detector (right). The probe is slid along the line to sample the voltage standing wave pattern point by point.

The VSWR is calculated from the voltage amplitudes read on the detector while φ is deduced from the position of the minima (or of the maxima) along the line.

---

[8] When the vector network analyzer (VNA) HP8410A was introduced in 1967 [Anderson and Dennison, 1967] by the Hewlett Packard Company (now Agilent), the catchword was "Stamp out slotted lines".

It can be demonstrated that the position z_{min} of a minimum voltage respective to a given reference plane (z_0) is related to the phase angle φ of Γ (representing the impedance at the reference plane) by:

$$\varphi = (2k+1)\pi - \frac{4\pi(z_{min} - z_0)}{\lambda}. \tag{2.12}$$

The positions of maximum voltage (z_{max}) are given by:

$$\varphi = 2k\pi - \frac{4\pi(z_{max} - z_0)}{\lambda}. \tag{2.13}$$

In Eqs. (2.12) and (2.13), λ is the wavelength in the line and k is a positive integer. The minima and maxima are separated by a quarter of wavelength ($\lambda/4$) and thus two consecutive minima (or maxima) are separated by half a wavelength ($\lambda/2$).

The measured impedance depends on the position (z_0) of the reference plane. It should obviously be chosen at the end of the line where the unknown impedance is connected. If the reference plane is chosen at any other position, the calculated impedance depends on the line length comprised between the load and the reference plane. This demonstrates the impedance transformation properties of transmission lines that will be addressed in a following paragraph.

The reference plane position can be accurately determined using standard impedances. The simplest one, and generally sufficient for slotted line measurements,[9] is the short ($Z = 0$). It is connected to the line in replacement of the unknown impedance, imposing a minimum voltage. The position of the subsequent $k = 1$ and $k = 2$ minima permit calibration of the wavelength and the position of the reference plane. Such a procedure requires a line of at least one wavelength. Hence, the slotted line can be reasonably used for frequencies above 1 GHz.[10]

The introduction of the Vector Network Analyzer (VNA) by the Hewlett Packard Company (now Agilent) profoundly changed the life of

---

[9] More elaborate calibration methods have been developed since the introduction of the vector network analyzers (VNAs). These will be described in Chapter 10.

[10] At this frequency λ is equal to 30 cm.

RF engineers and, despite their high costs, these instruments became very popular. A VNA directly reads the forward and reflected power at a given port of a network and, accordingly, is able to directly provide the complex impedance after some computations done in its embedded computer. These instruments cover the RF frequency range (typically 0.1–3000 MHz) and the microwave range (typically above 2 GHz, up to some hundreds of GHz) equally well. Most of the work regarding NMR probehead development is done under the GHz range and, although many measurements can be done without a VNA, this is a must have. Even if a VNA is not available in every lab, it is useful and formative to learn its language.

2.2.4 S-parameters

Linear networks are fully characterized by a set of parameters (Z, Y, S and others, see Chapter 3). These parameters, determined from measurements at the network ports, fully characterize its behavior in any environment without regard to the specific content of the network. This has been taken into consideration, for example, for simulation of an NMR probe [Lemdiasov *et al.*, 2011; Lemdiasov and Ludwig, 2012]. In this approach, the S-parameters of the network of conductors are determined. Any external component (capacitors, sources, etc...) can be connected and the resulting relevant properties of the probe can be calculated using a linear network simulator.

Among the various parameters defined at the network terminals, or ports, the S-parameters are probably the most useful and the most efficient. Every RF and microwave measurement instrument is now able to provide these standard parameters.

The S-parameters are defined from the backward and forward wave amplitudes and phases at each port of a network. This implies that some characteristic impedance of the measurement system should be defined.

The characteristic impedance of most RF and microwave systems is standard and equal to 50 Ohms.[11] This is commonly the impedance of the coaxial lines that transmit energy between the different modules of an RF system. Let this characteristic impedance be named Z_0 in the following:

$$Z_0 = 50\Omega. \quad (2.14)$$

The *S*-parameters can be defined for networks having any number of ports, but these are more easily explained considering a two-port network (Fig. 2.9).

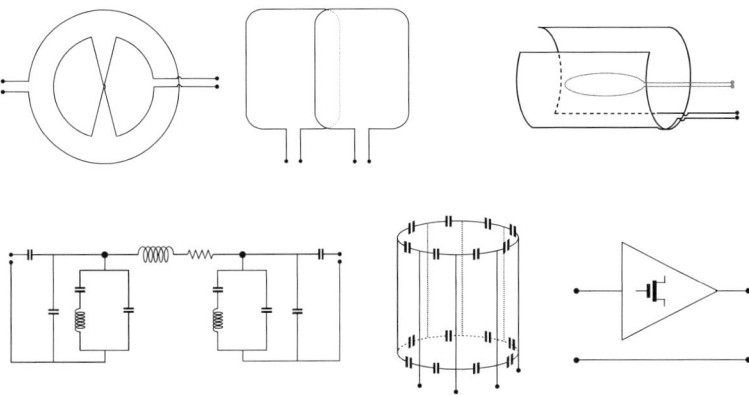

Fig. 2.9 Typical two-port network. From left to right and up to down: a quadrature surface coil, a two-element array coil, a transmit/receive coil configuration, a double tuned probe, a birdcage coil, a preamplifier.

A two-port network may be any "quadrature" probe, a two coil array, a transmit/receive probe system, a double resonant NMR probe, a circularly polarized birdcage coil, a preamplifier and so on. All of these devices will be considered in the rest of the book.

---

[11] TV systems use an impedance of 75 Ω instead. The choice of 50 Ω is a compromise between lower loss in a transmission line and higher power handling capability. 75 Ω is the optimum for lowest loss in the line [Lee, T. H., 2004, p.71].

If an excitation is applied to one port (while the other port is terminated in the characteristic impedance Z_0), part of the energy enters the network (forward) and another part is reflected back (backward), as already noted while measuring the impedance. Let a_1 be the voltage amplitude of the forward wave and b_1 the voltage amplitude of the reflected wave, respectively (Fig. 2.10).

One immediately sees that the ratio of b_1 to a_1 is the reflection coefficient already related to the VSWR [Eq. (2.10)]. The complex valued reflection coefficient is obtained if one measures not only the amplitudes, but also the relative phase between the forward and backward waves.

Then,[12]

$$\Gamma_{at\,port1} = \frac{b_1}{a_1}\bigg|_{a_2=0} = S_{11}$$

$$\Gamma_{at\,port2} = \frac{b_2}{a_2}\bigg|_{a_1=0} = S_{22}$$

(2.15)

A similar measurement done at port 2 (when port 1 is loaded by Z_0) gives S_{22}, Eq. (2.15).

The complete characterization of the two-port network requires two other measurements, the forward S_{21} and the reverse S_{12} transmission gains:

$$S_{21} = \frac{b_2}{a_1}\bigg|_{a_2=0}$$

$$S_{12} = \frac{b_1}{a_2}\bigg|_{a_1=0}$$

(2.16)

---

[12] When the output port of the network (port 2) is terminated in Z_0, the reflected wave from that load is zero. This is also the forward wave that enters port 2, hence $a_2 = 0$. Similarly, when port 1 is terminated in Z_0, $a_1 = 0$.

For passive components such as two-port NMR probes, S_{21} is equal to S_{12} and represents the isolation between the two ports. For an active device such as a preamplifier, S_{21} and S_{12} are certainly not equal. S_{21} is the gain of the amplifier while S_{12} is the isolation between the output and input ports.

The network behavior is now described in a matrix form as (refer to Fig. 2.10):

$$\begin{pmatrix} b_1 \\ b_2 \end{pmatrix} = \begin{bmatrix} S_{11} & S_{12} \\ S_{21} & S_{22} \end{bmatrix} \begin{pmatrix} a_1 \\ a_2 \end{pmatrix}. \quad (2.17)$$

The S-parameters are complex dimensionless values. The magnitudes can be expressed on a linear scale, or more frequently on a logarithmic scale as a power ratio in dB:

$$\left|S_{ij}\right|(dB) = 20 * \log_{10}\left(\left|S_{ij}\right|\right). \quad (2.18)$$

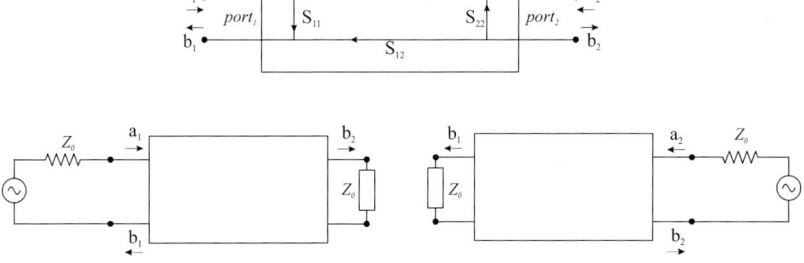

Fig. 2.10 Definition of the S-parameters (upper). Measurement of the S-parameters related to port 1 (S_{11} and S_{21}) is done while port 2 is terminated by Z_0 (bottom left). Similarly, the S-parameters related to port 2 (S_{22} and S_{12}) are measured with port 1 terminated by Z_0 (bottom right).

The magnitude of S-parameters can be directly measured with a Scalar Network Analyzer (SNA) giving access to the isolation, gain or mismatch in an easy (and cheap) way. But, to fully characterize the input or output impedance of a two-port network, the phase of S-parameters, as provided by a VNA, is required.

The input impedance of the network (the impedance seen by a given source) is obtained from S_{11} as:

$$Z_{in} = Z_0 \frac{1+S_{11}}{1-S_{11}} \qquad (2.19)$$

and, similarly, the output impedance (the source impedance for any circuit connected to the output port) is given by:

$$Z_{out} = Z_0 \frac{1+S_{22}}{1-S_{22}}. \qquad (2.20)$$

The conversion of the diagonal elements of the S matrix (or reflection coefficients) into impedances requires calculations in the complex plane. In the first days of microwave engineering, the engineers did not have access to the computers we see nowadays; rather they were used to the slide rule or other similar equipment. To simplify the above calculation using complex numbers, Philip H. Smith invented, and improved over a decade [Lee, 2004, Chapter 3, pp. 60 ff], a genius chart (the Smith chart) that is still in use today, despite the availability of powerful desk computers. Some reasons are, among others, that it is easy (eventually funny) to use it, it pictures quickly and accurately the behavior of any RF or microwave network, it naturally provides the transformed impedance through a given transmission line, it is an invaluable way to learn about RF and microwave circuits and, above all, it is an efficient tool for designing a matching network.

2.2.5 The Smith Chart

Many textbooks have chapters introducing and explaining in detail the Smith Chart. Here we will briefly describe, without demonstration, some of its properties that can be useful when designing NMR probes. Also the Smith chart may be used whenever possible to illustrate what the lucky designer can see on their own VNA display.

RF Components

The S-parameters can be displayed on a VNA in a variety of ways, either as a function[13] of frequency (Bode plot), or on a rectangular coordinate (imaginary *versus* real components, linear scale). In the latter case, an overlay is superimposed on the rectangular coordinate display as a polar plot or as the Smith Chart (Fig. 2.11).

The polar plot displays the magnitude and phase of S-parameters at a given frequency. The Smith Chart displays the resistive and reactive components of the normalized[14] impedance (or admittance) corresponding to the measured S-parameters [Eq. (2.19) or (2.20)].

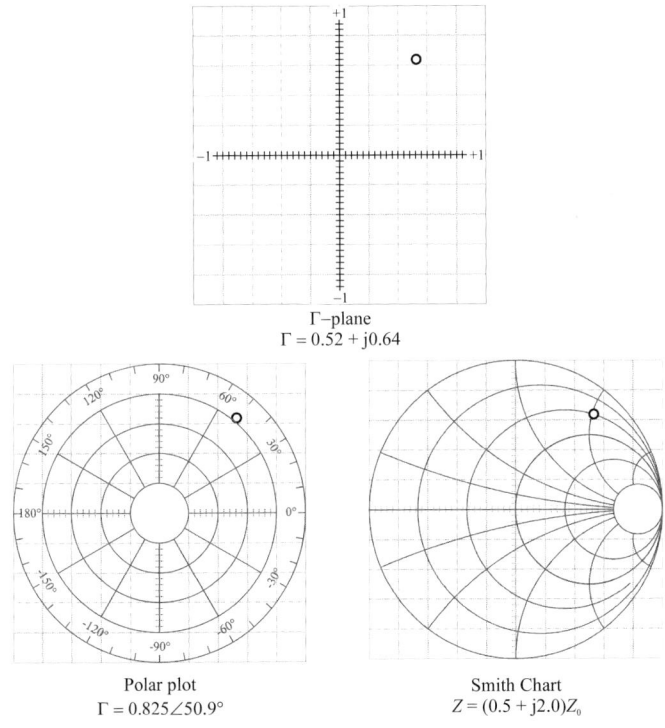

Fig. 2.11 Some ways to display the components of the complex reflection coefficient. The impedance can be directly obtained with a VNA using a Smith Chart overlay.

---

[13] The magnitude is plotted on a linear scale (between 0 and 1) or as a power ratio log scale ($-\infty$ and 0 dB).

[14] Respective to Z_0.

The phase reference is established by the location of the "reference plane" which is determined in the simplest case by replacing the impedance to be measured by a short.[15] Eventually, a "port extension" can be inserted to change the position of the reference plane. This corresponds to a phase rotation around the center of the chart (Fig. 2.11) equal to:

$$\Delta\varphi = 2\theta = \frac{4\pi l}{\lambda}, \qquad (2.21)$$

l being the distance by which the reference plane is displaced and λ the wavelength in the port extension line.[16] This provides the impedance transformation properties of any transmission line, another important application of the Smith Chart.

The center of the Chart corresponds to $\Gamma = 0$ (Fig. 2.12). At this point the measured impedance is equal to Z_0 and the normalized impedance is equal to unity. The horizontal line corresponds to a pure resistive component ranging from a short ($0 \angle 180°$)[17] to an open ($\infty \angle 0°$). Above this line, the reactive component is positive (inductance, $|Z| \angle \varphi$, where $0° < \varphi < 180°$). Below, the reactive component is negative (capacitance, $|Z| \angle \varphi$, where $-180° < \varphi < 0°$).

The mapping of the normalized impedance (Z-plane) into the Γ-plane is a direct application of an equation similar to Eq. (2.19), ensuring that if Γ is known then the normalized impedance Z/Z_0 is uniquely determined. This mapping results in contours of constant resistance and reactance [Fig. 2.13(a)] and contours of constant conductance and susceptance [Fig. 2.13(b)]. From the Γ-point, one can immediately obtain the unknown impedance (or admittance). The transformation of that impedance by a transmission line or any network (a matching network for example) can also be easily deduced. Many of these useful properties of the Smith Chart, relevant to NMR probe design, will be presented when necessary.

---

[15] More accurate calibration procedures will be discussed in Chapter 10.

[16] The forward and the backward wave are each phase shifted by θ, hence the factor of 2.

[17] The phasor notation (magnitude $\angle$ phase) is used here.

RF Components

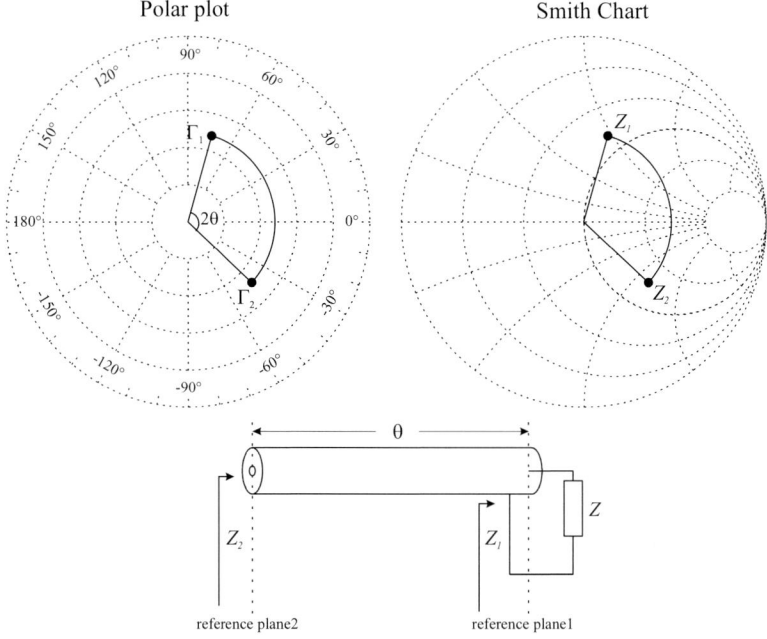

Fig. 2.12 Change of the reflection coefficient and, consequently, of the impedance when looking through a transmission line.

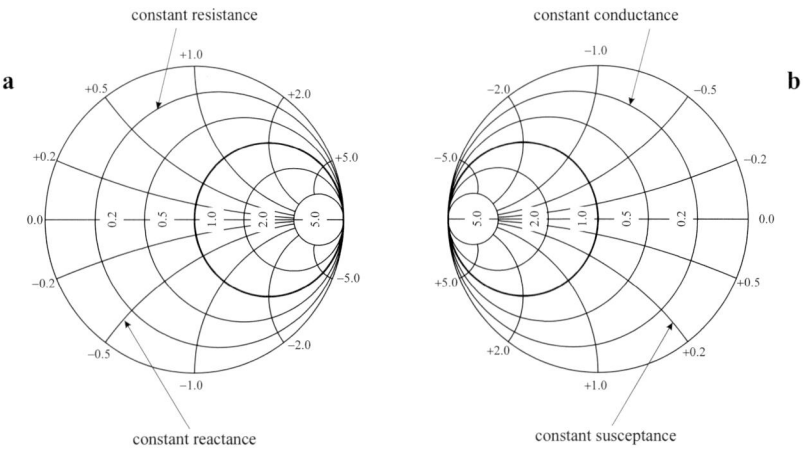

Fig. 2.13 Impedance and admittance mapping in the Γ-plane.

2.3 Electrical Properties of Materials

One of the crucial electrical characteristics of a material used when designing an NMR probe is related to the dissipation of energy as heat. Most of the time, this process leads to a degradation of the SNR.

The sources of energy loss or heating in probe components are essentially the conductor resistance and the dissipation of energy in the capacitor's dielectric or in any substrate supporting the circuit. These losses are represented in electrical circuits as a resistance.

High permeability materials (ferrite and powdered-iron substrate) are also present in RF and microwave devices. These magnetic materials are avoided in an NMR probe. Their properties will not be described in detail here, except occasionally when related to inductors.

The thermal losses in any sample interacting with the probe coil can also be represented by a resistance that adds to the resistance of the coil itself. These losses, already discussed in Chapter 1, result from eddy currents created in the sample by the oscillating magnetic field (B_1) and from the dissipation of energy due to the conservative electric field.

2.3.1 *Resistance of metal conductors*

A metal conductor is one of the main parts of an NMR probehead. Its resistance depends on the metal resistivity[18] (ρ), the conductor shape, the cross-sectional dimensions (area A) and on the metal magnetic susceptibility (μ). When driven by alternating current (AC) it also depends on the frequency (f).

The DC resistance (r_{DC}) of a piece of straight wire of length l and cross-sectional area A is given by:

$$r_{DC} = \frac{\rho l}{A}. \qquad (2.22)$$

---

[18] The resistivity ρ (ohms m) is the inverse of the conductivity σ (siemens m^{-1}).

Table 2.1 Resistivity (ρ) and conductivity (σ) of some materials at room temperature (295 K). The skin depth (δ) is calculated from Eq. (2.24).

Material	Resistivity (ρ) (ohms m)	Conductivity (σ) (siemens m^{-1})	Skin depth (μm) @ 1 MHz	1 GHz
Gold	2.45×10^{-8}	4.1×10^{7}	79	2.5
Copper	1.70×10^{-8}	5.9×10^{7}	66	2.1
Silver	1.60×10^{-8}	6.3×10^{7}	64	2.0

The AC resistance (r_{AC}) is larger than r_{DC} due to the "skin effect" and to "proximity" effects. Both effects result in a nonuniform current distribution in the conductor[19] cross-section. The corresponding decrease in the effective area (A_{eff}) increases the RF resistance compared to the DC resistance value. The resistance of a conductor of length l is:

$$r_{AC} = \frac{\rho l}{A_{eff}}. \qquad (2.23)$$

Due to the skin effect, the current density decreases exponentially from the surface of the conductor to the interior. The skin depth is defined as the depth where the current has decreased by 2.718 (e). This roughly represents the conductor thickness where the current can be assumed to be uniformly distributed. As indicated in Table 2.1, the thickness is very small at NMR frequencies (>40 MHz), in the range of 2–10 μm for copper conductors. The skin depth δ is given by:

$$\delta = \sqrt{\frac{2\rho}{\omega\mu}}, \qquad (2.24)$$

where ω is the angular frequency $2\pi f$ (rd/s), μ is the magnetic permeability of the material ($\mu = 4\pi \times 10^{-7}$ H m^{-1} for nonmagnetic materials) and ρ is the resistivity (ohms m).

For copper, the skin depth at room temperature (295 K) is:

---

[19] At RF, the current flows primarily on the surface of the conductor, hence the name "skin effect".

$$\delta = \frac{66}{\sqrt{f_{MHz}}} \, \mu m. \qquad (2.25)$$

At very low frequencies,[20] when the skin depth is greater than the wire radius, the resistance is independent of frequency and is equal to R_{DC}. At high frequencies, when the skin depth is smaller than the wire diameter, the resistance increases as the square root of frequency (Fig. 2.14, and Terman, 1943, p. 30).

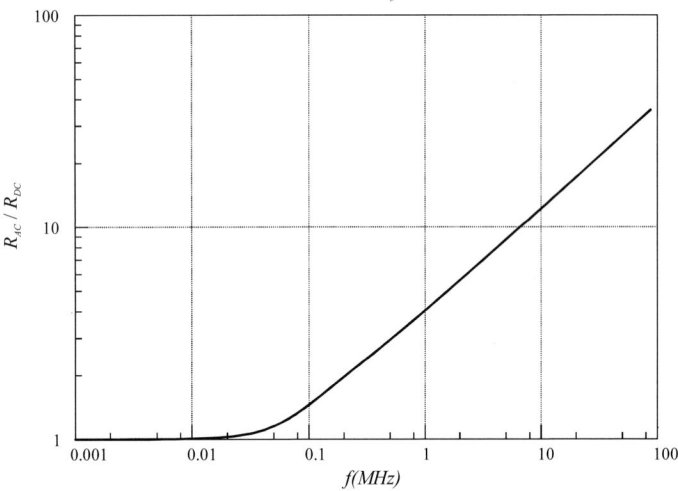

Fig. 2.14 AC to DC resistance ratio of a copper round wire as a function of frequency. The ratio values are calculated from Table 4 of the Radio Engineers Handbook [Terman, 1943, p. 31].

From the fit of the data plotted in Fig. 2.14, the AC resistance of a round isolated copper wire of diameter d (mm) is obtained as:

$$\frac{r_{AC}}{r_{DC}} = 3.78 d \sqrt{f_{MHz}} + 0.267. \qquad (2.26)$$

---

[20] Typically less than 10 kHz for a 1 mm copper isolated round wire, Fig. 2.14.

This equation is accurate better than 99% when $d\sqrt{f}$ is greater than 0.25 (d in mm and f in MHz). For a 1 mm wire diameter this corresponds to frequencies above 60 kHz.

In the RF domain, the constant term in Eq. (2.26) contributes less than 10% for frequencies around 1 MHz, and less than 3% for frequencies greater than 5 MHz, (again for a 1 mm wire diameter). This term can be neglected, so that, when the skin depth δ is much smaller than the wire radius, the resistance of a round copper wire can be approximated by [Terman, 1943, p. 35]:

$$r_{AC} = \frac{83.2\sqrt{f_{MHz}}}{d_{mm}} m\Omega/m. \qquad (2.27)$$

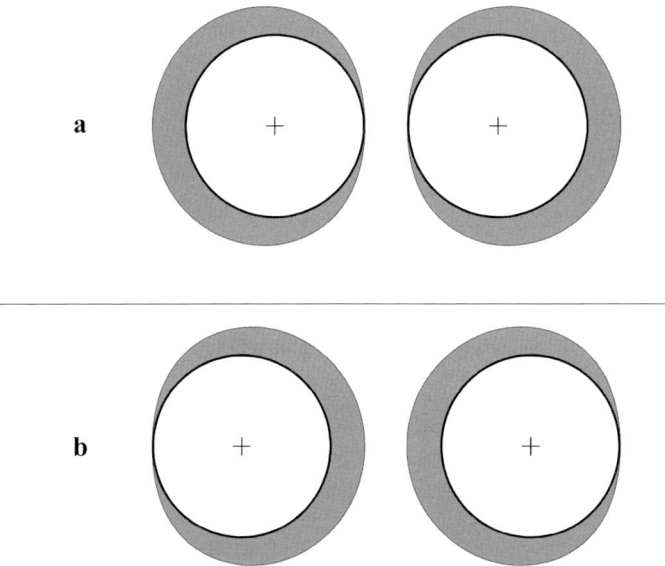

Fig. 2.15 Proximity effect generated by the currents flowing on two close round wire conductors. The current density is proportional to the grey area. (a) The currents in both conductors are flowing in the same direction. (b) The currents are flowing in opposite directions.

Eq. (2.27) is valid only for isolated conductors. If several conductors are close together the resistance increases due to the "proximity effect" (Fig. 2.15).

Currents of the same sign tend to repel each other. Similarly, currents of opposite sign tend to attract each other. Both interactions reduce the cross-section of the current flow, increasing the resistance. Contrary to the skin depth, this effect does not depend explicitly on the frequency[21] but on the ratio of the self and mutual inductances[22] of the proximate conductors.

The resistance for conductor shapes other than the round wire will be addressed specifically when required.

2.3.2 Losses in dielectric substrate

When a dielectric is immersed in an electric field, charges of opposed sign appear on the surface of the material (Fig. 2.16). This polarization mechanism comes from setting up and alignment of the electric dipoles in the medium through the influence of the applied electric field.

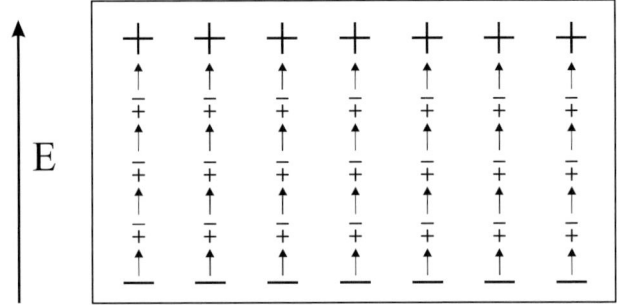

Fig. 2.16 Polarization mechanism of a dielectric material placed in an electric field E.

---

[21] This is true at frequencies above 1 MHz for typical wire conductors used in a probe, when the reactance of the conductor is much larger than its resistance.

[22] The self inductance depends slightly on the frequency due to the nonuniform current distribution (skin effect). However, the effect is only significant at very low frequency.

The total displacement electric field D in the dielectric material is given by:

$$\vec{D} = \varepsilon_0 \vec{E} + \vec{P} = \varepsilon \vec{E}. \tag{2.28}$$

The last equality in Eq. (2.28) assumes that the polarization is linearly related to the applied electric field. The proportionality constant is called the permittivity. It may be scalar (generally a complex number) if the medium is isotropic or a tensor. If the response of the dielectric to the electric field is not linear, ε is itself a function of E.

A dielectric material is commonly characterized by its relative permittivity (ε_r) and loss tangent (tan δ).

The complex permittivity can be written as:

$$\varepsilon = \varepsilon' - j\varepsilon'', \tag{2.29}$$

where ε' is the familiar lossless permittivity,

$$\varepsilon' = \varepsilon_0 \varepsilon_r, \tag{2.30}$$

ε_r is the relative permittivity referred as the dielectric "constant".

The imaginary component of the permittivity (ε'') originates from the time laps of the polarization establishment when an oscillating electric field is applied to the medium. This delay induces energy losses and heating of the material. The losses are characterized by the so-called loss tangent or phase angle.[23]

$$\tan \delta = \frac{\varepsilon''}{\varepsilon'} \tag{2.31}$$

ε' and tan δ depend on both the temperature and frequency in a wide range of ways depending on the prominent polarization mechanisms. Analogous to a friction mechanism, the response of the polarization to a rapidly varying electric field will be delayed, eventually canceled if the frequency is too high.

---

[23] For a low-loss dielectric material, tan $\delta \approx \delta$.

The upper limit can be as large as 10^{16} Hz for electronic polarization mechanisms and decreases to 10^{13} Hz for atomic polarization.

When the dominant mechanism is the orientation of pre-existing electric dipoles (orientational polarization[24]) the upper limit may cover a very wide range of frequencies, typically from 10^2 to 10^{10} Hz which are the characteristic frequencies of the Debye dielectric relaxation.

For materials used in electronic circuits or probe construction, tan δ tends to be frequency independent in the RF range. However, when approaching the characteristic frequency of dielectric relaxation, ε'' (and the losses) increases dramatically. This generally happens in the microwave domain.

In contrast, high loss dielectric materials (poly(methyl methacrylate) or Plexiglas®, some ceramic material of very high permittivity, some organic compounds and biological tissues) exhibit a frequency dependent tan δ, even in the RF range.

Table 2.2 shows some typical values of ε' and tan δ for dielectric materials used in NMR probe designs. These values are only indicative as they depend on many parameters. The frequency, the temperature, the orientation of the electric field with respect to crystal axes and the presence and nature of impurities in the material are parameters that are rarely specified in the data sources.

From the table, it can be seen that Plexiglas®, PVC and FR4 must be used with some caution. They could be convenient as a support for conductors but probably should be avoided as a dielectric in distributed capacitors or transmission lines.

An excellent review of dielectric properties of materials and their use in the context of MR appeared recently [Webb, 2011].

---

[24] In these cases, the Debye theory applies.

Table 2.2 Approximate dielectric "constant" and loss tangent for some materials encountered in probe designs.

Material	ε_r [1]	tan δ (10^{-4})	Usage
Air	1.0006		Transmission line standard
PTFE (Teflon®)	2.1	<2	Dielectric, TL [2], capacitor
Polyethylene	2.3	<5 [3]	Dielectric, TL [2]
Fused quartz	3.8	<2	Dielectric, capacitor
Sapphire	10	2	Dielectric, capacitor
FR4 (epoxy glass)	4.8	80 [4]	Dielectric, substrate
Plexiglas®	2.8	140 [5]	Substrate
PVC (polyvinyl chloride)	2.9	160 [5]	Substrate

Notes:
(1) Relative dielectric "constant", ε' can be obtained from Eq. (2.30).
(2) TL = Transmission Line.
(3) The cut-off frequency for this dielectric is quite low, around 10 GHz.
(4) At Ultra High Frequency (UHF).
(5) This is a very approximate value valid around 1 MHz.

2.4 Passive Linear Components at RF

2.4.1 *Conductors*

The most important part of an NMR probe is the conductive part which produces a near magnetic field in the sample space when driven by a given current. It can be made of different shapes, a single conductor of round or rectangular cross-section or eventually a multi-wire conductor.

Conductors are characterized primarily by their resistance and inductance. The impedance of an isolated conductor is:

$$Z = r + jL\omega. \tag{2.32}$$

The resistance r must be as small as possible in any case. It represents the intrinsic resistance of the wire and the additional losses due to inductive and capacitive couplings with the surrounding conductors and samples. The shape of the conductor may greatly influence this resistance. This will be considered in some detail in the following.

The reactive component of the impedance increases with the frequency. This has the consequence that, for a given current (a constant magnetic field amplitude), the near electric field (the voltage) increases, leading to a less efficient probe. Therefore, the inductance of the wire should be kept "reasonably" low in a good design.

Finally, the ratio of the length of the wire with respect to the wavelength should ideally be kept as small as possible ($\lambda/20$). Otherwise, one will be confronted with radiation losses and phase shift issues. One solution usually applied in NMR probehead design is the segmentation of the conductors by capacitor insertion.

2.4.1.1 Round wire

The resistance r of a round wire at RF (assuming that its diameter d is much greater than the skin depth δ) has been already addressed in paragraph 2.3.1. Eq. (2.27) is repeated here for completeness:

$$r = \frac{83.2\sqrt{f_{MHz}}}{d_{mm}} \; m\Omega/m. \qquad (2.33)$$

The resistance increases as the square root of frequency due to the skin effect. Note that Eq. (2.33) is only valid for an isolated wire. The presence of neighboring conductors may dramatically change this value (the proximity effect).

For obvious practical reasons, the diameter cannot be increased indefinitely in order to reduce the resistance. Typical diameter values used in many NMR probe constructions range from 0.6 mm to 2.5 mm. Lower wire diameters are used in "microcoils" while greater diameters are used for large body coils at low frequency. In those latter cases a plumbing tube is preferable than a solid wire, as it is lighter and easier to shape. The resistances of the tube and of the solid wire are identical at RF because the skin depth is much smaller than the thickness of the tube walls.

The inductance (nH) of a solid round wire, or of a tube, of diameter d and length l (mm) is given by [Terman, 1943, p. 17]:

$$L_{nH} = 0.2 l_{mm}\left[\ln(4l/d) - 1 + \mu\delta\right], \quad (2.34)$$

where μ is the permeability of the material and δ the skin depth. At low frequency, the last term $\mu\delta$ is 0.25 for nonmagnetic materials such as copper. At high frequency, this term can be dropped out [Grover, 2004, p. 269].

2.4.1.2 Ribbon conductor

The resistance r of an isolated conductor of any cross-section, can be written as [Terman, 1943, p. 35]:

$$r = K \frac{261\sqrt{f_{MHz}}}{P_{mm}} \; m\Omega/m. \quad (2.35)$$

P is the perimeter of the wire and K is a factor that depends on the uniformity of the current distribution around the cross-section of the wire.

The current on the cross-section of an isolated round wire is uniformly distributed [Fig. 2.17(a)]. In this case, $K = 1$ and one finds Eq. (2.33) quoted above.

In contrast, the current density on a wire of rectangular cross-section is not uniform [Fig. 2.17(b)], and K may increase to 2 or 3 depending on the ratio of the thickness (t) to the width (w) of the ribbon. It should be noted that the above expression with $K = 1$ and $P = w + t$ has sometimes been given in the literature, leading to an undervalued resistance.

From the graph of the function $K(w/t)$ shown by Terman [Terman, 1943, p. 36], it appears that K is proportional to $\log_{10}(w/t)$ when $w/t > 10$. Then, one may estimate the high frequency resistance of a ribbon as:

$$r \approx \frac{261\sqrt{f_{MHz}}}{2(w+t)_{mm}}\left(1 + 0.54\log_{10}\left(\frac{w}{t}\right)\right) \; m\Omega/m, \; w/t > 10. \quad (2.36)$$

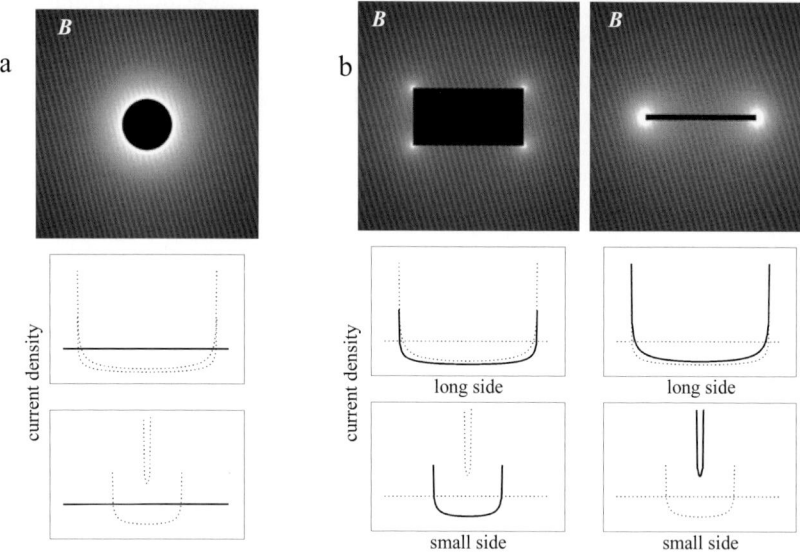

Fig. 2.17 Upper: magnetic field generated by conductors of different shapes when driven by the same total current. Lower: the current density is represented along the perimeter of the conductors.

Obviously this equation is only valid when the length of the ribbon is much smaller than the wavelength.

The inductance of a ribbon can be approximated[25] by [Grover, 2004, p. 35; Jin, 1999, p. 60]:

$$L_{nH} = 0.2 l_{mm} \left[\ln\left(2l/(w+t)\right) + 0.50 \right] . \qquad (2.37)$$

---

[25] The formula given by Terman (1943, p. 51) is, with the same units as Eq. (2.37): $L = 0.2l \left[\ln\left(2l/(w+t)\right) + 0.5 + 0.2235\left((w+t)/l\right) \right]$. The last term is usually negligible. More accurate and much more complicated formulae can be found, for example, in the sources of FastHenry [Kamon et al., 1994] or in the book by Niknejad [Niknejad, 2008, Eq. (6.46) on page 158].

The inductive properties of conductors and complex networks made up of round wires (of diameter d) or thin strip conductors (of width w) are comparable using the following relationship:

$$w = 2.241d \, . \tag{2.38}$$

This equation has been obtained by equating Eq. (2.34) and the Terman equation (see note 26) [Giovanetti et al., 2010].

Accurately calculating the inductance of a bar of rectangular cross-section for any frequency is a difficult task. After the pioneering works of Grover and others, the impedance estimation of ribbon conductors has a renewed interest due to its importance in the development of integrated circuits [Ruheli, 1972; Wu et al. 1992]. Recent papers [Matsuki and Matsushima, 2012; Piątek et al. 2012; Piatek and Baron, 2012] show that this topic is still relevant.

The calculation of the inductance of ribbons, and of a complex network of conductors, can finally be performed using a linear network analyzer such as FastHenry[26] [Kamon et al., 1994] or using integral methods [Jin, 1999]. These methods will be addressed in detail in Chapter 3.

When the sheet conductors are imbedded into striplines (i.e. including a dielectric substrate and a ground plane), accurate methods have been developed, most of which are well described in the book by Gupta et al. [Gupta et al. 1996].

2.4.1.3 Litz wire

Litz wires[27] were designed to increase the effective area of RF current flow compared to solid wires where the effective area is limited by the

---

[26] FastHenry is a so-called inductance extraction program. It performs a mesh analysis of the complex network of conductors after decomposition of each bar into several smaller interacting sections. The (partial) inductance of each bar is calculated from an elaborate formula due to Ruheli [Ruheli, 1972] and the mutual inductances between bars are estimated using the thin wire approximation and formulae given by Grover [Grover, 1946].

[27] Also termed Litzendraht conductors [Terman, 1943, p. 37].

skin depth effect. The litz wire is formed of a large number of strands of fine insulated wires, connected in parallel at the ends of the conductor so that the current is expected to be shared evenly.

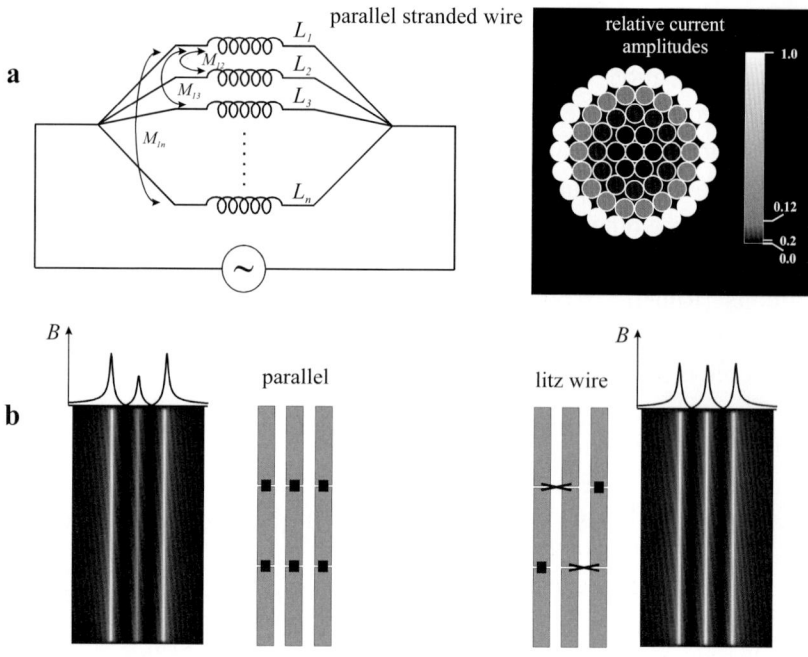

Fig. 2.18 (a) Model and current distribution in a multi-wire conductor. The current in each wire is calculated taking into account the magnetic interaction between all wires. If the wires are parallel, the current essentially flows in the external layer. (b) Demonstration of the litz wire effect. Right: the three conductors are straight and parallel. The current in the middle conductor is smaller, as shown by the corresponding magnetic field amplitude. Left: two crossovers have been inserted such that the current is evenly shared between the three conductors, as shown by the magnetic field amplitude around the conductors.

However, if simply done without special care, the litz wire will not function as expected. Assume all strands are straight and parallel, and the proximity effect tends to reject the current in strands at the periphery of the wire. This leads to an effective area comparable to the skin depth layer of a plain round conductor [Fig. 2.18(a)].

To be effective, the strands of the litz wire are woven, or transposed, in such a way that all strands intersect all points of the conductor cross-sectional area. In this way, the current is evenly distributed among each strand, increasing the effective area presented to the current [Fig. 2.18(b)]. The litz wire is very effective at frequencies below 0.5 MHz but it is generally known to lose its quality at frequencies greater than 2 MHz. The inter-strand capacitance is one of the parameters that limits the usable frequency range of litz wires.

At very low frequencies, the use of litz wires instead of solid copper wires of the same diameter is advantageous for NMR coil building [Croon *et al.*, 1999].

Grafendorfer *et al.* proved that a significant SNR increase is obtained up to 3.8 MHz and suggested that the litz cable could be optimized to extend its use at RF frequencies up to 10 MHz [Grafendorfer *et al.*, 2005 and 2006].

The concept of the litz wire is also the basis for the original design of the so-called "litz coil" invented and developed by Doty *et al.* and others [Doty *et al.*, 1999; Doty, 2000; Tang and Jelicks, 2002], extending its utility in the UHF range. The basic element of a litz coil is a "wire" that provides two parallel paths to the current (Fig. 2.19).

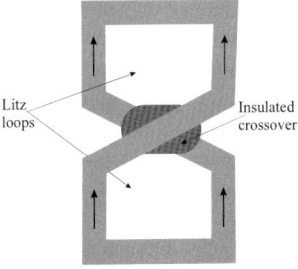

Fig. 2.19 The principle of the so-called litz coil elements.

An insulated crossover is inserted such that the current flow is evenly shared between the two paths (in other words, the inductances for the two paths are equal [Doty and Entzminger, 2012 p. 249]). In this way, they designed highly effective high frequency probes up to 600 MHz.

The well-known Doty Litzcage™ [Doty et al., 2007] is one member of these designs.

2.4.2 Cables and transmission lines

Cables or transmission lines are used to transport RF energy. These are constituted of two conductors arranged in asymmetric (coaxial, microstrip, stripline, coplanar waveguide), or balanced (linear or twisted bifilar wires) configurations (Fig. 2.20).

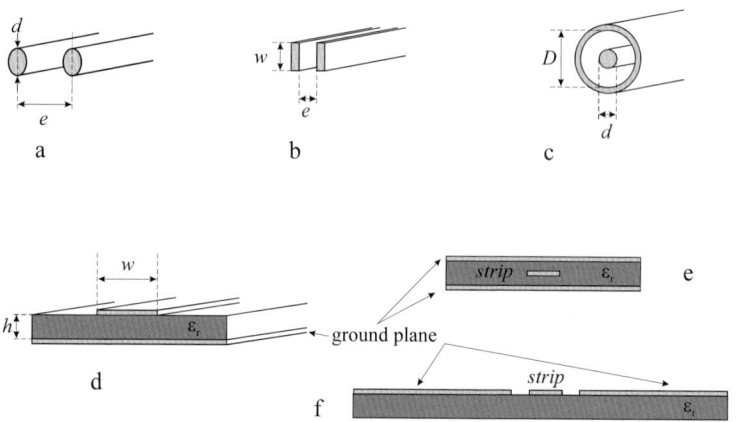

Fig. 2.20 Geometries of transmission lines. (a–c) Transmission lines using conductors separated by air or dielectric materials. (d–f) Stripline technology. It comprises a strip, a substrate and a ground plane.

Efficient transfer of energy is done when the transmission line functions in the so-called differential mode. In this mode, the currents in the two wires flow in opposite directions [Fig. 2.21(a)]. The electromagnetic field created outside the line is negligible.

However, an undesired mode may also be excited, the so-called common mode. In this mode, the currents in both conductors flow in the same direction [Fig. 2.21(b)]. Electromagnetic energy is radiated outside the cable, leading to losses, probe coil detuning and RF interference. It may also lead to patient burning in clinical MRI. This mode should obviously be avoided, particularly when concerned with energy transport. On the contrary, this mode may be used in the design of high frequency probe resonators (Chapter 9). In the following, only the differential mode will be addressed.

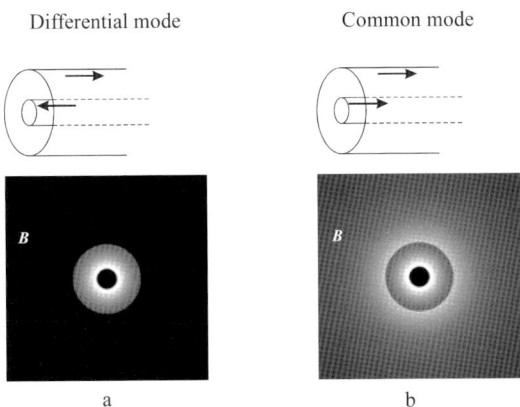

Fig. 2.21 Magnetic field generated around the conductors of a coaxial line functioning in the differential (a) or common mode (b). The current amplitudes in both cases are assumed to be equal.

2.4.2.1 *Wave propagation and characteristic impedance*

The electrical properties of a transmission line functioning in the differential mode are fully described by the distributed inductance and capacitance along the cable and their associated resistances, which represent the losses (Fig. 2.22).

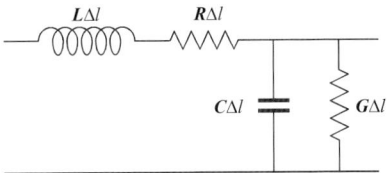

Fig. 2.22 Model for an elementary segment of a transmission line.

The inductance L per unit length of a transmission line is the inductance of the loop formed by the current flowing in one of the conductors (of unit length) and returning back *via* the other one. The capacitance C per unit length is the capacitance between the two conductors. It depends on the geometry of the conductors and on the dielectric material.

Due to the skin effect, L increases slightly at low frequency (typically <1 MHz). C is generally constant in the NMR frequency range, as far as the dielectric permittivity is itself frequency independent in this range.

The losses in the transmission lines are determined by the series resistance r of the conductors and by the conductance G of the dielectric medium (Fig. 2.22). r increases with the square root of frequency, as already noted [Eq. (2.26)]. The dependence of G on frequency may be more complicated, but if ε_r and tan δ are constant over the frequency range of interest, G is proportional to the frequency. This is usually the case for common coaxial cables operating in the NMR frequency range.

When the line is submitted to a given excitation, a wave propagates along its z axis (Fig. 2.23). If the line[28] is infinite the energy flows in the forward direction and, possibly, attenuates due to dissipation of energy in the line.

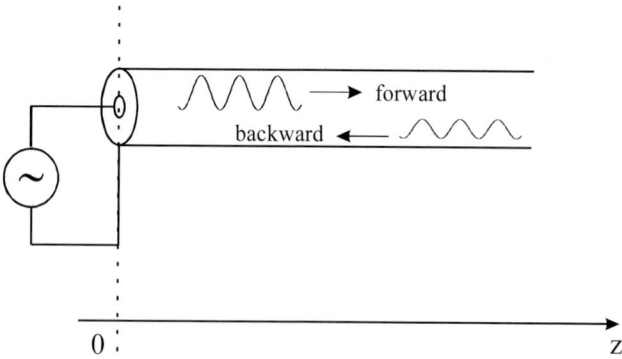

Fig. 2.23 Forward and backward waves propagating along a transmission line, excited by a source connected to one of its terminals.

---

[28] The line is also supposed to be uniform.

Any physical quantity[29] Λ associated with a propagating wave obeys the d'Alembert equation, which relates the space and time functional dependence of Λ. The one-dimensional propagation equation can be written as:

$$\frac{\partial^2 \Lambda}{\partial z^2} = \frac{1}{v_p^2} \frac{\partial^2 \Lambda}{\partial t^2}, \qquad (2.39)$$

where v_p is the propagation speed. Dissipative attenuation mechanisms, i.e. losses in the line, are accounted for with the assumption that $v_p \in \mathbb{C}$.

It is common in circuit theory to leave out the time dependence of Λ by considering a sinusoidal excitation and assuming that a steady state is reached. When the line is driven by a sinusoidal excitation, the steady state voltage $V(z)$ and current $I(z)$ are given by the solutions of the following wave equations [Pozar, 1998, p. 58]:

$$\frac{d^2 V(z)}{dz^2} - \gamma^2 V(z) = 0$$
$$\frac{d^2 I(z)}{dz^2} - \gamma^2 I(z) = 0 \qquad (2.40)$$

These equations can be straightforwardly deduced from the d'Alembert wave equation, defining the propagation constant γ as:[30]

$$\gamma^2 = \frac{-\omega^2}{v_p^2}, \qquad (2.41)$$

where ω is the pulsation frequency ($2\pi f$) of the sinusoidal excitation.

The propagation constant γ has the dimensions meter^{-1}. It is a complex number:

$$\gamma = \alpha + j\beta. \qquad (2.42)$$

---

[29] Such as voltage or current.

[30] The minus sign comes from the second derivative with respect to time of $\Lambda(z,t)$, assuming a sinusoidal time dependent solution.

Assuming an infinite line, the solution to Eq. (2.40) can be written as:

$$V^+(z) = V(0)e^{-\gamma z} = V(0)e^{-\alpha z}e^{-j\beta z}$$
$$I^+(z) = I(0)e^{-\gamma z} = I(0)e^{-\alpha z}e^{-j\beta z} \quad , \qquad (2.43)$$

where $V(0)$ and $I(0)$ are initial conditions[31] defined by the excitation. The + sign on V and I indicates that these are associated with the forward wave, originating at $z = 0$ and traveling in the $+z$ direction.

The physical meaning of α and β coefficients becomes evident from the above equations. α is the attenuation factor (the magnitude of A decreases exponentially with the distance z) and β is the propagation phase shift.

The propagation constant γ obviously depends on the distributed resistance R (Ω m^{-1}), inductance L (H m^{-1}), conductance G (Ω^{-1} m^{-1} or S m^{-1}) and capacitance C (F m^{-1}) of the transmission line. From the telegrapher's equations, which relate the current and voltage along the line, it can be demonstrated that [White, 2004, p. 95]:

$$\gamma = \sqrt{(R + j\omega L)(G + j\omega C)}. \qquad (2.44)$$

If the line is infinite and uniform, the propagation is in the forward direction. The ratio of voltage to current at any position of the line is then [Pozar, 1998, p. 58]:

$$\frac{V^+(z)}{I^+(z)} = \sqrt{\frac{R + j\omega L}{G + j\omega C}} = Z_c. \qquad (2.45)$$

The ratio has the dimensions of impedance and does not depend on the position along the line. It is the characteristic impedance Z_c[32] of the line.

---

[31] Generally V_0 and I_0 are complex numbers.

[32] The notation Z_c will be used whenever possible to denote the characteristic impedance of a given line. Z_0 will be used instead to name the characteristic impedance of the measuring system.

If the line is of finite length and terminated in a load of impedance different from Z_c, some energy is reflected back. The corresponding voltage $V^-(z)$ and current $I^-(z)$ of the backward wave obey Eq. (2.40). Then, similarly to Eq. (2.45):[33]

$$\frac{V^-(z)}{-I^-(z)} = \sqrt{\frac{R+j\omega L}{G+j\omega C}} = Z_c. \qquad (2.46)$$

The characteristic impedance Z_c is a complex number unless the terms R and G could be neglected in Eq. (2.45) or (2.46). In this case, the line is said to be low-loss.

2.4.2.2 Low-loss transmission lines

Lines that are used to transport RF energy are obviously designed to be low-loss, that is, the resistance R (of the conductors) and the conductance G (of the dielectric substrate) are much smaller than the reactance ($L\omega$) and the susceptance ($C\omega$), respectively:

$$\begin{aligned} R \ll L\omega \\ G \ll C\omega \end{aligned}. \qquad (2.47)$$

For standard 50 Ω coaxial cables, C is typically of the order of 100 pF m^{-1} and is a constant from very low frequency (1 kHz) up to the microwave range (>>1 GHz). Due to the skin effect, L decreases from low frequency to a plateau value reached at around 1 MHz. At this frequency and above, L is typically equal to 250 nH m^{-1}.

R is of the order of 1–5 Ω m^{-1} at 200 MHz (see below), depending mostly on the diameter of the cable. Because R is proportional to the square root of the frequency while $L\omega$ is proportional to f, the resistance and the reactance become of the same order of magnitude at around 100 kHz. The first condition of Eq. (2.47) is generally fulfilled above one or a few megahertz.

---

[33] The minus sign results from the fact that the current in the backward wave is opposed to the current in the forward wave.

G is less than $2.10^{-4}\,\Omega^{-1}\,m^{-1}$ at 200 MHz. Because G is proportional to the frequency, the second condition of Eq. (2.47) is easily fulfilled for all frequencies ($C\omega = 0.125\,\Omega^{-1}\,m^{-1}$ at 200 MHz).

Eqs. (2.44) and (2.45) can be rearranged as:

$$\gamma = j\omega\sqrt{LC}\sqrt{1 + R/j L\omega + G/jC\omega + RG/(j\omega)^2 LC}$$
$$Z_c = \sqrt{\frac{L}{C}}\sqrt{\frac{1 + R/jL\omega}{1 + G/jC\omega}} \qquad (2.48)$$

At frequencies higher than 1 MHz, when the conditions of Eq. (2.47) are fulfilled, Z_c can be approximated as:

$$Z_c = \sqrt{\frac{L}{C}}. \qquad (2.49)$$

The characteristic impedance of a low-loss line is real-valued and does not depend explicitly on frequency. This is generally true above 1 MHz and up to a few GHz. At lower frequencies, the characteristic impedance of a standard coaxial cable becomes complex and its magnitude increases from the nominal impedance 50 Ω to about 58–60 Ω at 1 kHz.

To the same degree of approximation, the propagation constant γ is [Pozar, 1998, p. 91]:

$$\gamma = \frac{1}{2}\left(R\sqrt{\frac{C}{L}} + G\sqrt{\frac{L}{C}}\right) + j\omega\sqrt{LC}. \qquad (2.50)$$

The speed of propagation or phase velocity (v_p) is, using Eq. (2.41) and the imaginary component of γ [Eq. (2.50)]:

$$v_p = \frac{1}{\sqrt{LC}}. \qquad (2.51)$$

It is independent of frequency above approximately 1 MHz. In this case, the line is said to be distortionless.[34]

The propagation speed is commonly characterized by the so-called velocity coefficient,

$$k = \frac{v_p}{c}, \qquad (2.52)$$

where c is the speed of light in a vacuum.

Finally, the velocity coefficient can be related to the relative permittivity of the dielectric medium in which the wave propagates. This is given by equating the Maxwell equations, which characterize the physical system, and the d'Alembert general propagation equations. It can be shown (for $\mu = \mu_0$) that:

$$k = \frac{1}{\sqrt{\varepsilon_r}}. \qquad (2.53)$$

2.4.2.3 Attenuation in low-loss lines terminated by Z_c

The real part of the propagation constant for a low-loss transmission line is, from Eq. (2.50):[35]

$$\alpha = \frac{1}{2}\left(\frac{R}{Z_c} + GZ_c\right). \qquad (2.54)$$

---

[34] At low frequencies (some hundreds of kilohertz and lower), the speed of propagation may depend on the frequency inducing some distortion in the transmitted signal. It can be shown, however, that if the line is designed such that $R/L = G/C$ (Heaviside conditions), the line becomes absolutely distortionless. Because G/C is generally smaller than R/L, Heaviside lines are constructed by inserting inductors at regular intervals, to increase L per unit length.

[35] The factor ½ in Eq. (2.54) results from Taylor series expansions of terms like $\sqrt{1+x}$ as $1 + x/2$, where x is $R/L\omega$ or $G/C\omega$.

The constant α represents the attenuation of the forward or backward wave which propagates along the line. It is expressed in *Neper* per unit length (Np m^{-1}). The ratio of the voltage $V(z = 1)$ at the end of a line of unit length to the excitation voltage $V(0)$ applied at the input is given, from Eq. (2.43), by:

$$\frac{V(z=1)}{V(0)} = e^{-\alpha}. \qquad (2.55)$$

Eq. (2.55) assumes that there is no reflection at the end of the line which is supposed to be terminated by its characteristic impedance. In this case, the attenuation of a low-loss transmission line is given by:

$$\alpha_{Neper/m} = -\ln\left|\frac{V(z=1m)}{V(0)}\right|. \qquad (2.56)$$

The attenuation is commonly expressed in dB. It represents the amount of power dissipated in the line. If the line is terminated with Z_c, and only in this case:

$$\begin{aligned}\alpha_{dB} &= 20 \cdot \log_{10}(e) \cdot \alpha_{Neper} \\ \alpha_{dB} &= 8.686 \cdot \alpha_{Neper}\end{aligned} \qquad (2.57)$$

This attenuation is sometimes designated as the "matched loss" *ML*. For a line of length *l*, it is written as:

$$ML_{dB} = 10\log_{10}\left(e^{2\alpha l}\right), \qquad (2.58)$$

where α is the attenuation constant (in NP m^{-1}) and *l* the length (in m) of the line.[36]

---

[36] The factor of two results from the fact that α is defined from the ratio of voltages [Eq. (2.56)] and *ML* is the ratio of power which is proportional to the square of the voltages ratio. Eq. (2.58) is consistent with Eq. (2.57).

The matched loss is generally less than 1 dB per meter at 1 GHz (Table 2.3) for currently used coaxial cables. It is mainly dependent on the cable diameter. The design of the shield and the quality of the dielectric material are also important parameters in this respect.

Table 2.3 Typical characteristics for some common coaxial cables having a nominal impedance of 50 Ω.

Standard reference [1]	Dielectric	Outer diameter (inches)	Inductance (nH m^{-1})	Capacitance (pF m^{-1})	Attenuation (dB m^{-1}) @ 1 MHz	1 GHz
RG316	PTFE [2]	0.1	237	95	0.028	0.95
RG58	PE [3]	0.2	252	101	0.014	0.64
RG223 [4]	PE [3]	0.2	252	101	0.013	0.49
RG214	PE [3]	0.34	252	101	0.006	0.26
RG405 [5]	PTFE [2]	0.086	235	94	0.022	0.74
RG402 [5]	PTFE [2]	0.141	235	94	0.012	0.41
RG401 [5]	PTFE [2]	0.25	235	94	0.007	0.25

Notes:
(1) The coaxial cables may have other references and their characteristics may slightly vary depending on the manufacturer.
(2) PTFE: Polyfluoroethylene or Teflon®, velocity 0.70.
(3) PE: Polyethylene, velocity 0.66.
(4) This cable has better performance than the similar RG58 due to double shielding.
(5) These microwave cables are the so-called "semi-rigid coax". The outer conductor is a thin copper tube. Only the RG401 is nonmagnetic. The smaller ones, in their standard versions, have a steel center conductor. These semi-rigid cables have some advantage with respect to mechanical (and consequently electrical) stability. Flexible, "formable", versions are also proposed by various manufacturers. They have similar characteristics (sometimes better in the very high microwave range) to the "old" reference.

The dielectrics of common coaxial cables (PE or PTFE) have very similar properties in the frequency range of interest (1–1000 MHz) with only a small advantage for PTFE. However, PTFE is much better than PE in the microwave range, especially above 10 GHz. PTFE is also more resistant than PE to the heat of the soldering iron. The permittivity of PTFE is slightly lower than that of PE (2.1 and 2.3, respectively). Hence, the capacitance per unit length is slightly lower, as indicated in Table 2.3. In practice, the capacitance varies with different manufacturers as it depends on the geometry and on the exact composition and physical nature of the dielectric.

Practically all the common (low-loss) coaxial cables exhibit a typical frequency dependence of the transmission parameter $S_{21}(f)$, as represented in Fig. 2.24 for the very common RG58 cable. The functional dependence fits the following equation:

$$-S_{21} = \alpha_{dB} = a\sqrt{f} + b \cdot f . \qquad (2.59)$$

The first term, which is proportional to the square root of frequency, represents the contribution of the conductors to the loss.

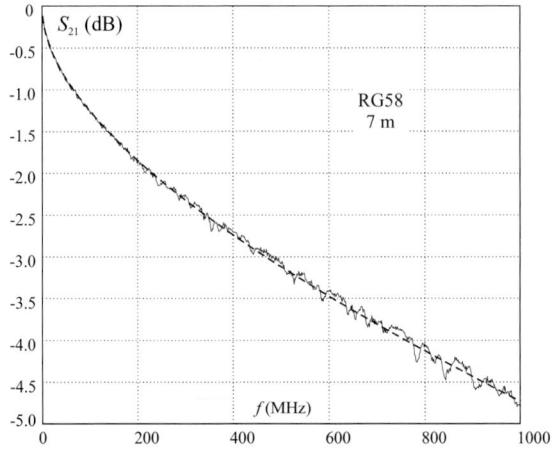

Fig. 2.24 Transmission measured for a 7m piece of RG58 coax as a function of frequency, fitted to Eq. (2.59).

This is usually the dominant contribution in the frequency range of interest (1 MHz to 1 GHz). As a result, the loss of a coaxial cable is mainly dependent on its diameter, as shown in Table 2.3. The larger the cable, the lower the attenuation is.

The second term in Eq. (2.59), proportional to the frequency, represents the contribution of the dielectric. It dominates at frequencies above the GHz.

From a and b, values given sometimes by the manufacturers or which can be deduced from measurement of the attenuation of a piece of cable,[37] one can calculate[38] the physical constants R and G at a given frequency.

Table 2.4 shows these parameters for some common 50 Ω coaxial cables. R and G are given at 200 MHz, but the values at any other frequency f can be obtained, remembering that R is proportional to the square root of f and G is proportional to f.

Table 2.4 Typical attenuation constants for a few common 50 Ω coaxial cables.

$$\alpha_{dB/m} = a\sqrt{f_{MHz}} + bf_{MHz}$$

Standard reference [1]	Dielectric	Outer diameter (inches)	a (dB m^{-1})	b (dB m^{-1})	R (Ω m^{-1}) @ 200 MHz	G (Ω^{-1} m^{-1}) @ 200 MHz
RG316	PTFE [2]	0.1	0.027	10×10^{-5}	4.4	9×10^{-5}
RG58	PE [3]	0.2	0.014	20×10^{-5}	2.3	18×10^{-5}
RG223 [4]	PE [3]	0.2	0.013	8×10^{-5}	2.0	7×10^{-5}
RG214	PE [3]	0.34	0.006	7×10^{-5}	1.0	6×10^{-5}
RG405 [5]	PTFE [2]	0.086	0.022	4×10^{-5}	3.6	4×10^{-5}
RG402 [5]	PTFE [2]	0.141	0.012	3×10^{-5}	2.0	3×10^{-5}
RG401 [5]	PTFE [2]	0.25	0.007	2.5×10^{-5}	1.1	2.5×10^{-5}

Notes: refer to Table 2.3

As already noted, the resistance is very dependent on the diameter of the cable. The roughness of the conductors (especially the shield) is another factor that may also control the resistance. For example, the resistance of RG401 is almost identical to that of RG214, although its diameter is smaller. In contrast, G is less dependent on the cable dimensions but is more sensitive to the coverage efficiency of the shield conductor and somehow to the dielectric material. PTFE appears to be slightly better than PE.

These data are approximate and may depend on the cable manufacturer, but it is obvious that an RG58 will always be worse than a double shielded RG223, but always better than an RG316 of smaller

---

[37] A length of 3–5 m is recommended for this measurement in the 1–1000 MHz frequency range.
[38] Using Eq. (2.54).

diameter, as far as one is concerned with one of the most important parameters, the attenuation.

2.4.2.4 Attenuation when the line is not terminated by its characteristic impedance Z_c

Consider, as an extreme case, a source with an internal impedance of 50 Ω (for example an NMR coil already matched with a lossless network) that is connected, using a small RG316 coaxial cable, to an LNA having a small input impedance[39] of 1 Ω real (Fig. 2.25). Let the cable length, depending on several geometric constraints, be one meter.

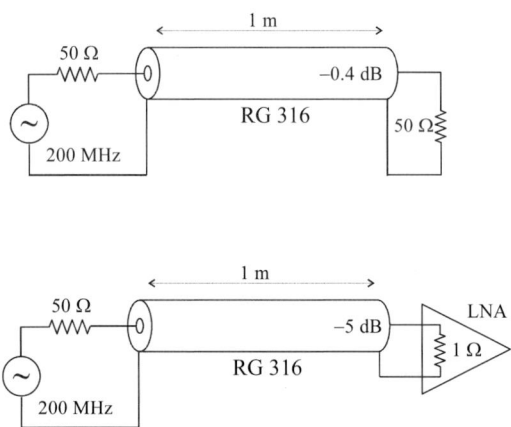

Fig. 2.25 Losses expected with a coaxial line (RG316) when terminated by its characteristic impedance (upper) and by the small impedance of a low noise amplifier (lower).

In these conditions, the noise factor (F) may be expected to be degraded by the additional loss in the cable of around 0.4 dB at 200 MHz. The *NF* of the LNA will probably be ruined, but the overall figure may still be close to 1 dB, an acceptable value.

---

[39] The use of a low noise preamplifier with small input impedance will be addressed elsewhere. Other typical cases of low impedance load will be encountered with transmission line matching networks.

In fact, we will observe that the loss in the cable is more than 5 dB, leading obviously to an unacceptable *NF*. Using a larger, but more rigid cable, like the RG223 could be a much better solution regarding its attenuation. The additional loss is indeed estimated to be only 0.2 dB at 200 MHz from the characteristics of the cable. In fact, we will get a loss of about 3.5 dB, still an unacceptable figure.

Let us consider another circuit (Fig. 2.26). We want to connect an inductive load (typically an NMR coil) having an impedance of 1.0 + j.100 Ω at 200 MHz (Q = 100) to a 50 Ω network.

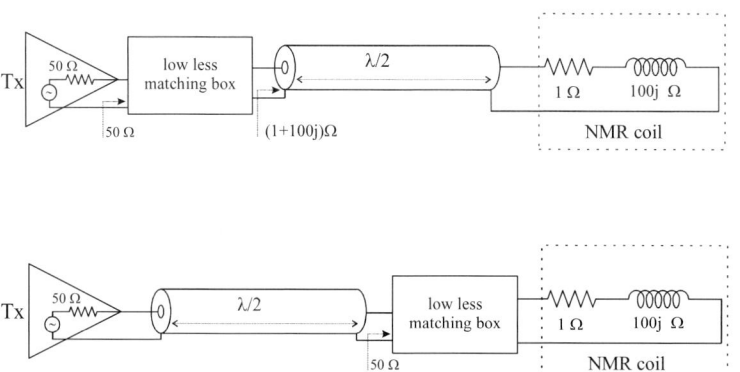

Fig. 2.26 The two circuits differ only by the position of the matching box which transforms the 1 + 100j Ω impedance into 50 Ω. The transmission line of the top circuit is not terminated by its characteristic impedance. A high standing wave ratio and consequently high losses are expected in this case. The losses in the transmission line of the bottom circuit are expected to be much lower.

For some reason,[40] the matching box cannot be put directly close to the coil but should be displaced at distance. The connection between the coil and the matching box will be made with a half wavelength piece of coaxial cable designed to copy the coil impedance to the input of the matching box exactly. At 200 MHz, the cable length will be about half a meter. The attenuation of the cable is expected be 0.1 or 0.2 dB for the RG223 or the RG316, respectively. Again, these figures seem to be

---

[40] This case, related to "remote matching" will be discussed in more detail in the next chapter.

acceptable. However, one will observe that the degradation of the sensitivity, due entirely to the loss in the cable, will be 6.0 dB with the RG223 and 8.6 dB with the RG316, respectively. The power delivered to the coil will be divided by a factor of 4 and 7(!), respectively, depending on the cable used.

These high losses are due to standing waves of high amplitude appearing in the line when not terminated by its characteristic impedance. For high VSWR, the loss in the line is much higher than the "matched" loss indicated in the data sheets [ARRL Handbook, 2009, p. 21–28].

To quantify this effect, let us consider a lossy, low-loss, line terminated by an arbitrary load impedance Z_L (Fig. 2.27).

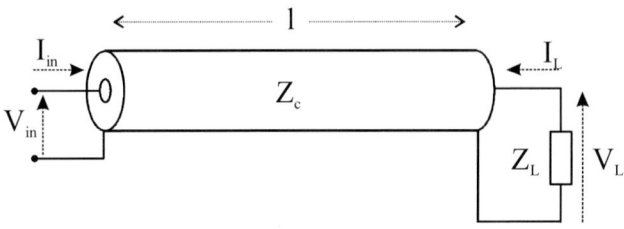

Fig. 2.27 Circuit used to simulate the transmission losses on a line of characteristic impedance Z_c when terminated by a load Z_L.

Let Γ_L be the reflection coefficient at the load:

$$\Gamma_L = \frac{Z_L - Z_c}{Z_L + Z_c}. \qquad (2.60)$$

The active power delivered to the load is given by:[41]

$$P_L = \mathrm{Re}\{V_L I_L^*\} = \mathrm{Re}\{V_L V_L^* / Z_L^*\}. \qquad (2.61)$$

---

[41] The voltage and current amplitudes are the Root Mean Square (RMS) values. Assuming a sinusoidal steady state and using the peak values, a factor of ½ should be inserted in Eq. (2.61).

V_L is the sum of the voltages of the backward (V^+) and reflected or forward (V^-) wave. Using the fact that the ratio of V^- to V^+ is the reflection coefficient Γ_L, P_L can be written as:

$$P_L = \text{Re}\left\{\frac{V_L^+ V_L^{+*}(1+\Gamma_L)}{Z_L^*}\right\} = |V_L^+|^2 \text{Re}\left\{\frac{(1+\Gamma_L)}{Z_L^*}\right\}. \quad (2.62)$$

From Eq. (2.60), and similarly to Eq. (2.19), one may write:

$$Z_L = Z_c \frac{1+\Gamma_L}{1-\Gamma_L}. \quad (2.63)$$

Finally,

$$P_L = |V_L^+|^2 \text{Re}\left\{\frac{1}{Z_c}\right\}(1-|\Gamma_L|^2). \quad (2.64)$$

The active power at the input of the line is:

$$P_{in} = \text{Re}\{V_{in} I_{in}^*\}. \quad (2.65)$$

The total voltage V_{in} and current I_{in} are, respectively, the sum of the voltage and current amplitudes of the backward and forward waves at the input. Using Eqs. (2.45) and (2.46) and rearranging, one obtains:

$$P_{in} = \text{Re}\left\{\frac{(V_{in}^+ V_{in}^{+*} - V_{in}^- V_{in}^{-*}) + (V_{in}^- V_{in}^{+*} - V_{in}^+ V_{in}^{-*})}{Z_c^*}\right\}. \quad (2.66)$$

The second term of the numerator is purely imaginary,[42] while the first term is purely real. At this point, one assumes that the line is low-loss so that the characteristic impedance is purely real. Then:

$$P_{in} = \frac{1}{Z_c}(|V_{in}^+|^2 - |V_{in}^-|^2). \quad (2.67)$$

---

[42] A complex number $AB^* - BA^*$ is purely imaginary for any $A, B \in \mathbb{C}$.

Let A be the "matched" power attenuation factor[43] of the line of length l:

$$A = e^{2\alpha l}. \qquad (2.68)$$

The voltage amplitude at the input of the forward wave V_{in}^+ is $\sqrt{A}$ times the voltage amplitude at the load V_L^+, while the backward wave is attenuated by the same factor:

$$\begin{aligned} |V_{in}^+| &= |V_L^+| \cdot \sqrt{A} \\ |V_{in}^-| &= |V_L^-|/\sqrt{A} \end{aligned}. \qquad (2.69)$$

Using Eq. (2.69) and the fact that the ratio of the backward to forward wave amplitudes is equal to $|\Gamma_L|$, one obtains:

$$P_{in} = \frac{|V_L^+|^2}{Z_c}\left(A - |\Gamma_L|^2/A\right). \qquad (2.70)$$

Finally, the ratio of the power dissipated in the load to the power delivered at the input of the line is:

$$\frac{P_L}{P_{in}} = \frac{\left(1 - |\Gamma_L|^2\right)}{\left(A - |\Gamma_L|^2/A\right)}. \qquad (2.71)$$

The above equation confirms that the efficiency of the line depends strongly on the reflection coefficient at the load. If the line is terminated by its characteristic impedance ($|\Gamma_L| = 0$), the attenuation is, as expected, the matched loss A. In contrast, if the reflection is high ($|\Gamma_L| = 1$), the ratio of power delivered to the load tends to 0 whatever the matched loss A of the line. All the available power is dissipated in the line. Obviously, if the line is lossless ($A = 1$) no power is dissipated in the line ($P_L = P_{in}$) whatever the loading conditions.

---

[43] If the line is lossless, $A = 1$, otherwise $A > 1$. The attenuation factor A is related to the matched loss ML_{dB} [Eq. (2.58)] by $A = 10^{ML/10}$.

It is worth noting that the magnitude of the reflection coefficient at the input is lower than that at the load, as deduced from Eq. (2.69):

$$|\Gamma_{in}| = |\Gamma_L|/A. \qquad (2.72)$$

Any load terminating a long line appears as a matched load(!). A long low-loss transmission line of about 100 m constitutes a good broadband resistive load, having the characteristic impedance of the line.

The power attenuation issue of lossy transmission lines (even low-loss) is of particular importance for communication applications where a long coaxial cable connects an antenna to a transmitter. The ARRL Handbook has a chapter[44] entirely dedicated to transmission lines, including a complete discussion on this topic.

This is also of importance when using transmission lines in NMR probehead circuits. It will be demonstrated in the following chapter that a short length of low-loss coaxial cable used as an impedance transformer may lead to a 10 dB loss of sensitivity. When used with some care, the impedance transformation property of transmission lines is however a useful tool in RF circuits.

2.4.2.5 *Impedance transformation*

The voltage $V(z)$ and current $I(z)$ at any point on the line is the sum of the voltages and currents associated with the backward and forward waves:

$$\begin{aligned} V(z) &= V^+(z) + V^-(z) \\ I(z) &= \frac{1}{Z_c}\left[V^+(z) - V^-(z)\right] \end{aligned} \qquad (2.73)$$

The ratio of $V(z)$ to $I(z)$ is the impedance to that point:

$$Z(z) = Z_c \frac{V^+(z) + V^-(z)}{V^+(z) - V^-(z)}. \qquad (2.74)$$

---

[44] See, for example, Chapter 21 in the 2009 edition.

It is independent of z if the line is terminated by its characteristic impedance (no reflected waves) and equal to Z_c. When the line is not terminated by Z_c the impedance varies periodically along the line.

Consider, for example, a line terminated by a short ($Z_L = 0$) (Fig. 2.28). A standing wave pattern will appear, with the voltage and current amplitude changing periodically along the line. At a distance from the load equal to half a wavelength, the total voltage is zero while the total current magnitude reaches a maximum value. The impedance at this position is a copy of the load impedance. Halfway between this point and the load (at $\lambda/4$ from the load) the total voltage is a maximum, the forward and reflected voltage amplitudes being equal. At this point, the forward and backward current amplitudes are also equal but of opposite sign, hence the total current is null and the impedance is infinite.

The corresponding reflection coefficient Γ, Eq. (2.8), changes from -1 ($Z = 0$) at the end of the line to $+1$ ($Z = \infty$) at $\lambda/4$, then returns to -1 at $\lambda/2$ and so on.

Assuming a lossless line, the magnitude of the reflection coefficient is constant over the line [Eq. (2.69), $A = 1$]. Only the phase of Γ rotates along the line with an angle equal to 2π radians for a $\lambda/2$ line. The direction of rotation can be estimated assuming a line length much smaller than a quarter of wavelength. In this condition, the shorted line obviously behaves as an inductor. Starting from the $-1,0$ coordinate on the polar plane, the Γ vector moves in the positive Γ – half plane, hence the rotation is clockwise when the length of line increases (Fig. 2.28).

The transformed impedance of any load by a line of electric length l can be predicted on a Smith Chart overlay, simply by moving along a circle,[45] centered on the chart, by an angle of $4\pi l/\lambda$, in the specified direction [Fig. 2.29(a)].

If the line is lossy, but low-loss, the propagation phase shift is almost equal to that of a lossless line [Eq. (2.50)]. The rotation angle is equal to that quoted above. However, the magnitude of Γ is not a constant. It

---

[45] If the line is lossless.

decreases[46] as the length of line increases (or as the frequency increases) [Eq. (2.72)], as shown on Fig. 2.29(b).

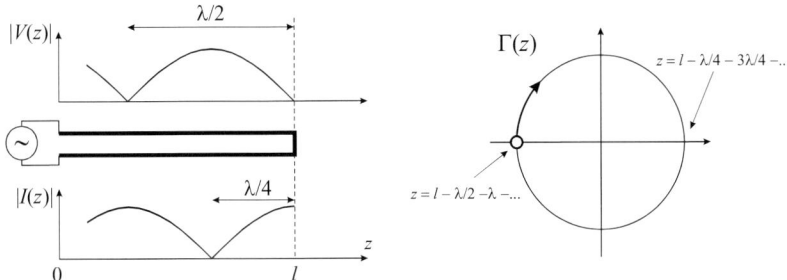

Fig. 2.28 Standing wave pattern on a lossless transmission line terminated by a short (left). Only the voltage and current magnitude are shown. The reflection coefficient along the line, when moving from the end of the line to the source, is represented in the Γ-plane (right). Γ is related to the standing wave amplitude and phase.

Quantitatively, rewriting Eq. (2.74) for $z = 0$ (the input) and expressing $V^+(0)$ and $V^-(0)$ as a function of the voltages $V^+(l)$ and $V^-(l)$ at the load, one straightforwardly obtains:

$$Z_{in} = Z_c \frac{V^+(l)e^{\gamma l} + V^-(l)e^{-\gamma l}}{V^+(l)e^{\gamma l} - V^-(l)e^{-\gamma l}}. \quad (2.75)$$

At the load, the reflection coefficient is:

$$\Gamma_L = \frac{V^-(l)}{V^+(l)}. \quad (2.76)$$

---

[46] As a result, a very long line loaded by any impedance (shorted, opened or else) behaves as a broadband load equal to the characteristic impedance of the line! ($\Gamma \to 0$ as $a \to \infty$).

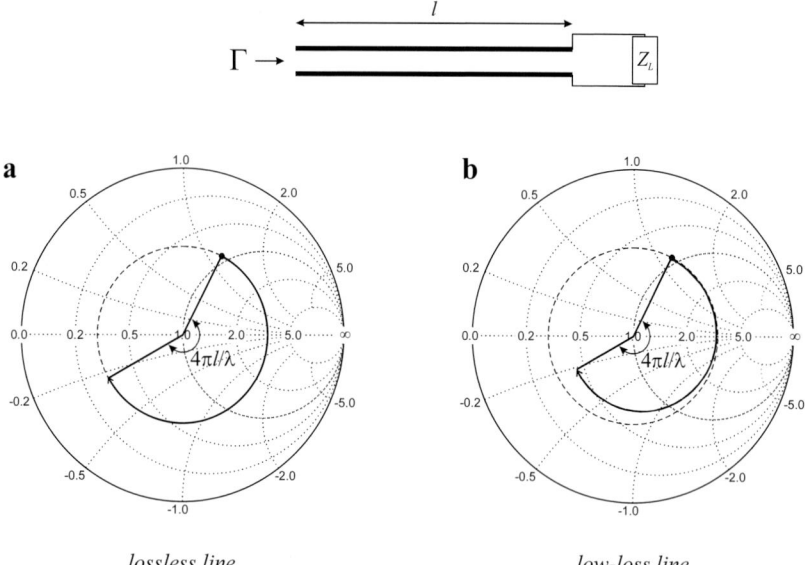

Fig. 2.29 Impedance transformation of a load Z_L by a transmission line of length l, represented on the Smith Chart. The center of the chart corresponds to a real impedance equal to Z_c, the characteristic impedance of the line. (a) The line is assumed to be lossless. The reduced impedance (Z_L/Z_c) in the Γ-plane moves on a circle as the length of line increases. The phase shift is determined by the length l of the line relative to the wavelength λ. (b) The line is assumed to be low-loss. Its characteristic impedance is still real-valued, equal to Z_c. The phase shift is identical to that of the lossless case. The magnitude of the Γ vector decreases as the line length (and the attenuation) increases.

Hence,

$$Z_{in} = Z_c \frac{1+\Gamma_L e^{-2\gamma l}}{1-\Gamma_L e^{-2\gamma l}} \quad (2.77)$$

where:

$$\gamma = \alpha + j\beta$$
$$\beta = 2\pi/\lambda \quad . \quad (2.78)$$

From Eq. (2.77), the reflection coefficient at the input of the line can be deduced immediately as:

$$\Gamma_{in} = \Gamma_L e^{-2\gamma l} = \left(\Gamma_L e^{-2\alpha l}\right) e^{-j2\beta l}. \quad (2.79)$$

The first term of the product describes the decrease of the magnitude of Γ_{in}, as already given by Eq. (2.72). The phase (second term of the product) describes the rotation of the Γ_{in} vector by an angle of $2\beta l$, in the clockwise direction (negative trigonometric sense of rotation) as already guessed. Eq. (2.79) illustrates how to use the Smith Chart to determine the impedance of any loaded line quickly and easily.

Using Eq. (2.8), one obtains the well-known equation of impedance transformation by a lossy line:

$$Z_{in} = Z_c \frac{Z_L + Z_c \tanh(\gamma l)}{Z_c + Z_L \tanh(\gamma l)}. \quad (2.80)$$

If the line is lossless, $\tanh(\gamma l)$ is purely imaginary and equal to $j\tan(\beta l)$:

$$Z_{in} = Z_c \frac{Z_L + jZ_c \tan(\beta l)}{Z_c + jZ_L \tan(\beta l)}. \quad (2.81)$$

Fig. 2.30 shows the typical impedance variation of a shorted and opened transmission line of a given length as a function of frequency. The line behaves as an inductor, a capacitor or a parallel or series resonator, depending on the ratio of its length to the wavelength and on the loading impedance. Such lines are used in matching networks (known as "stubs" in microwave engineering). They can also be used as a replacement for lumped RF components, as described in Section 2.4.4.

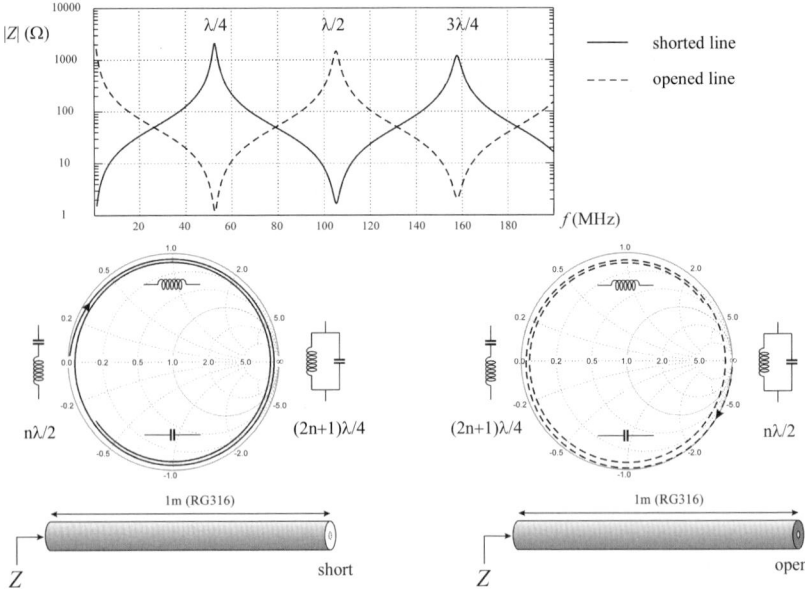

Fig. 2.30 Upper: magnitude of the impedance seen at the terminal of a line ending in a short or an open, as a function of frequency. A 1 m length of RG316 coaxial cable is assumed. The load impedance is copied when the frequency corresponds to a multiple of λ/2. The load impedance is reflected every odd multiple of λ/4. Note that the short impedance increases and the open impedance decreases as the frequency increases, due to the attenuation in the line. Lower: the phase of the complex impedance is represented on a Smith Chart overlay. The low impedance corresponds to a series resonant circuit, while the high impedance corresponds to a parallel resonant circuit. In between, the line is equivalent to a capacitor or an inductor, as depicted on the chart.

2.4.3 Discrete passive components

2.4.3.1 Resistors

As a lumped device in electronic circuits, the resistor has different shapes and sizes, depending essentially on the average power that it can tolerate. A real device is not a pure resistor at RF. A simplified model of a resistor (Fig. 2.31) includes a "parasitic" series inductance inducing phase shifts between the current and voltage. Stray capacitances between the device terminals and between the device and the surroundings add additional pathways to RF currents that modify the simple representation of a resistor.

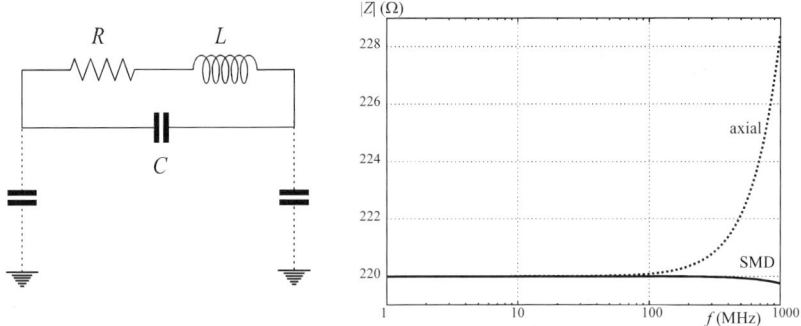

Fig. 2.31 High frequency model of a resistor (left). The resulting impedance (magnitude) varies with the frequency, as shown for two typical devices: an axial lead resistor and a Surface Mount Device (SMD). Both are metal coated.

Resistors are mainly[47] constituted of a thin metallic or carbon layer deposited on an insulating substrate. Due to the high resistivity of the material, the skin effect does not play a significant role on the resistance value which remains almost constant in the RF frequency domain. The series inductance is roughly equal to that of a conductor having the shape

---

[47] Other types of resistor exist. Carbon composition resistors were the original design. They are still available but less used. Wire-wound resistors are used in high power low frequency circuits. They are not suitable for RF applications.

of the resistor. The inductance of an axial lead resistor is typically of the order of 10 nH plus the inductance of the wire connections. The inductance of an SMD (Surface Mount Device) packaged resistor is one order of magnitude smaller.

Very broadband resistances from DC to several tens of GHz can be designed as a coaxial transmission line having a resistive central conductor. These devices are used as well-defined loads in RF circuits or as calibration standards for RF measurements.

2.4.3.2 Inductors (generalities)

Inductors are, by nature, the main parts for NMR detection by the Faraday induction method. The inductive parts of the probe interact with the sample, ideally through the magnetic near field components and possibly not through the electric near field components. The design of these inductors will be addressed throughout the book. In this paragraph we shall be concerned with the properties of inductors that are encountered in probe electrical circuits such as multiple frequencies resonant circuits, filters, diplexers, traps, chokes, etc.

Inductors have a resistance R and a reactive impedance X, which increases linearly with the frequency:

$$X = L\omega. \tag{2.82}$$

L is the inductance and ω the pulsation ($\omega = 2\pi f$). The ratio of X to R is the quality factor Q of the inductor, as defined by Eq. (2.5).

$$Q = \frac{L\omega}{R}. \tag{2.83}$$

Typically, the quality factor of inductors made of conventional conductive material, such as copper at room temperature, can be in the range of 10 to 1000. Inductors made of superconducting materials have a much higher Q factor, of the order of 10000 or more.

An essential property of inductors is that they are self-resonating at frequencies generally well in the RF range. It results that the above equations are only valid when f is well below the Self-Resonance

Frequency (*SRF*) of the inductor. Models that describe the behavior of inductors at higher frequencies are presented below for each specific kind of inductor.

2.4.3.3 *The air-solenoid (high frequency model, Q optimization)*

Most of the inductors encountered in probe circuits are shaped as a solenoid. It is the best component to efficiently store magnetic energy. Eventually, the NMR probehead itself may be a solenoid coil.

A solenoid is a helical winding of a round or flat conductor on a cylindrical mount (Fig. 2.32).

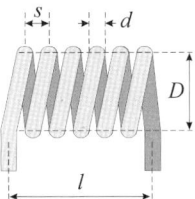

Fig. 2.32 A single layer solenoid coil and its dimensions: D – mean diameter; d – wire diameter; s – winding step; l – wire length from center to center.

The design of these inductors has been very well-known since the beginning of the last century and its optimization has been the subject of many studies [Terman, 1943].

Its impedance is characterized mainly by its inductance. It also includes a resistance, distributed capacitance between the windings and the capacitance between its terminals (Fig. 2.33).

The interwinding capacitance leads to the Self-Resonance (SR) of the coil. The SR generally occurs when the inductive impedance is between 600–1200 Ω. Additional magnetic and electric couplings of the inductor with the surroundings will modify its electrical characteristics.

For example, magnetic coupling will induce eddy currents in surrounding conductors that will modify the coil inductance L and eventually its resistance (resistive losses in the coupled conductors).

Electric coupling with the surrounding conductors will also add stray capacitances and pathways for RF currents [Grandi et al., 1999].

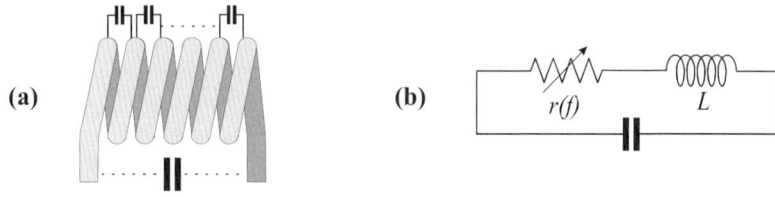

Fig. 2.33 (a) Parasitic capacitors in a solenoid coil. The capacitor between the terminals includes the parasitic with the circuit supporting the coil. (b) A simplified high frequency model of the solenoid coil. Its series resistance varies with frequency (mainly due to the skin effect on the conductor). All the parasitic capacitances are summed up in the parallel capacitor. The inductance value is assumed to be a constant (the high frequency external inductance).

The low frequency inductance of a cylindrical coil can be estimated with the following formula[48] [Raffin, 1979, p. 174] (dimensions are in mm, L is given in nH):

$$L = 0.4\pi^2 n^2 \frac{r_m^2}{l + e + r_{ext}} AB$$

$$A = \frac{10l + 12e + 2r_{ext}}{10l + 10e + 1.4r_{ext}}$$ (2.84)

$$B = 0.5 \log_{10} \left[100 + \frac{14 r_{ext}}{2l + 3e} \right]$$

The geometrical parameters (r_m, r_{ext} and l) are shown in Fig. 2.34. n is the total number of turns.

This formula proves to be accurate enough (1–5%) for any shape of cylindrical coil, from the single turn flat ring to the loop gap, as well as for the multilayer solenoid. If a greater accuracy is desired one may use

---

[48] This formula has been reported by R. A. Raffin as due to Morgan Brooks and H. M. Turner (University of Illinois).

specialized formulae; some of which are given in Chapter 10. Grover's book [Grover, 2004] is a useful reference for formulae of this kind.

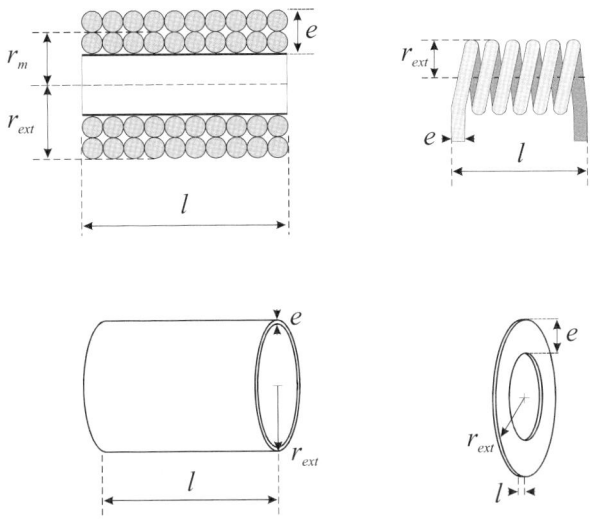

Fig. 2.34 Eq. (2.84) can be conveniently applied to the multilayer or single layer solenoid, to the single sheet coil (the loop-gap) and to the flat circular coil. The corresponding dimensions are shown for each type of coil.

The single layer solenoid coil inductance can be estimated with simpler formulae such as [Terman, 1943, p. 55] ($l > 0.8r$, $e \ll r$):

$$L_{nH} = 39.4 \frac{r_{mm}^2 n^2}{9 r_{mm} + 10 l_{mm}}. \quad (2.85)$$

This is an approximation of the popular Nagaoka formula, which gives the inductance of a thin single layer solenoid and high surface coverage[49] (uniform current) as (SI units):

$$L = \frac{\pi \mu_0 n^2 D^2}{4l} F(D/l). \quad (2.86)$$

---

[49] Correction for insulating space [Grover, 2004, p. 149] can be usually neglected.

$F(D/l)$ is a correction to take into account the finite length of the coil. F ranges from unity for an infinitely long solenoid to about 0.69 for a coil having its length equal to its diameter. It decreases to 0.2 as the length of the coil becomes one tenth of the diameter. Other corrections taking account of the finite thickness and shape of the wire could be added if a high accuracy is desired [Rosa, 1906].[50]

Values for $F(D/l)$ are tabulated, for example, in Grover's book [Grover, 1946, p. 144], or can be estimated from:

$$u = \frac{D}{l}$$

$$k = \sqrt{\frac{u^2}{1+u^2}} \qquad (2.87)$$

$$F\left(\frac{D}{l}\right) = \frac{4}{3\pi}\left[\frac{\sqrt{1+u^2}}{u^2}\left(\mathrm{K}(k) - \mathrm{E}(k)\right) + \sqrt{1+u^2}\,\mathrm{E}(k) - u\right]$$

where $\mathrm{K}(k)$ and $\mathrm{E}(k)$ are the complete elliptic integrals of the first and second kind, respectively. An algorithm to calculate these integrals from King's method [Fan et al., 1987] is described in Appendix A.

The inductance calculated from the three above formulae, Eqs. (2.84), (2.85) and (2.86), are compared together in Fig. 2.35, assuming a single layer coil of given diameter and wire of varying length.

In the present paragraph, Eq. (2.86) will be used to evaluate the quality factor Q of the coil, assuming the frequency is well below the SRF. Q is the most important parameter when designing an inductor that enters a probe circuit.

The resistance R of a single thin layer solenoid coil can be written as:

$$R = \rho \frac{n^2 D}{l\delta} p\xi. \qquad (2.88)$$

---

[50] Excellent papers concerning this topic are available on the web. For example (at the time of writing) [Weaver, 2012; Knight, 2013].

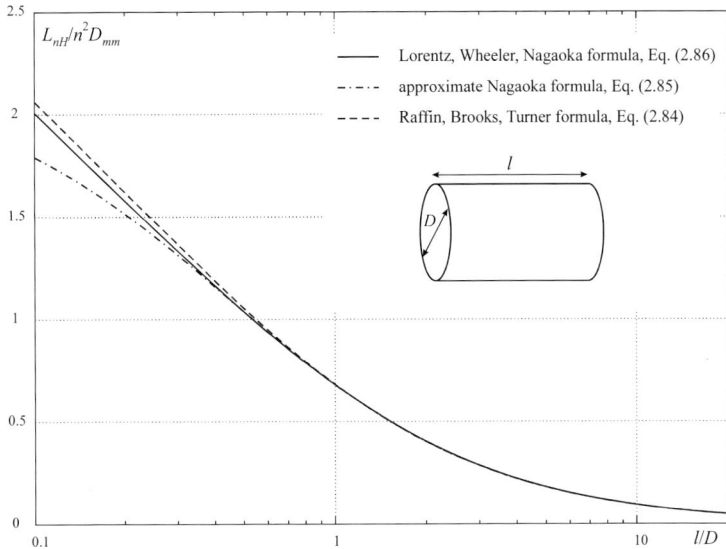

Fig. 2.35 Comparison between Eq. (2.84) and Eq. (2.86) for the calculation of the inductance of a single layer solenoid as a function of the ratio of its length (*l*) to diameter (*D*). It is remarkable that Eq. (2.84) gives very similar results to the "exact" Eq. (2.86). Note that Rosa's corrections have been neglected.

In this equation, ρ is the conductibility of the wire, δ is the skin depth (inversely proportional to the square root of the frequency), p is the "winding pitch" ($p = s/d$ where s is the distance between the centerlines of two consecutive turns and d is the wire diameter), and ξ is a "proximity factor" which increases the resistance as the wires come close together.

Combining Eqs. (2.86) and (2.88) with Eq. (2.83), one gets:

$$Q = \left[\frac{\pi\mu_0\omega\delta}{4pk\xi\rho}D\right]F(D/l). \qquad (2.89)$$

The Q factor of a single thin layer solenoid does not depend on the number of turns, for a given geometry. Well below the *SRF*, it increases as the square root of the frequency [Hoult, 1978] due to the term $\omega\delta$. Furthermore it increases with the coil diameter for a given D/l ratio.

For a given frequency and geometry, the Q value depends on the product $p\xi$, hence on the turn spacing. The optimum winding pitch p is of the order of 1.5.

The Medhurst formula [Medhurst, 1947a,b] provides a quantitative estimation at frequencies much lower than the *SRF*:

$$Q = 7.5 D_{mm} \psi \sqrt{f_{MHz}} . \qquad (2.90)$$

ψ is a function of D/l and of the winding ratio p. For optimum spacing and for coils having a number of turns greater than 3, ψ can be estimated from the following formula [Callendar, 1947; Medhurst, 1947c]:

$$\psi = \frac{1}{1.03 + 0.4 D/l} . \qquad (2.91)$$

The highest frequency to which these formulae [especially Eq. (2.91)] are valid essentially depends on the coil dimensions. As a rule, the frequency should be much lower than the *SRF*. Medhurst [Medhurst, 1947a,b] has shown that the resonant wavelength of a coil can be written as:

$$\lambda_{SRF} = \alpha_{SRF} l_w , \qquad (2.92)$$

where α_{SRF} (Table 2.5) is a dimensionless factor that depends only on the ratio D/l and l_w is the total length of wire. The formula applies to short and long solenoids but is valid only when one end of the coil is earthed. Combining all the above constraints, the best Q factor is obtained when the D/l ratio is between 0.7 and 2.0 [Terman, 1943].

The lowest ratio (a long solenoid) is the best choice when the coil is used as an NMR probe. A long coil improves the magnetic field homogeneity in a large portion of the coil volume (high filling factor). Solenoid microcoils [Webb, 2012] are typically very long solenoids (low D/l ratio). In these cases, the winding may be tighter to improve the Q factor. A detailed discussion on the optimization of these microcoils can be found in the literature [Webb, 1997; Minard and Wind, 2001; Webb, 2013].

Table 2.5 Factor α_{SFR} to estimate the self-resonant wavelength [Eq. (2.92)] of a cylindrical coil as a function of the ratio of its length (l) to diameter (D) (from Medhurst, 1947b, p.86).[51]

l/D	0.1	0.15	0.20	0.25	0.30	0.35	0.40	0.45	0.50	0.60	0.70
α_{SFR}	8.3	7.1	6.3	5.8	5.4	5.0	4.8	4.5	4.3	4.0	3.8
l/D	0.80	0.90	1.0	1.5	2.0	2.5	3.0	3.5	4.0	4.5	5.0
α_{SFR}	3.6	3.5	3.4	2.9	2.7	2.6	2.3	2.5	2.4	2.4	2.3
l/D	6.0	7.0	8.0	9.0	10	15	20	25	30	40	50
α_{SFR}	2.3	2.2	2.2	2.1	2.1	2.1	2.0	2.0	2.0	2.0	2.0

For applications in probe circuits, such as filters, traps or multiple tuning, the diameter of the coil should be about twice its length. Kan and Gonord [Kan and Gonord, 1992] demonstrated that the optimum ratio is 2.44 for these designs, corresponding to relatively short coils.

When the coil dimensions are chosen, the number of turns and wire spacing (or equivalently the wire diameter) remain to be determined. A usual rule is that the maximum of Q is obtained when the wire is "slightly smaller" than the one required to occupy all the available space [Terman, 1943], i.e. to cover almost all the solenoid coil surface. If the winding is tight ($p = 1$) the proximity effect dramatically increases the resistance by a factor of 3 or more. On the other hand, if the spacing between turns increases, the proximity parameter ξ diminishes, but p increases and the Q factor tends again to diminish [Eq. (2.89)]. As a rule, the optimum wire spacing between turns should be equal to the wire radius ($p = 3/2$). At this point, the number of turns remains to be fixed. It is ultimately determined by the coil inductance value, the coil diameter and the working frequency.

Efficient tools for the design of solenoid coils can be found on the web. For example, the calculator by Serge Stroobandt (ON4AA),[52] based on the theory of helical resonators, accurately predicts the electrical properties of solenoid coils (L, Q, SRF) over a large frequency range.

---

[51] Note: the table is presented here as the original one, i.e. α is given as a function of l/D.
[52] Available at: http://hamwaves.com/antennas/inductance.html.

Coils with a diameter larger than a few mm can be made easily and optimized starting from the formulae given here. On the other hand, coils with very small dimensions are commercially available as SMDs. These coils are wound on ceramic substrates and are nonmagnetic, as required for NMR. Coilcraft (Inc., Cary, Illinois, USA and Europe, Cumbernauld, Scotland) is a well-established company which proposes a large range of values, from the nH to some µH, suitable for NMR receiver probe circuits.

2.4.3.4 High frequency model of the air-solenoid

The high frequency model [Fig. 2.33(b)] shown in the previous section is valid up to frequencies just above the *SRF*. At higher frequencies, the distributed capacitance between each turn is best modeled by a distributed resonant system [Corum and Corum, 2001]. This is especially true when the number of turns is large.

As a matter of fact (Fig. 2.36), the impedance of a solenoid coil increases linearly with the frequency well below the SR, according to Eq. (2.82). When approaching the SR, its impedance increases considerably up to $QL\omega_{SRF}$. The maximum Q value is near ω_{SRF} then Q decreases at higher frequencies. Above the *SRF*, the impedance of the coil decreases continuously as RF current leaks through the interwinding capacitance. The current leakage is moderate anyway, especially for a long solenoid coil. Hence the impedance remains high over a broad frequency range up to a limit that depends on the design. This property will benefit the design of broadband choke inductance (Section 2.4.3.7).

The solenoid (and most inductors) therefore has two distinct frequency domains of functioning. At "low frequencies" (well below the *SRF*), the inductor has a well-defined inductance value. It can be used in filters, matching networks or other electronic circuits where a known inductance is required. *S*-parameters or SPICE models are provided by coil manufacturers up to a frequency just above the *SRF*, allowing the use of the inductor with confidence. At higher frequencies, the phase of Z becomes uncertain.

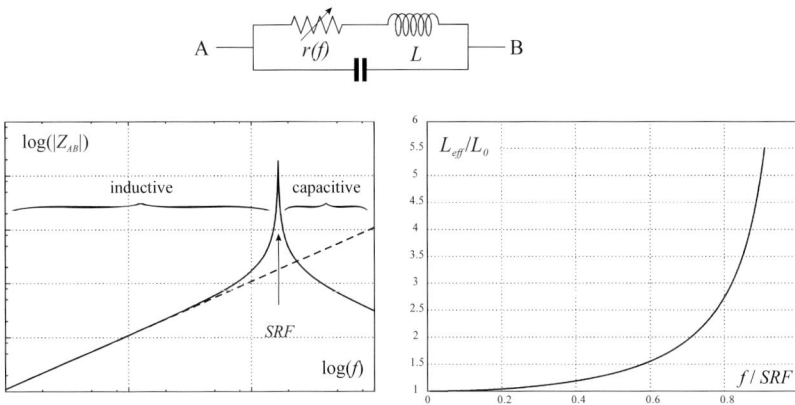

Fig. 2.36 Left: magnitude of the impedance of an inductor as a function of frequency up to, and slightly above, its *SRF*. The dashed line is the impedance of the ideal inductor. Right: the effective inductance as a function of the ratio of frequency to the *SRF*, in the domain of inductive impedance.

The parasitic couplings of the coil with the surroundings may significantly modify its electrical characteristic, making the in-circuit inductor modeling a difficult task. Evaluation of the behavior of the coil on a prototype board before the final realization is recommended.

2.4.3.5 *Toroidal inductors*

A toroidal coil concentrates the magnetic flux in its interior, minimizing the interaction with the surroundings. It is particularly simple to calculate the magnetic field amplitude inside the toroid by direct application of Ampere's law. As a result, the magnetic field per unit current is given, in SI units, by the following equation:

$$B/I = \frac{\mu n}{\pi D}, \qquad (2.93)$$

where n is the number of turns, D is the toroid diameter to the center line (Fig. 2.37) and μ is the magnetic susceptibility of the material of the toroid core.

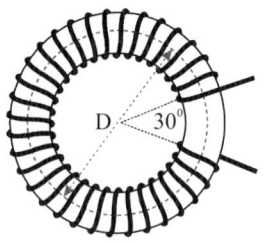

Fig. 2.37 Shape and dimensions of a toroid coil. An angle of 30° between the terminals of the winding is generally recommended for optimum characteristics.

The inductance can be deduced by a simple application of Faraday's law [first equality in Eq. (2.94)] and the definition of the inductance [second equality in Eq. (2.94)]:

$$e = -nA\frac{\partial B}{\partial t} = -L\frac{\partial I}{\partial t}, \qquad (2.94)$$

where e is the voltage induced by a time varying magnetic flux $(B.A)$ resulting from a given current (I) variation. A is the cross-sectional area of the toroid core. Assuming that the magnetic field B [Eq. (2.93)] is uniform inside the toroid, the inductance is obtained as:

$$L_{nH} = \frac{0.4\mu_r n^2 A_{mm^2}}{D_{mm}}. \qquad (2.95)$$

μ_r is the relative permeability of the core material, equal to unity for an air wound coil.

Toroidal coils are frequently formed on a high permeability core, increasing the inductance by a factor of hundreds or more. These designs have the advantages that, for a given inductance value, the coil size can be much smaller than that of an air coil, the interaction with the surroundings is minimized because of the efficiency of the core in concentrating the magnetic flux and the coil is easy to build. The inductance is less sensitive to the turn coiling arrangement than an air toroid coil, but to maximize the *SRF* it is recommended that the turns are spread evenly around the circumference of the core while leaving some separation between the two coil terminals.

The main disadvantage is that these coils are magnetic and cannot be used in close proximity to the NMR coil sensor. Also, the electric properties (permeability and losses) of the core are highly dependent on the nature of the material and on the frequency. Specific material should be chosen as a function of the operating frequency range. Furthermore, the magnetic core is sensitive to saturation, depending on its size and on its constitution.

The inductance of a toroid wound on a ferrite material is given by:

$$L_{nH} = n^2 A_L, \qquad (2.96)$$

where A_L is a numerical value, characteristic of the core material, given by the manufacturer[53] as a number representing the inductance in mH of a 1000 turn coil.

Eq. (2.96) applies only to ferrite cores. For powdered-iron cores, the A_L index is given as µH per 100 turns. In those cases, the inductance value in Eq. (2.96) should be divided by 10.

2.4.3.6 *Other inductors (single loop, printed spiral coil)*

Although the solenoid coils are probably the most widely used inductors, other designs are the best choices, depending on several constraints.

---

[53] A_L is often given in "units" of mH per 1000 turns. This may appear to be incorrect as the inductance is proportional to the square of the number of turns. In fact, A_L is a numerical value indicating that a coil wound with 1000 turns on the chosen core will have an inductance of A_L mH. It results that a one turn inductance on this core would be equal to A_L nH, as indicated by Eq. (2.96). The required number of turns to make a given inductance L is given as $n = 1000\sqrt{L_{mH}/A_L}$. This is valid for ferrite cores. For powdered-iron toroid cores, of smaller permeability, A_L is commonly given as µH per 100 turns. Hence, for a coil wound on a powdered-iron core, the required number of turns is $n = 100\sqrt{L_{\mu H}/A_L}$.

Any conductor, a round wire or a sheet of metal (Section 2.4.1), is an inductor. A straight round wire of copper with a diameter of 1 mm and length of 16 mm exhibits a high frequency inductance of the order of 10 nH.

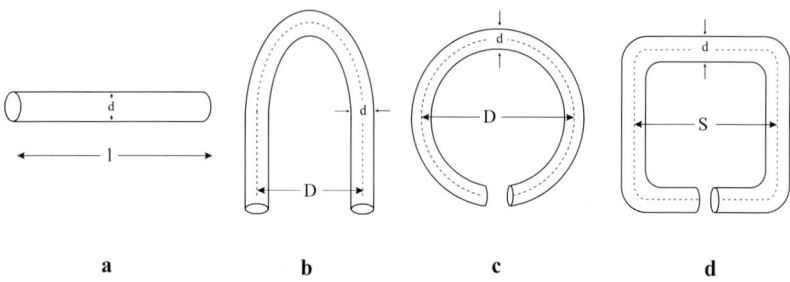

Fig. 2.38 Inductance of a length (l = 16 mm; d = 1 mm) of copper round wire shaped as: (a) straight wire (L = 10 nH), (b) hair pin (L = 9–5 nH, depending on the spacing between terminals), (c) circular coil (L = 5.5 nH), (d) square coil (L = 4.6 nH).

The high frequency inductance of a circular ring of tubular cross-section [Fig. 2.38(c)] is given by [Terman, 1943, p. 52]:

$$L_{nH} = (\pi/5) D_{mm} \left[\ln(8D/d) - 2 \right], \tag{2.97}$$

where D is the mean diameter of the loop (in mm) and d is the wire diameter.

The high frequency inductance of a square of round wire of side S and wire diameter d [Fig. 2.38(d)] can be estimated from [Terman, 1943, p. 52]:

$$L_{nH} = 0.8 S_{mm} \left[\ln(2S/d) + d/2S - 0.774 \right]. \tag{2.98}$$

The formula for a rectangular coil is much more complicated [Terman, 1943, p. 53]. An estimate of the inductance can be obtained assuming that loops of any shape but with equal area have about the same inductance (Fig. 2.39).

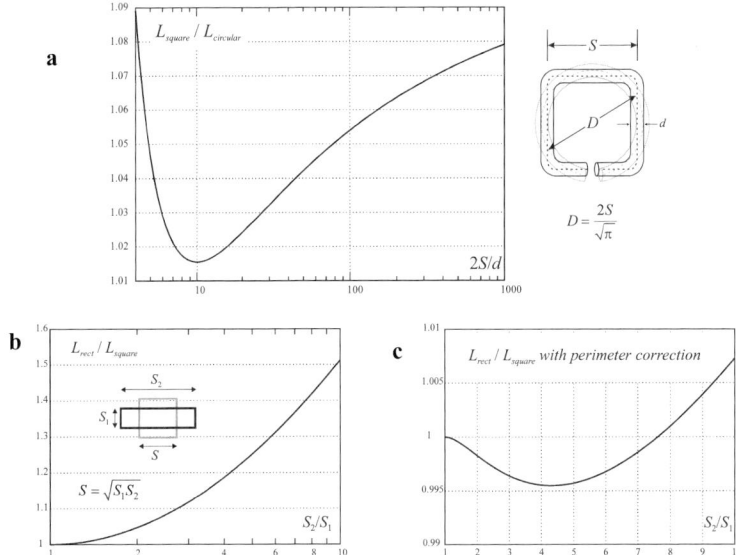

Fig. 2.39 Upper: comparison between the inductances of a square and a circular coil having the same area as a function of the wire diameter (a). Lower: comparison between the inductances of a rectangular and a square coil of the same area as a function of the shape factor (b). The inductances for both coils were calculated solving the inductance matrix of the network (simprobe). A ratio $2S/d = 20$ was assumed. An empirical correction has been applied as a function of the coil perimeter (c).

However, a correction must be added if the perimeters of the coils differ significantly [Fig. 2.39(b)]. As a result, the correction to the calculated inductance of a rectangular coil, compared to the inductance of a square coil of the same area, is given by the following empirical formula:

$$L_{rect}/L_{square} = x \cdot h(x)$$
$$h(x) = 1 - 0.1845(x-1). \qquad (2.99)$$
$$x = \frac{1+\alpha}{2\sqrt{\alpha}}$$

The ratio (x) of the perimeters of the rectangular coil and of the square coil of the same area is related to the aspect ratio (α) of the rectangular coil (the ratio of the long side to small side lengths), as indicated in the above equations. With this correction, the rectangular coil inductance accuracy is greatly improved [Fig. 2.39(c)].

The spiral loop (Fig. 2.40) is widely used on circuit boards. Its advantage is an easy and stable design.

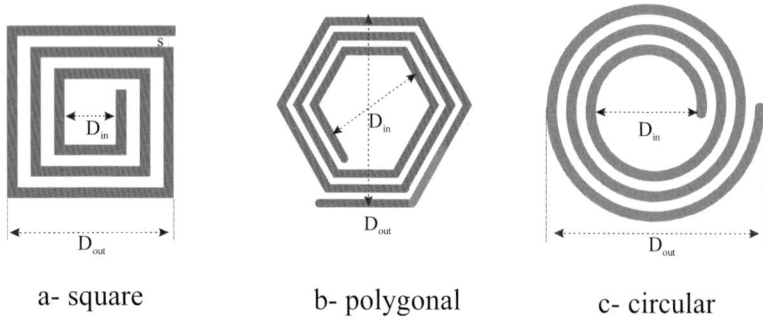

a- square   b- polygonal   c- circular

Fig. 2.40 Shapes and dimensions of spiral coils referred to in the text.

The inductance of a planar, square-shaped, spiral coil [Fig. 2.40(a)] can be estimated from [Mohan et al. 1999]:

$$L_{nH} = 2.94 \frac{n^2 D_{mm}}{1 + 2.75\eta}, \qquad (2.100)$$

where D is the mean diameter of the coil and η is a fill factor defined as [Lee, 2004, p. 232]:

$$\eta = \frac{D_{out} - D_{in}}{D_{out} + D_{in}} \qquad (2.101)$$

and

$$D = \frac{D_{out} + D_{in}}{2}. \qquad (2.102)$$

If the planar spiral coil is shaped as a polygon with N sides [$N > 4$, Fig. 2.40(b)], a general approximate formula for the inductance has been proposed by Thomas H. Lee [Lee, 2004, p. 232], which is rewritten here with the inductance in nH and dimensions in mm, as:

$$L_{nH} = 0.2\pi D_{mm} n^2 \left(\frac{4 A_{out}}{\pi D_{out}^2} \right) \times$$

$$\left[\ln\left(\frac{2.46 - 1.56/N}{\eta} \right) + \left(0.20 - \frac{1.12}{N^2} \right) \eta^2 \right]. \quad (2.103)$$

The first term in parentheses is the ratio of the surface of the circle enclosing the polygonal coil (A_{out}) to the surface of the circle of diameter D_{out} [Fig. 2.40(b)]. When the number N increases to infinity (the polygon tends to a circle), this term becomes equal to unity.

The inductance calculated by Eq. (2.103) is independent of the width (w) of the strip conductor and separation (s) between turns [Fig. 2.40(a)]. In fact, the error in L increases when the ratio s/w increases. As a result, the error in Eq. (2.103) is smaller than 8% when s is smaller than three times the conductor width w [Mohan, 1999, p. 1421].

Another estimate for the inductance of a circular spiral coil is:

$$L_{nH} = 9.84 \frac{n^2 D_{mm}}{4 + 11\eta}. \quad (2.104)$$

Finally, Eq. (2.84) can be used, for the circular spiral coil, when s/w is small. The length l of the coil is the diameter of the wire or the thickness of the strip conductor.

The inductance (L/Dn^2) estimated from the above expressions is compared in Fig. 2.41 for different geometries and as a function of the filling factor η.

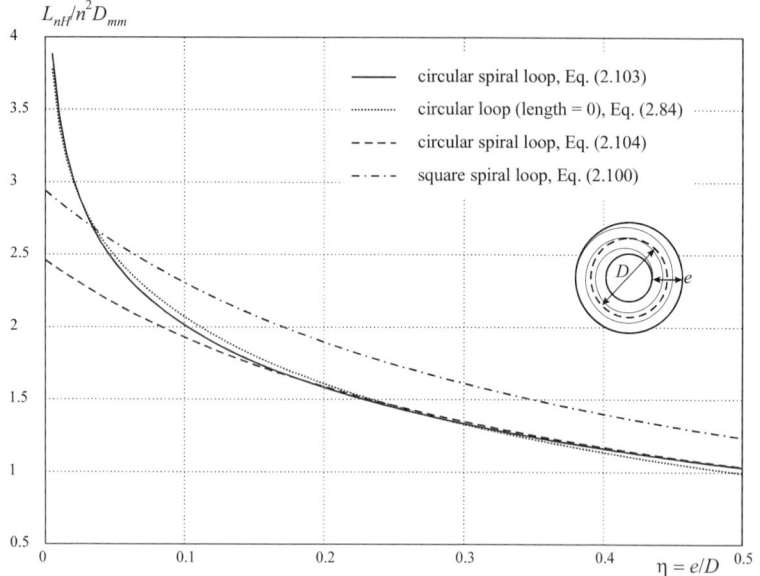

Fig. 2.41 Comparison of the inductances calculated for a thin circular spiral loop as a function of the fill ratio. Note that Eq. (2.84) still gives very similar results to the most accurate Eq. (2.103). For low fill ratios, the inductance calculated by the approximate Eq. (2.104) deviates significantly from above. The inductance of the square spiral coil of the same area is also shown. The inductance value is clearly overestimated. The inductance of such a coil is not exactly proportional to the square of the number of turns [Niknejad, 2007, p. 144]. This effect has not been considered here.

2.4.3.7 Choke inductor

A choke inductor is intended to present a high impedance at RF. It is required in DC bias lines to power up a preamplifier or to trigger commutation diodes in an active detuning circuit (Fig. 2.42). The required impedance at the operating frequency depends on the impedance value of the circuit where the inductor is inserted. For 50 Ω systems, the choke impedance must be of the order of some hundreds of Ohms, typically between 0.3 and 1 kΩ. An inductance of about 1 µH has an impedance $|Z|$ equal to 630 Ω at 100 MHz. It is a good candidate for an RF choke operating at this frequency.

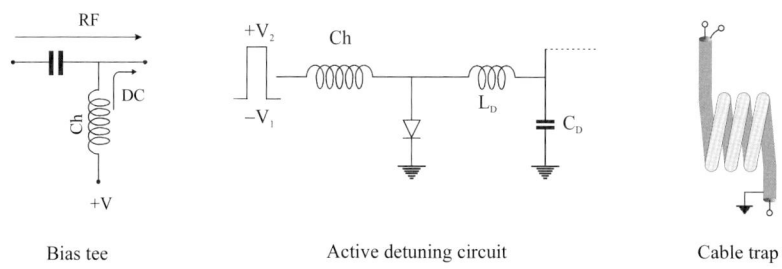

Fig. 2.42 Probe circuits where a choke inductance is used. Left: in the bias tee circuit, the choke provides a path to the DC power of a preamplifier and a high impedance to the RF. Middle: the RF path in the active detuning circuit is blocked toward the pulse power supply by the choke and toward the ground by the (very small) reverse capacitance of the PIN diode. Right: a cable trap is formed by coiling the coax as an inductor of high impedance. Cable "braid currents" are blocked in this way, ensuring that the cable functions in the differential mode.

The quality factor of choke inductors is of moderate importance and they still operate conveniently at frequencies much higher than the coil *SRF*. The RF impedance is high over a large frequency range but the DC resistance should be low. The evaluation of a given choke can conveniently be done by observing the transmission losses (S_{21} parameter, for example) of a bias tee circuit [Fig. 2.43(a)]. Using the same setup, S_{11} provides an estimation of the impedance of the choke inductor as a function of frequency [Fig. 2.43(b)].

Small size choke inductors supporting high DC current and having a *SRF* in the Very High Frequency (VHF) range but usable to the microwave domain are readily available commercially.

For example, Coilcraft provides wideband bias chokes (models 4310LC), capable of handling DC currents larger than about 2–3 A, with inductance values of 1.3 µH and 3.5 µH and *SRF* of 235 and 188 MHz, respectively.

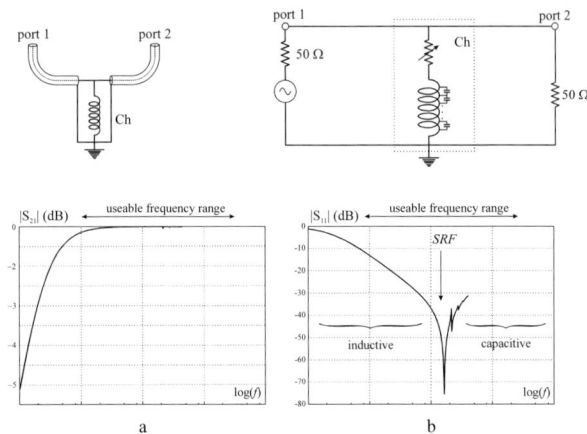

Fig. 2.43 An experimental setup to evaluate a choke inductor in a 50 Ω system (upper left). The circuit model (upper right) used to calculate the magnitude of S_{21} (transmission) and S_{11} (related to the impedance) as a function of frequency (lower).

These choke inductances are specified to exhibit a flat bandwidth with high impedance up to 6 GHz. These are small solenoid coils wound on a bar of ferrite material (Fig. 2.44).

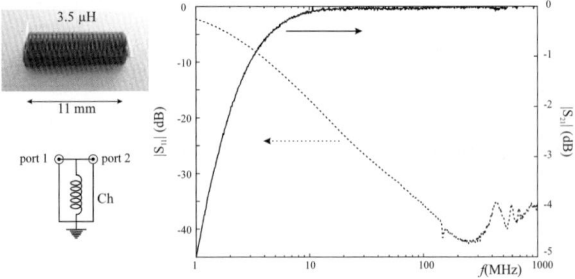

Fig. 2.44 Typical frequency behavior of a commercial very broadband choke (Coilcraft model 4310LC). In a 50 Ω system the transmission is almost perfect starting around 20 MHz. If used with greater input/output impedances, the operating frequency band will be reduced.

Although the ferrite material may lose some of its properties at very high frequencies, the design of the choke as a long solenoid coil contributes to maintaining high impedance up to the microwave frequencies. The lower frequency boundary is in the range where the choke still behaves as an inductor. In this frequency range, the reactive component is $L\omega$ [Eq. (2.82)], where L is the low frequency inductance. The ferrite material increases L, thus contributing to efficiently increasing the usable bandwidth of the choke towards the low frequencies.

The popular VK200 inductor (Fig. 2.45) is another choke usable from 5–15 MHz[54] up to the VHF range. These chokes are made of a few turns (one to three) of a wire threaded in holes drilled in a high permeability ferrite cylinder. These do not exhibit a clear resonant frequency. The ferrite material is essential for the functioning of these chokes. It results that the usable frequency range is moderate and some losses could be expected in a bias tee configuration.

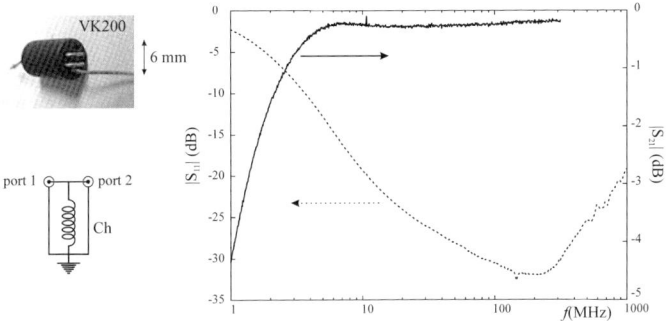

Fig. 2.45 The frequency behavior of the widely used VK200. The self-resonance is considerably broadened due to the ferrite core. The transmission characteristics are rather poor for use in circuits like those represented in Fig. 2.42. Also, the bandwidth tends to be reduced to frequencies lower than 500 MHz. Such a choke may be preferentially used in power supply filter circuits.

All these choke inductors, having a core of high magnetic permeability, are not usable in an NMR probe circuit in close proximity to the coil sensor.

A nonmagnetic choke inductor of moderate size can be made easily by coiling 22 turns of enameled copper wire (o.d. 0.3 mm, awg#29) on

---

[54] Depending on the number of turns.

the body of an axial-lead 1/2W old carbon composition resistor[55] (length 9.5 mm, diameter 4.0 mm, Fig. 2.46). Such a choke operates from about 100 MHz to above 1 GHz. Its low frequency inductance has been measured to be 880 nH. If a higher inductance is required, to operate at a lower frequency, two or three chokes can be connected in series. Eventually, a wire with a smaller diameter will permit an increase in the number of turns, and hence an increase in the inductance value (Fig. 2.46).

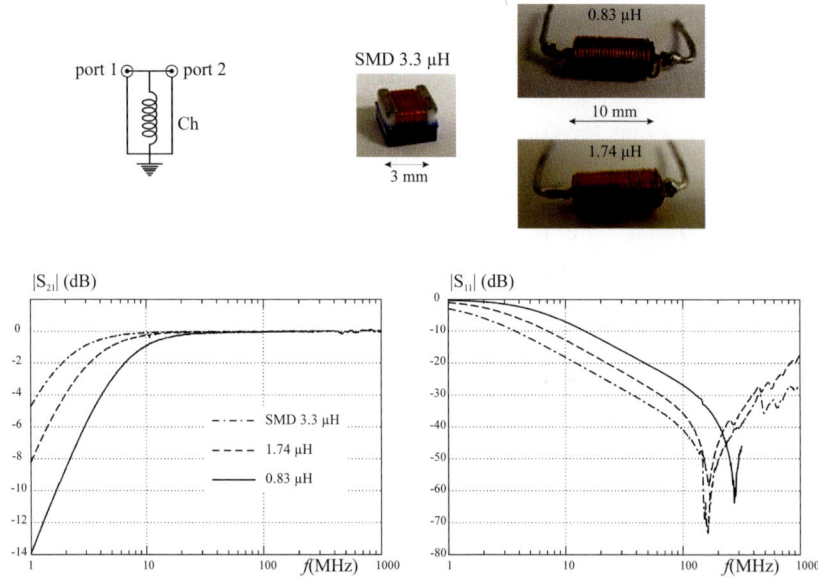

Fig. 2.46 Nonmagnetic choke inductors, commercial or homemade. The lower limit of the operating frequency range depends on the low frequency inductance value, as clearly visible on the S_{21} (left). The upper limit of the operating range is less dependent on the SRF (right), at least up to the GHz range.

Small SMD ceramic chip inductors of inductance above 1μH or more can also be used as a choke (Fig. 2.46) if the DC current is less than a few hundreds of mA.

---

[55] The resistance value does not matter. It can be anything greater than some kΩ. The resistor is only used as the coil mount.

2.4.3.8 Adjustable inductor

Adjustable inductors can be found in a variety of equipment, operating generally under GHz.

Depending on the available space, power requirements and frequency, the adjustable inductor can have different aspects and sizes.

Old variable inductors such as the roller inductor and the variometer are seldom used in NMR probe circuits operating nowadays at frequencies in the range of hundreds of MHz. However, these may be useful for very low frequency MR experiments, including NQR.

The roller inductor is a solenoid with a moving connection that allows the number of turns in the circuit to be changed, and hence the inductance value.

The variometer is constituted of a pair of coils connected in series and arranged in such a way that the angle between their axes can be changed from 0 to 90°. The mutual inductance between both coils adds or subtracts to the inductances of the isolated coils, changing the total inductance as a function of the angle between the coil axes.

Many RF instruments include small variable inductors. Basically, these are made of a solenoid coil and a movable threaded core that changes the permeability, hence the inductance.

The core is made of ferrite or powdered-iron material which increases the inductance as it fills in the interior of the solenoid coil.

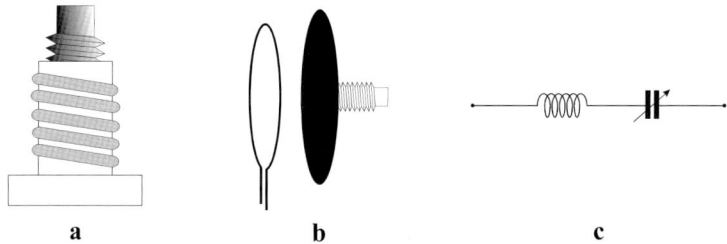

Fig. 2.47 Variable inductors. (a) A movable core (high μ or low ρ) more or less fills the region of high flux of the coil. (b) The extension of the magnetic flux created by the loop is controlled by a movable conductive disk. (c) A variable reactance is made up using an inductor and a variable capacitor in series.

It can be also made of a conductor, aluminum, brass or copper, which decreases the inductance as it fills the coil [Fig. 2.47(a)]. The latter is obviously best suited for an NMR probe. Any conductive material close to a coil disturbs the magnetic flux distribution, hence modifies the inductance. An example of a variable small inductor at very high frequency is shown in Fig. 2.47(b). Eddy currents reduce the inductance as a function of the distance between the movable piece of copper and the coil.

Last, but not least, a variable inductor, operating at a specific frequency, can be made with a fixed inductor in series with a variable capacitor [Fig. 2.47(c)].

2.4.3.9 *Capacitors (generalities)*

In this paragraph we describe the properties of capacitors that are used in probe electrical circuits (multiple frequency resonant circuits, filters, diplexers, traps, etc...). These, except variable capacitors, are mostly multilayer ceramic chip capacitors made of a stack of parallel conducting plates separated by a dielectric material.

A capacitor has reactive impedance X which decreases with the frequency as:

$$X = \frac{1}{C\omega}. \qquad (2.105)$$

The dielectric losses of a capacitor can be represented by a high resistance $r_{dielectric}$ in parallel with the capacitor. According to Eq. (2.7), the Q factor originating from these loss mechanisms is:

$$Q \approx r_{dielectric} C\omega. \qquad (2.106)$$

The contribution of the conductor losses (skin effect) is better represented by a resistance $r_{conductor}$ in series with the capacitor. The corresponding Q factor can be written from Eq. (2.5) as:

$$Q \approx \frac{1}{r_{conductor} C\omega}. \qquad (2.107)$$

Depending on the dominant loss mechanism, the Q factor of a capacitor can be independent of frequency[56] [Eq. (2.106)] or inversely proportional to the 3/2 power of the frequency[57] [Eq. (2.107)]. Above 30 MHz, the losses in multilayer ceramic chip capacitors are dominated by the conductor losses. Thus, at RF, all the losses are best represented by a series resistance, the so-called *ESR* (Equivalent Series Resistance).

The Q factor of a capacitor at RF is thus defined as:

$$Q = \frac{1}{ESR * C\omega}. \qquad (2.108)$$

The Q factor for capacitors used in NMR probe circuits is typically one order of magnitude greater than the Q for inductors.

The above equations are only valid well below the *SRF* of the capacitor. The high frequency model of RF capacitors is addressed in the following paragraph.

2.4.3.10 Capacitors (high frequency model)

Fig. 2.48 shows the circuit that models capacitors up to the microwave range. C_0 is the low frequency capacitance measured around 1 MHz and *ESR* is the series resistance representing the losses in the capacitor. L_s and C_p are parasitic components leading to series and parallel resonances of the device.

The first parasitic resonance (*SRF*) is due to the combination of the inductance L_s in series with C_0. In the capacitive domain this results in a change of the effective capacitance value as a function of frequency according to:

$$C_{\mathit{eff}} = \frac{C_0}{1-\omega^2 L_s C_0} = \frac{C_0}{1-\left(\omega/\omega_{SRF}\right)^2}. \qquad (2.109)$$

---

[56] The tanδ for most high quality dielectrics is a constant over a large frequency range. The corresponding parallel resistance is inversely proportional to the frequency.

[57] Due to the skin effect, the resistance of conductors is proportional to the square root of the frequency.

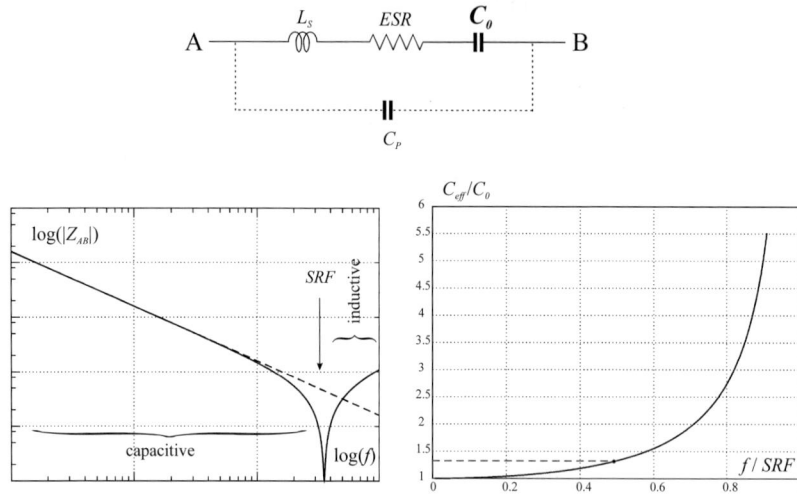

Fig. 2.48 Upper: high frequency model of a capacitor. Lower: magnitude of a capacitor impedance as a function of frequency up to and slightly above its *SRF* (left). The dashed line is the impedance of the ideal capacitor. The effective capacitance as a function of the ratio of frequency to the *SRF*, in the domain of capacitive impedance (right).

At half the *SRF*, the effective capacitance is increased by more than 30% of the nominal value! For multilayer ceramic chip capacitors, the series inductance of the component is approximately 0.3 to 2 nH, essentially depending on the case dimensions, not on the nominal capacitance value. The corresponding *SRF* for a 10 pF capacitor is typically between 1 and 3 GHz, just above the operating frequency range of most NMR experiments, but a nominal 100 pF capacitor of small size ($L_S = 0.7$ nH) exhibits a *SRF* of the order of 600 MHz, which results in an effective capacitance of 133 pF at 300 MHz.

Additional parasitic inductances and capacitances associated with the NMR probe circuit reduce the *SRF*. It should be kept in mind that a wire connection of only 10 mm[58] has an inductance of 5 nH, which adds to L_s, reducing *SRF* by a factor between two and three!

A parasitic capacitance C_p between the two terminals of the capacitor leads to parallel resonances, but generally at much higher frequencies than the *SRF*. This effect can be neglected at NMR frequencies.

2.4.3.11 Fixed capacitor

Unlike inductors, capacitors included in NMR probe designs are, most of the time, obtained commercially.

The preferred capacitors are microwave ceramic chip capacitors. The advantages are a relatively high *SRF*, very high *Q* at RF frequencies, good stability and high breakdown voltage.

The American Technical Ceramics (ATC) company maintains a line of microwave capacitors that are widely used in NMR probes. Typically, the capacitors are proposed in four different sizes. The so-called cases A, B, C and E are fabricated as rectangular boxes with roughly squared case size footprint dimensions of 1.5, 3, 6 and 10 mm, respectively. These are rated to working voltages up to 150, 500, 2500 and 3600 V, respectively.

Equations for the *SRF* and Q values are given below for such capacitors as a function of size, capacitance values and frequencies.

$$SRF_{MHz} = k_{SRF} / \sqrt{C_{pF}}, \qquad (2.110)$$

$$Q = \frac{k_Q}{C_{pF}^\alpha f_{MHz}^{1.5}}. \qquad (2.111)$$

Approximate constant values for k_{SRF}, k_Q and α are estimated from the manufacturer datasheets and are given in Table 2.6 for the ATC 100 series.[59]

---

[58] A sheet of copper of 3 mm width and 10 mm length has an inductance of 4.8 nH [Eq. (2.82)] and a round wire with a diameter of 1.2 mm and length of 10 mm has an inductance of 5.0 nH [Eq. (2.79)].

[59] The *SRF* and *Q* values obtained with these equations are only indicative. Reliable values are obtained when the capacitance is comprised between 10 and 100 pF and the

Table 2.6 Approximate constants for Eq. (2.110) and (2.111) as a function of the size of the ATC 100 microwave chip capacitors. L is the length (the distance between the terminals), W is the width and T the thickness of the capacitor.

Case	Size ($L \times W \times T$ mm)	k_{SRF} $(MHz/\sqrt{pF})$	k_Q $(p^{-a}MHz^{3/2})$	α
A	1.40 × 1.40 × 1.45	9500	1.2 × 10^7	0.62
B	2.79 × 2.79 × 2.59	7000	2.8 × 10^7	0.85
C	5.84 × 6.35 × 3.68	4750	0.56 × 10^7	0.61
E	9.65 × 9.65 × 4.32	3650	0.58 × 10^7	0.65

Nowadays, many manufacturers propose a line of microwave chip capacitors specifically designed for NMR, such as Voltronics Corporation, Temex Ceramics, Vishay, Murata Manufacturing Co.

These capacitors can be obtained nonmagnetic, at least without nickel barrier terminations, or constructed with compensation for the diamagnetic susceptibility of the materials.

2.4.3.12 Variable capacitor

Variable capacitors include a static and a movable electrode separated by a high quality dielectric (Teflon, quartz, glass, sapphire, air and vacuum).

These capacitors are generally bulkier than fixed capacitors for the same capacitance value. The Q factor is worse and the SRF lower. Another difficulty is keeping a reliable contact with low resistance between the movable electrode and the corresponding terminal.

Due to the challenge in fabricating a good quality variable capacitor, these should mostly be obtained commercially. Microwave models are again the best choices for NMR probes. They assure high Q factors at RF with minimum size. However, the adjustable range is small, typically not greater than about 30 pF. Larger ranges are available but at the expense of the breakdown voltage and/or of the size and of the quality factor.

The best quality and highest withstanding voltage are obtained with vacuum variable capacitors. These are however bulky, fragile and expensive. Teflon capacitors are more robust and less expensive.

---

frequency between 100 MHz and the SRF. Greater accuracy can be obtained from the datasheets from each manufacturer.

Many manufacturers provide nonmagnetic variable capacitors, specially designed for NMR and MRI. The best known company is Voltronics Corporation. Also, Polyflon (a Crane co. company) or Sprague Goodman are well established suppliers of components for the NMR probe builders.

2.4.3.13 Resonators

The NMR probe coil that couples with the magnetic moments of the spins is generally a resonator. This comes from impedance matching the coil inductor to the resistive system impedance of 50 Ω (following chapter).

Other type of resonators may be encountered in NMR probe circuits such as traps or multiple tuning circuits that involve coupled resonators. Inductors or capacitors behave as resonators near the *SRF*, as do transmission lines that are not terminated by the characteristic impedance.

Eventually, dielectric resonators have been used for NMR and EPR [Webb, 2011].

In this paragraph the general properties of electrical RLC resonators, constituting a resistor, an inductor and a capacitor, are described.

Depending on the way these components are assembled together, one gets a series resonator (Fig. 2.49) or a parallel resonator (Fig. 2.51).

The resistance of the series resonator represents the losses in both the capacitor and inductor. The losses in the parallel resonator are usually dominated by the conductor and radiation of the coil inductor. These losses are represented by the resistance r_L. The resistance associated with the capacitor (r_C) is usually negligible.

The impedance across the terminals A and B of the series RLC resonator is:

$$Z_{AB} = r + jL\omega + \frac{1}{jC\omega} = r + j\left(L\omega - \frac{1}{C\omega}\right). \qquad (2.112)$$

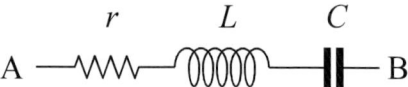

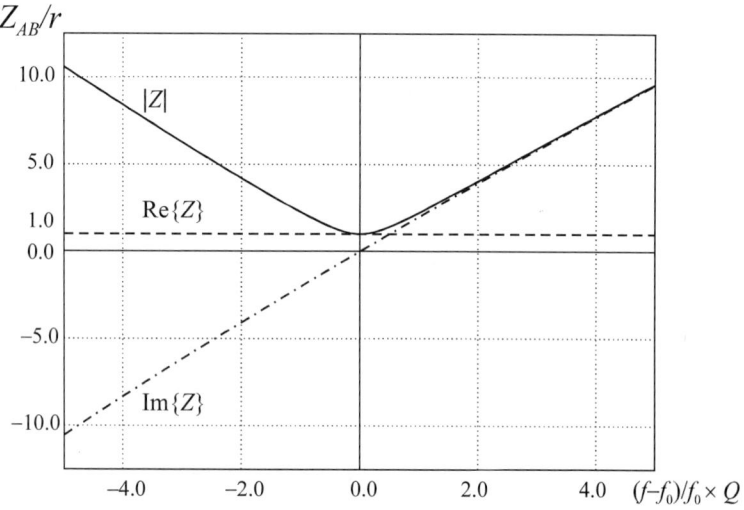

Fig. 2.49 Series RLC resonant circuit. The impedance components are shown on a reduced frequency scale around the resonant frequency.

The resonance occurs when the capacitance exactly compensates the inductor impedance:

$$L\omega_0 = \frac{1}{C\omega_0}. \qquad (2.113)$$

This happens for only one frequency, ω_0, defined by the well-known resonance condition:

$$LC\omega_0^2 = 1. \qquad (2.114)$$

When this condition is met, the energy exchanges back and forth, between the capacitor and the inductor, at each cycle, and dissipates slowly in the resistive component r. The dissipation rate (τ^{-1}) is related to the quality factor Q of the resonator:

$$Q = \frac{L\omega_0}{r} = \frac{1}{rC\omega_0}$$
$$\tau = \frac{2Q}{\omega_0}$$
(2.115)

On resonance, the impedance $|Z_{AB}|$ is a minimum (Fig. 2.49):

$$|Z_{AB}|(\omega_0) = \text{Re}\{Z_{AB}\} = r.$$
(2.116)

Thus, the current is a maximum and in phase with the voltage applied to the terminals. Around the resonance frequency it follows a rapidly varying function which can be represented by the so-called "universal resonance curve", Fig. 2.50 [Terman, 1943 p. 137]. The current (magnitude) decreases by a factor of 0.707 when the frequency of the applied voltage deviates by $\Delta f = f_0/Q$ around the resonant frequency.

The voltage across the capacitor or the inductor is maximum on resonance and is equal to the voltage applied to the terminals multiplied by Q. Around the resonance frequency, the voltage across the capacitor and inductor follows a similar function as the universal resonance curve.

The impedance $|Z_{AB}|$ across the terminals of the parallel resonant circuit similarly follows the universal resonance curve, provided Q is high (>10). The impedance Z_{AB} of the parallel circuit is given by:[60]

$$\frac{1}{Z_{AB}} = \frac{1}{r + jL\omega} + jC\omega,$$
(2.117)

---

[60] If Q is high (>10), the exact way in which the resistance is divided into the inductor or capacitor branches is of little importance [Terman, 1943, p. 144].

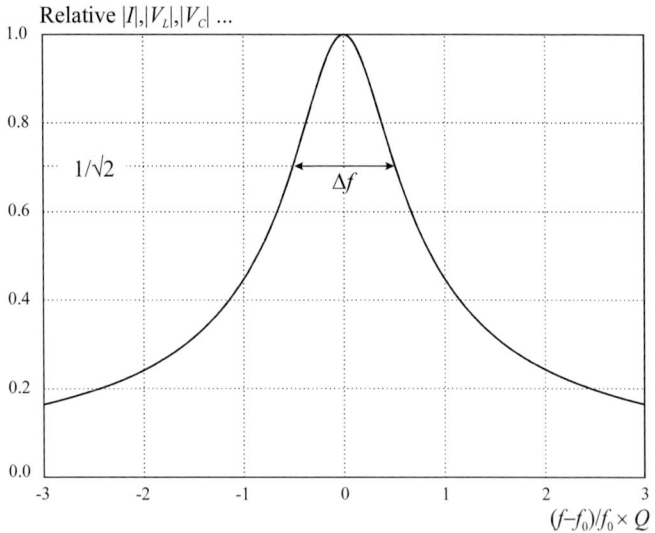

Fig. 2.50 Universal resonance curve. The current and voltages (magnitude) across the inductor and capacitor of a series resonant circuit are shown on a reduced frequency scale. A source of constant voltage is assumed. The curve is valid for $Q > 100$.

After some straightforward algebraic manipulations, one gets:

$$XL = L\omega$$

$$XC = \frac{1}{C\omega}$$

$$Z_{AB} = \frac{rXC^2}{r^2 + (XL - XC)^2} - jXC\frac{r^2 + XL(XL - XC)}{r^2 + (XL - XC)^2}$$
(2.118)

The resonant frequency of the parallel resonator can be defined in three different ways.

The first way to define the resonance is given by the condition that the impedance of the inductor (XL) is equal to the impedance of the capacitor (XC). This condition is equivalent to the series resonance condition Eq. (2.114). At this frequency, the impedance Z_{AB} is:

$$Z_{AB}(\omega_0) = \frac{XL^2}{r} - jXL = L\omega_0(Q - j).$$
(2.119)

The imaginary part of Z_{AB} is not cancelled out but is Q times smaller than the real part of the impedance. It may be neglected if Q is high. At this frequency, the real part of the impedance Z_{AB} is high, equal to the reactance of either the inductor or the capacitor, multiplied by Q. The magnitude $|Z_{AB}|$ reaches a maximum of similar value.

$$\operatorname{Re}\{Z_{AB}(\omega_0)\} = QL\omega_0$$
$$|Z_{AB}|(\omega_0) = L\omega_0\sqrt{1+Q^2} \approx QL\omega_0 \quad (2.120)$$

Typically, if the reactance of one of the branches of the resonator is about 100 Ω and the Q factor is 100, the impedance near resonance reaches a value of 10 kΩ.

The second way to define the resonance is given by the condition that the imaginary part of Z_{AB} cancels out:

$$r^2 + XL(XL - XC) = 0. \quad (2.121)$$

Rearranging Eq. (2.121), one gets

$$\delta = \frac{r^2 C}{2L}$$
$$LC\omega_{0'}^2 = 1 - 2\delta \quad (2.122)$$

The correction term 2δ is of the order of $1/Q^2$. Thus, the condition Eq. (2.122) is identical to Eq. (2.114) to better than 1% if Q is greater than 10. The real part and magnitude of Z_{AB} are equal:

$$\operatorname{Re}\{Z_{AB}(\omega_{0'})\} = |Z_{AB}|(\omega_{0'}) =$$
$$L\omega_{0'}^2 \left(Q + \frac{1}{Q}\right) \approx QL\omega_0 \quad (2.123)$$

The third way to define the resonance of the parallel resonator is given by the condition that the real part of Z_{AB} is a maximum. This occurs when:

$$LC\omega_{0''}^2 = 1 - \frac{r^2 C}{2L} = 1 - \delta. \quad (2.124)$$

Again, if the Q factor is high, the resonance condition Eq. (2.124) is identical to that of Eq. (2.122) or Eq. (2.114).

The real part and magnitude of Z_{AB} are given by:

$$|Z_{AB}|(\omega_{0''}) \approx \text{Re}\{Z_{AB}(\omega_{0''})\} = \frac{QL\omega_{0''}}{1 + \delta^2(1+Q^2) - 2\delta}. \quad (2.125)$$

The three resonance frequencies of the parallel resonator do not differ by an amount greater than 1% if Q is higher than 10. Hence the resonance condition of the series configuration will be commonly assumed.

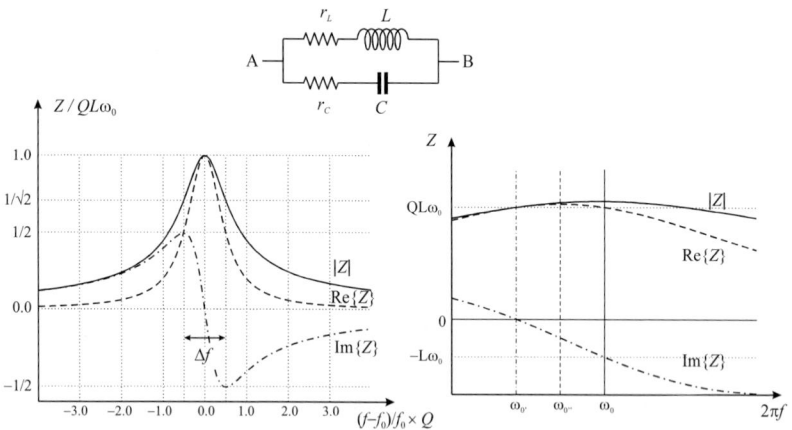

Fig. 2.51 Resonant parallel circuit. The impedance components are shown on a reduced frequency scale around the resonant frequency (left). The width Δf is defined as the bandwidth at half height of the real component of Z and at 0.707 of the magnitude of Z. The three resonant frequencies, as defined in the text, are shown to the right in an expanded view around to ω_0.

The functional dependence of the impedance Z_{AB} on frequency is shown in Fig. 2.51 for a high Q parallel resonator. The "bandwidth" of the circuit $\Delta\omega$, the Q factor and the resonant frequency are related by the following equation:

$$Q\Delta\omega = \omega_0. \qquad (2.126)$$

With this definition, $\Delta\omega$ is the width at half height of $\text{Re}\{Z_{AB}(\omega)\}$ or the width at 0.707 times the height of $|Z_{AB}|(\omega)$.

On resonance, the current through the terminals A and B is a minimum due to the high impedance $|Z_{AB}|$ of the circuit. The currents in the inductor and capacitor are equal and of opposed phase.

When the resonator is matched to a given impedance R_S (Fig. 2.52), the power gain is equal to unity, assuming the matching box is lossless. The current amplitude in each branch of the resonator is a maximum on resonance and equal to:[61]

$$|I_0| = \sqrt{\frac{2P_{in}}{r}}, \qquad (2.127)$$

where P_{in} is the *rms* power delivered by the matched source.

The voltage across the terminals of the resonator is:

$$|V_{AB}| = Q\sqrt{2rP_{in}} = \sqrt{2P_{in}QL\omega}. \qquad (2.128)$$

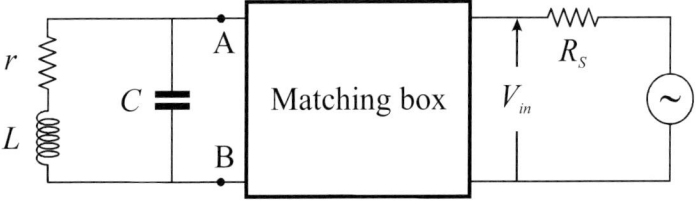

Fig. 2.52 The voltage across the parallel resonator is magnified by a factor given by Eq. (2.129), when power matched to a low impedance source.

---

[61] The factor of two assumes a sinusoidal excitation.

The on resonance voltage magnification is given by:

$$\frac{|V_{AB}|}{|V_{in}|} = \sqrt{\frac{QL\omega}{R_S}}. \qquad (2.129)$$

Assuming, for example, a 100 Ω inductor having a Q factor of 100, the voltage magnification when driven by a source with an internal impedance of 50 Ω will be equal to 14. The peak voltage across the capacitor will reach 4.5 kV if the resonator is driven by a 1 kW transmitter. Such figures should be taken into account when designing an NMR probe.

2.4.4 Distributed components made by transmission lines

In some designs it would be convenient to fabricate the required lumped elements of a probe circuit using transmission line sections. This would be the case, for example, for the design of flexible probeheads made up using specially modified ink-printers [Mager *et al.*, 2010] or possibly 3D printers.

A short section of opened transmission line behaves as a capacitor. If shorted at one end, the line behaves as an inductor.

The impedance presented by a length of opened line (Fig. 2.53) is, assuming a lossless transmission line [Eq. (2.81)], given by:

$$Z_{AB} = \frac{Z_c}{j\tan\theta}. \qquad (2.130)$$

If shorter than $\lambda/4$, the line behaves as a capacitance C:

$$C = \frac{\tan\theta}{Z_c \omega}. \qquad (2.131)$$

Similarly, the impedance presented by a length of shorted line (Fig. 2.53) is, again assuming a lossless transmission line, given by:

$$Z_{AB} = jZ_c \tan\theta. \qquad (2.132)$$

RF Components

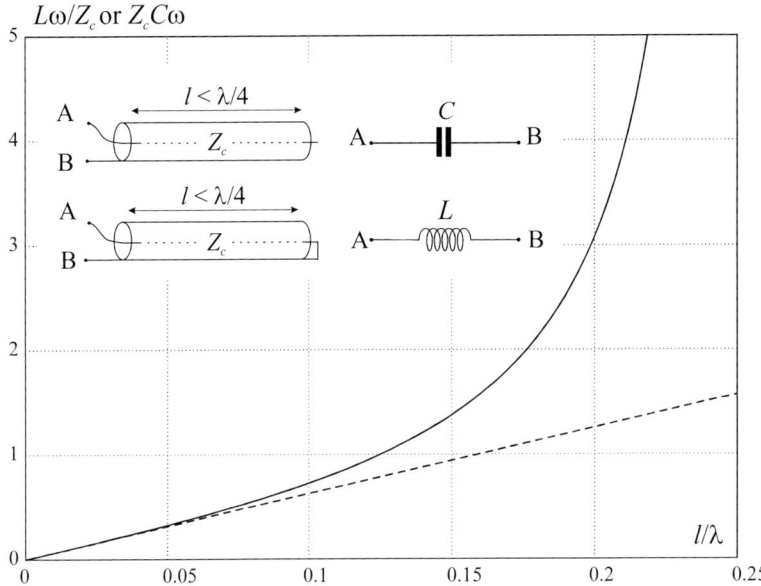

Fig. 2.53 Relative impedance of an opened (capacitance) or shorted (inductance) transmission line as a function of length. The dashed line shows the impedance of the corresponding pure capacitance or inductance.

If shorter than $\lambda/4$, the line behaves as an inductance L:

$$L = \frac{Z_c \tan\theta}{\omega}. \qquad (2.133)$$

Z_c is the characteristic impedance of the line and θ is related to the line length by:

$$\theta = 2\pi \frac{l}{\lambda}. \qquad (2.134)$$

In the equation above, l is the physical length of the line and λ is the wavelength in the line, which is related to the frequency f by:

$$\lambda = \frac{c}{f\sqrt{\varepsilon_r}}, \qquad (2.135)$$

where c is the speed of light, and ε_r the relative permittivity of the dielectric material of the line.

The capacitance C and the inductance L are frequency dependent, except for very short section of line ($\theta < 25°$). In this condition, the capacitance can be written as:

$$C \approx \frac{1}{Z_c} \frac{\sqrt{\varepsilon_r}}{c} l. \qquad (2.136)$$

In the above configuration the line is assumed to be functioning in the differential mode. The inductance L is given by:

$$L \approx Z_c \frac{\sqrt{\varepsilon_r}}{c} l = L_d, \qquad (2.137)$$

where L_d is the inductance of the differential mode.

Considering a standard 50 Ω line with a Teflon dielectric ($\varepsilon_r = 2.1$), the inductance is about 2.4 nH cm^{-1} and the capacitance is about 1.0 pF cm^{-1}. These values correspond to the inductance and capacitance per unit line length.

For longer lines, the effective capacitance and inductance are given by an equation similar to Eq. (2.109):

$$\begin{aligned} C_{\mathit{eff}} &= \frac{C}{1-\left(f/f_{\lambda/4}\right)^2} \\ L_{\mathit{eff}} &= \frac{L_d}{1-\left(f/f_{\lambda/4}\right)^2} \end{aligned}. \qquad (2.138)$$

In the equation above, $f_{\lambda/4}$ is the resonant frequency of a quarter wave line, which is given by:

$$f_{\lambda/4} = \frac{c}{4l\sqrt{\varepsilon_r}}. \qquad (2.139)$$

Another common configuration (printed capacitor) is sketched in Fig. 2.54(a). In contrast to the previous case, the line functions in the common mode. The opened section of line behaves as a capacitance in series with a "parasitic" inductance. The capacitance is given by Eq. (2.136). The series inductance is determined by the inductance of the line conductors joining A and B. For a symmetric stripline of length l (Fig. 2.55) the "parasitic" inductance L_c can be estimated using Eq. (2.37). It is related to the differential mode inductance L_d by the following equation:

$$L_c = \frac{1}{2}L_d + M, \qquad (2.140)$$

where M is the mutual inductance between the two parallel strips. L_c does not depend on the characteristic impedance of the line, while C and L_d do.

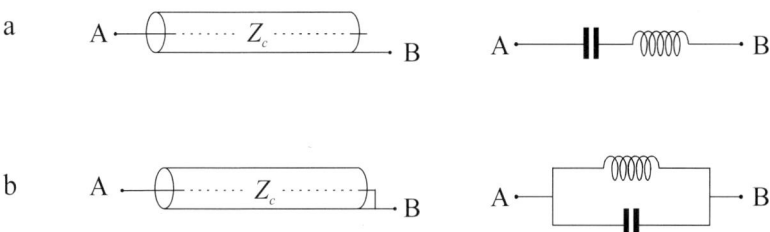

Fig. 2.54 (a) A capacitor made with a transmission line functioning in the common mode. The terminals of the capacitor are at opposed ends of the line. This is in contrast with Fig. 2.53 where the terminals were at one end of the line. The equivalent circuit is shown to the right. The inductance is given by the "common mode inductance" L_c given in Eq. (2.140). The capacitance is between the two conductors, given by Eq. (2.136). (b) An inductor made on the same principle, with only one end shortened. It is equivalent to a parallel resonant circuit. The inductance and capacitance of the circuit have the same values as above.

A realistic simulation of the printed capacitor can be done as shown in Fig. 2.55(a). The line is decomposed into elementary low-pass Π-filter cells. The cut-off frequency of each elementary cell must be much higher than the expected operating frequency.

The simulated response of this printed capacitor [Fig. 2.55(b)] is consistent with a series L–C resonant circuit with inductance and capacitance values as given by Eq. (2.37) and Eq. (2.136), respectively.

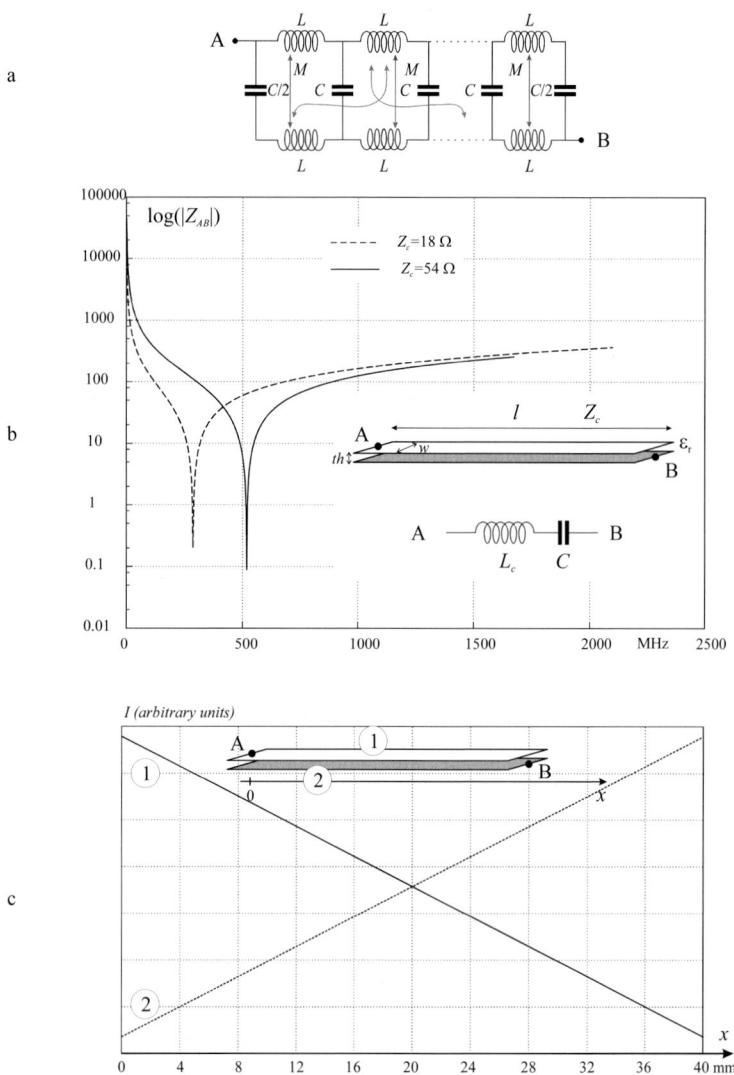

Fig. 2.55 (a) The line made of two parallel strips is decomposed into ten small cells. The mutual inductance between all conductors is taken into account in the simulation. (b) For a given length (l = 40 mm) and strip width (w = 5 mm) the resonant frequency of the printed capacitor depends on the characteristic impedance of the line (Z_c = 18 Ω, th = 1 mm or Z_c = 54 Ω, th = 1.6 mm). (c) The current amplitude in each strip varies linearly from one end to the other. Both current amplitudes are positive (common mode). The total current over the line is a constant.

The total current in the section of the transmission line is the same as in the equivalent series resonant circuit. In the upper conductor of the line [Fig. 2.55(c)] it decreases linearly from the connected point A to the open end. In the lower conductor, it increases linearly from the open end toward the connected point B. This current distribution is frequency independent, except for the amplitude of the total current which increases obviously when approaching the resonance frequency of the printed capacitor.

As an example, let us consider the resonant square coil of Fig. 2.56(a).

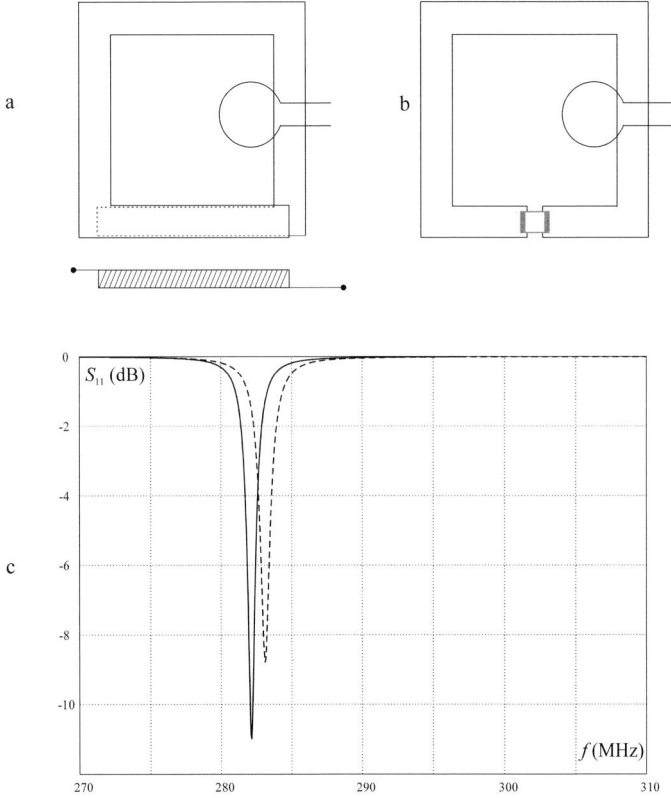

Fig. 2.56 A square coil tuned by a printed cap (a) and tuned by a discrete capacitor of the same capacitance (b). The two configurations resonate at about the same frequency (c).

One side of the coil is made of the superposition of two strips separated by a dielectric substrate. This part, behaving as a capacitor, resonates the coil at a frequency defined by the total capacitance of the section of line. The "parasitic" inductance of the section of line is absorbed into the total inductance of the coil. The result is that the square coil integrating a printed capacitor circuit [Fig. 2.56(a)] or a lumped capacitor of equal capacitance C [Fig. 2.56(b)] resonate at very similar frequencies, 282 MHz and 283 MHz, respectively [Fig. 2.56(c)].

The tuning capacitance C, given by Eq. (2.136) is represented in Fig. 2.57 as a function of the length of the section of line for different Z_c values and for different dielectric materials.

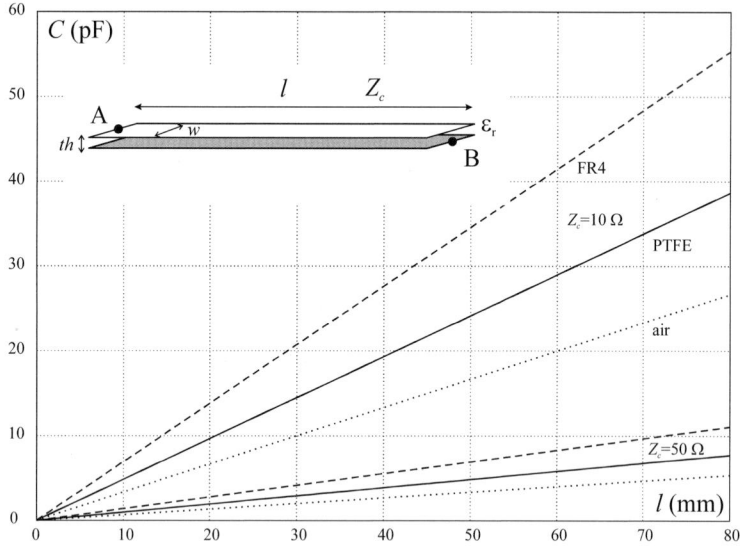

Fig. 2.57 Capacitance of a printed cap as a function of length, for typical dielectric materials and two characteristic impedances Z_c. The characteristic impedance is determined by the strip width (w) and the substrate thickness (th), for a given dielectric substrate.

A large range of capacitance values, up to 50 pF, can be obtained using the appropriate dielectric and characteristic impedance and using a reasonably long section of line. As a result, this technique can be helpful in many probe designs.

If the section of line is shorted at one end [Fig. 2.54(b)], it behaves as a trap (parallel resonant circuit, Fig. 2.58) when its length is equal to a quarter of the wavelength.

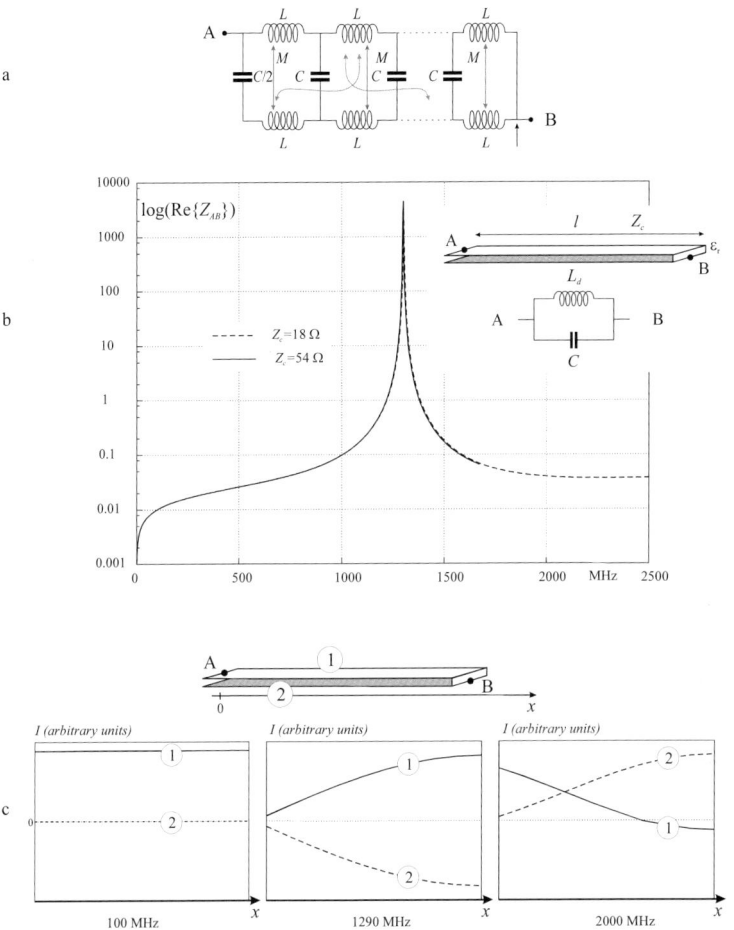

Fig. 2.58 (a) Model of a two strip transmission line. (b) Frequency response of the two strips ($l = 40$ mm, $w = 5$ mm) separated by a dielectric and connected as shown. It behaves as a parallel resonant circuit. The resonance frequency does not depend on the characteristic impedance of the line. (c) The current distribution on the strips depends on the frequency. At low frequency (left), current flows from A to B in the upper strip. At resonance (middle), the line functions in the differential mode. The resonance is determined by the line length ($\lambda/4$). The line functions in a mixture of differential and common mode above the resonance frequency, (right).

The resonance frequency is given by Eq. (2.139). In contrast with the open ended configuration, it does not depend on Z_c but solely on the permittivity of the dielectric material and on the length l of the line section.

Close to the resonance, the line functions in the differential mode. As a result, the total current on this section is null (Fig. 2.58), as it should be with a trap circuit. At lower frequencies, all the current passing in the circuit flows mostly in the upper conductor of the line (Fig. 2.58).

At higher frequency, the line functions in a mixture of differential and common modes, leading to a nonzero net current. The resonant frequency is so high that this trap circuit is hardly usable in an NMR coil circuit.[62] A high permittivity dielectric substrate could possibly be used to decrease its resonant frequency.

2.5 Non-linear Devices (Diodes)

Diodes are two-terminal components that can be encountered in probe circuits, in RF measuring instruments and many electronic modules. Diodes can be used as switching devices, RF small signal rectifiers, voltage controlled RF attenuators and, eventually, voltage controlled capacitors. Other diodes are able to emit light [Light Emitting Diodes (LEDs) and Laser diodes]; they will not be considered here.

Basically, a diode is formed by the junction between two regions (N and P) of a semiconductor crystal doped with different impurities. The N semiconductor is rich in negative charges (electrons) while the P semiconductor is rich in positive charge (holes). The electrons can flow easily from the N to P regions if the junction is properly polarized, leading to a small resistance, typically of the order of several ohms. If the voltage across the diode is reversed no current can flow, corresponding to a high resistance. When reverse-biased, the diode presents a small capacitance. The typical voltage–current characteristic of a diode is schematized in Fig. 2.59.

---

[62] For example, the section of line previously described as a printed capacitor on PTFE substrate resonates near 1290 MHz.

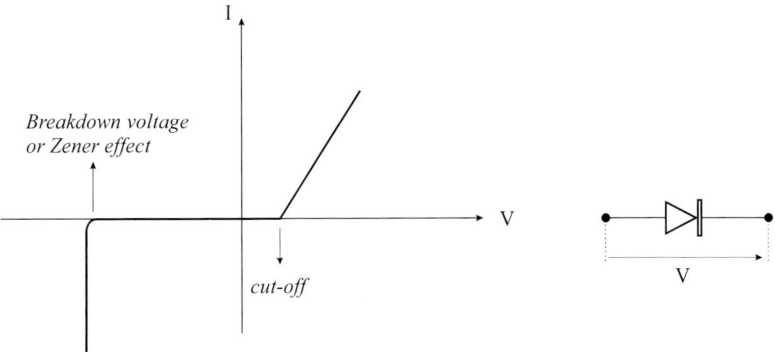

Fig. 2.59 Typical voltage–current characteristic of a semiconductor diode.

The cut-off voltage depends on the specific constitution of the diode. Germanium diodes typically exhibit a cut-off voltage of 0.2 V, while silicon diodes exhibit a cut-off voltage of 0.6 V. Schottky diodes that are used in high-speed switching power supplies and in RF and microwave applications exhibit a cut-off of about 0.4 V. The direct voltage of LEDs is much higher, of the order of 2 V.

2.5.1 Silicon and Schottky P–N diodes

Typical applications of these diodes are high-speed switching and RF detectors.

As a switching device, they can be used, for example, in the Transmit/Receive (TR) switching interface between the probehead and the NMR console or MRI scanner (Chapter 4). Passive decoupling circuits also make use of these commutation diodes. Very common part numbers are the 1N4118 (or 1N914) and 1N4148. Although many probe circuits still use these diodes, PIN diodes are nowadays preferred due to their much better switching characteristics in the RF and microwave frequency range.

As an RF detector, germanium diodes or Schottky diodes are the simplest and cheaper means to measure the amount of RF power delivered to or reflected from a circuit (Chapter 10). They permit characterization of frequency resonances, evaluation of probe efficiency, measurement of Q factors, etc... Such crystal detectors are generally extremely broadband. For example, the HP8470B low barrier Schottky diode detector operates from DC to 18 GHz. The typical response of such a crystal detector to an applied input power is shown in Fig 2.60.

The sensitivity is limited to an input power of around −30 dBm when connected to a standard oscilloscope. When the output voltage of the detector is lower than 10 mV (input power of around −17 dBm), the rectified voltage is proportional to the square of the input voltage (square law response range). When the output voltage is above 200 mV (corresponding to an input power of 10 dBm or 1 mW in 50 Ω), the response is linear.

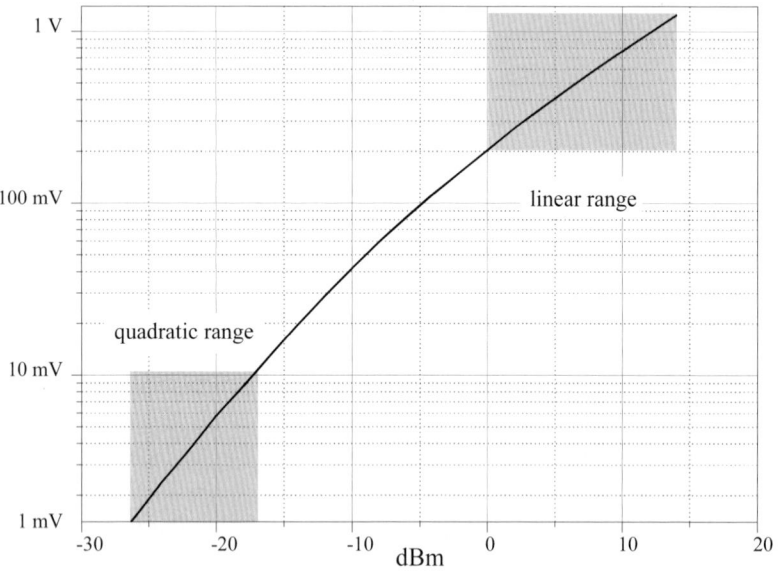

Fig. 2.60 Typical response of a broadband crystal detector, matched to 50 W (output rectified DC voltage as a function of the input RF power). 0 dBm corresponds to 1 mV in 50 W. The range of the horizontal scale corresponds to the power delivered by standard RF generators. The vertical scale corresponds to the standard oscilloscope sensitivity.

2.5.2 PIN diodes

PIN diodes include an "intrinsic" (I) layer between the P and N semiconductors. Accordingly, they exhibit different characteristics compared to a conventional diode. At low frequency, they behave similarly to a rectifier, but at high frequency they behave as a resistor. The transition frequency depends on the fabrication of the diode. Depending on the thickness of the I-region, PIN diodes can operate from the MHz to the microwave range.

The forward resistance of a PIN diode can be controlled over as much as four decades, making this component suitable for the design of voltage controlled switch or very wide band attenuators. When a DC current is applied to the diode, it exhibits a very low forward-biased[63] resistance (typically 0.1–1 Ω). The resistance increases to about 10 kΩ when no DC current is applied. The reverse-biased capacitance is also extremely small, of the order of 1 pF, typically much lower than the capacitance of conventional silicon diodes.

These diodes are supplanting the less expensive silicon diodes in passive and active probe circuits due to their much better RF characteristics. Common nonmagnetic PIN diodes suitable for NMR receiving circuits are the MA4P7470F from MACOM. The high currents required in transmit coils can be withstood using, for example, the nonmagnetic UM9415 or HUM2020 from Microsemi.

Remarkably, PIN diodes controlled by a small DC current of a few mA can handle a huge RF current in switching circuits.

2.5.3 Varactors

All diodes exhibit a capacitance when reverse-biased. The capacitance is due to the depletion region that separates the positive and negative charges of the P and N semiconductor layers. Because the thickness of the depletion region is proportional to the square root of the reverse

---

[63] The Microsemi HUM2020 has a typical resistance of 0.1 Ω at 4 MHz when the bias current is of the order of 500 mA.

voltage, the capacitance is inversely proportional to the square root of the applied voltage.

Varactors [Barnes *et al.*, 1961] are specially designed to present a relatively large capacitance (of the order of 10–100 pF) when reverse-biased. They are suitable to be used as voltage controlled tuning capacitance permitting compact probeheads to be built. However, distortions are to be expected in the presence of strong signals due to the non-linear characteristics of varactors. Accordingly, they should be used preferentially in receive-only probehead circuits.

The non-linear characteristics are however beneficial in applications such as harmonics multipliers, mixers and parametric amplifiers [Qian *et al.*, 2012; 2013].

2.6 Active Devices (Low Noise Amplifier or LNA)

The most critical part of the receiver chain of an NMR console is the preamplifier. Its role is to amplify the small signals from the NMR probe to a level that allows them to be conveniently treated by the receiver circuits, while adding a minimum of noise.

2.6.1 *Noise Factor (F) and Noise Figure (NF)*

The noise factor[64] F of an amplifier is defined as the ratio of the total output noise power to the noise output power arising from thermal noise[65] [Fish, 1994].

---

[64] The non-dimensional ratio F is conventionally named "noise factor" and the logarithm, in dB, is named "noise figure" [Fish, 1994, p. 103].

[65] At RF, the noise is mostly the "thermal noise" due to the random fluctuation of conduction electrons. The mean square voltage of the noise is proportional to the absolute temperature, to the bandwidth in which it is measured and is independent of the frequency. In addition, there is a "shot noise" resulting from the random passage of charge carriers across a potential barrier. Such a noise appearing in semiconductors has the same characteristics as the thermal noise.

The noise factor F determines how the SNR ratio is spoiled at the output of the amplifier.

$$F = \frac{total\ output\ noise\ power}{output\ noise\ power\ from\ source} = \frac{Input\ SNR}{Output\ SNR}. \quad (2.141)$$

The noise figure NF (in dB) is related to F by:

$$NF = 10\log_{10}(F). \quad (2.142)$$

For an ideal amplifier, $F = 1$ or $NF = 0$ dB. Inevitably, electronic devices add some noise, increasing F. Good RF amplifiers typically exhibit F ranging from 1.12 (0.5 dB) to 2 (3.0 dB).

An amplifier is generally made of a cascade of amplifier stages proving the required gain (Fig. 2.61). The F of a cascade of amplifiers is given by [Lee, 2004, p. 464]:

$$F = F_1 + \frac{F_2 - 1}{G_1} + \frac{F_3 - 1}{G_1 G_2} + \ldots, \quad (2.143)$$

where F_i is the noise factor of the ith stage and G_i its power gain.

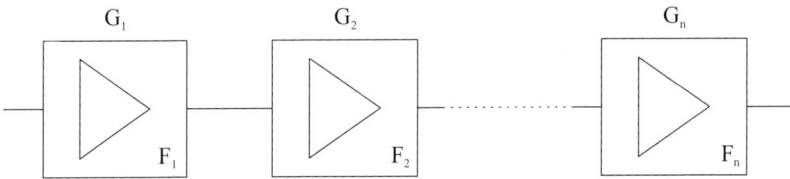

Fig. 2.61 An amplifier is constituted of n stages, each having a gain G and a noise factor F. In a well-designed amplifier chain, the first stage alone determines the F of the whole preamplifier. This requires that the gain G_1 of the first stage is sufficiently high. However, gains must be carefully distributed among each of the amplifier stages, avoiding any overload. The corresponding characteristic is the "1 dB compression" (P1dB) and/or "the third-order intercept" level (IP3). The 1 dB compression point is the output level at which a gain reduction of 1 dB is observed. The IP3 is the level producing third-order mixing products at the output having the same level as would be expected assuming a linear response of the amplifier. These parameters characterize the non-linearity of the amplifier. The NF and the IP3 (or the P1dB) determine the dynamic of the receiver which is particularly important for NMR instruments.

The F is primarily determined by the first stage if its gain (G_1) is sufficiently high. In this case, the output noise of the first stage exceeds the input noise of the second stage.

Typically, if the gain of the first stage is 15 dB ($G_1 = 30$) and the NF of the second stage is 3 dB ($F_2 = 2$), the F of the cascaded amplifier increases by 0.03. The NF becomes worse by about[66] 0.14 dB (if F_1 is close to unity) which could be acceptable.

The NF of an LNA typically ranges from 3 dB (Fig. 2.62) to less than 0.5 dB.[67]

Broadband RF amplifiers operating from 1 to 1000 MHz are common. They exhibit a gain of 25–30 dB and an NF of about 3–3.5 dB from 20–30 MHz to 1 GHz. Below 20 MHz, the NF generally increases to 4 dB at 10 MHz.

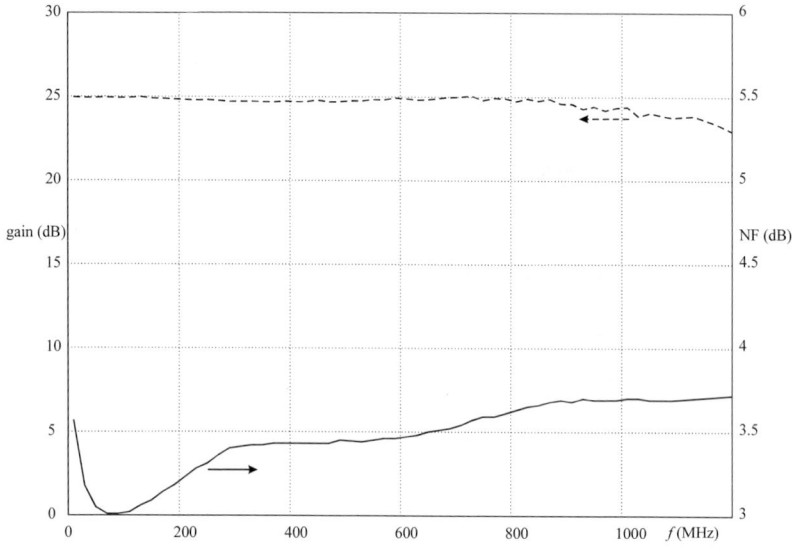

Fig. 2.62 Typical gain and NF of a small-signal, 50 Ω, broadband preamplifier operating from 10 MHz to 1 GHz.

---

[66] The added noise contribution of the second stage can be estimated as $1/(\ln(10)F_1)$ dB.

[67] In these cases, they are called "ultra low noise" amplifiers.

The NMR preamplifiers in the 1990s had an *NF* typically comprising 1.5–2.5 dB operating down to 10–20 MHz. They are generally broadband (Fig. 2.63).

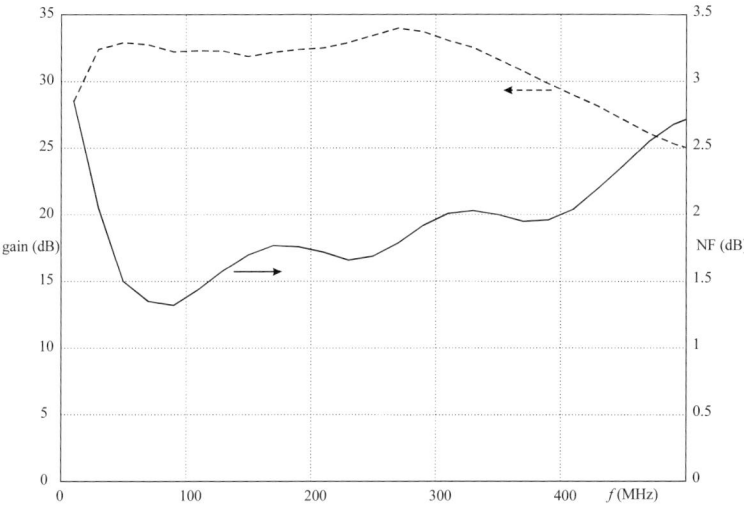

Fig. 2.63 Typical gain and *NF* of a broadband preamplifier of an NMR instrument of the 1980s–1990s, operating in the RF range.

Nowadays, the *NF* of these LNA has improved down to about 1–1.5 dB for standard preamplifiers, still broadband. Cooled preamplifiers, with improved *NF*, were also specially designed, principally for direct carbon-13 spectroscopy.

Ultra low noise preamplifiers, operating down to about 30 MHz, are now available with an *NF* of 0.5 dB or less. They are generally narrowband, nonmagnetic and specially designed for array coils used in MRI.

2.6.2 *LNA designs for NMR frequencies*

The primary requirement of a preamplifier for NMR is that it has the lowest possible *NF*. It must also support strong signals without distortion (high third-order intercept point or IP3), at least for spectroscopy on biological samples or for proton MRI. The total gain of the preamplifier should be about 25–30 dB.

A preamplifier (Fig. 2.64) comprises a first stage using an ultra low noise transistor and providing a gain of about 15–20 dB. A second stage is expected to provide a gain of around 10–15 dB. The second stage also improves the isolation between the input and the output of the LNA (S_{12} < −40 dB).

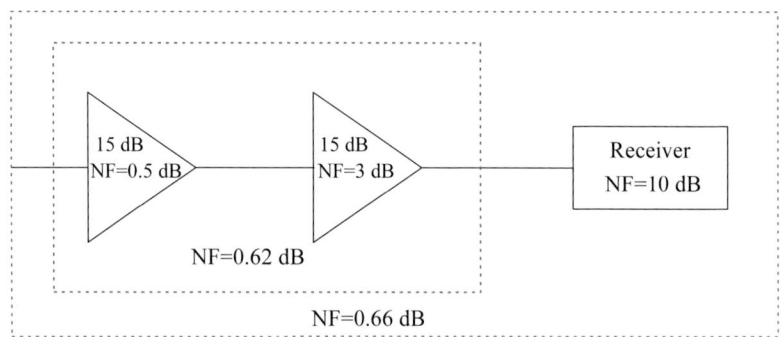

Fig. 2.64 A typical receiver channel of an NMR instrument. The preamplifier includes two stages with an ultra low noise front end transistor. With the values quoted here, the *NF* of the input transistor is brought to 0.62 dB. The preamplifier is designed to drive a receiver having a front end *NF* of 10 dB. In this case (the worst), the *NF* value is due to cabling and to a high-level front end passive mixer. Nevertheless, the total *NF* of the receiver chain would be 0.66 dB.

The *NF* of the second stage should not be worse than 3 dB[68] to keep the front end *NF*. Such a preamplifier is able to drive a mixer or a receiver having a front end *NF* as high as 10 dB, without significant degradation of the total *NF* (Fig. 2.64).

Some of the commonly available GaAsFET or recent HEMT (High Electron Mobility Transistor) transistors are usable in the RF range down to the MHz range, although they have been mainly designed for operation in the UHF and microwave domain.[69]

---

[68] A Monolithic Microwave Integrated Circuit (MMIC) such as the MAR series by Mini-Circuits can be used as a second stage.

[69] The manufacturers generally provide the characteristics of these transistors from 500 MHz up to several GHz, seldom from 100 MHz and extremely rarely below.

Many LNA designs[70] have been proposed during the last three decades using the widely known MGF1302 GaAsFET transistor. A single stage is expected to provide a gain of around 20 dB with *NF* under 0.7 dB in the VHF band. A generic preamplifier using this transistor is shown in Fig. 2.65.

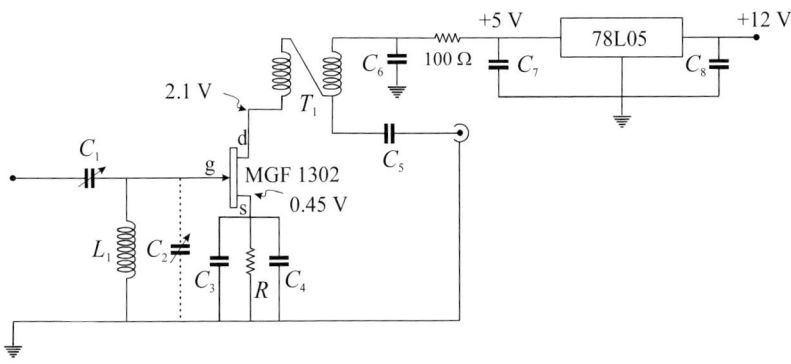

Fig. 2.65 A typical low noise preamplifier, widely used at RF and VHF. The input stage includes a matching *L*-network designed to transform the source impedance of 50 Ω into the noise matched impedance of the MGF 1302. The resistance *R* controls the bias current in the transistor. C_{3-8} are low impedance capacitors. In some designs, a capacitor C_2 is included, increasing the source impedance while using a moderately-sized inductor. The output is broadband (transformer T_1).

The output is broadband using a 4:1 transformer while the input is designed to match a standard 50 Ω source to the optimum noise source impedance of the transistor. The resistance in the transistor source connection is adjusted to bias the drain current to an optimum value. This design provides a good example of what could be the first stage of a good narrowband LNA for NMR. It has been used by manufacturers (for example Advanced Receiver Research, Burlington, USA) providing a low frequency (down to about 20 MHz) LNA readily usable for NMR.

Fig. 2.66 shows the gain and *NF* that have been obtained with a commercial LNA based on this design. The gain increases to 26 dB

---

[70] The radio-amateur literature is full of such designs for the 144 MHz, 432 MHz and 1296 MHz frequency bands.

around 20 MHz and the *NF* is a minimum (1 dB) around the same frequency.

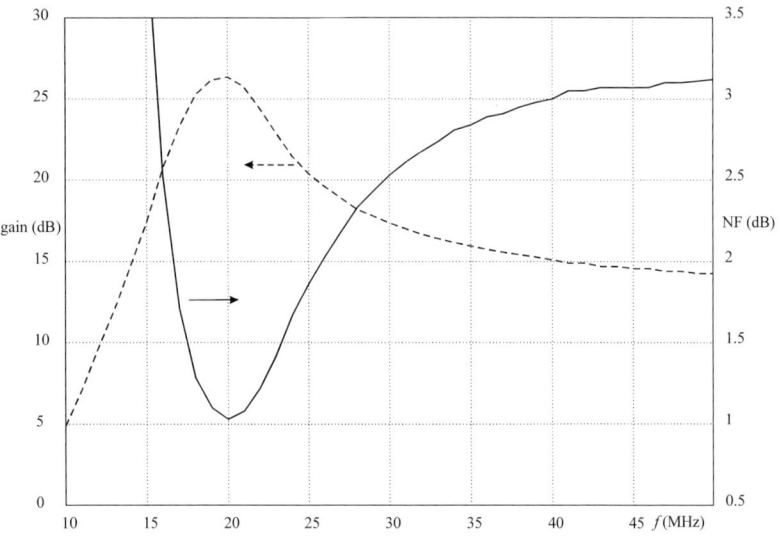

Fig. 2.66 Typical gain and *NF* obtained with the design shown in Fig. 2.65. The *NF* shows a sharp minimum around 20 MHz, the frequency at which the input matching network matches the 50 Ω source impedance to the optimum source impedance of the transistor at this frequency. The gain is a maximum around the same frequency due to the input network.

Above 30 MHz, the gain is roughly constant (15 dB) over a wide frequency band and the *NF* increases to around 3 dB. Such behavior is typical of a preamplifier based on a circuit such as that depicted in Fig. 2.65. The input LC circuit is designed to provide the required impedance transformation at 20 MHz for the best *NF*. Due to this circuit, the input impedance of the preamplifier is much lower than 50 Ω.

The more recent Enhanced mode pseudomorphic HEMT (E-p-HEMT) potentially has a minimum *NF* that can be as low as 0.1–0.3 dB at RF, allowing the design of an LNA with *NF* around 0.5 dB without too many difficulties. Again, these transistors are generally fully characterized at UHF and above.

Microwave Technology, Inc (Fremont, USA) provides low frequency[71] (from 43 to 127 MHz) LNAs, the MPM series, suitable for NMR and based on their own GaAs devices. These LNA have two amplification stages, provide a gain of 25–30 dB, and an *NF* less than 0.5. They are typically narrowband (Fig. 2.67). The input impedance is generally very low, and these preamplifiers are well suited for array coil applications.

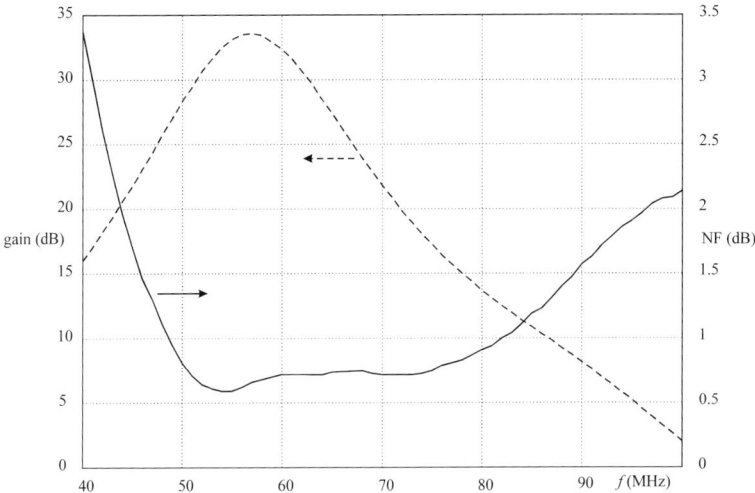

Fig. 2.67 Gain and *NF* of a preamplifier using a p-HEMT transistor. The preamplifier includes two stages with a tuned output network for impedance matching to 50 Ω. The *NF* and gain response are very dependent on the adjustment of the input *L*-network for noise matching and of the output circuits.

The first stage of an amplifier using an enhanced mode p-HEMT transistor is similar to the GaAsFET preamplifier shown in Fig. 2.65, except the biasing. The resistance *R* in the source connection of the transistor is not required. Instead, an active biasing circuit is recommended (Application Note AN5057, AVAGO Technologies). The input transformer is always an LC circuit. The corresponding

---

[71] They can be optimized to lower frequencies, down to 30 MHz, without significant degradation of their characteristics.

components should be of very high quality in order to minimize any losses which may be deleterious to the NF.

Due to the growing demand for ultra LNAs operating at the clinical MRI scanner (1.5 and 3.0 T, corresponding to a proton frequency of 64 and 128 MHz, respectively), the market proposes a lot of solutions. Among others, Agile MWT (Tampa Bay, USA) proposes two LNAs operating in the ranges 40–50 MHz and 60–70 MHz, with a gain of 30 and 29 dB, respectively, an *NF* of 0.3 dB and an input impedance of 2 Ω. Wan Tcom,[72] Inc (Chanhassen, USA) proposes narrow band LNAs designed for center frequencies of 64 MHz to 400 MHz with a gain around 28 dB, an *NF* of 0.4 dB and input impedance Z_{in} ranging from 0.1 to 1.5 Ω, depending on the specific model. This manufacturer also proposes a model operating at a lower frequency of 32 MHz with a gain of 28 dB, an *NF* of 0.7 dB and Z_{in} of 3 Ω.

Ultra low noise LNAs operating at low frequency (lower than 20–30 MHz) are possibly, although generally seldom, available. Advanced Receiver Research proposes[73] a line of LNA exhibiting an *NF* of around 0.5 dB down to 2.0 MHz. One example of such an LNA has already been described above. It exhibited an *NF* of about 1 dB at 20 MHz, which is not too bad, but that could possibly be improved after some careful adjustment.

Finally, integrated circuits are now available. The model TL5500 from Texas Instruments, for example, is specially designed for MRI operating at 64 or 128 MHz with a gain of about 28 dB and an *NF* lower than 0.5 dB.

The design of LNAs at very low frequency (below 10 MHz) seems to be problematic and probably requires different technologies. This may be due to the so-called "excess noise" that is generated in a resistor or a semiconductor submitted to a current. This noise adds to the thermal noise, independent of the current flowing in the device, and has already been solely considered in the previous discussion. The power spectrum of the excess noise has an approximate $1/f$ dependency. It could therefore dominate the thermal noise at low frequencies. This noise is known as

---

[72] A Wireless And Network Telecommunication Company.
[73] http://www.advancedreceiver.com/page12.html.

"low frequency" or "flicker" noise [Fish, 1994, pp. 84–88]. The equations that apply to the thermal noise and the previous definition of NF are no longer valid. An important difference is that the $1/f$ noise power available within a given frequency band (f_{min}, f_{max}) is dependent on the ratio f_{max}/f_{min} (proportional to $\log(f_{max}/f_{min})$) while the thermal noise power is proportional to the bandwidth $(f_{max} - f_{min})$.

2.7 Metamaterials (Flux Guides)

A Metamaterial (MM) is an artificial periodic structure composed of arrays of elements whose dimensions are small compared to the interacting wavelength. The purpose of creating such materials is that they may completely change the wave propagation properties compared to the natural media. Much of the historic research related to MMs is weighted from the view of antenna beam shaping within microwave engineering just after World War II. Furthermore, MMs appear to be historically linked to the research related to artificial dielectrics throughout the late 1940s, the 1950s and the 1960s. In 1967, Victor Veselago produced an often cited, seminal work on a theoretical material that could produce extraordinary effects that are difficult or impossible to produce in nature. At that time he proposed that a reversal of Snell's law, an extraordinary lens, and other exceptional phenomena can occur within MMs. This theory lay dormant for a few decades. There were no materials available in nature, or otherwise, that could physically realize Veselago's analysis. Not until 33 years later did the properties of such materials became a subdiscipline of physics and engineering.

However, at the end of the 20th century this description was expanded by John Pendry, a physicist from Imperial College in London. In the 1990s he was consulting for a British company, Marconi Materials Technology, as a condensed matter physics expert. The company manufactured a stealth technology made of a radiation-absorbing carbon for naval vessels. The company did not understand the physics of the material and asked Pendry if he could understand how the material worked. Pendry discovered that the radiation absorption property did not come from the molecular or chemical structure of the material, i.e., the carbon per se, but came from the long and thin physical shape of the carbon fibers. After successfully deducing and realizing the carbon fiber structure, Pendry further proposed that he try to change the magnetic properties of a nonmagnetic material, also by altering its physical structure. The material would not be intrinsically magnetic, nor inherently susceptible to being magnetized. He envisaged fabricating a nonmagnetic composite material which could mimic the movements of electrons orbiting atoms. However, the structures are fabricated on a scale that is magnitudes larger than the atom, yet smaller than the radiated wavelength. The radiation of interest is from radio waves and microwaves, through infrared to the visible wavelengths. Scientists view this material as "beyond" conventional materials. Hence, the Greek word "meta" was attached, and these are called metamaterials.

2.7.1 *Magnetic metamaterials at RF*

In the near field domain, the magnetic and electric field components of the electromagnetic radiation are decoupled. In this regime, both field components may be manipulated independently by appropriate materials having a specific permeability or permittivity. The focusing or the transfer of the near field magnetic component in an NMR experiment could be done using an MM having the appropriate permeability, irrespective of its permittivity.

Magnetic MMs have been specifically developed for NMR and MRI since the beginning of the 2000s by several laboratories that have proposed and demonstrated potential applications [Wiltshire *et al.*, 2001; Behr *et al.*, 2004; Allard and Henkelman, 2006; Freire *et al.*, 2008; Mosig *et al.*, 2009; Radu *et al.*, 2009; Freire *et al.*, 2010; Khennouche *et al.*, 2012; Xie *et al.*, 2012].

In contrast to the "field profiling" approaches that manipulate the spatial distribution of the near field magnetic component using a limited number of properly disposed coils, the MM is expected to behave as a homogeneous material having a non-static, magnetic permeability.

The relative[74] magnetic permeability μ_r can be either positive (paramagnetism) or negative (diamagnetism). Particular values of the effective permeability are of special interest for applications in NMR, namely $\mu_r = 0$ (reflector), $\mu_r = -1$ (lens) or $\mu_r \to \infty$ (flux guide).

For an MM having a discrete structure, an effective permeability, which is a characteristic of a continuous medium, can be defined [Pendry, 1999; Chen *et al.*, 2004; Smith *et al.*, 2005] after a "homogeneization" procedure [Baena *et al.*, 2008].

Close to the material, the discrete structure is apparent and the detailed spatial distribution of the near field component of the magnetic field will depend on the constitutive elements of the material. But, at some distance, the material may look like a homogeneous macroscopic medium. The sole restriction that material elements should fulfill in order to constitute the basic structure of an MM is that their dimensions and the distance between them in the lattice are much smaller than the wavelength. This is not an issue for the NMR frequencies in the RF domain as wavelengths are of the order of meters.

A magnetic MM can be viewed as a network of small magnetic dipoles that can be aligned coherently in such a way that the desired permeability is obtained. Ferromagnetic materials exhibit high permeability but are not usable in an NMR environment and do not provide an access to the negative permeability range, required for applications such as focusing or sub-wavelength imaging. A simple conductive ring exhibits, at RF, a diamagnetic susceptibility.[75] However, the permeability is moderate and is hardly tractable. On the contrary, a tuned loop (a resonator) exhibits, near resonance, a frequency dependent magnetic permeability that can cover a very large range from positive to negative values.

---

[74] The relative permeability is defined as $\mu_r = \mu/\mu_0$ where μ_0 is the permeability of the vacuum.

[75] The diamagnetic permeability is due to currents induced in the loop by varying magnetic field applied perpendicular to the rung plane. It should not be confused with the static magnetic susceptibility, which is characteristic of a so-called nonmagnetic material such as copper or alumina.

Quantitatively, the effective relative permeability of a (homogenized) periodic structure of small resonators is written as:

$$\mu_{eff}(\omega) = 1 - \frac{F\omega^2}{(\omega^2 - \omega_{0m}^2) + j\omega\gamma}. \qquad (2.144)$$

ω_{0m} is the "magnetic resonant frequency" of the material, γ is a dispersive factor representing the losses in the resonators. The resonators being very much smaller than the wavelength, the losses arise mostly from the resistance of the conductor and of the dielectric material forming the tuning capacitance. γ is related to the resonator Q factor by [Lipworth et al., 2014]:

$$\gamma = \frac{\omega_{0m}}{2Q}. \qquad (2.145)$$

F is a "filling factor" representing the fraction of material volume that is magnetically active. It is related to the ratio of the effective diameter (d) of the resonant elements and the lattice spacing between them (p) as:

$$F = \frac{\pi}{4}\left(\frac{d}{p}\right)^2. \qquad (2.146)$$

In practical designs, F is determined empirically.

The components (real and imaginary) of μ_{eff} are plotted as a function of frequency in Fig. 2.68, according to Eq. (2.144).

The effective permeability exhibits a strong variation around the "resonance frequency" of the material [Pendry et al., 1999].

The response of the MM around the resonant frequency allows one to define three narrow regions corresponding to the so-called $\mu_{eff} \to \infty$, $\mu_{eff} = -1$ and $\mu_{eff} = 0$ conditions that are of interest in the context of NMR/MRI.

Below the resonance, μ_{eff} is greater than 1 and is a maximum at resonance ($\mu_{eff} \to \infty$ case). Above the resonance, the permeability possibly becomes negative, if the losses in the material are small (Fig. 2.68 – high Q), providing access to the $\mu_{eff} = -1$ and $\mu_{eff} = 0$ cases.

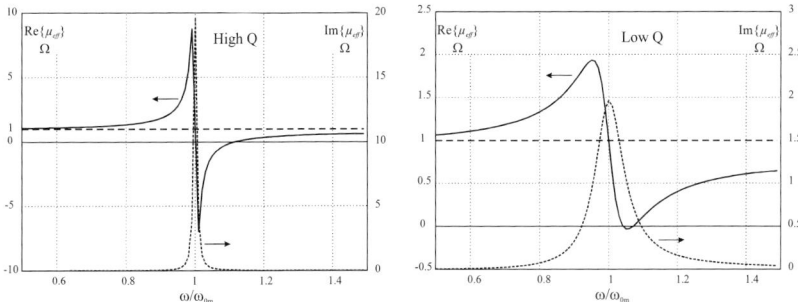

Fig. 2.68 Effective permeability of a metamaterial constituted of a lattice of resonators, as a function of frequency [Eq. (2.144)]. Right: using high Q ($Q = 50$) resonators. A negative permeability is easily obtained above resonance. Left: using low Q resonators ($Q = 5$). The effective permeability is always positive.

The material resonance ω_{0m} occurs when the "dipoles" representing the microstructure of the MM are collectively aligned by a given magnetic excitation. Both the amplitude and the phase of the response will determine the particular magnetic properties of the material.

In fact, the condition defined as either $\mu_{eff} \to \infty$ or $\mu_{eff} = -1$ or $\mu_{eff} = 0$ concerns different components of the complex permeability. In the ideal case of a lossless material:

$$\mu_{eff} = 0 \Leftrightarrow \mu' = 0, \mu'' = 0$$
$$\mu_{eff} = -1 \Leftrightarrow \mu' = -1, \mu'' = 0 \quad (2.147)$$
$$\mu_{eff} \to \infty \Leftrightarrow \mu' = 0, \mu'' \gg 1$$

The magnetic permeability of the MM can be evaluated by observing the change of the complex impedance of a finite-sized loop coupled with the material. The imaginary component of the loop impedance is sensitive to the real component of the permeability (μ'). Reciprocally, the real component of the loop impedance is dependent on the imaginary component of μ_{eff} (μ''). Such a property has been used, for example, to evaluate the effective permeability of the Swiss-Roll structure [Wiltshire *et al.*, 2001]. Another possibility is to observe the transmission properties (scattering parameter S_{21}) of the material.

The frequency range in which the effective permeability is negative (which provides the unusual magnetic properties of MM that are so attractive) is defined, in the limit of a low-loss material, as:

$$\omega_{0m} < \omega < \frac{\omega_{0m}}{\sqrt{1-F}}. \qquad (2.148)$$

The upper limit is called the magnetic plasma frequency, in analogy with dielectric plasmon resonance. At this frequency the permeability is null.

Due to the resonant nature of the MM elements, the desired permeability (and/or permittivity) is obtained only in a very narrow frequency band. As a result the MM are well adapted for NMR, this technique being naturally narrow-band.

Due to the discrete nature of the lattice, and even if the lattice constants are extremely small respective to the wavelength, the near field produced very close to the MM elements can be very different to that expected for a continuous medium. Although this can sometimes be an advantage for some NMR experiments, this should be taken into account in the design and evaluation of the MM in this context [Algarin *et al.*, 2011].

Finally, as a lens, an MM is particularly attractive due to its unusual property that the resolution is not limited by the imaging wavelength [Pendry, 2000]. This gives the possibility of performing sub-wavelength high resolution images of an object, using RF. This has already been demonstrated by MR images obtained without spatial encoding in an experiment using a lattice of Swiss-Rolls [Wiltshire *et al.*, 2003a; Smith *et al.*, 2004]. The resolution is limited here by the separation of the lattice elements.

2.7.2 Components of a magnetic metamaterial

While it was recognized, as early as 1952, that the magnetism of a conducting ring can be considerably increased by inserting a capacitor (forming a resonant circuit) [Shelkunoff and Friis, 1952], the development of magnetic MM components only started during the late 1990s. These structures rely on a ring resonator (Split Ring Resonator or SRR [Pendry et al., 1999]) that operates from the microwave range up to several THz (Fig. 2.69). At the opposite side of the frequency spectrum, capacitively loaded SRRs, tuned loop resonators or multi-turn configurations allow MM magnetic components resonating in the RF range, down to some MHz, to be designed.

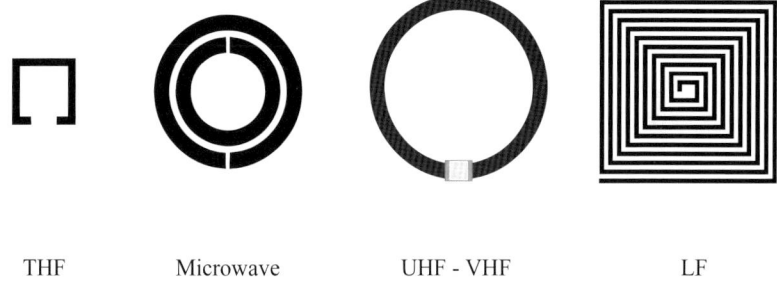

| THF | Microwave | UHF - VHF | LF |

Fig. 2.69 Examples of basic magnetic structures for metamaterials. They are resonant circuits exhibiting a large change in the permeability, near resonance frequency.

Magnetic components like the SRR are particularly well suited to applications in NMR, especially MRI. At this point, it is noteworthy that metallic wires may also exhibit magnetic properties [Pendry et al., 1999]. This has been exploited in the design of a "wire collimator" or endoscope for MRI [Radu et al., 2008; Radu et al., 2009].

MM elements for the manipulation of the electric component of the electromagnetic field have also been proposed by Smith et al. [Smith et al., 2000]. The elementary cell of these materials incorporates a magnetic resonant loop and small conductor rod. The rod gives rise to an effective negative permittivity, below the plasmon resonance. With this latter improvement a material with a negative refraction index has been obtained [Shelby et al., 2001] approaching the Perfect Lens [Pendry, 2000].

Up to now, the incorporation of dielectric components in the MM lattice has not been considered for NMR. It can however be envisaged that the electric near field components of an NMR coil could also be manipulated with some profit, permitting the negative effects of the E-field to be avoided, especially within biological samples.

In the following, only magnetic elements of MM, acting on the H-field, will be considered and the discussion will be limited to "structures" that have been already proposed for NMR.

2.7.2.1 *"Swiss-Rolls"*

These resonators have a finite dimension along their principal axis (Fig. 2.70). They are designed on concepts similar to the high frequency stripline resonator [Decorps and Fric, 1969, 1972] and loop-gap [Hardy and Whitehead, 1981; Froncisz and Hyde, 1982]. A higher inductance, which allows a high magnetic polarizability to be obtained at low frequency down to the RF range, is obtained with the Swiss-Roll[76] [Pendry et al. 1999; Wiltshire et al., 2001]. It consists of several turns of a metal/dielectric laminate onto a cylindrical form. The metal should ideally be plated directly on a good quality dielectric rather than using adhesive copper.

The resonance frequency of a Swiss-Roll can be estimated by the following equation [Pendry et al., 1999; Behr et al., 2004]:

$$\omega_0 = \sqrt{\frac{sc^2}{2\pi^2 \varepsilon_r r_{\mathit{eff}}^3 (N-1)}}, \qquad (2.149)$$

where (Fig. 2.70) s is the distance between turns, ε_r is the relative permittivity of the dielectric substrate, N is the number of turns, c is the speed of light in free space and r_{eff} is an effective radius which takes account of the finite thickness of the roll.

---

[76] This resonator has also been proposed later as the Scroll-Coil resonator [Stringer et al., 2005].

An approximate estimation for r_{eff} is given by the mean radius of the roll ($r_0 + Ns/2$), where r_0 is the inner diameter. Remarkably, the resonant frequency does not depend on the coil length.

Swiss-Roll cells have been assembled together to form an MM flux guide [Wiltshire et al., 2001], a device transferring the image of a planar magnetic source with a resolution limited by the roll diameter [Wiltshire et al., 2003b] and a magnetically isotropic $\mu = -1$ lens capable of sub-wavelength imaging [Wiltshire et al., 2006]. As a flux guide in a classical MRI experiment, only one Swiss-Roll element may transmit the RF magnetic flux from one end to the other [Behr et al., 2004].

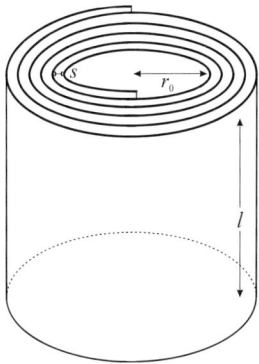

Fig. 2.70. The Swiss-Roll structure, also known as the scroll-coil. It is constituted of a metal/dielectric laminate rolled on a cylinder. The structure is self-resonant due to the distributed inductance/capacitance.

2.7.2.2 Ring planar resonators

The SRR, proposed by Pendry et al. [Pendry et al., 1999] is composed of two concentric rings split at opposed positions [Fig. 2.71(b)]. The two rings form a relatively high impedance transmission line.[77] The line is forced in the common mode due to the position of the gap. This configuration provides a lower resonant frequency than that possibly obtained with a single split ring [Fig. 2.71(a)].

---

[77] It has a high inductance to capacitance ratio.

A very approximate formula for the resonance frequency of a planar SRR [Fig. 2.71(b)] is [Pendry et al. 1999; Radkovskaya et al., 2005]:

$$\omega_0 = \frac{c}{\pi}\sqrt{\frac{3s}{r^3}}, \qquad (2.150)$$

where r is the mean radius of the SRR and s is the width of the gap between the two rings.

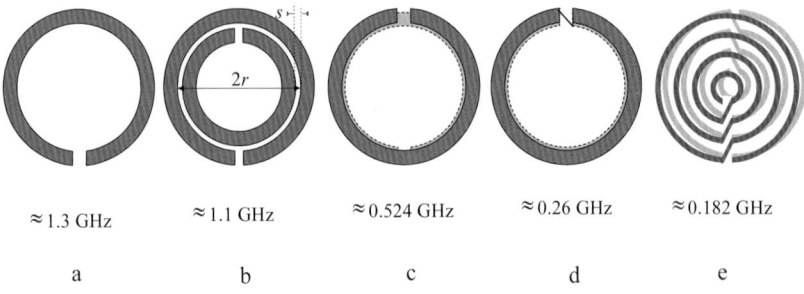

≈ 1.3 GHz   ≈ 1.1 GHz   ≈ 0.524 GHz   ≈ 0.26 GHz   ≈ 0.182 GHz

a   b   c   d   e

Fig. 2.71 A few split ring resonators characterized by equivalent external dimensions and of decreasing resonant frequency. (a) a simple split ring resonates when its perimeter is roughly equal to half the wavelength. (b) A second split ring adds some capacitance, decreasing the resonant frequency. (c) The capacitance between the two split rings is increased compared to (b), decreasing the resonant frequency further. (d) The inductance of the common mode is increased by the interconnect across the gap. (e) A multi-turn split ring resonator.

Typically, an SRR having a mean diameter of about 1 cm and a 0.2 mm gap between the two rings resonates at a frequency of about 3 GHz. If the two rings are etched on the opposed faces of a dielectric substrate, like in the parallel-plate split conductor (PPR) [Gonord et al., 1988] or the Twin Horseshoe Resonator (THR) [Gonord and Kan, 1994] designs, the electric field is confined in between the two rings, providing a better decoupling between the electric and magnetic field components. With such designs the resonance frequency can easily be decreased into the VHF–UHF range.

The resonance frequency of the THR (also known as the broadside-coupled SRR [Wiltshire, 2009, pp. 14–5; Freire et al., 2013]) can be estimated by the following equation [Haziza et al., 1997]:

$$\omega_0 \approx 4\sqrt{\frac{cZ_c}{L_t l_t \sqrt{\varepsilon_r}}}. \qquad (2.151)$$

In this expression, Z_c is the characteristic impedance of the line, l_t is the circumference of the resonator and L_t is the series inductance of the two rings, taking into account the mutual between the rings ($L_t = 2(L + M)$).
Z_c can be estimated from formulae given by Wheeler [Wheeler, 1965]:

$$Z_c = \frac{120\pi}{\sqrt{\varepsilon_r}} \left\{ \begin{array}{l} \dfrac{w}{h} + 0.44127 + \dfrac{\varepsilon_r + 1}{2\pi\varepsilon_r}\left[\ln\left(\dfrac{w}{h} + 0.94\right) + 1.45158\right] \\ + \dfrac{0.08226(\varepsilon_r - 1)}{\varepsilon_r^2} \end{array} \right\}^{-1} \qquad (2.152)$$

which is valid in the "wide strips" approximation. In practice it can be used when $w/h > 1$. w is width of the strip, h is the dielectric thickness and ε_r its relative permittivity.

If the strips are narrow ($w/h < 1$), the characteristic impedance can be estimated by the "narrow strips approximation", again given by Wheeler as [Wheeler, 1964]:

$$Z_c = \frac{120}{\sqrt{\varepsilon_{\mathit{eff}}}} \left[\begin{array}{l} \ln\left(\dfrac{4h}{w}\right) + \dfrac{1}{8}\left(\dfrac{w}{h}\right)^2 \\ -\dfrac{1}{2}\dfrac{\varepsilon_r - 1}{\varepsilon_r + 1}\left(0.45158 + \dfrac{0.24156}{\varepsilon_r}\right) \end{array} \right]. \qquad (2.153)$$

$$\varepsilon_{\mathit{eff}} = \frac{\varepsilon_r + 1}{2}$$

The resonance frequency of a THR is typically in the VHF–UHF domain. For example, the THR shown in Fig. 2.71(c) (mean diameter 30 mm, width of traces 4 mm, Duroid dielectric substrate thickness 1.6 mm and $\varepsilon_r = 2.4$) resonates at 524 MHz.

Resonance frequencies below 100 MHz can be obtained with a capacitively loaded SRR or with a simple loop tuned by a chip capacitor.

However, the inductance of the ring of the MM particle is an important parameter in the final μ_{eff}. At low frequencies, a high inductance should be preferred. Also a small granularity of the MM is desired, especially in the domain of NMR/MRI where the sensing coils should be close to the sample. The elementary cells of the MM should therefore be as small as possible. A multi-turn helical coil is well suited to exhibiting a high inductance in a small unit cell. For example a so-called Multi-turn Split-ring Resonator (MSTR) [Serfaty et al. 1997] of similar overall dimensions to the THR in Fig. 2.71(c) exhibits a resonance frequency of 182 MHz with four turns.

The THR or the MSTR resonators are built on a transmission line model, operating in the common mode. This mode is induced by the opposed position of the gaps in the SRR. Another configuration, which exhibits a higher inductance, is when the transmission line is forced in the common mode, as shown in Fig. 2.71(d). In this design, the gaps face each other so that the transmission line formed by the two conducting traces would function in the differential mode unless the common mode is forced by a crossover connection through the dielectric substrate.

As a result, the resonant frequency is divided by a factor of two compared to the equivalent THR [Fig. 2.71(c) and 2.71(d)]. A recently published design [Lipworth et al., 2014] using such multi-turn coils connected in series exhibits a resonance frequency of around 12 MHz with an overall dimension of less than 2 cm. This resonator is constituted of two multi-turn helical coils (17 turns each) etched on the two sides of a thin substrate (thickness about 0.254 mm). The width of the conductor is 0.2 mm and separated by 0.2 mm.

2.7.2.3 Transmission line network

Other magnetic resonators have been proposed for operation at RF with possible applications in NMR/MRI. These rely on the so-called Composite Right/Left-Handed (CRLH) MM transmission line resonator [Caloz and Itoh, 2005].

In the design by Mosig et al. [Mosig et al., 2009], the unit cell is constituted of a multilayer stack of a thin dielectric substrate with metallization on both faces and another thicker dielectric with a bottom ground plane (Fig. 2.72). The intermediate metallization is split along the y axis in order to form two capacitances (C_a) with the upper layer. An inductor (L_b) made of a thin conducting trace connects the middle of the upper metallization to the ground plane through a via. The equivalent circuit shown in Fig. 2.72 includes the "parasitic" elements (L_a and C_b) completing the resonant circuits.

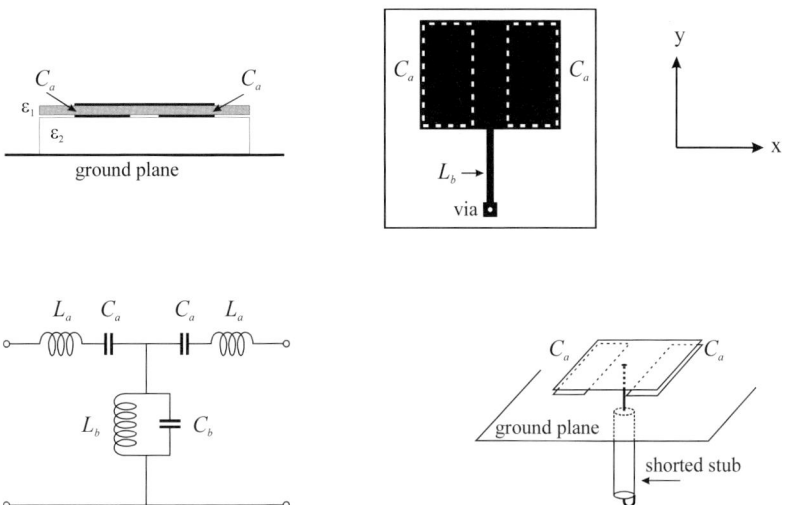

Fig. 2.72 CRLH MM transmission line resonator as designed by Mosig *et al.* (upper) [Mosig *et al.*, 2009] and Rennings *et al.* (lower right) [Rennings *et al.*, 2013]. The equivalent circuit is shown lower left. The main difference between the two designs is the L_b, C_b resonant circuit. In the design by Mosig *et al.* it is constituted of a thin strip relating the upper plate to the ground plane through a via. The circuit is tuned by the parasitic capacitance C_b between the strip and the ground plane. In the original design by Rennings *et al.*, the resonant circuit L_b, C_b is done by a shorted coaxial stub.

These elements are aligned in series forming a so-called Zeroth-Order Resonant Antenna (ZORA) capable of producing a near field magnetic component for NMR/MRI [Mosig *et al.*, 2009; Rennings *et al.*, 2013]. In these conditions, the "metamaterial" itself is used as a transmit/receive coil.

2.7.3 RF metamaterial designs

The MM resonant particles are combined together to form 2D or 3D lattices (Fig. 2.73), depending on the desired final applications.

Typically, flux compressors or flux guides are highly anisotropic. The resonant magnetic components are combined in an axial or planar configuration giving rise to effective permeability as:

$$\mu_\| \gg 1, \ \mu_\perp = 1, \quad (2.154)$$

where the $\|$ axis corresponds to the principal axis of symmetry of the MM lattice.

On the contrary, the construction of an MM lens focusing the magnetic field should be an isotropic 3D structure [Freire et al., 2008].

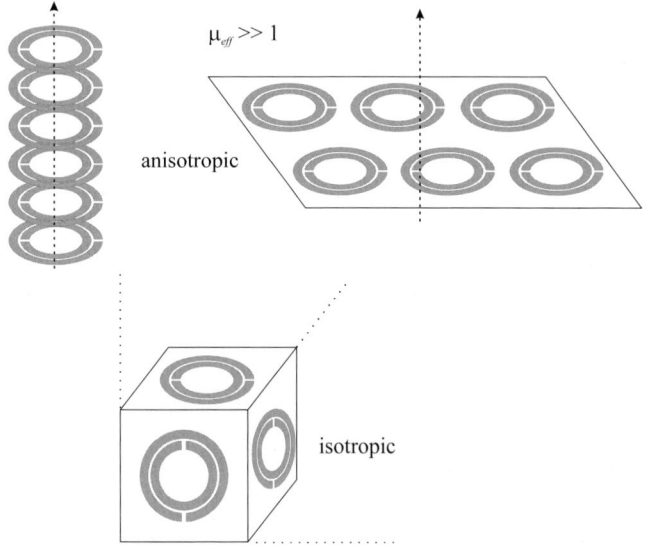

Fig. 2.73 Upper: arrangement of resonant microstructures forming a highly anisotropic magnetic material. Lower: the micro resonators are arranged on the faces of a cube forming a metamaterial unit cell. This cubic cell is repeated to form a final lattice having isotropic magnetic properties.

2.7.3.1 Flux compressors and flux guides

A tuned solenoid coil placed in a magnetic field oscillating at the same frequency would act as a cylindrical material of high permittivity. Such a "flux concentrator" could be used as an NMR sensor, placing the sample inside the solenoid. This design, not strictly an MM, is described separately in Chapter 4 (Fig. 4.27).

A tuned solenoid, shaped as a cone, or a set of tuned loops shaped on the same former, will act as a "flux compressor" [Wiltshire et al., 2004]. Such a material can be the particle of an MM assembly but can also be a simple "compressor" collecting the signal in a large area on one face and outputting it to a smaller one, or *vice versa*.

The Swiss-Roll is another device well suited for constructing a flux guide. This has been already demonstrated either as a 2D array of resonators [Pendry et al., 1999; Wiltshire et al., 2001] or as a single "bar" of high magnetic permeability [Behr et al., 2004].

Swiss-Rolls can also be assembled to form a magnetic yoke [Allard and Henkelman, 2006], concentrating the magnetic component of the RF field in the desired pathway.

2.7.3.2 Lenses

The ideal MM lens is a material that should exhibit a negative refraction index, $n = -1$ [Pendry, 2000]. For NMR/MRI, the effective component of the near field being solely magnetic, a $\mu = -1$ MM could be a "magnetic lens". Applications of this $\mu = -1$ MM have already been reported in the domain of MRI [Freire et al., 2008; Freire et al., 2010; Algarin et al., 2011]. In these cases, the lens manipulates the magnetic component of the RF field but the imaging is done using the classical spatial frequency encoding. Here, the lens acts as a magnetic field focuser.

Experiments have also been reported demonstrating that such MMs can be used for sub-wavelength imaging [Wiltshire et al., 2003; Smith et al., 2004]. The resolution remains relatively poor but could be probably improved. Combining such an imaging technique with classical spatial encoding techniques, improvement in the acquisition time could be envisaged.

Magnetic RF lenses have been the subject of numerous developments during the last decade. It appears that a thin planar 2D configuration cannot be a good lens. Better, it can be a reflector, possibly combined with a magnetic lens [Lopez et al., 2011].

The recently developed (magnetic) RF lenses generally have the following characteristics:
- the unit cell is constituted of "ring" resonators disposed on the faces of a cube, giving rise to an isotropic effective permeability.
- The "ring" resonators have very small dimensions with respect to the wavelength but have a high inductance.

2.7.4 Dielectric resonators

A high permittivity material can be efficient in manipulating and producing an electromagnetic field [Webb, 2011]. Strictly speaking this is not an MM as already defined, but it has some similarities with respect to its use in MR (EPR and NMR) as it is able to create and manipulate the magnetic near field inherent in these experiments.

Recent convincing demonstrations, among others, of the applicability of dielectric materials in the domain of high field MRI have been published. In these applications, the dielectric is used either as a pad tailoring the electromagnetic field within the sample [Haines, et al., 2010], as a high homogeneity volume coil producing a circularly polarized B_1^+ field [Aussenhofer and Webb, 2012] or as a transmit/receive coil [Aussenhofer and Webb, 2014].

The capability of a dielectric resonator to produce an (evanescent) magnetic field has been well-known for some time [Wen et al., 1996]. The first resonant modes that appear at the low end of the frequency spectrum are of interest for MR applications. These are the $TE_{01\delta}$ mode, producing an evanescent magnetic field in the axial direction and very close to the surface of the dielectric resonator,[78] and the $HEM_{11\delta}$, which

---

[78] This mode can easily be demonstrated by a simple experiment, as shown by Andrew Webb during his lectures. A simple bottle containing water constituted a cylindrical dielectric resonator. The resonant $TE_{01\delta}$ mode can be excited with a coupling loop (electrically balanced) connected to the S_{11} port of a network analyzer. The S_{11} presents a

produces a transverse magnetic field in the equatorial plane [Aussenhofer and Webb, 2012].

The resonant frequency of the first TE_{01} mode can be estimated with the following empirical formula:

$$f_{MHz} = \frac{3.4 \times 10^4}{a_{mm}\sqrt{\varepsilon_r}}\left(\frac{a}{l} + 3.45\right), \qquad (2.155)$$

where a is the radius of the dielectric cylinder, l is the length (Fig. 2.74) and ε_r is the relative permittivity.

The dielectric material can be a solid cylinder or a water suspension of dielectric particles.[79] The latter is relatively easy to make and to shape in the desired aspect.

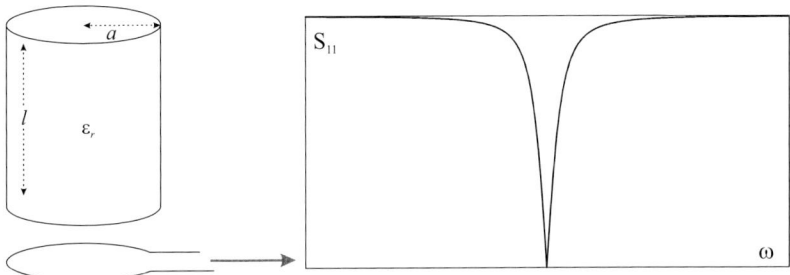

Fig. 2.74 Evidence of a dielectric resonant TE01 mode by magnetic coupling.

The use of these dielectric materials for NMR can circumvent many of the issues encountered with a conventional RF coil at high fields (radiation losses, SR, increased resistance, conservative electric field).

---

dip at the resonant frequency demonstrating the exchange of energy with the dielectric resonator. The resonant frequency can be moved by adding amounts of water in the bottle.

[79] For example, the widely available and cheap calcium titanate ($CaTiO_3$) [Haines et al., 2010].

Another advantage of the dielectric resonator is that its size is much lower than a resonant cavity of the same frequency.

During recent years, the number of applications of dielectric resonators for MR, and especially for MRI at very high field, has considerably grown. Yet, the association with magnetic MM has not been envisaged. However, thanks to the possibility of magnetic lenses to manipulate the evanescent magnetic field, this association offers a new approach for NMR and EPR experiments on large objects and at very high frequency.

Chapter 3

Introduction to Linear Network Analysis

An NMR probe is an assembly of linear passive[1] components like resistors, inductors, and capacitors connected to a set of conductors. The conductors, which couple with the magnetic moments of the nuclear spins from which the NMR signal is obtained, are the most important parts of the probe.

A numerical simulation, even approximate, is invaluable to understand the electromagnetic properties of an NMR probehead and to evaluate its performance in the final NMR experiments. To do this, a linear network analysis is fast and reliable. This approach does not replace the 3D full electromagnetic simulation that is obviously required in a final evaluation step, especially if the probe is to be used in a clinical environment. However, this last step requires high performance computing and a lot of time. Three-dimensional simulations will be briefly described in Chapter 7.

In the present chapter, a linear network analysis which can be made routinely at each step of the probe design is proposed. First, a short presentation of network theory introduces three complementary matrix formalisms: the impedance and admittance matrices, the chain matrix, and the scattering matrix. Some applications of the corresponding approaches are described. Second, the linear network analysis applied to the NMR probe will be described with particular emphasis on a simple

---

[1] Non-linear passive components such as diodes or active components such as preamplifiers are out of the purpose of this chapter. However, the outputs of the network analysis described here can be used with any network simulators based on SPICE if desired.

linear network analysis program that has been used for most of the simulations shown in the book.

3.1 Introduction to Network Theory

The ports of a given circuit (network) allow the communication with the external world. A voltage (V) and a current (I) entering the device exist at each port of the network (Fig. 3.1).

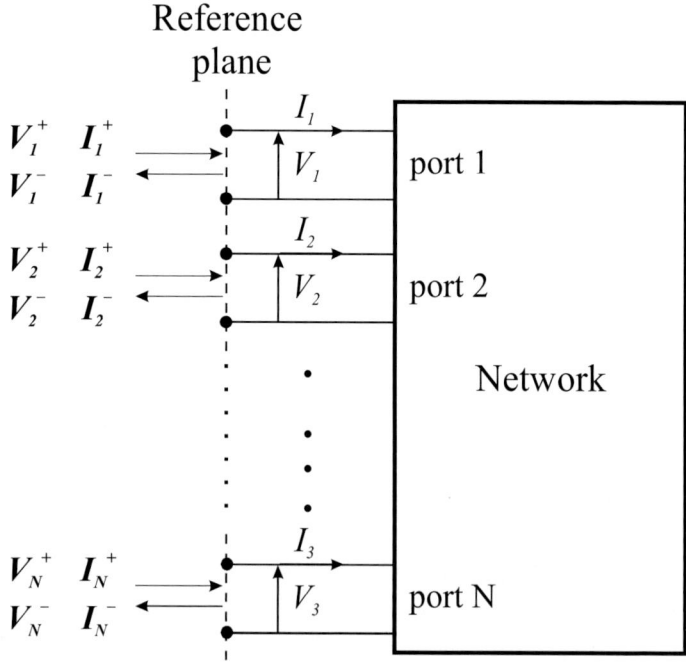

Fig. 3.1 N-ports general network. The voltage V and current I at each port are the result of the combination of the incident (V^+, I^+) and the reflected (V^-, I^-) wave according to Eq. (3.1).

The electrical properties of the circuit, when connected to sources and loads, can be described by the relationships that exist between these quantities. They can be defined as the ratio of voltage to current, impedance, or admittance, leading to the so-called impedance or admittance matrix representation. Another possibility, that still completely characterizes the network, is to express the (V,I) vector at one port as a function of the corresponding vector at another port. This is the transmission (or chain or ABCD) matrix representation.

The most efficient way to characterize an RF circuit is via a network analyzer that is able to give full information over a large frequency range. It characterizes the relative amplitudes and phases of the incident, reflected and transmitted waves for an arbitrary structure, or what is commonly called the scattering matrix or S-matrix. The term S-parameters generally describes the elements of the S-matrix. The S-parameters are important in RF design because they bear relevant information about the circuit to be analyzed and, most importantly, are easily measured. This approach is especially useful when circuits are connected together by transmission lines of given characteristic impedance Z_0, as is the case for an NMR spectrometer. All its RF parts – the probe, transmitters, receivers, filters – are connected together through transmission lines of constant characteristic impedance of 50 Ω. The physical quantities that are considered at each port of a given network are the complex (amplitude and phase) incident and reflected waves. The corresponding matrix is the so-called scattering matrix.

All these representations are able to fully characterize any network. Depending on the circumstances, one representation or another is best suited. Fig. 3.2 provides the correspondence between the different matrix representations in the case of a two-port network.

The relationships between the quantities describing the network as impedance, admittance, and transmission matrices are straightforward. The relation with the scattering matrix makes use of the following relationships:

$$V = V^+ - V^-$$
$$I = I^+ - I^-$$
(3.1)

The voltage (or current) at each port is decomposed into two quantities that are finally related to the reflection coefficient [Eq. (3.2)] and ultimately to the impedance Z, appearing at the reference plane of the considered port (Fig. 3.1).

	S	Z	Y	ABCD
S_{11}	S_{11}	$\dfrac{(Z_{11}-Z_0)(Z_{22}+Z_0)-Z_{12}Z_{21}}{\Delta Z_0}$	$\dfrac{(Y_0-Y_{11})(Y_0+Y_{22})+Y_{12}Y_{21}}{\Delta Y}$	$\dfrac{A+B/Z_0-CZ_0-D}{A+B/Z_0+CZ_0+D}$
S_{12}	S_{12}	$\dfrac{2Z_{12}Z_0}{\Delta Z}$	$\dfrac{-2Y_{12}Y_0}{\Delta Y}$	$\dfrac{2(AD-BC)}{A+B/Z_0+CZ_0+D}$
S_{21}	S_{21}	$\dfrac{2Z_{21}Z_0}{\Delta Z}$	$\dfrac{-2Y_{21}Y_0}{\Delta Y}$	$\dfrac{2}{A+B/Z_0+CZ_0+D}$
S_{22}	S_{22}	$\dfrac{(Z_{11}+Z_0)(Z_{22}-Z_0)-Z_{12}Z_{21}}{\Delta Z}$	$\dfrac{(Y_0+Y_{11})(Y_0-Y_{22})+Y_{12}Y_{21}}{\Delta Y}$	$\dfrac{-A+B/Z_0-CZ_0+D}{A+B/Z_0+CZ_0+D}$
Z_{11}	$Z_0\dfrac{(1+S_{11})(1-S_{22})+S_{12}S_{21}}{(1-S_{11})(1-S_{22})-S_{12}S_{21}}$	Z_{11}	$\dfrac{Y_{22}}{|Y|}$	$\dfrac{A}{C}$
Z_{12}	$Z_0\dfrac{2S_{12}}{(1-S_{11})(1-S_{22})-S_{12}S_{21}}$	Z_{12}	$\dfrac{-Y_{12}}{|Y|}$	$\dfrac{AD-BC}{C}$
Z_{21}	$Z_0\dfrac{2S_{21}}{(1-S_{11})(1-S_{22})-S_{12}S_{21}}$	Z_{21}	$\dfrac{-Y_{21}}{|Y|}$	$\dfrac{1}{C}$
Z_{22}	$Z_0\dfrac{(1-S_{11})(1+S_{22})+S_{12}S_{21}}{(1-S_{11})(1-S_{22})-S_{12}S_{21}}$	Z_{22}	$\dfrac{Y_{11}}{|Y|}$	$\dfrac{D}{C}$

Fig. 3.2 Relationships between elements of the S, Z, Y and ABCD matrices.

	S	Z	Y	ABCD
Y_{11}	$Y_0 \dfrac{(1-S_{11})(1+S_{22})+S_{12}S_{21}}{(1+S_{11})(1+S_{22})-S_{12}S_{21}}$	$\dfrac{Z_{22}}{|Z|}$	Y_{11}	$\dfrac{D}{B}$
Y_{12}	$Y_0 \dfrac{-2S_{12}}{(1+S_{11})(1+S_{22})-S_{12}S_{21}}$	$\dfrac{-Z_{12}}{|Z|}$	Y_{12}	$\dfrac{BC-AD}{B}$
Y_{21}	$Y_0 \dfrac{-2S_{21}}{(1+S_{11})(1+S_{22})-S_{12}S_{21}}$	$\dfrac{-Z_{21}}{|Z|}$	Y_{21}	$\dfrac{-1}{B}$
Y_{22}	$Y_0 \dfrac{(1+S_{11})(1-S_{22})+S_{12}S_{21}}{(1+S_{11})(1+S_{22})-S_{12}S_{21}}$	$\dfrac{Z_{11}}{|Z|}$	Y_{22}	$\dfrac{A}{B}$
A	$\dfrac{(1+S_{11})(1-S_{22})+S_{12}S_{21}}{2S_{21}}$	$\dfrac{Z_{11}}{Z_{21}}$	$\dfrac{-Y_{22}}{Y_{21}}$	A
B	$Z_0 \dfrac{(1+S_{11})(1+S_{22})-S_{12}S_{21}}{2S_{21}}$	$\dfrac{|Z|}{Z_{21}}$	$\dfrac{-1}{Y_{21}}$	B
C	$\dfrac{1}{Z_0} \dfrac{(1-S_{11})(1-S_{22})-S_{12}S_{21}}{2S_{21}}$	$\dfrac{1}{Z_{21}}$	$\dfrac{-|Y|}{Y_{21}}$	C
D	$\dfrac{(1-S_{11})(1+S_{22})+S_{12}S_{21}}{2S_{21}}$	$\dfrac{Z_{22}}{Z_{21}}$	$\dfrac{-Y_{11}}{Y_{21}}$	D
	$|Z|=Z_{11}Z_{22}-Z_{12}Z_{21}$	$Y_0=1/Z_0$	$|Y|=Y_{11}Y_{22}-Y_{12}Y_{21}$	$\Delta Y=(Y_{11}+Y_0)(Y_{22}+Y_0)-Y_{12}Y_{21}$ $\Delta Z=(Z_{11}+Z_0)(Z_{22}+Z_0)-Z_{12}Z_{21}$

Fig. 3.2 (continued)

$$\Gamma = \frac{V^-}{V^+} = \frac{Z - Z_0}{Z + Z_0}.\qquad(3.2)$$

The above equation and the concept of reflection coefficient imply that the S-parameter matrix is defined in a particular system having a specific characteristic impedance Z_0.

3.1.1 Impedance and admittance matrices

The impedance matrix **Z** of a general network (Fig. 3.1) relates the voltages and currents as:

$$\begin{bmatrix} V_1 \\ V_2 \\ \cdot \\ \cdot \\ \cdot \\ V_N \end{bmatrix} = \begin{bmatrix} Z_{11} & Z_{12} & \cdot & \cdot & Z_{1N} \\ Z_{21} & Z_{22} & \cdot & \cdot & Z_{2N} \\ \cdot & & & & \cdot \\ \cdot & & & & \cdot \\ Z_{N1} & Z_{N2} & \cdot & \cdot & Z_{NN} \end{bmatrix} \begin{bmatrix} I_1 \\ I_2 \\ \cdot \\ \cdot \\ \cdot \\ I_N \end{bmatrix}.\qquad(3.3)$$

The admittance matrix is defined in a similar way, by inverting the above equation.

$$\begin{bmatrix} I_1 \\ I_2 \\ \cdot \\ \cdot \\ \cdot \\ I_N \end{bmatrix} = \begin{bmatrix} Y_{11} & Y_{12} & \cdot & \cdot & Y_{1N} \\ Y_{21} & Y_{22} & \cdot & \cdot & Y_{2N} \\ \cdot & & & & \cdot \\ \cdot & & & & \cdot \\ Y_{N1} & Y_{N2} & \cdot & \cdot & Y_{NN} \end{bmatrix} \begin{bmatrix} V_1 \\ V_2 \\ \cdot \\ \cdot \\ \cdot \\ V_N \end{bmatrix}.\qquad(3.4)$$

In general, each Z_{ij} or Y_{ij} element is complex. For an N-port network, the impedance (or admittance) matrices are of size $N \times N$.

As an example, let us consider a two-port T-network (Fig. 3.3).

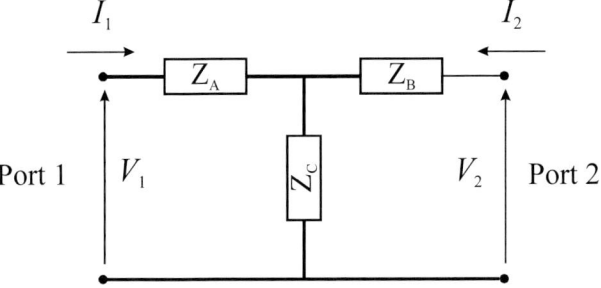

Fig. 3.3 A general two-port T-network, with voltage and current definitions.

From Eq. (3.3), Z_{11} is the input impedance at port 1 when port 2 is open:

$$Z_{11} = \left.\frac{V_1}{I_1}\right|_{I_2=0} = Z_A + Z_C, \qquad (3.5)$$

and similarly

$$Z_{22} = \left.\frac{V_2}{I_2}\right|_{I_1=0} = Z_B + Z_C. \qquad (3.6)$$

The transfer impedance Z_{12} can be determined measuring the open-circuit voltage when a current I_2 is flowing through port 2:

$$Z_{12} = \left.\frac{V_1}{I_2}\right|_{I_1=0} = Z_C \qquad (3.7)$$

and

$$Z_{21} = \left.\frac{V_2}{I_1}\right|_{I_2=0} = Z_C. \qquad (3.8)$$

The Z-matrix is always symmetric for any passive network.

Similarly the impedance matrix of a Π network (Fig 3.4) can be obtained as:

$$Z_{11} = \frac{Z_A(Z_B + Z_C)}{Z_A + Z_B + Z_C}, \tag{3.9}$$

$$Z_{22} = \frac{Z_B(Z_A + Z_C)}{Z_A + Z_B + Z_C}, \tag{3.10}$$

$$Z_{12} = Z_{21} = \frac{Z_A Z_B}{Z_A + Z_B + Z_C}. \tag{3.11}$$

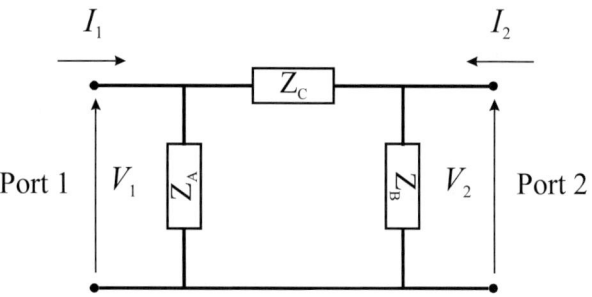

Fig. 3.4 Two-port Π network.

3.1.2 The transmission (ABCD) matrix

When the network is constituted of a cascade connection of two or more two-port networks, the ABCD matrix is well suited to describing the behavior of the complete system. The transmission (chain or ABCD) matrix relates the input and output (V, I) vectors (Fig. 3.5) by:

$$\begin{pmatrix} V_{in} \\ I_{in} \end{pmatrix} = \begin{bmatrix} A & B \\ C & D \end{bmatrix} \begin{pmatrix} V_{out} \\ I_{out} \end{pmatrix}. \tag{3.12}$$

Introduction to Linear Network Analysis 163

Fig. 3.5 Voltage and current definition in the ABCD matrix representation. Note the change of sign of the output current as compared with the Z or Y representation.

The transmission matrix of a cascade of N networks (Fig. 3.6) is simply given by a matrix multiplication:

$$\begin{pmatrix} V_0 \\ I_0 \end{pmatrix} = \mathbf{A}_1 \mathbf{A}_2 \ldots \mathbf{A}_N \begin{pmatrix} V_N \\ I_N \end{pmatrix}. \qquad (3.13)$$

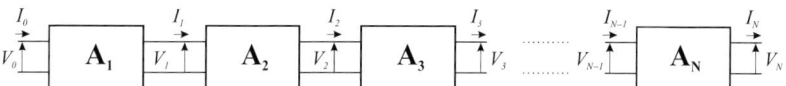

Fig. 3.6. A cascade of two-port networks is easily represented by the chain matrix formalism as the product of the chain matrices of each elementary network.

The chain matrix elements for three basic two-port networks are shown in Fig. 3.7.

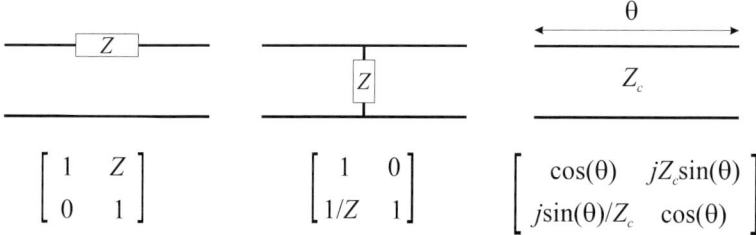

$$\begin{bmatrix} 1 & Z \\ 0 & 1 \end{bmatrix} \qquad \begin{bmatrix} 1 & 0 \\ 1/Z & 1 \end{bmatrix} \qquad \begin{bmatrix} \cos(\theta) & jZ_c\sin(\theta) \\ j\sin(\theta)/Z_c & \cos(\theta) \end{bmatrix}$$

Fig. 3.7 ABCD matrix of basic two-port networks: a series impedance (left), a shunt impedance (middle) and a section of lossless line (right). For a lossy line, $\cos(\theta)$ and $\sin(\theta)$ are replaced by $\cosh(\gamma l)$ and $\sinh(\gamma l)$, respectively.

A useful application of the chain matrix formalism is the design of networks built with lumped elements that replace transmission lines, for a narrow frequency band.

A transmission line is frequently used as a phase shifter or an impedance transformer in the design of a network, but the line is sometimes too bulky to fit in the available space. This will be the case at low frequencies (of the order of 100 MHz) as well as in the microwave frequency range when designing monolithic integrated circuits. On the other hand, the required characteristic impedance may not be a standard value. In those cases, it is sometimes preferable to replace the line by its equivalent network realized with lumped components. This concept was developed in a very easy manner, especially for microwave technology (see, for example, Parisi, 1989), using the transmission or chain matrix formalism.

The electrical properties of a transmission line can be entirely described by its ABCD chain matrix:[2]

$$\mathbf{A} = \begin{pmatrix} \cos\theta & jZ_C \sin\theta \\ j\sin\theta/Z_c & \cos\theta \end{pmatrix}, \qquad (3.14)$$

where Z_c represents the line characteristic impedance [Pozar, 1998, p. 208].

The replacement circuit takes the form of either a Π low-pass or a T high-pass filter (Fig. 3.8). Eventually, the circuit can be a band-pass or stop band filter. In this case, the equivalent line will be dual band (see, for example, the dual band 180° hybrid described in Chapter 5).

The chain matrices of the low-pass filters [Fig. 3.8(a)] are easily obtained after multiplication of the three chain matrices describing the components of the circuit (refer to Fig. 3.7 to write the matrix corresponding to each component). For the Π low-pass filter, one obtains:

---

[2] We assume a lossless line here.

$$\mathbf{A}_{LP} = \begin{pmatrix} 1 & 0 \\ jC_1\omega & 1 \end{pmatrix} \begin{pmatrix} 1 & jL_1\omega \\ 0 & 1 \end{pmatrix} \begin{pmatrix} 1 & 0 \\ jC_1\omega & 1 \end{pmatrix}$$

$$\mathbf{A}_{LP} = \begin{pmatrix} 1 - L_1C_1\omega^2 & jL_1\omega \\ jC_1\omega(2 - L_1C_1\omega^2) & 1 - L_1C_1\omega^2 \end{pmatrix} \quad (3.15)$$

and for the T low-pass filter [Fig. 3.8(a)]:

$$\mathbf{A}_{LP} = \begin{pmatrix} 1 - L_2C_2\omega^2 & jL_2\omega(2 - L_2C_2\omega^2) \\ jC_2\omega & 1 - L_2C_2\omega^2 \end{pmatrix}. \quad (3.16)$$

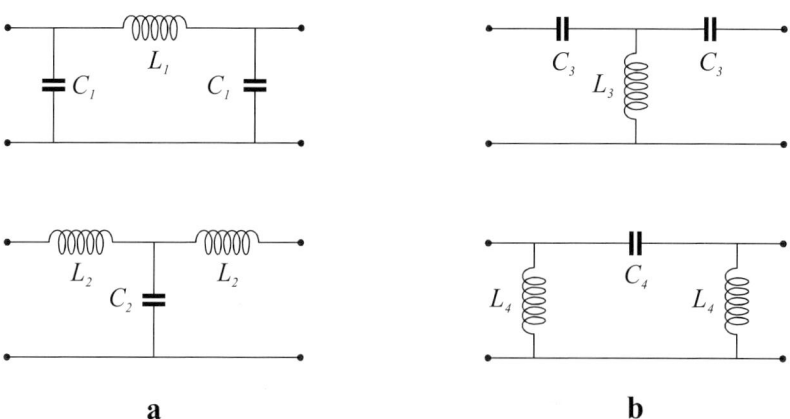

Fig. 3.8 Low-pass (a) and high-pass (b) filter cells that are used to model a transmission line at a given frequency, or very close to it.

The component values for L and C are given by equating matrices $\mathbf{A}$, Eq. (3.14), and $\mathbf{A}_{LP}$, Eq. (3.15) or (3.16). Thus:

$$L_1\omega = Z_c \sin\theta$$
$$C_1\omega = \frac{1}{Z_c} \frac{\sin\theta}{1 + \cos\theta} \quad (3.17)$$

and

$$L_2\omega = Z_c \frac{\sin\theta}{1+\cos\theta}$$
$$C_2\omega = \frac{\sin\theta}{Z_c}$$
(3.18)

The component values for L and C should be positive, hence:

$$0 \le \theta \le \pi$$
$$k\lambda \le l \le (2k+1)\lambda/2 \quad k \in \mathbb{N}$$
(3.19)

Similarly the component values for the high-pass filter [Fig. 3.8(b)] are given by equating the corresponding chain matrices.

The matrix of the T high-pass filter is given by:

$$\mathbf{A}_{HP} = \begin{pmatrix} 1 & -j/C_3\omega \\ 0 & 1 \end{pmatrix} \begin{pmatrix} 1 & 0 \\ -j/L_3\omega & 1 \end{pmatrix} \begin{pmatrix} 1 & -j/C_3\omega \\ 0 & 1 \end{pmatrix}$$

$$\mathbf{A}_{HP} = \begin{pmatrix} 1 - \dfrac{1}{L_3 C_3 \omega^2} & \dfrac{-j}{C_3\omega}\left(2 - \dfrac{1}{L_3 C_3 \omega^2}\right) \\ \dfrac{-j}{L_3\omega} & 1 - \dfrac{1}{L_3 C_3 \omega^2} \end{pmatrix},$$
(3.20)

and for the Π high-pass filter [Fig. 3.8(b)]:

$$\mathbf{A}_{HP} = \begin{pmatrix} 1 - \dfrac{1}{L_4 C_4 \omega^2} & \dfrac{-j}{C_4\omega}\left(2 - \dfrac{1}{L_4 C_4 \omega^2}\right) \\ \dfrac{-j}{L_4\omega} & 1 - \dfrac{1}{L_4 C_4 \omega^2} \end{pmatrix},$$
(3.21)

from which one gets:

$$L_3\omega = -\frac{Z_C}{\sin\theta}$$
$$C_3\omega = -\frac{1}{Z_c}\frac{1+\cos\theta}{\sin\theta}$$
(3.22)

and

$$L_4\omega = -Z_c \frac{1+\cos\theta}{\sin\theta}$$
$$C_4\omega = -\frac{1}{Z_c \sin\theta}.$$
(3.23)

The component values should be positive, hence the phase angle of the lines modeled by the high-pass circuit should verify:

$$\pi \leq \theta \leq 2\pi$$
$$(2k+1)\lambda/2 \leq l \leq (k+1)\lambda \quad k \in \mathbb{N}.$$
(3.24)

3.1.3 *The scattering matrix*

The S-matrix formalism has already been introduced in Chapter 2 for a two-port network. It was shown that the S-parameters[3] are related to the incident and reflected wave amplitudes at each port of the network.

For the N-port network shown in Fig. 3.1, the S-matrix is a generalization of Eq. (2.17). It is defined as follows:

$$[V^-] = [S][V^+],$$
(3.25)

or explicitly as:

$$\begin{bmatrix} V_1^- \\ V_2^- \\ \cdot \\ \cdot \\ \cdot \\ V_N^- \end{bmatrix} = \begin{bmatrix} S_{11} & S_{12} & \cdot & \cdot & \cdot & S_{1N} \\ S_{21} & S_{22} & \cdot & \cdot & \cdot & S_{2N} \\ \cdot & & & & & \cdot \\ \cdot & & & & & \cdot \\ \cdot & & & & & \cdot \\ S_{N1} & S_{N2} & \cdot & \cdot & \cdot & S_{NN} \end{bmatrix} \begin{bmatrix} V_1^+ \\ V_2^+ \\ \cdot \\ \cdot \\ \cdot \\ V_N^+ \end{bmatrix}.$$
(3.26)

---

[3] The S-parameters are the elements of the S-matrix.

The reflected or output wave at each port is a function of the input waves at all ports. The scattering parameters, S_{ij}, are defined as:

$$S_{ij} = \left. \frac{V_i^-}{V_j^+} \right|_{V_k^+ = 0\ \forall k \neq j} . \quad (3.27)$$

The scattering matrix formalism was initially developed during the 1960s as an invaluable tool for characterizing electronic devices in the newly developing field of microwave technology. The Hewlett-Packard Company (now Agilent) was one of the leaders in this development. Nowadays, the *S*-parameters are a must for characterizing any components at any frequencies. *S*-parameters have already been mentioned in Chapter 2 for two-terminal components such as resistors, capacitors, and inductors, for which the manufacturers generally provide the *S*-parameters.

3.2 Linear Network Simulation of Probe Coil Circuits

Many commercial or free network simulators are based on SPICE (Simulation Program with Integrated Circuit Emphasis), initially developed at the University of California, Berkeley and released as the 2g.6 version[4] in 1983.

SPICE can do much more than is required to simulate a probe circuit, which generally include only linear passive devices. It does transient or time-domain large signal analysis, linear small signal frequency domain analysis and noise analysis.

The last versions of SPICE, freely distributed, are based on a modified nodal analysis that allows simulating probe circuits including inductors, resistors, and capacitors. For example, SPICE has previously been used to predict the resonance of probes such as the "free elements" birdcage resonator [Fakri *et al.*, 1996].

---

[4] The 2g.6 version is written in Fortran. The version 3f.4, written in the C language, was released in 1993. An open source version is in continuing development as ngspice (http://ngspice.sourceforge.net/presentation.html).

3.2.1 NMRP: a simple tool for probe evaluation

At the time of the first edition, programs had been specifically developed for probe simulation. Since that time, the software has been improved with respect to speed of calculation and the number of tools available to create the required inductance matrix and to analyze the outputs.

NMRP[5] is written in C and the critical parts use the GNU Scientific Library (GSL) [Galassi *et al.*, 2009]. Accordingly, the sources are freely distributed under the GNU license.

The use of GSL for matrix manipulation[6] improved the calculation speed by a factor of thousands compared to the initial version. It becomes reasonable, using a desk computer to simulate a network with up to 10000 components (conductors, capacitors, resistors) with 4 GB of RAM memory.

The software, written with little knowledge of advanced numerical methods, is however still limited but may be helpful to demonstrate many characteristics of a probe, from the accurate prediction of the resonance spectrum to the calculation of current density on conductors and evaluation of the magnetic field distribution in free space.

The software still lacks an efficient memory management, a parallelized algorithm, and a graphical interface.

3.2.1.1 Simprobe: a linear probe network analyzer

Simprobe, which is part of the NMRP package, is a linear network analyzer designed to calculate the current and voltage in each branch of a network as a function of frequency. It is specifically devoted to the calculation of currents in the conductive parts of a probe. We are therefore mainly concerned with passive components (Resistance, Inductance and Capacitance, abbreviated as RLC) that can be inductively coupled together, and eventually with external linear sources (the transmitter amplifiers or the sample spin magnetization).

---

[5] NMRP means « NMR probeheads ». This is a package of utilities designed to simulate NMR probe properties.

[6] Vector and matrix calculations are based on the Blas and Lapack libraries.

Knowing the current, the next step is to estimate the magnetic field distribution created by the probe using the segment method (Chapter 7). Briefly, the probe is decomposed into linear segments of complex valued currents, each creating a magnetic field in space. The field components are easily calculated. The total field is obtained by summing up the contributions of all the current segments. Furthermore, the behavior of the probe at any frequency can be simulated and compared with the measurements done on the RF workbench.

The probe network analysis is done in two steps, using two fundamental laws: Kirchhoff's laws and Ohm's law. The first step involves Kirchhoff's laws for node currents and loop voltages. At this stage, the topology of the network is described as a set of branches,[7] nodes, and loops (Fig. 3.9).

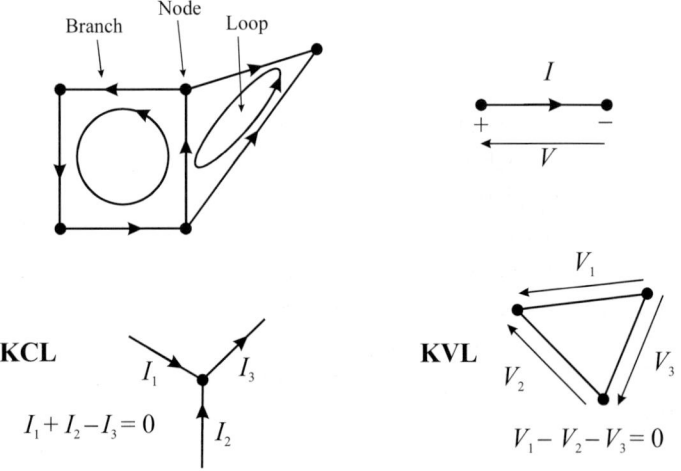

Fig. 3.9 Upper: definitions of a network topology in graph elements and sign conventions for currents and voltages. Lower: a representation of Kirchhoff's laws for currents at node (KCL) and for voltages in loop (KVL).

The second step involves Ohm's law, written for each branch of the network and relating the branch current to the branch voltage. The

---

[7] In graph theory, the branches are named edges [Deo, 1974].

external voltage and current sources and the branch impedances are specified at this stage.

The graph describing an electrical network is a directed graph. An arbitrary direction for the current in each branch is firstly defined. The voltage sign across an element of the network is determined according to the usual convention that the current enters the branch (component) at the positive edge (Fig. 3.9).

The first Kirchhoff law states that at each node of the network the sum of currents is equal to zero. The second law states that the sum of voltages in each loop of the network is also equal to zero.

Let n be the number of nodes and b the number of branches. For node k, the Kirchhoff current law (KCL) is expressed as:

$$\sum_{j=1}^{b} a_{jk} I_j = 0. \qquad (3.28)$$

The coefficients a_{ik} are equal to +1, −1 or 0, depending on whether the current I for branch j is directed away from (+1) or towards (−1) the kth node. a_{ik} is equal to 0 if branch j is not connected to node k. The matrix $\mathbf{A} = [a_{ik}]$ is called the branch-node incidence matrix. Because one of the node equations can always be written as a linear combination of the others, the number of KCL independent equations is $n - 1$ (also equal to the number of trees in the graph).

Similarly, for each independent loop s, the Kirchhoff voltage law, or KVL, is expressed as:

$$\sum_{j=1}^{b} b_{js} V_j = 0, \qquad (3.29)$$

where the coefficients b_{js} are equal to +1, −1 or 0, depending on the loop direction[8] with respect to the branch (or current) orientation. Specifically, b_{js} is equal to +1 if the jth current is in the same direction as the sth loop,

---

[8] The loop direction is initially chosen arbitrarily.

and equal to -1 if the jth current is in the opposite direction. Obviously, b_{js} is equal to 0 if branch j is not part of loop s.

The matrix $\mathbf{B} = [b_{js}]$ is called the mesh-branch incidence matrix. The number m of independent KVL equations is equal to the number of branches minus the number of trees, thus:

$$m = b - n + 1. \tag{3.30}$$

Ohm's law relates the voltage and current for each branch of the network. For the most general branch, including a complex impedance Z_k, a current source G_k and a voltage source E_k (Fig. 3.10), the voltage V_k across the branch k can be expressed as:

$$V_k = -E_k + Z_k G_k + Z_k I_k. \tag{3.31}$$

In the case of an inductive coupling (mutual inductance) between different branches (frequently encountered with an NMR probehead), the voltage source E_k includes the electromotive force created by the current I_j carried by branch j which is coupled to branch k:

$$E_k = Z_{kj} I_j, \tag{3.32}$$

where Z_{kj} is the coupling impedance between branches k and j. Accordingly, Ohm's law can be rewritten as:

$$V_k = \left[-E_k + Z_{kk}(G_k + I_k) \right] + \sum_{j=1}^{b} Z_{kj} I_j, \tag{3.33}$$

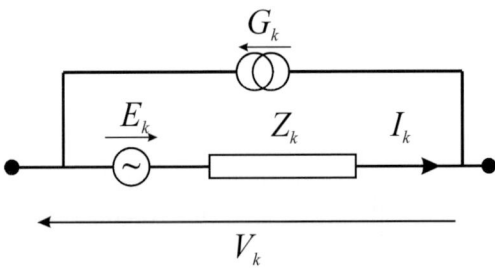

Fig. 3.10 The general branch includes a voltage source (E_k), a current source (G_k) source, and an impedance Z_k. All quantities are complex.

where E_k and G_k are external independent voltage and current sources. In Eq. (3.33), Z_{kk} is the (self-)impedance of branch k and Z_{kj} ($k \neq j$) is the mutual impedance between branches j and k.

Solving the network equations is necessary for only one of the unknown vectors **V** or **I**. This is usually done by the so-called nodal or mesh analysis, making use of Ohm's equations together with the appropriate Kirchhoff Law (KVL or KCL).

In the nodal analysis, the unknown is the node voltage vector $\mathbf{V}_{node}$. The voltages are referenced to the potential of a particular node, usually the ground node. Thus the number of unknowns is $n - 1$, consistent with the number of KCL independent equations. The network equations are formed by combining the KCL and Ohm's law which are written in a matrix form as:

$$\mathbf{Y}_{node} \cdot \mathbf{V}_{node} = \mathbf{I}_{sources}. \qquad (3.34)$$

External voltage sources are eventually converted into current sources by means of Norton's theorem. $\mathbf{Y}_{node}$ is the admittance matrix formed from the known branch admittance matrix $\mathbf{Y}_{branch}$ using the following relationship:

$$\mathbf{Y}_{node} = \mathbf{A} \cdot \mathbf{Y}_{branch} \cdot \mathbf{A}^T, \qquad (3.35)$$

where **A** is the branch-node incident matrix, Eq. (3.28).

In the mesh analysis, the unknowns are the mesh currents $\mathbf{I}_{mesh}$. The number of unknowns is the number of loops l, which is also the number of independent KVL equations which combine with the Ohm's law equations to give:

$$\mathbf{Z}_{mesh} \cdot \mathbf{I}_{mesh} = \mathbf{V}_{sources}. \qquad (3.36)$$

External current sources are eventually converted into voltage sources by means of Thevenin's theorem. $\mathbf{Z}_{mesh}$ is the mesh impedance matrix, which is related to the known $\mathbf{Z}_{branch}$ matrix, appearing in the Ohm's law equations, as

$$\mathbf{Z}_{mesh} = \mathbf{B} \cdot \mathbf{Z}_{branch} \cdot \mathbf{B}^T, \qquad (3.37)$$

where **B** is the mesh-branch incidence matrix, Eq. (3.29).

The branch currents are the desired outputs. They are given by the elements of the vector $\mathbf{I}_{branch}$, of dimension b. This vector is directly obtained from $\mathbf{I}_{mesh}$ (of dimension l), and *vice versa*, by the following relations:

$$\mathbf{I}_{mesh} = \mathbf{B} \cdot \mathbf{I}_{branch}, \qquad (3.38)$$

$$\mathbf{I}_{branch} = \mathbf{B}^{\mathrm{T}} \cdot \mathbf{I}_{mesh}. \qquad (3.39)$$

The most popular and common approach is probably based on nodal formulation due to its simplicity in writing the incidence matrix **A**. However, due to the mutual inductive coupling between parts of the NMR probe, the mesh analysis is most appropriate[9] for solving the network equations.

Hence, the vector $\mathbf{I}_{mesh}$ is obtained from the solution of Eq. (3.36) and the required $\mathbf{I}_{branch}$ is calculated from Eq. (3.39).

Specifically, Ohm's law for NMR probes, not involving external current sources G_k, is written in matrix form from Eq. (3.33) as:

$$\mathbf{V} = -\mathbf{E} + \mathbf{Z} \cdot \mathbf{I}_{branches}, \qquad (3.40)$$

where

$$\mathbf{E} = \left[E_k \right]_{k=0,b-1} \qquad (3.41)$$

is the vector of independent (external) branch sources[10] and

$$\mathbf{Z} = \left[Z_{kj} \right]_{k,j=0,b-1} \qquad (3.42)$$

is the branch impedance matrix.

---

[9] A modified nodal analysis (referred to in the literature as Mesh-Nodal-Analysis, or MNA) can also be implemented, as in SPICE, retaining the advantages of the nodal approach, based solely on matrix **A**.

[10] The external sources are mainly the pulse and decoupling power transmitters.

To solve Eq. (3.40) for the branch currents, the branch voltages vector **V** must be eliminated. This is done, using KVL, in the following manner. First, Ohm's law, Eq. (3.40), is written as a function of $\mathbf{I}_{mesh}$ as:

$$\mathbf{V} = -\mathbf{E} + \mathbf{Z} \cdot \mathbf{B}^T \cdot \mathbf{I}_{mesh}. \tag{3.43}$$

Multiplying by **B** and using $\mathbf{BV} = 0$ (KVL), one obtains a set of m linear equations with m unknown mesh currents:[11]

$$\left[\mathbf{B} \cdot \mathbf{Z} \cdot \mathbf{B}^T\right] \cdot \mathbf{I}_{mesh} = \mathbf{B} \cdot \mathbf{E}, \tag{3.44}$$

where the "mesh impedance" matrix, in brackets, is formed from the known branch impedance matrix **Z** (including all self and mutual inductances) and the mesh-branch incidence matrix **B**. Finally, the current vector $\mathbf{I}_{branch}$ of dimension b is obtained from the mesh current vector $\mathbf{I}_{mesh}$, of dimension m ($<b$), using Eq. (3.38).

From a practical point of view, Eq. (3.44), which involves complex numbers, needs to be rewritten in terms of real numbers to be numerically calculable on a standard computer. This problem has been the subject of numerous works producing efficient and complex methods. We choose to use the simplest and most understandable, although less efficient, approach.

Any complex valued matrix **Z** can be decomposed into real-valued matrices as:

$$\mathbf{Z} = \mathbf{Z}^R + i\mathbf{Z}^I, \tag{3.45}$$

where $i = \sqrt{-1}$. The elements of $\mathbf{Z}^R$ and $\mathbf{Z}^I$ are, respectively, the real and imaginary part of each complex element of **Z**. Using this simple matrix decomposition, the equation $\mathbf{ZI} = \mathbf{V}$ in $\mathbb{C}$ is:

$$\left(\mathbf{Z}^R + i\mathbf{Z}^I\right)\left(\mathbf{I}^R + i\mathbf{I}^I\right) = \mathbf{V}^R + i\mathbf{V}^I. \tag{3.46}$$

---

[11] Eq. (3.44) differs from Eq. (3.36) in the definition of the external source voltage vectors. In Eq. (3.36), the voltage source vector has the dimension m (the number of meshes). In Eq. (3.44), E has the dimension b (the number of branches). Both vectors are related by the mesh-branch incidence matrix.

Equating the real part and imaginary part, Eq. (3.46) separates into two *coupled* equations in $\mathbb{R}$:

$$\begin{aligned} \mathbf{Z}^R \mathbf{I}^R - \mathbf{Z}^I \mathbf{I}^I &= \mathbf{V}^R \\ \mathbf{Z}^I \mathbf{I}^R + \mathbf{Z}^R \mathbf{I}^I &= \mathbf{V}^I \end{aligned}. \qquad (3.47)$$

This set of coupled equations is finally written as a unique set of linear *independent* equations:

$$\mathsf{Z}\mathsf{I} = \mathsf{V}, \qquad (3.48)$$

where Z is the impedance super matrix $\in \mathbb{R}$

$$\mathsf{Z} = \begin{pmatrix} \mathbf{Z}^R & -\mathbf{Z}^I \\ \mathbf{Z}^I & \mathbf{Z}^R \end{pmatrix} \qquad (3.49)$$

and I, V are the current and voltage super vectors $\in \mathbb{R}$

$$\mathsf{I} = \begin{pmatrix} \mathbf{I}^R \\ \mathbf{I}^I \end{pmatrix} \quad \mathsf{V} = \begin{pmatrix} \mathbf{V}^R \\ \mathbf{V}^I \end{pmatrix}. \qquad (3.50)$$

Eq. (3.48) is finally inverted to give, separately, the real and imaginary parts of the unknown currents.

3.2.1.2 Building the Z-matrix

In this formulation, the probe circuit is fully defined by two matrices: the impedance super-matrix and the KVL matrix.

The diagonal elements of each sub-matrix of Z describe, separately, the real and imaginary components of the self-impedance of each branch. The off-diagonal elements of the sub-matrices in Z describe the coupling between each pair of branches.

The probe is decomposed into the simplest possible elementary components (the branches). Each branch is thus described by a series RLC circuit (Fig. 3.11).

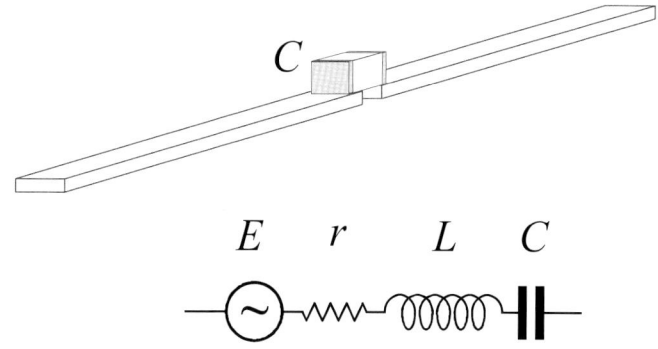

Fig. 3.11 Components of a basic probe and their equivalence as lumped elements (E – external source, r – the conductor resistance and possibly the electron spin resonance of the capacitor, L – the partial conductor inductance).

The resistor, inductor, or capacitor can be lumped elements or a conductor (of partial inductance L) having a resistance (r) and eventually incorporating a lumped capacitor (C). The coupling between branches is limited to the mutual inductance between the conductors (Fig. 3.12). Stray capacitance can eventually be included as lumped capacitor.[12]

The purpose of the building utilities is to calculate the elements of the matrix Z. In the present software version, the frequency dependence of the elements of Z is calculated in the simprobe engine from Eqs. (2.82) and (2.105), assuming ideal L and C components[13] (Fig. 3.13). The resistance of the conductor is assumed to be proportional to the square root of frequency. Hence, the inputs to simprobe are the values of

---

[12] These capacitors must be part of a loop to have a significant effect on the electrical behavior of the network.

[13] This simplification can be overridden. The Z-matrix can be calculated at a desired frequency by an external program and converted into the appropriate r, L, and C.

inductance, capacitance, and resistance[14] for each branch. The frequency at which the resistance is defined must also be provided.

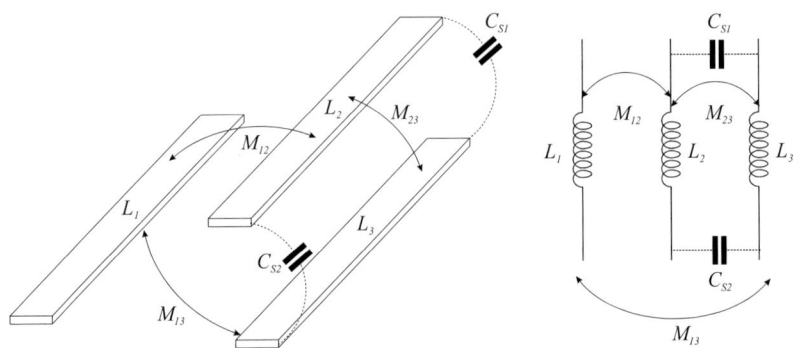

Fig. 3.12 Description of the probe conductor network, including the partial inductance of each element, the mutual inductance between them, and eventually the stray capacitances. The resistances are not shown here but they are obviously included in a complete probe description.

For the off-diagonal elements representing the mutual inductance between each pair of branches, the frequency dependence is the same as for an ideal inductance.

The calculation of the inductance matrix is done from the equations already given in Chapter 2 for the round wire [Eq. (2.34)] and for the flat strip [Eq. (2.37)]. The mutual inductance is also calculated from formulae[15] as given in Grover's book [Grover, 2004[16]]. These formulae are identical to those implemented in the FastHenry software [Kamon *et al.*, 1994].

Any other algorithm can be use to perform the calculation of the inductance matrix. For example the integration of the Neumann

---

[14] Broadband resistors are automatically recognized if they are associated with L = C = 0 in the branch impedance description.

[15] Assuming the thin wire approximation.

[16] This edition is a facsimile of the original book which appeared in 1946.

equations is another way to calculate the inductance, which has been proved to be efficient and accurate [Jin, 1999; Giovanetti, *et al.*, 2002].

The KVL matrix (incidence matrix **B**) provides the topology of the network. It can be written straightforwardly for simple circuits. For complex networks, the present approach is to write the matrix, case by case.

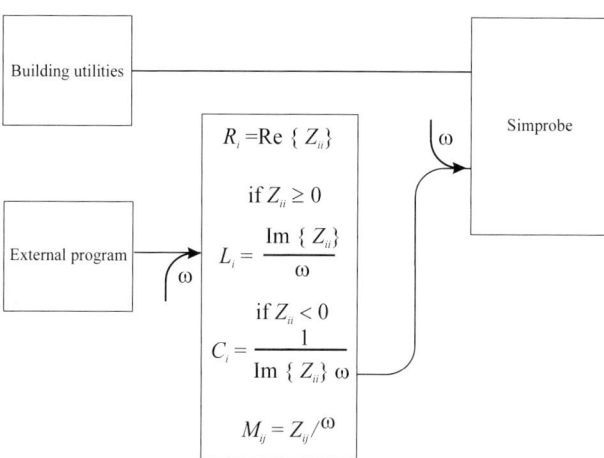

Fig. 3.13 Two different methods to pass the *r*, *L*, *C* values to simprobe. The building utilities directly provide the required values. Alternatively, an external program calculates the impedance values at a given frequency ω which are easily converted into the required *r*, *L*, *C* values, as shown.

A general algorithm could be elaborated using the graph theory [Deo, 1974], such as implemented in FastHenry [Kamon *et al.*, 1994]. Yet this has not been included in the present version of the software.

3.2.1.3 *Analysis of simprobe outputs*

The outputs of simprobe are the vectors $\mathbf{I}_{branch}$ and $\mathbf{V}_{branch}$ calculated from the vector $\mathbf{I}_{mesh}$, solution of Eq. (3.44), and using Eq. (3.39) and Eq. (3.33), respectively.

The magnetic field distribution created by the network of conductors can be calculated from $\mathbf{I}_{branch}$ as described in Chapter 7. All matrix elements (Z, Y, A, S) related to any ports of the network can also be calculated from the vectors $\mathbf{I}_{branch}$ and $\mathbf{V}_{branch}$.

For example, the impedance of a one port network (Fig. 3.14) is simply given by:

$$Z_k = \frac{V_k}{I_k}. \tag{3.51}$$

The reflection coefficient at the same port is:

$$\Gamma = \frac{Z_k - Z_0}{Z_k + Z_0}. \tag{3.52}$$

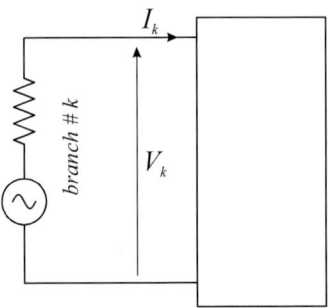

Fig. 3.14 A one-port network is powered by a source having any internal impedance described by a branch k. The impedance Z_k of the network is given by the ratio of V_k to I_k.

The S-parameters S_{11} and S_{21} of a two-port probe circuit [Fig. 3.15(left)] are obtained, according to Eqs. (2.15) and (2.16), when port 1 is powered and port 2 is terminated by 50 Ω.

Hence:

$$S_{11} = \frac{Z_1 - Z_0}{Z_1 - Z_0}$$
$$S_{21} = \frac{V_2 - I_2 Z_0}{V_1 + I_1 Z_0}. \tag{3.53}$$

The S-parameters S_{22} and S_{12} are obtained similarly when port 2 is powered and port 1 is terminated by 50 Ω [Fig. 3.15(right)].

$$S_{22} = \frac{Z_2 - Z_0}{Z_2 - Z_0}$$
$$S_{12} = \frac{V_1 - I_1 Z_0}{V_2 + I_2 Z_0}$$
(3.54)

In Eqs. (3.53) and (3.54), the indices 1 and 2 denote the branches, including the source and load connected to these ports (Fig. 3.15).

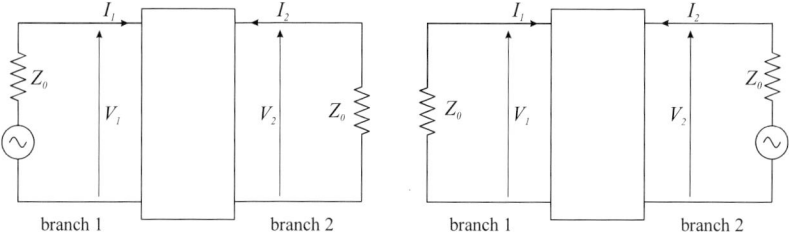

Fig. 3.15 Circuits used to calculate S_{11}, S_{21} (left) and S_{22}, S_{12} (right) for a two-port network.

As an example, let's tune, match, and predict the amplitude of the magnetic field created by the simple probe circuit shown in Fig. 3.16. A 10 W power transmitter with 50 Ω internal impedance is assumed. This setup corresponds to a typical small transmit–receive surface coil operating at 200 MHz on a 4.7 tesla spectrometer/scanner.

The circuit includes a square coil made of a round wire, tuning and matching capacitors (see next chapter), and the RF source.

The first step is to calculate the inductance of the rectangular coil (its dimensions are given in Fig. 3.16). This can be done by Eq. (2.98) or simulating the circuit of Fig. 3.17 and extracting the inductance.[17] As a result, these two methods give $L = 75.8$ and 75.3 nH, respectively.

The resistance of the coil can be calculated using Eq. (2.27) but it is underestimated in practice. A more realistic value can be obtained from

---

[17] The inductance is given by $L = Z_k/\omega$, where Z_k is the impedance seen across the source branch k and ω the frequency at which the circuit has been simulated.

impedance measurements done on the coil or from an initial guess of the Q factor. A value of $Q = 100$ is standard for such a coil operating at 200 MHz.

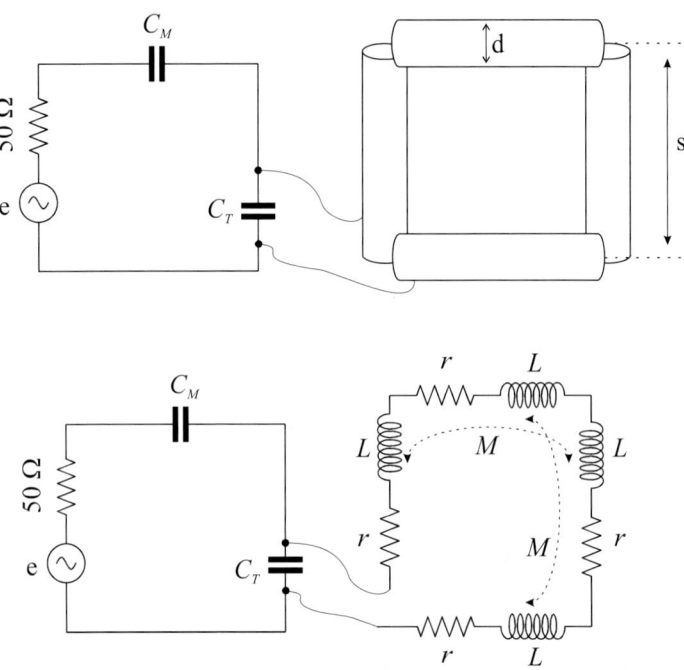

Fig. 3.16 Upper: description of a square coil made up of round wires and tuned/matched to 50 Ohms. Lower: its representation as lumped components, including the partial inductance of each conductor and the mutual inductance between them. $s = 30$ mm, $d = 1.2$ mm. The tuning and matching capacitor are equal to $C_T = 7.25$ pF and $C_M = 1.16$ pF, assuming $Q = 100$ at 200 MHz.

The tuning and matching capacitors are calculated as described in the next chapter and the complete probe circuit simulated. The calculated S-parameter (S_{11}) confirms that the probe is perfectly tuned and matched (Fig. 3.18).

The tuning and matching capacitors are calculated as described in the next chapter and the complete probe circuit simulated. The calculated S-parameter (S_{11}) confirms that the probe is perfectly tuned and matched (Fig. 3.18).

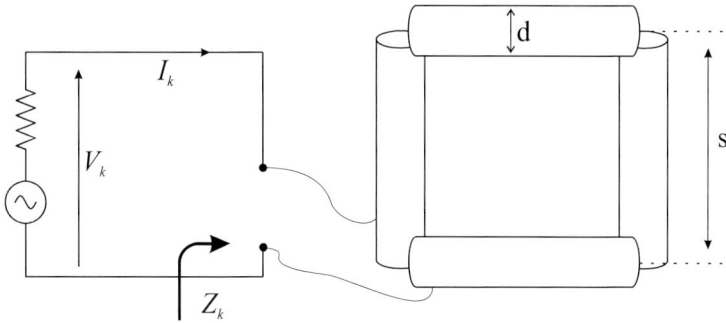

Fig. 3.17 Description of the circuit allowing calculation of the inductance of the square loop. A source with any internal impedance is used. The inductance is deduced from the imaginary component of the impedance Z_k, calculated at a given frequency ω. If the resistance of the loop is negligible compared to $|Z_k|$, the calculated inductance does not depend on frequency.

The magnetic field amplitude at the coil center is obtained from the measurement of the pulse duration corresponding to a rotation of 90° of the macroscopic spin magnetization. This can be measured [Keifer, 1999] with a small spherical sample containing doped[18] water. The magnetic field amplitude can also be calculated from the current flowing in the probe conductors. The comparison between the measured and calculated values allows evaluation of the probe design. The 90° pulse width (PW_{90}) is related to the NMR efficient magnetic field amplitude by the following equations:

---

[18] The sample solution should be doped with paramagnetic impurities to speed up the relaxation rate of the water proton. To avoid overloading the receiver, the sample could be made with a mixture of heavy and light water. With a large amount of heavy water, almost all the NMR-detectable water molecules are partially deuterated (HDO). A mixture of a similar amount of H_2O and D_2O is not recommended as it leads to a doublet.

$$PW_{90} = \frac{\pi}{2\gamma_N B_1^+}, \qquad (3.55)$$

where γ_N is the gyromagnetic ratio of the nucleus.

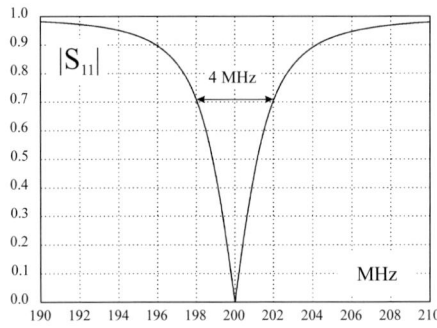 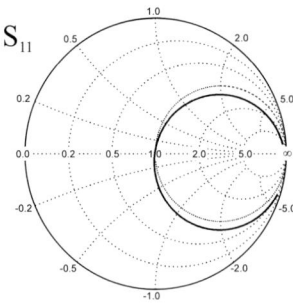

Fig. 3.18 Calculated reflection coefficient (S_{11}) of the tuned/matched probe of Fig. 3.16. Left: $|S_{11}|$ is represented on a linear scale as a function of frequency. The magnitude of the reflection coefficient is null at 200 MHz indicating a perfect match at this frequency. The width at $|S_{11}| = 0.707$ is equal to 4 MHz in agreement with an unloaded Q equal to 100. Right: the complex valued S_{11} is represented on a Smith Chart overlay as given by a Vector Network Analyzer.

For proton nuclei:

$$PW_{90} = \frac{58.72}{B_1^+} \; \mu s / Gauss. \qquad (3.56)$$

With the test coil described above, one expects a 90° pulse length of 68 µs with a 10 W transmitter ($B_1^+ = 0.87$ Gauss[19]).

The measured value can be compared with the calculated NMR effective field component (Chapter 7), by:

$$B_1^+ = B_{1\,calculated}^+ \sqrt{\frac{P_{measured} Q_{measured}}{1000}}, \qquad (3.57)$$

---

[19] 1 Gauss = 10^{-4} tesla.

where it is assumed that the calculation has been done with $Q = 100$ and $P = 10$ W. The transmitter power and Q factor can be measured with adequate means, usually available in an RF lab.

3.2.1.4 Calculating inductances

The calculation of the square coil inductance proved to be accurate when compared to standard formulae. In this calculation, the coil was decomposed into four linear conducting wires. For more complicated structures, the coil should be decomposed into an appropriate set of linear segments[20] to fit its geometry as closely as possible.

Fig. 3.19(a) shows the decomposition of a single layer thin solenoid.

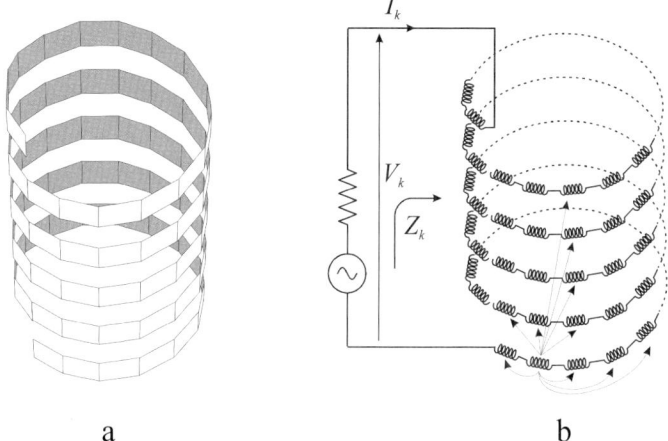

Fig. 3.19 (a) Description of the thin single layer solenoid coil as a series of small strips ($N_{seg} = 16$ per turn in the figure, $N_{turn} = 5$). The strip width is 2 mm and the center to center turn distance is 3.2 mm. The coil diameter is 10 mm. (b) Inductive representation of the solenoid coil.

Each segment is a strip conductor having the length of a turn divided by N_{seg}. The width of each segment conductor is the width of the ribbon

---

[20] Some parts of coils can be more efficiently decomposed using circular segments. In the present version of the building utilities, they are limited to segments contained in a plane perpendicular to the z-axis.

wire used to build the solenoid coil.The calculation of the inductance of the coil modeling the solenoid is straightforward using the procedure already described for the square coil. The network is built on $N_{turn} \times N_{seg}$ straight ribbon conductors. The diagonal elements of the matrix Z are obtained from the inductance of the small elementary ribbon conductor. The non-diagonal elements are calculated from the mutual inductance between each pair of segments, assuming the thin wire approximation. Closed-form formulae are used as described in the previous chapter.

The inductance is obtained after solving Eq. (3.44) for the network shown in Fig. 3.19(b) and applying Eq. (3.51). The result is 146 nH, which is significantly larger than the inductance calculated (120 nH) from the Nagaoka formula, Eq. (2.85). This difference is attributable to the approximations made here. The elementary strips are shorter than their width giving rise to an inaccurate partial inductance value and the calculation of the mutual inductance assumes a uniform current distribution on the solenoid conductor, which is certainly not the case.

3.2.1.5 Calculating current densities

To take into account the nonuniformity of current distribution, each ribbon segment is decomposed into a set of N_{strip} small parallel conductors [Fig. 3.20(a)].

The width of the elementary strip is the width of the ribbon divided by N_{strip}, becoming smaller than the strip length. One may therefore expect a better accuracy for the diagonal elements of Z.

The network is now described by $N_{turn} \times N_{seg} \times N_{strip}$ linear segments. The current in each ribbon is decomposed into $(N_{strip} - 1)$ current loops. The connection between all the segments forming the complete solenoid coil is finally described by a single loop including the source.

The calculated inductance of the solenoid represented in Fig. 3.20 is 129 nH, significantly lower than the value obtained above (146 nH) with the assumption of a uniform current on the strip conductor, but close to the value given by the Nagaoka formula (120 nH).

A calculation for a solenoid of similar dimensions and round wire (Fig. 3.21) results in an inductance value of 132 nH, which is comparable with the inductance predicted by Eq. (2.84) (137 nH).

Introduction to Linear Network Analysis 187

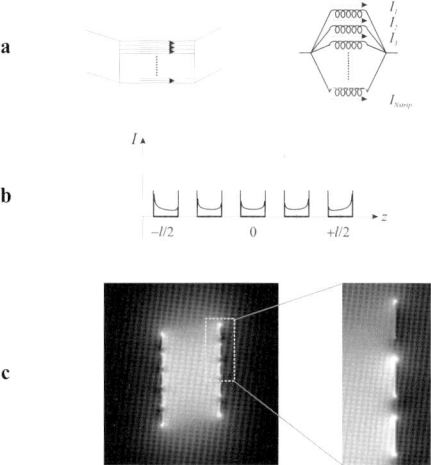

Fig. 3.20 (a) Each strip element of the solenoid in Fig. 3.19 is decomposed into 64 strips ($N_{strip} = 64$). (b) The calculation of the current density in each turn. (c) The generated magnetic field in a plane containing the coil axis, zoom in on the area in the proximity of the strips.

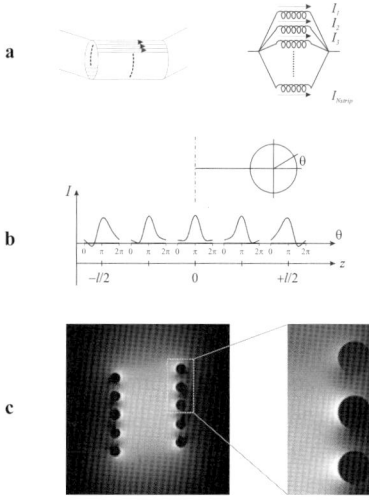

Fig. 3.21 (a) Decomposition of the surface of a round wire into small strips. (b) The calculation of the surface current density. (c) The magnetic field generated by a solenoid of the same dimensions as in Fig. 3.20 (internal diameter 10 mm, wire diameter 2 mm), zoom in on the area in the proximity of the wires.

The current density shows the proximity effect between the solenoid turns [Figs. 3.20(b) and 3.21(b)]. It is best visualized by the magnetic field distribution close to the conductors [Figs. 3.20(c) and 3.21(c)].

3.2.1.6 Calculating resonant spectrum

As an example, the resonant spectrum of a planar low-pass ladder structure composed of nine strips tuned with 200 pF capacitors (Fig. 3.22) is calculated using the linear network analyzer described here and the results are compared with previous calculations and measurements reported by Jin [Jin, 2012].

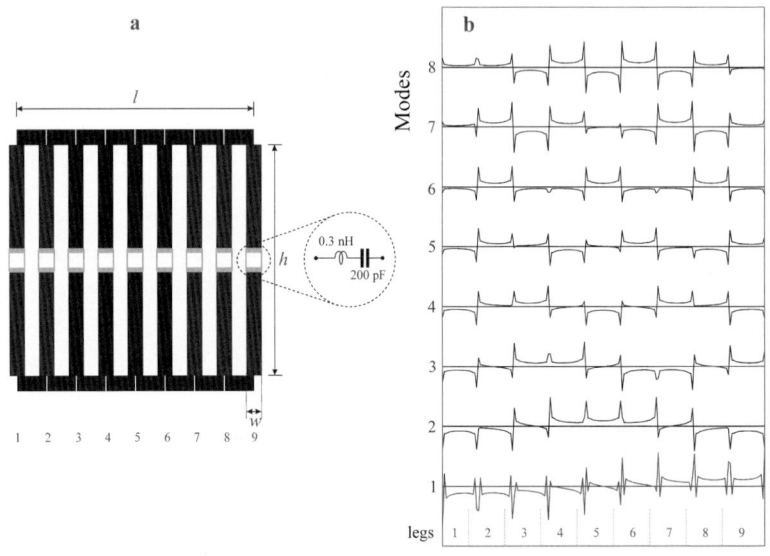

Fig. 3.22 Left: representation of the ladder filter from Jin [Jin, 2012], as calculated here (l = 22.8 mm, h = 10.685 mm, w = 1.43 mm). A simplified high frequency model of the capacitor takes into account the self-resonance frequency of 650 MHz. Right: representation of the current density (arbitrary amplitudes) calculated on the nine rungs for each resonant mode of the filter (mode 1 is the lowest frequency mode).

The circuit comprises a network of conductors, defining the mutual inductance matrix, and a set of lumped capacitors. The results are summarized in Table 3.1.

Table 3.1 Comparison of the measured and calculated resonant frequencies of the low-pass ladder network (Fig. 3.21).

C = 200 pF			$C^{(2)}$ = 200 pF + 0.3 nH		$C_{eff}^{(3)}$ = 188.4 pF	
Measured[1]	Calculated	Δ(%)	Calculated	Δ(%)	Calculated	Δ(%)
48.2	46.3	−4.1	47.4	−1.69	47.8	−0.84
75.5	73.0	−3.4	75.5	0.00	75.3	−0.40
93.8	91.0	−3.1	94.3	+0.53	93.9	+0.10
107.2	104.1	−3.0	107.8	+0.56	107.4	+0.19
117.5	114.1	−3.0	117.9	+0.34	117.7	+0.17
125.1	121.5	−3.0	125.4	+0.24	125.4	+0.24
130.2	126.7	−2.8	130.6	+0.31	130.7	+0.38
133.7	129.8	−3.0	133.7	0.00	133.9	+0.15

Notes:
(1) The measured frequencies have been reported by Jin [Jin, 2012].
(2) A gap (2.1 mm) is inserted in the rungs.
(3) The self-resonance frequency of the capacitor is assumed to be 650 MHz (consistent with case A of ATC100 porcelain multilayer capacitors).

Using the geometry of Fig. 3.22, the calculated resonant frequencies are systematically lower than the measured ones. The agreement is improved assuming an effective capacitance of 188.4 pF (third column of Table 3.1). Alternatively, a good agreement is also obtained (second column of Table 3.1) inserting a gap in the rungs and assuming the high frequency model of the capacitor, as described in Chapter 2, Section 2.4.3.3.

The calculated resonant spectrum did not improve when taking into account the current distribution on the rungs even assuming their nonuniform distribution [Fig. 3.22(b)].

This result is in contrast with the previous calculation of the inductance of a thin single layer solenoid. There are several reasons that may explain this discrepancy. The ratio of the distance between strip centers and strip widths is larger in the ladder filter than for the solenoid. The spread of the resonant spectrum is highly dependent on L/M of the rungs in the ladder case while the effective self inductance depends mainly on $L - M$ for the solenoid case. Finally, for the ladder case, the widths of the strip elements are much shorter than their length.

3.2.1.7 Mutual inductance and current distribution

The calculation of the mutual inductance, from closed-form formulae, is generally performed assuming the thin wire approximation. In this approach the distance between the current filaments differs from the centered geometrical space between the conductors (Fig. 3.23). This distance is called the Geometric Mean Distance or GMD [Grover, 2004, pp. 17ff].

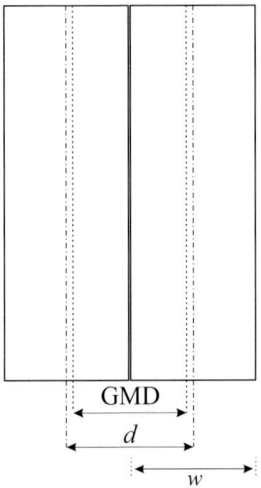

Fig. 3.23 Representation of the Geometrical Mean Distance (GMD) compared to the center to center distance d, for two edge-touching planar thin strip conductors.

If the conductors are thin coplanar strips, touching together, as considered here, the GMD is $e^{-0.1137}$ (0.89) times the center to center distance d of the two strips. The correction decreases rapidly with the ratio d/w. If d is two times the strip width, the GMD is $0.98d$. The GMD assumes, however, that the current density is identical on both coupled strips.

To investigate the effect of the current distribution on the mutual inductance, consider two planar strip conductors of varying width and connected in different ways. In the first two networks (Fig. 3.24), the two strips are connected in series with the source assuring that the total current amplitudes in both conductors are equal.

Introduction to Linear Network Analysis 191

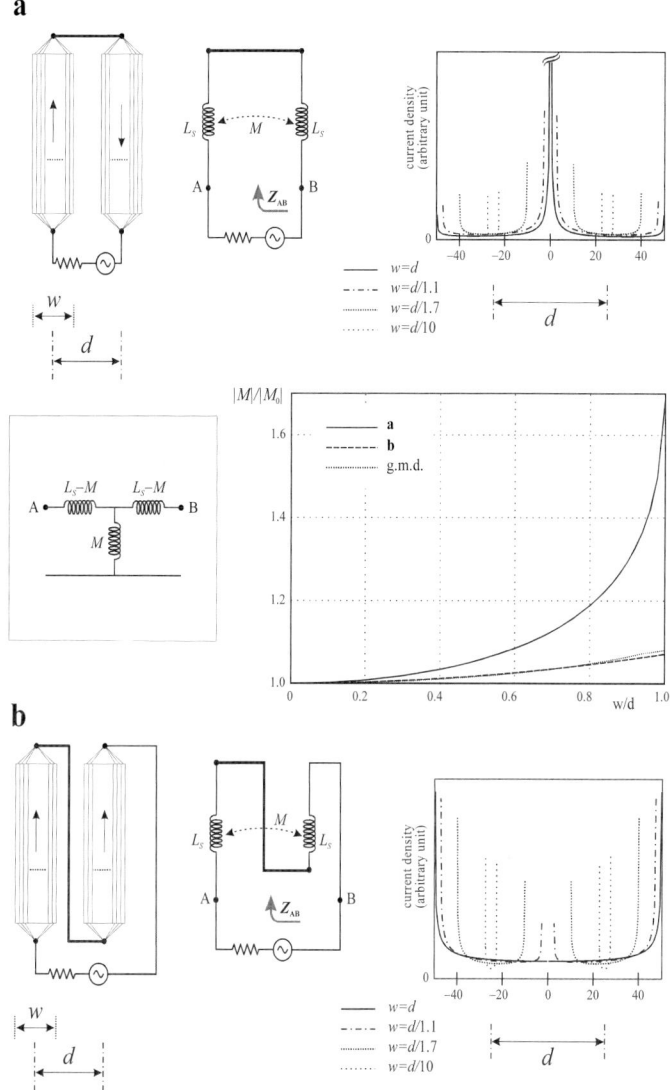

Fig. 3.24 Strip decomposition, circuit representation and current densities for two parallel strip conductors connected in series as shown (a) and (b). Middle left: equivalent circuit for two coupled inductors. Middle right: ratio of the calculated mutual inductances M/M_0 for circuits (a) and (b) and using the GMD as a function of w/d. M_0 is the mutual inductance calculated using the centered thin wire approximation.

The series connection is such that the current directions are opposed [Fig. 3.24(a)] or identical [Fig. 3.24(b)]. Each strip is decomposed into a large number (100) of parallel strip filaments in order to take into account the current distribution, which is not uniform but symmetrically shaped on the strips, as shown in Fig. 3.24.

According to the equivalent circuit representing two coupled inductors, the mutual inductance M is calculated from the following equation:

$$M\omega = L_S\omega - \frac{Z_{AB}}{2}, \qquad (3.58)$$

where L_S is the partial self inductance of each separated strip and Z_{AB} is the impedance calculated at frequency ω.[21]

When the currents in both strips have opposed directions [Fig. 3.24(a)], the current density is higher on the facing side of the strips, according to the proximity effect. The calculated mutual inductance appears strongly dependent on the ratio w/d (Fig. 3.24) when compared to the centered thin wires approximation. In contrast, when the currents in both strips have the same direction the high current density appears on the opposed sides of the strips [Fig. 3.24(b)]. Accordingly, the calculated mutual inductance appears to be very close to that predicted by the GMD. Thus, the centered thin wire approximation gives accurate results (within 5%) when $w/d < 0.85$ if the currents in both strips have the same direction. When the currents are opposed, the centered wire approximation is accurate to within 5% only when $w/d < 0.5$.

In a second network (Fig. 3.25), one strip is excited by a source thus inducing a voltage in the second strip. The resultant induced current is dissipated into a load resistance of 50 Ohms connected to the second strip. The current amplitudes are different in each strip, as is the current distribution (Fig. 3.25). The mutual impedance is obtained from the calculated S-parameters of this two-port network. According to the equivalent circuit representing the two coupled inductors and using Eqs.

---

[21] Note that the inductance does not depend on frequency in these calculations, provided the resistance of the conductor is much smaller than the inductive impedance.

(3.7) and (3.8) and the matrix coefficients transformation table (Fig. 3.2), the mutual inductance is obtained from:

$$M\omega = \text{Im}\{Z_{12}\} = \text{Im}\left\{Z_0 \frac{2S_{12}}{(1-S_{11})(1-S_{22})-S_{12}S_{21}}\right\}. \quad (3.59)$$

When the strip conductors are touching together, the calculated mutual inductance still differs significantly (by 38%) from the mutual calculated from the centered thin wires approximation. The discrepancy is nevertheless smaller when the strips are connected in series as in Fig. 3.24(a). As a result, the centered thin wire approximation gives accurate results within 5% when the ratio w/d is smaller than about 0.6.

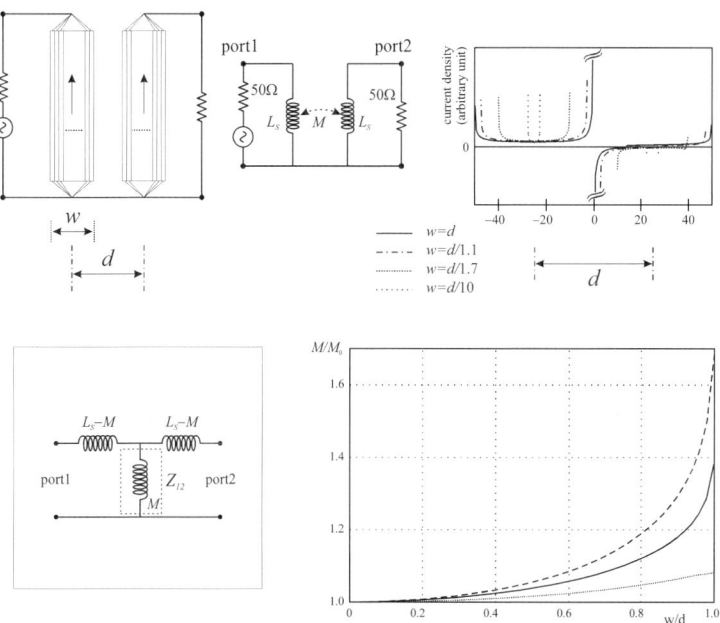

Fig. 3.25. Upper: strip decomposition, circuit representation, and current densities for two parallel strip conductors mutually coupled as shown. Lower left: equivalent circuit for the two coupled inductors. Lower right: ratio of the calculated mutual inductances M/M_0 for the coupled inductor circuit (full line), using the GMD (dotted line) and for the circuit shown in Fig. 3.24(a) (dashed line) as a function of w/d. M_0 is the mutual inductance calculated using the centered thin wire approximation.

3.2.2 Other dedicated linear circuit analyzers

In this section, we briefly review some probe simulators that allow a linear network analysis of the RF circuit. This analysis is invaluable for an understanding of the behavior of the device on the RF workbench as well on the NMR spectrometer or MRI scanner and to track issues of detuning or to identify the desired resonant mode. This can be done using the currently available 3D full electromagnetic simulators, but this is time-consuming and requires high performance computational resources.

Any probe network can be modeled by lumped elements as resistor, capacitor, and inductor as described above for NMRP. The impedance determination of these components is the initial step for the calculation of the electrical properties of the circuit and of the currents flowing in the conductors. In NMRP, only the inductance matrix is calculated. This is done using mainly closed-form formulae.

Another approach has been described by Giovannetti *et al* [Giovannetti *et al.*, 2002; Giovannetti *et al.*, 2007] in which the self and mutual inductances are obtained from an integration method which proves to be accurate.[22] The resistances of conductors are evaluated from closed-form formulae and the sample-induced resistances are calculated, using, for example, FDTD[23] [Giovannetti *et al.*, 2011]. Therefore, the method of Giovannetti *et al.* may be considered as a mixed algorithm, with the aim to compute the elements of the Z-matrix of the probe network as precisely as possible.

For a long time, Doty *et al.* have designed and used a specific linear network analyzer to predict the component value required by a given probe design, to analyze its electrical properties, and to understand many aspects of its coupling with the external world. To our knowledge, the code (either source or compiled) of this analyzer has not been released. But the reader can find some examples in the literature such as, for example, the design of birdcage resonators [Doty *et al.*, 2007].

The Doty *et al.* analyzer is not intended to calculate the magnetic field distribution as the NMRP or Giovannetti *et al.* simulator can do

---

[22] The inductance calculator is written in IDL language and is available from the authors.
[23] Finite Difference Time Domain.

because it treats only lumped elements. However, the elements that can be included in the circuit model allow an accurate simulation, taking account of both the magnetic and electric couplings in an elegant manner. For example, the rungs of a birdcage coil (or those of the ladder filter presented before) are modeled by an ideal transformer (magnetic coupling) completed by a piece of transmission line (extra inductance, capacitive coupling, and losses). Other classical elements such as an ideal inductor or capacitor can be included in the network. This offers a complete model for a given probe circuit that could subsequently be simulated using, for example, a SPICE-like analyzer. As an example, the ladder filter described above can be modeled as shown in Fig. 3.26.

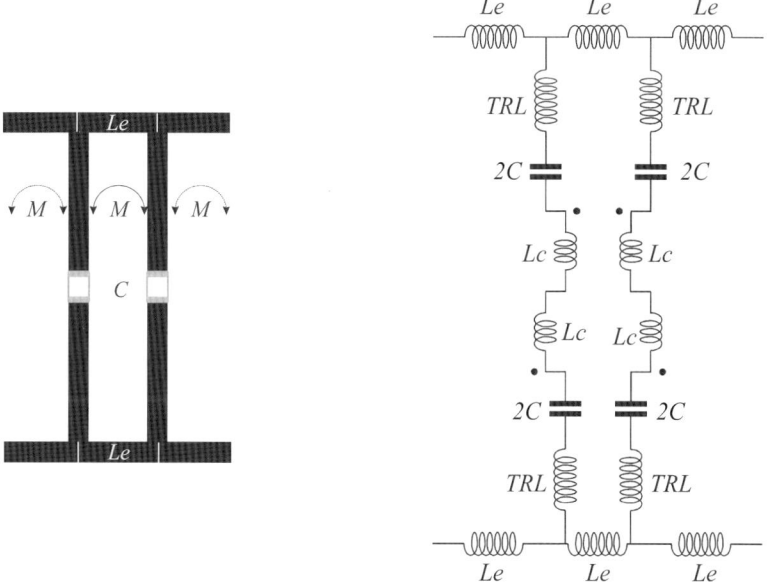

Fig. 3.26 Part of the ladder filter model shown in Fig. 3.22 using elements as in the Doty *et al.* analyzer. *Lc* are parts of ideal transformers, *TRL* are parts of transmission lines and *Le* are the partial inductances of the end connections.

The model can also be refined, including stray capacitances or any other parasitic couplings, allowing an analysis of undesired common modes.

Another general approach for probe simulation, based on the S-matrix, has recently been introduced [Kozlov and Turner, 2009; Lemdiasov et al., 2011]. This method provides a bridge between a time consuming 3D full electromagnetic simulation and efficient RF simulators. Instead of calculating the Z-matrix of the probe network, as in NMRP or in the Giovannetti et al. simulator, the S-matrix is obtained at only one frequency from a 3D full electromagnetic simulation of the multi-port network of probe conductors, including the coupling with the surroundings (sample). The obtained S-parameters are extrapolated assuming that the effective inductance and resistance are nearly constant in a narrow frequency range around the operating frequency [Lemdiasov and Ludwig, 2012]. In this way, the tuning and matching components can be determined easily from a simple linear network analysis. The slowest step (the 3D electromagnetic simulation) is limited to one calculation but the inherent accuracy is retained. The efficiency of an RF network simulation, using lumped elements derived from the S-matrix is also retained.

The Birdcage Builder [Chin et al., 2002] is a simulator specifically designed for estimating the capacitor needed to tune a birdcage coil (low-pass, high-pass or band-pass) at a given frequency. The trick of the calculation resides in assuming an effective inductance of a specific branch that takes into consideration the mutual contribution of the other branches as well as the ideal current distribution for a given resonant mode. In this way, using closed-form formulae for partial and mutual inductance, the tuning capacitors can be quickly estimated, but at a specified resonant mode. The software, which is freely accessible, proved to be accurate enough to provide good starting values for building a convenient birdcage resonator designed for a desired operating frequency.

Chapter 4

Interfacing the NMR Probehead

The basis of the NMR signal detection by the Faraday inductive method consists of coupling an inductor to the nuclear spin magnetization. Interfacing this "coil" to the Radio Frequency (RF) electronics of the NMR spectrometer or the MRI scanner requires impedance matching of the coil inductance and resistance to the system impedance, which is generally a pure 50 Ω resistance.

As outlined in Chapter 2, probes can be used as both transmitter and receiver, as transmit-only, or as receive-only. In all these cases, the probe coil, or the probe resonator, must be matched to 50 Ω.

Impedance matching allows an optimum energy transfer when the probe is used as a transmit coil to excite the spin magnetization. In this case, the best efficiency is obtained when the power delivered by the transmitter is entirely dissipated in the probe resistance.[1] In other words, the probe should be viewed as a 50 Ω load impedance at the operating frequency ω_S [Fig. 4.1(a)].

Due to the Principle of Reciprocity [Hoult, 2000], the best efficiency of the probe as a receiver would be obtained under the same conditions, as far as the optimum power transfer is concerned.

In the receive mode, the probe is usually connected to a Low Noise Amplifier (LNA). The best signal-to-noise ratio is generally not obtained when the optimum power transfer condition is fulfilled. Rather, the probe should be "noise matched" to the LNA input transistor. As a matter of fact, the preamplifier includes its own matching network that transforms the standard 50 Ω impedance into the required optimum impedance (Z_{opt})

---

[1] Including all the losses (sample, coil conductors, radiation loss, etc…).

for the best noise factor. In other words, the probe should be viewed as a source with an internal impedance equal to 50 Ω at the operating frequency ω_S [Fig. 4.1(b)].

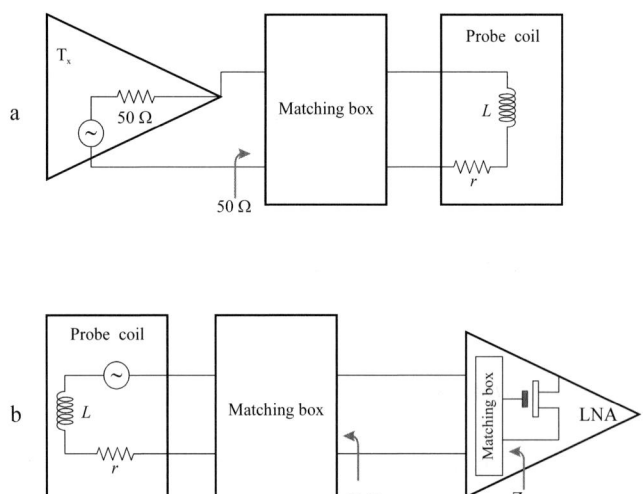

Fig. 4.1 (a) Transmit mode: the probe coil is viewed as a 50 Ω load for optimum power transfer. (b) Receive mode: the probe is viewed as a source with internal impedance equal to 50 Ω used by the LNA matching box for an optimum noise figure.

It should be noted here that, in these conditions, the preamplifier input impedance differs from 50 Ω, implying that a mismatch exists between the probe and the preamplifier (Chapter 2). While power transfer is not optimum, the best noise factor (the best sensitivity) is expected [Rohde and Bucher, 1988]. This mismatch should however be taken into account, especially when a lossy long cable is used to connect the probe to the LNA.

In this chapter, impedance matching circuits, specifically designed for NMR probeheads will be described. Remote matching, through transmission lines or "wireless" coupling will be introduced with particular emphasis on the advantages and drawbacks of the method.

Exact expressions for the component values of the matching networks will be given whenever possible. This would be useful in a computer-aided design program. Exact solutions are also invaluable for subsequent simulation of the probe electromagnetic properties.

The necessity of electrically balancing the NMR coil will be demonstrated. Balanced matching circuits, derived from the general ones, will subsequently be described.

Transmit/Receive (TR) switches, damping circuits, baluns and cable traps are accessories for interfacing a matched probe to the NMR console. They are described in a specific section. Interfacing a coil to an LNA is addressed with particular emphasis on the case when it is a component of a receiving probe array.

Finally, the principle of broadband and non-resonant probe designs will be described in the last section.

4.1 Impedance Matching

Whatever the operating mode of the probe, transmitting or receiving, it must be impedance matched to 50 Ω.

The inductance of the probe coil is typically of the order of 100 Ω and its resistance is of the order of 1 Ω (Q = 100). Matching such impedance to a pure resistance of 50 Ω is necessarily realized in a narrow band of frequency, according to the Bode–Fano criteria [Bode, 1945; Fano, 1948]. It results that the probe will be mostly[2] a resonator.

4.1.1 The Bode–Fano limit

The Bode–Fano relationship gives an upper limit to the frequency bandwidth for which one has an "acceptable" impedance matching, in fact an acceptable "return loss" [Pozar, 1998, pp. 295–297]. In its original formulation, this criterion states that the impedance matching of a resistance R shunted with a capacitance C can be made using any lossless matching network with the following constraint [Bode, 1945, pp. 363–368; Fano, 1948]:

---

[2] Broadband probe designs will be considered at the end of this chapter.

$$\int_0^\infty \ln\left(\frac{1}{|\Gamma(\omega)|}\right) d\omega \leq \frac{\pi}{RC}. \tag{4.1}$$

In the equation above $\Gamma(\omega)$ is the reflection coefficient defined as:

$$\Gamma = \frac{Z - Z_0}{Z + Z_0}, \tag{4.2}$$

where Z is the impedance of the parallel RC circuit to be matched to Z_0. A similar relation holds for a series rL circuit:

$$\int_0^\infty \ln\left(\frac{1}{|\Gamma(\omega)|}\right) d\omega \leq \frac{\pi r}{L}. \tag{4.3}$$

Because the above integral is limited to a defined positive value, a perfect match can be made for only one frequency (or for a discrete set of frequencies). Let Γ_m be the upper limit of acceptable reflection coefficient within a given frequency band $[\omega_1, \omega_2]$. Assuming an ideal case where Γ is constant and equal to Γ_m in the bandwidth $\Delta\omega = \omega_2 - \omega_1$,

$$\begin{aligned}\Gamma(\omega) &= \Gamma_m \quad \begin{cases}\omega \geq \omega_1 \\ \omega \leq \omega_2\end{cases} \\ \Gamma(\omega) &= 1 \quad \begin{cases}\omega < \omega_1 \\ \omega > \omega_2\end{cases}\end{aligned} \tag{4.4}$$

the Bode–Fano criterion for the inductance L in series with a resistance r [Eq. (4.3)] can be written as:

$$\ln\left(\frac{1}{|\Gamma_m|}\right) \Delta\omega \leq \frac{\pi r}{L}. \tag{4.5}$$

Improving impedance matching ($\Gamma_m \to 0$) implies a smaller bandwidth $\Delta\omega$. Furthermore, when r/L diminishes (i.e., the Q factor increases), the maximum achievable bandwidth diminishes for a given Γ_m. It appears therefore that a high Q circuit (that is the requirement for a good sensitive probe) is harder to match on a broad frequency band. (Fig. 4.2).

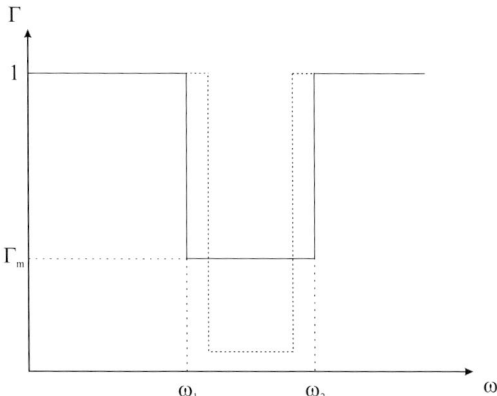

Fig. 4.2 The integral Eq. (4.3) is bounded. An improvement of Γ in the pass-band (Γ_m) results in a narrowing of the bandwidth.

4.1.2 Connected matching circuits

4.1.2.1 Basic L-matching circuits

Impedance matching is very well documented in the RF and microwave literature. The simplest matching circuit, which could be the best choice for NMR probe design, is an L-network formed from two conjugate reactive impedances, as shown in Fig. 4.3.

The component values, assuming that R_1 and R_2 are pure resistive impedance, are given by [RSGB Handbook, 1982, pp. 12.41]:

$$\begin{aligned} p &= R_2/R_1 \\ X_1 &= \pm R_1\sqrt{p-1} \\ X_2 &= \mp R_2/\sqrt{p-1} \end{aligned} \quad . \quad (4.6)$$

R_2 must be always greater than R_1. R_2 can be assigned to Z_0, or to the high impedance ρ of a resonator. R_1 can be assigned to Z_0 or to the resistance r of the inductor. X_1 and X_2 being conjugate, if X_1 is an inductance, then X_2 must be a capacitance and *vice versa*.

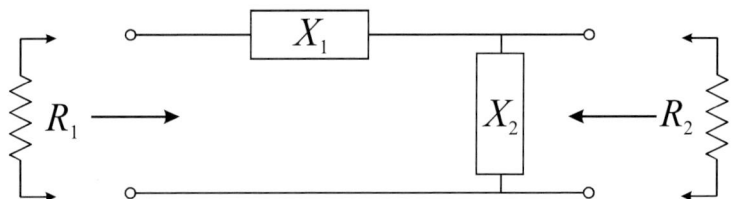

Fig. 4.3 Basic L matching network. The network is made out of pure, conjugate, reactances X_1 and X_2 ($X > 0$ means an inductor, $X < 0$ means a capacitor). The reactance values, as a function of R_1 and R_2, are given in Eq. (4.6).

Among the various possibilities based on the L-matching circuit in Fig. 4.3, one retains four circuits (Fig. 4.4) which give rise to the principal sketches used for NMR probe designs. The circuits shown in Fig. 4.4(a) are defined for $r < Z_0$, while the circuits in Fig. 4.4(b) are defined for $\rho > Z_0$.

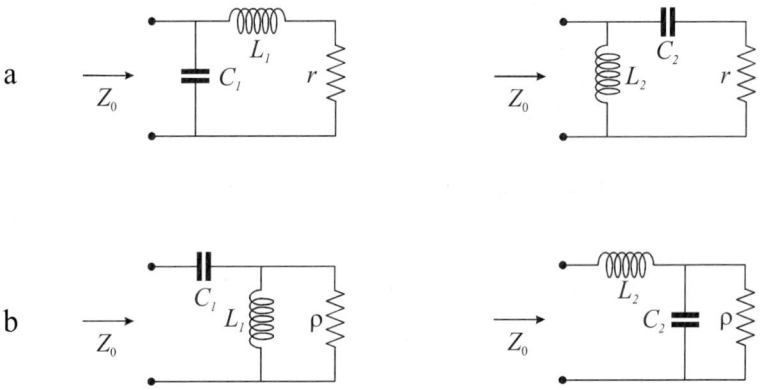

Fig. 4.4 Basic L-networks designed to match resistive loads. (a) $r < Z_0$. (b) $\rho > Z_0$. The input impedance can be either capacitive (left), or inductive (right).

The component values are derived from Eq. (4.6). For the circuits shown in Fig. 4.4(a) one gets:

$$L_1\omega_S = \frac{1}{C_2\omega_S} = \sqrt{r(Z_0-r)}$$
$$\frac{1}{C_1\omega_S} = L_2\omega_S = Z_0\sqrt{\frac{r}{Z_0-r}}$$
(4.7)

and, for the circuits shown in Fig. 4.4(b):

$$L_1\omega_S = \frac{1}{C_2\omega_S} = \rho\sqrt{\frac{Z_0}{\rho-Z_0}}$$
$$\frac{1}{C_1\omega_S} = L_2\omega_S = \sqrt{Z_0(\rho-Z_0)}$$
(4.8)

The equations above are valid only when the impedances to be matched (r or ρ) are pure resistances. Prior to matching the complex impedance of the probe, including its inductance and resistance, it must be transformed into a pure resistance.

4.1.2.2 Series tuned/parallel matched

In this configuration, the coil inductance is cancelled out (series tuned) by a capacitance[3] C_0 (Fig. 4.5). The small resistance r of the probe coil is then matched to Z_0 using one of the circuits shown in Fig. 4.4(a). The cancellation of the coil inductance by C_0 implies that the matching is narrow band.

In the first circuit (Fig. 4.5), the capacitance C_0 and the inductance L_1 of the matching network can be combined together into a single capacitance[4] C_T, the so-called tuning capacitor. The parallel capacitance C_1 [Fig. 4.4(a)] is identified as the matching capacitor and is named C_M accordingly.

The component values are derived straightforwardly from Eq. (4.7):

---

[3] C_0 is defined by $LC_0\omega_S^2 = 1$.
[4] If $1/C_0\omega_S$ is larger than $L_1\omega_S$, which is usually the case.

$$\frac{1}{C_M \omega_S} = Z_0 \sqrt{\frac{r}{Z_0 - r}}$$
$$\frac{1}{C_T \omega_S} = L\omega_S - \sqrt{r(Z_0 - r)}$$
(4.9)

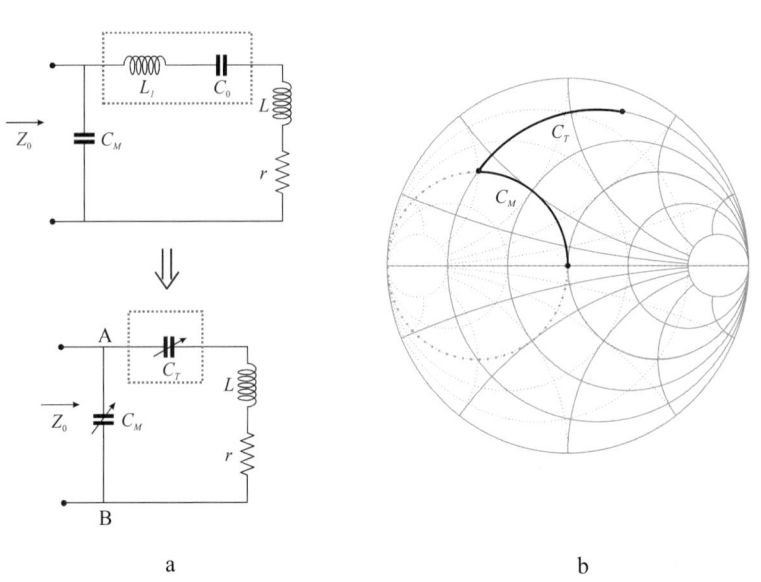

Fig. 4.5 (a) Series tuned/parallel matched circuit derived from the L-network shown in Fig. 4.4(a, left). The coil inductance is tuned at ω_S by the capacitance C_0. Then, the coil resistance r is matched to 50 Ω by the (L_1, C_M) circuit. L_1 is associated with C_0, into the final tuning capacitance C_T ($C_T > C_0$). (b) The tuning (C_T) and matching (C_M) pathways are depicted on the Smith Chart.

As depicted on the Smith Chart [Fig. 4.5(b)], the matching procedure is done as follows:

- First, the tuning capacitor C_T moves the coil impedance (including its resistance and inductance) to the unit admittance circle. At this point, the admittance looking across the C_M terminals is $Y_{AB} = (Y_0 + jB)$ Ω^{-1}.
- Second, C_M moves the admittance Y_{AB} to the center of the plot where Γ = 0 or Y/Y_0 = 1.0. In other word, the admittance $C_M \omega_S$ cancels out the residual admittance B.

The series combination of C_0 and L_1 implies a slight detuning of the resonator to the lower frequencies (C_T is larger than C_0).

The second matching network shown in Fig. 4.4(a, right) leads to the circuit shown in Fig. 4.6(a). It can obviously be derived from the Smith Chart [Fig. 4.6(b)]. Extending the previous solution, the tuning capacitor C_T is decreased further until Y_B moves to the symmetric position on the unit admittance circle. At this point, the center of the plot can be reached using the proper inductance L_2 which is identified here as L_M. Because C_T is smaller than C_0 [Fig. 4.6(b)], the resonator is slightly detuned toward higher frequencies.

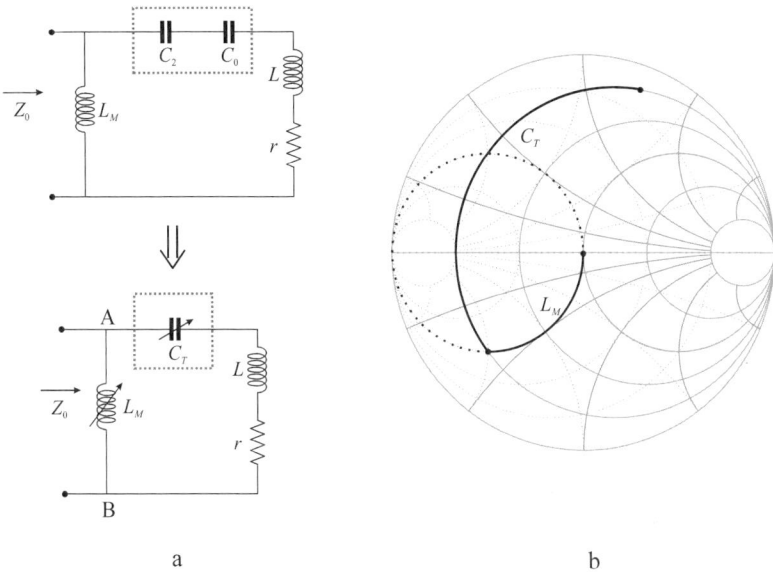

Fig. 4.6 (a) Series tuned/parallel matched circuit derived from the L-network shown in Fig. 4.4(a, right). The coil inductance is tuned at ω_S by the capacitance C_0. Then, the coil resistance r is matched to 50 Ω by the (C_2, L_M) circuit. C_2 is combined with C_0 into the final tuning capacitance C_T ($C_T < C_0$). (b) The tuning (C_T) and matching (L_M) pathways are depicted on the Smith Chart.

The corresponding component values are again straightforwardly derived from Eq. (4.7):

$$L_M \omega_S = Z_0 \sqrt{\frac{r}{Z_0 - r}}$$

$$\frac{1}{C_T \omega_S} = L\omega_S + \sqrt{r(Z_0 - r)} \tag{4.10}$$

The circuit of Fig. 4.5, including only two capacitances, is best suited for NMR probe designs, as capacitors usually have a better Q factor than inductors.

In contrast, the inductance L_M in the circuit of Fig. 4.6 must be adjusted according to the probe loading (change of r). It is usually preferable to use a fixed inductance. Matching the circuit to various coil loading conditions implies that a variable matching capacitor should be added as shown in Fig. 4.7.

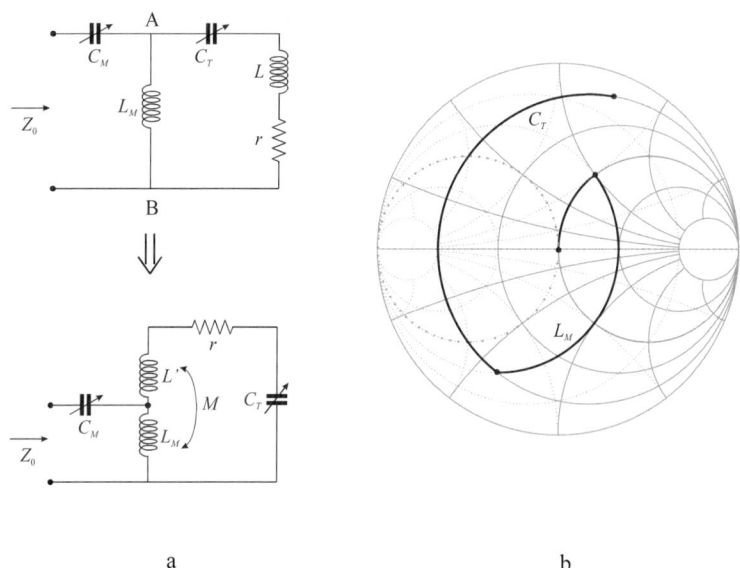

Fig. 4.7 (a) Circuits derived from the circuit shown in Fig. 4.6. The variable matching inductance is replaced by a fixed inductance and a variable capacitor. The coil and matching inductance can be combined, leading to a circuit in which the matching capacitance is connected to a tapping point on the coil inductor. (b) As shown on the Smith Chart, compared to Fig. 4.6 the tuning capacitance (C_T) and the matching inductance (L_M) are both decreased until the impedance reaches the unit circle.

In this case, a higher resonant frequency than that predicted by Eq. (4.10) must be provided, as is shown on the corresponding Smith Chart [Fig. 4.7(b)]. Furthermore, the fixed inductance L_M must be higher than the value required for the maximum loading of the coil.

Eventually, the inductance L_M may be part of the coil itself, the connection being made at an appropriate tapping point.

4.1.2.3 Parallel tuned/series matched

In the series tuned/parallel matched circuits, the coil inductance was cancelled out by a series resonant configuration. It can also be cancelled out by a parallel resonant circuit. In this case, the impedance across the coil becomes a large resistance which can be matched by the circuits in Fig. 4.4(b).

Fig. 4.8(a) shows the corresponding circuit. Let C_0 be the capacitance used to tune the probe coil at the operating frequency ω_S. The inductance L_1 in parallel to C_0 is equivalent to a single capacitance C_T, provided $1/C_0\omega_S$ is smaller than $L_1\omega_S$, which is usually the case. The capacitance C_T being smaller than C_0 [Fig. 4.8(b)], the coil resonance is slightly shifted toward higher frequencies. The impedance across the terminals A and B is positive (inductance). It is cancelled out by the matching capacitance C_M.

Only approximate component values can be easily obtained from Eq. (4.8) due to the approximate definition of the resistance ρ of the (L, C_0) resonant circuit:

$$\rho \approx QL\omega_0 \approx QL\omega_S. \qquad (4.11)$$

Using the first order expansion in r/Z_0 (or Z_0/ρ), one obtains the following approximate solution:

$$\begin{aligned} C_T\omega_S &= \frac{1}{L\omega_S}\left(1 - \sqrt{\frac{r}{Z_0}}\right) \\ C_M\omega_S &= \frac{1}{\sqrt{Z_0 QL\omega_S}} \end{aligned} \qquad (4.12)$$

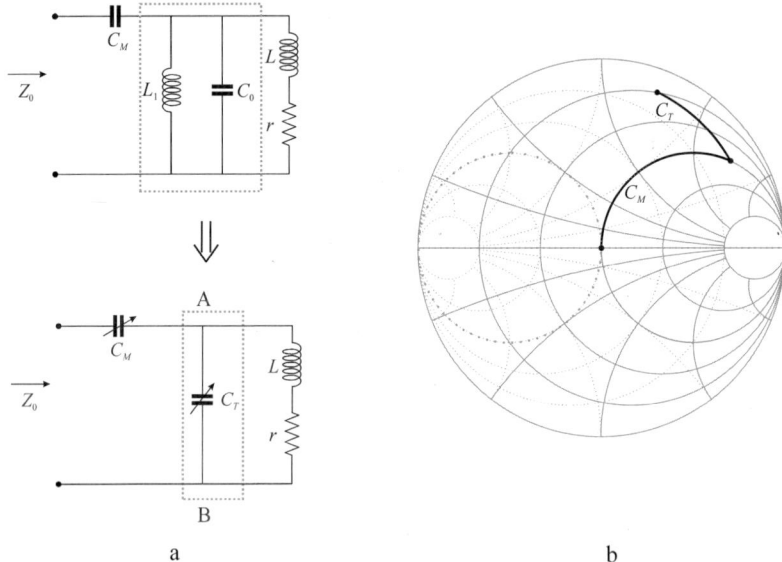

Fig. 4.8 (a) Parallel tuned/series matched circuit derived from the L-network shown in Fig. 4.4(b, left). The coil inductance is tuned at ω_S by the capacitance C_0. The high resistance ($\rho \approx QL\omega_S$) of the coil resonator is matched to 50 Ω by the (L_1, C_M) circuit. L_1 is associated with C_0, into the final tuning capacitance C_T ($C_T < C_0$). (b) The tuning (C_T) and matching (C_M) pathways are depicted on the Smith Chart.

Another matching solution (Fig. 4.9) is obtained by increasing the capacitor C_T such that the impedance Z_{AB} reaches the symmetric point on the unit resistance circle [Fig. 4.9(b)]. The resonance of the coil is shifted to lower frequencies. The impedance Z_{AB} is negative and is cancelled out by the matching inductor L_M.

The corresponding approximate expressions are:

$$C_T \omega_S = \frac{1}{L\omega_S}\left(1 + \sqrt{\frac{r}{Z_0}}\right). \tag{4.13}$$

$$L_M \omega_S = \sqrt{Z_0 Q L \omega_S}$$

Exact expressions for the matching components can be derived from the solution for C_T of the following equation:

$$Z_0 = \frac{rXC_T^2}{r^2 + (XL - XC_T)^2},$$ (4.14)

$$XC_T = \frac{1}{C_T \omega_S} \quad XL = L\omega_S$$

where XC_T is the absolute value of the impedance of the tuning capacitor C_T at the operating frequency ω_S and XL is the reactance of the coil inductor.

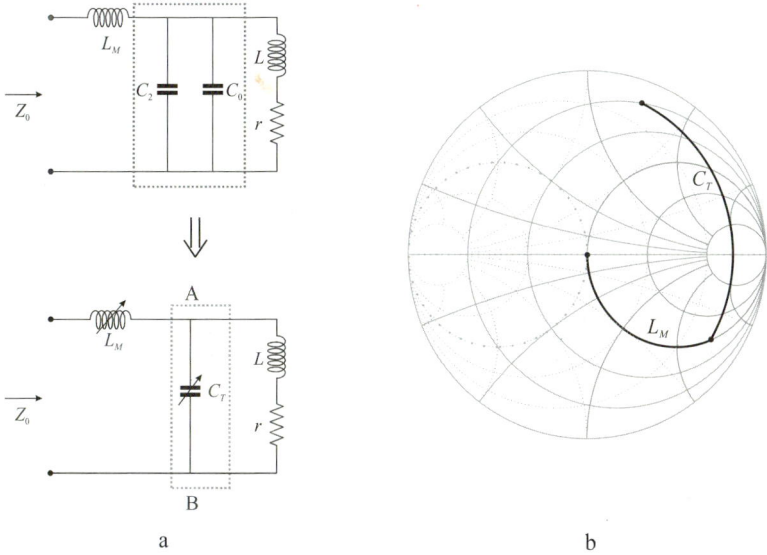

Fig. 4.9 (a) Parallel tuned/series matched circuit derived from the L-network shown in Fig. 4.4(b, right). The coil inductance is tuned at ω_S by the capacitance C_0. Then, the high resistance ($\rho \approx QL\omega_S$) of the coil resonator is matched to 50 Ω by the (C_2, L_M) circuit. C_2 is combined with C_0 into the final tuning capacitance C_T ($C_T > C_0$). (b) The tuning (C_T) and matching (C_M) pathways are depicted on the Smith Chart.

Eq. (4.14) shows that the real component of the impedance across the tuning capacitor C_T must be equal to Z_0. Solving the equation for C_T, one obtains:

$$C_T\omega_S = \frac{Q \pm A}{B}, \tag{4.15}$$

where

$$Q = \frac{L\omega_S}{r} \tag{4.16}$$

$$A = \sqrt{\frac{B}{Z_0} - 1} \tag{4.17}$$

$$B = r(1+Q^2). \tag{4.18}$$

The residual imaginary component of the impedance should be compensated by a series matching capacitance or inductance (Fig. 4.10). The imaginary component of the impedance Z_{AB} across the tuning capacitor C_T is:

$$\operatorname{Im}\{Z_{AB}\} = -Z_0 \frac{1 + Q\left(Q - \frac{XC_T}{r}\right)}{\frac{XC_T}{r}}. \tag{4.19}$$

From Eqs. (4.15) and (4.18):

$$\frac{XC_T}{r} = \frac{1+Q^2}{Q \pm A}. \tag{4.20}$$

Substituting Eq. (4.20) into Eq. (4.19) and after some algebraic manipulation, one obtains the reactive component of the probe resonator at ω_S:

$$\operatorname{Im}\{Z_{AB}\} = \mp Z_0 A. \tag{4.21}$$

The plus sign in Eq. (4.21) above corresponds to the minus sign in Eq. (4.15), as expected. In that case, the resonance frequency of the probe resonator is shifted to high frequencies. The reactive impedance at

ω_S is inductive and will be compensated by a matching capacitor C_M [Fig. 4.10(a)]. The corresponding value is given by:

$$C_M \omega_s = \frac{1}{Z_0 A}. \qquad (4.22)$$

If the plus sign is chosen in Eq. (4.15), the resonance frequency of the probe resonator is shifted to lower frequencies. The reactive impedance at ω_S is capacitive and will be compensated by a matching inductor L_M [Fig. 4.10(b)]:

$$L_M \omega_s = Z_0 A. \qquad (4.23)$$

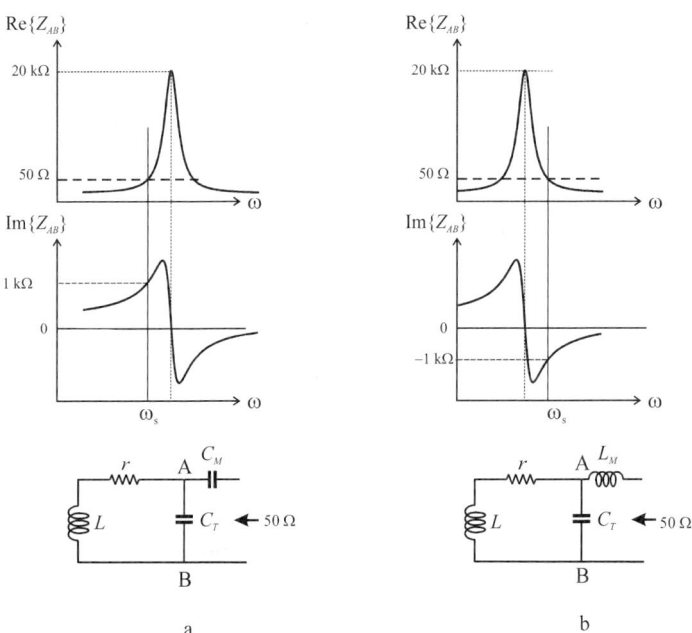

Fig. 4.10 Sketches illustrating the tuning/matching procedure in the case of the parallel tuned/series matched circuit. (a) The coil inductance L is tuned to a higher frequency than ω_0 ($LC_T\omega_0^2 = 1$), such that the real component of the impedance, Re{Z_{AB}}, is equal to 50 Ω. Then the residual, positive, impedance Im{ Z_{AB}} is cancelled out by C_M. (b) The coil inductance L is tuned to a lower frequency than ω_0, such that Re{Z_{AB}} = 50 Ω. Then the residual, negative, impedance Im{Z_{AB}} is cancelled by L_M.

4.1.2.4 *Efficiency of the capacitive matching circuits*

The capacitors used in a matching circuit are generally of much better quality than the probe coil itself in terms of Q factors. Accordingly, the losses in the matching box are usually neglected.

Practical case:

Let the coil reactive impedance be equal to 100 Ω with a Q factor of 100 at 200 MHz ($r = 1$ Ω). The tuning and matching capacitors are assumed to be commercial multilayer microwave capacitors with known Q values estimated from the formulae given in Chapter 2.

Both circuits, the parallel tuned/series matched and the series tuned/parallel matched, are considered. The capacitance values are first calculated, assuming a lossless matching network. Their resistances (equivalent series resistance *ESR*, defined in Chapter 2) are estimated from the Q value of Eq. (2.111) as:

$$ESR = \frac{1}{QC\omega}. \qquad (4.24)$$

Finally, the power lost in a resistance[5] r driven by a current i_C is:

$$P = r|i_C|^2. \qquad (4.25)$$

Taking into account the resistance of the capacitors a small mismatch of the order of −25 to −30 dB is observed. Possibly, the capacitance values can be slightly readjusted to improve the matching better than −40 dB (Table 4.1).

The component values estimated for the circuits in Fig. 4.11 are summarized in Table 4.1.

---

[5] The resistance r is either the coil resistance or the *ESR* of the capacitors.

Table 4.1 Components values and simulated electrical characteristics of the circuits shown in Fig. 4.11. The coil inductance is $(1 + j100)\,\Omega$ at 200 MHz.

Circuit	C_T (pF)	r_{CT} (Ω)	C_M (pF)	r_{CM} (Ω)	S_{11} (dB)	$|i_2|/|i_1|$	E
Parallel tuned series matched	6.8	0.062	1.2	0.08	−43	6.9	−0.1 dB
Series tuned parallel matched	8.6	0.058	106	0.04	−45	6.7	−0.3 dB

Note: the reflection coefficient S_{11}, the current ratio, and the efficiency can be calculated using any linear network analyzer.

The losses in the matching circuits arose essentially from the current that circulates in the resonator (i_2), the current (i_1) in the feed line being about one order of magnitude smaller. Therefore, the undesired losses in the parallel tuned/series matched circuit [Fig. 4.11(a)] originate mostly from the resistance of the tuning capacitor. In contrast, both the tuning and the matching capacitors in the series tuned/parallel matched circuit [Fig. 4.11(b)] contribute to the losses.

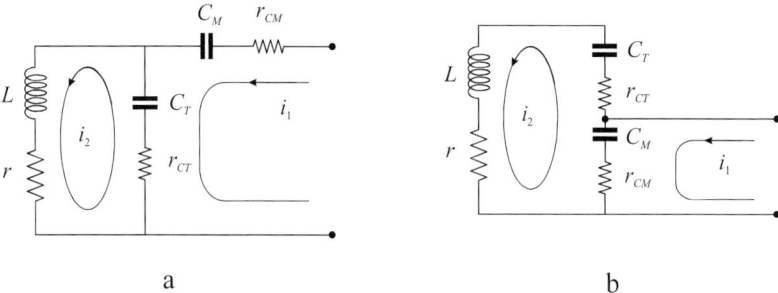

Fig. 4.11 Parallel tuned/series matched (a) and series tuned/parallel matched (b) circuits, including the resistance (*ESR*) of the capacitors, used to estimate the efficiency of the circuits (Table 4.1).

Because the *ESR* values of the contributing capacitors are comparable, the losses in the parallel tuned/series matched circuit are about half the losses in the series tuned/parallel matched circuit.

The efficiency (in dB) of the matching circuit can be defined as:

$$E_{dB} = 10\log_{10}\left(\frac{P_{coil}}{P_{total}}\right), \qquad (4.26)$$

where P_{coil} is the power dissipated in the coil resistance r and P_{total} is the power entering the circuit. Both circuits having a return loss smaller than −40 dB, the total incident power is equal to the power delivered by the source (power matching). The power dissipated in the probe coil is calculated from the current i_2 flowing in the resonator. As expected, the current i_2 is the same for both circuits when the matching components are lossless, but differs slightly when realistic resistances have been included in the capacitor's model. As a result, the efficiency of −0.1 dB, calculated for the parallel tuned/series matched circuit, is close to an acceptable value. However, the series tuned/parallel matched circuit has an efficiency of −0.3 dB which becomes comparable to the noise figure (*NF*) of a very good preamplifier. This indicates that, without special care or particular needs, the parallel tuned/series matched circuit may be preferred.

4.1.2.5 *Transmission line matching (remote matching)*

Transmission line matching is widely used in microwave engineering [Pozar, 1998, pp. 258–295; Lee, 2004, pp. 87–107]. Matching a probe coil using a transmission line may be suitable when the variable tuning/matching components cannot be placed close to the coil.

The matching of an inductive load by a single line is briefly described below. The required line characteristic impedance is not standard and is generally hard to build, except at very high frequencies, using microstrip technology or high quality impedance tuners.

Thus, matching circuits using transmission lines of standard characteristic impedances will be considered. This easy to implement approach will require additional lumped elements connected to the line terminals and may therefore be described as remote matching.

A lossless transmission line transforms a load impedance Z_L (Fig. 4.12) according to the following equation (Section 2.4.2.5):

Interfacing the NMR Probehead

$$Z_{AB} = Z_C \frac{Z_L + jZ_C \tan(\theta)}{Z_C + jZ_L \tan(\theta)}, \quad (4.27)$$

where Z_C is the characteristic impedance of the line and θ is the phase angle:[6]

$$\theta = 2\pi \frac{l}{\lambda} = 2\pi \frac{l_e}{\lambda_{vacuum}}. \quad (4.28)$$

l is the physical length of the line and λ is the wavelength in the cable. l_e is the so-called electrical length of the line, i.e., the length of an air-line leading to the same phase shift. The wavelengths are related to the frequency f by:

$$\lambda = \frac{c}{f\sqrt{\varepsilon_r}} = k\lambda_{vacuum}, \quad (4.29)$$

where ε_r is the dielectric constant. k is the velocity factor which depends on the dielectric material of the line and c is the speed of light.

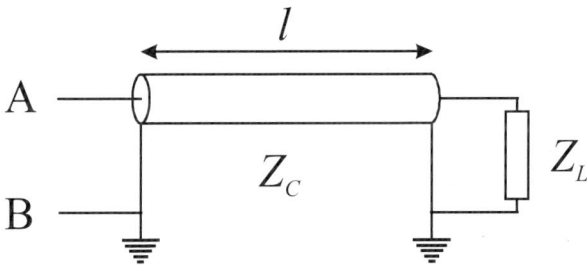

Fig. 4.12 Circuit using the transformation properties of a transmission line to match a load Z_L to $Z_{AB} = Z_0$. The required impedance transformation is realized for a given length (l) and characteristic impedance Z_C of the line, according to Eqs. (4.30) and (4.31). If Z_L is purely resistive, the impedance matching is realized by the so-called quarter wave transformer, according to Eq. (4.32).

---

[6] The rotation angle in the Γ-plane is twice this value.

Both the characteristic impedance Z_C and the length of the line should be adjusted for matching any impedance $r + jXL$ to Z_0. They are given by [White, 2004, pp. 109]:

$$Z_C = \sqrt{\frac{Z_0 r - r^2 - XL^2}{1 - \dfrac{r}{Z_0}}} \qquad (4.30)$$

$$\tan(\theta) = \frac{Z_C(Z_0 - r)}{Z_0 XL}. \qquad (4.31)$$

Because Z_C should be a real number, the possibilities of matching an inductive load are very restricted. The condition that XL^2 should be smaller than $r(Z_0 - r)$ is fulfilled only for either very small Q coils or small inductances.

If $XL = 0$, the matching circuit shown in Fig. 4.12 is the well-known quarter-wave transformer which matches purely reactive loads.

$$\begin{aligned} Z_C &= \sqrt{Z_0 r} \\ \theta &= (2n+1)\frac{\pi}{2} \quad n \in \mathbb{N} \end{aligned} \qquad (4.32)$$

It is worth noting that broadband matching of pure resistances can be achieved by successive impedance transformations through sections of quarter-wave transformers. Similar wide band tuning/matching networks can be built using an adjustable segmented transmission line such as in a state-of-the-art design by Qian and Brey [Qian and Brey, 2009]. In their design, the tuning/matching is done by sliding two $\lambda/4$ dielectric slugs in a high quality air-line. The setup is equivalent to four connected transmission lines and the dielectric slugs control the forward and backward waves in the lines until the input reflection is cancelled out.

The matching of an inductive load with a transmission line can be done simply using a line with standard characteristic impedance Z_0. The Smith Chart (Fig. 4.13) shows how this process can be realized in two steps. First, a line of the appropriate length transforms the coil impedance $r + jL\omega_S$ into $Z_0 + jL_m\omega_S$, which is located on the unit impedance circle. Second, the residual inductance L_m is cancelled out by a series matching capacitor C_M. This circuit is very similar to the parallel tuned/series matched circuit already discussed. Here the length of transmission line is equivalent to the tuning capacitor.

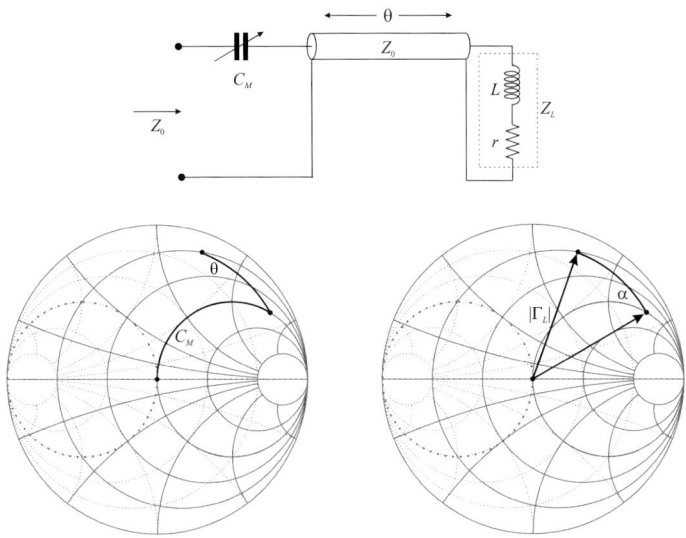

Fig. 4.13 Matching an inductive load (L,r) using a line of standard characteristic impedance Z_0 is realized firstly by a phase shift θ which moves Z_L to a point on the unit circle. Then, impedance matching is completed by the series capacitor C_M (left). The angle α describing the motion of the impedance (right) is equal to twice the phase shift θ. Note that θ is related to the physical line length by Eq. (4.28).

The length of line is given by the condition that the real part of the impedance Z_{AB}, Eq. (4.27), is equal to Z_0. Assuming a lossless line of characteristic impedance equal to Z_0, this condition can be written as:

$$\operatorname{Re}\left\{\frac{Z_L + jZ_0 \tan(\theta)}{Z_0 + jZ_L \tan(\theta)}\right\} = 1, \qquad (4.33)$$

where

$$Z_L = r + jL\omega_S. \qquad (4.34)$$

Eq. (4.33) is easier to solve geometrically in the Γ-plane than analytically. According to Fig. 4.13, the solution is given by the intercept of the circle of radius Γ_L centered in the Γ-plane with the unit impedance circle. Γ_L is the reflection coefficient at the load.

The coordinates of the intercept point are given by:

$$\begin{aligned} \Gamma_x &= |\Gamma_L|^2 \\ \Gamma_y &= |\Gamma_L|\sqrt{1-|\Gamma_L|^2} \end{aligned} \quad . \tag{4.35}$$

The required length of the line is obtained from the angle α between the vectors representing the initial reflection coefficient Γ_L and the target corresponding to the intercept. The angle α is given by the scalar product of these two vectors:

$$\alpha = \cos^{-1}\left(\operatorname{Re}\{\Gamma_L\} + \operatorname{Im}\{\Gamma_L\}\sqrt{\frac{1}{|\Gamma_L|^2}-1} \right) = 2\theta, \tag{4.36}$$

where the phase shift θ is defined by Eq. (4.28).

Eq. (4.36) assumes that the line is lossless. In fact, even the better lines are lossy. It results that the line length l obtained from Eqs. (4.36) and (4.28) is approximate, but provides a starting point[7] for the design.

In the case of a lossy line, a numerical solution for the physical length l of line is obtained, assuming again that the real impedance at the input of the line must be equal to Z_0:

$$\operatorname{Re}\{Z_{in}(l)\} = Z_0. \tag{4.37}$$

Z_{in} is calculated using Eq. (2.80), where the propagation constant γ and the characteristic impedance Z_C are complex numbers given, in the most general case, by Eqs. (2.44) and (2.45). The line parameters, R, L, G, and C can be obtained for any real line as described in Chapter 2 (see Section 2.4.2.1).

---

[7] This gives rise to an overestimated line length.

When connected to a high Q inductive load, the line exhibits a set of sharp resonances as its length is varied (Fig. 4.14). The solutions to Eq. (4.37) can be determined efficiently in three steps.

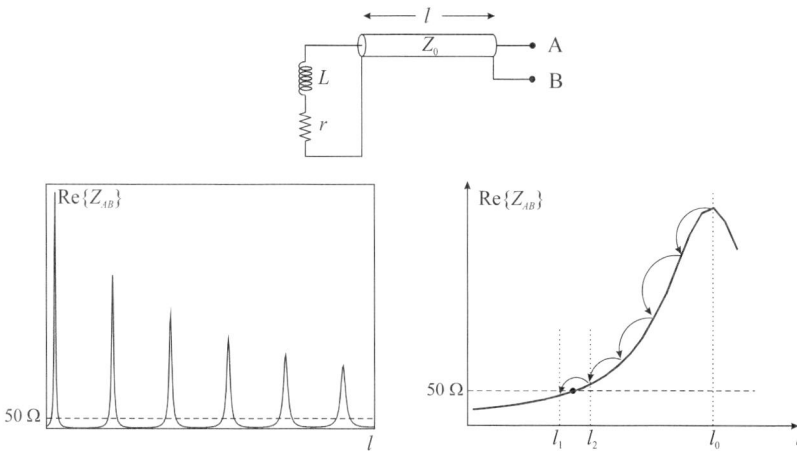

Fig. 4.14 Numerical determination of the length of line of characteristic impedance Z_0 that allows matching of an inductive load (L,r) to the resistive impedance Z_0. First, the real component of the impedance Z_{AB} (Re{ Z_{AB} }) is calculated as a function of line length, at ω_S, using Eq. (2.80) and a R,L,C,G model of the real line. The circuit exhibits a number of resonant modes, as shown on the left. The first mode (corresponding to the shortest line length) is selected. Second, starting from this resonance, the solution for l of the equation Re{$Z_{AB}(l)$} = 50 Ω is bracketed downward in an interval $[l_1,l_2]$ by a steepest descent procedure, shown on the right. Third, a root-finding algorithm is used to find the solution in this interval.

First, the length of the line is varied until the sign of Im{Z_{in}} changes (at resonance). The corresponding length l_0 is close to the resonance of the line terminated by the inductive load. Second, one solution of Eq. (4.37) around the resonance is bounded in an interval $[l_1, l_2]$ (Fig. 4.14). The interval is determined by a simple steepest descent procedure by bracketing the Z_0 value. Third, the solution is refined in the previously determined interval using a conventional root-finding algorithm.

The required length to transform an inductive load of 100 Ω at 200 MHz using a 50 Ω transmission line is shown in Fig. 4.15, as a function of the coil Q factor.

The lossless approximation is valid for very good low-loss lines such as the semi-rigid cables RG402 or RG401, but becomes significantly inaccurate for a low-loss line such as RG316.

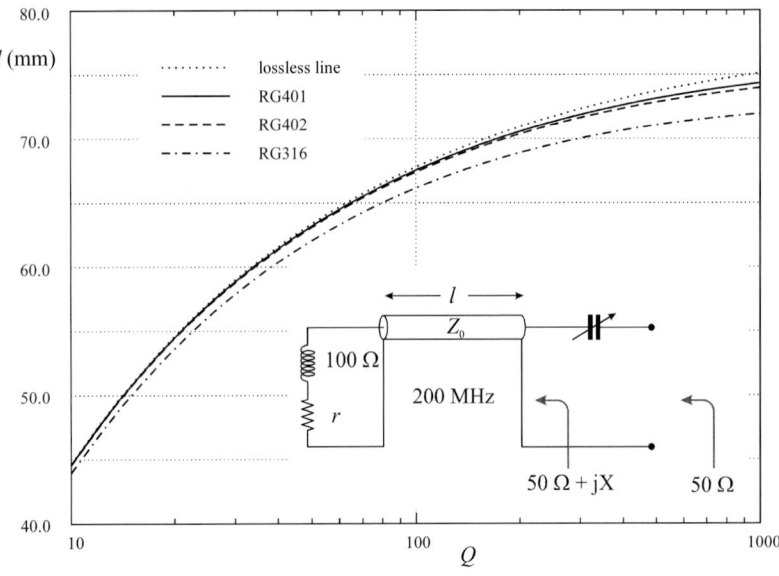

Fig. 4.15 The length of different coaxial lines transforming an inductive impedance ($r + j100$) Ω into ($50 + jX$). Ω is represented at 200 MHz as a function of $Q = 100/r$. The residual reactance X will be compensated by the matching capacitor to complete impedance matching.

The line length is obviously a function of the probe coil loading. In practice, the length cannot be changed easily. Preferably, it will be fixed to a value smaller than the one required when the load of the probe is at a maximum. Then, the effective length of the line can be virtually increased by a variable (tuning) capacitor connected in parallel to the line terminals [Fig. 4.16(a)]. This configuration can be viewed as a remote matching circuit. The line length is short: only a few centimeters, as in the above numerical example. Such a length may be too small to fit the requirements of a remote control. Thus, interposing a $\lambda/2$ line[8] between the coil and a matching circuit may be another attractive solution [Fig.

---

[8] The $\lambda/2$ line simply copies the impedance at a longer distance.

4.16(b)]. However, this alternative should be used with extreme care as the losses in the line can very quickly destroy the probe efficiency.

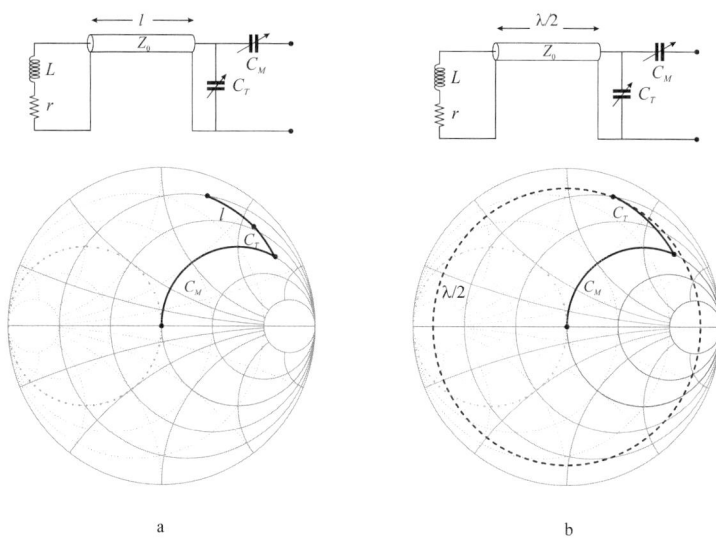

Fig. 4.16 Two remote matching circuits using a transmission line. (a) A small piece of line pre-tunes the coil inductance L. Tuning is completed by the addition of the capacitance C_T (equivalent to increasing the length of the line). (b) the $\lambda/2$ transmission line copies the coil impedance which is tuned/matched by a classical parallel tuned/series matched circuit.

4.1.2.6 *Efficiency of transmission line matching networks*

The problem of matching a reactive load through a transmission line has been already discussed in the case of communication equipment [for example, ARRL Handbook, 2009, pp. 21.6–21.11] as well as in the NMR literature [Martin and Daly, 1986; MacLaughlin, 1989; Walton and Conradi, 1989; Rath, 1990a; Villa *et al.*, 1999; Kodibagkar and Conradi, 2000; Qian and Brey, 2009]. It is shown that the losses depend critically on the standing wave ratio of the line. A high Q reactive load, such as an NMR probe, leads to very high standing wave ratios, thus possibly to high losses. This will be considered quantitatively here with some details.

As already outlined in Section 2.4.2.4, the losses in a line not terminated by its characteristic impedance may be much higher than the losses expected from the manufacturer's data sheet. Even if using a so-called low-loss line, the efficiency of the remote matching circuit may decrease dramatically.

Consider the circuit shown in Fig. 4.17 in which a transmission line is interposed between a lossless matching circuit and the probe coil.

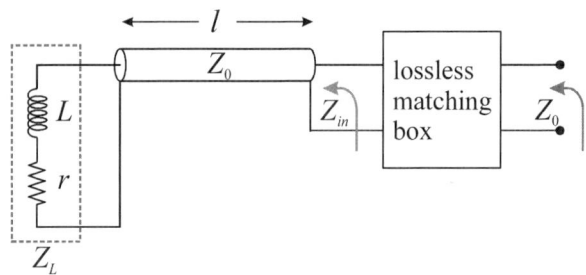

Fig. 4.17 Circuit to evaluate the efficiency of a remote matching circuit using a transmission line.

One will consider two cases. In the first one, the line length is adjusted to the value just required to tune the coil. In the second one, the line length is set to $\lambda/2$ in order to copy the coil impedance at a remote position, terminated by a classical parallel tuned/series matched circuit.

The probe efficiency is evaluated as the ratio of the effective power dissipated in the coil resistance to the power delivered by a transmitter with internal source impedance equal to Z_0. If the probe is used as a receiver, the same efficiency will be expected as far as The Principle of Reciprocity is applicable.

The ratio of the active power dissipated in the load to the active power delivered at the input of the line is given by Eq. (2.71), which is repeated here:

$$\frac{P_L}{P_{in}} = E = \frac{\left(1-|\Gamma_L|^2\right)}{\left(A-|\Gamma_L|^2/A\right)}. \tag{4.38}$$

Γ_L is the reflection coefficient at the load, which is very close to unity if the coil has a high quality factor. A is the "matched" power attenuation of the line. Because there is a match between the source and the input impedance of the line, the power applied to the line is equal to the power delivered by the source. Hence, this ratio is the probe efficiency.

Practical case

The inductive load is assumed to be an inductance of 100 Ω at 200 MHz with a Q factor ranging from 10 to 1000.

The required length line is calculated as outlined in the previous paragraph from which one deduces the "matched" attenuation A. Obviously, the efficiency is unity for any lossless line ($A = 1$) whatever the load impedance.

Three different lines used practically will be considered. The RG316 is a low-loss cable of moderate quality which has the advantages that it is flexible and of small diameter (about 3 mm with its sheath). The RG402 has about the same external diameter but is rigid and of better quality. The RG401 is of even better quality but is bulkier (external diameter larger than 6 mm) and also rigid. All these cables have a PTFE[9] dielectric. Their characteristics are given in Tables 2.3 and 2.4, allowing Eq. (4.37) to be solved numerically and the efficiency ratio to be calculated from Eq. (4.38). The efficiency is shown in Fig. 4.18 for the three cables, as a function of the quality factor Q of the inductive load. It is clear that the efficiency decreases as the coil Q increases! Losses can be as large as 4 dB for standard Q values ($Q = 200$) with less than 7 cm of RG316 cable. The losses are even more catastrophic when using a $\lambda/2$ line, as is obvious from Fig. 4.19.

Similar results would be obtained even if the coil was pre-tuned because the magnitude of the reflection coefficient at the load would not change significantly.

These rather deceiving results have been verified experimentally on the RF workbench with a simple setup. A loop of about 30 mm diameter

---

[9] Polytetrafluoroethylene or Teflon®.

is etched on a printed circuit board ($L = 56.3$ nH, $r = 0.34$ Ω, $Q = 210$ at 200 MHz).

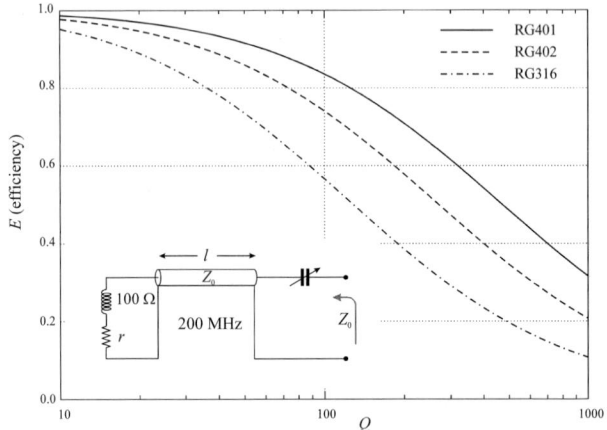

Fig. 4.18 Efficiency of a remote matching using the minimum line length to tune the coil inductance at 200 MHz, as a function of Q ($Q = 100/r$).

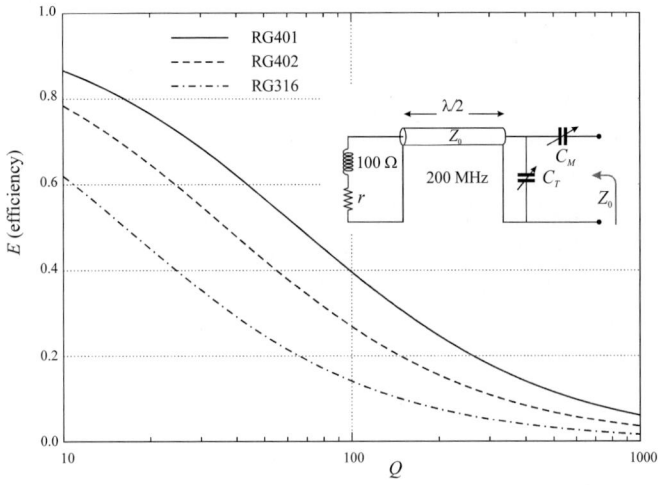

Fig. 4.19 Efficiency of a remote matching using a $\lambda/2$ transmission line to copy the coil impedance at 200 MHz, as a function of Q ($Q = 100/r$).

A small balanced sniffer coil (10 mm diameter) is fixed at about 10 mm from the coil plane (coupling between both coils of about −7.5 dB). The loop is carefully tuned and matched at 200 MHz with high quality components soldered close to the coil.Its Q factor is evaluated from S_{11} and can be changed by loading the coil using samples containing water with different salt concentrations. The S_{21} transmission parameter between the loop and the sniffer coil directly gives an evaluation of the power dissipated in the equivalent resistance of the probe coil. A small piece of about 10 cm of RG316 cable is interposed between the loop and the tuning/matching circuit which is again carefully adjusted at 200 MHz using high quality variable capacitors. S_{21} is measured again and compared with the value obtained when the matching circuit was connected directly to the coil. The difference represents the efficiency of the probe coil. Values of 2 to 3 dB were currently measured. The difference increased to more than 10 dB when a $\lambda/2$ line is interposed instead of the 10 cm piece of line.

The only way to decrease the losses in the line is to reduce $|\Gamma_L|$ before connecting the coil to the line. This can be done with a lossless pre-matching circuit interposed between the coil and the line but this considerably reduces the tuning/matching range of the remote control.

Another possibility is to use a line of higher impedance. Two straight parallel conductors are probably the best, provided the coil is pre-tuned close to the operating frequency. A remote capacitor will add the necessary capacitance to reach the desired frequency and a matching capacitor will remove any residual reactance as in the classical parallel tuned/series matched circuit.

Finally, the adjustable segmented transmission line design of Qian and Brey [Qian and Brey, 2009] based on a high quality impedance tuner is a good solution in certain circumstances.

4.1.3 *Flux coupled matching circuits*

It is assumed here that the coil probe has already been tuned at a given frequency ω_0, possibly close to the operating frequency ω_S. The principle of inductive matching is to couple the probe resonator with a "sniffer" coil. The transformed impedance at the coupler coil terminals is that of the parallel resonator, reduced in amplitude to a level depending on the coupling amount (Fig. 4.20).

Matching consists of moving the sniffer coil in the magnetic field created by the resonator in such a way that the real part of the transformed impedance is equal to the desired target Z_0. Possibly, the frequency of the resonator should be shifted in order that this condition is met at ω_S. Finally, as with the previous circuits, a residual reactance should be cancelled out by its conjugate impedance.

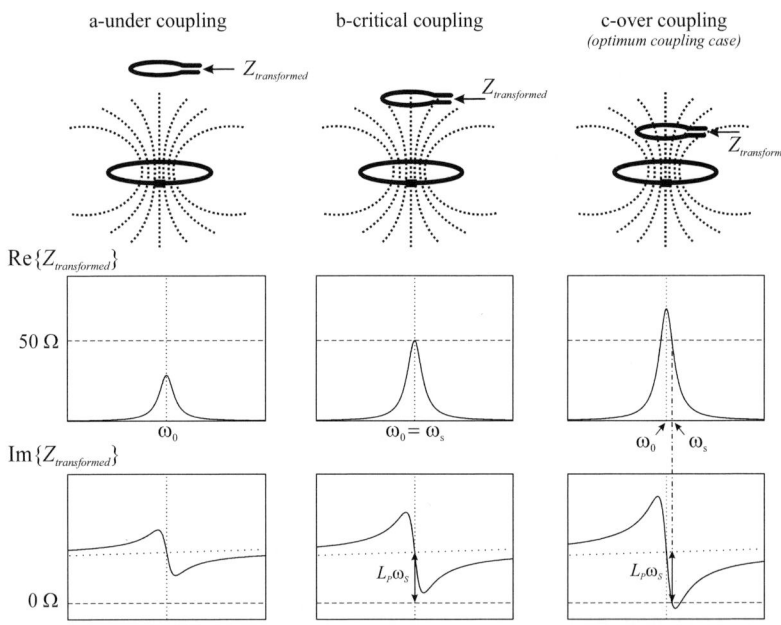

Fig. 4.20 Transformed impedance of a resonator, tuned to ω_0, by a coupling loop of inductance L_P. (a) The coupling is too weak to allow impedance matching. Re$\{Z_{transformed}\}$ is always smaller than 50 Ω. (b) The coupling is such that Re$\{Z_{transformed}\}$ = 50 Ω at ω_0. To complete the impedance matching, the residual Im$\{Z_{transformed}\}$ can be cancelled out by a capacitor. (c) The coupling is slightly increased such that the two conditions Re$\{Z_{transformed}\}$ = 50 Ω and Im$\{Z_{transformed}\}$ = 0 are simultaneously fulfilled at ω_S.

Three different conditions of coupling can be distinguished. When the coupling loop is placed far away from the resonator (under coupling), the transformed impedance is much lower than Z_0 and impedance matching is not possible. As the sniffer coil is brought closer to the resonator, the real component of the transformed impedance becomes exactly equal to Z_0 at the resonant frequency for a particular position of the coupling coil (critical coupling). The residual reactive component is exactly[10] that of the coupling loop. It can be cancelled via a series capacitance which tunes the sniffer coil at ω_S. Further decreasing the distance between the coupling loop and the resonator, the resistive component of the transformed impedance becomes higher than Z_0 at resonance (over coupling). This situation is similar to the capacitive coupling scheme presented in Section 4.1.2 with the difference that the impedance value at ω_0 is reduced by the inductive coupling. Consequently, the difference between ω_0 and ω_S is much smaller than for the capacitive matched circuits and the tuning need of the resonator is diminished (Fig. 4.21).

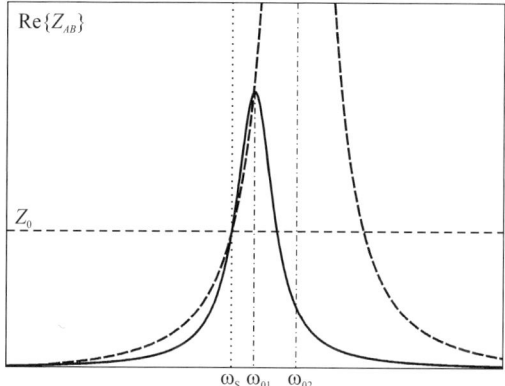

Fig. 4.21 When the resonator is inductively coupled, the real component of the impedance is reduced in amplitude (full line) compared to the impedance across the tuning capacitor (dashed line). As a consequence, the frequency shift required to fulfill the matching condition $\text{Re}\{Z_{AB}\} = 50\ \Omega$ is smaller for the inductive coupling case than for the capacitive coupling circuit.

The reactive component of the transformed impedance will also be compensated by its conjugate impedance. For one particular coupling condition (optimum coupling), the residual reactance is "self-cancelled"

---

[10] The imaginary component of the resonator impedance, and of its image by the coupling loop, is zero.

by a judicious balance between the transformed reactance of the resonator and the inductance of the coupling loop [Fig. 4.20(c)].

4.1.3.1 Theory of inductive matching

The inductive coupling circuit[11] and its equivalent description are shown in Fig. 4.22. The circuit comprises the resonator (L, C_T), the coupling loop (L_P) and a "matching" capacitor C_P which will compensate any positive residual reactance. We deliberately ignore the possibility of compensating a negative reactance by a variable inductor.

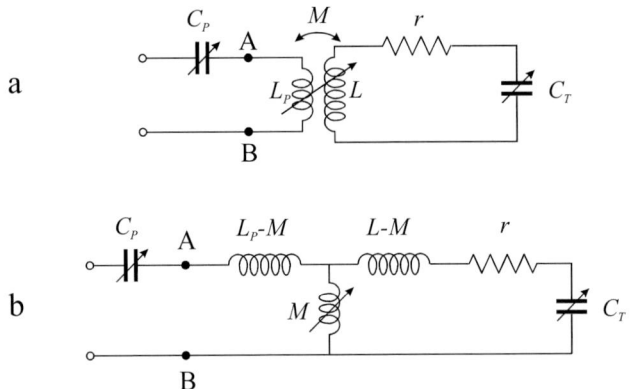

Fig. 4.22 (a) The inductive matching circuit and its adjustable parameters. C_T is the tuning capacitor. M is the mutual inductance between the coil probe inductance L, L_P is the coupling loop inductance and C_P a matching capacitor. (b) The equivalent circuit representation.

The transformed impedance of the resonator, looking at the terminals A and B of the coupling coil, is easily obtained from the equivalent circuit [Fig. 4.22(b)] as:

$$Z_{AB}(\omega) = \frac{rM^2\omega^2}{r^2 + XS^2} + jM\omega\frac{r^2 + XS^2 - M\omega XS}{r^2 + XS^2} \\ + j(L_P\omega - M\omega)$$ (4.39)

---

[11] The inductive coupling has been extensively described in the NMR literature [Decorps et al., 1985b; Froncisz et al., 1986a; Hoult and Tomanek, 2002].

XS represents the residual reactance of the probe resonator:

$$XS = L\omega - \frac{1}{C_T \omega}. \tag{4.40}$$

Matching the resonator to Z_0 at ω_S is fulfilled when the real part of Z_{AB} is equal to Z_0 and its imaginary part is compensated by the conjugate impedance (the capacitance C_P). The first condition is given by:

$$\frac{rM^2\omega_S^2}{r^2 + XS^2} = Z_0 \tag{4.41}$$

and the second condition by:

$$M\omega_S \frac{r^2 + XS^2 - M\omega_S XS}{r^2 + XS^2} + (L_P\omega_S - M\omega_S) = \frac{1}{C_P\omega_S}. \tag{4.42}$$

Let X_P be the reactive impedance of the primary circuit including the coupling loop and the matching capacitor C_P:

$$XP = L_P\omega_S - \frac{1}{C_P\omega_S}. \tag{4.43}$$

The tuning capacitor is given by Eq. (4.41) which may be rearranged as:

$$XS^2 = r\left(\frac{M^2\omega_S^2}{Z_0} - r\right). \tag{4.44}$$

When the tuning condition of Eq. (4.44) is realized, the matching condition, Eq. (4.42), simplifies to:

$$XP = XS\frac{Z_0}{r}. \tag{4.45}$$

Eqs. (4.44) and (4.45) allow the design of various inductive matching schemes according to the value of the three adjustable parameters, M, X_S and X_P.

4.1.3.2 Inductive matching with tuned coupling loop

The minimum mutual inductance M, allowing impedance matching is given by:

$$M\omega_S = \sqrt{rZ_0}. \qquad (4.46)$$

If M is lower than this value (under coupling), Eq. (4.44) has no real-valued solution.

When M is set to the value defined by Eq. (4.46), corresponding to the critical coupling condition, the resonator must be tuned at the operating frequency ($X_S = 0$, $\omega_0 = \omega_S$) and the coupling loop must be series tuned with the capacitor C_P ($X_P = 0$). The capacitances of the circuit [Fig. 4.23(a)] are given by:

$$L_P C_P \omega_S^2 = 1, \qquad (4.47)$$

$$L C_T \omega_S^2 = 1. \qquad (4.48)$$

It is noticeable that the tuning [Eq. (4.48)] and matching [Eq. (4.46)] adjustments are independent. The capacitance C_P is independent of the probe loading.

Fig. 4.23(b) shows the representation in the Z-plane of the transformed impedance Z_{AB}, around the resonant frequency.[12] In the critical coupling condition, the diameter of the circle representing Z_{AB} is equal to Z_0. The real component of the impedance Z_{AB} fulfills Eq. (4.41) at the resonance frequency of the probe resonator. The imaginary component, which is equal to $L_P \omega_S$, is compensated by C_P in series with L_P.

---

[12] The graph of the impedance $Z(\omega)$ of a parallel resonator in the Z-plane is a circle. The diameter of the circle is equal to the real-valued impedance at resonance ($QL\omega_0$). The circle is spanned in the clockwise direction as the frequency increases. At resonance, the point representing the impedance is located on the horizontal axis (Im$\{Z\} = 0$). The two opposite points on the circle lying on a diameter parallel to the vertical axis correspond to the frequencies equal to $\omega_0(1 \pm 1/2Q)$. Almost all the circle is spanned in a small frequency range around the resonance, of the order of $\pm 3\omega_0/Q$.

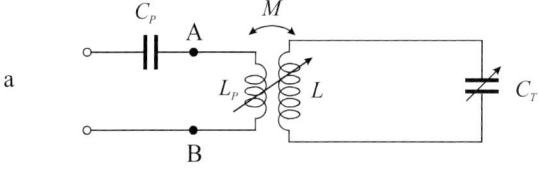

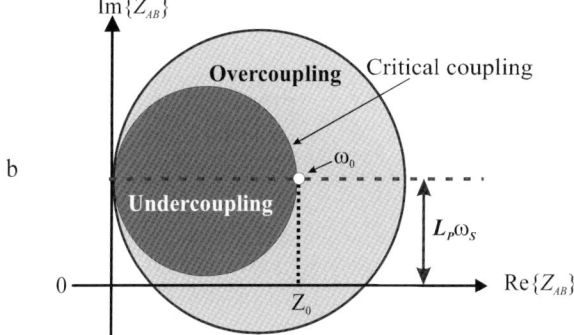

Fig. 4.23 Representation of the impedance Z_{AB} of the circuit (a) in the Z-plane (b). Matching is realized when the mutual inductance is adjusted so that $\text{Re}\{Z_{AB}\} = 50 \; \Omega$ at $\omega_0 = \omega_S$. The imaginary component $\text{Im}\{Z_{AB}\}$ is equal to the coupling loop impedance. It is cancelled out by the capacitor C_P.

From this plot, it is obvious that in under coupling conditions, there is no matching solution.

4.1.3.3 Inductive matching with a non-tuned coupling loop

By increasing the mutual inductance it is possible to get the matching conditions using solely the coupling loop without additional components (Fig. 4.24).

In this case, the mutual inductance is adjusted such that the residual reactive impedance is "self-cancelled." The transformed reactive impedance of the resonator becomes equal and opposite to the impedance of the coupling coil. The primary probe coil resonance must be shifted so that Eq. (4.44) is fulfilled. This condition may be called the "optimum coupling". Because the capacitor C_P is assumed to be a short (its impedance is zero), XP is:

$$XP = L_P\omega_S. \tag{4.49}$$

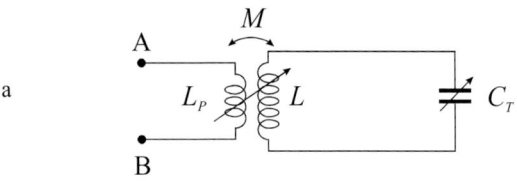

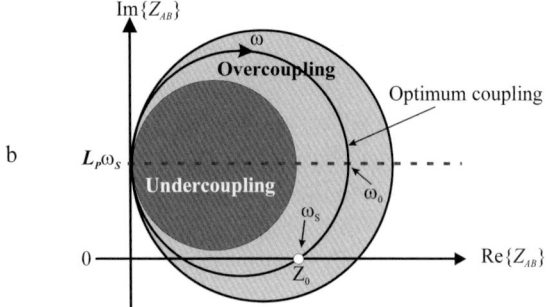

Fig. 4.24 The mutual inductance is adjusted (optimum coupling) such that the circle representing the impedance Z_{AB} in the Z-plane crosses the real axis at Z_0. The tuning frequency ω_0 of the coil resonator is adjusted so that the frequency is equal to ω_S at the crossing point. In this case, the residual impedance of the coupling loop is self-cancelled.

Using Eq. (4.44),[13] Eq. (4.45) can be rewritten as:

$$L_P\omega_S = \frac{Z_0}{r}\sqrt{r\left(\frac{M^2\omega_S^2}{Z_0} - 1\right)}, \tag{4.50}$$

from which the mutual inductance corresponding to the optimum coupling solution is obtained:

---

[13] Only the positive solution of Eq. (4.44) is acceptable here.

$$M\omega_S = \sqrt{rZ_0\left[1+\left(\frac{L_P\omega_S}{Z_0}\right)^2\right]}. \qquad (4.51)$$

The tuning frequency of the resonator is given by Eq. (4.45) which is rewritten as:

$$XS = \frac{rL_P\omega_S}{Z_0}. \qquad (4.52)$$

The optimum coupling corresponds to the case when the diameter of the circle representing Z_{AB} in the Z-plane intercepts the horizontal axis at Z_0 [Fig. 4.24(b)]. At this point, Im$\{Z_{AB}\}$ is zero; there is no need for a matching capacitor. However, this point corresponds to a frequency slightly higher than the resonant frequency of the probe resonator [Fig. 4.22(b)]. In other words, the resonator must be tuned at a frequency slightly lower than ω_S. The tuning capacitor C_T should be slightly increased, accordingly to Eq. (4.52).

4.1.3.4 Inductive matching with fixed mutual and variable capacitor

All the previous circuits require adjustment of the mutual impedance between the coupling loop and the probe resonator for every loading condition. This requires some mechanical means in order to move the coupling coil relative to the resonator. This design can be bulky and complicates the probe mechanics.

In practice it is sometimes better to fix the position of the coupling loop and to add a variable capacitor to remove any residual reactive impedance, as shown in the most general circuit shown in Fig. 4.22(b).

Once the mutual M has been fixed, the tuning and matching can be adjusted with variable capacitors. The corresponding capacitances are obtained from Eqs. (4.44) and (4.45) which are expressed as:

$$\frac{1}{C_T\omega_S} = L\omega_S \pm \sqrt{\frac{r}{Z_0}\left(M^2\omega_S^2 - rZ_0\right)} \qquad (4.53)$$

and

$$\frac{1}{C_P \omega_s} = L_P \omega_s \pm \sqrt{\frac{Z_0}{r}\left(M^2 \omega_s^2 - rZ_0\right)}. \quad (4.54)$$

In over coupling conditions, the mutual impedance is always larger than rZ_0, so there is always at least one acceptable solution. If the coupling is comprised between the critical and optimum conditions, the square root term in Eq. (4.54) is lower than $L_P\omega_S$. There are two possible matching solutions, depending on the chosen sign in Eqs. (4.53) and (4.54). If the plus sign is chosen, the resonator frequency must be shifted to a lower frequency than ω_S. If the coupling is higher than the optimum matching condition, defined by Eq. (4.51), only the + sign is acceptable. Fig. 4.25 shows the Z-plane representations corresponding to these two cases.

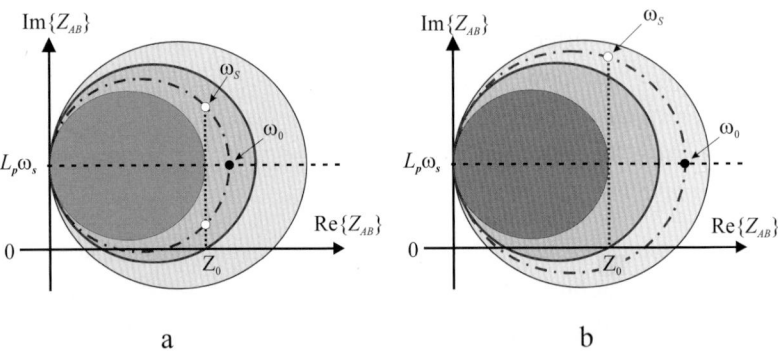

Fig. 4.25 (a) The coupling strength is between the critical and optimum coupling. There are two possible matching solutions (represented by white points) that can be completed with a matching capacitance in series with the coupling loop. (b) The coupling strength is larger than the optimum coupling. There is only one matching solution ($\text{Im}\{Z_{AB}\} > 0$) with $\omega_0 > \omega_S$.

From a practical point of view, the position of the coupling loop (without a series capacitor) is firstly determined experimentally when the probe coil is maximally loaded (minimum Q value). Then, the loop is slightly moved away from the probe, so that the right term of Eq. (4.54) becomes positive for every practical loading condition. From this point, impedance matching can always be performed by the matching capacitor in series with the coupling loop.

4.1.3.5 Efficiency of the inductive coupling

In the previous discussion we have neglected the resistance r_P of the primary loop. This has only little importance on the matching performance because this resistance contributes only to the real part of the impedance which adds to Z_0. The matching conditions must be modified by replacing Z_0 by $Z_0 - r_P$ in order to take into account this resistance.

The resistance r_P is, however, the source of undesired losses. An estimation of the corresponding lost power is given by:

$$P_{loss} = r_P |i_P|^2, \qquad (4.55)$$

where $|i_P|$ is the effective current circulating in the primary circuit. Assuming a perfect power match, the sum of the power dissipated in the probe coil and in the primary loop is equal to the power delivered by the source (of internal impedance Z_0).

Hence,

$$Z_0 |i_P|^2 = r_P |i_P|^2 + P_{probe}. \qquad (4.56)$$

The efficiency is defined as:

$$E = \frac{P_{probe}}{Z_0 |i_P|^2}. \qquad (4.57)$$

From these two equations, the power efficiency of the inductive coupling circuit is:

$$E = 1 - \frac{r_P}{Z_0}. \qquad (4.58)$$

The efficiency does not depend on the inductive coupling scheme. It depends solely on the ratio of the coupling loop resistance to Z_0. If Z_0 is equal to 50 Ω, a 0.5 Ω resistance[14] leads to an efficiency of 0.99 (−0.04 dB). The inductive coupling is potentially an efficient coupling method. In addition the probe resonator is self-balanced. However, it has some disadvantages compared to the connected matching circuits: it is bulkier and it is somehow coupled to the sample, possibly leading to distortions of the magnetic field distribution created by the probe resonator [Hoult and Tomanek, 2002].

4.1.3.6 Coupled resonators

The coupling methods described so far require that the resonator should be tuned as a function of its loading. However, in case access to the tuning capacitor C_T is not possible, one may want to remotely "tune" the coil using an external circuit [Schnall et al., 1986ab; Kuhns et al., 1988; Silver et al., 2001; Volland et al., 2010].

This remote "wireless" method of coupling has some similitude with remote matching through a transmission line. The isolated probe resonator is coupled to a second resonator incorporating its own tuning/matching components (Fig. 4.26).

In the same way that a piece of line, terminated by any impedance, could be matched to Z_0 using an appropriate L-matching network, a loop coupled to any resonator may be tuned and matched without great difficulties. Intuitively, if the probe resonator is tuned at a frequency far away from the operating frequency ω_S, and is "moderately" coupled, most of the incident power will be dissipated in the tuned/matched coupling circuit rather than in the probe resonator. To efficiently share the available power, the two resonators must be strongly coupled [Kuhns et al., 1988]. Two resonant modes will appear, depending on the relative phase of currents in both loops. The frequencies for these modes depend on the coupling strength and on the isolated resonant frequency of each tuned coil. Changing the tuning of the external coupling loop, a resonant mode can be brought to the desired frequency ω_S (remote tuning).

---

[14] A small coupling loop of about 50 nH has a resistance of about 0.5 Ω at 200 MHz.

Interfacing the NMR Probehead 237

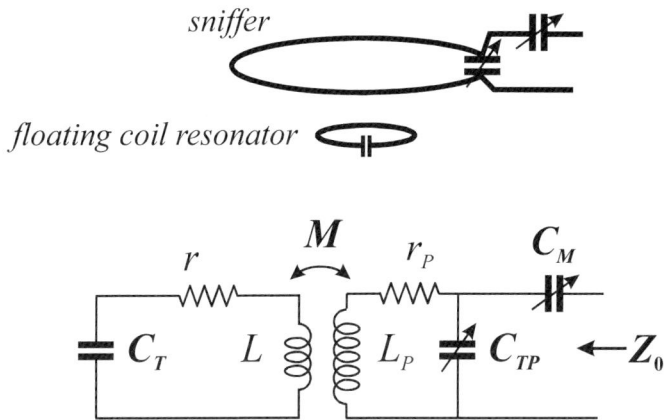

Fig. 4.26 A remote tuning/matching circuit used to collect the signal of a coil tuned at a fixed frequency resonance.

The efficiency of such a coupling scheme depends on many factors such as the Q of both resonators, the difference between their tuning frequencies, the difference between the probe resonator frequency and the operating frequency, the mutual coupling inductance, etc...

In practice, the best results would obviously be obtained when the frequency of the probe resonator is set as close as possible to the operating frequency. For a given mutual inductance M, the resonant frequency ω_P ($L_P C_{TP} \omega_P^2 = 1$) of the coupling resonator (Fig. 4.26) could be very different than ω_S. In the limit that ω_P is very high ($C_{TP} \to 0$), the circuit becomes identical to inductive matching with fixed mutual and variable capacitor (Section 4.1.3.4) and therefore exhibits a good efficiency. Furthermore, the magnetic field created by the floating resonator is moderately perturbed by the field created by the external coupling coil.

In contrast, if the probe resonator is tuned at a frequency far away from ω_S, the mutual inductance must be increased in order to become efficient. The perturbation of the resonator magnetic field will increase and more power will be dissipated in the coupling circuit.

4.1.3.7 Flux concentrators

Another kind of coupled resonator has recently been proposed in which the sensitivity of a large probe coil is considerably improved [Sakellariou et al., 2007] when the "filling factor" is an issue. In other words, the best sensitivity for a given sample is obtained when the probe coil fits as closely as possible to the sample.

If a small sample occupies only a small region of a large coil, the sensitivity is poor. If the sample fills the interior of a specifically designed resonator almost completely, the sensitivity can be high. To interface this small resonator to the spectrometer or scanner, it is not always possible, or simple, to use one of the previously described connected matching circuits.

One possibility is to use an external large probe coil[15] to couple the small resonator containing the sample. In this case, the floating resonator is maximally coupled to the large coil and the achieved sensitivity can be close to that of the small coil connected in the usual way.

The small floating resonator acts as a "flux concentrator" that locally amplifies the magnetic field created by the larger coil. This is illustrated in Fig. 4.27. A small solenoid coil is strongly coupled to a larger saddle coil interfacing the floating solenoid to the spectrometer or scanner console.

Such a large coil is always a part of a conventional NMR instrument. Setting up such an experiment is therefore very easy. In the original work mentioned at the beginning of the paragraph, the large coil was a commercial probe already designed for solid state NMR experiments and the small solenoid was put inside the large fixed coil. The great advantage here was that the small coil could be rotated at high speed, as is usual in solid state NMR experiments, without the issues of stable connections. In addition, the filling factor is optimized, whatever the size of the large coupling resonator.

Such a setup has also been employed in liquid NMR spectroscopy experiments [Tang and Jerschow, 2010]. It can also be envisaged for MRI with similar improvements.

---

[15] Possibly pre-existing in the NMR instrument.

Interfacing the NMR Probehead

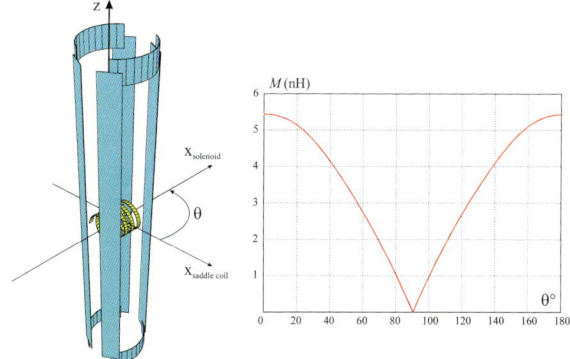

Fig. 4.27 A small solenoid (yellow) containing the sample is coupled to a conventional saddle coil (cyan). The mutual inductance between both coils is shown to the right, as a function of the angle between their magnetic axes.

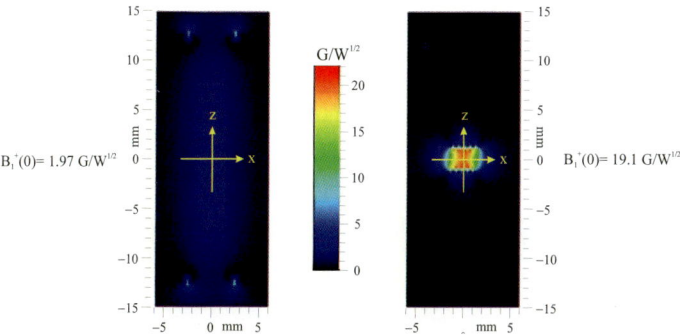

Field amplitude (B_1^+) in the xz plane created by the saddle coil (left) and by the solenoid (right) when they are tuned and matched to 50 Ω at 400 MHz and powered by 1 W. The Q factor of each resonator is assumed to be equal to 200.

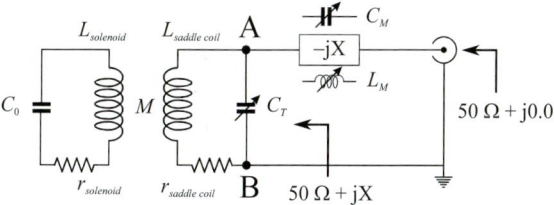

The coupled resonator circuit under investigation.
$L_{solenoid} = 30.2$ nH ($r_{solenoid} = 0.38$) ; $L_{saddle\ coil} = 70$ nH ($r_{saddle\ coil} = 0.88$).

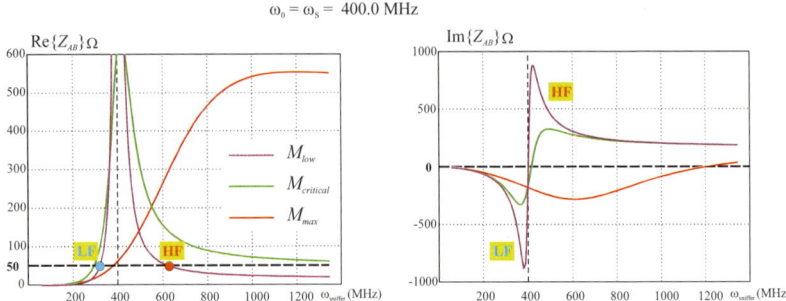

Fig. 4.27 (continued). The solenoid is assumed to be tuned exactly to the operating frequency of 400 MHz (C_0 = 5.245 pF). Three coupling conditions are investigated.

Maximum coupling ($\theta = 0$, M = 6.33 nH). There is only one possibility to match the saddle coil (Re$\{Z_{AB}\}$ = 50 Ω, right), corresponding to a tuned frequency lower than 400 MHz (Low Frequency or LF solution, C_T = 2.53 pF). In this case, Im$\{Z_{AB}\} < 0$ (left). It can be cancelled by an inductor L_M = 63.25 nH. Critical coupling (M = 1.733 nH, $\theta = 72°$). There is also only one LF solution (C_T = 4.2 pF). Im$\{Z_{AB}\}$ being negative, it can be cancelled by an inductor (L_M = 69.27 nH). Under coupling ($\theta = 80°$, M = 1.0 nH). There are two solutions, to LF and to HF. The HF solution (C_T = 0.926 pF) is preferred as Im$\{Z_{AB}\}$ is positive and can be cancelled using a capacitor (C_M = 1.354 pF) instead of an inductor.

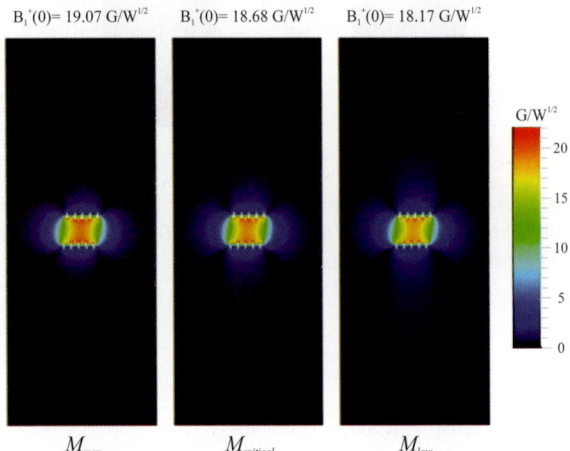

Amplitude of the effective magnetic field at the center of the solenoid coupled by the saddle coil tuned and matched at 400 MHz. The probe configuration is most efficient for the maximum coupling. Note that the field amplitude M_{low} does not depend on the chosen solution, HF or LF.

Interfacing the NMR Probehead

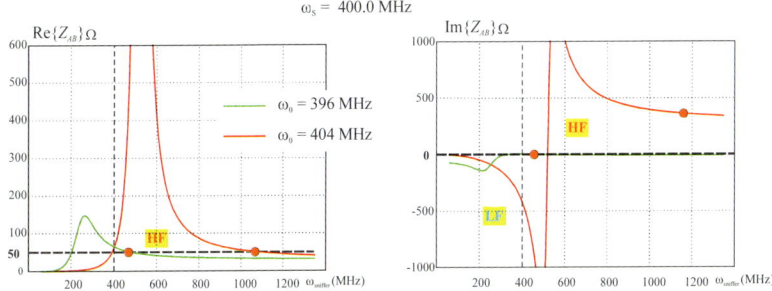

Fig. 4.27 (continued). The solenoid is assumed to be tuned close to ω_S, to 404 MHz (C_0 = 5.142 pF) and 396 MHz (C_0 = 5.352 pF), respectively. For each case, there are two solutions (LF and HF) leading to Re$\{Z_{AB}\}$ = 50 Ω. The HF solution is preferred, allowing the remaining reactance Im$\{Z_{AB}\}$ to be cancelled with a capacitor. If ω_0 = 404 MHz, the HF solution may be impracticable because the saddle coil should be tuned to around 1 GHz (C_T = 0.317pF and C_M = 1.05 pF). The LF solution (C_T = 2.39 pF) would require a matching inductor (L_M = 150 nH). If ω_0 = 396 MHz, the HF solution is practically realizable with C_T = 1.561 pF and C_M = 5.7 pF.

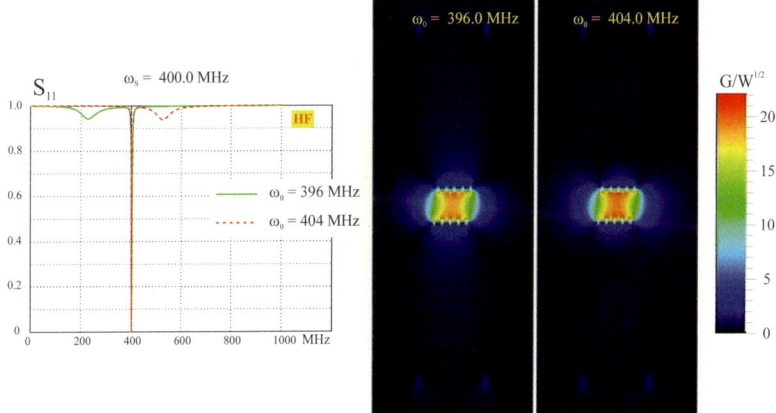

The S_{11} spectra are shown for both tuning conditions of the solenoid and using the HF solution. The magnetic field amplitude is slightly higher when the solenoid coil is tuned to 404 MHz ($B_1^+(0)$ = 19.1 G W$^{-1/2}$) rather than to 396 MHz ($B_1^+(0)$ = 18.5 G W$^{-1/2}$). Again, the magnetic field amplitude does not depend on the chosen solution (LF or HF). The magnetic field amplitude is almost identical to the reference field value, indicating that a detuning of the solenoid coil by about 1% has no influence on the probe efficiency.

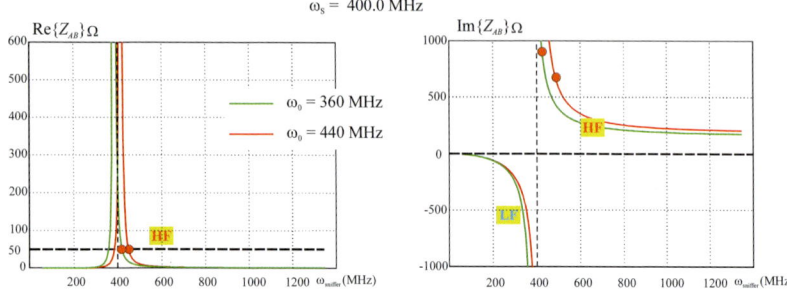

Fig. 4.27 (continued). The solenoid is assumed to be tuned with a difference of 10% relative to the operating frequency, to 440 MHz (C_0 = 4.335 pF) and 360 MHz (C_0 = 6.475 pF), respectively. For each case, there are two solutions (LF and HF) leading to Re{Z_{AB}} = 50 Ω. The HF solution is preferred, allowing the remaining reactance Im{Z_{AB}} to be cancelled with a capacitor. The HF solutions are practicable for both cases. If ω_0 = 440 MHz, the HF solution leads to (C_T = 1.8 pF and C_M = 0.32 pF). If ω_0 = 396 MHz, the HF solution is leads to C_T = 2.06 pF and C_M = 0.38 pF.

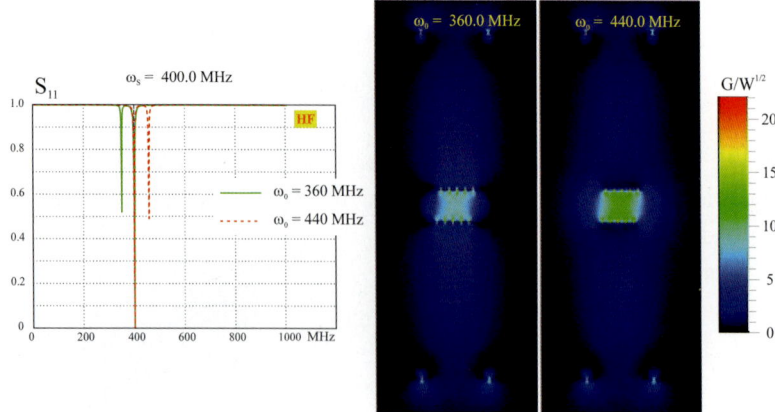

The S_{11} spectra are shown for both tuning conditions of the solenoid and using the HF solution. The magnetic field amplitude is higher when the solenoid coil is tuned to 440 MHz ($B_1^+(0)$ = 11.1 G W$^{-1/2}$) rather than to 360 MHz ($B_1^+(0)$ = 8.4 G W$^{-1/2}$). The magnetic field amplitude does not depend on the chosen solution (LF or HF). However, the magnetic field amplitude is significantly lower than the reference field value ($B_1^+(0)$ = 19.1 G W$^{-1/2}$). This indicates that the tuning frequency of the solenoid coil should not differ by more than a few percent from the desired operating frequency and, preferably, slightly higher.

The use of a smaller coil improves the sensitivity and this coupling scheme avoids many connection issues. Since the pioneering work by Schnall *et al*. [Schnall *et al.,* 1986a], recent applications in this direction have been published [Silver *et al.,* 2001; Volland *et al.,* 2010] and are still actively being developed [Ginefri *et al.,* 2012].

4.2 Balancing the Probehead

The detection of the NMR signal by the "inductive method" is associated with the magnetic component of the "near field." The probe should essentially be sensitive to the magnetic part of the electromagnetic field. The electric field component of the near, source of additional energy losses, and sample heating should be as low as possible. Furthermore, any conductors submitted to time varying currents radiate energy through the "far field" components of the electromagnetic field ("antenna" effect). The radiated energy is useless[16] and definitively lost. This results in a degradation of the sensitivity, as well as undesired strong couplings of the probe with the surroundings.

The design of an NMR probe should primarily take care of two factors: the minimization of the amplitude of the conservative electric field in the sample and, at the same time, the maximization of the magnetic field component. The radiated field ("antenna" effect) should be avoided by all means.

These constraints can be fulfilled in part by proper design of the circuits interfacing the probe with the NMR console. This will be discussed in this paragraph showing the advantages of balancing the probe respective to both the reduction of the near electric field and the radiated far field.

The dielectric losses, associated with the near field component of the electric field, are due to the probe capacitive coupling with the sample which usually has a poor dielectric quality. The corresponding energy loss is proportional to the square of the amplitude of the electric field $|E|$.

---

[16] New MRI methods use true antennas to excite travelling wave modes in the magnet bore [Brunner *et al.*, 2009; Webb *et al.*, 2010].

Balancing the probe divides |E| by a factor of two, hence divides the losses by a factor of four. Another complementary way of minimizing the electric fields inside the probe is using a "high current probe" (low inductance) design.

The electromagnetic energy radiation off the probe comes into play when the probe dimensions reach about one tenth, or more, of their proper wavelength. With constantly increased static magnetic fields of the imagers/spectrometers, one will be confronted with this problem even for small probe dimensions (of the order of some centimeters). Again, equilibrating the probe potential respective to the ground has the effect of decreasing the "antenna effect" while reducing the radiated energy. In addition, an electromagnetic "shield" contributes to eliminating the radiating field, but it will be demonstrated later that the shield could result in a strong degradation of the sensitivity in certain circumstances.

4.2.1 Evidencing the electric losses effect

Electric losses are due to energy dissipated by currents in the sample. These currents are induced through the parasitic capacitors formed between the coil and the sample. These capacitors appear in parallel with the probe tuning capacitor (Fig. 4.28).

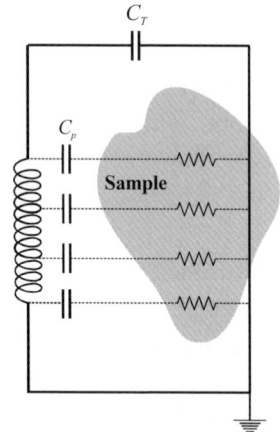

Fig. 4.28 The parasitic capacitance introduced into the probe electrical circuit by a conductive sample.

Their importance can be evaluated by the decrease of the intrinsic resonance frequency when the probe is loaded by a dielectric sample.

The losses, and ultimately the sensitivity, can be evaluated by comparing the corresponding Q values obtained for the different load conditions.

4.2.1.1 Experimental setup

We consider two cylindrical coils of the same internal volume (sample dimensions) but of different inductances. These coils are expected to create the same magnetic field amplitude for the same Q factor, but different conservative electric field amplitudes. The characteristics for both coils will be compared in a balanced and unbalanced configuration.

The first coil is a solenoid coil with four turns (Fig. 4.29). Its inductance is equal to 130 nH, hence an impedance of 163 Ω at 200 MHz.

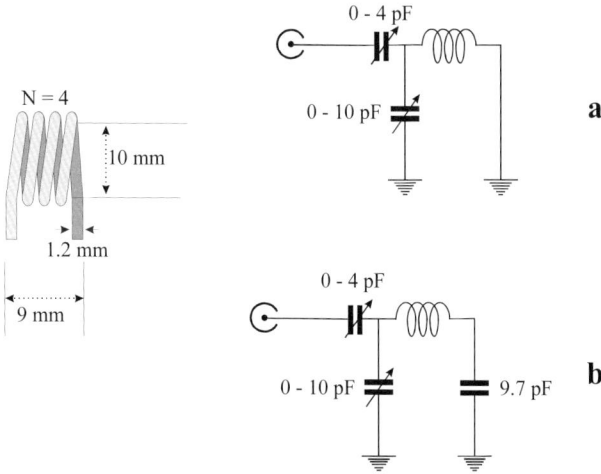

Fig. 4.29 Experimental circuit used to illustrate the dielectric losses effect. Geometry of the solenoid coil and its impedance matching circuits, at 200 MHz, (a) for non-symmetrical and (b) symmetrical design.

The second one is a single turn "loop gap" coil (Fig. 4.30) with a much smaller inductance of 7.6 nH (impedance of 9.6 Ω at 200 MHz).

The two coils have a sample volume able to accept a 10 mm diameter glass tube that is either empty or filled with distilled water or a 9 g L^{-1} NaCl aqueous solution. The distilled water sample mimics the effect of a moderately lossy sample, while the salt solution mimics a typical biological tissue that will induce both electric (due to its poor dielectric quality) and magnetic (due to its conductivity) losses.

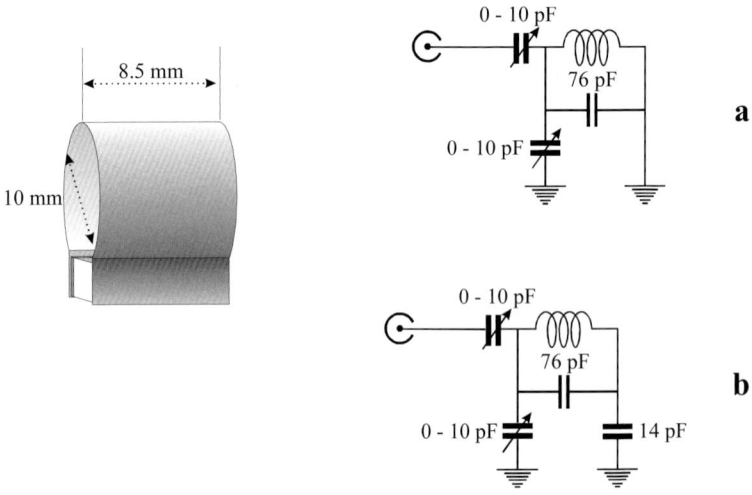

Fig. 4.30 Experimental circuit used to illustrate the dielectric losses effect. Geometry of the loop gap coil and its impedance matching circuits, at 200 MHz, (a) for non-symmetrical and (b) symmetrical design.

4.2.1.2 Expected sensitivity

The ultimate sensitivity can be written as (Chapter 1):

$$S = K \frac{B_1}{I} \frac{1}{\sqrt{r}}, \qquad (4.59)$$

where B_1/I is the effective RF magnetic field amplitude created by the coil per unit current, r is the coil resistance and K is a constant including the frequency, the geometry of the coil, and other physical parameters (temperature, receiver bandwidth, number of spins, nucleus magnetic properties…). Note that the ratio $B_1/I\sqrt{r}$ is expressed in gauss $W^{-1/2}$ in the non-SI[17] units system.

It will be shown here that the relative sensitivity for the two coils, having similar dimensions, is expected to be solely proportional to the square root of the Q factor.

Assuming a perfect impedance matching, the transmitter power P is entirely dissipated in the coil resistance r. In this case, the current $|I|$ in a probe coil of inductance L is given by

$$|I| = \sqrt{\frac{P}{r}} = \sqrt{\frac{PQ}{L\omega}}. \qquad (4.60)$$

The magnetic field amplitude created at the center of the cylindrical coil carrying the current $|I|$ is:

$$B_1 = F_G n |I|, \qquad (4.61)$$

where F_G is a geometric factor depending only on the coil dimensions and n is the number of turns.

The geometric factor for the solenoid coil is given by (Chapter 8):

$$F_G = \frac{4\pi}{\sqrt{d^2 + l^2}}, \qquad (4.62)$$

where d is the mean coil diameter and l its length. With the dimensions quoted in Fig 4.29, one gets $F_G = 0.46$ gauss A^{-1} per turn.

For the loop gap configuration, the corresponding geometrical factor is $F_G = 0.44$ gauss A^{-1}, taking into account the current distribution on the copper foil[18] (Chapter 8).

The sensitivity S, Eq. (4.59), can finally be rewritten as:

$$S = K \frac{nF_G}{\sqrt{L\omega}} \sqrt{Q}. \qquad (4.63)$$

---

[17] 1 gauss = 10^{-4} tesla.

[18] Eq. 4.61 overestimates F_G by about 10%.

It results that the relative[19] sensitivity of the two coils is $0.15\sqrt{Q}$ and $0.14\sqrt{Q}$ gauss $W^{-1/2}$ for the solenoid and loop gap coil, respectively. This demonstrates that the sensitivity depends solely on Q when the geometry has been fixed.

4.2.1.3 Frequency shifts

Table 4.2 summarizes the results obtained with the two coils, in an unbalanced and balanced configuration.

Table 4.2 Electrical characteristics of the circuits shown in Fig. 4.29 (solenoid, $L = 130$ nH) and Fig. 4.30 (loop gap, $L = 7.6$ nH) in the unbalanced and balanced configuration and in different loading conditions. S (in gauss $W^{-1/2}$) is proportional to the sensitivity (assuming $K = 1$).

		Unbalanced			Balanced		
		$\Delta\nu$	Q	$S(G\ W^{-1/2})$	$\Delta\nu$	Q	$S(G\ W^{-1/2})$
Solenoid	Empty	–	181	1.95	–	246	2.28
	Distilled water	10 MHz	168	1.88	6 MHz	234	2.22
	NaCl solution	12 MHz	40	0.92	8 MHz	72	1.23
Loop gap	Empty	–	350	2.62	–	336	2.65
	Distilled water	1.4 MHz	330	2.55	1.0 MHz	326	2.53
	NaCl solution	1.4 MHz	110	1.47	1.2 MHz	134	1.62

For the unbalanced case, one end of the coil is connected to the ground and a "classical" capacitive coupling network [Figs. 4.29(a) and 4.30(a)] is used to match the impedances. For the balanced case, the matching device has been designed, as explained in the forthcoming

---

[19] The proportionality constant K is identical for both coils.

section, in order to equilibrate the coil potentials respective to the ground [Figs. 4.29(b) and 4.30(b)]. Using the same sample, a 10 mm diameter tube filled with distilled water (ε_r about 80 at 295 K), the shift for the solenoid coil is 10 MHz and only 1.4 MHz for the loop gap coil, in the non-symmetric configuration. Although the parasitic capacitance is expected to be slightly larger with the loop gap coil than with the solenoid, the resonant frequency shift clearly indicates that the contribution of the parasitic capacitors to the current path is much larger with the solenoid than with the loop gap. The tuning capacitor is indeed about 5 pF for the solenoid and 83 pF for the loop gap coil and dominates largely in the latter case.

These results confirm that with a lower inductance coil, creating a lower electric field, the current flows preferentially in the lossless tuning capacitor than in the parasitic lossy capacitors associated with the sample.

In the symmetric configuration, the shift is reduced, for both coils, by a factor of roughly √2. In this case, the parasitic capacitor is in parallel with the tuning capacitor, which is double the value of the non-symmetric case (Fig. 4.31). The current flowing through the parasitic capacitor is thus diminished and the corresponding losses are expected to be smaller.

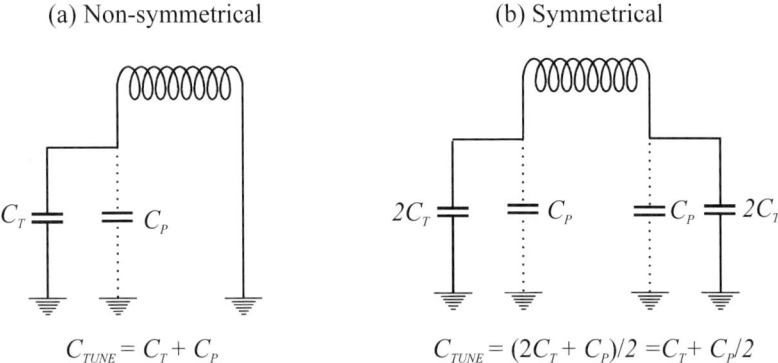

Fig. 4.31 Equivalent impedance matching circuit for a (a) non-symmetrical and (b) symmetrical design, evidencing the contribution of the parasitic capacitance C_P.

Equivalently, the electric field in the sample is halved in the symmetric configuration, resulting in an expected decrease by a factor of four of the corresponding electric losses.

Using a sample containing a salt solution at a concentration similar to the biological samples (9 g of NaCl per liter) the observed shifts are roughly the same (Table 4.2) as observed with the distilled water sample. The dielectric constant is roughly the same for both samples, but, as shown in the next paragraph, both the electric and magnetic losses will contribute to a decrease in the Q factor.

4.2.1.4 Q factors

The Q factors are measured as explained in Chapter 10 and allow an estimation and comparison of the expected sensitivities for the two coils, in different loading conditions.

First of all, it is interesting to compare the Q factor obtained with the sample-free coils. For the solenoid coil, one observes a significant increase of the Q factor when the coil is in a symmetric configuration compared with the non-symmetric one. This effect is most probably due to the fact that the length of the wire approaches a non-negligible fraction of the wavelength ($\lambda/10$ at 200 MHz). The coil begins to radiate electromagnetic energy resulting in corresponding losses and thus increasing the resistance. This "antenna effect" will be more clearly demonstrated in the following paragraph, using an open structure (a single loop) of similar inductance. For the loop gap coil, the Q factor is independent of the configuration, symmetric or not, and is higher compared to the solenoid coil due to its intrinsic low impedance.

Using the distilled water sample, one observes only a small decrease of the sensitivity for both configurations. The dielectric losses are moderate, as would so be for most organic solutions used in chemistry analysis. At this point, it is interesting to note that the loop gap low inductance coil is expected to provide the best sensitivity (Table 4.2) even for the low-loss dielectric solution.

On the contrary the conductive biological samples, containing salts, imply a dramatic decrease of the Q factors. Part of the losses arise from the magnetic losses due to the currents induced in the sample that are proportional to the magnetic field created by the coil, B_1. Of course, this field is needed for NMR experiments, so these losses cannot be reduced without paying a price in the sensitivity. The other losses come from the poor dielectric properties of the solution. The worst sensitivity is obtained in the unbalanced solenoid coil configuration. With the symmetric configuration, the Q factor increases by 80% and the sensitivity by 34%. As expected, the dielectric losses are smaller with the loop gap and the gain in sensitivity is moderate, but still significant, with the symmetric configuration (about 10%).

These results are definitively in favor of the symmetric configuration for probe design. In this configuration the conservative electric field is halved and, consequently, the losses are divided by four. Both configurations create the same magnetic field amplitude, for the same Q factor, but the low inductance design is better, again because the electric field is minimized.

4.2.2 *Evidencing the radiation losses: the antenna effect*

To illustrate the consequences of the radiated energy on the Q factor and consequently on the sensitivity of an NMR coil, we shall now consider a simple loop of 60 mm diameter tuned to 200 MHz (Fig. 4.32). The length of coil wire is about $\lambda/8$ at this frequency.

When the coil is not electrically balanced respective to the ground, for example by connecting one end directly to the ground [Fig. 4.32(a)], it exhibits two modes of operation. One corresponds to a "true loop," sensitive only to the magnetic component of the radio frequency field, in perfect agreement with the NMR experiment needs. The second mode of operation is that of a non-directional, vertical antenna of small dimension, known as the "antenna effect" [*ARRL Handbook*, 1988, ch. 39].

The radiation resistance of a balanced small loop can be estimated from [Ramo *et al.*, 1994, p. 206]:

$$R_r = \frac{\pi}{6}\left(\frac{\mu}{\varepsilon}\right)^{\frac{1}{2}}\left(\frac{l}{\lambda}\right)^4, \tag{4.64}$$

where l is the perimeter of the loop and λ the wavelength. With the dimensions quoted in Fig. 4.32, the radiation resistance of the loop is estimated to be equal to 0.05 Ω. It is negligibly small compared to the resistance of the coil conductor which is estimated to be about 0.4 Ω. In contrast, the radiation resistance of a grounded straight wire dominates the coil resistance when it is not balanced.

As a result, the measured Q factor of the non-symmetric configuration is about 80, much higher than the best value measured in a balanced configuration ($Q = 400$, Table 4.3).

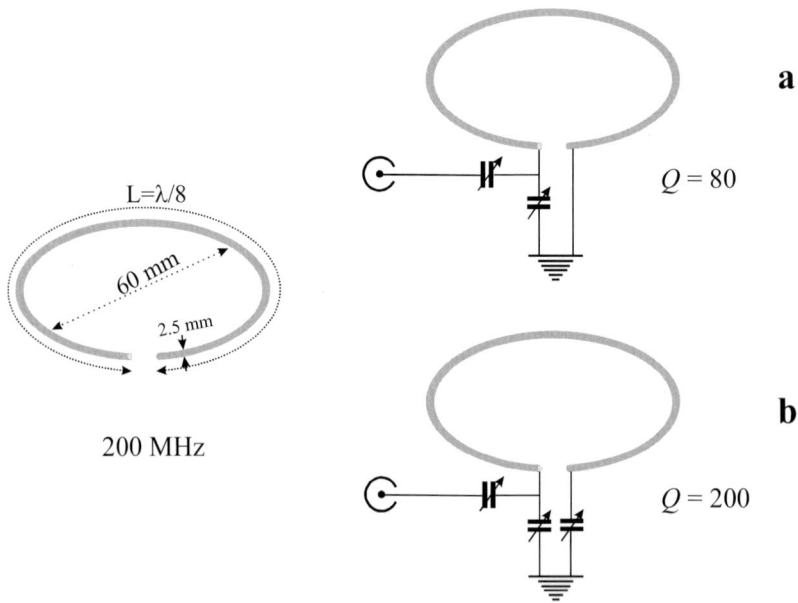

Fig. 4.32 Experimental setup used to illustrate the radiation losses effect. The loop coil design in a (a) non-symmetrical and (b) symmetrical configuration.

When a symmetric coupling scheme [Fig. 4.32(b)] is used, the Q factor increases to 260, indicating that the antenna effect is considerably reduced.

Another possibility to cancel out the radiating energy is to place the coil into a closed shield. Although this solution is indispensable for large resonators to minimize the interaction with neighboring metal pieces, there are still problems. The phase shift of the RF current on long conductors induces large field heterogeneities.

To avoid these problems, a "segmentation" of the coil wire is desirable. This is done by soldering capacitors from place to place thus allowing a maximum conductive length of up to $\lambda/20$. This also has the advantage of distributing the electric field all around the coil and of diminishing the contribution of the coil antenna mode.

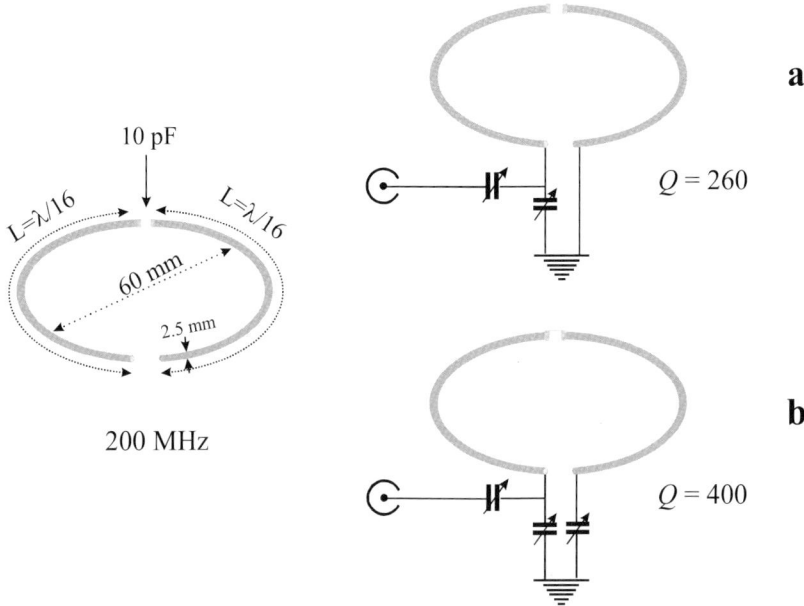

Fig. 4.33 Experimental setup used to illustrate the radiation losses effect. Segmentation design for the loop coil in a (a) non-symmetrical and (b) symmetrical configuration.

As a result, the segmentation of the loop coil in two halves, separated by a 10 pF capacitor (Fig. 4.33), increases the Q factor to 200 in the non-symmetric coupling scheme (Table 4.3). The antenna effect is effectively reduced by the capacitor segmentation as evidenced by the doubling of the Q factor. An additional improvement by a factor of two ($Q = 400$, Table 4.3) is finally obtained with the balancing of the segmented coil.

An example of solenoid segmentation is given by Cook and Lowe [Cook and Lowe, 1982].

Table 4.3 Q factor measured at 200 MHz for a tuned/matched probe coil made of a loop of wire (diameter 60 mm, length of the order of $\lambda/8$), in various coupling configurations. The probe is empty.

	Unbalanced coupling	Balanced coupling
Non-segmented (Fig. 4.5)	80	260
Segmented (Fig. 4.6)	200	400

4.2.3 Symmetrical capacitive coupling networks

The necessity of using a balanced matching network has been well demonstrated for *in vivo* experiments by Murphy-Boesch and Koretsky [Murphy-Boesch and Koretsky, 1983]. They provide a spectacular sensitivity improvement with such a configuration for ^{31}P spectroscopy of rat organs, using implanted coils. Due to the strong coupling of the NMR probe with the rat tissues, it was particularly important to reduce the electric field near to the coil. Since the success of this approach, almost all the NMR probe designs, especially those dealing with animal or human experiments use their coupling scheme.

4.2.3.1 *Splitting the matching capacitor*

The idea given by Murphy-Boesch and Koretsky consists of splitting the matching capacitor of the standard circuit shown in Fig. 4.8(a) into two capacitors connecting both ends of the coil, one to the ground and the other one to the cable central connection (Fig. 4.34). Provided the two capacitors have twice the capacitance given by Eq. (4.22), the impedance matching is still realized (the two matching capacitors being in series, the resulting capacitance is just halved).

The same current flows through both capacitors, and, because they have the same impedance, the voltages respective to the ground at the coil ends have almost equal amplitudes but opposed phases, realizing the required electrical balancing. Furthermore, the voltage amplitude is halved compared to the unsymmetrical case, providing the expected reduction in electrical losses. The virtual ground point is located near the center of the coil, also resulting in a reduction in radiation losses due to the antenna effect.

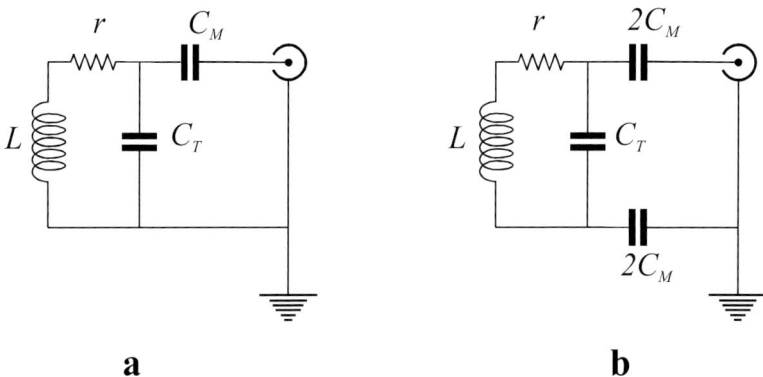

Fig. 4.34 Conversion of the basic capacitive matching circuit (a) into a simple, balanced matching circuit (b).

Ideally, the voltages (respective to the ground) would be of exactly opposed phase and of the same amplitude. In fact, the symmetry is only approximate due to the finite resistance of the coil. As a result, the voltages exhibit small, in phase, real components and large, out-of-phase, imaginary components as well as different amplitudes.

Fig. 4.35 gives an example, for a typical coil, assuming a Q value ranging from 10 to 1000. Clearly, the asymmetry increases for the lower Q values, but this procedure still significantly improves the balancing properties of the probe.

In practice, these coupling schemes should be preferred, whenever it is possible. Calculation of the components is very easy and can be done using the formulae already given (Section 4.1.2.3). The matching capacitance should be simply doubled. Adjusting the symmetric matching network is not very critical in practice. It is usually sufficient to use one fixed and one variable matching capacitor. Provided the fixed one is chosen slightly greater than the calculated value (assuming the lowest expected Q value), the network can adapt any loading condition with satisfactory equilibration of the probe. In any case, the results will be better than those obtained without any balancing.

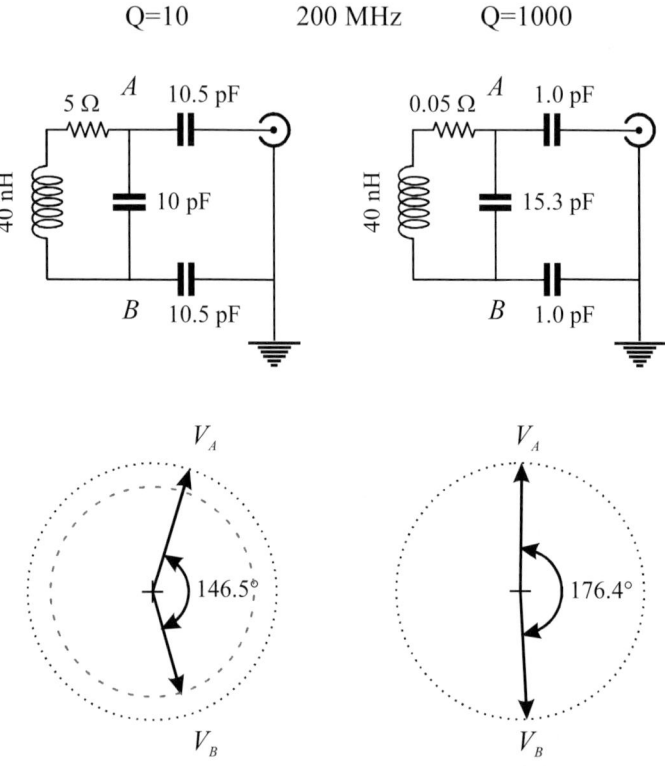

Fig. 4.35 Upper: example of a simple balanced matching circuit, at 200 MHz, for low and high Q values. Lower: voltages at points A, B represented in the complex plane. For the low Q case there are differences in amplitudes between the corresponding voltages, as well as an important departure from π angle value for the corresponding phases.

4.2.3.2 A versatile capacitive balanced matching network

The circuit presented in Fig. 4.36 is a capacitive matching network that incorporates almost all the possible impedance matching techniques using capacitive networks.

It also allows the probe coil to be precisely balanced, permitting the equilibration of the voltage amplitudes. Only the phase shift resulting from the probe coil resistance cannot be compensated.

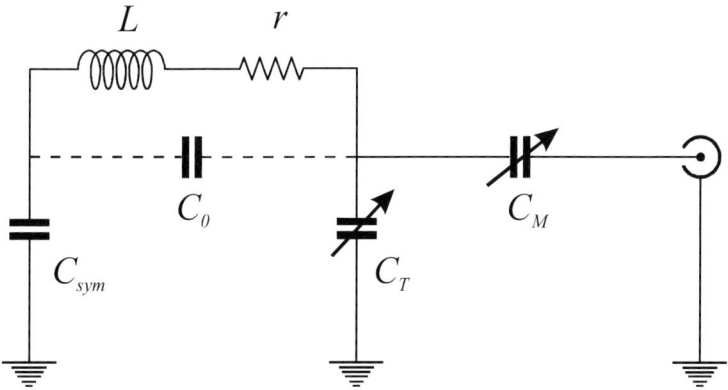

Fig. 4.36 A balanced capacitive network. C_{sym} is the symmetrization capacitor, C_T the tuning capacitor, and C_M the matching one. C_0 is the optional pre-tuning capacitor.

Furthermore, it has the advantage that the fine tuning capacitor has its rotor connected to the ground, avoiding pathways for parasitic currents through the mechanical components that may connect the capacitor to the operator. Because one of the capacitors should be kept constant, only two adjustable capacitors are required, (C_T and C_M, Fig. 4.36). Finally, the possibility to "pre-tune" the coil with a capacitor C_0 permits the concentration of the current where it should be, in the probe coil.

The calculation of the components is straightforward. First, we consider the circuit shown in Fig. 4.36 without the pre-tuning capacitor C_0, as shown in Fig. 4.37. The coil inductance is split into two equal parts. Tuning half the inductance with the capacitor C_{sym}, the voltage across this part of the circuit is zeroed, leading to a virtual ground close to the center of the probe coil. But, due to the resistive component $r/2$ (Fig. 4.37), there remains a small voltage that cannot be compensated.

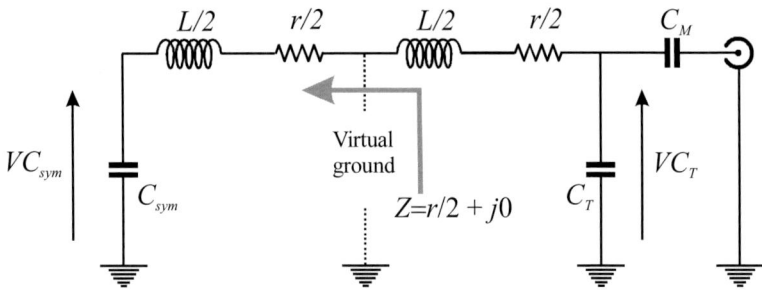

Fig. 4.37 Splitting the balanced capacitive network. When $L/2$ is tuned on resonance by C_{sym}, there remains a small voltage to the ground that cannot be compensated.

The principle of operation outlined above leads to an approximate solution for equilibrating the probe, especially for a low Q coil. In order to improve the balancing, one would like to set the voltages across C_{sym} and C_T equal in amplitude and opposed in phase. Writing VC_{sym} as a function of VC_T (Fig. 4.38), one gets:

$$VC_T + jL\omega_S I + rI - VC_{sym} = 0, \qquad (4.65)$$

where I is the current flowing in the circuit.

The condition $VC_T = -VC_{sym}$ gives:

$$2VC_{sym} = jL\omega_S I + rI = j2I/C_{sym}\omega_S . \qquad (4.66)$$

The last equality of Eq. (4.66) cannot be satisfied unless r is equal to zero, as expected.

But, it is still possible to equate the magnitude of VC_{sym} and VC_T. This is obtained when:

$$\left| jL\omega_S + r \right| = \frac{2}{C_{sym}\omega_S} . \qquad (4.67)$$

Or, after some algebraic manipulations:

$$XC_{sym} = \frac{1}{C_{sym}\omega_S} = \frac{\sqrt{r^2 + L^2\omega_S^2}}{2}. \qquad (4.68)$$

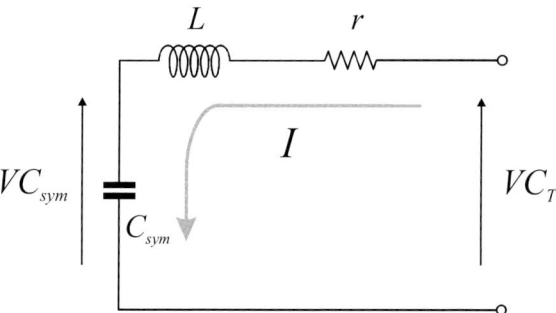

Fig. 4.38 Relevant voltages and current in the balanced circuit network.

In the limit when r tends to zero, half of the inductance is tuned to ω_S by the capacitor C_{sym}. It is only slightly dependent on the probe loading if Q is high (r is small as compared to $L\omega_S$). In practice, the capacitor C_{sym} should be kept fixed.

C_T and C_M can now be calculated to match the circuit to 50 Ω at the spectrometer frequency ω_S. This is easily done by the procedure already outlined (Section 4.1.2.2) bearing in mind the equivalent circuit near the frequency of interest (Fig. 4.39).

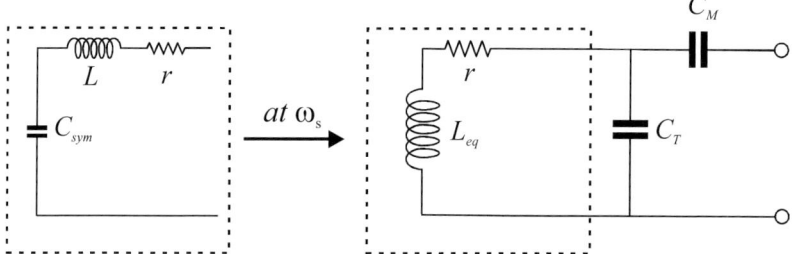

Fig. 4.39 The balanced coil (left) is tuned and matched at ω_s via C_T, C_M (right). L_{eq} is the equivalent inductance of L in series with C_{sym}.

The reactance of the impedance formed by L in series with C_{sym} is always positive[20] at ω_S. It is equivalent to an inductance L_{eq} in series with the resistance r of the coil. For a high Q coil, L_{eq} is close to $L/2$, otherwise L_{eq} is:[21]

$$L_{eq} = L - \frac{L}{2}\sqrt{1 + \frac{1}{Q^2}}. \qquad (4.69)$$

Using the equations from Section 4.1.2.3, one immediately obtains all the component values for the balanced circuit shown in Fig. 4.36, in the case where $C_0 = 0$.

$$C_T \omega_S = \frac{Q_{eff} - A}{B}, \qquad (4.70)$$

$$C_M \omega_S = \frac{1}{Z_0 A}, \qquad (4.71)$$

where Q_{eff} is the effective Q factor of the equivalent inductance L_{eq}:

$$Q_{eff} = \frac{L_{eq} \omega_S}{r}. \qquad (4.72)$$

A and B are given by equations similar to Eqs. (4.17) and (4.18), replacing Q with Q_{eff}:

$$A = \sqrt{\frac{B}{Z_0} - 1}, \qquad (4.73)$$

$$B = r\left(1 + Q_{eff}^2\right). \qquad (4.74)$$

Finally, when C_0 is different from zero, the component values can be derived from similar equations as above.

---

[20] Remember that C_{sym} compensates only half of the total inductance.
[21] Using Eq. (4.68).

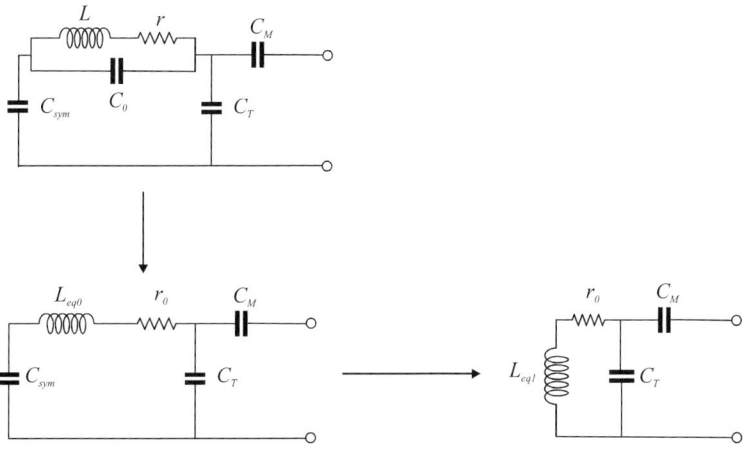

Fig. 4.40 Sequential transformation of the balanced circuit incorporating C_0 into the basic parallel tuned/series matched capacitive matching network.

The parallel circuit L, r, C_0 (Fig. 4.40) is equivalent to an inductance L_{eq0} in series with a resistance r_0. C_{sym} is still calculated from Eq. (4.68) where r and L are replaced by their equivalent values. The circuit including L_{eq0}, r_0 and C_{sym} is now equivalent to an inductance L_{eq1} that finally allows calculation of C_T and C_M. This is summarized in the following equations.

Let $XC_0 = 1/C_0\omega_s$ and $XL = L\omega_s$ be the impedance of C_0 and L, respectively, at the spectrometer ω_s. From Eq. (2.118):

$$r_0 = \frac{rXC_0^2}{r^2 + (XL - XC_0)^2} \tag{4.75}$$

and

$$XL_{eq0} = L_{eq0}\omega_s = -XC_0 \frac{r^2 + XL(XL - XC_0)}{r^2 + (XL - XC_0)^2}. \tag{4.76}$$

Similarly with Eq. (4.68):

$$XC_{sym} = \frac{1}{C_{sym}\omega_S} = \frac{\sqrt{r_0^2 + XL_{eq0}^2}}{2}. \qquad (4.77)$$

The equivalent inductance L_{eq1} (Fig. 4.40) is given from:

$$XL_{eq1} = L_{eq1}\omega_S = XL_{eq0} - \frac{1}{C_{sym}\omega_S}. \qquad (4.78)$$

These equations define a new effective Q factor as:

$$Q_{eff1} = \frac{XL_{eq1}}{r_0}. \qquad (4.79)$$

The capacitance C_T and C_M are finally obtained from equations similar to Eqs. (4.70), (4.71), (4.73), and (4.74) replacing Q_{eff} by Q_{eff1}.

More generally, the complex impedance at any feeding point on a resonator can still be matched in a balanced condition using the same circuit. Let R and X be the resistive and reactive component at the feed point, respectively (Fig. 4.41). Provided X is positive (otherwise the capacitive matching network would not work), the reactive impedance appears as an inductor in series with a resistance and the above equations can be applied directly.

The circuit provides equality of the VC_{sym} and VC_T voltage amplitudes whatever the loading conditions. However, as already noted the coil resistance makes the phase difference between the two voltages less than 180°.

Considering the circuit shown in Fig. 4.38 and Eq. (4.65) again, one obtains the phase difference φ between the two voltages (VC_T leads the voltage VC_{sym}).

When Eq. (4.67) is fulfilled (i.e., the voltage magnitudes are equal), φ is given by:

$$tg\frac{\varphi}{2} = Q_{app}. \qquad (4.80)$$

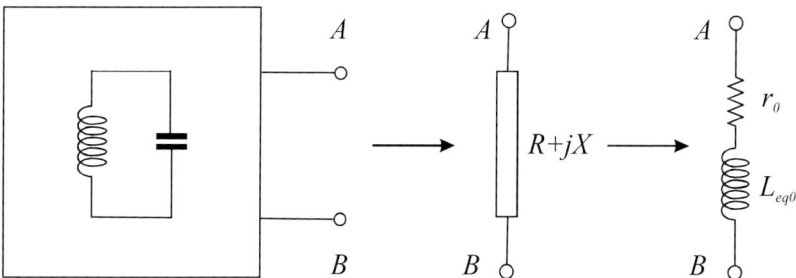

Fig. 4.41 Equivalent circuit for the feed point impedance of a resonator, at a frequency lower than its self-resonance. In this condition, the reactance X is positive as required. The resonator can be tuned and matched at the operating frequency ω_S using the balanced circuit with the component values as described in the text. The values of r_0 and Le_{q0} appearing in Eq. (4.77) should be identified with R and X/ω_S, as shown in the figure.

Q_{app} is the quality factor for the impedance appearing at the feed points (not to be confused with the Q_{eff} already defined).

For the circuit in Fig. 4.36, the apparent quality factor is equal to the quality factor Q of the coil if not using the pre-tuning capacitor C_0. For the circuit using the pre-tuning capacitor C_0, it is equal to $L_{eq0}\omega_S/r_0$ (Fig. 4.40). For the circuit in Fig. 4.41:

$$Q_{app} = \frac{X}{R}. \qquad (4.81)$$

In practice, the capacitance C_{sym} is almost independent of the probe loading and should be kept fixed once it has been adjusted for a good balance of the probe. Both the matching (C_M) and tuning (C_T) capacitors can be adjusted as for the classical circuit for any probe loading. In these conditions, the magnitudes of the voltages VC_{sym} and VC_T are equal, while their phase difference is kept in a narrow range close to the ideal 180° value, as a function of Q. This is shown in Fig. 4.42 for a typical example.

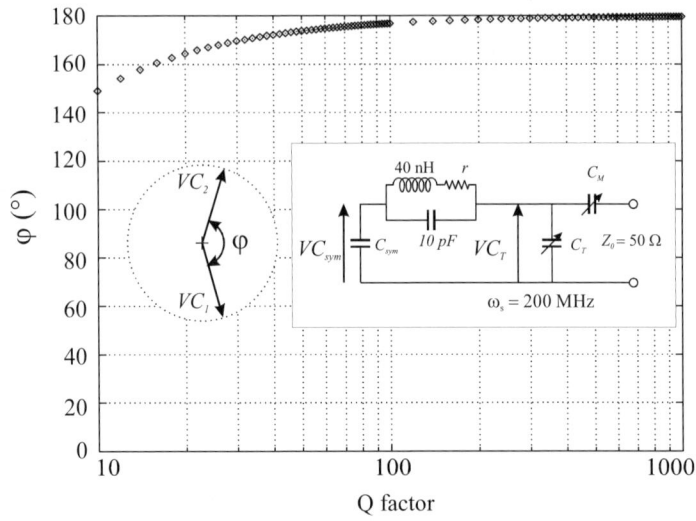

Fig. 4.42 Q dependence of the phase difference between the end coil voltages for a specific pre-tuned coil.

4.2.3.3 Capacitive bridge

Another circuit that provides coil balancing whatever the load conditions is the circuit shown in Fig. 4.43. This circuit has been already described by Chen and Hoult [Chen and Hoult, 1989]. It represents a clever extension of the Murphy-Boesch and Koretsky symmetric configuration [Fig. 4.34(b)] as is shown on the corresponding Smith Chart [Fig. 4.43(b)].

The difficulty in the original balanced circuit was maintaining the equality of the two matching capacitances $2C_M$ [Fig. 4.34(b)] assuring the balance of the coil. In the circuit shown in Fig. 4.43(a), this condition is always fulfilled as the two capacitances C are equal by construction. As a result, the tuning capacitor C_T should be set lower than the one required by the classical circuit. Then, the capacitance C ($C/2$ on the Smith Chart) moves the impedance to the unit admittance circle. At this point, the capacitance C_M adds the necessary admittance to move the impedance to the center of the chart, *i.e.* impedance matching to Z_0.

A detailed and comprehensive discussion on how to adjust this circuit has been given by Alecci [Alecci et al., 2006].

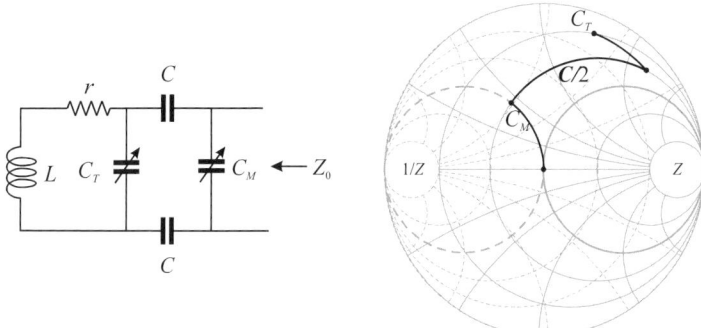

Fig. 4.43 The circuit is intermediate between the parallel tuned/series matched and series tuned/parallel matched circuits already described. In the former circuit, the tuning is realized with C_T that moves the impedance to the unit circle of the chart. The impedance matching is realized with C/2 which moves the admittance to the unit circle. The impedance matching is realized with C_M.

4.3 Summary of useful matching circuits

The first part of Table 4.4 shows five basic circuits designed to match the coil impedance to the purely resistive impedance Z_0.

The second part of the table shows three circuits that are derived from the basic ones and designed to balance the probe coil. Note that the inductive coupling circuits are auto-balanced.

The choice between different circuits, capacitive or inductive coupling, parallel matched or series matched, depends on a few constraints.

If ample space is offered, an inductive coupling could be preferred due to its balancing properties and simple implementation.

Array coils designed according to Roemer et al.'s circuit [Roemer et al., 1990] require that the parallel matched circuit should be preferred. In most other cases, the series tuned circuit and its derivatives could be a better choice due to its good efficiency and more accessible component values.

Table 4.4 Some basic coupling circuits and the corresponding formulae allowing component characterization. ω_S is the operating frequency.

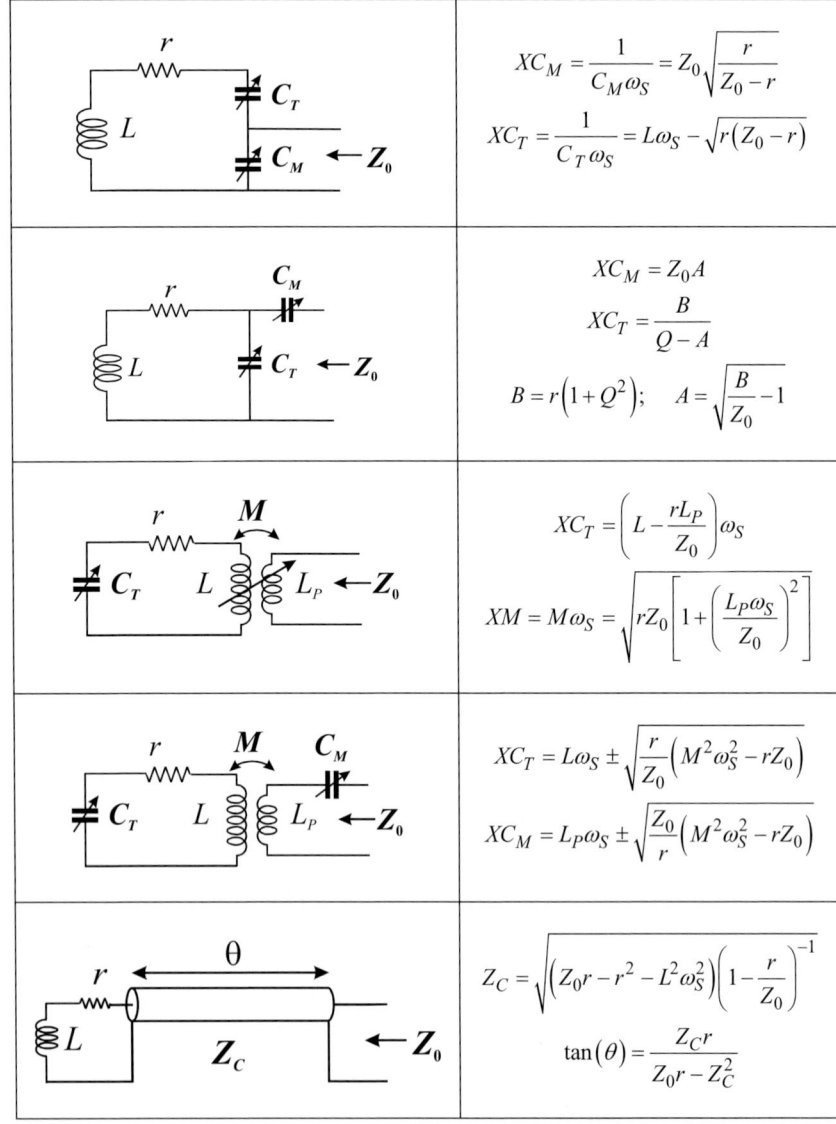

Table 4.4 (continued). Symmetric matching networks.

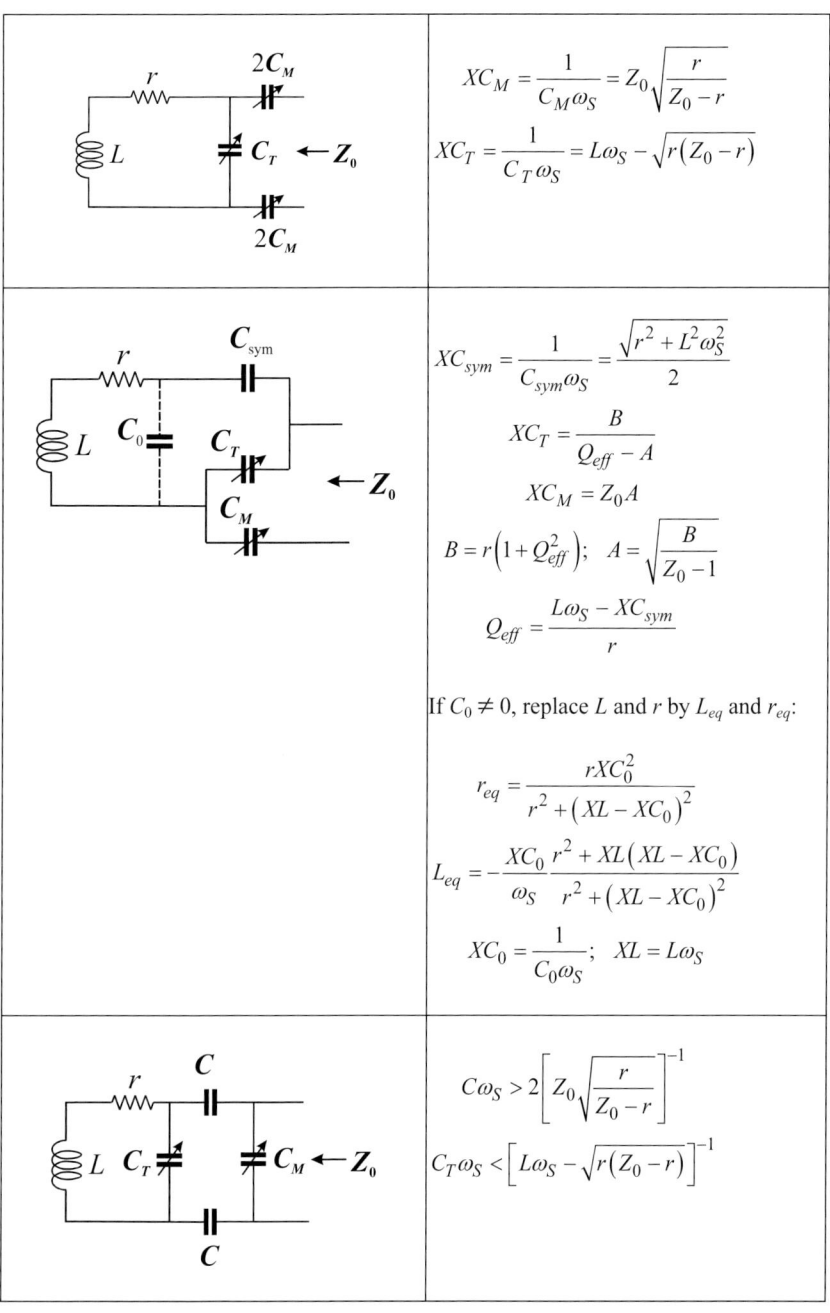

4.4 Accessories

Interfacing the probe resonator to the NMR/MRI console requires some additional components. Some of these devices are briefly presented here.

4.4.1 *Transmit/Receive (TR) switches*

When the same coil is used as transmitter and receiver, a switching device allows the dual functioning of the probe, connected to the transmitter amplifier during the short period of the excitation pulses or to the first stage of the receiver preamplifier while collecting the NMR signal.

The simplest switching circuit (Fig. 4.44) that is still in use after about half a century is made by means of a pair of crossed diodes and one $\lambda/4$ line [Lowe and Tarr, 1968]. During the receiving period, the crossed diodes in series with the transmitter isolate the probe signal from getting dissipated in the transmitter source impedance and isolate the receiver from the transmitter noise. The signal runs from the probe to the preamplifier without significant attenuation.

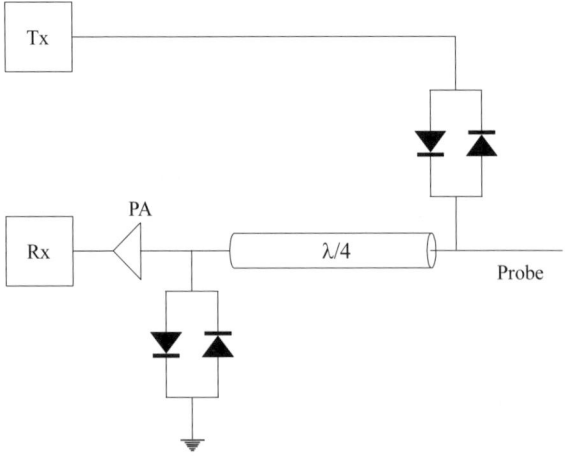

Fig. 4.44 TR passive switching network (Tx – transmitter, Rx – receiver, PA – preamplifier).

During the transmitting period, the high pulse current activates the diodes. The crossed diodes parallel to the preamplifier input act as a short circuit. The $\lambda/4$ transmission line being terminated in a short shows infinite input impedance directing the transmitter power to the probe.

The crossed diodes, placed between the transmitter and the probe, induce a loss of power

$$P_{loss\_diodes} = r_d I^2, \qquad (4.82)$$

where r_d is the forward resistance of the diodes and I the current provided by the amplifier. The incident power is given by

$$P = Z_0 I^2, \qquad (4.83)$$

where Z_0 is the characteristic impedance of the cable (assuming impedance matching). The loss, in dB, is given by

$$loss = -10 \log_{10}\left(\frac{P - P_{loss\_diodes}}{P}\right) = -10 \log_{10}\left(\frac{Z_0 - r_d}{Z_0}\right). \qquad (4.84)$$

Typically, r_d is of the order of one ohm, thus the loss is less than 0.1 dB, which is an acceptable value.

From the receiving point of view, the diodes present high resistive impedance and a small capacitance. Using proper diode components, the parasitic capacitance of the diode is sufficiently low at the usual NMR frequencies to neglect the corresponding loss.

In practice, standard silicon switching diodes such as the very common 1N4148 are used for isolating transmit power amplitudes of a kilowatt. The crossed diodes are made of pairs of two or four diodes connected in parallel.

PIN diodes are now becoming very common; they may be preferably used as they exhibit a very low direct resistance when conducting and very high impedance when reverse-biased.

The design of the quarter-wave transmission line is not very critical. In practice, a single line is sufficient within a bandwidth close to one octave. These lines can also be replaced by their equivalent lumped LC network [McLachlan, 1980].

Active circuits, using PIN diodes and $\lambda/4$ transmission lines can be designed as well (Fig. 4.45). The $\lambda/4$ isolating cells can be doubled to improve the receiver protection during transmitting. The advantage of such circuits is essentially a versatile timing control of the switching period and possibly a better TR isolation. However, they require additional components (choke inductors, blocking capacitors) and additional driving circuits in order to control the PIN diodes.

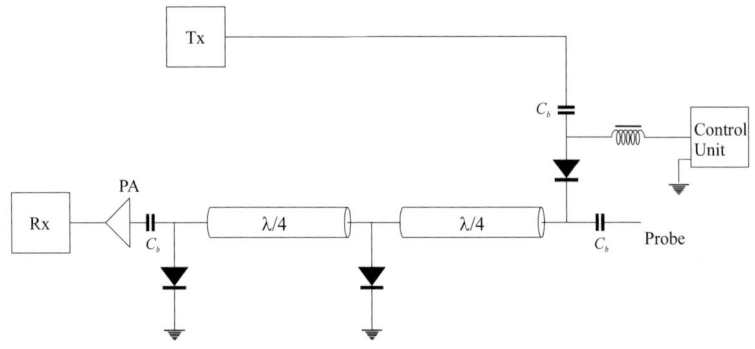

Fig. 4.45 Active TR switching circuit. The diodes are activated by an external control unit. C_b are blocking capacitors which isolate the transmitter and preamplifier (PA) input from DC currents controlling the PIN diodes. Several $\lambda/4$ cells can be connected in series to achieve any desired attenuation of the high power transmitter RF pulses.

4.4.2 *Damping circuits*

The sensitivity of the probe resonator is proportional to the square root of Q. As a consequence, the transient response of a high sensitivity probe could be an issue, especially when fast recovery is desired for collecting the very small signal from the nuclear spins after a high power RF pulse. The decay time constant[22] τ is proportional to the Q factor and inversely proportional to the operating frequency:

---

[22] The time constant of Eq. (4.85) assumes an exponential decrease of the voltage across the probe coil proportional to $e^{-t/\tau}$.

$$\tau = \frac{2Q}{\omega_S}. \tag{4.85}$$

This is particularly an issue at low field NMR, with NMR of low gamma nuclei, with Nuclear Quadrupolar Resonance (NQR) experiments or low field EPR [Devasahayam et al., 2000].

Assuming an average probe having a Q factor equal to 200 at 10 MHz operating frequency, the time constant τ is of the order of 5 µs. The voltage across the coil should decrease by a factor of at least 10^9 or more before the data acquisition starts.[23] Thus, the required delay between the end of the pulse gating and the start of the receiving period should be at least 20τ, i.e., 0.1 ms in the present example. This is generally too long, especially for fast relaxing nuclear species. To improve the recovery time, the Q factor of the probe should therefore be considerably reduced. Except in cases where the sensitivity is not a problem (which is rarely the case with NMR!) a permanent damping of the probe is probably not desirable.[24] Spoiling the Q factor during the pulse can be done with a passive circuit, as shown in Fig. 4.46(a). The damping resistance R comes in parallel to the resonant circuit when the diodes are activated by the RF pulse. Accordingly, the Q factor drops to a value that can be controlled with the value of R. During the receiving period, the Q factor of the resonator is recovered, leading to an optimum sensitivity.

Due to the change of the Q factor, the probe matching differs during the receiving and transmitting periods. Some additional circuits, similar to the passive TR switches already described, can be inserted in order to select the appropriate matching networks [Fig. 4.46(b)].

---

[23] The peak voltage during the RF pulse is typically of the order of 1 kV and the minimum voltage before acquisition should be less than the signal, assumed to be 1 µV.

[24] It should be noted that the signal to noise is generally not evaluated in terms of the uncompromising reduction of the intrinsic sensitivity of the probe when any of the Q spoiling methods described here are applied (permanent or controlled energy dissipation, over coupling). In contrast, it is generally admitted that the sensitivity is improved. NMR signals always being exponentially decaying, a reduction of the recovery time may compensate, possibly overpass, the decrease of the intrinsic sensitivity which would be obtained without damping the Q factor.

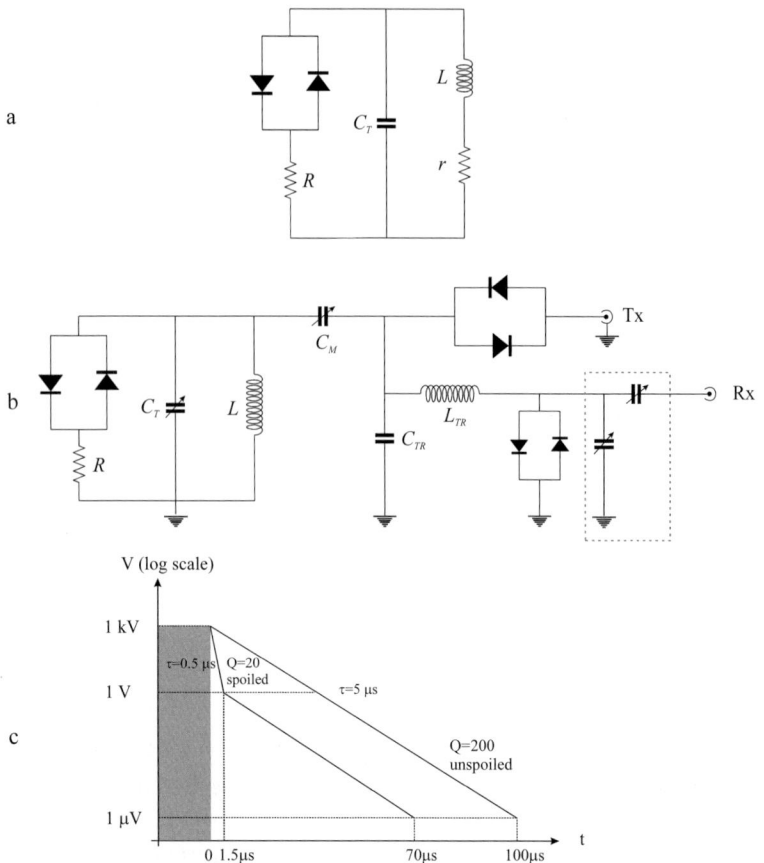

Fig. 4.46 (a) A simple passive Q-spoiling circuit. (b) The TR switch associated to this circuit. The L_{TR}-C circuit has the same role as the $\lambda/4$ TR switch shown in Fig. 4.44. During receiving, a matching box comprising C_{TR} and C_{MR} is inserted in the receive channel to match the probe to 50 Ω assuring that the source impedance, as seen by the preamplifier, is optimum for the best noise figure. (c) The diodes are conducting during the RF pulse and during recovery as far as the voltage across the C_T capacitor is greater than about 1 V. During this period, the rLC resonator is efficiently damped by the resistance R. When the voltage decreases below 1 V, the diodes become inoperative. This limits the recovery time improvement. However, damping during the pulse has the advantage of limiting the amplitude and phase distortions that may occur during the nuclear spin manipulation period.

During the dead time period, when the voltage across the resonator is higher than about 1 V (diode threshold voltage), the accumulated energy in the probe resonator is efficiently dissipated in R, leading to a significant decrease of this period of time.[25] However, when the voltage across the diodes becomes lower, they are no longer activated and the Q factor returns to the unspoiled value. The recovery time increases accordingly. The total recovery time is improved but still remains high due to the last period during which the diodes are inoperative [Fig. 4.46(c)]. To speed up the recovery during this last period, Hoult proposed a clever method which benefits from the fact that the input impedance of the LNA is generally different from Z_0 [Hoult, 1984]. Using the appropriate length of cable, the input impedance of the preamplifier can be transformed into the required damping impedance, without a significant degradation of the noise factor of the receiving chain. This design has the advantage that it does not require important modifications of the existing circuitry. In this case, the chosen tuning/matching circuit, either series tuned/parallel matched or parallel tuned/series matched, may have an influence on the damping efficiency [Miller *et al.*, 2000].

Over coupling the probe to the receiver or transmitter is another way to reduce the recovery time [Chingas, 1983]. Although the damping is permanent, it has less impact on the probe efficiency than using a dissipating resistance.

To improve the recovery, especially during the low-voltage period, it is probably better to activate the diodes with an external control circuit. A number of circuits have been proposed, allowing a five-fold reduction of the recovery time [for example, Andrew and Jurga, 1987; Jurga *et al.*, 1992]. In addition, these circuits improve the isolation of the preamplifier, avoiding receiver overload during transmit time and minimizing noise leakage from the transmitter during receive time.

An efficient dissipation of the accumulated energy in the probe tank circuit can also be done through a 1:1 transformer inserted in the probe circuit, as shown in Fig. 4.47. A reduction of the recovery time by an

---

[25] This is referred to in the literature as the high-voltage portion of the ringdown.

order of magnitude has been obtained with such a circuit [Peshkovsky et al., 2005].

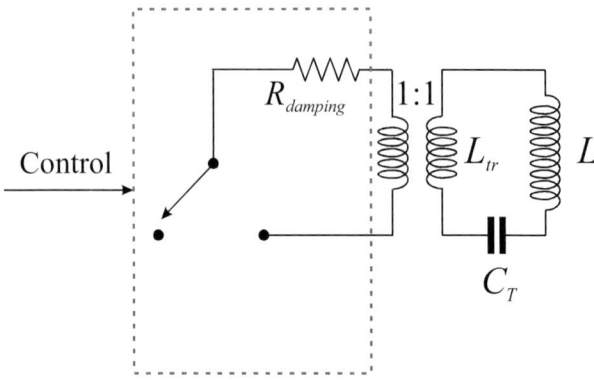

Fig. 4.47 Damping circuit using a 1:1 transformer inserted in the probe coil circuit. The controlling circuit switches the damping resistor across the secondary of the transformer, allowing efficient dissipation of energy. This circuit is appropriate at low frequency when the inductance L_{tr} can be made much smaller than L.

Very recently, a simple and efficient circuit has been proposed [Aissaini et al., 2014]. The circuit was designed for low frequency NQR experiments but the requirements are exactly the same as for low field NMR. However, the active switching device is a Complementary Metal Oxide Semiconductor (CMOS) integrated circuit, limiting the usable RF power. Q-switching is obtained by a dissipating resistance inserted in series with the C_{sym} capacitor of the balanced circuit (Fig. 4.36). The advantage is that the degree of damping can be controlled with precision by a judicious choice of the dissipating resistance r_{low} and of the ratio of C_{S1}/C_{S2} (Fig. 4.48).

All these circuits aim to diminish the recovery time after the RF pulse. In fact, a high Q probe resonator also has deleterious consequences during the transmitting period, introducing amplitude and phase distortions of the RF pulses. This leads to difficulties in accurately manipulating the spins as required in many pulse sequences. A comprehensive theory has already been presented [Barbara et al., 1991] that helps in understanding the experimental response of the coupled spin-probe system and is unavoidable to optimize the final "damping" circuit. Some of the methods already described here can help in reducing these issues, but other techniques using adapted pulse shaping (amplitude- and phase-controlled) have recently been described [Takeda et al., 2009; Tabuchi et al., 2010]. They rely on the design of specific pulse sequences which could be applied whenever possible[26] to improve both the accuracy of the spin manipulation and the reduction of the receiver dead time.

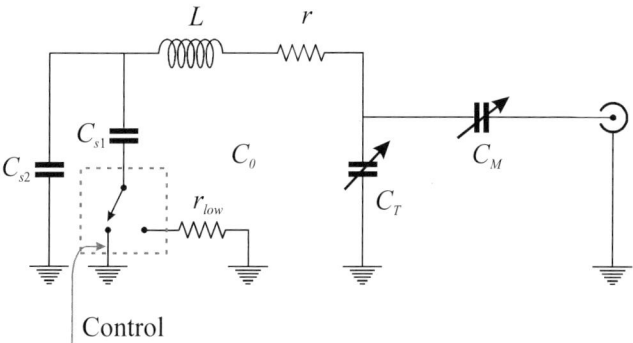

Fig. 4.48 The damping circuit designed by Aissaini et al. is based on a clever modification of the balanced matching circuit shown in Fig. 4.36. The C_{sym} capacitor is split and a damping resistor r_{low} is inserted in series with one capacitor using a CMOS switching circuit. The adjustment of r_{low} and of the ratio C_{s2}/C_{s1} allows a fine optimization of the recovery of the circuit.

4.4.3 Baluns and cable traps

The electrically-balanced probe is expected to be connected to a coaxial asymmetric transmission line. In principle, a "balanced to unbalanced" transformer (balun) is required. The role of this device is to avoid currents flowing on the external conductor of the coaxial cable, leading to the deleterious "common mode" [Peterson, 2012].

At high frequency, when the skin depth is much smaller than the thickness of the cable shield, the coaxial cable is constituted of three conductors: the center conductor, the inner surface of the shield

---

[26] Modern designs of pulse-programming and pulse-shaping circuits allow an easy implementation of these techniques.

conductor and the outer surface of the shield conductor [Ott, 2009, pp. 44–105]. The interior of the cable is subjected to the differential mode, but currents can be induced on the third conductor (common mode) through the so-called "common impedance." The common mode arises, in part, from the coupling of the cable shield (i.e., the external side of the outer conductor) to the electromagnetic field created by the probe. The corresponding currents circulating on the external side of the cable outer conductor induce additional loss sources, modify the tuning and matching of the probe coil, pick up extra noise from the surroundings, and radiate energy. Common modes are also responsible for undesired cross-talk between the ports of a multiple-probe network. More importantly, when dealing with animal or human experiments, the common mode currents may cause serious burns.

Common modes related to NMR probe circuits are frequently broad resonant modes (Fig. 4.49).

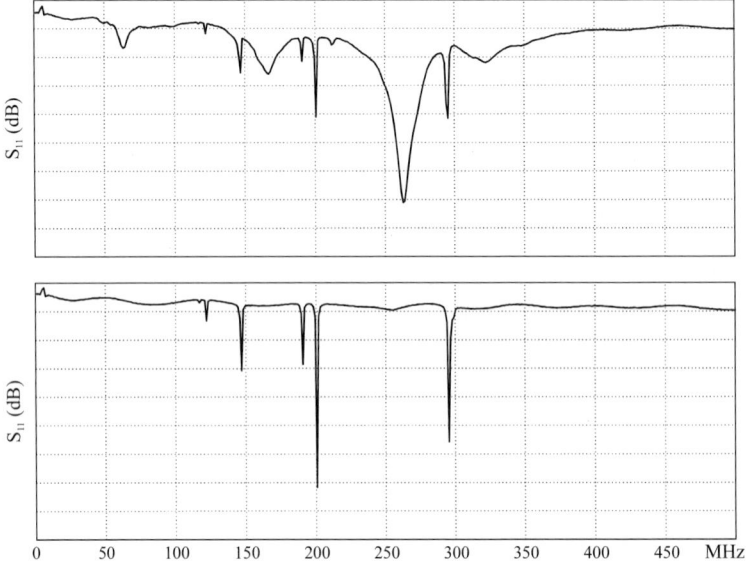

Fig. 4.49 A typical S_{11} spectrum of an NMR probe (a high-pass linear birdcage coil in this example). Upper: common modes are evidenced by broad resonant modes. Lower: the common modes have been suppressed after insertion of a broadband cable trap.

These modes may couple with the resonant modes of the probe, absorbing energy and modifying the electrical characteristics of the probe. The common mode is frequently evidenced by the so-called "hand effect" that affects coil tuning when the operator approaches the hand or touches the insulating jacket of the cable. It is also sensitive to the position of the cables respective to the surrounding metal conductors.

The parasitic common modes should be avoided by using an appropriate balun inserted between the electrically-balanced probe[27] and the asymmetric coaxial cable.

An efficient protection against common mode is in fact provided by a so-called "common mode choke" or "cable trap," which is designed to suppress currents on the outer conductor of the cable. These devices are good baluns, but they can also be used in circumstances not necessarily involving balanced circuits.

4.4.3.1 *LC-balun*

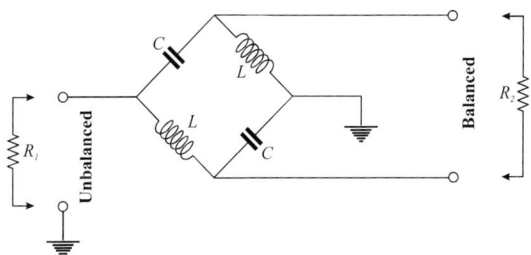

Fig. 4.50 An *LC*-balun circuit for matching purely resistive impedances.

The circuit shown in Fig. 4.50 has a double role. It is a matching circuit and a balun. It can match virtually any resistive load, but the component values may become unmanageable for extreme R_1/R_2 impedance ratios.

---

[27] An electrically-balanced probe reduces the electric near field in the sample, reduces the radiated field energy, and reduces the electromagnetic interaction with the electric field components of the far field. But, asymmetric couplings (both inductive and capacitive) with the surroundings may ruin the initial balanced configuration of the probe circuit.

The balun is formed of a bridge, with a capacitance and an inductance in opposed branches. The component values are given by:

$$L\omega_S = \frac{1}{C\omega_S} = \sqrt{R_1 R_2}. \qquad (4.86)$$

Although the impedance match is realized on a large frequency bandwidth when both resistances are equal, the balun is operative only at the designed frequency ω_S.

4.4.3.2 The 4:1 λ/2 balun transformer

The circuit shown in Fig. 4.51 shows a balun is which the feed points are driven by the voltage existing at the ends of a λ/2 transmission line. These voltages are opposed in phase due to the insertion of the half-wave transmission line.

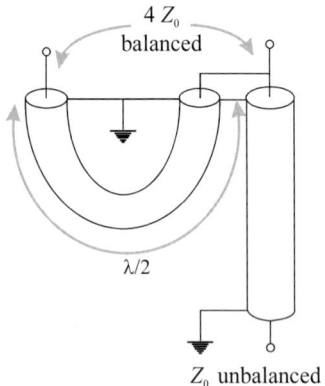

Fig. 4.51 The λ/2 balun 4:1 transformer. The λ/2 transmission line introduces a 180° phase shift leading to balanced voltage at the feed points. As a consequence, the impedance looking at the feed points is four times the input impedance. The characteristic impedance of the transmission lines used here should be equal to Z_0 for best performance.

The impedance at the feed points is four times the input impedance, Z_0, which should also be the characteristic impedance of the transmission lines used in the design.

The impedance at the feed points may be an advantage in some probe designs as the currents flowing in the connections to the probe would be halved compared to the currents that would flow in the connections to a probe matched to Z_0. However, the voltage is doubled, leading to a higher conservative electric field and potentially higher dielectric losses.

The $\lambda/2$ balun operates at the frequency for which the line is cut to half-wave, similar to the LC balun.

4.4.3.3 Broadband balun transformers

The Guanella balun [Guanella, 1944; Ruthroff, 1959] schematically represented in Fig. 4.52 is a 1:1 impedance transformer which can operate on a very broad frequency range, depending on its design.

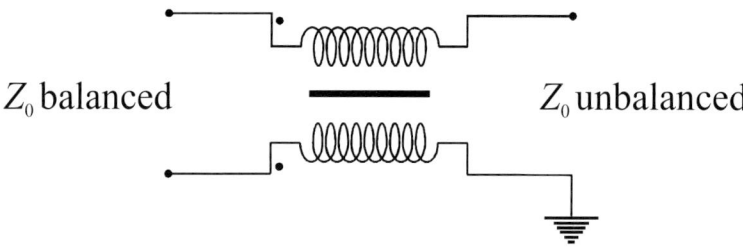

Fig. 4.52 The Guanella balun is a 1:1 transmission line impedance transformer. The net current (common mode) is ideally zero. The currents flowing in both "windings" are opposed in phase (differential mode).

The lowest operating frequency of a broadband transformer is limited by the coil inductance, while the high frequency limit is imposed by the coupling constant. Thus the design of broadband transformers, especially using air cored inductances, is quite difficult. In contrast, the large inductance and strong coupling conditions can be better fulfilled when employing ferrite cores. The power capability of these flux coupled transformers is limited by the saturation properties of the ferrite material that can give rise to a non-linear response, hence harmonic distortions. However, when broadband transformers are made up of coaxial, or bifilar,[28] transmission lines wound on ferrite cores [Fig. 4.53(a)] they can still handle high power in their functioning band [Guanella, 1944; Ruthroff, 1959].

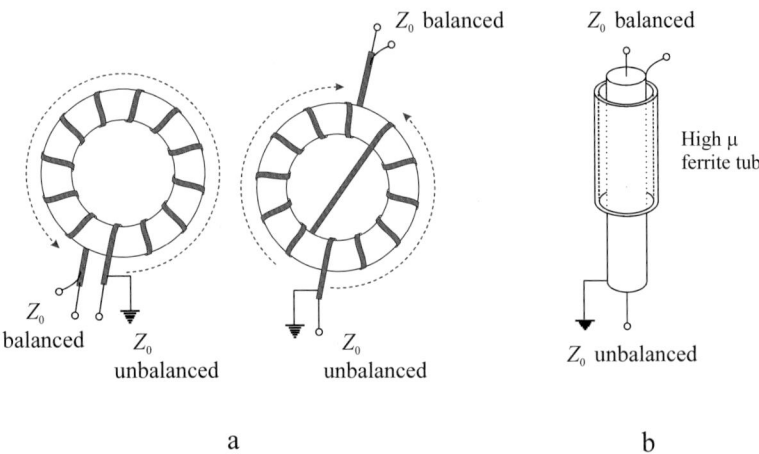

a   b

Fig. 4.53 Broadband baluns built on the Guanella balun concept [Sevick, 1994]. (a) The baluns are built as a toroidal coiling of a transmission line. The turns can be in the same direction or in opposed directions on half the coil (Reisert balun). (b) A coaxial line is slit into a stack of high permeability tubes efficiently blocking the common mode currents on the external conductor of the line.

In this case, the magnetic flux in the ferrite core is due to the common mode currents which are expected to be much lower than the efficient differential current modes. The technology is well documented [Sevick, 1994; 2001] and is probably the better choice for designing devices on any central frequency from a few MHz to the GHz. But the presence of the ferrite core is the limiting factor that may render these devices useless in close proximity to the probe coil. They are however valuable for the construction of testing devices that can be used on the RF workbench.

Another version of a broadband transmission line balun transformer is shown in Fig. 4.53(b). Instead of coiling the cable on a magnetic core

---

[28] The Ruthroff design uses a three wire winding and improves the bandwidth of the circuit characteristics.

(or possibly on a ferrite cylindrical bar), the cable is inserted into a high ferrite material. Again, the external conductor of the cable exhibits a high inductance, hence a high impedance, efficiently blocking the common mode currents.

In close proximity to the probe, the cable should preferentially be air coiled. This constituted an efficient nonmagnetic broadband Guanella's balun anyway, operating at frequencies higher than about 100 MHz (Fig. 4.54). For example, an RG316 cable coiled by five turns on a cylindrical form (diameter 20 mm) constitutes a choke inductor having an impedance of about 250 Ω at 100 MHz.

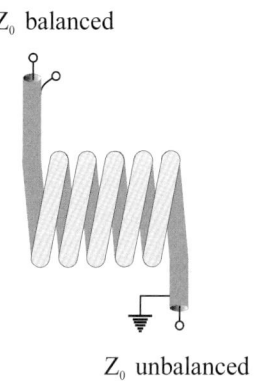

Z_0 balanced

Z_0 unbalanced

Fig. 4.54 A broadband balun or cable trap is made by coiling a transmission line. The external conductor of the line constitutes a choke inductor. The inner space of the line is subjected to the non-attenuated differential mode. This balun is the nonmagnetic version of the Guanella 1:1 transformer shown in Fig. 4.53.

4.4.3.4 *Tuned cable traps*

Tuned cable traps operating on a narrow band can be very efficient, but sometimes difficult to tune. The $\lambda/4$ balun [Fig. 4.55(a)], sometimes called the "bazooka" balun, uses the properties of a short circuited quarter-wave transmission line that presents an infinite impedance at the input. This balun is a 1:1 transformer. The design is not very critical due to their large bandwidth respective to the frequency range. The bazooka balun can be shortened and its electrical length can be adjusted using a variable capacitor located at its input [Fig. 4.55(b)]. The simple trap balun shown in Fig. 4.55(c) provides another way to block the cable braid currents.

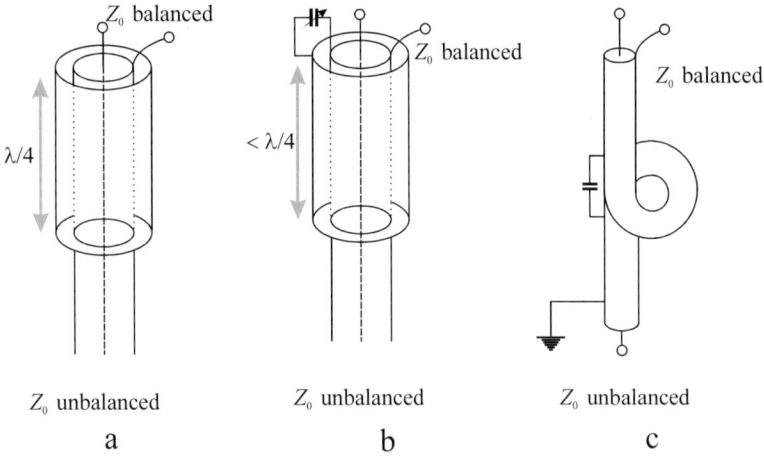

Fig. 4.55 Narrow band baluns (cable traps). (a) The l/4 or "bazooka" balun. Note that the braid of the inner cable is soldered to the bottom of the external conductor of the l/4 line. (b) A shortcut version of the bazooka balun. The l/4 line is adjusted by a capacitor, allowing the balun to be accurately tuned. (c) The common choke inductance is made up by a tuned trap circuit formed with the coaxial braid conductor.

A high impedance circuit is formed by the cable loop inductance tuned at the frequency of interest by a capacitor. This balun is also narrow band.

An ingenious device was proposed [Seeber *et al.*, 2004] as a versatile trapping circuit for elimination of the common mode currents. Among others, its main advantages are that it does not connect or solder to the cable and that it can fit over any cable assemblies without affecting the coil parameters (in contrast to a standard trapping circuit). It can also solve difficult problems such as the case of common mode currents that are not induced by the probe itself and for which a balanced match is not a solution. This could be, for example, the case of a receiving coil placed inside a larger resonator (see Chapter 9). In this case, the large emitter probe can induce currents on the cable shield connecting the small receiver probe to its preamplifier. Only a trap, conveniently placed on the cable, can reject the undesired currents.

The floating trap [Fig. 4.56(a)] is a sort of transmission line formed by two concentric cylinders separated by a dielectric of low permittivity (recommended). The line is short circuited at one end and has a length shorter than a $\lambda/4$. It is tuned by capacitors soldered at the other end between the inner and outer cylinder, looking like a capacitive tuned bazooka balun. The main point here is that the coaxial cable that carries the common mode current acts as the central conductor of a transmission line formed with the inner cylinder of the floating trap balun. In this way the undesirable common mode currents that flow on the coaxial cable shield are effectively blocked by the floating trap.

Another ingenuous solution allows tuning of the trap in a very easy way. It consists of separating the assembly into two parts [Fig. 4.56(b)]. The gap between each part can be adjusted (modifying the mutual inductance, hence the inductance of the trap) to tune the high Q trap to the exact frequency desired. In addition, the slit trap can be easily removed, put on or displaced on any cable assemblies according to each particular common mode current problem encountered.

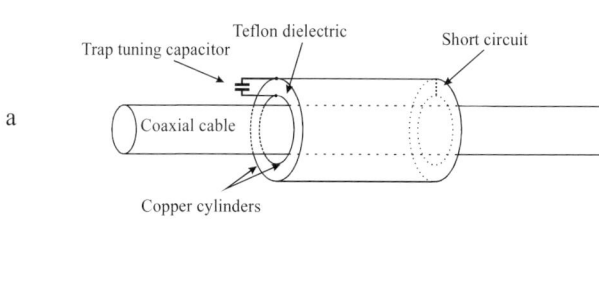

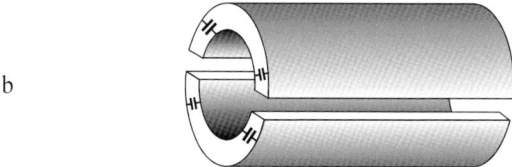

Fig. 4.56 (a) Floating trap. (b) The trap fine tuning is done by adjusting the distance between the two parts of the split cylinder.

The position of the floating trap on the cable shield should be determined for best common mode rejection. The corresponding currents vary indeed along the cable shield [Beck *et al.*, 2000]. They present points of minimum and maximum amplitudes having a pattern of characteristic standing waves. The best position for the trap is thus where the current amplitude is maximum. The corresponding points can be located empirically or determined using current sensors placed around the cable assembly.

Finally, in cases of pronounced difficulties with cable common mode currents, a recommended practice [Peterson *et al.*, 2002; 2003] is that the probe is balanced and connected to the coaxial cable through a balun, that can be a cable trap (preferably shielded). Another trap (the floating trap for example) can be placed at a $\lambda/4$ distance away from the probe connections, where the undesirable currents are expected to be at a maximum.

4.5 Interfacing the Probe to a Low Noise Amplifier (LNA)

The LNA has already been presented in Chapter 2. As outlined at the beginning of this chapter [Fig. 4.1(b)], the input of the LNA is impedance matched for better noise performance and not for better power transfer. The complete circuit, including the probe coil, is shown in Fig. 4.57(a) and the impedance transformations are illustrated on the Smith Chart shown in Fig. 4.57(b).

The matching network at the input of the LNA is designed to transform the internal impedance (assumed to be $Z_0 = 50\ \Omega$) of a given probe into the optimum source impedance, characterized by the so-called optimum reflection coefficient Γ_{opt}.

$$\Gamma_{opt} = \frac{Z_{opt} - Z_0}{Z_{opt} + Z_0};\ \ \Gamma_{opt}, Z_{opt} \in \mathbb{C}\ . \tag{4.87}$$

It results that the input impedance of the LNA differs significantly from the Z_0 impedance depending on the particular design of the preamplifier, especially on the type of front end transistor (Fig. 4.58).

Interfacing the NMR Probehead   285

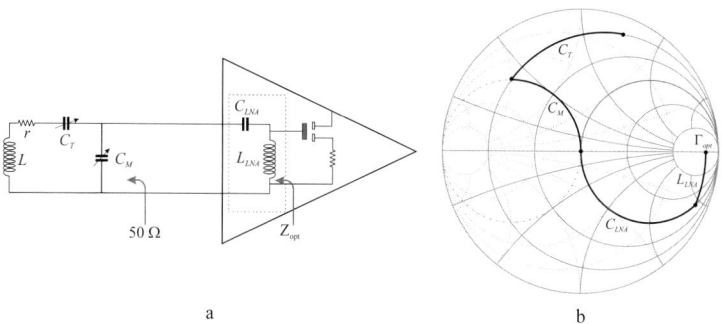

a   b

Fig. 4.57(a) Circuit showing a probe coil (L,r) tuned (C_T) and matched (C_M) to 50 Ω and connected to an LNA. The network at the input of the LNA is designed to transform the probe 50 Ω impedance into the optimum noise source impedance Z_{opt}. (b) Smith Chart showing the pathway to obtain Γ_{opt}.

The first stage of the ultra low noise preamplifiers used nowadays in NMR/MRI is generally a microwave MOSFET[29] transistor. These devices typically present very high input impedance at RF frequencies.

The optimum source impedance for the best noise performance (lowest *NF*) of these transistors increases at RF frequencies to about 1–2 kΩ with a low reactive component (few ohms). The optimum reflection coefficient is typically around $\Gamma_{opt} \approx 0.95 \angle 0°$ at NMR frequencies. The L-network described in Fig. 4.4(b) is well suited to performing the impedance transformation of the probe (50 Ω) into the optimum impedance (≈ 1.5 kΩ).

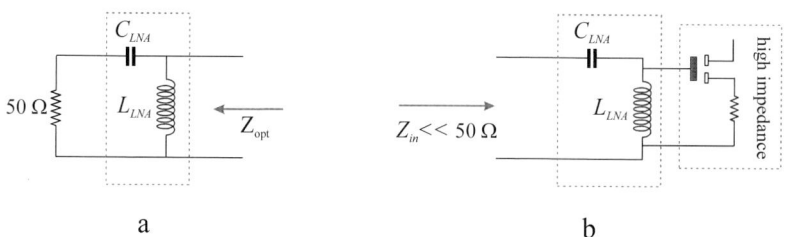

a   b

Fig. 4.58(a) Impedance at the output of the L-matching network of the LNA. (b) Impedance at the input of the LNA.

---

[29] Specifically pseudomorphic High-Electron-Mobility Transistor (pHEMT).

The impedance ratio Z_{opt}/Z_0 being quite large, the L-network [L_{LNA} and C_{LNA}, Fig. 4.57(a)] is resonant close to the operating frequency ω_S. The input impedance of the LNA can therefore be very low at ω_S, in the order of 1 Ω. Connecting the probe coil[30] to this preamplifier using a short section of a good coaxial cable should not be an issue. Toward the probe coil circuit, the cable is properly terminated assuring that the source impedance seen by the front end transistor is at an optimum. Accordingly, the best *NF* could thus be obtained even if the cable is not terminated in its characteristic impedance toward the LNA. We are not concerned here with optimum transfer of energy. However, attention should be paid to extra resistance in series with the low input impedance of the LNA that could generate noise and destroy the preamplifier *NF*. A modern transistor may exhibit a minimum *NF* as low as 0.1 dB at NMR frequencies. In practice, the final *NF* of the completed circuit could be around 0.5–0.6 dB without special care, which is not too bad a value. It can sometimes be decreased to around 0.3 dB after careful adjustments.

The low input impedance of the LNA is also one of the keys for the design of array coils. The invention of coil arrays for MRI is due to Roemer *et al.* [Roemer *et al.*, 1990]. Since that time, this concept has been extensively used in the design of receive-only probes, mostly for MRI. In this approach, the signals delivered by each coil in the array are combined to offer the high sensitivity and resolution of a small surface coil, but on a larger volume.

The main problem arising when designing an array is given by the coupling between coils. This has been elegantly solved by overlapping the adjacent coils to give zero mutual inductance (Chapter 9). But distant coils are still coupled together. In their original paper, Roemer *et al.* [Roemer *et al.*, 1990] showed that decoupling between nearest neighbors as well as distant coils can be further improved by connecting each coil to a low impedance preamplifier.

To understand how this works, let us consider the circuit describing one cell in the array, as shown in Fig. 4.59(a). The probe coil is assumed to be an inductive load of (1.0 + j100.0) Ω. The coil is matched to 50 Ω using a series tuned/parallel matched circuit. The choice of this matching

---

[30] Already tuned and matched to 50 Ω.

circuit allows the benefit from the low input impedance of the LNA to minimize the current in the coil circuit during reception.

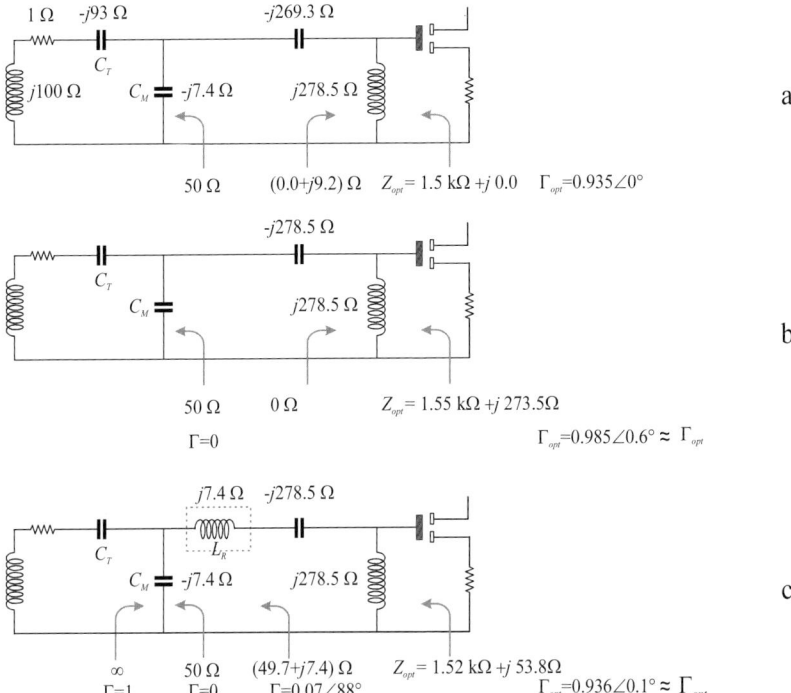

Fig. 4.59 (a) Probe circuit and input circuit of the LNA, adapted for the best noise figure. The input impedance of the LNA is low, slightly inductive. The resistance of the inductor has been neglected and the input impedance of the front end transistor is assumed to be very high. (b) The input circuit of the LNA is made resonant at the operating frequency after slight adjustment of the input capacitance. The source impedance seen by the front end transistor is still close to the optimum for the best noise figure. (c) The inductor L_R is inserted to tune the capacitor C_M of the probe circuit. The source impedance seen by the front end transistor is still optimum.

In this way, the coupling of noise and NMR signal to other coils is eliminated. The noise decoupling between coils is an important aspect of the efficiency of the "NMR phase array."[31]

The optimum source impedance for the best noise performance of the front end transistor is assumed to be about 1.5 kΩ ($\Gamma_{opt} \approx 0.935\ \angle 0°$). The impedances of the L-matching network components are calculated using Eq. (4.8). Assuming that the impedance at the gate of the MOSFET is very high, it could be neglected. Therefore, the input impedance of the LNA is estimated to be about $(0.0 + j9.2)\ \Omega$. In practice, the impedance can be made almost zero[32] at the operating frequency after a small modification of the capacitor C_{LNA} or possibly adding a conjugate impedance to cancel out the excess of reactance. It will have very little influence on the final source impedance which will still be very close to the optimum for the best NF [$\Gamma \approx \Gamma_{opt}$, Fig. 4.59(b)].

Roemer et al. suggested the insertion of an inductor L_R in the circuit including the probe matching capacitor C_M and the low input impedance of the LNA [Fig. 4.59(c)]. This circuit forms a choke that blocks the current in the series resonant probe coil circuit, improving (and controlling) noise and signal decoupling between the coils in the array. The L_R inductance should be set equal to the complex conjugate of C_M in order to make a parallel resonant circuit including L_R, C_{LNA}, L_{LNA} and C_M. From Eq. (4.9), the required impedance for L_R is equal to 7.4 Ω in the present example. The insertion of this small inductance will have little influence on the NF as the optimum Γ_{opt} at the gate of the transistor will again be closely reached [Fig. 4.59(c)].

The required series inductance can also be realized using the appropriate length of coaxial cable connecting the probe matching inductor to the LNA. Possibly, the length could be adjusted by a capacitor placed in parallel to one end of the line.

---

[31] As it was originally named by Roemer et al. [Roemer et al., 1990] in reference to phased array radar.

[32] The resistance of L_{LNA}, the impedance of the MOSFET in parallel to L_{LNA} and some additional loss contributions would make the input impedance not zero but still very small, of the order of 1 Ω.

The most important condition that an array must fulfill is that the interaction between coils is minimized while maintaining the best *NF*. This can be obtained using an LNA with low impedance and a probe coil already carefully matched to 50 Ω. Decoupling between coils can be optimized on the workbench using a small inductor as L_R or a short length of cable. The effective cable length can be adjusted using a variable capacitor in front of the LNA or, possibly, by slightly readjusting the capacitance C_M. In this way, the decoupling between coils can be optimized without degrading the *NF* of the receiver channels.

4.6 Ultra-broadband and Ultrafast Recovery Probes

In many magnetic resonance experiments, the short duration of the NMR signal after excitation requires a probe that recovers very quickly after application of high power pulses. Using a probe resonator, the decay time constant of the "ringing" transient after the pulse is proportional to its *Q* factor [Eq. (4.85)]. This problem is particularly crucial at low frequency. Some circuits have already been presented in Section 4.4.2 to overcome, in part, this issue. They were designed to momentarily increase the bandwidth by spoiling the *Q* factor of the probe resonator.

Because some energy is dissipated in the damping resistor, some loss in sensitivity is to be expected. In fact, the advantages obtained while increasing the probe bandwidth can sometimes largely overcome the sensitivity issue, especially for any experiment requiring the acquisition of the signal very early after the end of the pulses. This possibility is particularly interesting for solid state NMR, NQR, and experiments on very fast relaxing nuclei.

In addition, an ultra-broadband probe is able to record the signals from all the NMR nuclei in a given sample, without multiple tuning the probe resonator (Chapter 6). For some configurations, especially microcoils, the multiple tuning circuits may be the source of important losses. The corresponding loss of sensitivity could be of the same order or even higher than the losses that can be expected using an ultra-broadband probe. Recently, it was demonstrated that a broadband non-resonant microcoil is efficient for recording multinuclear (from the deuterium to the proton) NMR spectra of micrograms of a small molecule [Fratila et al., 2014].

Ultra-broadband probes were initially designed using lumped element delay lines. Recently, a new class of broadband coils, using transmission lines, is emerging. This will possibly lead to new designs and concepts far away from a resonant probe. These concepts are briefly described in the following.

4.6.1 Delay line ultra-broadband NMR probe

The earlier ultra-broadband NMR probes were designed as lumped element delay lines, as proposed by Lowe and Engelsberg [Lowe and Engelsberg, 1974]. This kind of probe has, curiously, been the subject of only a few studies. But, lately, its interest has been renewed by Kubo and Ichikawa [Kubo and Ichikawa, 2003].

The circuit is sketched in Fig. 4.60. The probe itself is made out of Π or T low-pass filter sections connected in series. In practice, a solenoid is wound on the sample and capacitors are soldered along the windings and connected to the ground.

The corresponding setup constitutes an approximation of a transmission line, characterized by its characteristic impedance Z_c,

$$Z_c \approx \sqrt{\frac{L}{C}}, \tag{4.88}$$

having a cutoff frequency of the order of:

$$f_{cutoff} \approx \frac{1}{\pi\sqrt{LC}}. \tag{4.89}$$

L and C are the effective inductance[33] and the capacitance, respectively, of the elements constituting the delay line.

The theoretical analysis of the delay line probe showed that it has sensitivity equivalent to a resonator having a Q factor equal to:

---

[33] The inductances constituting the delay line are mutually coupled.

$$Q_{eq} = \omega_S \tau_D, \qquad (4.90)$$

where τ_D is the delay time of the line. The decay time constant of the probe is given by:

$$\tau_R \approx 2/\omega_{cutoff}. \qquad (4.91)$$

If the operating frequency ω_S is lower than ω_{cutoff}, the delay line is non-dispersive and the pulse travels down the line without much distortion, for a wide frequency range. If ω_{cutoff} is high, the transient time may be very short and independent of ω_S. A compromise should be found if one expects a Q_{eq} not too low, hence a useful sensitivity.

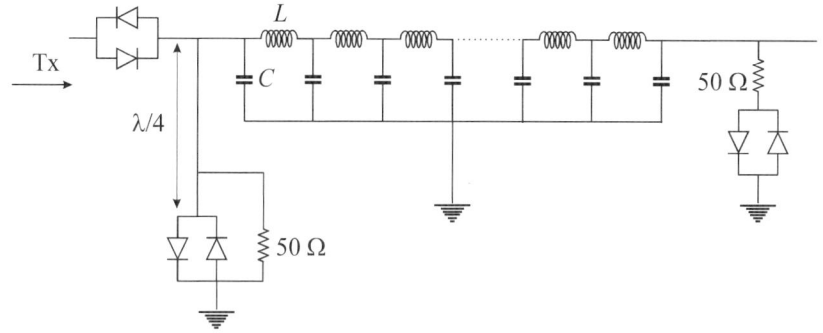

Fig. 4.60 A schematic representation of the delay line probe in its environment as designed by Lowe and Engelsberg [Lowe and Engelsberg, 1974]. The required termination (50 Ω) of the line during the transmitting and receiving periods is assured by the PIN diode commutation. During transmission, the termination on the left side of the delay line is short circuited and isolated by the quarter-wave line. On the right side, the delay line is effectively terminated in 50 Ω. A quarter-wave line in series with the preamplifier and terminated by additional two PIN diodes (not shown) protects the receiver chain as usual. During receiving, all diodes are opened, hence only the left side of the delay line is terminated by the 50 Ω load. The delay line appears as a source with an internal impedance of 50 Ω. The probe itself is made out of a solenoid with capacitors soldered between the ground and points along the solenoid winding.

The key of this probe design is the pulse travelling the delay line and dissipating in the load resistance (50 Ω) that terminates the line. During the receiving period, the line should also be terminated in its characteristic impedance. The classical interface to the spectrometer has to be modified accordingly (Fig. 4.60).

4.6.2 *Transmission line ultra-broadband NMR probe*

The design of the delay line probe has recently been extended to a new class of ultra-broadband probe using transmission lines [Murphree *et al.*, 2007 and 2012; Scott *et al.*, 2012].

In these designs, the conductors of a "classical" NMR probe are replaced by sections of transmission lines shaped as a saddle coil [Murphree *et al.*, 2007 and 2012] or as a solenoid [Scott *et al.*, 2012].

The transmission line constituting the probehead is properly terminated in its characteristic impedance by a resistive load avoiding, or limiting, standing waves associated with resonant modes. The impedance of the probe becomes independent of frequency.

The conductive parts of the transmission line probes are subjected to currents able to create magnetic fields in their close proximity, allowing detection of NMR signals by the inductive method as a conventional probehead.

The conductor pattern of Murphree *et al.*'s probe is that of a saddle coil. The saddle coil is etched on one side of a flexible insulating substrate. Additional conductors are etched on the other side forming a microstrip transmission line. The characteristic impedance of the line can be controlled by the geometry of the conductors, the thickness, and the permittivity of the insulator [Pozar, 1998].

In the design by Scott *et al.* the probehead is shaped as a solenoid coil. The second conductor of the transmission line is conductor paint containing silver. The paint surface is extended over the solenoid until the desired characteristic impedance is obtained. A copper strip is attached to the silver paint on the solenoid and soldered to the probe grounded connection.

The terminating resistor plays an important role on the desired frequency band. It can be adjusted as required and possibly completed by some matching components, but this will somehow reduce the probe bandwidth.

The concept of a transmission line used in a broadband configuration, i.e., properly terminated, can be applied to any "classical" probe configuration. Examples have been given here for the solenoid and the saddle coils.

There is however a price to pay with respect to the sensitivity, especially due to the necessity to terminate the probe in a resistive load that absorbs part of the energy travelling in the probe coil. Also, the transmission lines are generally working in the differential mode. This fact limits the extension of the sensitive volume close to the conductors.

In any case, sensitivity losses that can be expected using certain conventional configurations (multiple tuned microcoils, experiments on fast relaxing nuclei, broad solid state spectra, NQR) can probably be overcome using the concept of a non-resonant probe.

4.6.3 Non-resonant probe circuit

These probes are simply made out of a traditional coil inductance which is directly connected to the transmission line. To be functional, the probe must fulfill some conditions. The Q factor of the coil inductance should be low. It may possibly be lowered by additional broadband resistors, which seems counterintuitive.

This concept proved to be well adapted to the design of microcoils [Fratila et al., 2014]. The probe is made out of a micro spiral coil etched on an insulating material.

Fratila et al. showed, from simulations in the supplementary material, that the complex impedance of the coil is dominated by its resistance up to at least 500 MHz. The real part of the impedance (about 5 Ω) is frequency independent up to 1 GHz and higher than the reactive component which becomes detectable above 100 MHz. The distributed capacitance between the windings of the coil probably contributes to increase the bandwidth of the coil in a similar manner as observed for broadband choke inductances (Chapter 2). Such a behavior is typical for microcoils with many turns. In addition, the circuit is damped by adding a termination resistor and, possibly, the matching is improved with a resistor connected in series with the feed coaxial line.

Matching a low Q coil with additional resistances is expected to be deleterious with respect to the sensitivity. In fact, the authors reported a very similar sensitivity compared to a proton-tuned microcoil.

This illustrates the difficulty in tuning and matching very small coils and demonstrates that the non-resonant circuit may be useful with the additional advantage of being naturally multinuclear.

Chapter 5

Quadrature Driving

NMR involves a macroscopic magnetization that rotates around the static magnetic field in a defined direction. To create this magnetization, a rotating magnetic field B_1 is applied to the ensemble of microscopic spins, leading to a certain degree of coherence in their precession motion. The rotation direction of the applied B_1 field must coincide with the direction of the precession motion of spins.

The oscillating magnetic field created by a single coil is intrinsically linearly polarized. It can be decomposed into two counter-rotating components, only one of these components being efficient for NMR. Therefore, the probe should preferentially produce a circularly polarized magnetic field. In this case, the RF power will be more efficiently used to create the B_1 field and hence the macroscopic magnetization. In this case, from the Principle of Reciprocity, the best receiving sensitivity is expected.[1]

A probe able to produce a rotating magnetic field may be constituted of two independent coils or by two uncoupled resonant modes producing orthogonal field components.

It requires that the probe has two feeding ports excited with sources having equal amplitudes and 90° phase difference. This is done using a "quadrature splitter" which splits the transmitter excitation power. The same circuit, behaving as a "quadrature combiner," adds the two 90°

---

[1] One can expect that driving the probe in a "quadrature" mode allows, ideally, a sensitivity gain by a factor of $\sqrt{2}$ and halves the transmitter RF power required to rotate the nuclear magnetization [Chen *et al.*, 1983; Glover *et al.*, 1985] compared to the single port drive case.

components of the electromagnetic force induced by the rotating magnetization into a unique signal fed to the preamplifier (Fig. 5.1).

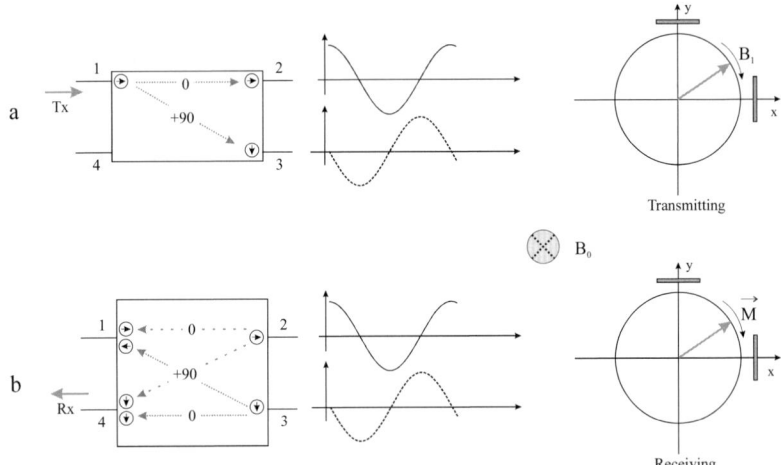

Fig. 5.1 Principles of driving the probe in a quadrature mode by a 90° hybrid device. (a) The transmitter is connected to port 1, port 4 being isolated from port 1. The pulse power is split in quadrature mode to ports 2 and 3, creating a rotating magnetic field. (b) The voltages induced by the rotating magnetization into coil x and coil y are out-of-phase by 90°. They are fed into ports 2 and 3 of the hybrid. Ports 2 and 3 are isolated from each other. The resulting signals that appear at ports 1 and 4 are phase shifted by the hybrid so that the two quadrature signals appearing in phase at port 4 are summed up, while they appear out-of-phase at port 1 and thus cancel out, assuring that no signal leaks toward the transmitter load.

In this chapter we will describe how to insert the quadrature "splitter/combiner" circuit between the probe and the transmitter and receiver channels of the console. Some circuits which are able to drive the probe in the quadrature mode will be described, making use of the well documented microwave technology [Pozar, 1998; Andrews, 2006].

5.1 Interfacing the Quadrature Probehead to the Console

Interfacing the quadrature probe to the console (Fig. 5.2) can be realized on any standard instruments which are most frequently wired for using a single coil probe. It is generally sufficient to disconnect the transmitter cable from the preamplifier box and reconnect it to the "input" port of a quadrature hybrid. The "isolated" port 4 of the hybrid is wired to the preamplifier input which normally includes the $\lambda/4$ line and the protecting crossed diodes. The Tx and Rx ports of the hybrid are intrinsically isolated. The crossed diodes in series with the power amplifier can be wired or not depending on the availability of the corresponding connections in the instrument. The outputs, named I and Q, are connected to the two ports of the quadrature probe.

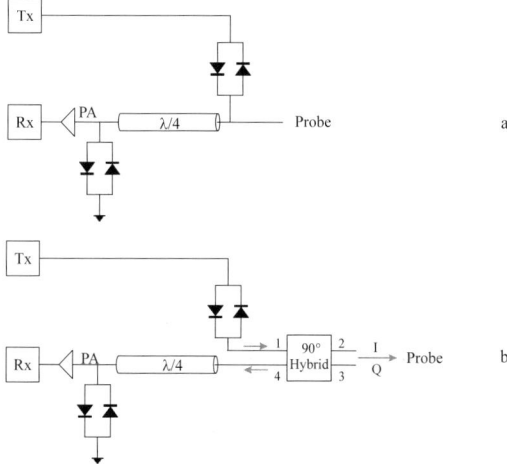

Fig. 5.2 (a) Standard setup for interfacing a single coil probe to the console (as already described in Chapter 4). (b) Connecting a quadrature probe only requires the insertion of a 90° hybrid as shown. Note that the labeling I and Q of the probe ports is arbitrary. If these ports are not assigned properly, the B_1 field will rotate in the opposite direction compared to the nuclear magnetization. The sense of rotation depends on the direction of the static magnetic field B_0, which is generally unknown, and on the sign of the nuclear gyromagnetic ratio. In practice, the assignment of the I and Q ports could be tested for the best NMR signal.

At this point it should be noted that the probe must be properly connected to the "quadrature splitter/combiner," in order to excite the spins and to get a signal.[2] It is possible to predict *a priori* the correct way to connect the I and Q ports to the appropriate ports of the probe. However, the direction of the static magnetic field B_0 is usually unknown. It can be determined in a variety of ways,[3] but in practice, the assignment of the hybrid ports can be done very simply by observing the signal, or the image, while exchanging the connections I and Q to the probe or exchanging the hybrid connections to the receiver (Rx) and the transmitter (Tx).

The quadrature "splitter/combiner" can be either a so-called 90° hybrid [Fig. 5.3(b)] or a 180° hybrid associated with a λ/4 transmission line [Fig. 5.3(c)].

These hybrids can be derived from a general four-port network as shown in Fig. 5.3(a). The hybrid junction, derived from this network, has specific properties leading to the following S-matrix:

$$S = \begin{bmatrix} 0 & S_{12} & S_{13} & 0 \\ S_{21} & 0 & 0 & S_{24} \\ S_{31} & 0 & 0 & S_{34} \\ 0 & S_{42} & S_{43} & 0 \end{bmatrix} \quad (5.1)$$

The four-port hybrid is assumed to exhibit a perfect match at all ports. It results that the diagonal elements of S are null. The circuit is assumed to be passive and reciprocal. It results that each non-diagonal element S_{ij} is equal to S_{ji}. Also, the non-diagonal elements S_{14}, S_{41}, S_{23} and S_{32} are

---

[2] Note that all nuclear spins do not rotate in the same direction. For example, nitrogen-15 spins rotate in the opposite sense to protons or carbon-13. Thus, in multinuclear experiments involving such nuclei, the probe connections should be done according to the sign of the nuclear gyromagnetic ratio. In practice, it will be easy to determine the rotation direction for protons; the correct connections for nitrogen-15 could be deduced accordingly.

[3] A compass or, preferably, a non-ferromagnetic loop driven by a DC current can help in determining the orientation of B_0.

equal to zero, meaning that ports 1 and 4 are isolated, as well as ports 2 and 3.

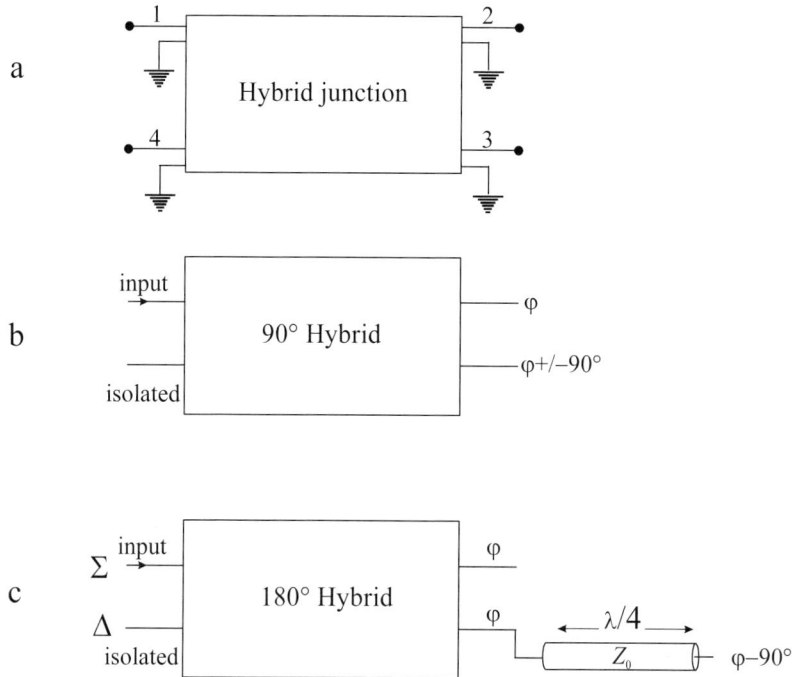

Fig. 5.3 Four-port hybrids. (a) The general hybrid junction with port numbering. (b) The 90° (or quadrature) hybrid splits the input signal into two signals of equal amplitude having a phase difference of ±90°, depending on its specific design. (c) The 180° hybrid splits the signal fed to the Σ input into two signals of equal amplitude and phase. The output signals are out-of-phase if the input signal is fed to the Δ port. In quadrature NMR, a λ/4 line is inserted in one of the outputs to introduce an additional phase shift of 90°. The corresponding device is similar to a 90° hybrid.

The role of the hybrid junction is to evenly split a signal applied to port 1 (or 4) between ports 2 and 3. If the network is lossless and reciprocal, the magnitude of the corresponding matrix elements fulfills:

$$|S_{12}|=|S_{13}|=|S_{42}|=|S_{43}|=\frac{1}{\sqrt{2}}. \quad (5.2)$$

It can be demonstrated that, under these conditions, there are only two possible designs leading to the so-called 90° hybrid and 180° hybrid, described, respectively, by the following *S*-matrices:

$$S_{90} = \frac{1}{\sqrt{2}} \begin{bmatrix} 0 & 1 & j & 0 \\ 1 & 0 & 0 & j \\ j & 0 & 0 & 1 \\ 0 & j & 1 & 0 \end{bmatrix} e^{j\varphi} \quad (5.3)$$

and

$$S_{180} = \frac{1}{\sqrt{2}} \begin{bmatrix} 0 & 1 & 1 & 0 \\ 1 & 0 & 0 & -1 \\ 1 & 0 & 0 & 1 \\ 0 & -1 & 1 & 0 \end{bmatrix} e^{j\varphi}. \quad (5.4)$$

The phase of the transferred signal is offset by a constant φ, depending on the particular design of the hybrid.

For the 90° hybrid, a signal fed to port 1 (or 4) is split into quadrature signals at port 2 and 3. Reciprocally, quadrature signals fed to ports 2 and 3 are summed up (in-phase) or subtracted (out-of-phase) at ports 1 or 4, depending on the relative phase of the input signals.

For the 180° hybrid, a signal fed to port 1 is split into in-phase signals at ports 2 and 3, while a signal fed to port 4 is split into out-of-phase signals at ports 2 and 3. Reciprocally, signals fed to ports 2 and 3 are summed up at port 1 (Σ port) and subtracted at port 4 (Δ port).

Due to the symmetry of the device, ports 1 and 4 can be exchanged with ports 2 and 3, respectively.

Some of the devices that can be used for connecting the probe in a quadrature mode are reviewed in the following paragraphs. The design providing the quadrature splitting, recombining, and transmitter/receiver isolation uses either broadband transformers[4] (well suited to low frequency operations) [Hoult *et al.*, 1984], or devices that are well-known from microwave technology, as branch lines or ring hybrids. At microwave frequencies, these devices are built using waveguides [Poole, 1967] but more frequently now they take the simple form of planar transmission lines built on a microstrip or on stripline technologies. The microstrip lines are formed by copper traces etched on one face of a dielectric sheet while the other face is covered by a conductive plane allowing the return current path. The dimensions of the line (width, thickness, and permittivity of the dielectric) determine its characteristic impedance. The striplines are similar except that the central conductor is embedded into the dielectric coated on both faces by a conductive plane (Chapter 2).

As the frequency domain used in NMR is approaching the GHz, microstrip technology becomes applicable. However, *in vivo* experiments are still mostly performed at lower frequencies where standard transmission lines, or their lumped element equivalents, may preferably be used.

Commercial hybrids are available in many packages and technologies, either magnetic or nonmagnetic. Their characteristics are frequently established on a large bandwidth covering an octave or more. This is usually obtained at the price of some ripple in the pass-band which could generally be acceptable. Anyway, the sensitivity improvement of a quadrature probe is still important compared to the single port drive experiment.

As the NMR experiment on a given nucleus is typically narrowband, simple, and very efficient, hybrids can be constructed in the lab using available components such as coaxial lines, inductors, and capacitors.

---

[4] These are constructed as Guanella or Ruthroff balun transformers [Guanella, 1944; Ruthroff, 1959; Sevick, 2001].

5.2 Quadrature Hybrids

These four port devices are used in a large number of high frequency circuits related to radio communications and measurement instruments (power splitters and combiners, mixers, reflection bridges, etc.).

5.2.1 *λ/4 transmission line hybrid*

The branch-line hybrid shown in Fig. 5.4 is a popular element for microwave engineering [Pozar, 1998]. It consists of four $\lambda/4$ transmission lines that assure the required 90° phase shifts.

The two lines that connect port 1 with port 2, and port 3 with port 4, have a characteristic impedance of $Z_0/\sqrt{2}$. This impedance is required to match each port to the characteristic impedance Z_0.

The two other lines connecting port 1 with port 4 and port 2 with port 3 have a characteristic impedance of Z_0.

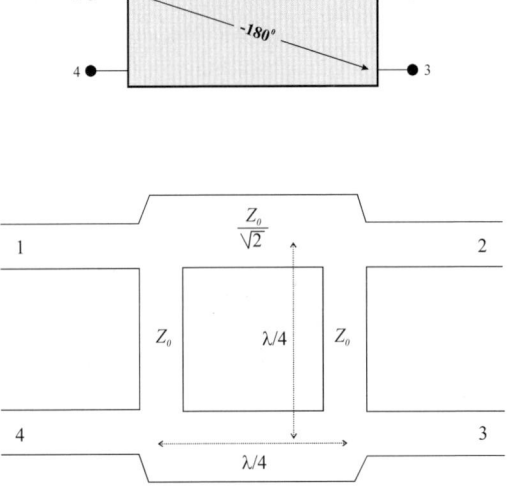

Fig. 5.4 A simple narrowband quad hybrid used in microwave circuits. Its central frequency is determined by the length ($\lambda/4$) of the striplines. Note that each port has a "return" path to the ground, not shown in the figure. When using stripline technology, the return path is the ground plane on which the circuit is build. Each port is driven through coaxial connectors.

The hybrid is a symmetrical four-port network that can conveniently be analyzed using the odd–even analysis method [Reed and Wheeler, 1956; Pozar, 1998]. The aim of this analysis is to superimpose two different modes of driving the circuit that results in the desired one. In the present case, one assumes that port 1 is driven by a voltage source of amplitude V and zero phase, while port 4 is not driven at all ($V_4 = 0$).

This driving condition is obtained by the sum of an "even" mode, in which port 1 and 4 are excited with sources of $+V/2$ volt each, and an "odd" mode, in which port 1 and 4 are driven by sources of $\pm V/2$ volts, respectively (Fig. 5.5).

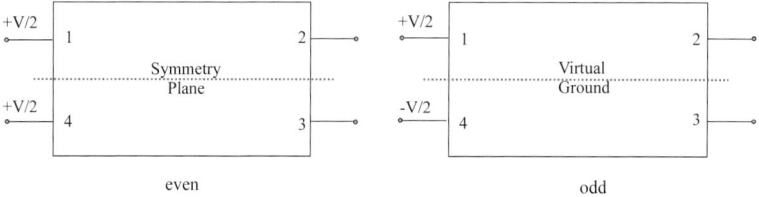

Fig. 5.5 The even–odd decomposition of a symmetrical four-port device. In the odd mode, the symmetry plane is a virtual ground (V = 0).

Now, the symmetry properties of the circuit allow one to define a symmetry plane that lies midway between the two $\lambda/4$ lines connecting port 1 with port 2 and port 4 with port 3, respectively. This plane separates the circuit into two identical halves constituted of a $\lambda/4$ line and two $\lambda/8$ stubs connected in parallel with each port (Fig. 5.6). The difference between the odd and even modes resides solely in the stub termination.

In the even mode [Fig. 5.6(a)], the $\lambda/8$ stubs are terminated by an open connection lying on the symmetry plane. From the transmission line equation,[5] Eq. (2.81), the $\lambda/8$ open-ended stub behaves as a capacitor of impedance $XC = -jZ_0$. In the odd mode [Fig. 5.6(b)], the stubs are terminated by a short circuit at the virtual ground. The $\lambda/8$ stub becomes an inductor of impedance $XL = jZ_0$.

---

[5] One assumes lossless lines.

The analysis of the half-hybrid circuit (a two-port network) is greatly simplified using the ABCD chain matrix formalism (Chapter 3).

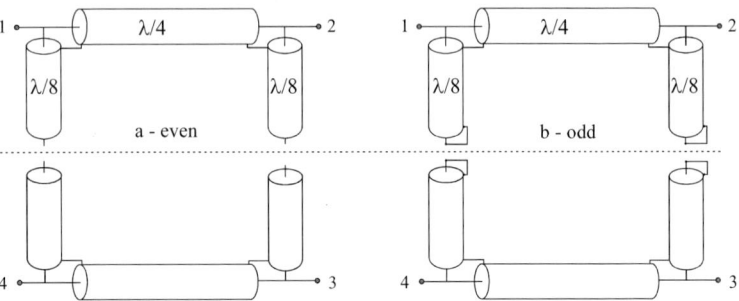

Fig. 5.6 Odd–even mode decomposition of the branch-line quadrature hybrid into two equivalent sub-circuits. The dashed line indicates the position of the symmetry plane. In the even mode, ports 1 and 4 are excited by in-phase incident waves. In the odd mode, ports 1 and 4 are excited by opposed-phase incident waves.

In this formalism, the output (voltage, current) vector is related to the corresponding input vector by a simple matrix relationship:[6]

$$\begin{pmatrix} V_{in} \\ I_{in} \end{pmatrix} = \mathbf{A} \begin{pmatrix} V_{out} \\ I_{out} \end{pmatrix}, \qquad (5.5)$$

where $\mathbf{A}$ is a 2 × 2 matrix describing the circuit. If the circuit can be decomposed into elementary quadrupoles of known chain matrix, then the whole matrix is simply obtained from the product of each elementary matrix $\mathbf{A}_i$, in order, from the input to the output:

$$\mathbf{A} = \mathbf{A}_1 \mathbf{A}_2 ... \mathbf{A}_n . \qquad (5.6)$$

Knowing the chain matrix, one can easily calculate the input (or output) impedance, when the circuit is terminated in a given impedance Z_L, and the output voltage as a function of the input vectors. These

---

[6] The "direct" or "reflected" equations that relate, respectively, the output to the input voltages or the reverse, can be encountered as well. The corresponding chain matrices are the inverse of each other.

quantities (input impedance and output voltage) are also completely described by the scattering matrix associated with the network and can be measured with the appropriate instruments (Chapter 10).

The chain matrix for the half-hybrid corresponding to the even mode is thus obtained from the product of the chain matrices describing, respectively, the open $\lambda/8$ stub at the input port, the $\lambda/4$ transmission line of impedance $Z_0/\sqrt{2}$ and the $\lambda/8$ stub in parallel with the output port. The chain matrix for the open $\lambda/8$ stub of the transmission line of impedance equal to Z_0 is given by:

$$\mathbf{A}_1 = \mathbf{A}_3 = \begin{pmatrix} 1 & 0 \\ Y & 1 \end{pmatrix} = \begin{pmatrix} 1 & 0 \\ j/Z_0 & 1 \end{pmatrix}, \tag{5.7}$$

where Y is the admittance of the section of line, connected in parallel to the quadrupole ports. The chain matrix for the $\lambda/4$ transmission line of characteristic impedance $Z_c = Z_0/\sqrt{2}$ is:

$$\mathbf{A}_2 = \begin{pmatrix} \cos\theta & jZ_c \sin\theta \\ j\sin\theta/Z_c & \cos\theta \end{pmatrix} = \begin{pmatrix} 0 & jZ_0/\sqrt{2} \\ j\sqrt{2}/Z_0 & 0 \end{pmatrix}, \tag{5.8}$$

Z_0 being the system characteristic impedance, usually 50 Ω.

The chain matrix for the even circuit is thus:

$$\mathbf{A}_{even} = \mathbf{A}_1 \mathbf{A}_2 \mathbf{A}_3 = \begin{pmatrix} -1/\sqrt{2} & jZ_0/\sqrt{2} \\ j/Z_0\sqrt{2} & -1/\sqrt{2} \end{pmatrix}. \tag{5.9}$$

Similarly, the chain matrix for the shorted $\lambda/8$ stub is:

$$\mathbf{A}_1 = \mathbf{A}_3 = \begin{pmatrix} 1 & 0 \\ Y & 1 \end{pmatrix} = \begin{pmatrix} 1 & 0 \\ -j/Z_0 & 1 \end{pmatrix}, \tag{5.10}$$

and the chain matrix describing the odd mode circuit is:

$$\mathbf{A}_{odd} = \begin{pmatrix} 1/\sqrt{2} & jZ_0/\sqrt{2} \\ j/Z_0\sqrt{2} & 1/\sqrt{2} \end{pmatrix}. \tag{5.11}$$

Assuming the circuit terminated in a matched load (50 Ω), one may calculate the input impedance and the output voltage from the **A** matrix.

The input impedance, assuming that the output termination is $V_2/I_2 = Z_0$, is given by:

$$Z_{in} = \frac{V_1}{I_1} = \frac{a_{11}Z_0 + a_{12}}{a_{21}Z_0 + a_{22}}, \qquad (5.12)$$

where a_{ij} are the elements of matrix **A** given by Eqs. (5.9) and (5.11).

The input impedance of the half-hybrid, for the even mode is:

$$Z^{in}_{even} = \frac{-Z_0/\sqrt{2} + jZ_0/\sqrt{2}}{j/\sqrt{2} - 1/\sqrt{2}} = Z_0, \qquad (5.13)$$

and for the odd mode:

$$Z^{in}_{odd} = \frac{Z_0/\sqrt{2} + jZ_0/\sqrt{2}}{j/\sqrt{2} + 1/\sqrt{2}} = Z_0. \qquad (5.14)$$

Hence, the input impedance is matched to Z_0 for both the even and odd half circuit, resulting from the impedance characteristic chosen for the $\lambda/4$ line.

From the superposition principle, the reflection coefficient at ports 1 and 4 of the whole circuit is, respectively, the sum and difference of the reflection coefficients for the even and odd modes. Because the reflection coefficient is 0 for both modes, the impedance is obviously matched for the whole circuit as well.

The output voltage (V_2) of the half-hybrids can also be obtained from the chain matrix coefficients. Assuming that the input voltage is V_1 (= $V/2$) and that the load impedance is Z_0, the output voltage is (Fig. 5.7):

$$V^{out}_{even} = \frac{V_1}{a_{11} + a_{12}/Z_0} = -V\frac{(1+j)\sqrt{2}}{4} \qquad (5.15)$$

for the even-mode, and:

$$V_{odd}^{out} = \frac{V_1}{a_{11} + a_{12}/Z_0} = \pm V \frac{(1-j)\sqrt{2}}{4} \quad (5.16)$$

for the odd-mode. The +/− sign corresponds to the output voltage at port 2 ($V/2$ at input port 1) and port 3 ($-V/2$ at input port 4), respectively.

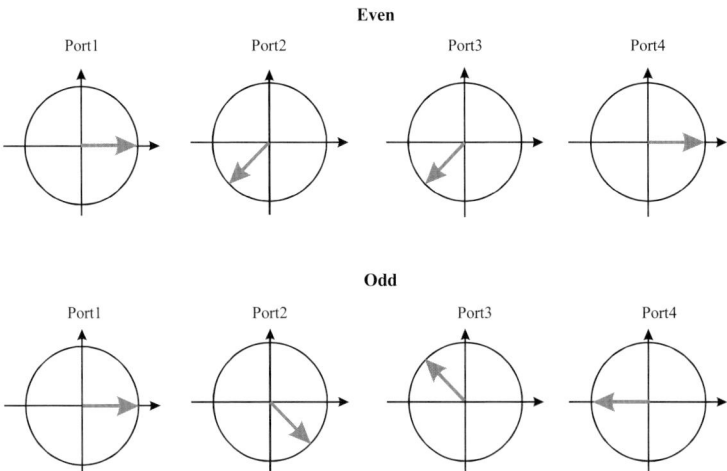

Fig. 5.7 Complex plane representation of the voltages at each port of the hybrid, for the even mode (upper) and the odd mode (lower). Ports 1 and 4 are excited by in-phase (respectively out-of-phase) incident waves in the even (respectively odd) mode. Ports 2 and 3 are the output ports.

The voltage at the output ports of the whole circuit is given by the superposition of the even and odd mode voltages calculated above. Thus, assuming $+V$ is applied at port 1, the output voltages are:

$$V_2 = -j \frac{V\sqrt{2}}{2} \quad (5.17)$$

and

$$V_3 = -\frac{V\sqrt{2}}{2}. \quad (5.18)$$

The voltages have the same amplitudes and a 90° phase difference. Port 3 is shifted by 180° respective to the input, while the voltage at port 2 lags by 90° (Fig. 5.8).

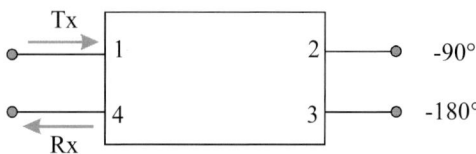

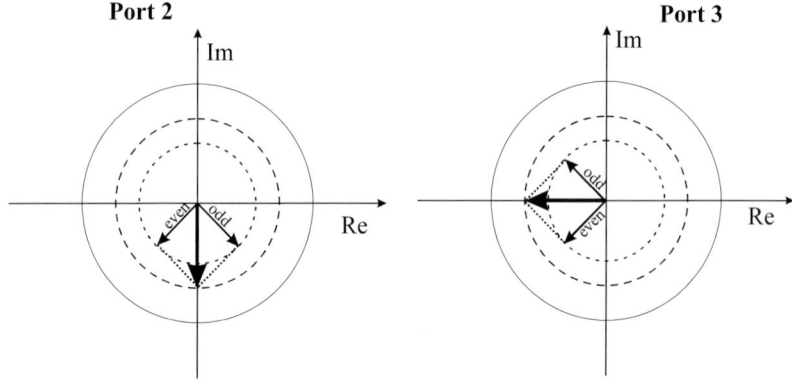

Fig. 5.8 Superposition of the odd–even modes. The voltages at output ports 2 and 3 are obtained, assuming that port 1 is driven by a 1 volt incident wave (transmitting period).

During the transmitting period, the output voltage amplitudes are reduced by a factor of √2 compared to the input. This is an obvious result because half the input power is dissipated evenly in each load, corresponding to a −3 dB loss. The effective rotating magnetic field created by the probe is higher, by the same factor of √2, compared with the effective field that would have been created by the same power source in a single port driving configuration. Indeed, in the latter case the effective field amplitude would be proportional to $V/2$.

During the receiving period, the nuclear magnetization induces two voltages shifted in phase by 90°. These two voltages are fed to ports 2 and 3 and combined to give a nonzero voltage at the Rx port (port 4, Fig. 5.8) and zero voltage at the Tx port (port 1, Fig. 5.8).

It is noticeable that a rotating magnetic field is created after an excitation to port 1 while the voltages created by a magnetization rotating in the same direction add to port 4.

The quadrature hybrid described so far is very easy to build for microwave frequencies. The transmission lines of any characteristic impedance can be realized by microstrip technology. At NMR radio frequencies, the required line dimensions may become more difficult to handle. Although the hybrid can be realized using coaxial transmission lines, the nonstandard characteristic impedance of $50/\sqrt{2}$ Ω for the $\lambda/4$ may also cause a problem. This can be solved using a lumped element equivalent network, as shown later in Section 5.2.3.

5.2.2 *λ/8 transmission line hybrid*

This hybrid [Chen and Hoult, 1989; Fitzsimmons *et al.*, 1993] is similar to the branch-line hybrid but is less bulky. It is therefore more appropriate at low frequencies, around 200 MHz.

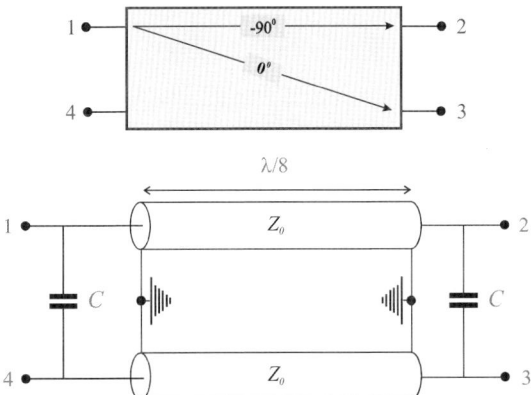

Fig. 5.9 A 90° hybrid built on transmission lines with standard characteristic impedance $Z_0 = 50$ Ω.

310   NMR Probeheads for Biophysical and Biomedical Experiments

Two of the $\lambda/4$ lines of the branch-line hybrid are replaced by a shorter section of a line having standard characteristic impedance. The two other lines are replaced by capacitors (Fig. 5.9).

An understanding of the circuit behavior can be achieved by the odd–even analysis method as outlined above. The circuit is split into two identical parts incorporating a capacitor $2C$ at each port [Fig. 5.10(a)]. Each capacitor is connected to the (virtual) ground symmetry plane in the odd mode and is kept unconnected in the even mode.

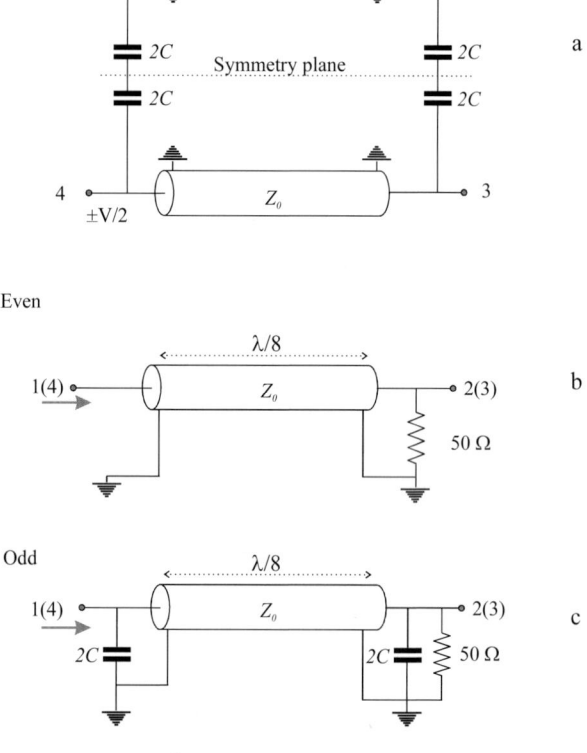

Fig. 5.10 (a) Odd–even decomposition of the hybrid. In the even mode, (b), the sub-circuit is simply a section of the $\lambda/8$ transmission line. In the odd mode, (c), the same line is terminated by a capacitance $2C$ at each end.

In the even mode, the network is simply a line terminated by its characteristic impedance [Fig. 5.10(b)]. The input impedance is equal to Z_0 (50 Ω). The voltage appearing at the output port has the same amplitude as the input voltage ($V/2$) but lags by a $-45°$ phase shift due to the propagation in the $\lambda/8$ line.

In the odd mode, the network is a Π circuit formed by two capacitors (of capacitance $C' = 2C$) and the $\lambda/8$ transmission line of characteristic impedance Z_0. The chain matrix of the circuit is:

$$\mathbf{A}_{odd} = \begin{pmatrix} 1 & 0 \\ jC'\omega & 1 \end{pmatrix} \frac{\sqrt{2}}{2} \begin{pmatrix} 1 & jZ_0 \\ j/Z_0 & 1 \end{pmatrix} \begin{pmatrix} 1 & 0 \\ jC'\omega & 1 \end{pmatrix}$$
$$= \sqrt{2}/2 \begin{pmatrix} 1 - Z_0 C'\omega & jZ_0 \\ jC'\omega(2 - Z_0 C'\omega) + j/Z_0 & 1 - Z_0 C'\omega \end{pmatrix} \quad (5.19)$$

When the circuit is terminated by 50 Ω, the input impedance of the odd mode circuit, is given by:

$$Z_{in} = Z_0 \frac{1 - Z_0 C'\omega + j}{jZ_0 C'\omega(2 - Z_0 C\omega) + j + 1 - Z_0 C'\omega}. \quad (5.20)$$

Z_{in} is equal to Z_0 (50 Ω) when the impedance of C' is $Z_0/2$ (25 Ω). Thus, the capacitance value for the complete hybrid must be:

$$XC = \frac{1}{C\omega} = 50\Omega. \quad (5.21)$$

The output voltage for the odd mode is:

$$V_{odd}^{out} = \frac{V_1}{a_{11} + a_{12}/Z_0} = \mp V \frac{(1+j)\sqrt{2}}{4}. \quad (5.22)$$

The amplitude of the output voltage is again equal to that of the input voltage ($V/2$). The $-$ sign corresponds to the output voltage at port 2 when $+V/2$ is applied at port 1. In this case, the output voltage lags by $-135°$. The $+$ sign corresponds to the output voltage at port 3 when $-V/2$ is applied at port 4. In this case, the output voltage lags by $+45°$.

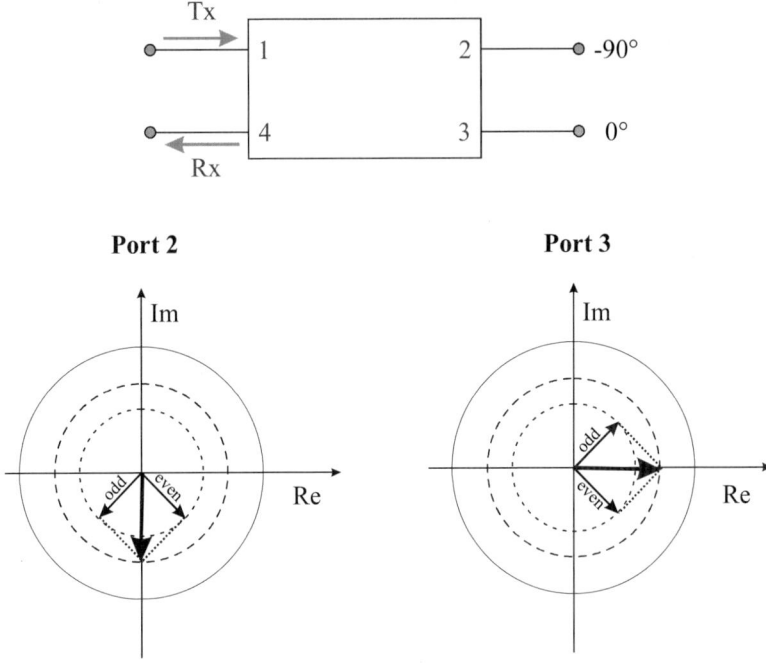

Fig. 5.11 Superposition of the odd–even modes for the $\lambda/8$ transmission line hybrid. The voltages at output ports 2 and 3 are obtained, assuming that port 1 is driven by a 1 volt incident wave.

Addition of the even and odd modes leads again to equal output voltage amplitude ($\sqrt{2}/2$) at ports 2 and 3. The phase difference is also equal to 90° as with the $\lambda/4$ branch-line hybrid. But, in this case, port 3 is leading compared to port 2. The results of this analysis are summarized in Fig. 5.11.

The adjustment of the hybrid circuit is not very critical, especially concerning the length of lines, provided the symmetry is carefully respected. The hybrid is narrowband but its bandwidth is much greater than the frequency range required by an NMR experiment (Section 5.2.4).

An example of layout is shown in Fig. 5.12, designed for 200 MHz. The lines should be carefully paired. The required length ($\lambda/8$) can be precisely determined by measuring the resonant frequencies for a long piece of a sample of line (Chapter 10).

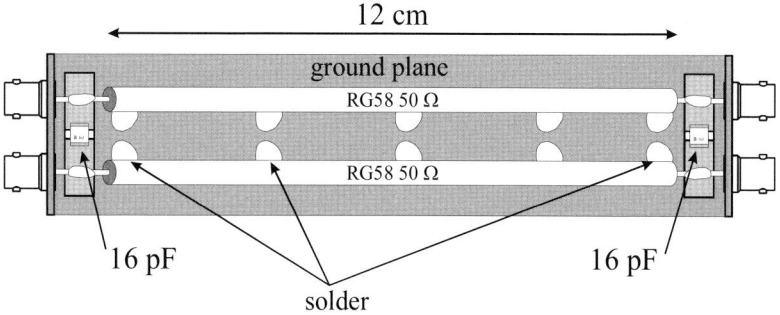

Fig. 5.12 Layout of a 90° hybrid designed for 200 MHz. The standard coaxial cable used here has a velocity factor of 0.66. The electric length of $\lambda/8$ thus corresponds to 12.375 cm. The slightly smaller length takes into account the phase shift introduced by the connections (that should be kept as short as possible). The capacitors are mounted on small sheets of printed circuit board glued on the main board that constitutes the ground plane. Fine adjustments (impedance matching and balancing) can be done using adjustable capacitors as a replacement for the 16 pF.

A small error in the line length can be compensated for by using variable capacitors to adjust the hybrid at the desired frequency, the imbalance being the most sensible characteristic.

Simulations of the circuit using perfectly cut lossless transmission lines indicate that it can possibly be adjusted over a wide frequency band (100 MHz or more!) without significant degradation of the isolation [Fig. 5.13(a)]. The impedance matching [Fig. 5.13(b)] also remains very good and essentially depends on the output load impedance (not shown). The phase difference between the output ports is also kept constant and equal to 90°, as required [Fig. 5.13(c)]. However, the imbalance in the output amplitude voltages is dependent on the capacitance values [Fig. 5.13(d)]. The imbalance is acceptable (<5%) over a 10 MHz range around the frequency for which the device has been designed. This indicates that the lines should be cut with a precision of about ±5%.

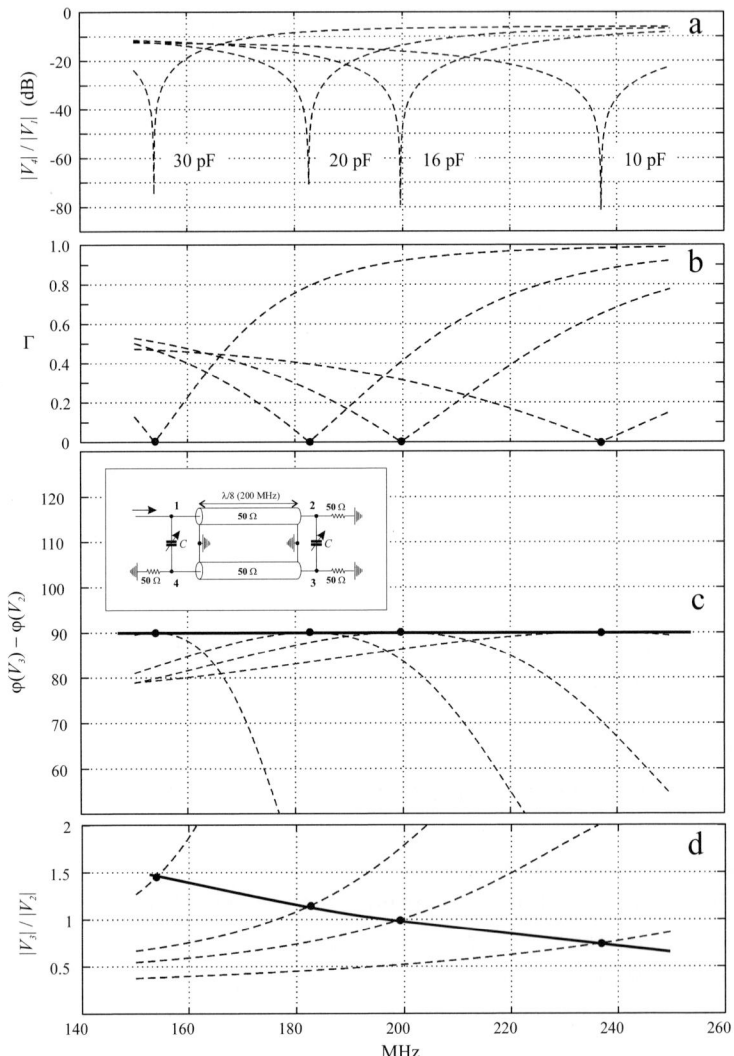

Fig. 5.13 Performance of the $\lambda/8$ quad hybrid as a function of frequency for different capacitance C. (a) Isolation between ports 1 and 4. (b) Reflection coefficient at port 1. (c) Phase difference between ports 2 and 3. (d) Voltage ratio at ports 2 and 3. Note that the balance between ports 2 and 3 is only perfect at the frequency for which the lines have been cut.

A small difference between the electric length of the lines has no dramatic influence on the hybrid performance, except for the relative phase outputs. As a result, assuming a difference in length of ±0.025 λ (±20%) the output voltage amplitude imbalance follows a dependence on the capacitance values as for the case of perfectly equilibrated lines. The isolation does not depend on the capacitors and remains close to –50 dB. The reflection coefficient is a little bit worse (about –25 dB) over the whole capacitive range indicating a slight mismatch. The phase shift, still independent of the capacitance values, is either lower or higher than the optimum 90° by as much as 10°. A 1° error in the phase shift would therefore require a good mechanical design and an accurate line cutting. The sign of the phase error provides an indication of the line length discrepancies (which line is longer).

Finally, the difference in the two capacitance values essentially introduces an important degradation of the isolation properties at port 4, as well as an important impedance mismatch. Two variable capacitors should be used in this case. They can easily be adjusted by looking at the reflection coefficient and/or at the isolation on port 4. All in all, building the device is easy.

The length of lines is not very critical, but they should be carefully sized to be as equal as possible since the phase difference depends on it. Two variable capacitors allow the device to be tuned and matched precisely at the desired frequency. The cable braid should be soldered along the ground plane constituted by the walls of the box enclosing the hybrid (Fig. 5.12). Provided the mechanical parts are well constructed (in particular taking care with the symmetry of the device), an acceptable hybrid characterized by small phase errors, a small leakage through the "isolated" port, and a small impedance mismatch would be very easy to get.

5.2.3 *Lumped element quadrature hybrids*

The two hybrid circuits described above can be replaced by their lumped element equivalents, facilitating the construction in some circumstances, especially if there is no room for using the transmission line technique.

5.2.3.1 Quarter-wave hybrid equivalent

The *LC* circuit of Fig. 5.14 is the Π-network equivalent of a transmission line (Chapter 2).

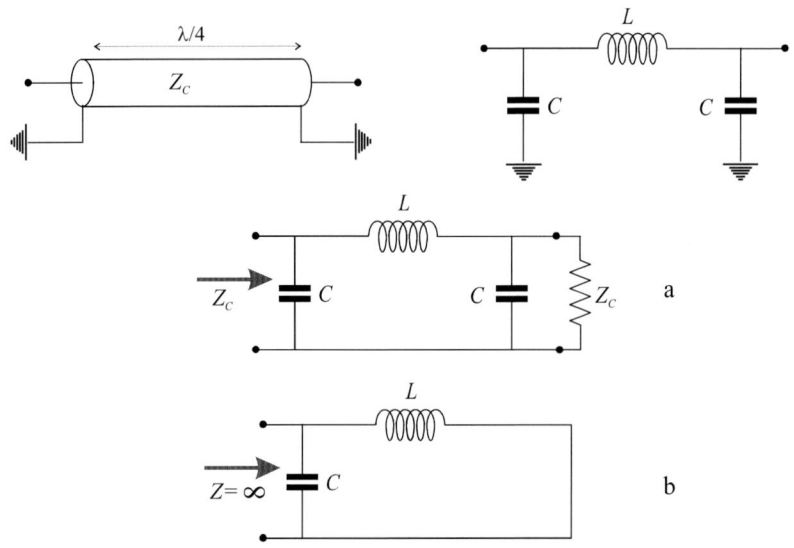

Fig. 5.14 Π filter (upper left) equivalent of a transmission line. The circuit replacing a quarter-wave transmission line should realize the following conditions: (a) the filter should present the same impedance Z_c at its input port as the load; (b) the filter input impedance should be infinite when terminated by a short circuit.

From Eqs. (3.17) and (3.18), the component values of the Π-filter modeling a quarter-wave transmission line ($\theta = \pi/2$) of characteristic impedance Z_c are given by:

$$L\omega = \frac{1}{C\omega} = Z_c. \qquad (5.23)$$

It can be easily verified that the characteristics of a λ/4 transmission line [Figs. 5.14(a) and 5.14(b)] are fulfilled.

The microwave quadrature hybrid of Fig. 5.4 can then be replaced by the lumped element hybrid shown in Fig 5.15.

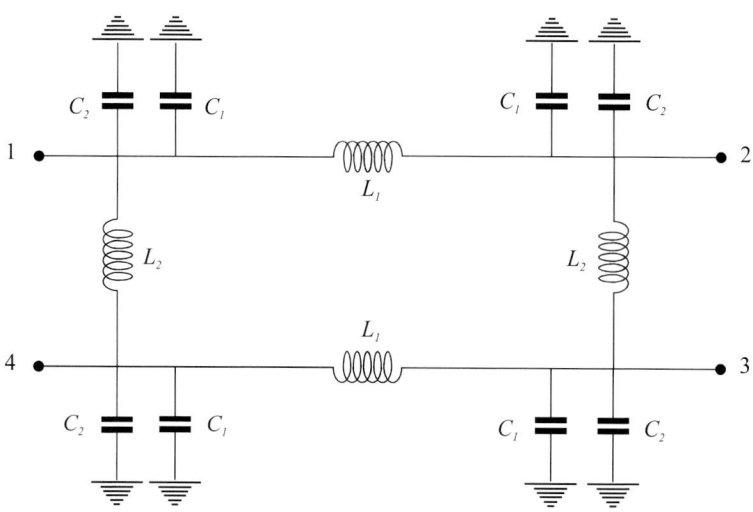

Fig. 5.15 The complete lumped element equivalent of the branch-line quad hybrid built on quarter-wave lines.

The component values, for a load impedance Z_0 and operating frequency ω_S, are given by:

$$L_1 = \frac{Z_0}{\omega_S \sqrt{2}}, \qquad (5.24)$$

$$L_2 = \frac{Z_0}{\omega_S}. \qquad (5.25)$$

The value for C is given by the sum of the capacitances C_1 and C_2 of the lumped elements that are in parallel (Fig. 5.15).

C_1 is the capacitance of the equivalent circuit for the $\lambda/4$ transmission lines of characteristic impedance $Z_0/\sqrt{2}$:

$$C_1 = \frac{\sqrt{2}}{Z_0 \omega_S}. \tag{5.26}$$

C_2 is the capacitance of the equivalent circuit for the Z_0 $\lambda/4$ lines:

$$C_2 = 1/Z_0 \omega_S. \tag{5.27}$$

The equivalent capacitance C is:

$$C = C_1 + C_2 = \frac{1+\sqrt{2}}{Z_0 \omega_S}. \tag{5.28}$$

An evaluation of the hybrid performances has been made for a device designed for a central frequency of 200 MHz and an impedance of 50 Ω. Simulation of the output voltages (ports 2 and 3) was done at 200 MHz, assuming a random variation of the component values in a 5% range (±2.5%) around their nominal values. This variation corresponds to fairly good components. The evaluation was done on 500 simulated circuits.

Figure 5.16 shows the complex plane representation of the two output voltages (port 2 and 3) when 1 V is applied at the input port 1. The phase difference and the voltage amplitude are around 90° and 0.707 V, respectively, as already expected. The deviations from the optimum value, due to component defaults, appear to be of the same order as the error in the component values.

Assuming the same spreading in component values, the reflection coefficient is constantly of the order of −30 dB, indicating a slight mismatch. The isolation between ports 2 and 3 is of the order of 35 dB, representing an acceptable figure.

It should be noted that these results are obtained on totally uncorrelated component values. They can thus be considered as the worst of cases. Provided the components can be carefully paired in order to keep a good symmetry of the device, the practical results could be much better.

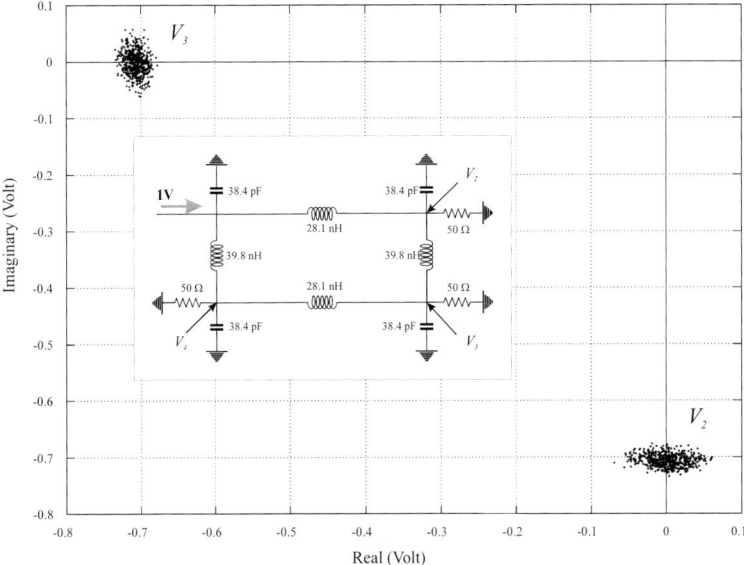

Fig. 5.16 Complex plane representation of the output voltage values, assuming an input of 1 V and 500 randomly chosen component values within a 5% interval. The indicated nominal values are calculated for a 200 MHz quad hybrid.

5.2.3.2 $\lambda/8$ hybrid equivalent

Similarly, the $\lambda/8$ hybrid circuit can be replaced by the lumped element network represented in Fig. 5.17. The capacitance C_1 is equal to that of the original circuit [Fig. 5.9 and Eq. (5.21)], i.e.,

$$C_1 = \frac{1}{Z_0 \omega_S}. \tag{5.29}$$

The other components, namely the inductance L and the capacitance C_2, are given by the Π-network equivalent of the $\lambda/8$ transmission line. From Eqs. (3.17) and (3.18), the component values of the Π-filter modeling this line ($\theta = \pi/4$) of characteristic impedance Z_c are given by:

$$L = \frac{Z_0 \sqrt{2}}{2\omega_S}, \qquad (5.30)$$

$$C_2 = \frac{\sqrt{2}-1}{Z_0 \omega_S}. \qquad (5.31)$$

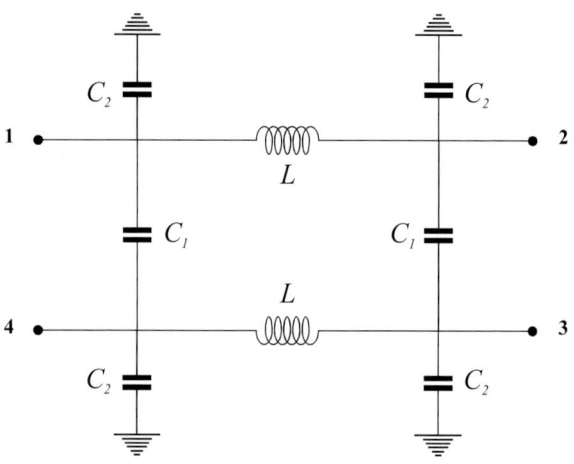

Fig. 5.17 Lumped element circuit equivalent to the $\lambda/8$ quad hybrid.

Obviously, the same precautions with respect to the symmetry of the design as discussed in the previous paragraph should be taken here too. Adjusting the components may, however, be easier, and the lumped element approach appears to be a convenient way to build compact hybrid devices. However, attention should be paid to avoid any magnetic coupling between the inductors that would perturb the functioning of the hybrid.

5.2.4 *Frequency response of the quad hybrids*

A $\lambda/4$ (Fig. 5.18) and a $\lambda/8$ (Fig. 5.19) quad hybrid, built on transmission lines and on lumped elements, have been simulated in order to compare their performances on a frequency scale, assuming a design for a central frequency of 200 MHz.

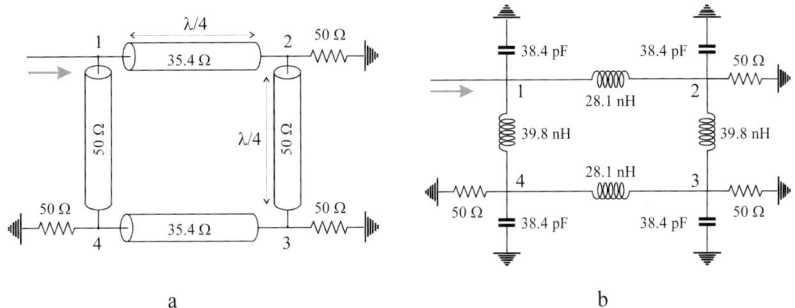

Fig. 5.18 (a) Transmission line and (b) lumped element versions of a quarter-wave quad hybrid designed for a central frequency of 200 MHz.

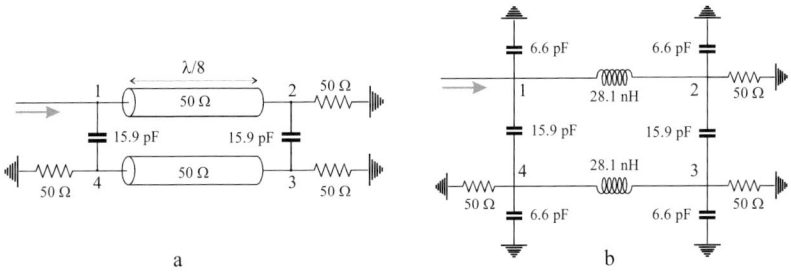

Fig. 5.19 (a) Transmission line and (b) lumped element versions of a $\lambda/8$ quad hybrid designed for a central frequency of 200 MHz.

The isolation properties (V_4), the input impedance (the reflection coefficient), and the phase difference and amplitude imbalance between the two output ports are presented in Figs. 5.20 to 5.23 as a function of frequency.

All hybrids behave similarly with frequency. Because the simulated circuits assume perfect components, the isolation is infinite at 200 MHz. Figure 5.20 represents only the range from –10 dB to –60 dB, the latter limit being considered to be the best in the real world. Interestingly, as long as the hybrid is well constructed, the isolation can be quite good (between –35 to –40 dB) over a large frequency range (5 MHz, 2.5%) compared to the typical bandwidth of NMR experiments.

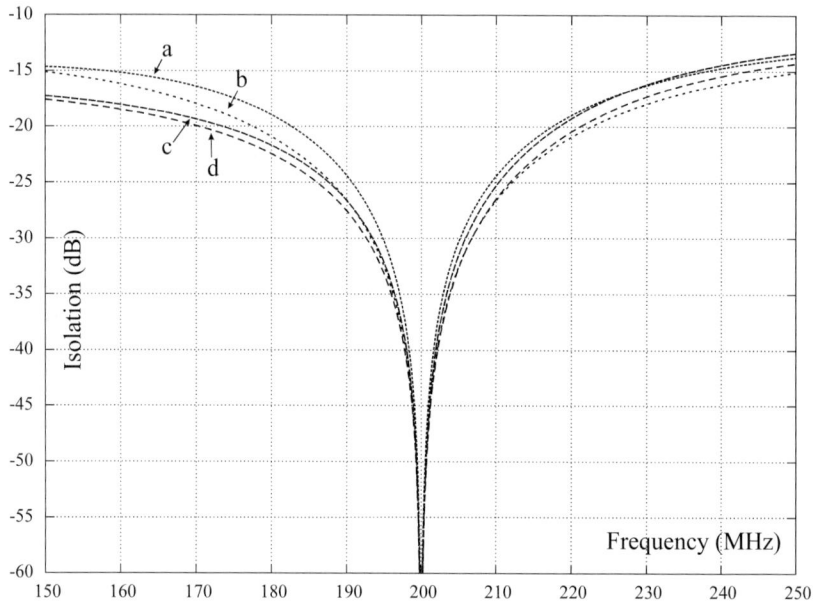

Fig. 5.20 Isolation properties of quad hybrids designed for a 200 MHz central frequency. (a) and (b) refer to the $\lambda/4$ hybrid (the lumped element and transmission line versions, respectively). (c) and (d) refer to the $\lambda/8$ hybrid (the lumped element and transmission line versions, respectively).

The relative broadband property of these devices is also demonstrated when examining the reflection coefficient Γ that represents the deviation of the input impedance from the characteristic 50 Ω (Fig. 5.21). The narrow response around 200 MHz [Fig. 5.21(e)] represents the reflection coefficient curve that would be obtained with a moderate Q factor (100), NMR probe matched at 200 MHz. The response from the quad hybrid is much larger, indicating that the probe circuit will dominate the impedance mismatching rather than the hybrid.

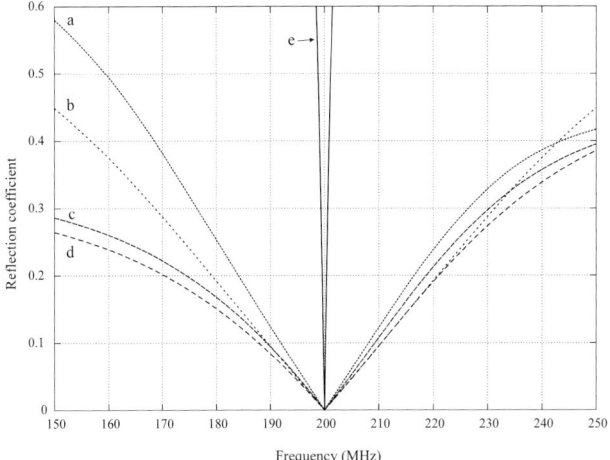

Fig. 5.21 Input (port 1) reflection coefficient [defined by Eq. (5.47)] of the quad hybrids designed for a 200 MHz central frequency. (a) to (d) are the same as in Fig. 5.20. For comparison purposes, the reflection coefficient for a $Q = 100$ resonant circuit is given in (e).

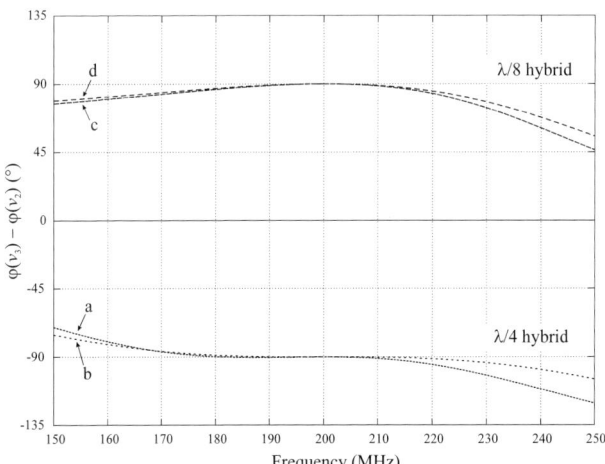

Fig. 5.22 Phase difference between the two output ports (2, 3) for the four hybrids (see legend of Fig. 5.20 for details). Note the different phase shifts expected for the quarter-wave (a, b) and the $\lambda/8$ (c, d) hybrids.

Another parameter of interest is the phase difference between the two output ports (V_2 and V_3, Fig. 5.22). Note, as already described, that the quarter-wave and the $\lambda/8$ hybrid have different phase properties. The output V_2 is in advance by 90° from V_3 in the case of a $\lambda/4$ hybrid, while it is the reverse for a $\lambda/8$ hybrid. The phase variation is relatively smooth near the central frequency for both hybrid types. Finally, balance of the voltage amplitude value of the two output ports is shown in Fig. 5.23.

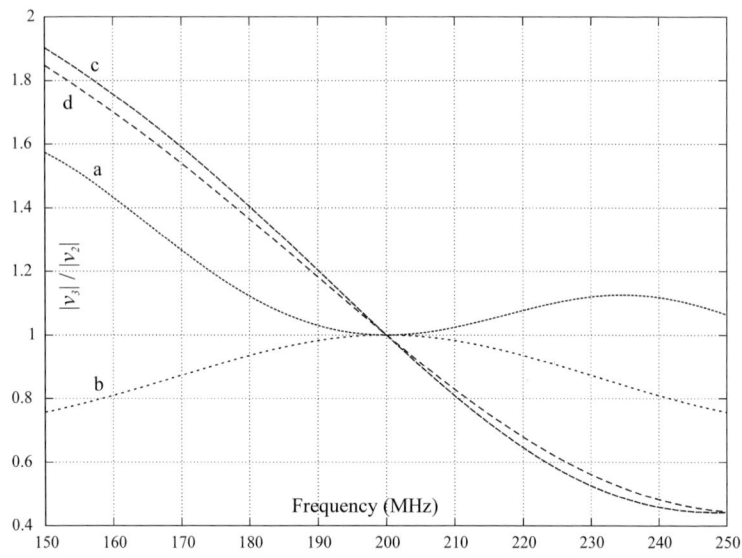

Fig. 5.23 Balancing properties of the output ports (2, 3) of the quad hybrids designed for 200 MHz. (a) to (d) are the same as in Fig. 5.20.

Clearly, the branch-line quarter-wave hybrid appears to be better than the $\lambda/8$ hybrid, from this point of view. The voltage ratio varies smoothly around the central frequency while it is much more sensitive with the $\lambda/8$ variant. Considering again that the NMR experiment works on a very narrow frequency band, the imbalance observed with the $\lambda/8$ hybrid may be acceptable for this application.

5.3 180° Hybrid

The ring hybrid, or rat-race, is a 180° hybrid that can also be used for quadrature mode NMR. At microwaves it is built using waveguides[7] or using the microstrip or stripline technologies (Fig. 5.24).

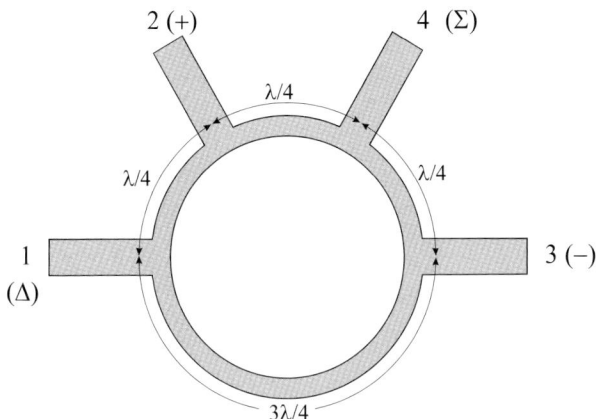

Fig. 5.24 The rat-race or ring hybrid. The picture represents the copper sheets, located on one face of a double sided printed board. It constitutes the main current path. The return path is made out by the uniform copper sheet on the other side of the dielectric material.

At most NMR frequencies, coaxial transmission lines will preferably be used. At the lower frequency range, or if no sufficient place is available, a lumped element equivalent circuit may be used instead.

The circuit is basically a 180° hybrid that can function as a splitter or as a combiner. It has four ports, a "sum" port, a "difference" port, and two output ports (Fig. 5.24). As a splitter, there is a 0° or 180° phase difference between the output ports depending on the excited input port. If the input is applied to port 4 ("sum" port), the signal is evenly split on ports 2 and 3 with no phase difference, and port 1 is isolated ($V = 0$). If the input is applied to port 1 ("difference" port), the output signals at ports 2 and 3 have a 180° phase difference and port 4 is isolated. Reciprocally, when two signals are applied to ports 2 and 3 (the hybrid

---

[7] Known as the magic-T [see, for example, Poole, 1967].

functions as a combiner), the sum and difference of inputs are formed at ports 4 and 1, respectively, hence the name for these ports.

When used in conjunction with an NMR probe operating in the quadrature mode, one of the output signals of the 180° hybrid must be shifted in order to get a net 90° phase shift difference. This is done simply by adding a λ/4 transmission line in series with one of the output ports, as shown in Fig. 5.25.

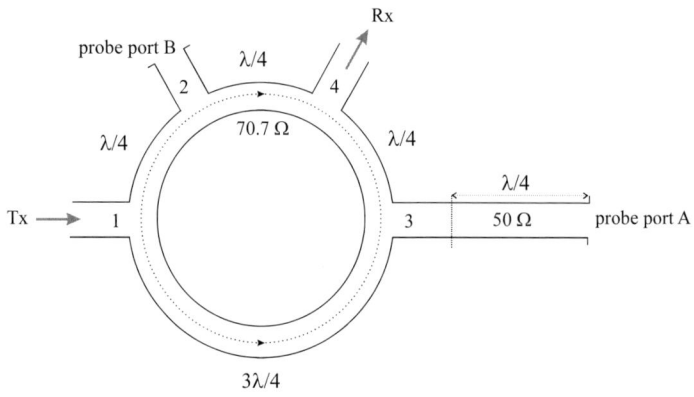

Fig. 5.25 Adding a 90° phase shifter in one of the output ports transforms the 180° hybrid into a quad hybrid usable for quadrature mode NMR.

5.3.1 *The 180° rat-race hybrid*

The functioning of the 180° hybrid circuit can be understood as follows [Chen and Hoult, 1989]. Assuming that a current enters the ring at the input port 1 (transmitter), it will split and travel in the ring, both clockwise and counterclockwise (Fig. 5.25). When reaching port 3, the two currents have travelled the same length of 3λ/4 and hence add together. The total signal at this point lags the input current by 270° (3λ/4). Similarly, the ring currents add together at port 2 with the same phase (one has travelled by λ/4 and the other by 5λ/4). At this point the output signal has a phase difference of 90° respective to the input. It results that the two output ports 2 and 3 differ in phase by 180°. Finally, if the currents arrive opposed in phase at port 4 (one has travelled by λ/2 and the other one by λ) it results in a cancellation of the output signal (isolation).

A complete analysis of this device can be done again using the odd-even analysis. The symmetry plane now lies between ports 1, 2 and 3, 4 allowing the circuit to decompose into two identical ones (Fig. 5.26).

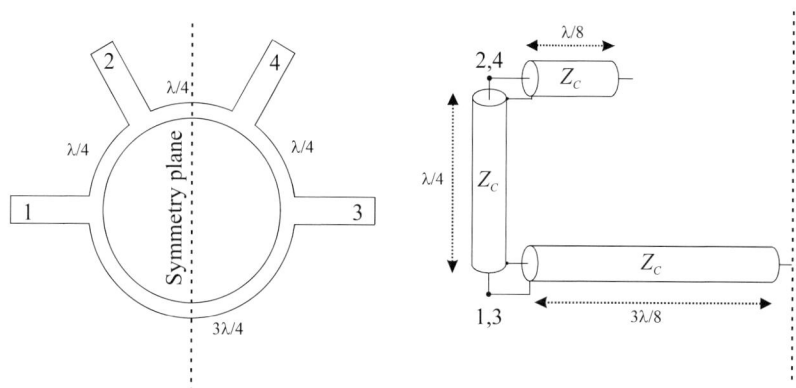

Fig. 5.26 Even–odd mode decomposition of the ring hybrid into two sub-circuits made of transmission lines with a characteristic impedance Z_c. The two $\lambda/8$ and $3\lambda/8$ lines are terminated either by a short (odd mode) or open (even mode) connection.

The stubs of each circuit are either open or short circuited, depending on the mode considered (even or odd). But, in contrast to the previous case, the sub-circuits are not symmetric respective to their corresponding input–output ports. Hence, for a complete description of the circuit properties we consider two cases separately. In one case, a unit amplitude incident wave is applied to port 1 (the difference port), where a $3\lambda/8$ stub is connected in parallel to the input port. In the second case, it is applied to port 4 (the sum port) where a $\lambda/8$ stub is connected.

Firstly considering the case where the incident wave is applied to port 1 [Fig. 5.27(a)], the ABCD chain matrices for the circuits corresponding to the odd and even modes are given by:

$$\mathbf{A}_{even} = \mathbf{A}_{open}^{3\lambda/8} \mathbf{A}^{\lambda/4} \mathbf{A}_{open}^{\lambda/8}, \qquad (5.32)$$

$$\mathbf{A}_{odd} = \mathbf{A}_{shorted}^{3\lambda/8} \mathbf{A}^{\lambda/4} \mathbf{A}_{shorted}^{\lambda/8}. \qquad (5.33)$$

The chain matrices corresponding to the $\lambda/8$ and $3\lambda/8$ stubs are obtained from Eq. (5.8) with $\theta = \pi/4$ and $3\pi/4$, respectively. Hence:

$$\mathbf{A}_{open}^{\lambda/8} = \mathbf{A}_{shorted}^{3\lambda/8} = \begin{pmatrix} 1 & 0 \\ j/Z_c & 1 \end{pmatrix}, \qquad (5.34)$$

$$\mathbf{A}_{open}^{3\lambda/8} = \mathbf{A}_{shorted}^{\lambda/8} = \begin{pmatrix} 1 & 0 \\ -j/Z_c & 1 \end{pmatrix}, \qquad (5.35)$$

where Z_c is the characteristic impedance of the lines.

Similarly, the chain matrix of the $\lambda/4$ transmission is:

$$\mathbf{A}^{\lambda/4} = \begin{pmatrix} 0 & jZ_c \\ j/Z_c & 0 \end{pmatrix}. \qquad (5.36)$$

Hence, the chain matrices for the even and odd mode are:

$$\mathbf{A}_{even} = \begin{pmatrix} -1 & jZ_c \\ 2j/Z_c & 1 \end{pmatrix} \qquad (5.37)$$

and

$$\mathbf{A}_{odd} = \begin{pmatrix} 1 & jZ_c \\ 2j/Z_c & -1 \end{pmatrix}. \qquad (5.38)$$

Given the above chain matrices and Eq. (5.12), the input impedance for the even and odd mode half circuits, when terminated by the impedance Z_L, are:

$$Z_{even}^{in} = Z_L k \frac{-1 + jk}{k + 2j} \qquad (5.39)$$

and

Quadrature Driving 329

$$Z_{odd}^{in} = Z_L k \frac{1+jk}{-k+2j}, \tag{5.40}$$

where $k = Z_c/Z_L$.

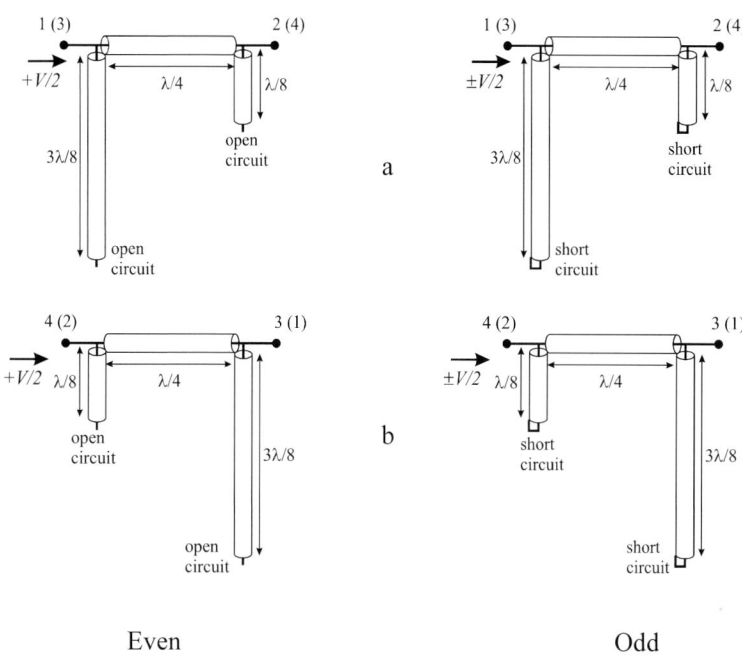

Even   Odd

Fig. 5.27 The different modes of excitation (a and b) of the half-hybrid circuits corresponding to the even and odd mode. In (a) excitation is applied to ports 1 and 3 while in (b) it is applied to ports 2 and 4. In the even mode, all ports are excited in phase. In the odd mode, port 3 (a) and port 2 (b) are excited by incident waves of opposed phase respective to ports 1 and 4. This corresponds to excitation applied to the whole hybrid, either to port 1 (a) or port 4 (b).

Similarly, viewing from port 4 [Fig. 5.27(b)] the even and odd chain matrices are:

$$\mathbf{A}_{even} = \mathbf{A}_{open}^{\lambda/8} \mathbf{A}^{\lambda/4} \mathbf{A}_{open}^{3\lambda/8}, \tag{5.41}$$

$$\mathbf{A}_{odd} = \mathbf{A}_{shorted}^{\lambda/8} \mathbf{A}^{\lambda/4} \mathbf{A}_{shorted}^{3\lambda/8}. \tag{5.42}$$

Hence:

$$\mathbf{A}_{even} = \begin{pmatrix} 1 & jZ_c \\ 2j/Z_c & -1 \end{pmatrix}, \quad (5.43)$$

and

$$\mathbf{A}_{odd} = \begin{pmatrix} -1 & jZ_c \\ 2j/Z_c & 1 \end{pmatrix}. \quad (5.44)$$

The input impedance of the sub-circuits is:

$$Z_{even}^{in} = Z_L k \frac{1+jk}{-k+2j} \quad (5.45)$$

and

$$Z_{odd}^{in} = Z_L k \frac{-1+jk}{k+2j}. \quad (5.46)$$

In contrast with the previous cases considered, the input port impedance for the even and odd modes differ from each other. Furthermore, they cannot be matched independently to a real-valued impedance of 50 Ω, as far as the impedance Z_L and k are real-valued. In other words, when applying an incident wave to the even–odd sub-circuits, it will be partially reflected. Fortunately, for a given k value, these reflected waves cancel out when superimposing the two odd and even modes. In this case, the input port is correctly matched to the desired 50 Ω impedance.

The amount of the reflected wave is defined by the reflection coefficient Γ, which can be written in terms of the input impedance Z_{in} and source impedance Z_0 as:

$$\Gamma = \frac{Z_{in} - Z_0}{Z_{in} + Z_0}. \quad (5.47)$$

Due to the superposition principle, the input port reflection coefficient of the total circuit is obtained as the sum of the even and odd mode reflection coefficients of the corresponding sub-circuit:

$$\Gamma = \Gamma_{even} + \Gamma_{odd}. \tag{5.48}$$

Input impedance matching ($\Gamma = 0$) implies that:

$$\Gamma = \frac{Z^{in}_{even} - Z_0}{Z^{in}_{even} + Z_0} + \frac{Z^{in}_{odd} - Z_0}{Z^{in}_{odd} + Z_0} = 0. \tag{5.49}$$

This condition is fulfilled if:

$$Z^{in}_{even} Z^{in}_{odd} = Z_0^2. \tag{5.50}$$

Assuming that the output ports are loaded by the system impedance Z_0 ($Z_L = Z_0 = 50\ \Omega$) the condition above [Eq. (5.50)] is realized for both cases (a) and (b) of Fig. 5.27, if:

$$k = \sqrt{2} \tag{5.51}$$

hence

$$Z_c = Z_0 \sqrt{2}. \tag{5.52}$$

The analysis of the response of the hybrid to an excitation voltage applied to port 1 (difference port) or to port 4 (sum port) must take into account the fact that the input impedance of all ports of each sub-circuit of Fig. 5.27 is different from Z_0. Consequently, the voltages for the whole circuit will be obtained by the superposition of the reflected and incident waves corresponding to the even and odd modes separately.

The reflection coefficient, Eq. (5.47), describes the amount of reflected waves at a junction point where there is an impedance change. Similarly, one can define a transmission coefficient, T, which describes the amount of energy effectively transferred to the load:

$$T = 1 + \Gamma = \frac{2Z_{in}}{Z_{in} + Z_0}. \tag{5.53}$$

The "voltages", in fact the scattered wave amplitude W at all ports of the network, are obtained by the superposition of the even–odd reflection and transmission coefficients. Thus, assuming that port 1 is driven with an incident wave of unit amplitude, one obtains:

$$\begin{aligned} W_1 &= \frac{1}{2}(\Gamma_{even} + \Gamma_{odd}) = 0 \\ W_2 &= \frac{1}{2}(T_{even} + T_{odd}) = \frac{j\sqrt{2}}{2} \\ W_3 &= \frac{1}{2}(\Gamma_{even} - \Gamma_{odd}) = -\frac{j\sqrt{2}}{2} \\ W_4 &= \frac{1}{2}(T_{even} - T_{odd}) = 0 \end{aligned} \quad (5.54)$$

$W_1 = 0$ means that the input port is matched to 50 Ω, as required. $W_4 = 0$ means that port 4 is isolated, with no voltage appearing at this port. The values for W_2 and W_3 indicate that for a unit volt applied at port 1 one obtains equal voltage amplitudes at both ports and a 180° phase difference. The incident wave is evenly split between the two ports with amplitude of $\sqrt{2}/2$ corresponding to half the input power, as it should be.

The results are summarized in Fig. 5.28(a). Reverting the wave propagation it becomes evident that when applying voltages of equal amplitude and opposed phase at ports 2 and 3, one should obtain a unit amplitude voltage at port 1 and a zero at port 4.

Similarly, when the incident wave is applied to port 4, after combining the odd and even modes one obtains:

$$\begin{aligned} W_1 &= \frac{1}{2}(T_{even} - T_{odd}) = 0 \\ W_2 &= \frac{1}{2}(\Gamma_{even} - \Gamma_{odd}) = -\frac{j\sqrt{2}}{2} \\ W_3 &= \frac{1}{2}(T_{even} + T_{odd}) = -\frac{j\sqrt{2}}{2} \\ W_4 &= \frac{1}{2}(\Gamma_{even} + \Gamma_{odd}) = 0 \end{aligned} \quad (5.55)$$

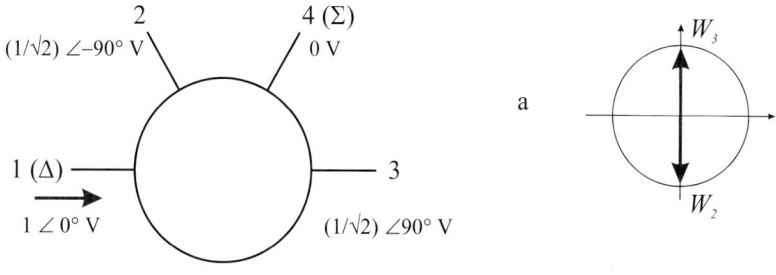

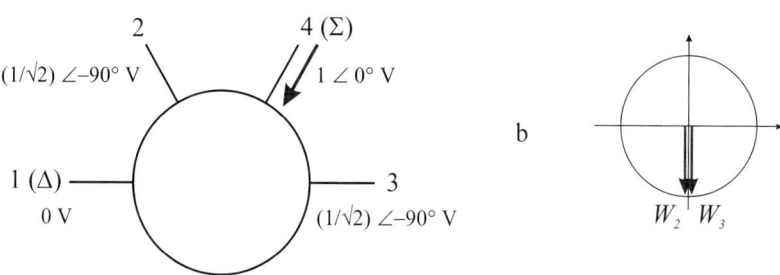

Fig. 5.28 Summary of output voltages obtained when exciting the hybrid either to port 1 (a) or port 4 (b) by an incident "wave" of 1 volt. Note that in both cases the input impedance is matched, provided all other ports are terminated in 50 Ω. Both the phasor notation (left) or the complex plane representation (right) are used.

Port 4 is matched ($W_4 = 0$), while port 1 is isolated ($W_1 = 0$). Again, the incident wave is evenly split but this time the two voltages appearing at ports 2 and 3 have the same phase. Reciprocally, applying in phase voltages of equal amplitude to ports 2 and 3, the sum will appear as a unit amplitude voltage at port 4 and zero at port 1 [Fig. 5.28(b)].

At this point, one may pay attention to some of the difficulties that could be encountered when trying to define voltages and currents (or similarly impedance or admittance) of "high frequency" circuits. This problem, very well-known for microwave circuits, is completely eliminated using the scattering matrix formalism. The matrix describes physical and measurable quantities, the amplitude and phase of waves travelling in a given direction. From these data (incident, reflected, and transmitted waves) one can finally obtain impedances or admittances and eventually voltages and currents. Although the latter appears to be somewhat of an abstraction to the microwave engineer [Pozar, 1998, p. 196], it is particularly important for the NMR experimentalist because the oscillating magnetic field that rotates the macroscopic magnetization depends on it. The scattering matrix formalism allows an elegant and easy procedure for calculating the behavior of complex networks. The reader is left to himself to apply this particularly efficient formalism to all the cases discussed so far in this chapter. The formalism is briefly presented in Chapter 3 together with the relations that exist between various matrix representations of networks (impedance matrix, chain matrix etc.) allowing calculation of all the properties of the circuit that concern the NMR experimentalist.

5.3.2 *Using the rat-race 180° hybrid in quadrature NMR*

In a 50 Ω system, the line characteristic impedance of the hybrid should be $50\sqrt{2}$ Ω, Eq. (5.52). The corresponding value (70.7 Ω) is, within less than 10%, close to the characteristic impedance (75 Ω or, eventually, 72 Ω in some cases) of standard coaxial lines. Such standard coaxial cable can then be used to build the hybrid [Yoda *et al.*, 1989], but at the expense of a slight degradation of the impedance matching (Section 5.3.4).

Another possibility is to use standard 50 Ω coaxial cables [Sank *et al.*, 1986; Chen and Hoult, 1989]. In this case, the output ports 2 and 3 must be loaded by a 25 Ω impedance, and the sum and difference ports 1 and 4 by the standard 50 Ω load.

The 25 Ω impedance required at ports 2 and 3 is realized using two 50 Ω cables in parallel, each connected to a 50 Ω load. In this way, the probe needs four ports, the two pairs of quadrature connections being driven by out-of-phase signals [Fig. 5.29(b)]. The probe electric field is thus balanced for each quadrature connection port. The 180° phase difference is simply obtained by an additional length of $\lambda/2$, 50 Ω, coaxial cable.

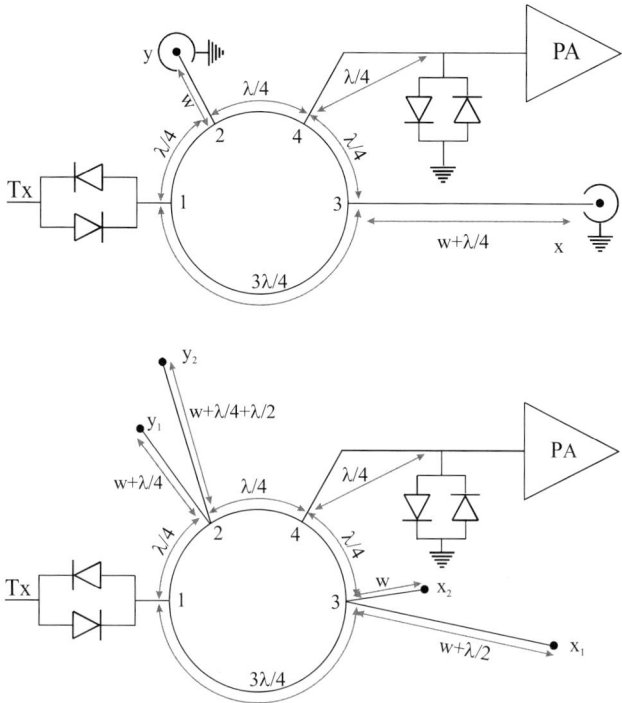

Fig. 5.29 Two coax ring hybrid configurations in an NMR environment. The Yoda *et al.* [Yoda *et al.*, 1989] circuit is built using 75 Ω cables (upper), while the Sank *et al.* scheme [Sank *et al.*, 1986] uses 50 Ω cables (lower).

The final configurations of the 90° hybrid, based on a rat-race hybrid using either 75 Ω or 50 Ω cables, are shown in Fig. 5.29. The length of wire w that appears at all ports is arbitrary. This extra length of coax is obviously required to connect the circuit and the probe.

5.3.3 Lumped element equivalent of the 180° hybrid

The rat-race hybrid can also be built replacing the transmission lines with their lumped element equivalent. Such a circuit is proposed in Fig. 5.30 that simulates the 180° hybrid.

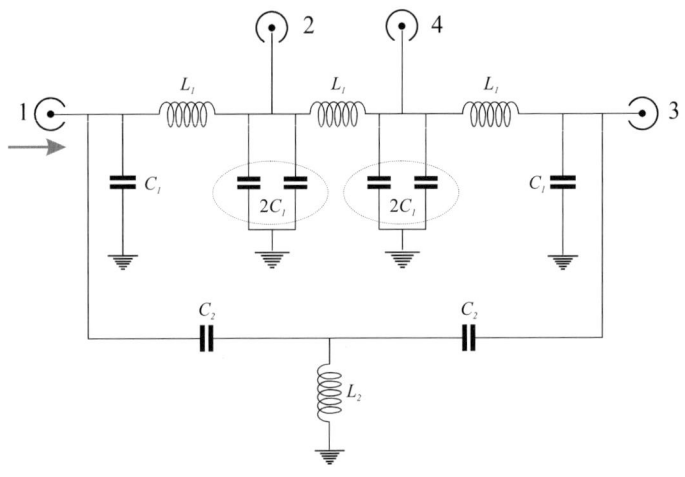

Fig. 5.30 Lumped element realization of the 180° ring hybrid.

Each of the $\lambda/4$ lines is replaced by its equivalent low-pass filter, while the $3\lambda/4$ line is simulated by a high-pass filter. In this way, a minimum of components is used for building the device. This configuration presents a frequency response that is quite different than the coax made counterpart (next paragraph). A similar frequency response can be recovered by replacing the high-pass filter by a cascade of three low-pass filters identical to those used in the upper branch. Because the hybrid should be used at only one frequency, the choice between the two circuits is just a matter of preference.

The component values are given by the following formulae:

$$L_1 = L_2 = \frac{Z_C}{\omega_S} = \frac{Z_0 \sqrt{2}}{\omega_S} \tag{5.56}$$

and

$$C_1 = C_2 = \frac{1}{Z_c \omega_S} = \frac{1}{\omega_S Z_0 \sqrt{2}}, \quad (5.57)$$

where Z_c is the characteristic impedance of the simulated line and Z_0 is the system impedance (50 Ω). A 180° hybrid designed for a frequency of $\omega_S = 200$ MHz and impedance of $Z_0 = 50$ Ω is shown in Fig. 5.31.

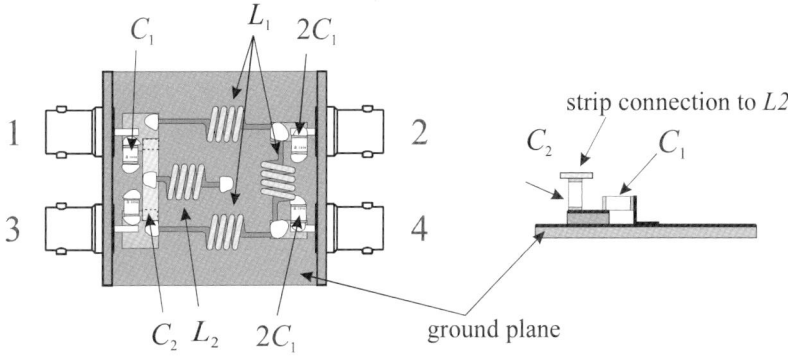

Fig. 5.31 Layout of a lumped element 180° hybrid designed for 200 MHz central frequency. All inductors are equal (56.27 nH, four turns of 1 mm diameter wire wound on air with an internal diameter of 5.3 mm. The spacing between each turn should be such that the coil length is 7 mm). The capacitors are 10 pF (C_1 and C_2) and 22 pF ($2C_1$) ATC chip capacitors. The actual capacitances should be $C_1 = C_2 = 11.25$ pF and $2C_1 = 22.5$ pF, respectively (Fig. 5.30). The small pieces of printed circuit board that are glued on the copper ground plane serve as a support for connections and provide the required additional capacitances to C_1 and $2C_1$. The performances measured at 200 MHz for a hybrid prototype were: $S_{12} = -3.1$ dB, $S_{13} = -3.07$ dB, $\Delta\varphi_{23} = 181°$ and $S_{14} = -35$ dB. The input (port 1) impedance was slightly different from 50 Ω (64.0 − j6.0), but it could be easily matched using a simple matching network (Chapter 4), thus improving the performances.

To connect the hybrid to the probe, it should be remembered that an additional 90° shift should be added to obtain the quadrature required for the NMR experiment. This can be done using the appropriate length of lines connecting the device to the probe.

5.3.4 Frequency response of the 180° ring hybrids

The performances of three hybrids have been simulated as a function of frequency and are compared in Figs. 5.32 to 5.35. Two hybrid circuits are built from coax lines with characteristic impedance of 70.7 Ω and of 75 Ω. A third hybrid is built on lumped elements, using the high-pass filter $3\lambda/4$ replacement version (Fig. 5.30).

Replacement of the coax line of optimum characteristic impedance by lines with standard impedance of 75 Ω has practically no influence on the phase difference, balance, and isolation properties of the hybrid.

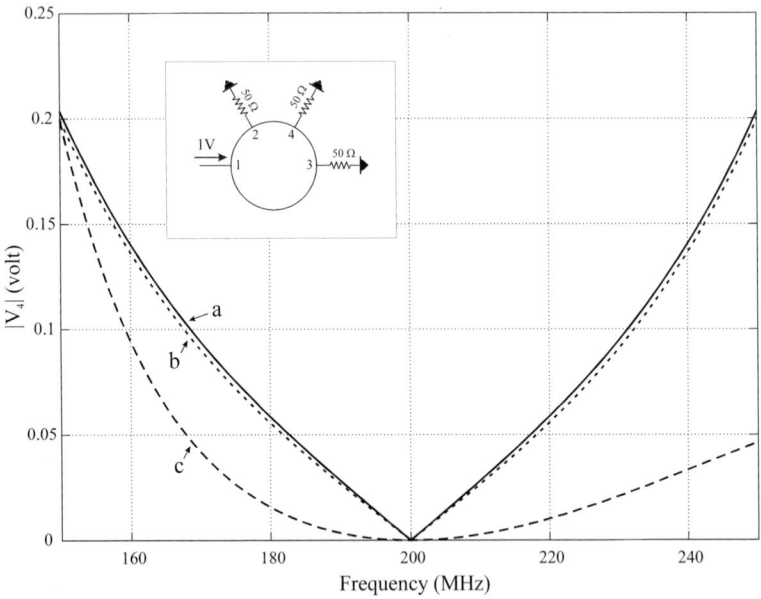

Fig. 5.32 Frequency dependence of the isolation properties of a 180° ring hybrid designed for a central frequency of 200 MHz, built on (a) 70.7 Ω or (b) 75 Ω transmission lines and (c) on the lumped element circuit of Fig. 5.30.

In contrast, the input port presents a significant impedance mismatch [Fig. 5.33(b), reflection coefficient near –25 dB] that may be acceptable, or possibly compensated by an appropriate network.

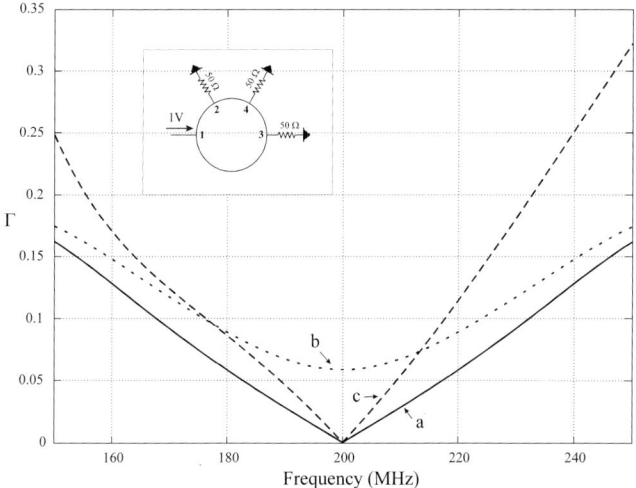

Fig. 5.33 Frequency dependence of the reflection coefficient at the input port of a 180° ring hybrid designed for 200 MHz central frequency, built on (a) 70.7 Ω or (b) 75 Ω transmission lines and (c) on the lumped element circuit of Fig. 5.30. Note the mismatch when using a 75 Ω coax cable.

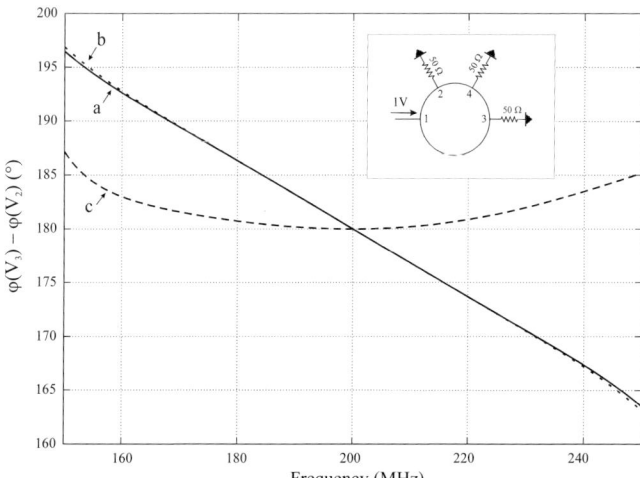

Fig. 5.34 Frequency dependence of the phase properties of a 180° ring hybrid designed for a central frequency of 200 MHz, built on (a) 70.7 Ω or (b) 75 Ω transmission lines and (c) on the lumped element circuit of Fig. 5.30.

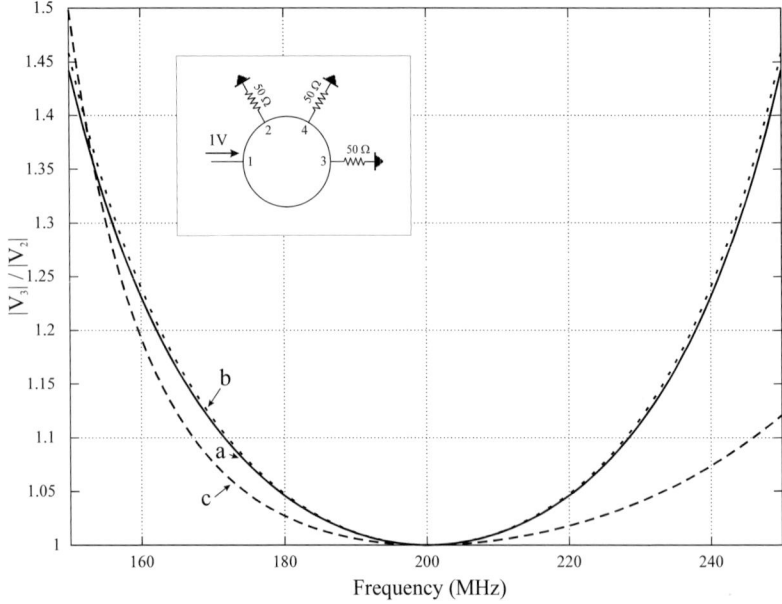

Fig. 5.35 Frequency dependence of the balancing properties of the output port of a 180° ring hybrid designed for 200 MHz central frequency, built on (a) 70.7 Ω or (b) 75 Ω transmission lines and (c) on the lumped element circuit of Fig. 5.30.

The lumped element hybrids may be used on a broader frequency range compared to their coax versions, except for the reflection coefficient. Despite its broadband characteristics, the device should be built carefully as its performance depends critically on its component values (should be better than 5%).

Although the lumped element device has several advantages, the coax version is probably much easier to build with satisfactory performances. But, the latter has the drawback that it requires cables that have nonstandard characteristic impedance. In fact, the choice will essentially depend on the availability of places to put the device.

5.4 Other 90° Hybrids

To our knowledge, the NMR literature essentially mentions the use of $\lambda/8$ branch-line or ring hybrids. Other designs are still possible, mainly those which can be found in the microwave literature. In particular, a design based on the principle of coupled lines can be envisaged. Some devices of this type are already used in NMR probe technology, for example to drive a single port probe simultaneously by two transmitters. This variant is required, for example, for "homonuclear" decoupling. Similar devices are also used to sample incident or reflected power in a transmission line. Generally, the coupled line devices have the drawback that the coupling is too loose to achieve the 3 dB coupling factor required for the splitter/combiner used for quadrature NMR probes. One solution to increase the coupling factor between the lines is known as the Lange coupler. The coupler is made of lines that are disposed in an "interdigital" configuration, increasing the coupling efficiency between the lines. Since its invention in 1969, this coupler has been the subject of many improvements. Such a design, difficult to realize at low frequency, is probably more useful at the frequencies above 400 MHz that are now more and more frequently used in NMR applications.

Broadband couplers can also easily be designed, using the principle of coupled lines.[8] A method frequently used to enlarge the bandwidth of coupling devices is to use a cascade of hybrids, each designed for a progressively increasing frequency associated with a progressive change of line characteristic impedances. Such a principle is used when designing broadband matching networks (a cascade of quarter-wave transformers of progressively increasing characteristic impedances). Using microstrip line technology, the design is particularly simple. For example, a 180° hybrid circuit can take the form shown in Fig. 5.36. The variation in the line widths allows control of its characteristic impedance and the spacing between them allows control of the coupling in between.

---

[8] At least at the microwave frequencies [Pozar, 1998].

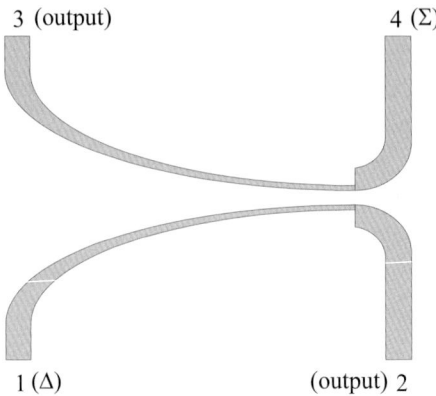

Fig. 5.36 A microwave broadband 180° hybrid coupler designed on microstrip technology. The drawing represents the etching pattern on one side of a double sided printed circuit board. The other side is left non-etched and constitutes the return path (ground plane).

Also based on the principle of cascaded devices, simple solutions have been proposed to increase the bandwidth of the quarter-wave branch hybrid. A corresponding design including two cascaded branch-line hybrids is shown in Fig. 5.37 [Andrews, 2006].

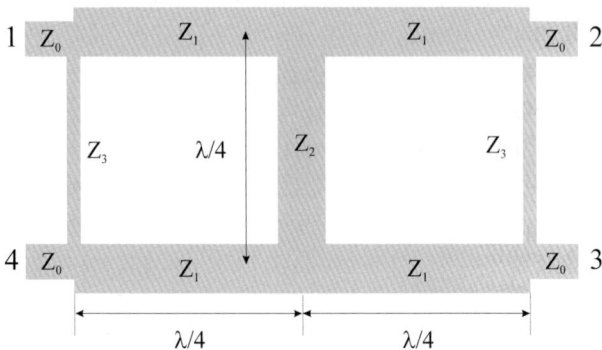

Fig. 5.37 Cascading two 90° hybrids increases the bandwidth of the device. The line characteristic impedances are given by: $Z_1 = Z_2 = Z_0/\sqrt{2}$; $Z_3 = Z_0(1+\sqrt{2})$.

The required line characteristic impedance values impair building the device using standard coax lines for use on a routine NMR frequency band but a lumped element circuit can easily be designed using the low-pass filter equivalents. Fig 5.38 shows one circuit designed for a central frequency of 200 MHz.

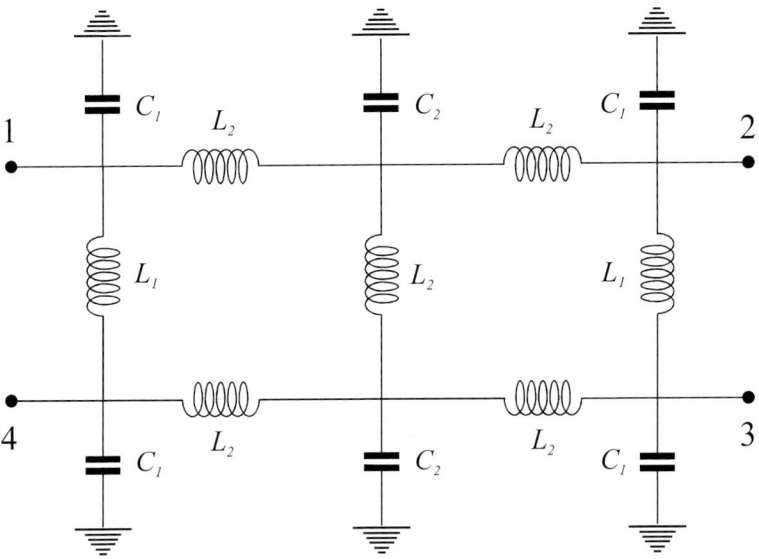

Fig. 5.38 Lumped element version of the double hybrid of Fig. 5.37. The component values are given by Eq. (5.23) using the characteristic impedances quoted in Fig. 5.37. For a 200 MHz design: L_1 = 96.06 nH, L_2 = 28.1 nH, C_1 = 29.1 pF, C_2 = 67.5 pF.

The frequency dependence of its characteristics clearly shows an increase in bandwidth. In particular, the quadrature phase is kept constant over more than a 50 MHz range (Fig. 5.39).

The design of quadrature hybrids, constructed with lumped elements and having broadband characteristics is extensively described in the book by David Andrews [Andrews, 2006]. These hybrids are designed on different concepts than the simple replacement of lines by lumped elements and can constitute a source of valuable designs for the NMR experimentalist.

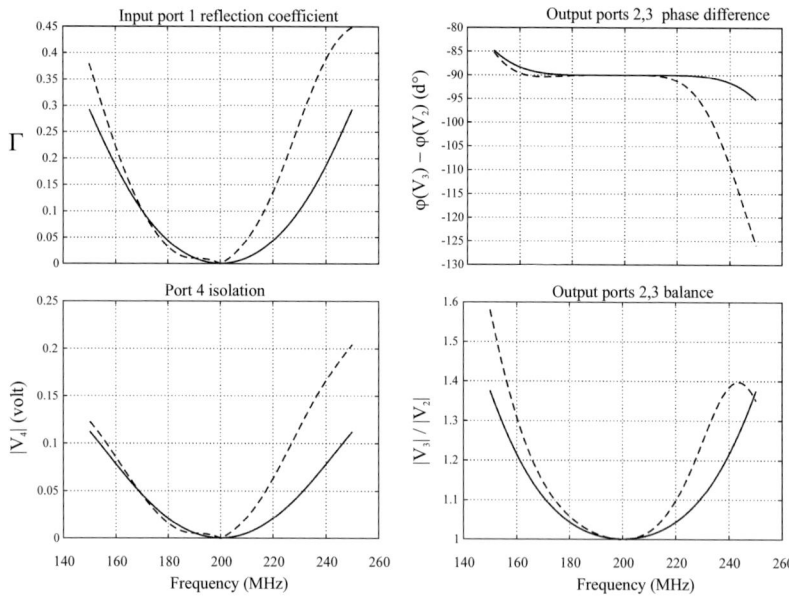

Fig. 5.39 Frequency dependence of the characteristic properties of double hybrids (Fig. 5.37 and 5.38), designed on transmission lines (full line) and on lumped elements (dashed line), for a central frequency of 200 MHz.

Chapter 6

Multiple Frequency Tuning

For many NMR experiments, the probe needs to excite and/or receive the signal from different types of nuclei, hence it should be impedance matched to the spectrometer transmitters and receiver(s) for different resonant frequencies.

For example, experiments involving X-nuclei (^{31}P, ^{13}C, ^{15}N, etc.) require, in addition, a proton channel that is generally used either for decoupling or for receiving. Eventually, a deuterium lock channel is added to synchronize the magnetic field to the frequency of the spectrometer reference oscillator, leading to a three channel[1] probe design.

In the localized spectroscopy experiments on X-nuclei, a proton channel is required for both shimming and localization purposes through the images. In this case, it is desirable that the same coil is used for both nuclei (X and ^{1}H) to ensure the observation of an identical volume. This is also best done using a single coil tuned to both frequencies.

As already shown in Chapter 4, matching a given coil inductance can only be achieved for a discrete number of frequencies at which the coil may be "tuned." In other words, tuning the coil is a process that cannot be dissociated from the impedance matching process. As the magnetic field used in NMR, and hence the frequency, increases the probe "coil"

---

[1] In a high resolution spectrometer designed for protein structural determination, the probe has at least four channels. One channel is devoted to the most important nucleus to be observed (proton) and is fed to a carefully designed coil. Two channels are devoted to decoupling (^{13}C and ^{15}N) and are fed to another, double tuned, single coil that is orthogonal to the proton coil and of greater diameter. Finally, the lock channel (deuterium) is fed to the proton coil, meaning it is double tuned.

tends to be self-resonant.[2] Thus, multiple frequency experiments imply, in this case, that the probe is a multiple mode resonator.

Figure 6.1 shows, from a general point of view, different ways to connect a probe coil or resonator at two or more frequencies. This can be done either through a single port [Fig. 6.1(a)] or through multiple ports [Fig. 6.1(b)], depending on the experimental needs.

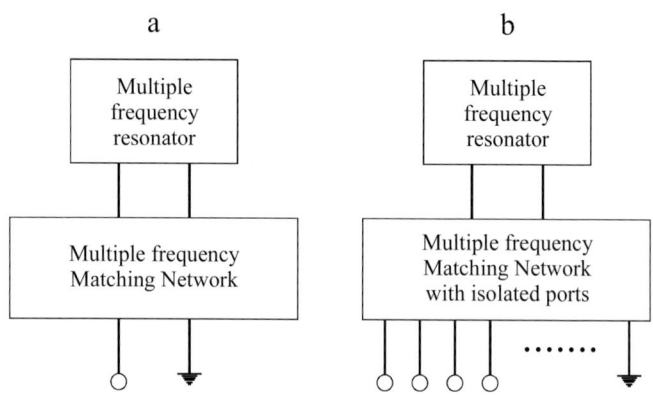

Fig. 6.1. Different multiple frequency matching schemes: (a) single port and (b) multiple ports.

The isolation between the different channels should also be provided, in order to avoid undesirable leakage. Leakage from a receiver port into a transmitter port would result in a loss of signal. Conversely, leakage from a transmitter port into an observing port will increase the noise level on the corresponding receiver. In a case where the probe is driven through a single port [Fig. 6.1(a)], isolation will be provided by an external diplexer (or multiplexer) that connects the single probe port to different channels.[3] In the multiple port case, the isolation will be

---

[2] This is frequently required at high frequencies, but also has the advantage that the current is concentrated in the device producing the B_1 magnetic field, even at lower frequencies.

[3] In this case, connections can also be made "manually" providing perfect isolation between channels!

provided internally via the matching circuits and/or using filters, or using mutually decoupled multiple coils.[4]

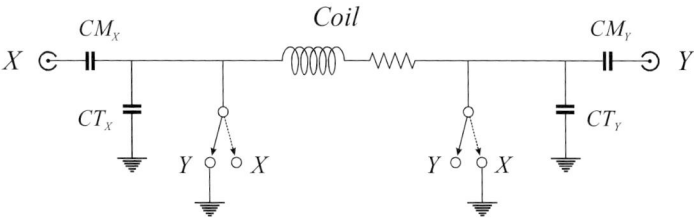

Fig. 6.2 Principle of the shunting method to double tune and match an NMR probe to nuclei X and Y. The probe coil is connected to the respective matching networks using frequency-dependent switches. These switches connect one end of the coil to the ground at the respective resonant frequencies of the nuclei (respectively v_X or v_Y). The opposite end is isolated from the ground and connected to the corresponding classical capacitive matching network. In this way, the ports are well isolated from each other and the switches do not interfere with the matching network components. These switches are realized using LC components or transmission lines as described in paragraph 6.1.

Figures 6.2–6.4 show different methods used to build an NMR probe operating at two frequencies. These methods can be easily extended for three or more frequencies (some examples will be considered later).

In the first case (Fig. 6.2), the sample coil is tuned and matched by separate networks that are connected to the probe by means of "frequency-dependent switches" [Stringer and Drobny, 1998].

In the second case (Fig. 6.3), the probe resonator is designed to work at the required frequencies prior to being connected to the matching network. The different resonant frequencies are obtained either using a tuning reactive network that presents one or more poles [Fig. 6.3(a),(b)] or using the properties of coupled resonators [Fig. 6.3(c),(d)]. The first approach is best suited to remote resonant frequencies, while the second approach is more suited to close resonant frequencies. Depending on the coupling strength, a splitting of the resonance modes can be obtained and adjusted to the desired frequencies.

---

[4] A typical configuration is to use two orthogonal coaxial coils, one for decoupling and one for receiving. In this case, the filling factor is generally not optimized for both coils.

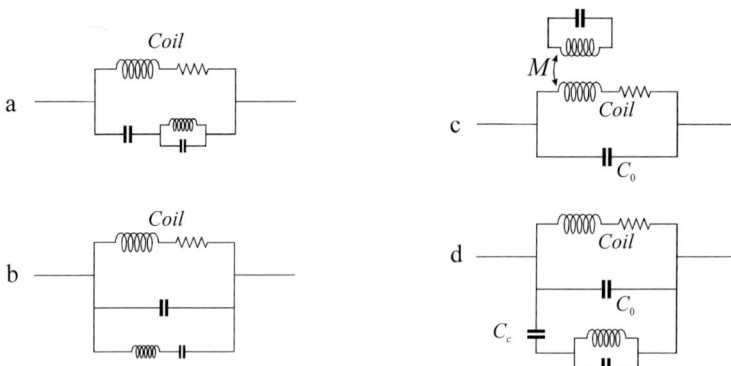

Fig. 6.3 Multiple-pole technique. (a) and (b) The probe coil is resonated by a complex reactive component. The corresponding network, including a resonant LC circuit, has a pole for a finite frequency. In this way, there are two frequencies at which the network impedance is the complex conjugate of the coil impedance, realizing the resonance. The corresponding circuits are described in Section 6.2. (c) and (d) The coil is already resonated with C_0 at a frequency intermediate between the two desired frequencies. The resonance response is split to the desired frequencies by coupling a "tank" circuit, either inductively (c) or capacitively (d). The final frequencies depend on the resonant frequencies of the coil resonator, including C_0, on the resonant frequency of the trap, and on the coupling strength characterized by M (mutual inductance) or C_c (coupling capacitance). The corresponding circuits are described in Section 6.3.

Finally, in the third case (Fig. 6.4), the probe is constituted of an assembly of two physically separated resonators [Fig. 6.4(a), (b)] or is a multi-mode single "coil" [Fig. 6.4(c)]. Isolation between the separated coils can be achieved by a crossed-coil configuration [Fig. 6.4(a)] or inserting a high frequency trap circuit[5] in the low frequency resonator [Fig. 6.4(b)]. Another possibility is to optimize the geometry of both coils in order to get a predetermined "field profile" (Chapter 9).

The multi-mode resonator configuration is generally based on a volume resonator having several resonant frequencies. It is therefore

---

[5] If the active volume of both coils should be the same, they probably could be strongly magnetically coupled. Then, the resonator having the lowest resonant frequency "screens" the high frequency resonator. A trap circuit, tuned at the high frequency and inserted in the low frequency resonator, precludes the screening effect.

especially suited to the Birdcage resonator [Hayes *et al.*, 1985] as described in Chapter 8.

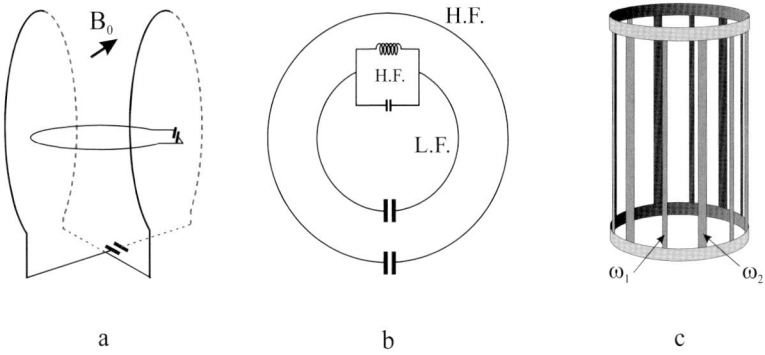

a   b   c

Fig. 6.4 Multiple coil and multi-mode configurations. (a) Two crossed coils are tuned to their own resonance. Due to geometric decoupling between both coils, the whole circuit can be tuned and matched independently. (b) The two coils are strongly coupled. To avoid interference between the two coils, a trap circuit tuned to the high frequency is inserted in the low frequency resonator. (c) A birdcage exhibits various modes of resonance. A multiple-resonant probe can be obtained by alternating resonant legs tuned at different frequencies (ω_1 and ω_2). Alternatively, the intrinsic resonant modes of a conventional birdcage can be tailored to obtain the required resonance frequencies.

This third class of methods will be addressed elsewhere, but it is worth mentioning at least three papers in the literature that are representative of these approaches. The paper by Alecci *et al.* [Alecci *et al.*, 2006] describes in detail, with useful practical hints, a sodium/proton probe constituted of two independent surface coils. The paper by Pang *et al.* [Pang *et al.*, 2012] provides a good introduction to a multi-resonant birdcage coil using alternating resonant legs. Finally, a paper by Webb *et al.* [Webb *et al.*, 2011] describes an original method using two resonant modes of a "single tuned" birdcage coil. The lowest homogeneous mode ($k = 1$, Chapter 8) is tuned to ^{31}P and a higher mode ($k > 1$) is tuned to the proton. The trick is to benefit from the dielectric effects at high frequency (300 MHz) that increase, near the center of the coil, the intrinsically heterogeneous B_1^+ field corresponding to $k > 1$.

In this chapter, we will consider some of the circuits that are used in NMR probe designs belonging to the two first approaches described here, i.e., switching (or shunting) the radio frequency (RF) current into the single probe coil through appropriate networks or providing the coil with multiple resonances using appropriate multiple-pole circuits.

6.1 Shunting Methods

These methods consist of driving a single coil probe at the desired frequencies through appropriate matching networks.

6.1.1 *Dual frequency switching circuits*

In a dual frequency configuration, one end of the single coil probe is tuned and matched by a simple capacitive network (Chapter 4) at one frequency, while the opposite end is shorted to the ground for the second frequency. This is realized via frequency-dependent switches.

A simple "*on*" switch is a series LC circuit that is a short circuit at resonance [Fig. 6.5(a)]. Its impedance, as a function of frequency is:

$$Z_{series}(\omega) = jL\omega\left(1 - \frac{\omega_{on}^2}{\omega^2}\right). \tag{6.1}$$

It may be also written as:

$$Z_{series}(\omega) = \frac{1}{jC\omega}\left(1 - \frac{\omega^2}{\omega_{on}^2}\right), \tag{6.2}$$

where ω_{on} is the resonance frequency.

Similarly, an "*off*" switch can be realized using a parallel resonant LC circuit [Fig. 6.5(b)]. Its impedance is given by:

$$Z_{parallel}(\omega) = jL\omega\left(1 - \frac{\omega^2}{\omega_{off}^2}\right)^{-1}, \tag{6.3}$$

that may be also written as:

$$Z_{parallel}(\omega) = \frac{1}{jC\omega}\left(1 - \frac{\omega_{off}^2}{\omega^2}\right)^{-1}, \qquad (6.4)$$

where ω_{off} is the resonant frequency. If $\omega = \omega_{off}$, the impedance is infinite, opening the connection. When $\omega < \omega_{off}$, the circuit appears as an inductance (the preferred path for the current), while at $\omega > \omega_{off}$, the impedance is a capacitance.

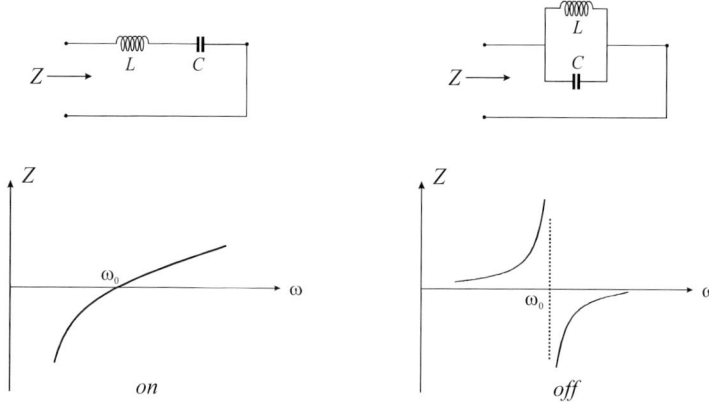

Fig. 6.5 Simple frequency-dependent "switches" (upper) and their respective frequency-dependent impedances (lower).

Dual frequency *on/off* switches can be realized by combining the simple circuits described above, but they can be better designed as shown in Fig. 6.6.

At a given frequency (ω_{on}), the network is a series resonant circuit presenting a null impedance, while at another frequency (ω_{off}) it behaves as a parallel high impedance resonant circuit. At this point, it should be remarked that these circuits can also be used as tuning elements of the main probe coil [Fig. 6.3(a),(b)], making use of their frequency-dependent impedance properties. In these cases, one uses the properties of the circuit to exhibit a given capacitive impedance at the required frequencies.

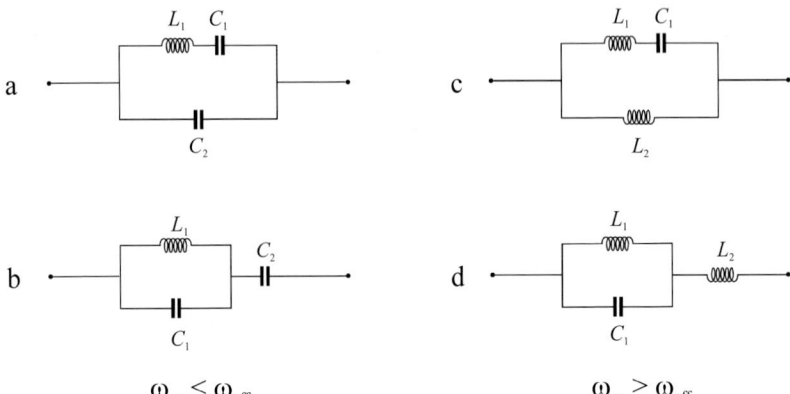

Fig. 6.6 *On/off* frequency-dependent switches. The circuits (a) and (b) present an *on* state at a frequency lower than the *off* state. The *on* and *off* state frequencies for the circuits (c) and (d) are in the reverse order. Note that these last two circuits include two inductors.

The component values required for each switch should now be determined as a function of ω_{on} and ω_{off}. Because there are three unknown component values, one of them could be freely chosen. With the circuits shown in Fig. 6.6(a),(b), it is natural to choose a given inductance value for L_1. Hence, for the circuit in Fig. 6.6(a), C_1 and C_2 are given by:

$$C_1 = \frac{1}{L_1 \omega_{on}^2}$$
$$C_2 = \frac{1}{L_1 \left(\omega_{off}^2 - \omega_{on}^2 \right)}, \quad (6.5)$$

and for the circuit in Fig. 6.6(b) by:

$$C_1 = \frac{1}{L_1 \omega_{off}^2}$$
$$C_2 = \frac{\omega_{off}^2 - \omega_{on}^2}{L_1 \omega_{on}^2 \omega_{off}^2}. \quad (6.6)$$

For these networks, the capacitance values are related to the *on* and *off* frequencies (ω_{on} and ω_{off}) by the following relationship:

$$\omega_{off}/\omega_{on} = \sqrt{1+C_2/C_1} . \tag{6.7}$$

With the circuits shown in Fig. 6.6(c),(d), it is natural to choose a given value for C_1. Then, for the circuit in Fig. 6.6(c), L_1 and L_2 are given by:

$$\begin{aligned} L_1 &= \frac{1}{C_1 \omega_{on}^2} \\ L_2 &= \frac{\omega_{on}^2 - \omega_{off}^2}{C_1 \omega_{on}^2 \omega_{off}^2} \end{aligned}, \tag{6.8}$$

and for the circuit in Fig. 6.6(d) by:

$$\begin{aligned} L_1 &= \frac{1}{C_1 \omega_{off}^2} \\ L_2 &= \frac{1}{C_1 \left(\omega_{on}^2 - \omega_{off}^2\right)} \end{aligned}. \tag{6.9}$$

A relation similar to Eq. (6.7) holds for L_1 and L_2:

$$\omega_{on}/\omega_{off} = \sqrt{1+L_2/L_1} . \tag{6.10}$$

The impedance of the switching circuits of Fig. 6.6(a) is given, as a function of frequency, by (Fig. 6.7, circuit A):

$$\left|Z_{switch}^A(\omega)\right| = \frac{L_1}{\omega} \frac{\left(\omega_{off}^2 - \omega_{on}^2\right)\left(\omega^2 - \omega_{on}^2\right)}{\left(\omega^2 - \omega_{off}^2\right)}, \tag{6.11}$$

and that of the switching circuits of Fig. 6.6 (b) by (Fig. 6.7, circuit B):

$$\left|Z_{switch}^{B}(\omega)\right| = \frac{L_1}{\omega} \frac{\left(\omega_{off}^2\right)^2 \left(\omega^2 - \omega_{on}^2\right)}{\left(\omega_{off}^2 - \omega_{on}^2\right)\left(\omega^2 - \omega_{off}^2\right)}. \quad (6.12)$$

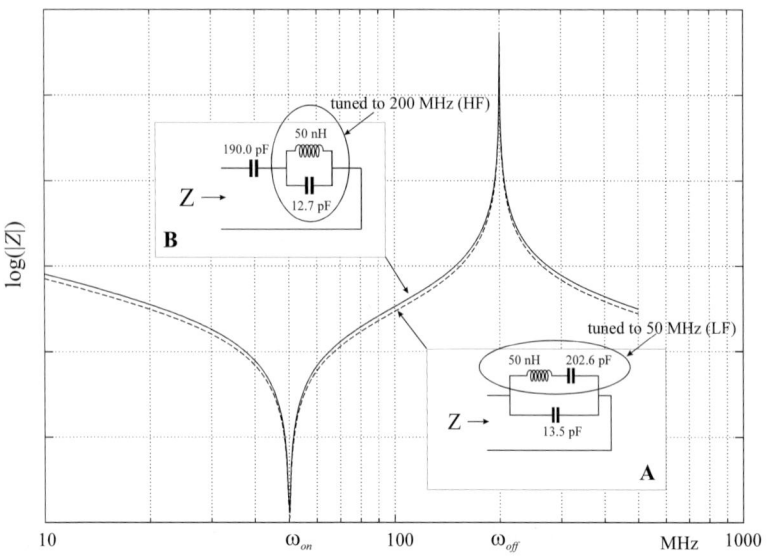

Fig. 6.7 Frequency dependence of the impedance of two *on/off* switches presenting the *on* and *off* state at 50 and 200 MHz, respectively. They are designed, after choosing a 50 nH inductance, using Eqs. (6.5) and (6.6).

Circuits A and B of Fig. 6.7 include only one inductor. Similar circuits can be built on the switches shown in Figs. 6.6(c) and 6.6(d), but involving two inductors. They are preferably avoided in NMR probe designs because the additional inductor L_2 leads to unnecessary resistive losses. The corresponding impedance formulae are given anyway for completeness.

For a circuit built on the switch shown in Fig. 6.6(c):

$$\left|Z_{switch}^{C}\right| = \frac{\omega}{C_1} \frac{\left(\omega_{on}^2 - \omega_{off}^2\right)\left(\omega_{on}^2 - \omega^2\right)}{\left(\omega_{on}^2\right)^2 \left(\omega^2 - \omega_{off}^2\right)}, \quad (6.13)$$

and for the circuit built on the switch shown in Fig. 6.6(d):

$$\left|Z_{switch}^{D}\right| = \frac{\omega}{C_1} \frac{\left(\omega^2 - \omega_{on}^2\right)}{\left(\omega_{on}^2 - \omega_{off}^2\right)\left(\omega^2 - \omega_{off}^2\right)}. \tag{6.14}$$

6.1.2 Practical double tuned circuits

Figure 6.8(a) shows a completed double tuned, single coil probe operating at the frequencies ω_{HF} and ω_{LF}. The corresponding tuning and matching component values can be obtained directly from the equations given in Chapter 4. Equations (4.15) and (4.22) give the tuning and matching capacitance values, respectively [approximate values can be obtained from Eq. (4.12)]. Due to the inevitable losses introduced by the switch components (mainly the inductance L_1), the matching capacitance, and eventually the tuning one, might need to be slightly readjusted. Practically, the switching circuits are seldom used in their complete configurations. Rather, they are most frequently simplified as shown in Fig. 6.8(b),(c). The first simplification [Fig. 6.8(b)] consists of keeping only the series resonant circuits. This still provides a good isolation between the two ports due to the low impedance of these circuits but adds a "parasitic" component in parallel to the tuning capacitors. The lumped element LC circuits could be, in principle, replaced by transmission lines, as shown in Fig. 6.8(c). In this case, one uses the properties of a $\lambda/4$ line that presents a null impedance when terminated by an open connection, thus behaving as a series resonant circuit. However, an important limitation that may render the circuit useless is that, for a particular ω_{HF}/ω_{LF} ratio, the longer low frequency (LF) $\lambda/4$ line may become an odd number of quarter-waves at high frequency (HF). In that case, the HF tuning capacitor is shorted to ground. In contrast, it could work perfectly when the ratio is an even number as for the $^1H/^{13}C$ pair. In this case, the carbon LF quarter-wave line appears as a λ line at the proton frequency.

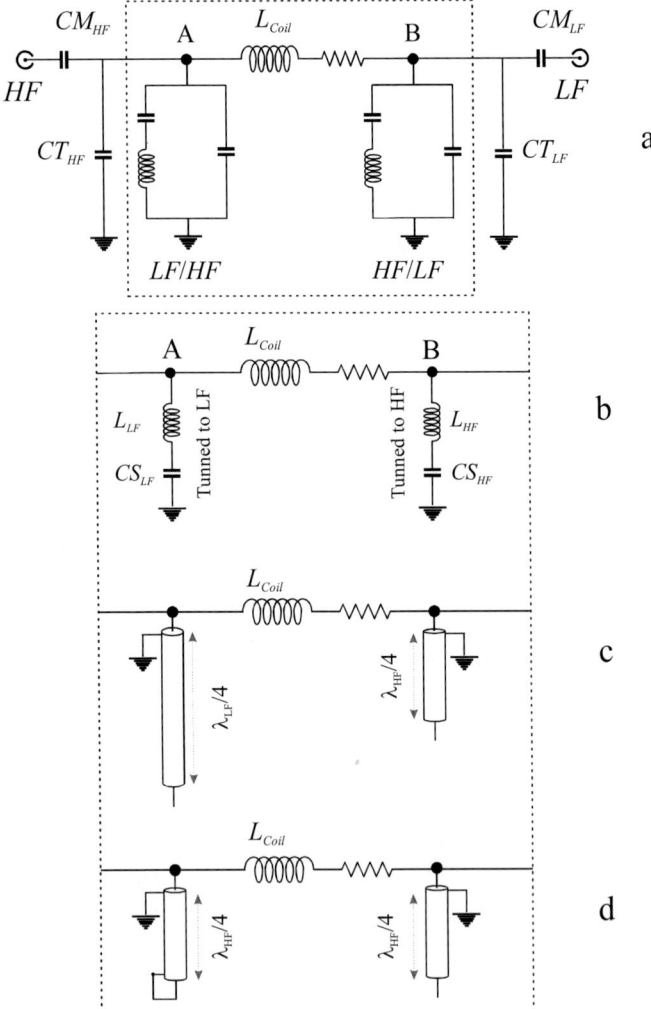

Fig. 6.8 A completed double tuned probe circuit based on the shunting method (a). The part enclosed by dashed lines is usually simplified as shown in the circuits (b), (c) and (d).

The infinite impedance of the opened end line is reported to the coil end connected to the *HF* tuning/matching network. Hence, the line does not introduce any parasitic impedance assuming the line is lossless. A significant degradation of the signal-to-noise ratio might be observed at the proton frequency,[6] if the lines are lossy. Furthermore, the geometrical line length can also be a limitation. Assuming a 4.7 T spectrometer (200 MHz for the proton and 50 MHz for the carbon), the required length for the *LF* switch is about 90 cm when using RG58 cable (characterized by a velocity coefficient of 0.6). This cable length may not easily fit in a narrow bore environment and may favor undesired common mode currents. Thus this circuit should be avoided, if possible. It is presented here only to draw the reader's attention to some of the pitfalls that could be encountered while designing RF circuits. The circuit shown in Fig. 6.8(d), which still makes use of transmission lines, was probably one of the first proposed in the literature [Cross *et al.*, 1976]. This clever circuit uses two $\lambda/4$ lines cut for resonating at the highest frequency. The line connected to the sample coil end near the *LF* matching network is opened, providing the required short to the ground at *HF* and preventing current leakage through the *LF* port.

On the opposite coil side, the shorted line presents a high impedance at *HF*, and a low, but nonzero, impedance at *LF*. The isolation of *LF* currents to the *HF* port is therefore limited in this configuration and may require, if necessary, an additional circuit (trapping circuits or filters) in the *HF* matching network. With these simplifications, the component values of the matching networks, as given in Chapter 3, should be modified taking into account the additional impedance that comes in parallel with the tuning capacitors, due to the finite impedance of the switching components [Fig. 6.9(b)–(d)].

For the circuit in Fig. 6.8(b), the tuning capacitance values corrected from the parasitic switch impedances[7] [Eq. (6.2)] are given by:

$$CT_{LF} = CT_{LF0} - CS_{HF} \left(1 - \frac{\omega_{LF}^2}{\omega_{HF}^2}\right)^{-1} \quad (6.15)$$

for the *LF* tuning/matching network, and

---

[6] In the case when the proton port is used only for decoupling, this configuration can still be used.

[7] Neglecting the resistive losses.

$$CT_{HF} = CT_{HF0} - CS_{LF}\left(1 - \frac{\omega_{HF}^2}{\omega_{LF}^2}\right)^{-1} \qquad (6.16)$$

for the *HF* tuning/matching network. CT_{HF0} and CT_{LF0} are the uncorrected tuning capacitance, as given by Eq. (4.15).

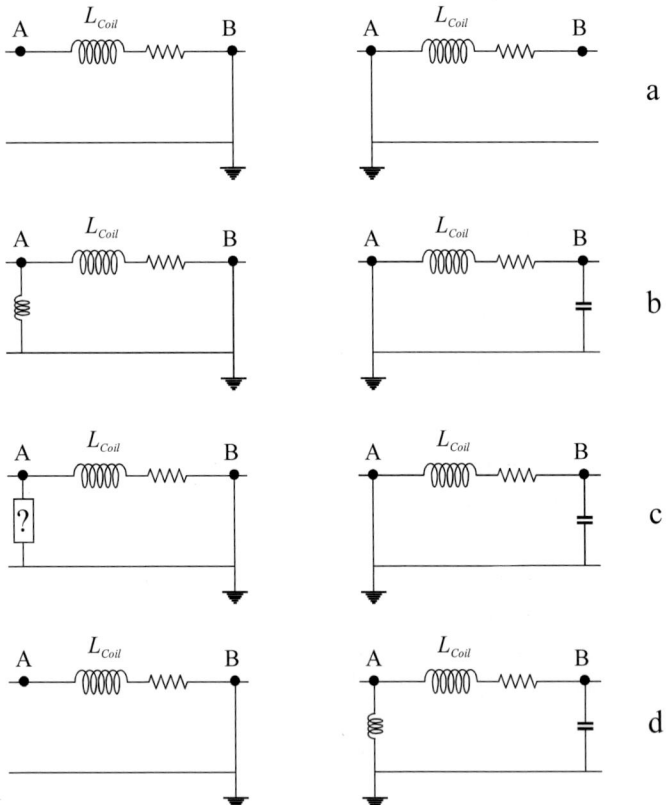

Fig. 6.9 Equivalent network of the circuits shown in Fig. 6.6 at *HF* (left) and *LF* (right). (a) One end of the probe coil is directly connected to the ground by the switching circuit. (b) The *LF* switching circuit has a positive reactive impedance at *HF* (inductor). The *HF* switch has a negative impedance at *LF* (capacitor). (c) The impedance of the *LF* circuit ($\lambda/4$ line at *LF*) can be positive, negative or null at *HF*, depending on the ratio of ω_{LF}/ω_{HF}. At *LF*, the *HF* switch has a positive impedance (capacitor). (d) The *HF* switching circuits act only at *LF*. They appear as an inductor (shorted line) at point A and as a capacitor at point B [as in (c)].

If the *HF* and *LF* frequencies are well separated, the circuits can be designed such that the corrections are negligible. For example, in the case of the ^{1}H/^{13}C pair, the frequency ratio ω_{HF}/ω_{LF} is roughly equal to four. Hence, the correction term in parentheses, Eq. (6.15), is close to unity. The parasitic impedance that appears in parallel with the *LF* tuning capacitor is thus roughly equal to that of CS_{HF}, which is expected to be much smaller than CT_{LF0}.[8]

On the *HF* side of the coil, the parasitic impedance of the *LF* switch in parallel to the tuning capacitor is positive [inductive, Eq. (6.1)]. In other words, the impedance, at *HF*, of the *LF* series resonant circuit is dominated by LS_{LF}. Hence, its impedance should be compensated by increasing the tuning capacitance [Eq. (6.16)]. The correction term is proportional to CS_{LF}, which is expected to be large respective to CT_{HF0}. But, being divided by approximately the square of the frequency ratio (a factor of 16 for the ^{1}H/^{13}C pair), the corrective capacitance may still be very small. This essentially depends on the chosen component values used to build the series resonant circuits, which are entirely at the designer's choice.

The matching capacitance values are simply given by Eq. (4.22), but should be slightly corrected again, in order to take account of the losses due to the switching circuits (at the experimental setup stage).

For the circuit shown in Fig. 6.8(c) the parallel parasitic impedance of the switches can be estimated as follows.

From Eq. (4.27), the impedance of an opened line of length *l* and of characteristic impedance Z_c is:

$$Z_{switch}(\omega) = \frac{Z_c}{j \tan(\theta)}, \tag{6.17}$$

where $\theta = (2\pi l)/\lambda$, and λ is the wavelength in the line at ω.

When *l* is an odd number of quarter-wavelength ($l = (2k+1)\lambda/4$), the impedance Z_{switch} is equal to zero, as expected.

---

[8] For typical designs, the *LF* tuning capacitance will be 10 to 100 times greater than the tuning capacitance of the *HF* series resonant circuit.

At any frequency ω, the angle θ of a line cut at $\lambda/4$ for ω_0 is given by:

$$\theta = \frac{2\pi}{\lambda_\omega}\frac{\lambda_{\omega_0}}{4} = \frac{\pi}{2}\frac{\omega}{\omega_0}. \qquad (6.18)$$

Hence, the *HF* $\lambda/4$ opened line, connected to the *LF* matching network, looks like a capacitor [Fig. 6.9(c)] at the *LF* frequency ($\theta < \pi/2$). Taking into account this correction, the final *LF* tuning capacitance value is given by:

$$CT_{LF} = CT_{LF0} - \left(Z_c \omega_{LF}\right)^{-1} \tan\left(\frac{\pi}{2}\frac{\omega_{LF}}{\omega_{HF}}\right). \qquad (6.19)$$

On the *HF* side of the coil, the $\lambda/4$ *LF* line has an impedance that can be either positive, negative or, eventually, null depending on the ratio ω_{HF}/ω_{LF}. The final *HF* tuning capacitance value is given by:

$$CT_{HF} = CT_{HF0} - \left(Z_c \omega_{HF}\right)^{-1} \tan\left(\frac{\pi}{2}\frac{\omega_{HF}}{\omega_{LF}}\right). \qquad (6.20)$$

If the ratio ω_{HF}/ω_{LF} is close to an odd number, $\tan(\theta)$ tends to $\pm\infty$ and the $\lambda/4$ *LF* line resonates near *HF*. In this case, tuning the sample coil to *HF* may become impossible.

In the circuit presented in Fig. 6.8(d), the tuning capacitor of the *HF* matching network is in parallel with the shorted $\lambda/4$ line cut for ω_{HF} (that presents infinite impedance at *HF*).

Hence no correction should be applied to this network, except eventually on the matching capacitor that should again be slightly adjusted to compensate the lines losses.

The calculation of the tuning capacitance value for the *LF* matching network is a little bit more complicated. It should take into account both the small inductance due to the shorted *LF* line that comes in series with the probe coil and the capacitance due to the open *LF* line.

Hence:

$$CT_{LF} \approx CT_{LF0}\left(1 - \frac{Z_c \tan\left(\frac{\pi}{2}\frac{\omega_{LF}}{\omega_{HF}}\right)}{\omega_{LF} L_{coil}}\right) - (Z_c \omega_{LF})^{-1} \tan\left(\frac{\pi}{2}\frac{\omega_{LF}}{\omega_{HF}}\right). \quad (6.21)$$

The first term in Eq. (6.21) takes into account the reduction of the tuning capacitance due to the increase of the total inductance. The second term subtracts the contribution due to the $\lambda/4$ *HF* opened transmission line.

Considering again the pair $^1H/^{13}C$ in a 4.7 T NMR spectrometer, the contribution of opened and shorted $\lambda/4$ proton 50 Ω lines are, respectively, of 26.4 pF and 66 nH at the carbon frequency. These values appear to be non-negligible and clearly would contribute significantly in the *LF* matching network. The lines must be of very good quality in order not to introduce extra losses.

6.1.3 *Multiple tuning of a single coil*

The scheme presented above can be extended quite easily to more than two frequencies. As an example, an ingenuous circuit (Fig. 6.10) has been proposed [Kan and Courtieu, 1980] as a triple resonance, single coil probe (^1H, ^{13}C, ^2H).

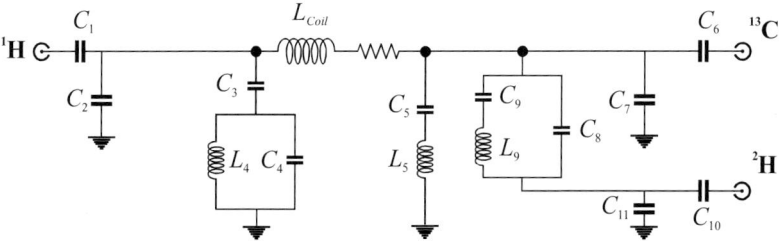

Fig. 6.10 An optimized triple tuned single coil probe as suggested by Kan and Courtieu.

The circuit has been optimized in order to add a minimum of components to the conventional capacitive matching networks required for each channel.

The probe circuit is made out of two principal parts. One part is a double tuned ^{1}H/^{13}C single coil circuit incorporating a dual frequency switching circuit (proton/carbon) provided by $C_3L_4C_4$ on the proton port, and a simplified switch made of a series circuit L_5C_5 (tuned to proton) on the carbon port. C_7 is part of the tuning carbon network. The second part is a ^{2}H matching network grafted onto the carbon side of the coil through the $L_9C_9C_8$ network. The latter circuit is a dual frequency switch characterized by zero impedance at the deuterium frequency and an infinite impedance at the carbon frequency.

Let us consider first the double tuned ^{1}H/^{13}C part of the circuit. Due to the L_5C_5 network on one hand and to the $L_9C_9C_8$ network on the other hand, it is completely isolated from the ^{2}H network at both proton and carbon frequency. Accordingly, it can be designed independently of the deuterium network and provides an example of a well-designed double tuned probe.

The tuning/matching capacitances can be estimated, referring to the equivalent circuits for each frequency of interest shown in Fig. 6.11(a),(b).

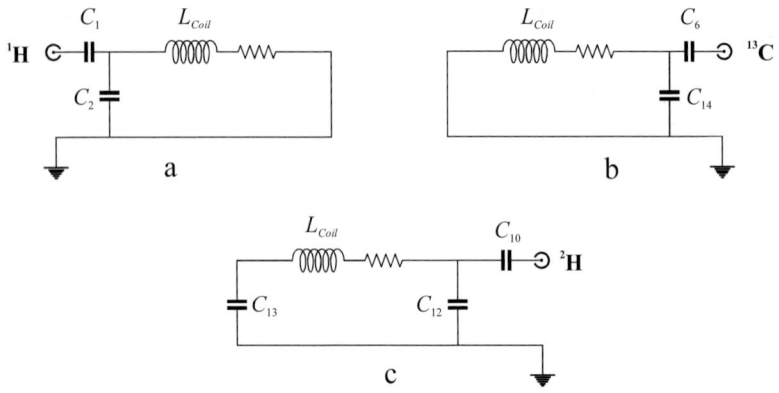

Fig. 6.11 Equivalent circuits of the triple tuned probe (Fig. 6.10) at the proton (a), carbon-13 (b) and deuterium (c) frequencies.

The proton matching network component values (C_1, C_2) are directly obtained by Eqs. (4.15) and (4.22) because none of the switching circuits ($L_4C_4C_3$ and L_5C_5) introduces a "parasitic" impedance. In contrast, the carbon tuning capacitor C_{14} of Fig. 6.11(b) is made out of C_7 in parallel with the impedance of the series circuit L_5C_5 resonating at ω_H. Hence, similarly with Eq. (6.15)

$$C_7 = C_{14} - C_5 \left(1 - \frac{\omega_C^2}{\omega_H^2}\right)^{-1}. \tag{6.22}$$

The required value for C_{14} is again calculated from Eq. (4.15). Usually, the C_5 capacitance at the input of the high frequency (proton) switch will be much lower than the required value for C_{14}, hence allowing a real, positive solution for C_7.

As a practical example, let us consider a sample coil with an inductance of 150 nH that should be triple tuned in a 4.7 T spectrometer operating at approximately 200, 50, and 31 MHz for the proton, carbon and deuterium, respectively. The chosen inductance is quite large for a 200 MHz probe, but it also represents a compromise for the carbon frequency at which it is expected to get a good signal-to-noise ratio.

Following the authors' comments on the design [Kan and Courtieu, 1980], the pairs L_4, C_4 and L_5, C_5, are resonating at ω_H. In addition, L_4 and L_5 have equal or smaller inductance compared with the sample coil and the best possible Q factor to minimize losses in the corresponding trap circuits. A value for $L_4 = L_5 = 50$ nH should fit all these criteria. A solenoid made of four turns of 8/10 mm copper wire wound on a 4.2 mm diameter support can be adjusted to a total length of 4.7 mm, giving the required inductance. Then $C_4 = C_5 = 12.7$ pF will tune the trap coils to resonance. Hence, C_3 is obtained from Eq. (6.6) as:

$$C_3 = \frac{\omega_H^2 - \omega_C^2}{L_4 \omega_H^2 \omega_C^2}, \tag{6.23}$$

where the *off* and *on* state are set to the proton and carbon frequency, respectively. It results that $C_3 = 190.0$ pF.

It is now time to build and align the dual proton/carbon probe part. The different steps are summarized in Fig. 6.12. Assuming a Q factor of the order of 200 at 200 MHz, the tuning and matching capacitances should be C_2 = 3.64 pF and C_3 = 0.6 pF, respectively. These are quite small values, resulting from a sample coil impedance at the highest acceptable limit (about 190 Ω). Thus the sample coil circuit is tuned and matched to ω_H with a short circuit replacing L_5C_5 [Fig. 6.12(a)].

Now inserting the circuit L_5C_5 (Fig. 6.12b), pre-tuned at ω_H, the probe is adjusted again for the proton frequency to optimum matching using essentially C_1. Only little changes should be observed due to the losses introduced by the series tuned circuit. This completes the proton channel alignment.

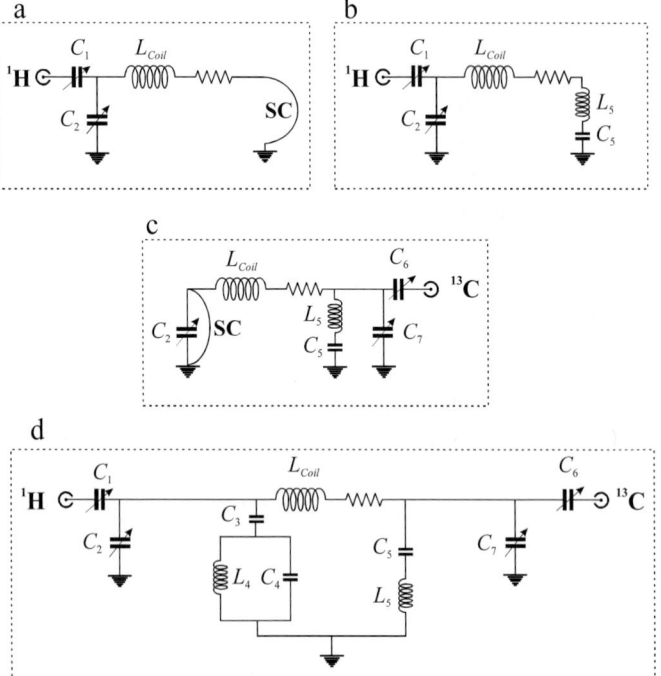

Fig. 6.12 Different steps of alignment of the triple tuned probe in Fig. 6.10 (see text for details). The circuit (d), obtained before drafting the deuterium channel, is a fully operational double tuned proton–carbon probe. SC is a "short circuit" that is temporarily connected. The switching circuits L_5C_5 and $L_4C_4C_3$ should be adjusted separately before being connected to the probe circuit.

The carbon channel is now aligned as follows. C_2 is temporarily short circuited and the probe is tuned and matched to ω_C, using C_7 and C_6 [Fig. 6.12(c)]. The corresponding values may be predicted from Eqs. (6.22) and (4.22) as $C_7 = 47.5$ pF and $C_6 = 6.6$ pF, assuming a Q factor of 100 at 50 MHz [C_{14} is 61.0 pF as given by Eq. (4.15)]. Replacing the short circuit by the $L_4C_4C_3$ network, which is already pre-tuned, the matching carbon capacitor C_6 should only be slightly readjusted, again due to the switch losses. This completes the dual proton–carbon part of the probe [Fig. 6.12(d)] whose isolation between all ports can be made higher than 50 dB [Kan and Courtieu, 1980] due to the use of series tuned circuits.

The second part, including the ^{2}H/^{13}C switching circuit in series with the deuterium matching network, can now be inserted to obtain the third resonance. The equivalent circuit at the deuterium frequency is given in Fig. 6.11(c).

This circuit is more complicated than the other ones because the two proton/carbon switches contribute to "parasitic" impedance at the deuterium frequency. On one end of the coil, C_{13} is the equivalent capacitance of the ^{1}H/^{13}C switching circuit $L_4C_4C_3$ at ω_D, in parallel with C_2. It is given by Eqs. (6.11) and (6.5):

$$C_{13} = C_2 + C_3\left(1 - \frac{\omega_D^2}{\omega_H^2}\right)\left(1 - \frac{\omega_D^2}{\omega_C^2}\right)^{-1}. \qquad (6.24)$$

On the other coil end, C_{12} [Fig. 6.11(c)] is the parallel combination of the deuterium and carbon tuning capacitors C_{11} and C_7, and the equivalent capacitance of the L_5C_5 network. Hence, using Eq. (6.15) again, one obtains:

$$C_{12} = C_{11} + C_7 + C_5\left(1 - \frac{\omega_D^2}{\omega_H^2}\right)^{-1}. \qquad (6.25)$$

The tuning capacitance C_{11} can be calculated from Eq. (6.25) with C_{12}, obtained as usual from Eq. (4.15) but taking account of the fact that the sample coil inductance is apparently decreased by the C_{13} series capacitance.

Hence, the inductance value to be used in Eq. (4.15) should be:

$$L' = L_{coil} - \frac{1}{C_{13}\omega_D^2}, \tag{6.26}$$

and the resulting quality factor becomes

$$Q' = \frac{L'\omega_D}{r} = \frac{L'}{L_{coil}}Q, \tag{6.27}$$

where r and Q are the resistance and quality factor, respectively, of the probe sample coil at the deuterium frequency.

From Eq. (6.24), C_{13} is estimated as 304.8 pF. The tuning/matching components for the deuterium channel can then be calculated from Eq. (4.15) (C_{12} = 381.27 pF) and Eq. (4.22) (C_{10} = 37.8 pF) assuming L' = 63.5 nH [Eq. (6.26)] and a resistance of the sample coil of 0.365 Ω at 31 MHz [Q = 80, then Q' = 33.9 from Eq. (6.27)]. C_{11} is deduced from Eq. (6.25) as 320.8 pF.

The adjustment of the deuterium matching network is done by connecting the capacitors C_{11} and C_{10} directly to C_6 (the $L_9C_9C_8$ network being temporarily short circuited) and connecting a 50 Ω load to the carbon port. The impedance connected to the carbon port may have a slight influence on the deuterium tuning adjustment. For a 50 Ω load at the *LF* port (carbon) the corresponding matching capacitor C_6 comes in parallel with C_7 and C_{11}. In this case C_{11} must be decreased accordingly. In contrast, if the load at the carbon port is high at ω_D, inserting for example a ^{2}H band stop filter, C_6 does not contribute to the deuterium channel tuning. Finally, the pre-tuned $L_9C_9C_8$ network is inserted and C_{10} will be slightly readjusted due to the loss introduced by L_9. This ^{2}H/^{13}C switching network can be designed around the inductance L_9 that should not be self-resonant, at least below the carbon frequency and preferably at all frequencies for which the probe is designed. As a rule of thumb, the impedance of the inductance to be used should not be greater than roughly 500 Ω at the highest frequency (200 MHz). Hence, the corresponding inductance should be equal to or less than 400 nH. A solenoid coil made of eight turns of 8/10 mm wire wound on an 8.8 mm diameter support and adjusted to a total length of 9.3 mm has the required inductance to be tuned to ω_D with a 65.9 pF capacitor (C_9). One gets $C_8 = 41.1$ pF, assuming the *on* state at ω_D and the *off* state at ω_C, respectively. The final circuit of the completed triple resonant probe is shown in Fig. 6.13.

Note that, due to the isolation scheme used here, the tuning/matching process at any ports will have no influence on the proton or carbon channels. In contrast, tuning the carbon channel (and to a much lesser extent, tuning the proton channel) would have little influence on the deuterium matching network adjustment.

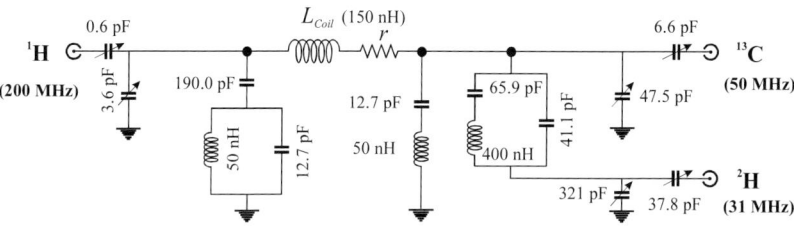

Fig. 6.13 The completed triple tuned probe with component values, designed for a 4.7 T spectrometer.

6.1.4 Balancing the shunting configurations

The methods described above have one main drawback. Due to the switching circuit, one end of the probe coil is connected to the ground at the frequency of interest, leading to a non-symmetrical configuration. On the other hand, attempts to equilibrate the coil probe imply that a voltage should exist at both ends of the coil, eliminating the circuit isolation advantage of this simple configuration. Thus, the shunting method appears to be of interest at low frequencies and for "small" sample coils, when the dielectric losses are expected to be negligible. In most cases however, it is highly desirable that the coil should be balanced. We propose here a balancing scheme based on the equilibrated configuration of Section 4.2.1.2.

6.1.4.1 *Approximate balanced circuit*

Let us consider again the example of a sample coil (150 nH) that should be tuned and matched at 200 MHz (^{1}H in a field of 4.7 T) and 50 MHz (^{13}C). As presented in the previous section, the sample coil inductance has the highest acceptable limit (188 Ω) for working at 200 MHz. This is why balancing the circuit is indispensable for optimizing the design. At the carbon frequency, the coil impedance has a much lower value (about 47 Ω), making the balancing less critical. This is justified even more if it is used as a decoupling channel. A similar argument applies if the second channel is used as a ^{2}H lock channel instead, where sensitivity is not of concern. Hence, one may accept that the probe is not equilibrated at the ^{13}C (or ^{2}H) frequency and still remains effective.

The circuit of Fig. 6.14 fulfills the above requirements in a very simple way. Let us start from the well-designed, equilibrated, and pre-tuned, proton circuit consisting of L_{coil}, C_0 (pre-tuning capacitor), C_1 (matching), C_2 (tuning) and C_3 (symmetrizing). The carbon circuit is grafted on the proton circuit adding an isolating trap circuit (L_6C_6) tuned to the proton frequency and a shunting circuit (L_7C_7) to close the probe coil circuit to the ground at ω_C.

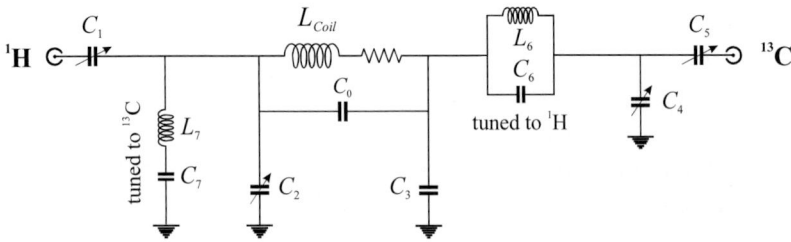

Fig. 6.14 A double tuned probe circuit that provides electrical balancing at the highest frequency mode (proton). Note that this mode is furthermore optimized by connecting a pre-tuning capacitor C_0 in parallel to the sample coil (Chapter 4). This circuit can be used efficiently for a proton-observe carbon-decouple probe configuration.

The L_6C_6 network can be replaced by a dual frequency switching circuit such as that shown in Fig. 6.6(a) or 6.6(b), but the impedance of L_6C_6 at ω_C is small enough[9] that such a complication may be prevented.

Again assuming a Q factor of 200 at 200 MHz, the component values (Fig. 6.12) can be easily obtained from Eq. (4.77) (C_1 = 1.16 pF), Eq. (4.70) (C_2 = 0.685 pF) and Eq. (4.71) (C_3 = 1.84 pF), assuming a pre-tuning capacitance of 3.3 pF. The value obtained for C_2 should however be corrected from the parasitic impedance of L_7C_7 at ω_H:

$$C_2 = 0.685 + C_7 \left(\frac{\omega_H^2}{\omega_C^2} - 1 \right)^{-1} pF . \qquad (6.28)$$

In order to make small corrections, C_7 should be as small as possible, corresponding to a large inductance L_7. But, L_7 should not be self-resonating below the proton frequency. Hence, we choose an inductance of 400 nH (as in the previous section) that represents a reasonable limit. The tuning capacitance at 50 MHz is therefore C_7 = 25.33 pF. The correction value is 1.69 pF. Hence, the L_7C_7 circuit contributes significantly to the proton tuning (C_2 = 2.38 pF). As far as L_7C_7 is made with good components, losses will be minimized, but the tuning range of the proton channel will be limited accordingly.

The calculation of the tuning and matching capacitance values at the carbon frequency requires the impedance estimation of the total circuit at ω_C, seen from the carbon matching network C_4C_5 (Fig. 6.15). It includes the circuit formed by the sample coil in parallel with $C_0 + C_3$ and in series with the proton trap C_6L_6. The latter is assumed to be designed as in the previous section, with L_6 = 50 nH and C_6 = 12.7 pF. The probe circuit including the proton trap is equivalent to a 215 nH inductance in series with a 0.55 Ω resistance at 50 MHz. It is worth noting at this point that the proton trap contributes to more than 30% of the reactive impedance. This has an important consequence respective to the apparent Q factor of the circuit. In the present case, assuming a lossless trap circuit, the apparent Q factor at the carbon port is 123 (215/0.55) instead

---

[9] This is allowed by the fact that the frequency ratio ω_H/ω_C is large. For closest frequencies, a complete switching circuit would have been desired.

of the "true" quality factor of the sample coil (100). This is a general behavior observed in the case of multiple tuned circuits where the Q factor does not represent the expected sensitivity. A better way to evaluate the probe is to simulate the circuit, predict the B_1 magnetic field (hence the 90° pulse), and compare the results thus obtained with the experiment. The simulation also allows optimization of the probe circuit through an analysis of the repartition of currents in the different parts of the circuits, localizing the losses. As a result, the losses should be preferably concentrated in the sample coil rather than in the trapping and/or shunting circuits.

The capacitance values required to tune and match the circuit to 50 Ω at ω_C are obtained from the equivalent inductance impedance shown in Fig. 6.15 and Eqs. (4.15) and (4.22), as $C_4 = 42.08$ pF and $C_5 = 4.95$ pF.

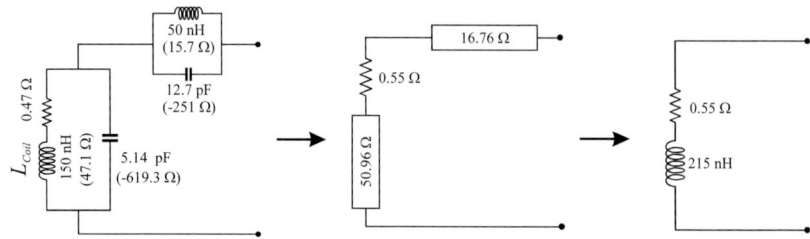

Fig. 6.15 Equivalent circuit of the dual frequencies probe (Fig. 6.14) as it appears at ω_C. The equivalent final L,r circuit is used to calculate the tuning and matching component values using the formulae of Chapter 4.

6.1.4.2 Multiple-frequency full balancing circuit

Figure 6.16 shows a circuit that provides the equilibration of the sample coil *at both frequencies*. The circuit is again based on the balanced configuration described in Section 4.2.1.2. Balancing is provided at both frequencies commuting the two tuning *LF* capacitors by an appropriate frequency-dependent switch. Isolation between both channels is provided by blocking circuits.

These blocking circuits can be a simple high impedance trap circuit. They are required in the *LF* connection in order to isolate the *HF* tuning components from the low impedance *LF* matching capacitor.

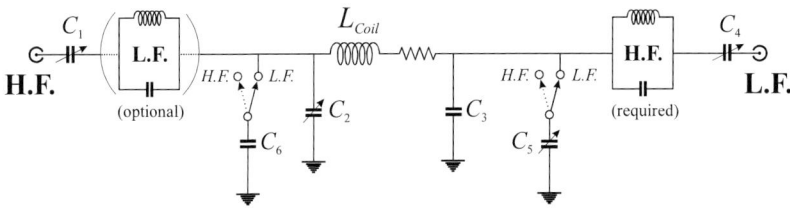

Fig. 6.16 A general double tuned single coil probe circuit that provides electrical balancing at both *HF* and *LF* resonant frequencies.

In contrast, the *LF* blocking circuit could be omitted in the *HF* connection as the matching capacitor C_1 presents a high impedance at *LF*. The protection of the *HF* port from *LF* energy leakage can be improved anyway by an external band stop filter.

The *HF* tuning capacitors are not switched in order to avoid any additional lossy components in the *HF* circuit. This has little influence on the *LF* tuning component adjustment because C_2 and C_3 have high impedance at *LF*. Only a slight imbalance will be introduced at *LF* by C_3 in parallel with C_5. Note that the symmetrizing capacitors are almost independent of the probe loading when the probe is balanced (Section 4.2.1.2). Hence, only one adjustable capacitor is required for tuning each channel, as indicated in Fig. 6.16.

The switches are realized according to the circuit shown in Fig. 6.6(b). In this case, the switch capacitors [C_2 in Fig. 6.6(b) and C_5 (or C_6) in Fig. 6.16] are in series and are replaced by only one component [Fig. 6.17(a)].

The calculation of the value for the components entering the circuit is straightforward.

First, the capacitances entering the proton circuit [Fig. 6.18(a)] are immediately obtained from Eqs. (4.70), (4.71) and (4.77). As the frequency-dependent switches have infinite impedance at 200 MHz, no correction is required.

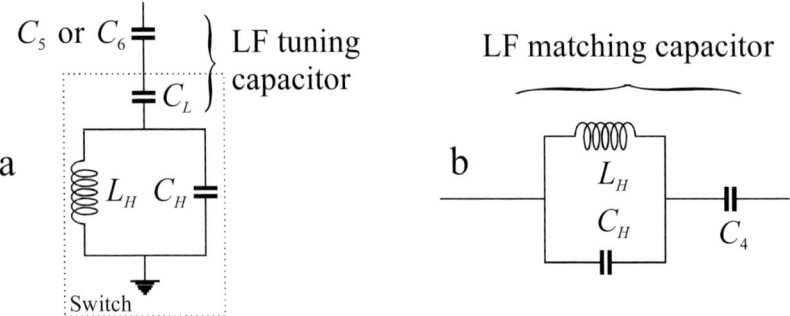

Fig. 6.17 Contribution of the switching circuits used in the probe design of Fig. 6.16 connected to the *LF* tuning and symmetrizing capacitors (a) and to the *LF* port matching capacitor (b). Note that the switches are identical and designed for $\omega_{on} = \omega_{LF}$ and $\omega_{off} = \omega_{HF}$, allowing a complete independence of the *HF* circuit alignment.

Second, the tuning, matching, and symmetrizing capacitances for the carbon circuit [Fig. 6.18(a)] are calculated similarly. Third, a correction is applied to the calculated values to take account of the parasitic elements originating from the proton circuit and from the switch impedance at 50 MHz (Fig. 6.17).

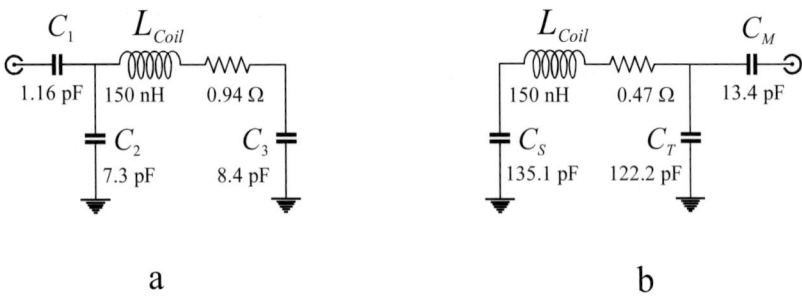

Fig. 6.18 (a) Component values required to tune and match the probe coil at 200 MHz, assuming a Q value of 200. (b) Component values required to tune and match the same probe coil at 50 MHz, assuming a Q factor of 100.

The final circuit is shown in Fig. 6.19, together with the simulated scattering parameters.

Multiple Frequencies Tuning

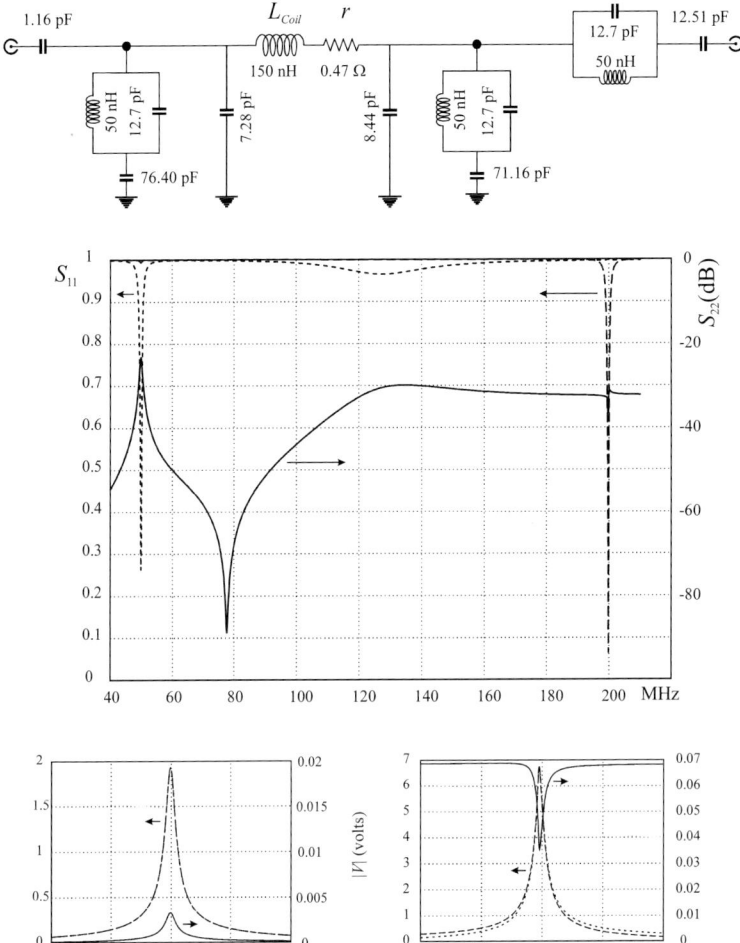

Fig. 6.19 Upper: the final double tuned (200/50 MHz) fully balanced probe circuit. Note that r is equal to 0.47 Ω at 50 MHz and 0.94 Ω at 200 MHz. Middle: the simulated scattering parameters for the above circuit including traps with realistic losses. Note that the port isolation is down to −40 dB at 200 MHz. Lower: voltages between the coil terminals and ground (dashed lines) and at its middle point (full line) when driven at the carbon port (left) and the proton port (right). Note that the midvoltage is two orders of magnitude lower (right scale) than at the coil terminals, indicating a good coil balancing.

6.2 Multiple-pole Circuits

This method is also generally preferred for relatively well separated resonant frequencies. It consists of replacing the tuning capacitor(s) by a network having one or more poles, similar to the circuits already described in the previous section as dual frequency switches (Fig. 6.6).

First, let us consider a simple resonator $L_{coil}C$ that has a given resonant frequency ω_0 (for example, a probe that is already designed for low frequency (ω_{LF}) nuclei like ^{39}K, ^{13}C, ^{31}P, etc.). Sometimes the probe must also be tuned to the proton frequency (ω_{HF}) in order to optimize the main field homogeneity, to localize a volume of interest on an image, or to irradiate the protons.

A well-known method for obtaining a new resonance is to insert another resonant circuit ($L_1 C_1$) into the initial circuit, as shown in Fig. 6.20. This resonant circuit is sometimes called a "trap," referring to its usage in radio communication antenna technology. The radiating wire (a resonant dipole) is split in different resonating sections by inserting traps, tuned for the desired frequencies. These traps block the current at their location, limiting the antenna length. In our case, the resonant circuit is better called a tank circuit, as it stores the electromagnetic energy.

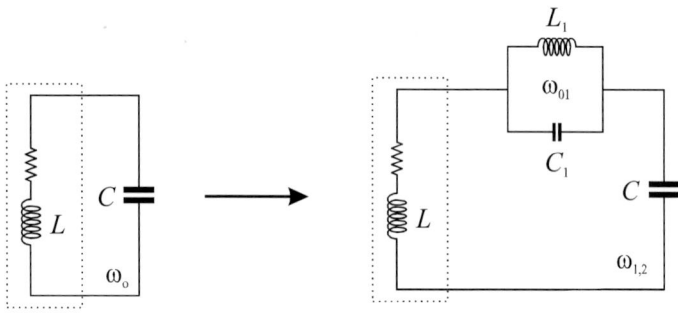

Fig. 6.20 A given resonator (left) is double tuned by inserting another resonant circuit (right) which provides a second pole.

After insertion of the trap, the circuit exhibits two resonances, as graphically shown in Fig. 6.21 and quantitatively given by:

$$\omega_{1,2} = \left[\frac{\Omega^2 \pm \sqrt{\Omega^4 - 4\omega_0^2 \omega_{01}^2}}{2}\right]^{1/2}, \quad (6.29)$$

where

$$\Omega^2 = \frac{1}{LC} + \frac{1}{L_1 C_1} + \frac{1}{LC_1} \quad (6.30)$$

and ω_0 and ω_{01} are the resonant frequencies of the LC and $L_1 C_1$ circuits:

$$\omega_0^2 = \frac{1}{LC} \quad \omega_{01}^2 = \frac{1}{L_1 C_1}. \quad (6.31)$$

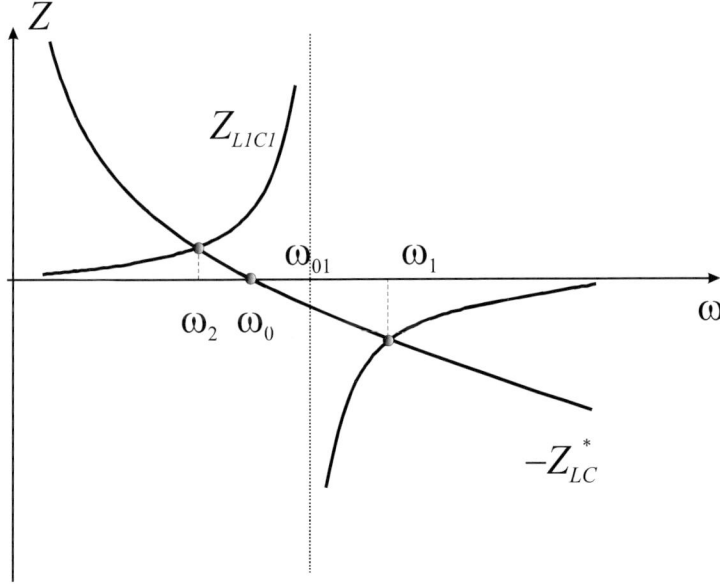

Fig. 6.21 Graphical representation of the solutions that give the resonant frequencies of the circuit in Fig. 6.20. Z_{L1C1} is the impedance of the trap that should be equal to the complex conjugate of the impedance of the series LC circuit (equal to $-Z_{LC}$). ω_0 is the "zero" of the LC circuit impedance and ω_{01} is the "pole" of the trap $L_1 C_1$ circuit impedance. ω_1 and ω_2 are the new resonant frequencies of the whole circuit.

To build the double tuned circuit, the resonance frequency ω_{01} of the tank circuit[10] can be freely chosen, but the choice has important consequences on the probe efficiency at ω_{LF} and ω_{HF}.

Usually, if the two desired resonances are far apart, the added tank circuit is tuned close to *HF*. Its impedance at *LF* is equivalent to a small inductance which only slightly depends on the trap resonance frequency. If this inductance, added to the sample coil, is small enough and well-designed, the losses associated with the additional coil are also minimized. On the contrary, the tank circuit tuning has a great influence on the *HF* resonant frequency. Its impedance should be equal to the small capacitance required to tune the sample coil at *HF* (if $\omega_{LF} \ll \omega_{HF}$, the capacitor *C* has a negligible low impedance at *HF*).

Hence, the tank circuit is the essential part that tunes the sample coil to *HF* and its loss contributions accordingly become of prime importance to the overall performance of the probe at *HF*. In this way, the probe efficiency is favored at *LF*. In other words, if resonating close to *HF*, the tank circuit stores a large part of the electromagnetic energy when the probe is excited at *HF* and consequently may dominantly contribute to the losses. When excited at *LF*, only a small part of the energy is concentrated in the tank circuit (low *L*), hence it is mostly dissipated in the sample coil.

Alternatively, the efficiency can be optimized at *HF* rather than at *LF* by reversing the role of the two resonating circuits (Fig. 6.22). Obviously, if the tank circuit coil, tuned close to *HF*, is the sample coil, then the *HF* electromagnetic energy stored in this circuit is efficiently dissipated in the sample. In contrast, the *LF* electromagnetic energy is dissipated in the added series resonant circuit $L_1 C_1$ that is assumed to be tuned close to *LF*. The efficiency of the probe at *LF* will consequently be lowered.

General rules for double tuning a probe resonator can be deduced from this discussion.

---

[10] The role of the trap circuit is different here than in the shunting circuits of the previous section. Here it is used as a tuning component, thus its design is more critical than if used as a high impedance device.

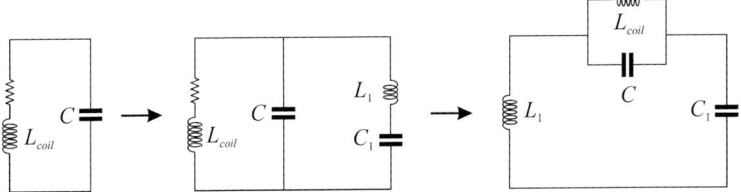

Fig. 6.22 An alternative method to double tune a resonator by connecting a series resonant circuit $L_1 C_1$ in parallel (upper). The advantage is that the initial resonator circuit is not opened. The final circuit is equivalent to the circuit in Fig. 6.20, where the role of L_1 and L_{coil} are interchanged (lower).

Starting from a previously designed single frequency resonator, a higher or lower frequency pole can be added to the circuit by either inserting a parallel resonant circuit (Fig. 6.18) or shunting the resonator by a series resonant circuit (Fig. 6.20). If the added circuit is tuned close to the desired new frequency, the performance of the probe will mostly be kept at the initial resonant frequency. But an efficiency compromise at both frequencies can be chosen by adjusting the tuning frequency of the added circuit in the range between *LF* and *HF*. A few relationships for approximating the efficiency are given in Section 6.5.

A quantitative approach to describing the multiple-poles resonating structure discussed so far is to consider that the sample coil inductance is tuned by a given frequency-dependent reactance connected in parallel (Fig. 6.23).

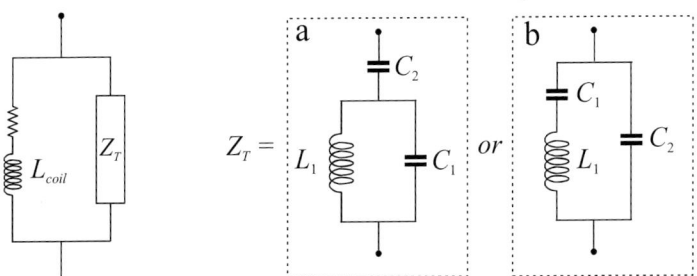

Fig. 6.23 A general procedure to multiple tune a sample coil L_{coil} by connecting in parallel a complex impedance (Z_T) that has different poles and zeroes. The networks (a) and (b) have a pole at ω_{off} and a zero at ω_{on}, as described in Section 6.1. They can be used to double tune a given sample coil.

When the coil inductance is equal to the complex conjugate of the parallel reactance, a resonance will be observed. Considering the case of double tuning, the circuits shown in Fig. 6.6 can be effectively used as candidates for the tuning reactance. For practical reasons, the circuits of Fig. 6.6(a) and 6.6(b) would be preferred because they use only one lossy inductance. The equivalence with the circuits described above becomes obvious (Fig. 6.23). Quantitatively, the required capacitive impedance to tune the coil at a given frequency (assuming a given resistance) is still obtained by the equations already given in Chapter 4 (see Table 4.4).

6.2.1 *Approximate double-pole network component values*

The components of the tuning circuits can be obtained by solving Eq. (6.11) or Eq. (6.12) for ω_{on} or ω_{off}.

Let C_{LF} and C_{HF} be the required capacitances to tune the probe coil inductance, L_{coil}, at ω_{LF} and ω_{HF},[11] respectively. Using either the circuit in Fig. 6.6(a) (hereafter referred to as circuit A) or in Fig. 6.6(b) (hereafter referred to as circuit B), the following equations should be solved in order to obtain the network component values:

$$\frac{1}{C_{HF}\omega_{HF}} = L_{coil}\omega_{HF} = Z^{A,B}(\omega_{HF}) \qquad (6.32)$$

and a similar equation for ω_{LF}:

$$\frac{1}{C_{LF}\omega_{LF}} = L_{coil}\omega_{LF} = Z^{A,B}(\omega_{LF}). \qquad (6.33)$$

The impedance $Z^A(\omega)$ and $Z^B(\omega)$ are given by Eq. (6.11) and Eq. (6.12), respectively, involving three unknowns, L_1, ω_{on} and ω_{off}. As the previous discussion suggests, it is natural to choose the resonant

---

[11] These are the "free running" resonant frequencies, which should, in a real probe circuit, be slightly different from the spectrometer frequencies as a result of coupling with the impedance matching networks (Chapter 4). An exact formulation is provided at the end of this section.

frequency of the tank circuit, ω_{on} (circuit A) or ω_{off} (circuit B), as a free parameter.

The solutions of Eqs. (6.32) and (6.33), giving the tuning network component values, are:

$$C_1 = \frac{1}{L_{coil}} \frac{\left(\omega_{on}^2 - \omega_{LF}^2\right)\left(\omega_{HF}^2 - \omega_{on}^2\right)}{\omega_{on}^2 \omega_{LF}^2 \omega_{HF}^2}, \qquad (6.34)$$

$$C_2 = \frac{1}{L_{coil}} \frac{\omega_{on}^2}{\omega_{LF}^2 \omega_{HF}^2}, \qquad (6.35)$$

$$L_1 = \frac{1}{C_1 \omega_{on}^2}, \qquad (6.36)$$

when using a series resonant circuit in parallel with a tuning capacitor C_2 (Fig. 6.22), and

$$C_1 = \frac{1}{L_{coil}} \frac{\omega_{off}^2}{\left(\omega_{off}^2 - \omega_{LF}^2\right)\left(\omega_{HF}^2 - \omega_{off}^2\right)}, \qquad (6.37)$$

$$C_2 = \frac{1}{L_{coil}} \frac{\omega_{off}^2}{\omega_{LF}^2 \omega_{HF}^2}, \qquad (6.38)$$

$$L_1 = \frac{1}{C_1 \omega_{off}^2}, \qquad (6.39)$$

when using a parallel tank circuit in series with a tuning capacitor C_2 (Fig 6.20).

As expected, the resonance frequency of the added circuit should be chosen in-between ω_{LF} and ω_{HF} (C_1 should be positive). On the other hand, when approaching both limits, the value for L_1 (circuit A) or C_1 (circuit B) diverges.

As a practical application, let us again consider the sample coil of 150 nH inductance, that should be tuned to both 50 and 200 MHz. The component values for the tuning network are given in Fig. 6.24 as a function of the resonant frequency of the "conjugate" resonator L_1C_1 (ω_{on}, circuit A or ω_{off}, circuit B) that is strongly (maximally) coupled with the circuit $L_{coil}C_2$.

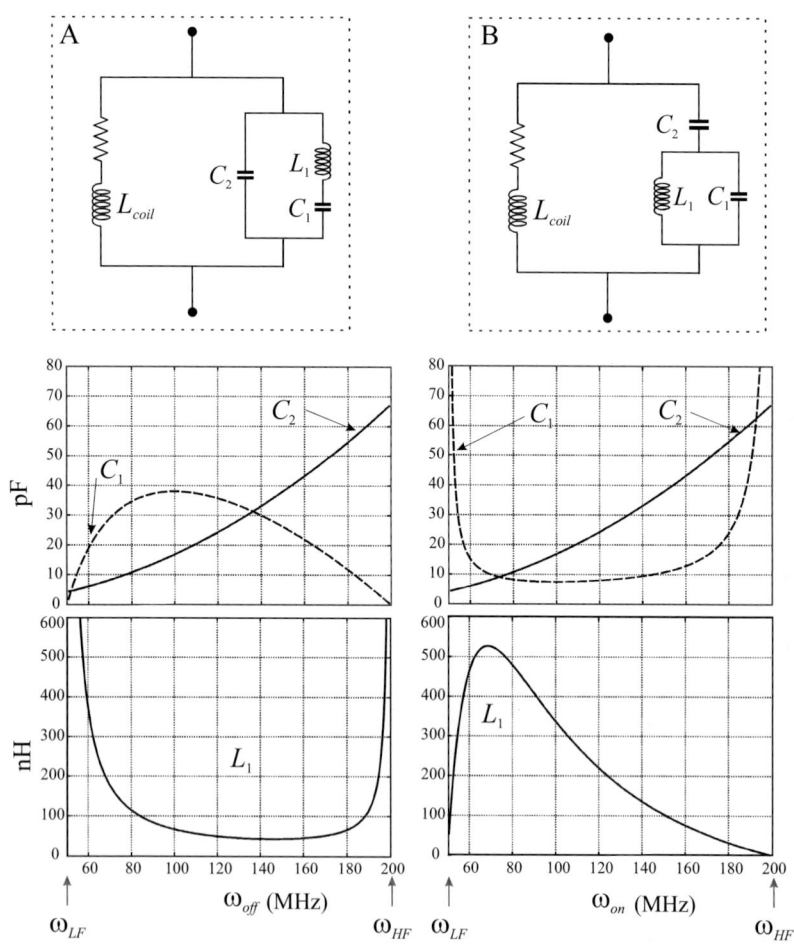

Fig. 6.24 Component values that double tune a 150 nH sample coil to 50 MHz (ω_{LF}) and 200 MHz (ω_{HF}), as a function of the designer choice for ω_{on} ($L_1C_1\omega^2_{on} = 1$, circuit A) or ω_{off} ($L_1C_1\omega^2_{off} = 1$, circuit B).

One obvious result is that C_2, which tunes L_{coil} to a given frequency, has an identical variation for the two circuits. Its value ranges from 4.2 pF to 67.5 pF enabling tuning of L_{coil} to 200 MHz and 50 MHz, respectively. In these extreme cases, the conjugate resonator L_1C_1 should be tuned to the "complementary" frequency (50 MHz or 200 MHz, respectively) and should be "transparent" to the $L_{coil}C_2$ resonance. Hence, the impedance of L_1C_1 should be either infinite (circuit A) or null (circuit B) at 50 MHz or 200 MHz, depending on the choice made for C_2. This can be achieved only when the inductance L_1 is either infinite and C_1 is zero (infinite impedance) or the capacitance C_1 is infinite and L_1 is null (zero impedance), as shown in Fig. 6.24 (circuit A or B). This is the obvious reason for the diverging values for L_1 or C_1 near the resonance frequencies ω_{LF} and ω_{HF}.

Between these two extreme conditions, the L_1C_1 resonance can be chosen, in principle, anywhere between the two desired frequencies ω_{LF} and ω_{HF}. As shown in Fig. 6.24, there is always a physically acceptable solution for both L_1 and C_1 and for both circuits. But, from a practical point of view, the range of possibilities will be somewhat limited by the need to use high quality components, especially respective to the inductance values.

We have already seen that the inductance entering a circuit functioning at 200 MHz should not exceed 400 nH. This value limits the acceptable lower value for ω_{off} to about 90 MHz. The upper limit for the ω_{off} frequency will be given by the lowest acceptable inductance value.

A value in the range of 10–15 nH appears reasonable in practice, still allowing a good quality coil to be built. Hence the upper limit of the available frequency range for ω_{off} will be near 190 MHz.

The possible frequency range for ω_{on} (circuit A) is greater compared with the corresponding one for circuit B, starting from about 60 MHz and ending close to 195 MHz. The lowest required inductance is always greater than 50 nH, which can easily be built. Hence there is a large range of possible choices to double tune a given coil. In fact, the ultimate criteria will be dictated by the efficiency compromise for both resonant frequencies (Section 6.5).

6.2.2 Exact solutions

Equations (6.34) to (6.39) are approximate. They are derived as they facilitate the prediction of the general behavior of this kind of circuit. The approximation comes from the fact that the reactive impedance of the tuning network is assumed to tune the sample coil to ω_{HF} and ω_{LF} precisely. However, matching the probe to the 50 Ω impedance using a capacitive network requires, as already seen in Chapter 4, the resonator to be slightly detuned by a quantity that depends on the coil Q factor. In this case, the impedance of the tuning capacitor given by Eq. (4.15), differs slightly from $L_{coil}\omega_{HF}$ or $L_{coil}\omega_{LF}$, as was assumed by Eq. (6.32) and Eq. (6.33).

In order to obtain the exact values for C_1, L_1 and C_2, the tuning components of the real circuit, incorporating the matching components, one has to solve Eqs. (6.32) and (6.33) using the values for C_{HF} and C_{LF}, as calculated from Eq. (4.15). The results will be given here only for the case of circuit B (Fig. 6.20) which is the most frequently used. Hence, from Eq. (6.12), the two equations to be solved are:

$$L_1 \frac{\omega_{off}^4 \left(\omega_{HF}^2 - \omega_{on}^2\right)}{\left(\omega_{off}^2 - \omega_{on}^2\right)\Omega_1^2} = \frac{1}{C_{HF}}, \qquad (6.40)$$

$$L_1 \frac{\omega_{off}^4 \left(\omega_{on}^2 - \omega_{LF}^2\right)}{\left(\omega_{off}^2 - \omega_{on}^2\right)\Omega_2^2} = \frac{1}{C_{LF}}, \qquad (6.41)$$

where Ω_1 and Ω_2 are defined by:

$$\Omega_1^2 = \omega_{HF}^2 - \omega_{off}^2 \qquad (6.42)$$

and

$$\Omega_2^2 = \omega_{off}^2 - \omega_{LF}^2 . \qquad (6.43)$$

As discussed above, ω_{off} is "freely" chosen between ω_{LF} and ω_{HF} and C_{HF} and C_{LF} are known quantities. Thus, Eqs. (6.40) and (6.41) will be solved for L_1 and ω_{on}. Using Eqs. (6.6) and (6.7), the capacitance values C_1 and C_2 will finally be obtained. The results are:

$$\omega_{on}^2 = \omega_{off}^2 - \frac{\Omega_1^2 \Omega_2^2 (K-1)}{K\Omega_1^2 + \Omega_2^2}, \tag{6.44}$$

where K is defined as:

$$K = \frac{C_{LF}}{C_{HF}}. \tag{6.45}$$

The component values are then given by:

$$L_1 = \frac{1}{\omega_{off}^4} \left(\frac{1}{C_{HF}} - \frac{1}{C_{LF}} \right) \frac{\Omega_1^2 \Omega_2^2}{\Omega_1^2 + \Omega_2^2}, \tag{6.46}$$

$$C_1 = \frac{1}{\omega_{off}^2 L_1}, \tag{6.47}$$

$$C_2 = \frac{C_1}{\omega_{off}^2} \left[\frac{K\Omega_1^2 + \Omega_2^2}{\Omega_1^2 \Omega_2^2 (K-1)} - 1 \right]^{-1}. \tag{6.48}$$

These equations assume that the L_1 coil is of high quality ($Q > 10$). The matching capacitor is still given by Eq. (4.22) assuming also that L_1 has a high Q factor. When a significant amount of energy is lost in L_1 (poor efficiency) the matching capacitor has to be significantly increased, even if the *apparent* Q factor of the matched circuit appears to be high.

6.2.3 *Multiple-pole tuning for more than two frequencies*

Tuning the probe coil to more than two frequencies can be done by further extending the previous method [Schnall, 1992], as shown for example in Fig. 6.23. The tuning reactance is itself built on a combination of coupled resonant circuits, providing additional poles.

The design of the multiple frequency tuning reactance should again be done taking into account the unavoidable efficiency compromises imposed by the experimental needs.

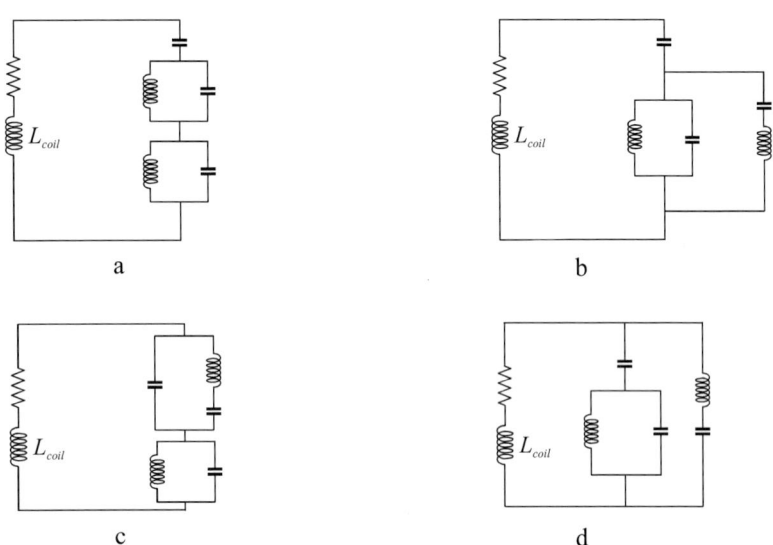

Fig. 6.25 A few reactive impedance circuits that can be used to triple tune a given sample coil. Each has three poles and/or zeroes. Other combinations are still possible, provided the reactive impedance has the required number of poles/zeroes.

6.2.4 Balancing the multiple-pole circuits

6.2.4.1 Inductive coupling

One obvious method to balance the coil is to drive it by an inductive coupling (Fig. 6.26).

Each channel is fed to the coupling loop at one end, the other end being terminated by the appropriate switching circuit. The switching circuit connected in parallel with the *HF* (*LF*) port should be off at *HF* (*LF*) and on at *LF* (*HF*). The switching circuits of Fig. 6.6(a),(b) and of Fig. 6.6(c),(d), respectively, fulfill this requirement.

In fact, a series resonant circuit in parallel to each port could be sufficient, if the *HF* and *LF* frequencies differ significantly. The series resonant circuit has two distinct roles. It presents null impedance closing the current path at its design frequency and it contributes to the isolation between both channels.

The impedance of the *LF* series resonant circuit is high at *HF* due to its inductance. Reciprocally, the impedance of the *HF* series resonant circuit is high at *LF* due to its capacitance. The introduction of these circuits in the coupling port will therefore have little influence on the matching for both channels.

Finally, the coupling loop should be over coupled (Section 4.1.3.4) to the resonator coil for both frequencies. In this case, the final matching can be realized by an adjustable capacitor inserted in series with the coupling loop (Fig. 6.26).

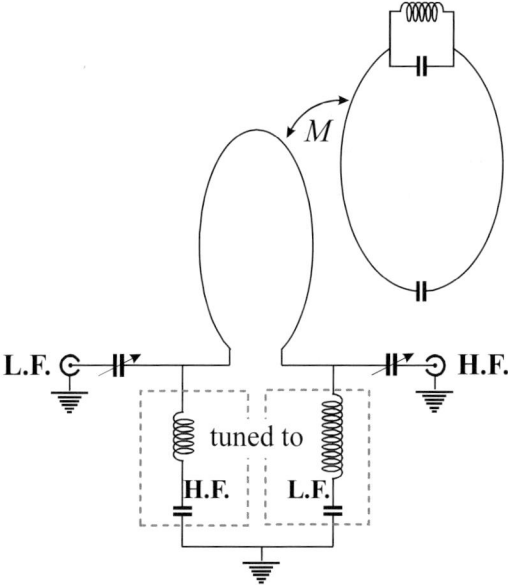

Fig. 6.26 An inductive coupling permits a natural balancing of the coil at both frequencies. The series resonant circuits close up the coupling loop circuit and provide isolation between ports.

6.2.4.2 The null-point method

The "null-point method" recently proposed by Wetterling *et al.* [Wetterling *et al.*, 2012] is a nice solution that provides a balancing of the coil at both frequencies and isolation between ports. The circuit can be applied when both frequencies differ significantly.

The double tuning of the coil is obtained using a circuit similar to Fig. 6.23(b), except that the coil is "pre-tuned" to a frequency intermediate between the high (*HF*) and low (*LF*) operating frequencies. The key to the method is to establish a virtual ground in the middle of the coil by splitting the pre-tuning capacitor equally and physically connecting the mid-point to the ground (Fig. 6.27). Then, the *LF* and *HF* ports are connected to virtual grounds that are created at *HF* and *LF*, respectively. In this way, the coil remains balanced and the ports are isolated from each other.

To understand the circuit functioning, let us first consider the circuit tuned at *HF* [Fig. 6.27(a)]. To shift the resonant frequency of the (L_{coil}, C_0) resonator, a trap circuit tuned at a frequency higher than *HF* is added in parallel to the coil. This trap circuit is part of the reactance that will be used to double tune the coil. The mid-point of the trap coil L_1 is a virtual ground (null-point).

Similarly, to shift the resonance of the circuit (L_{coil}, C_0) to *LF*, a capacitor C_2 should be added in parallel to C_0. This capacitance is split in two halves to create a null-point at *LF* [Fig. 6.27(b)].

In the double tuned circuit [Fig. 6.27(c)], the capacitance C_2 will not significantly move the *HF* null-point. C_2 indeed has negligible impedance at *HF*, provided there is a large difference between *HF* and *LF*. In contrast, the trap circuit L_1C_1 introduces a substantial reactive impedance at *LF*. This impedance can be compensated by changing the ratio between the two capacitors C_{2a} and C_{2b}[12] [Fig. 6.27(d)]. In this way, a null-point at *LF* is maintained at the point connecting C_{2a} and C_{2b}.

The matching capacitors at *HF* and *LF* can now be connected to the *LF* and *HF* null points, respectively [Fig. 6.27(b)]. Because the *LF* null

---

[12] C_{2a} should be increased to compensate for the inductive impedance of L_1C_1 at *LF*. C_{2b} should be decreased to maintain the total capacitance of C_{2a} in series with C_{2b}.

point is probably not exactly at the mid-point of L_1, a trap circuit is inserted in series with the *LF* matching capacitor improving the port isolation.

Finally, to fine tune the circuit at *HF*, the authors [Wetterling *et al.*, 2012] propose adjustment of the inductor L_1 [Fig. 2.47(a)]. Fine tuning at *LF* is done with a variable capacitor connected in parallel to C_{2a}, C_{2b}.

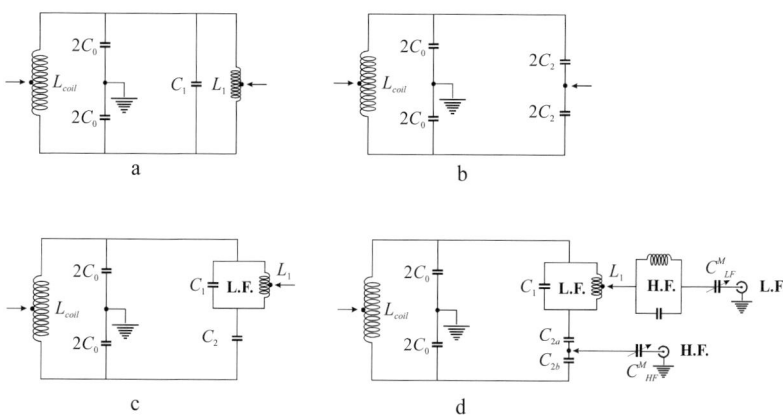

Fig. 6.27 (a) The resonator L_{coil}, C_0 is tuned to a frequency intermediate between *HF* and *LF*. The circuit $L_1 C_1$ ($\omega_{off} > HF$) decreases the effective inductance of L_{coil} leading to a resonance at *HF*. The reference ground is established by the physical connection to the mid-point of C_0. This established a virtual ground in the middle of L_{coil} (balancing) and another one in the middle of L_1. The arrows indicate these virtual grounds or "null-points". (b) The circuit L_{coil}, C_0 is tuned to LF by adding a capacitance C_2 in parallel to C_0. C_2 is split into two equal capacitances. The mid-point between the two capacitances (arrow) is at a zero potential with respect to the ground, creating a null-point. (c) Double tuning is obtained by inserting the trap circuit $L_1 C_1$ in series with the capacitance C_2. The mid-point of L_1 is still a null-point, but only at *HF* and if the impedance of C_2 can be neglected at this frequency. (d) A null-point at *LF* is created by splitting the C_2 capacitance into two different capacitances C_{2a} and C_{2b} to take account of the impedance of the $L_1 C_1$ circuit at *LF*. The *HF* port is connected at this point. Similarly, the *LF* port is connected to the *HF* null-point. To improve the isolation between the *LF* and *HF* channels, a trap circuit, tuned at *HF*, is inserted into the *LF* connection. Fine tuning at *LF* is realized with a capacitor connected to C_{2a}, C_{2b} and the inductor L_1 is made variable to fine tune at *HF*.

Approximate values for the circuit design can be obtained by solving the following equations:

$$\begin{aligned} Z_R(\omega_{HF}) &= Z^B(\omega_{HF}) \\ Z_R(\omega_{LF}) &= Z^B(\omega_{LF}) \end{aligned}, \quad (6.49)$$

where Z_R is the impedance of the resonant circuit $L_{coil}C_0$ and $Z^B(\omega)$ is the impedance of the circuit of Fig. 6.23(b). $Z^B(\omega)$ is given by Eq. (6.12).

The capacitances C_{2a} and C_{2b} are given by the following equations:

$$C_{2a} = 2C_2 - \frac{1}{|Z_T(\omega_{LF})|\omega_{LF}}$$

$$C_{2b} = \left[\frac{1}{C_2} - \frac{1}{C_{2a}}\right]^{-1} \quad (6.50)$$

C_2 is the solution of Eq. (6.49) and $Z_T(\omega_{LF})$ is the impedance of the L_1C_1 circuit at ω_{LF}.

The resonant frequency of $L_{coil}C_0$ can be freely chosen between the *HF* and *LF* [Wetterling et al., 2012]. A convenient value for C_0 could be around:

$$C_0 = \frac{1}{L_{coil}\omega_{HF}\omega_{LF}}. \quad (6.51)$$

Finally, the resonant frequency ω_{off} of the trap circuit L_1C_1 should be chosen higher than the *HF*. Intuitively, the inductance L_1, solution of Eq. (6.49), should be chosen much smaller than L_{coil} to favor efficiency at *LF*. However, too small a value for L_1 could significantly degrade the efficiency at *HF*.

6.3 Coupling Tank Circuits

The method described so far, using two (or more) maximally (strongly) coupled resonators, is well adapted for resonant frequencies that are far apart but is practically unusable for those close together. In contrast, by controlling the coupling between the two resonators, a fine adjustment of the resonant curve could be obtained.

The two resonators (the probe coil resonator and a "tank circuit") can be coupled together in a large variety of ways, some of which are shown in Fig. 6.28 [Terman, 1943; Haase *et al.*, 1998].

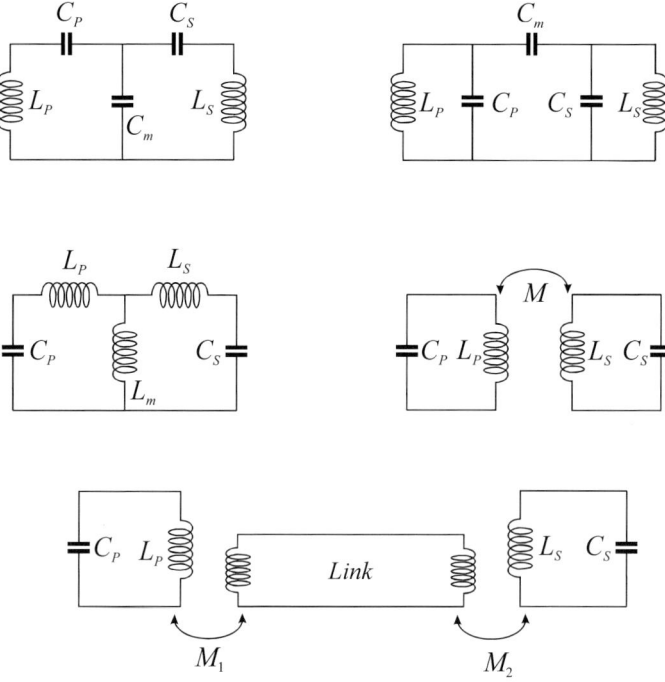

Fig. 6.28 A few coupled resonator configurations providing two resonant modes. Resonant frequencies can be adjusted by the coupling reactance value (C_m or L_m) or by the mutual inductance M.

As the coupling strength increases, the resonance curve changes from a simple broadening in the case of "under coupling" to the appearance of two resonances in the case of "over coupling" (Fig. 6.29). The technique has been well-known since the early days of radio communication [Terman, 1943] and was, among other applications, extensively used in the design of band-pass filters. It has also been applied to the design of double tuned NMR probes [Schnall *et al.*, 1986b; Fitzsimmons *et al.*, 1987].

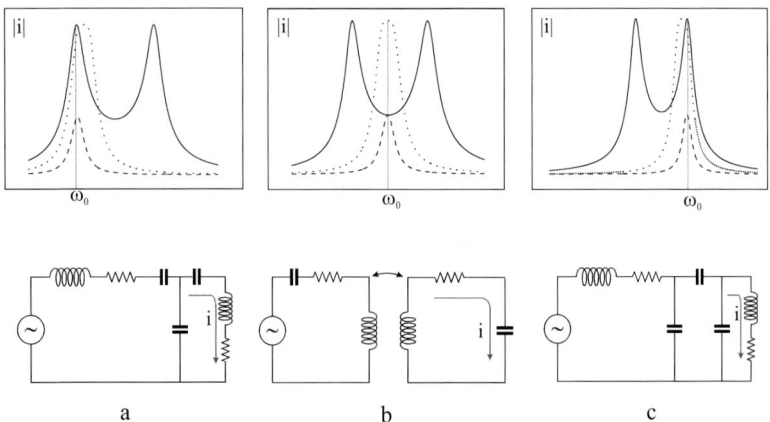

Fig. 6.29 Universal resonant curves obtained for three different coupling schemes of two identical resonators (same resonant frequency ω_0 and same L/C ratio). When loosely coupled, only one resonance is obtained, at ω_0 (dashed lines). If the coupling is increased to the critical value, the resonance curve broadens and becomes flat at its maximum (dotted lines). Increasing the coupling strength above the critical coupling, two resonant peaks appear (full lines) depending on the coupling scheme.

For any coupled schemes, two resonant modes[13] appear that differ principally in the current phase which flows in the probe resonator and the tank circuits. This fact may have important consequences for the magnetic field distribution if the tank circuit coil "sees" the sample volume. In this case the tank circuit magnetic field should be preferentially shielded such that the same field distribution is produced by the sample coil for both resonant modes.

---

[13] The two resonators can initially be tuned to the same resonant frequency (degenerate case). The coupling splits the degenerate state into two distinct resonant modes (symmetric and anti-symmetric combinations of the degenerate modes). In case the resonators are initially tuned to different frequencies, the coupling will mix the two modes to provide two new resonant modes characterized by different frequencies and (coil) currents. This behavior is sometimes compared to the behavior of two mechanically coupled pendulums, sound wave resonators (musical instruments), or even to coupled eigenstates of any quantum system.

The resonant mode frequencies of the coupled circuits depend essentially on their respective isolated resonant properties and on the coupling strength. For simplicity, one will neglect the resistances of both coils. Consequently, three cases should be considered.

The first case involves two identical resonators having the same resonant frequency ω_0 and the same L/C ratio. The second case is when the two resonators have the same ω_0 but differ in the L/C ratio. Finally, the general case where the two resonators have different characteristics will be considered. Furthermore, the coupled circuits can be described either as a T- or a Π-network. The former will be considered first as it is simpler from an analytical point of view. The formulae related to the latter will then be deduced, using the concept of a dual, or reciprocal, network.

The general coupled circuit is represented in Fig. 6.30 as a T-network. Z_m is the coupling impedance and Z_P and Z_S represent the "primary" and "secondary" impedance, respectively.

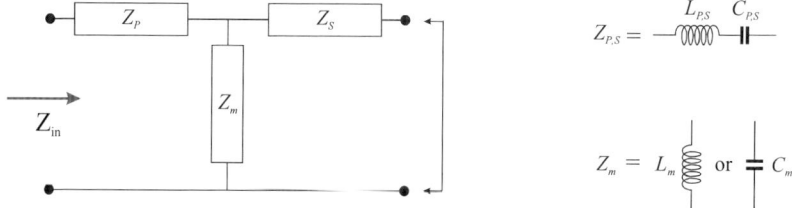

Fig. 6.30 T-network describing one type of coupled resonator circuit. Z_P and Z_S are the two resonators that are coupled through Z_m. The coupling device can be an inductor or a capacitor. The network is terminated by a short circuit and is driven at its input port.

To find out the resonant frequencies of the whole circuit (its zeroes), the following equation should be solved for ω:

$$Z_{in}(\omega) = 0, \quad (6.52)$$

where Z_{in} is the input network impedance when it is terminated by a short circuit.

Using, for example, the chain matrix formalism (Chapter 3), Z_{in} is given by:

$$Z_{in} = \frac{B}{D} = \frac{Z_P + Z_S + \frac{Z_P Z_S}{Z_m}}{1 + \frac{Z_S}{Z_m}}, \qquad (6.53)$$

from which one immediately deduces the following resonance equation:

$$Z_P Z_m + Z_S Z_m + Z_P Z_S = 0. \qquad (6.54)$$

6.3.1 Coupled identical resonators

In this case, the two resonators have the same resonant frequencies and the same L/C ratio. Hence, the impedances Z_P and Z_S are equal for all ω:

$$Z_P(\omega) = Z_S(\omega) = Z(\omega) \quad \forall \omega. \qquad (6.55)$$

Equation (6.51) becomes:

$$(2Z_m + Z)Z = 0 \qquad (6.56)$$

having two solutions:

$$Z = 0, \qquad (6.57)$$

$$Z = -2Z_m. \qquad (6.58)$$

The first solution corresponds to the zero of Z, i.e., the resonance frequency ω_0 of the two identical resonators. The frequency of the second mode, obtained from Eq. (6.58), can be adjusted by the coupling impedance without affecting the first resonant mode frequency. This is an interesting property that makes this kind of coupling attractive when designing dual closely spaced frequency NMR probes.

Assuming Z is a series resonant circuit, having ω_0 of zero, $Z(\omega)$ is a continuous function of frequency, having a negative value below ω_0 and a positive value above ω_0 (Fig. 6.31).

Hence, if the coupling impedance is an inductance ($Z_m > 0$ for all ω), Eq. (6.58) has only one solution for ω_2 less than ω_0 [Fig. 6.31(a)]. On the contrary, if the coupling impedance is a capacitance ($Z_m < 0$ for all ω), Eq. (6.58) has a solution for a frequency ω_2 greater than ω_0 [Fig. 6.31(b)].

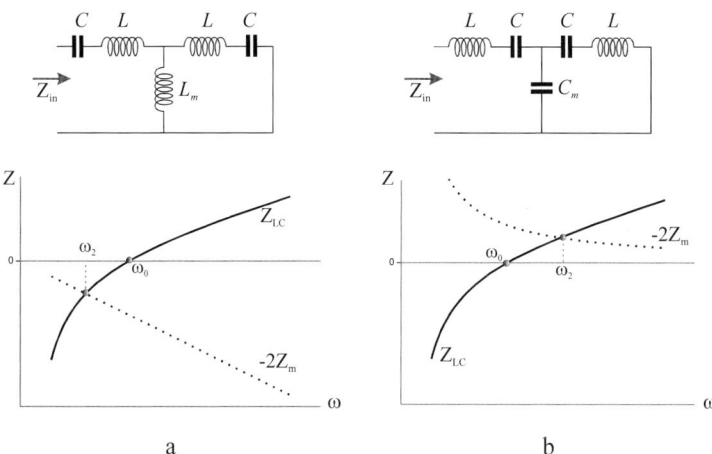

Fig. 6.31 Graphical representation of the resonance equation [Eq. (6.56)] when the resonators are assumed equivalent, having a resonant frequency equal to ω_0. With an inductive coupling ($-Z_m < 0$, increasing in absolute value with ω), one obtains a second resonance frequency, ω_2, lower than ω_0 (a). With the capacitive coupling ($-Z_m > 0$), one obtains a second resonant frequency higher than ω_0 (b).

Furthermore, the frequency difference between the two modes will increase with the coupling impedance (L_m increases or C_m decreases), resulting in either a decreased (inductive coupling) or increased (capacitive coupling) resonant frequency.

The solution of Eq. (6.58) is given by:

$$\omega_2 = \omega_0 \sqrt{1 + 2C/C_m} \qquad (6.59)$$

for the case of capacitive coupling, and

$$\omega_2 = \omega_0 \sqrt{\frac{1}{1+2L_m/L}} \qquad (6.60)$$

for the case of inductive coupling through the impedance L_m.

The above results are in agreement with those already obtained in an elegant manner by Haase [Haase et al., 1998], considering the equivalent circuit for each resonant mode.

It is usual to express the resonant frequencies as a function of the so-called "coupling coefficient" k. For the circuits considered here, it is given by:

$$k = \frac{C}{C+C_m} \qquad (6.61)$$

for the capacitive coupling, and

$$k = \frac{L_m}{L+L_m} \qquad (6.62)$$

for the inductive coupling case.

From Eqs. (6.59) and (6.60), the resonant frequencies are given by:

$$\omega_1 = \omega_0$$
$$\omega_2 = \omega_0 \sqrt{\frac{1+k}{1-k}} \qquad (6.63)$$

for the capacitive coupling case, and

$$\omega_1 = \omega_0$$
$$\omega_2 = \omega_0 \sqrt{\frac{1-k}{1+k}} \qquad (6.64)$$

for the inductive coupling case.

A more general equation that relates the resonant mode frequencies ω_1 and ω_2 of the two coupled identical resonators has already been proposed [Hong and Lancaster, 1996]:

$$\frac{|\omega_2^2 - \omega_1^2|}{\omega_2^2 + \omega_1^2} = k. \tag{6.65}$$

The above equation can also be derived from Eqs. (6.63) and (6.64).

6.3.2 Coupled resonators having the same resonance frequencies but different L/C ratio

In this case, the frequency dependence of Z_S and Z_P are functionally different, although the corresponding polynomials have the same zeroes.

Again assuming a series resonant LC circuit, its impedance as a function of frequency can be written as:

$$Z_i = \frac{1}{jC_i\omega}\left(1 - \frac{\omega^2}{\omega_i^2}\right) \quad i = P, S \tag{6.66}$$

or

$$Z_i = jL_i\omega\left(1 - \frac{\omega_i^2}{\omega^2}\right) \quad i = P, S, \tag{6.67}$$

where ω_i is the resonant frequency of the L_iC_i circuit (Fig. 6.28).

In the capacitive coupling case, Eq. (6.67) may be rewritten, conveniently using Eq. (6.66) for Z_P and Z_S and letting $\omega_P = \omega_S = \omega_0$.

Hence:

$$\frac{1}{C_m}\left(\frac{1}{C_P} + \frac{1}{C_S}\right)\left(1 - \frac{\omega^2}{\omega_0^2}\right) = -\frac{1}{C_PC_S}\left(1 - \frac{\omega^2}{\omega_0^2}\right)^2, \tag{6.68}$$

from which one gets the solutions for ω:

$$\omega_1 = \omega_0,$$
$$\omega_2 = \omega_0\sqrt{1 + \frac{C_P + C_S}{C_m}}. \tag{6.69}$$

In the case of inductive coupling, Eq. (6.67) is more convenient to express the impedance Z_P and Z_S. The resonance equation becomes:

$$L_m(L_P + L_S)\left(1 - \frac{\omega_0^2}{\omega^2}\right) = -L_P L_S \left(1 - \frac{\omega_0^2}{\omega^2}\right)^2, \qquad (6.70)$$

giving the resonant frequencies

$$\omega_1 = \omega_0$$
$$\omega_2 = \omega_0 \bigg/ \sqrt{1 + \frac{L_m(L_P + L_S)}{L_P L_S}}. \qquad (6.71)$$

This coupled resonator circuit therefore exhibits very similar properties to those already discussed when the resonators are assumed to be identical. The only difference is that the second mode resonant frequency is not simply related to the coupling constant k, as was the case in Eqs. (6.64) and (6.65).

6.3.3 General case (different coupled resonators)

In order to discuss the properties of the general case, it is more convenient to write Eq. (6.54) in a form that allows a graphical representation of the solutions. One obtains:

$$\frac{1}{Z_P} + \frac{1}{Z_S} = -\frac{1}{Z_m}. \qquad (6.72)$$

Figure 6.32 shows that the two resonant modes are shifted to lower or higher frequencies when the resonators are coupled by an inductor or a capacitor, respectively.

Quantitatively, Eq. (6.72) is rewritten making use of Eq. (6.66) or Eq. (6.67) in the case of capacitive or inductive coupling, respectively. The solutions for ω are obtained from the following quadratic equation:

Multiple Frequencies Tuning

$$X = \frac{\omega}{\sqrt{\omega_P \omega_S}}$$

$$X^4 - X^2 \left[\left(1 + \frac{C_P}{C_m}\right)\frac{\omega_P}{\omega_S} + \left(1 + \frac{C_S}{C_m}\right)\frac{\omega_S}{\omega_P} \right] + \left(1 + \frac{C_S}{C_m} + \frac{C_P}{C_m}\right) = 0 \quad (6.73)$$

in the case of capacitive coupling, and

$$X = \frac{\sqrt{\omega_P \omega_S}}{\omega}$$

$$X^4 - X^2 \left[\left(1 + \frac{L_m}{L_P}\right)\frac{\omega_S}{\omega_P} + \left(1 + \frac{L_m}{L_S}\right)\frac{\omega_P}{\omega_S} \right] + \left(1 + \frac{L_m}{L_S} + \frac{L_m}{L_P}\right) = 0 \quad (6.74)$$

in the case of inductive coupling.

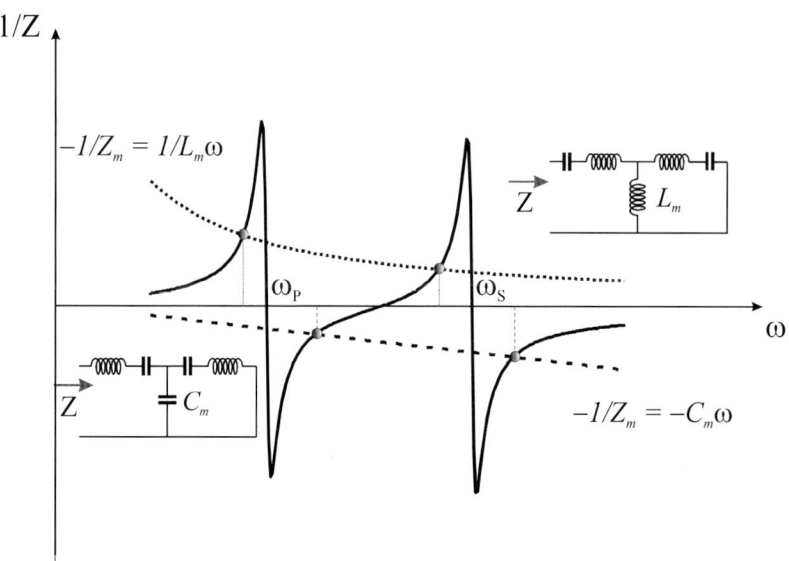

Fig. 6.32 Graphical representation of the resonant equation [Eq. (6.72)] for two different resonators (full line) resonating at ω_P and ω_S, respectively. The resonant frequencies are increased when capacitively coupled (dashed line), while they are decreased when inductively coupled (dotted line).

Note that the solutions are normalized to the geometric mean of the two free resonant frequencies ($\sqrt{\omega_P \omega_S}$) and that the direct ω coordinate is used for the capacitive coupling case. In the case of inductive coupling, the inverse ($1/\omega$) coordinate is used instead. This is a usual practice of network theory, especially used for the LC filter designs.

Finally, the currents in both coils are out-of-phase again for one mode and in-phase for the other one. This should be taken into account in case the tank coil sees the sample volume.

6.3.4 Flux coupled resonators

The equivalent circuit of the two coupled resonators is represented in Fig. 6.33.

The coupled coils are replaced by a T-network, similar to the coupling case already analyzed except that the primary, L_P, and secondary, L_S, coils are replaced by $L_P - M$ and $L_S - M$, respectively. M represents the mutual inductance, equivalent to the L_m in Fig. 6.31(b).

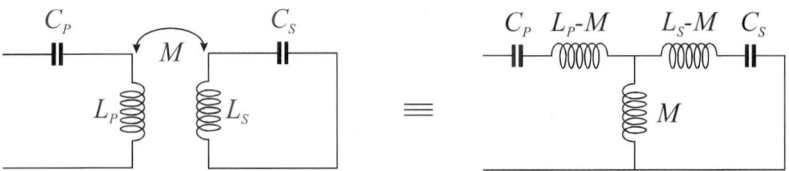

Fig. 6.33 Flux coupled resonator circuit (left) and its equivalent T-network (right).

The previous equations, discussed above, still apply, providing one replaces L_P, L_S, and L_m with the corresponding values quoted above.

In particular, when both resonators are identical ($L_P = L_S = L$; $C_P = C_S = C$), the upper frequency mode resonates at:

$$\omega_1 = \frac{1}{\sqrt{(L-M)C}}, \qquad (6.75)$$

i.e., it is displaced to HF compared to the resonant frequency ω_0 (= $1/\sqrt{LC}$) of the isolated resonators. This mode corresponds, as in the previous case, to out-of-phase currents circulating in each coil. The lower frequency mode (corresponding to in-phase currents in both coils) is still given by Eq. (6.60), as:

$$\omega_2 = \omega_1 \sqrt{\frac{1}{1 + 2M/(L-M)}}, \qquad (6.76)$$

which can finally be written as

$$\omega_2 = \frac{1}{\sqrt{(L+M)C}}. \qquad (6.77)$$

Expressed as a function of the coupling constant $k = M/L$, one obtains the well-known result:

$$\omega_{2,1} = \frac{\omega_0}{\sqrt{1 \pm k}}, \qquad (6.78)$$

where ω_0 is the resonant frequency of the isolated resonators.

If the resonators have the same resonant frequency ω_0 but have different L/C ratios, Eq. (6.78) is still valid with the coupling constant k defined as:

$$k = \frac{M}{\sqrt{L_P L_S}}. \qquad (6.79)$$

In case the resonant frequencies of the isolated circuits are not equal, the corresponding equations derived in the previous paragraph for the inductive case [Eq. (6.74)] can also be applied, provided one again replaces L_S and L_P by $L_S - M$ and $L_P - M$, respectively.

Finally, the case of coupling by an inductive link can be treated similarly, using the equivalent circuit shown in Fig. 6.34.

At this point, it should be mentioned that the flux coupled configuration may lead to a dependent RF magnetic field distribution mode. For example, if the sample is placed in-between two identical planar circular coils, the RF magnetic field will be homogeneous for the *LF* mode (Helmoltz coil configuration) while it will be a gradient for the *HF* resonant mode (Fig. 6.35). The latter may be useful when NMR experiments require RF field gradients [Friedrich and Freeman, 1988]. On the other hand, when using a "surface coil," i.e., the sample is placed outside the volume comprised between the two coils, the sensitivity regions for the *HF* and *LF* modes of the coil system may differ slightly.

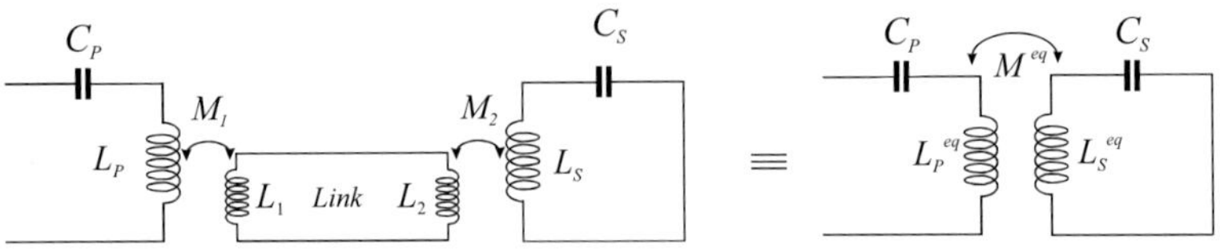

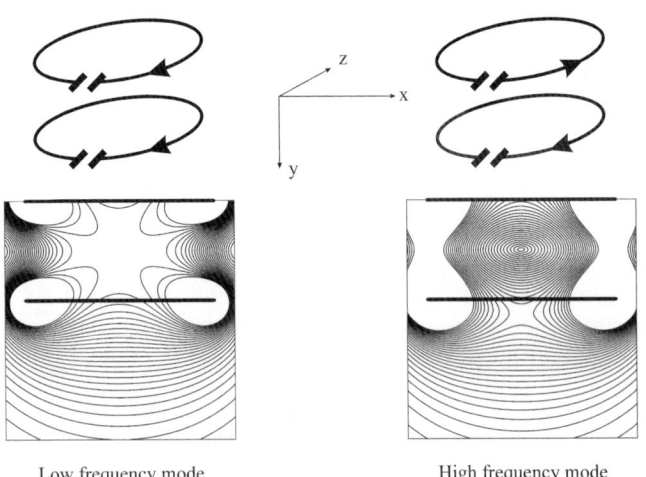

$$M^{eq} = \frac{M_1 M_2}{L_1 + L_2} \quad L_s^{eq} = L_s - \frac{M_2^2}{L_1 + L_2} \quad L_s^{eq} = L_s - \frac{M_2^2}{L_1 + L_2}$$

Fig. 6.34 Inductively coupled resonators through a link (right). The equivalent circuit and corresponding component values (right) are given by Terman [Terman, 1943].

Low frequency mode   High frequency mode

Fig. 6.35 NMR efficient magnetic field (B_1^+, Chapter 7) created by two identical inductively coupled resonators made out of tuned circular loops. The final resonator has two resonant modes, one with *in-phase* current (low frequency mode, homogeneous mode) and one with *out-of-phase* current (high frequency mode, gradient mode).

Until now, we have neglected the effect of the coil losses on the behavior of the coupled resonator. This approximation is justified, provided the Q factors are high enough.

However, if one is interested in the shape of the resonant curve, the resonator Q factors should be taken into account. This concerns the band-pass filter designs as well as shaping the resonant curve in order to control the probe bandwidth (its time response). Such a procedure has already been proposed for solid state NMR applications. The relevant well-known equations, giving the ratio of voltage across the secondary capacitor to the voltage applied to the primary circuit and the peak voltage frequencies across the secondary capacitor, are given by [for example, Terman, 1943]:

$$\frac{E_S}{E_P} = \frac{k}{\gamma^2 \left[k^2 + 1/Q_P Q_S - \left(1 - 1/\gamma^2\right)^2 + j\left(1 - 1/\gamma^2\right)\left(1/Q_P + 1/Q_S\right) \right]} \times \sqrt{\frac{L_S}{L_P}} \quad (6.80)$$

and

$$\omega_{2,1} = \frac{\omega_0}{\sqrt{1 \pm k\left[1 - \frac{k_c^2}{2k^2}\left(\frac{Q_P}{Q_S} + \frac{Q_S}{Q_P}\right)\right]^{1/2}}}. \quad (6.81)$$

In the equations above, γ represents the normalized frequency ($\gamma = \omega/\omega_0$) and k_c is the critical coupling constant. It is given by:

$$k_c = \frac{1}{\sqrt{Q_P Q_S}}. \quad (6.82)$$

These equations are only valid if the two isolated resonators are tuned to the same frequency ω_0. If slightly detuned, some approximations can be made to apply the same equations. In particular, if the Q factors of both resonators are equal, the detuning effect is equivalent to a slight increase of the coupling coefficient:

$$k_{app} = \sqrt{k^2 + \frac{(\omega_P - \omega_S)^2}{\omega_0^2}},\qquad(6.83)$$

where ω_0 is the frequency midway between primary and secondary resonant frequencies, ω_P and ω_S.

6.3.5 *Special case of a short circuited coil*

Apart from double tuning a given probe, interesting applications of a coupled resonator circuit may be to broaden the frequency response of a given probe, and eventually, to tune it remotely. The last application is of particular interest for superconducting probe coils and also for any resonator that cannot be directly connected to a tuning component.

A technique that is generally used to remotely tune a resonator is to couple it inductively with a coil that is short circuited (Fig. 6.36). In this case, the inductance of the resonator coil is decreased, thus increasing the resonance frequency.

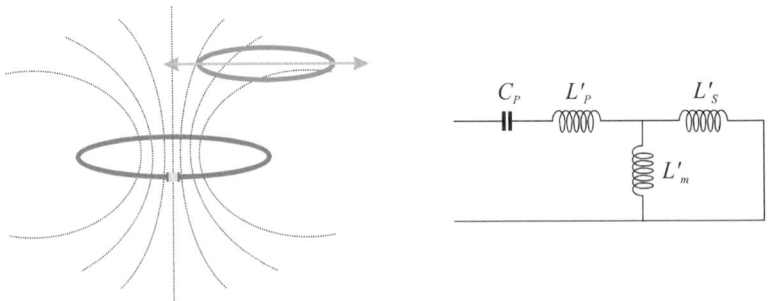

Fig. 6.36 Tuning a resonator by moving a short circuited loop in its magnetic flux (left), and the equivalent circuit (right).

The new resonance frequency is given by:

$$\omega_2 = \frac{1}{\sqrt{L_{eq} C_P}},\qquad(6.84)$$

where L_{eq} is the equivalent inductance made up with $L_P - M$ in series with the parallel combination of M and $L_S - M$, leading to:

$$L_{eq} = L_P(1 - k^2). \tag{6.85}$$

Hence,

$$\omega_2 = \frac{\omega_P}{\sqrt{1-k^2}}, \tag{6.86}$$

where ω_P is the resonant frequency of the isolated resonator

$$\omega_P = \frac{1}{\sqrt{L_P C_P}}. \tag{6.87}$$

This result can also be derived from Eq. (6.74) in the limit when ω_S tends to zero, corresponding to short circuiting the secondary coil. Hence:

$$-\frac{\omega_P'^2}{\omega^2}\left(1 + \frac{L_m'}{L_S'}\right) + \left(1 + \frac{L_m'}{L_P'} + \frac{L_m'}{L_S'}\right) = 0, \tag{6.88}$$

where the primed variables are obtained from the T-network equivalent of the flux coupled coil circuit as:

$$L_m' = M \quad L_P' = L_P - M \quad L_S' = L_S - M, \tag{6.89}$$

$$\omega_P'^2 = 1/(L_P - M)C_P. \tag{6.90}$$

6.3.6 *Π-network configuration*

Excepting the flux coupled case, the coupled resonators in a T-network configuration involve opening the resonant circuits in order to insert the coupling component.

In contrast, the Π-network configuration may have some advantages. It is very easy to add a second resonant mode to a previously existing resonator, without the drawback of the flux coupled mode which always leads to a resonant mode dependent on the magnetic field distribution.

The adjustment of the resonant mode frequencies will only require adjustment of the coupling impedance value. Usually it is easer to adjust a capacitance rather than an inductance, leading to the capacitive coupling scheme [see Fig. 6.38(a)] which will be preferred.

Quantitatively, the previous equations obtained for the T-network configuration may still be used. It is sufficient to transform the Π-coupling network into its equivalent T-circuit, using the new component values together with the equations derived in the previous section.

Consider the general Π- and T-networks in Fig. 6.37.

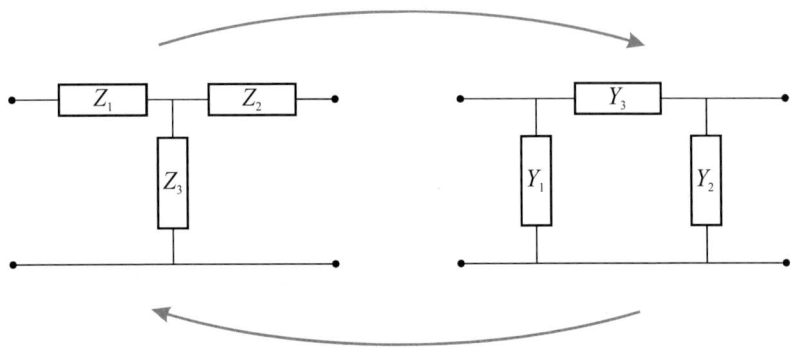

Fig. 6.37 A two terminal T-network and its equivalent dual Π-network. The relations that hold between the relative impedances or admittances of both networks are given by Eqs. (6.92) and (6.93).

They are equivalent (i.e., have the same transmission and impedance characteristics) if the admittance **Y** and impedance **Z** matrices (Chapter 3) fulfill the following condition:

$$\mathbf{YZ} = \mathbf{ZY} = 1. \tag{6.91}$$

Then the following relations hold:[14]

$$Z_{22} = \frac{Y_{11}}{Y_{11}Y_{22} - Y_{12}^2} = Z_2 + Z_3$$

$$Z_{11} = \frac{Y_{22}}{Y_{11}Y_{22} - Y_{12}^2} = Z_1 + Z_3 \qquad (6.92)$$

$$Z_{12} = \frac{-Y_{12}}{Y_{11}Y_{22} - Y_{12}^2} = Z_3$$

or

$$Y_{22} = \frac{Z_{11}}{Z_{11}Z_{22} - Z_{12}^2} = Y_2 + Y_3$$

$$Y_{11} = \frac{Z_{22}}{Z_{11}Z_{22} - Z_{12}^2} = Y_1 + Y_3 \qquad (6.93)$$

$$Y_{12} = \frac{-Z_{12}}{Z_{11}Z_{22} - Z_{12}^2} = Y_3$$

The coupling coefficient k is now defined as:

$$k = \frac{Z_{12}}{\sqrt{Z_{11}Z_{22}}} = \frac{Y_{12}}{\sqrt{Y_{11}Y_{22}}}. \qquad (6.94)$$

Now consider the capacitive coupling Π-network in Fig. 6.38(a).

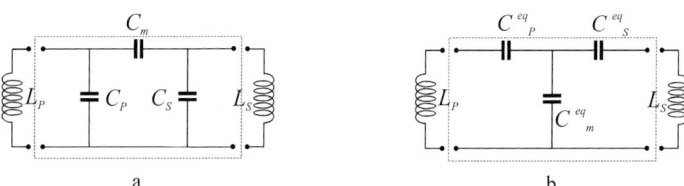

a   b

Fig. 6.38 (a) Π-network capacitive coupling and (b) its equivalent T dual network. The equivalent capacitance values are given by Eqs. (6.95) to (6.97).

---

[14] These can be obtained, for example, by equating the chain matrix elements for both networks.

It is equivalent to the capacitive T-network of Fig. 6.38(b) with the following transformations:

$$C_P^{eq} = \frac{\Delta}{C_S}, \tag{6.95}$$

$$C_S^{eq} = \frac{\Delta}{C_P}, \tag{6.96}$$

$$C_m^{eq} = \frac{\Delta}{C_m}, \tag{6.97}$$

where Δ is given by

$$\Delta = C_P C_S + C_m (C_P + C_S). \tag{6.98}$$

Furthermore, from Eq. (6.94), the coupling coefficient is given by:

$$k = \frac{C_m}{\sqrt{(C_m + C_P)(C_m + C_S)}}. \tag{6.99}$$

At this point, all the results described above for the T-network coupling schemes can be applied. In particular, Eqs. (6.63) and (6.69) can be used to predict the resonant frequencies of the coupled circuit.

Let us consider, for example, the case of two identical (LC) coupled resonators [Fig. 6.38(a)]. It results that $C_S = C_P$ and $C_S^{eq} = C_P^{eq} = C^{eq}$. The upper resonant mode frequency is given by [from Eq. (6.63)]:

$$\omega_2^2 = \omega_0'^2 \left(\frac{1+k}{1-k} \right), \tag{6.100}$$

where

$$\omega_0'^2 = \frac{1}{LC^{eq}}, \tag{6.101}$$

corresponding to the lower frequency mode. One obtains:

$$\omega_0'^2 = \frac{1}{LC}\left(\frac{1-k}{1+k}\right), \qquad (6.102)$$

where $1/LC$ is the square of the resonant frequency ω_0 of the isolated circuit. Replacing in Eq. (6.100), it results that the upper frequency mode is now the resonant frequency of the isolated circuits. The resonant frequencies are finally given by:

$$\omega_1 = \omega_0 \sqrt{\frac{1-k}{1+k}}. \qquad (6.103)$$

$$\omega_2 = \omega_0$$

6.3.7 *Summary of coupled resonator properties*

Depending on the coupling mode, one can obtain practically any resonant frequencies, starting from a given resonator. The behavior of the different methods discussed above is summarized in Figs. 6.39, 6.40, and 6.41.

When coupling two circuits having the same resonant frequency, two modes are created, one having the isolated resonance unchanged [Fig. 6.39(a),(b)] except when the flux coupled scheme is chosen.

When the coils are spatially coupled together two resonant modes appear, one higher and the other one lower compared with the isolated circuit resonance [Fig. 6.39(c)]. The current flow in each coil has different directions depending on the modes and on the coupling scheme. This is also shown in Figs. 6.39, 6.40 and 6.41.

For the T-coupling configuration, equal currents flow in opposite directions in the coupling component. Thus both circuits are virtually isolated from each other and the resonant frequency remains unchanged. When the currents flow in the same direction in the coupling component, a voltage appears at the junction, with the phase depending on the component type.

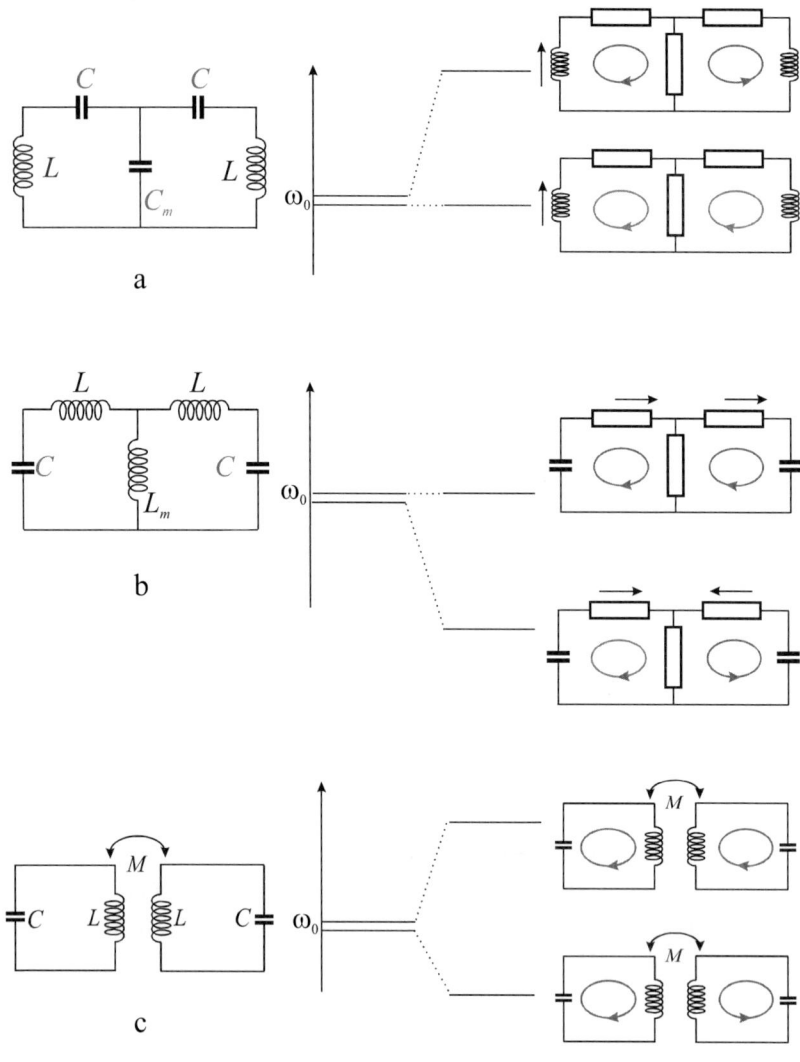

Fig. 6.39 Summary of resonant mode properties (frequency splitting and current phase) for the three principal configurations of identical resonators coupled through a T-network.

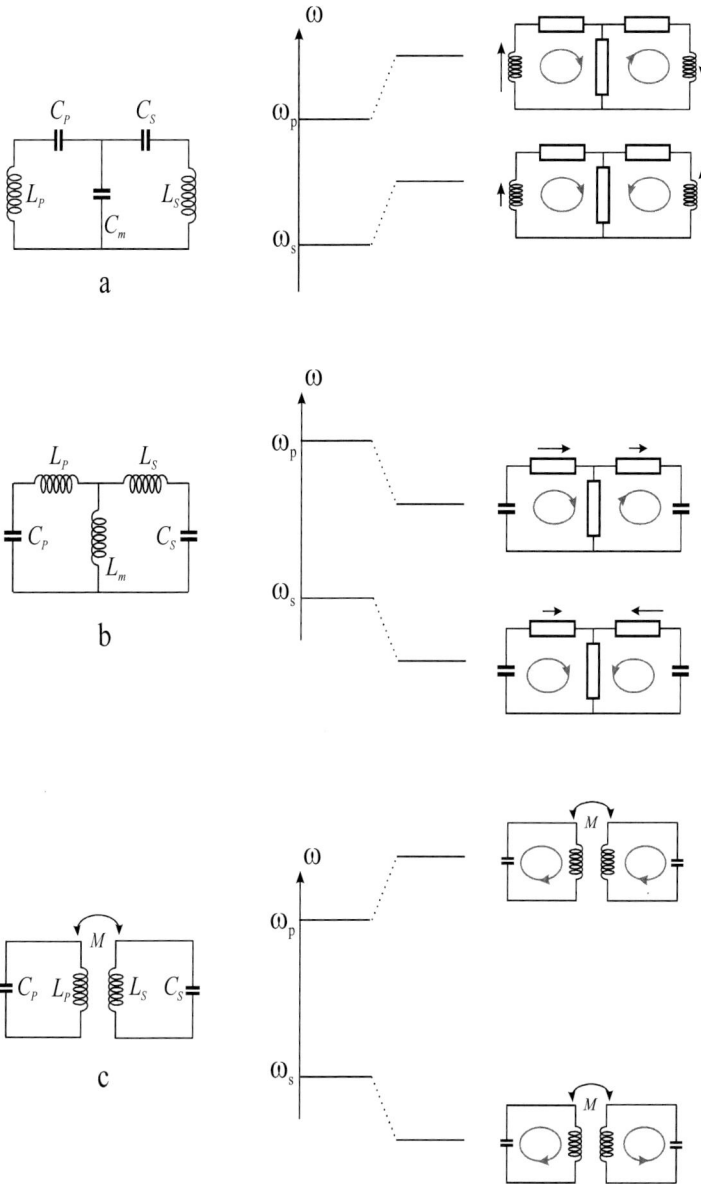

Fig. 6.40 Summary of resonant mode properties (frequency splitting and current phase) for two configurations of different resonators coupled through a T-network.

As a result, the corresponding mode has a resonance frequency that is shifted respective to the isolated resonance mode either to a higher or a lower frequency when the coupling component is a capacitor [Fig. 6.39(a)] or an inductor [Fig. 6.39(b)], respectively.

The Π-coupling scheme leads to the reverse scheme. When the currents flow in the same direction within both resonators, no voltage appears across the coupling component, making the resonators behave as isolated. The corresponding mode has an unchanged resonance frequency. When the currents flow in opposite directions, a voltage appears across the coupling component, leading to an effective coupling of both resonators. The frequency mode is shifted lower in the case of capacitive coupling.

Finally, in the case of a flux coupled coil, the equivalent T-network has reduced inductances of both resonator circuits. Hence, the frequency of the out-of-phase current mode is higher than that of the isolated circuit. On the contrary, it is lower for the resonant mode that has the currents in phase.

From a practical point of view, the coupling capacitive Π-network (Fig. 6.41) is particularly attractive.

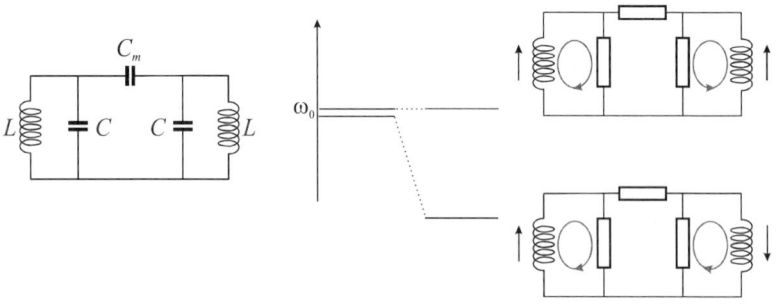

Fig. 6.41 Summary of resonant mode properties (frequency splitting and current phase) for two identical resonators coupled through a capacitive Π-network.

It can be applied simply to a given resonator, already optimized, to obtain a second resonant mode that can be finely adjusted to any frequency below the resonant frequency of the main resonator, using only one adjustable component. This method is especially recommended for very close frequencies like the proton and fluorine nuclei, for example (i.e., 200 MHz and 188 MHz).

6.4 Efficiency of a Multiple Tuned NMR Probe

When designing the coil probe for two, or more, resonant frequencies, one should take care that the "probe efficiency" will invariably be degraded with respect to an optimized single resonant probe. Additional losses come out mainly from the added inductor(s). As this inductor is being subjected to a current that depends heavily on the resonant mode, the corresponding losses depend on it accordingly. Generally, a multiple frequency probe should be designed by choosing the component values in order to preferably optimize the efficiency for a given mode of resonance. Another consequence of energy storing in the added resonant circuit(s) is that the apparent Q factor of the whole circuit will not represent the real sensitivity, but will be rather more or less dominated by the Q factor of the added resonator(s). In extreme cases, the sensitivity may become inversely proportional to the apparent Q factor.

The probe efficiency may be defined as:

$$E = \sqrt{\frac{Power\ dissipated\ in\ the\ probe\ coil}{Total\ power\ dissipated\ in\ the\ whole\ circuit}}. \quad (6.104)$$

The sensitivity (S/N ratio, or equivalently the pulse angle rotation at a given power level) is directly proportional to E.

Although each case should be analyzed individually, some general rules with respect to the efficiency of a dual frequency probe can be derived. This will provide a few guidelines to optimize a particular design.

6.4.1 *Efficiency of shunting methods*

Let us consider, as an example, the circuit of Fig. 6.42(a) which is a basic, and the simplest, double tuned probe circuit using the shunting method. The shunting circuits, assumed to be built with lumped elements, are tuned to the high (*HF*) and low (*LF*) frequency, respectively.

The efficiency of the circuit for both frequencies ω_{HF} and ω_{LF} can be derived from the equivalent circuit drawn in Fig. 6.42(b). Let ω_1 be the tuning frequency of the probe circuit (either ω_{HF} or ω_{LF}) and ω_2 the second working frequency (either ω_{HF} or ω_{LF}, respectively). In the following we shall calculate the efficiency at frequency ω_1.

The average power dissipated in a complex impedance Z is given by:

$$P_{av} = \text{Re}\{v \cdot i^*\}, \qquad (6.105)$$

where i^* is the complex conjugate value of current i that flows in Z. In the case of a series circuit of the type $Z = r + jX$, the power P_{av} is simply given by:

$$P_{av} = \text{Re}\{(r + jX)i \cdot i^*\} = r|i|^2. \qquad (6.106)$$

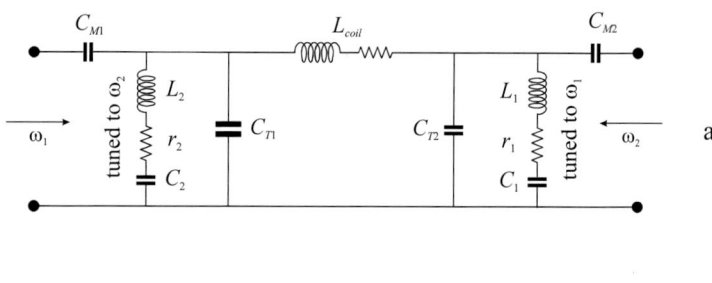

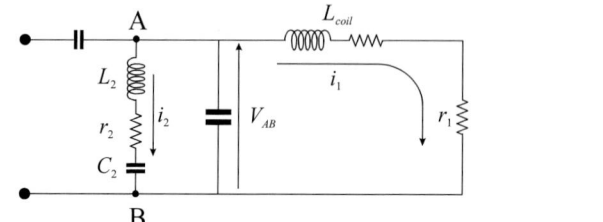

Fig. 6.42 (a) A double tuned single coil probe and (b) its equivalent circuit for one of the resonant mode frequencies (ω_1).

Hence, from Eq. (6.94), the efficiency can be obtained as:

$$E^2(\omega_1) = \frac{r|i_1|^2}{(r+r_1)|i_1|^2 + r_2|i_2|^2} = \frac{r}{r+r_1 + r_2\left(\frac{|i_2|}{|i_1|}\right)^2}, \quad (6.107)$$

where i_1 represents the current that flows in the coil and the series switch, tuned to the resonant frequency ω_1. i_2 represents the current that flows, at ω_1, in the switch tuned to ω_2. r represents the coil resistance, r_1 and r_2 are the switch resistances at ω_1.[15]

Let V_{AB} be the voltage at point A with respect to the ground [Fig. 6.42(b)]. The current i_1 can be expressed as:

$$|i_1|^2 = \frac{V^2}{XL_{coil}^2 + (r+r_1)^2}, \quad (6.108)$$

where XL_{coil} is the sample coil impedance at ω_1. Assuming that the L_1C_1 circuit is resonant at ω_1, i_2 is given by:

$$|i_2|^2 = \frac{V^2}{(XL_2 - XC_2)^2 + r_2^2}, \quad (6.109)$$

where XL_2 and XC_2 are the inductor L_2 and capacitor C_2 impedances at ω_1 of the series circuit tuned at ω_2 ($L_2C_2\omega_2^2 = 1$).

One finally obtains:

$$E(\omega_1) = \sqrt{\frac{r}{r+r_1 + \alpha r_2}}, \quad (6.110)$$

where the contribution α of the "parasitic" switch circuit tuned to ω_2 is given by

---

[15] The coil wire resistances are frequency dependent. They increase roughly with the square of the frequency. However, the sample coil resistance r may vary in a more complicated manner due to the complex frequency dependence of the sample losses.

$$\alpha = \frac{|i_2|^2}{|i_1|^2} \approx \left(\frac{L_{coil}}{L_2}\right)^2 \frac{1}{\left(1-\omega_2^2/\omega_1^2\right)^2}. \quad (6.111)$$

The above expression assumes that the coil resistance is much lower than its impedance ($Q > 10$) which is usually the case.

The best efficiency supposes that the extra "parasitic" contribution is negligible. Because r_1 and r_2 are obviously of the same order of magnitude, an optimum efficiency is obtained when α is much lower than unity.

Considering the case of a proton–carbon double tuned probe (ω_{HF} is four times ω_{LF}), the LF efficiency ($\omega_1 = \omega_{LF}$, $\omega_2 = \omega_{HF}$) can be approximated as:

$$E_{LF} \approx \sqrt{\frac{1}{1+\dfrac{L_{LF}}{L_{coil}}\dfrac{Q_{coil}}{Q_{LF}}}}. \quad (6.112)$$

The third term in the denominator of Eq. (6.110) is negligible because the term α is small (the frequency-dependent term in Eq. (6.111) is equal to $1/15^2$). At HF the corresponding contribution is no longer negligible:

$$E_{HF} \approx \sqrt{\frac{1}{1+\dfrac{L_{HF}}{L_{coil}}\dfrac{Q_{coil}}{Q_{HF}}+\dfrac{L_{coil}}{L_{LF}}\dfrac{Q_{coil}}{Q_{LF}}}}. \quad (6.113)$$

An optimum efficiency at LF requires that the inductance of the switching circuit, tuned to ω_{LF} (L_{LF}), is much lower than L_{coil}. This condition clearly degrades the efficiency at HF [Eq. (6.113)]. Thus the efficiency for both frequencies is a compromise. An attempt to improve the carbon efficiency would result in a dramatic degradation of the proton channel properties. A solution to this difficulty has been already proposed (Section 6.1.3). It consists of minimizing the parasitic current i_2. This can be obtained by replacing the simple series tuned circuit by a complete switch having ω_{on} equal to ω_{LF} and ω_{off} equal to ω_{HF}. On the LF side, this solution is not required because the power dissipated in the parasitic components belonging to the series resonant circuit at ω_{HF} is negligible. The circuit in Fig. 6.10 was designed taking into account all these requirements.

Although there is a large freedom to choose the switch components (in particular the value of the inductance L_{LF} or L_{HF}), the efficiency optimization implies that L_{LF} or L_{HF} should preferably be smaller than L_{coil}. The final choice will depend, however, on the possibility of building a high Q inductor having a self-resonant frequency higher than the working frequencies of the probe. In addition, the capacitors of the switching circuits must have a high Q (see Chapter 2). Hence, the range of available inductance is somewhat reduced. An analysis of each particular case should be performed prior to any design. An example was already given and fully discussed in Section 6.1.3.

6.4.2 *Efficiency of multiple-pole circuits*

In this case, the efficiency can be written as:

$$E = \sqrt{\frac{r}{r + \xi r_T}}, \qquad (6.114)$$

where r_T is the effective resistance of the complex tuning impedance at the frequency of interest, and ξ is the ratio of the squared *module* of currents flowing in the sample coil and in the tuning network [Fig 6.43(a)]. As already noted, the ratio ξ will determine the efficiency which will again depend on the resonant modes.

As an example, let us consider the tuning circuit made out of a capacitor in series with a tank circuit [Fig. 6.23(b)]. Assuming that the losses only come out from the coil resistance belonging to the tuning reactance, Eq. (6.114) may be rewritten as:

$$E = \sqrt{\frac{r}{r + \alpha r_1}} = \sqrt{\frac{1}{1 + \alpha \frac{L_1}{L_{coil}} \frac{Q_{coil}}{Q_1}}}, \qquad (6.115)$$

where r_1 is the tuning coil resistance and α is the ratio of the squared current modules flowing in the sample coil and the tank coil [Fig. 6.43(b)].

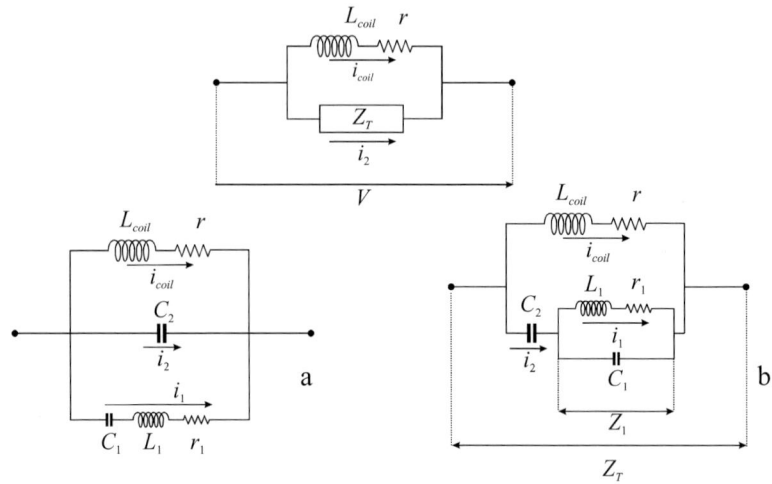

Fig. 6.43 (a) Currents, voltages and impedance for the general multiple tuned circuit and (b) and (c) double tuned circuits.

Assuming high Q coils ($Q > 10$), α can be written as a function of the reactive components of the circuit as:

$$\alpha = \frac{|i_1|^2}{|i_{coil}|^2} = \frac{L_{coil}^2}{L_1^2}\left(\frac{Z_1}{Z_T}\right)^2, \qquad (6.116)$$

where L_1 is the trap coil inductance, Z_1 its impedance at the frequencies of interest (ω_{HF} or ω_{LF}) and Z_T the impedance of the total tuning reactance. In other words, the frequency dependence of α determines the efficiency balance between LF and HF.

Usually, the sample coil is essentially tuned to LF by the series capacitor C_2 [Fig. 6.23(b)], while the tank circuit L_1C_1 behaves as a low inductance ($\approx L_1$) that comes in series with L_{coil}. Hence, the following approximations hold: $Z_T \approx L_{coil}\omega_{HF}$, $Z_1 \approx L_1\omega_{HF}$ implying $\alpha \approx 1$.

The *LF* efficiency becomes:

$$E_{LF} \approx \sqrt{\frac{1}{1+\dfrac{L_1}{L_{coil}}\dfrac{Q_{coil}}{Q_1}}}. \qquad (6.117)$$

At *HF*, the tank circuit tunes the sample coil, while the capacitor C_2 represents a low, negligible, impedance. Hence, $Z_T \approx Z_1$ and efficiency becomes:

$$E_{HF} \approx \sqrt{\frac{1}{1+\dfrac{L_{coil}}{L_1}\dfrac{Q_{coil}}{Q_1}}}. \qquad (6.118)$$

Improving the *LF* efficiency requires that L_1 is smaller than L_{coil}, but this has the side effect of degrading the *HF* efficiency. Optimization of the efficiency is again a compromise between the two resonant modes.

More generally, the efficiency can be expressed as a function of the trap circuit resonant frequency (ω_{off}) that can be freely chosen for the design of the double tuned probe (Fig. 6.6b). Using Eqs. (6.37) to (6.39) as a function of ω_{off}, as well as Eqs. (6.115) and (6.116), the efficiency at ω (either ω_{HF} or ω_{LF}) can be written as:

$$E = \sqrt{\frac{1}{1+\beta\dfrac{Q_{coil}}{Q_1}}}, \qquad (6.119)$$

where

$$\beta(\omega) = A\left(\frac{\omega_{off}^2/\omega_{HF}\omega_{LF}}{A+1-\omega_{off}^2/\omega^2}\right)^2 \qquad (6.120)$$

and

$$A = \frac{\left(\omega_{off}^2-\omega_{LF}^2\right)\left(\omega_{HF}^2-\omega_{off}^2\right)}{\omega_{HF}^2\omega_{LF}^2}. \qquad (6.121)$$

Obviously, A is always positive because ω_{off} is, at the same time, always higher than ω_{HF} and lower than ω_{LF} (see Section 6.2).

The dependence of $E(\omega_{HF})$ and $E(\omega_{LF})$ as a function of ω_{off} is given in Fig. 6.44 for the typical case of a carbon–proton (50–200 MHz) double tuned probe. Obviously, the efficiency is better if the quality factor (Q_1) of the coil tank circuit is better than that of the sample coil. In addition, as already discussed, the efficiency for both frequencies is a compromise depending on the choice of the resonant frequency ω_{off} of the tank circuit. As a general rule, the efficiency is degraded at the working frequency closest to ω_{off}. In addition, ω_{off} should not be chosen too close to the desired frequency as the efficiency drops rapidly in this range (Fig. 6.44).

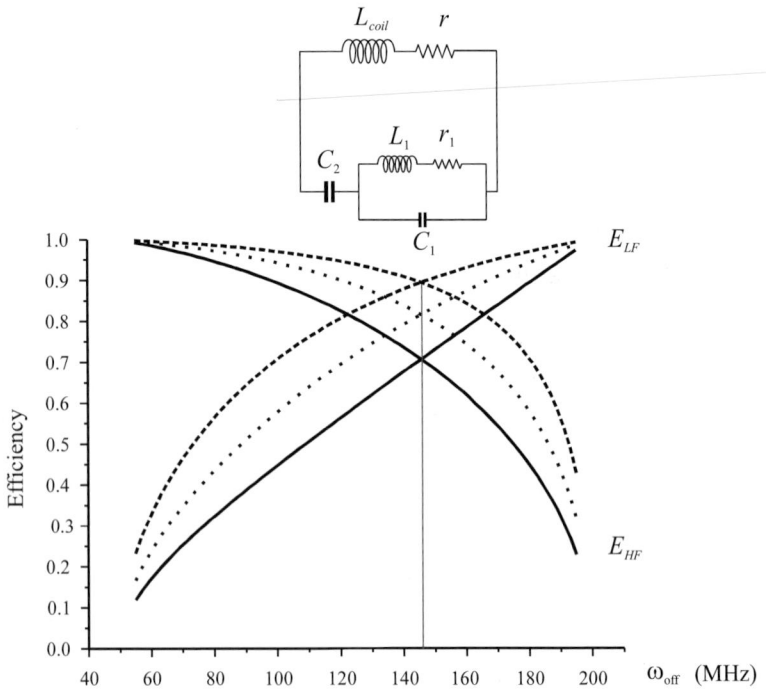

Fig. 6.44 Efficiency as a function of ω_{off} for a 50–200 MHz double tuned 150 nH coil [circuit, Fig. 6.43(b)] and for three different Q_{coil}/Q_1 ratios (1 - full line, 0.5 - dotted lines and 0.25 - dashed lines). The mark on the frequency scale corresponds to Eq. (6.122).

Finally, an equal compromise for both resonant modes is obtained when ω_{off} is given by:

$$\omega_{off}^2 = \frac{\omega_{HF}^2 + \omega_{LF}^2}{2}. \tag{6.122}$$

In that case, $\beta(\omega_{HF}) = \beta(\omega_{LF}) = 1$. The efficiencies for both modes are equal and given by:

$$E_{HF} = E_{LF} = \sqrt{\frac{1}{1+Q_{coil}/Q_1}}. \tag{6.123}$$

If the Q factor of the tank circuit is at least three or four times greater than Q_{coil}, the efficiency is acceptable for both resonant modes. A similar reasoning can be applied when using the complementary type of circuit involving a series resonant parallel circuit in parallel with a capacitor to tune the sample coil [circuit shown in Fig. 6.23(b)]. In this case, the roles of L_1 and L_{coil} are interchanged. This circuit may be used when one desires to optimize the *HF* efficiency at the expense of the *LF* efficiency, without interrupting the *HF* resonator circuit. This is justified, for example, when the proton channel is used as a detection channel while the carbon (or any other intrinsically low sensitivity X-nucleus) is used as a decoupling channel. It can also be used for adding a deuterium channel (lock) to a higher frequency probe (carbon, phosphorus, fluorine or proton). In these cases, the electromagnetic energy is now concentrated at *HF* in the tank circuit constituted by the sample coil, and consequently efficiently dissipated in it, as is required in the NMR experiment, provided the resonant "tank" circuit is tuned closest to ω_{LF}.

Quantitatively, the efficiency is still given by Eq. (6.115) with α given by [Fig. 6.43(b)]:

$$\alpha = \frac{|i_3|^2}{|i_1|^2} = \frac{L_{coil}^2}{L_1^2}\left(\frac{1}{1-\frac{\omega_{on}^2}{\omega^2}}\right)^2. \tag{6.124}$$

Using the component values given by Eqs. (6.34) to (6.36), one obtains the following expression for the efficiency:

$$E = \sqrt{\frac{1}{1+\eta\frac{Q_{coil}}{Q_1}}}, \qquad (6.125)$$

where

$$\eta(\omega) = C\left(\frac{\omega^2}{\omega^2-\omega_{on}^2}\right)^2 \qquad (6.126)$$

and

$$C = \frac{\left(\omega_{on}^2 - \omega_{LF}^2\right)\left(\omega_{HF}^2 - \omega_{on}^2\right)}{\omega_{LF}^2 \omega_{HF}^2}. \qquad (6.127)$$

The efficiency plots as a function of ω_{on} are shown in Fig. 6.45. It appears that the design of the circuit is more critical than the previous one, especially respective to the HF efficiency. As a result, ω_{on} should preferably be chosen close to LF, providing a good efficiency for both HF and LF. For a circuit including a parallel resonant circuit, the efficiency can be equally optimized (Fig. 6.45) for both frequencies (i.e., $\beta(\omega_{LF}) = \beta(\omega_{HF}) = 1$) at ω_{on} verifying the following relationship:

$$\frac{1}{\omega_{on}^2} = \frac{1}{2}\left(\frac{1}{\omega_{HF}^2} + \frac{1}{\omega_{LF}^2}\right). \qquad (6.128)$$

Because of the critical dependence of E on ω_{on}, it is therefore recommended that ω_{on} should be chosen close to the optimum one, given by Eq. (6.128).

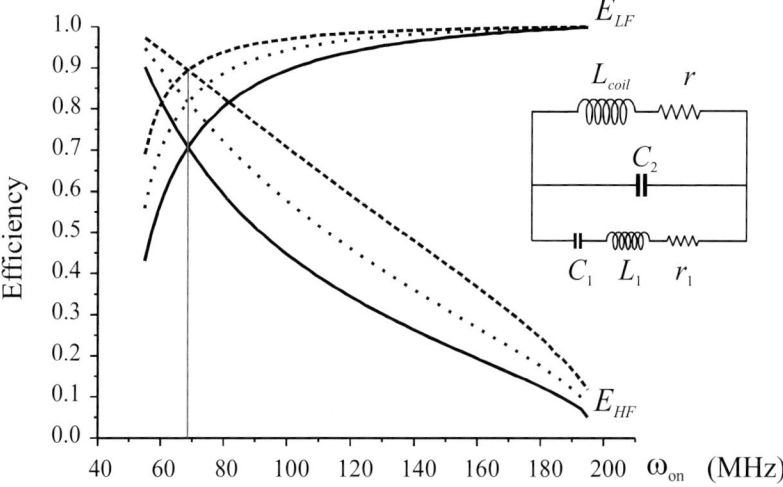

Fig. 6.45 Efficiency as a function of ω_{on} for a 50–200 MHz double tuned 150 nH coil [circuit shown in Fig. 6.39(c)] and for three different ratios of Q_{coil}/Q_1 (1 - full line, 0.5 - dotted lines and 0.25 - dashed lines). The mark on the frequency scale corresponds to Eq. (6.125).

6.4.3 *Efficiency of coupled resonant circuits*

As in the previous cases, where the resonators are very tightly coupled together, the efficiency will also depend on the resonant mode and on the proportion of current that flows in each resonator. Assuming that the coupling component is lossless, the efficiencies for both modes are still given by equations similar to Eqs. (6.117) and (6.118), where Q_1 and L_1 are the characteristics of the coupled tank circuit. This circuit is used when the resonant mode frequencies are close. Therefore, it is suitable that L_1 should have a similar value to L_{coil} providing the optimum efficiency for both *LF* and *HF*. The tank circuit should, however, be carefully designed in order to obtain its Q much higher than Q_{coil}.

6.5 Interfacing the Multiple Frequency Resonator to the Spectrometer

Only a few words will be given here, because examples are provided elsewhere in practical descriptions of some double tuned designs [McNichols *et al.*, 1999]. Furthermore, examples have been already given in this chapter for designs involving the shunting method (Section 6.1) and for the case of a double resonant mode carbon–proton probe (Section 6.2).

The design of the matching network should take account, as far as possible, of the leakage that may exist from one port to another, especially to avoid any losses of the receiver channel into the resistive load of the transmitter channel. On the other hand, noise leakage from a transmitter or decoupling channel into the receiver port should be avoided too. A way to avoid these undesired phenomena is to incorporate "trapping" circuits into the matching network, such as those extensively described here. Finally, the isolation between the ports can ultimately be performed using external high-pass, low-pass, or band-pass filters. The design and positioning of these filters depends on each particular case.

In the particular case where two channels are fed to the unique port of a double tuned probe the frequency-dependent switching circuits described in Section 6.1 can be used efficiently [Gonnella and Silverman, 1989; Ton That *et al.*, 1997; Hu *et al.*, 1998; Tadanki *et al.*, 2012]. When the two frequencies are close (such as proton and fluorine [Berkowitz and Ackerman, 1987]) a diplexer is well suited to interface the probe with the two spectrometer channels (Fig. 6.46). These circuits are widely used in many communication devices and are relatively simple to build. They can be designed via band-pass filters. Most usually encountered is the use of a combination of high- and low-pass filters [Matthaei *et al.*, 1980, pp. 991–1000].

Figure 6.47 gives an example of a simple diplexer designed for simultaneous proton–fluorine experiments that can be used on a 4.7 T apparatus (200–188 MHz).

The design is straightforward. It uses the switching circuit of Figs. 6.6(b) and 6.6(d), which can be identified, respectively, as a low-pass filter ($\omega_{on} < \omega_{off}$) and a high-pass filter ($\omega_{on} > \omega_{off}$). The simulated transmission coefficient (Fig. 6.47) clearly demonstrates the dual role of the diplexer for each design frequency.

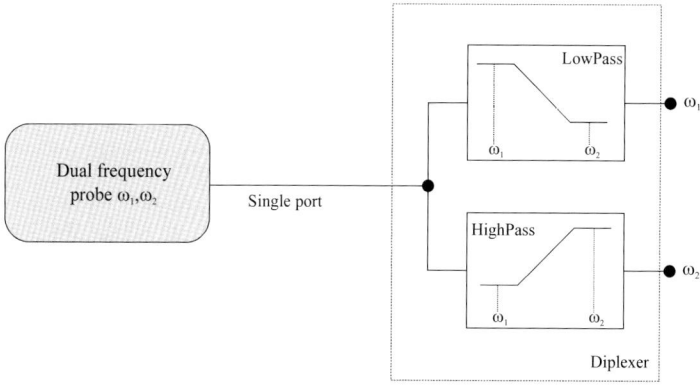

Fig. 6.46 Connecting a double tuned, single port probe to two different frequency channels through a diplexer.

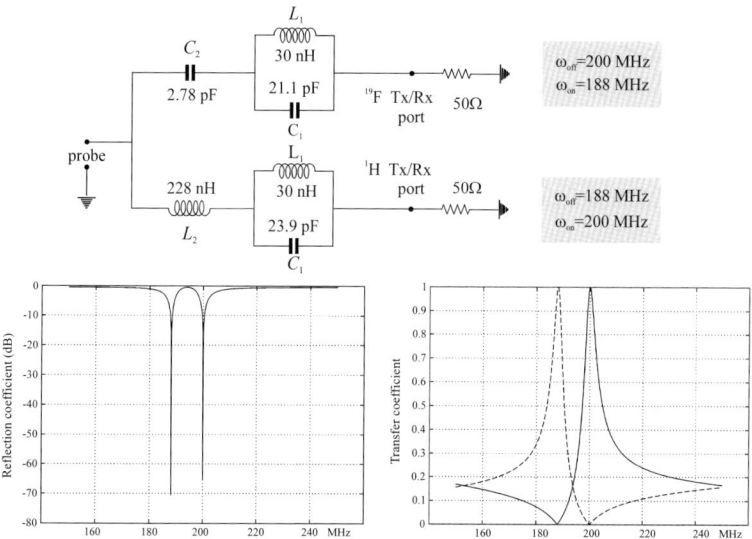

Fig. 6.47 A simple, narrowband, proton–fluorine (200–188 MHz) diplexer (upper) and its electrical performance (lower). The reflection and transmission coefficients are simulated when the network is driven at the probe port and is terminated by a 50 Ω load at the spectrometer (TX/RX) ports.

6.6 Is the Q Factor Representative of the Sensitivity?

Commonly, the probe sensitivity for a given resonant mode is quickly evaluated by the Q factor, which is easy to calculate from the frequency response of the circuit. The Q factors measured by this method are not representative of the sensitivity, especially for multiple tuned probes.

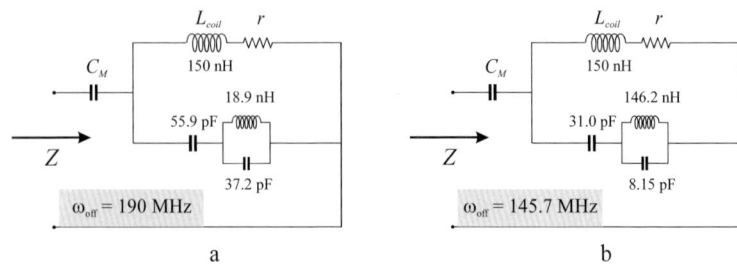

Fig. 6.48 Examples of double tuned probe circuits used to simulate the electrical properties (apparent Q factor): (a) ω_{off} is close to ω_{HF} (200 MHz); (b) ω_{off} is optimum, efficiency at ω_{HF} and ω_{LF} is equal (50 MHz).

Let us consider again the case of a proton–carbon (200–50 MHz) probe made out of a coil (150 nH) tuned by a reactive network incorporating a capacitor in series with a tank resonant parallel circuit. The circuit is matched to 50 Ω (Fig. 6.48) at each frequency of interest and the reflection coefficient is simulated as a function of frequency in order to estimate the apparent Q factor from the reflected half power points ($\Gamma = 0.707$ or -3dB).

The results are presented in Fig. 6.49 and compared to the reflection curves that would be obtained assuming the same coil is single tuned. Two cases have been considered, when the trap circuit is firstly tuned close to ω_{HF} ($\omega_{off} = 190$ MHz) and secondly to the frequency that gives equal efficiencies for both resonant modes [ω_{off} given by Eq. (6.122)].

The resonant curve of the double tuned coil always appears sharper than the one corresponding to the single tuned probe coil.

The increase in Q depends mainly on the position of ω_{off} relative to ω_{HF} or ω_{LF}. The largest increase is observed for the resonant mode that has its frequency closest to ω_{off} [ω_{HF} for example in Fig. 6.49(d)]. When ω_{off} is chosen close to the value which gives equally optimum efficiency for both modes, the apparent Q factor is roughly doubled for each mode. Note that in the simulations the Q factor of the tank circuit L_1C_1 is assumed to be very high (infinite). With finite Q values, the resonant curve will broaden slightly, but the main features will be kept anyway.

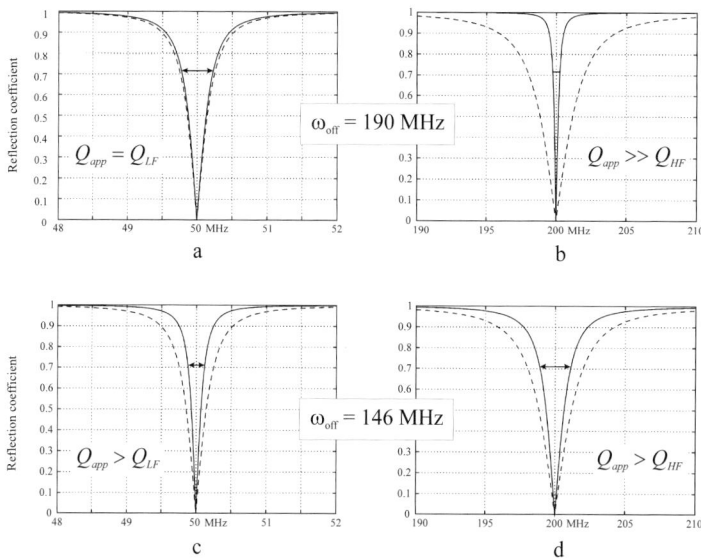

Fig. 6.49 Reflection coefficient, as a function of frequency, for the two designs of the double tuned circuit shown in Fig. 6.48 (full lines) compared to the reflection coefficient obtained for the single tuned probe at LF or HF.

Quantitatively, an estimate of the apparent Q factor can be obtained as follows. From the definition of the power factor (assuming reasonably high $Q > 10$), one can write:

$$Q_{app} = \frac{L_{coil}\omega|i_{coil}|^2 + L_1\omega|i_1|^2}{r_{coil}|i_{coil}|^2 + r_1|i_1|^2}. \qquad (6.129)$$

The numerator and the denominator represent the electromagnetic energy stored in the inductances and the energy lost in their resistances, respectively. Assuming the case of a circuit including a trap in series with a capacitor, the apparent Q factor is given by:

$$Q_{app} = Q_{coil} \frac{\alpha^{-1} L_{coil}/L_1 + 1}{\alpha^{-1} L_{coil}/L_1 + Q_{coil}/Q_1}, \qquad (6.130)$$

where α is the square of the current ratio i_1/i_{coil} as given, for example, by Eq. (6.116). Assuming high Q coils again, one obtains:

$$Q_{app} = Q_{coil} \frac{\beta^{-1} + 1}{\beta^{-1} + Q_{coil}/Q_1}, \qquad (6.131)$$

where β is given by Eq. (6.120). If one now assumes that the Q factor of the trap circuit is much greater than Q_{coil}, as it should be, the apparent Q factor becomes simply:

$$\lim_{Q_1 \to \infty} (Q_{app}) = Q_{coil} \frac{\beta^{-1} + 1}{\beta^{-1}}. \qquad (6.132)$$

Two practical cases are interesting to consider. The first one is when the trap circuit is tuned close to *HF*. In this case, the *LF* efficiency is favored. Hence $\beta(LF)$ is small while $\beta(HF)$ is high, leading to:

$$Q_{app}(LF) \approx Q_{coil} \qquad (6.133)$$

and

$$Q_{app}(HF) \approx \beta Q_{coil}. \qquad (6.134)$$

The apparent Q factor is close to the "true" value at *LF*, while it is much higher at *HF* [Fig. 6.49(a), (b)]. The multiplication factor, equal to β, in fact represents the degradation of the sensitivity (proportional to the efficiency) which is also dependent on the Q_{coil}/Q_1 ratio.

The second case of interest is when the circuit is designed for equal optimum efficiencies for both resonant modes [ω_{off} given by Eq. (6.122)]. In this case, $\beta(\omega_{HF}) = \beta(\omega_{LF}) = 1$ (see above), hence the apparent Q factor is *doubled* for each resonant mode [Fig. 6.49(b), (c)]. Obviously, all these results can be extrapolated to circuits tuned to more than two frequencies, the sensitivity depending essentially on the proportion of energy stored in the various resonators in the probe circuit.

Chapter 7

Magnetic Field Amplitude Estimation

An important step in designing an NMR probe is the knowledge of the resulting magnetic field configuration, in terms of uniformity and amplitude. The probe magnetic field is generated by all the constitutive conductive parts, hence the calculated field map, prior to any kind of NMR experiment, may give very important hints for the practical realization. All conductive parts of the NMR probe may be decomposed in elementary geometrical segments carrying currents. The elementary magnetic field vectors thus generated add together, based on the principle of superposition. The magnetic field vectors generated by the geometrical elements of the probe are easy to calculate using the Biot–Savart law. In the following, we recall some basic formulae for calculating the magnetic field generated by currents in a few simple geometrical configurations used later for calculating probe field maps.

7.1 The Biot–Savart Approximation

The Biot–Savart expression giving the magnetic field $d\vec{B}$ produced at a point P, distance r from the current I flowing in an arbitrary wire configuration (Fig. 7.1) is given by:

$$d\vec{B} = \frac{\mu_0}{4\pi} I \frac{d\vec{s} \times \vec{r}}{|r^3|}, \qquad (7.1)$$

where $d\vec{B}$ is the magnetic field produced by a small section of wire, $d\vec{s}$, r is the positional vector to the point where the magnetic field is calculated, I the current in the wire and μ_0 (= $4\pi \times 10^{-7}$) is the permeability of free space.

The module of the expression above gives the amplitude of the generated magnetic field:

$$dB = \frac{\mu_0}{4\pi} \frac{|ds||r|\sin\theta}{|r^3|}. \qquad (7.2)$$

The principle of superposition enables one to calculate the magnetic field generated by all sections of the wire and to add all the corresponding vectors, as shown in Fig. 7.1.

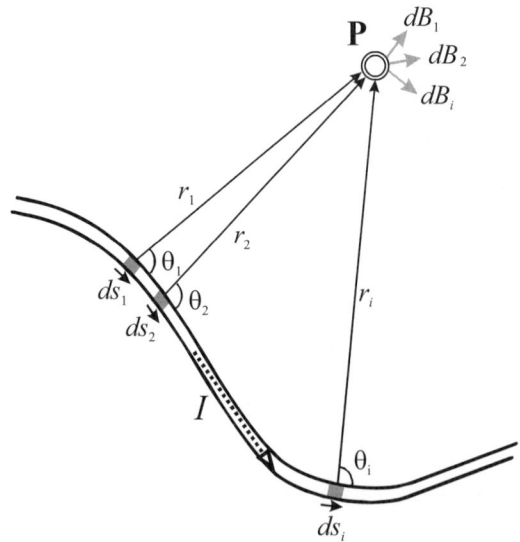

Fig. 7.1 Biot–Savart law for a few segments of a current carrying wire.

Eventually, the magnetic field $\boldsymbol{B}$ generated by the whole wire will be

$$\vec{B} = d\vec{B}_1 + d\vec{B}_2 \ldots + d\vec{B}_n = \sum_i d\vec{B}_i = \frac{\mu_0}{4\pi} I \sum_i \frac{d\vec{s}_i \times \vec{r}_i}{r_i^3}. \qquad (7.3)$$

The above expression shows that the final result depends on the number of segments describing the wire and their relative positions; in other words on the geometrical factor. As a direct consequence of Eq. (7.2), the magnetic field equations of some simple geometries will be presented.

7.1.1 *Magnetic field produced by straight wires*

The amplitude of the magnetic field generated by a thin *straight wire*, in the geometry shown in Fig. 7.2 is given by:

$$dB = \frac{\mu_0}{4\pi} I \left(\frac{\sin \theta}{a} \right)^2 \sin \theta dx, \tag{7.4}$$

where *dB* is the magnetic field generated by the segment *dx*, which is in the *x* direction, at the distance *a*.

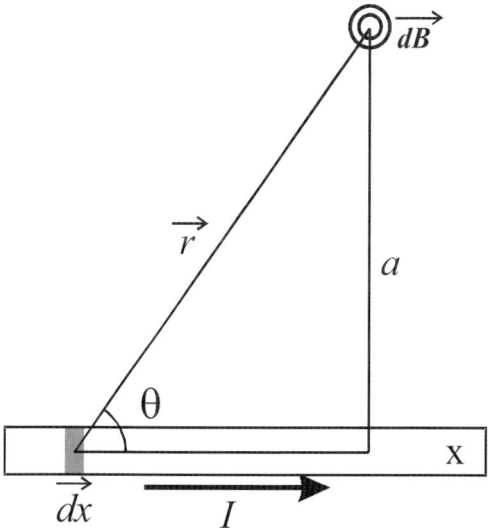

Fig. 7.2 The geometry of a straight wire. *dB*, perpendicular to the figure plane, points upward.

For the whole wire, considered infinitely long, the integration of Eq. (7.3), taking into consideration that it is much easier to express everything in terms of θ $\left(x = -\dfrac{a}{\tan(\theta)}\right)$ gives the well-known result:[1]

$$B = \dfrac{\mu_0}{4\pi a} I \int_0^\pi -\sin\theta d\theta = \dfrac{\mu_0}{2\pi a} I. \qquad (7.5)$$

Another straightforward case where Eq. (7.2) may be easily integrated is the magnetic field generated at a distance a from the *middle of the wire of length L*, Fig. 7.3.

$$B = \dfrac{\mu_0}{4\pi a} I \dfrac{L}{\left((L/2)^2 + a^2\right)^{1/2}}. \qquad (7.6)$$

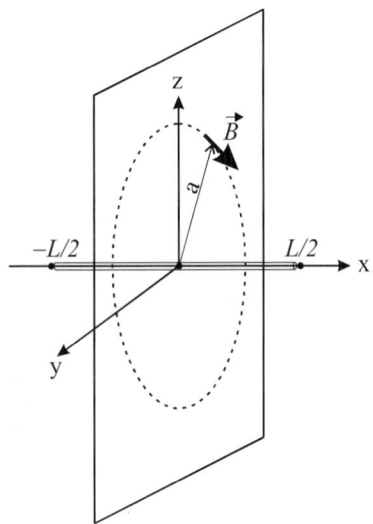

Fig. 7.3 Generated magnetic field in a plane passing through the middle point of a straight wire of length L.

---

[1] Ampere's theorem.

Generally, for a segment of thin wire of length L, the expression of the magnetic field induced at an arbitrary point P, depends on the distance vectors of both ends of the segments, as illustrated in Fig. 7.4.

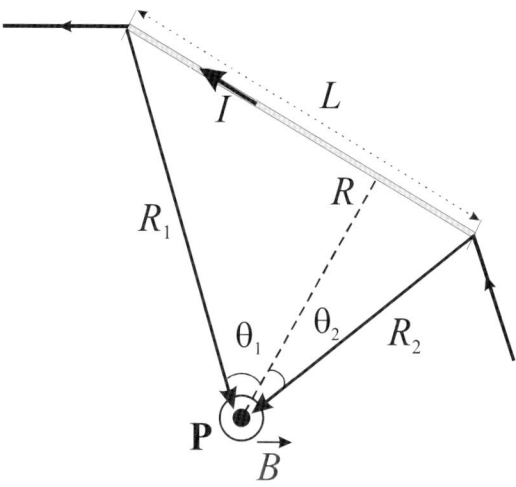

Fig. 7.4 Geometry for the calculation of the magnetic field created by a current flowing in a finite segment of straight wire, at an arbitrary point P.

In this case, the expression of the magnetic field induced at point P is given by:

$$\vec{B} = \frac{\mu_0 I}{4\pi} \frac{\vec{L} \times \vec{R}_1}{L^2 R_1^2 - (\vec{L}\vec{R}_1)^2} \left(\frac{\vec{L}\vec{R}_1}{R_1} - \frac{\vec{L}\vec{R}_2}{R_2} \right) \qquad (7.7)$$

or, in module

$$B = \frac{\mu_0}{4\pi} \frac{I}{R} (\sin\theta_1 + \sin\theta_2). \qquad (7.8)$$

Again, even for the most general case of a segment of wire, the magnetic field obtained for an arbitrary point P in space is dependent on the geometrical coordinates of the segment.

7.1.2 Magnetic field produced by a loop

The expressions of the field generated by other simple geometries are useful to mention, since a probe electrical design can always be reduced to the sum of elementary current segments and loops.

For the case of the magnetic field generated by a current loop, two situations will be of interest; the simplest one is considering the field in the very center of the current loop and the second one is deriving the expression for any point on the center axis. Both geometrical configurations are shown in Fig. 7.5.

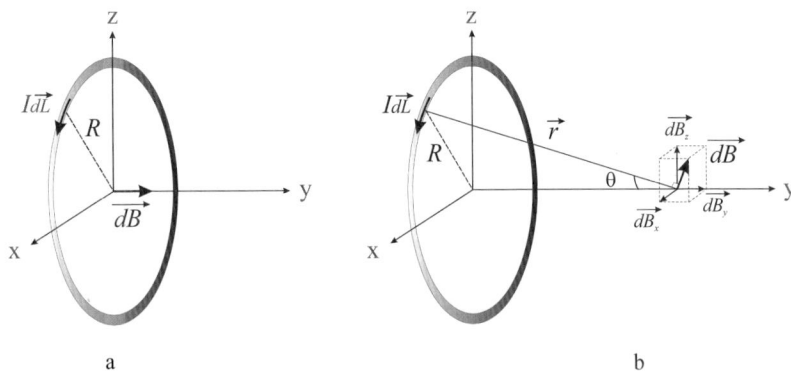

Fig. 7.5 The magnetic field generated by a current loop segment (a) in the very center of the loop and (b) on an arbitrary point on the center axis.

Applying the general form of the Biot–Savart law to calculate the magnetic field generated anywhere on the axis passing through the center of the current loop, for the element of current one gets:

$$d\vec{B} = \frac{\mu_0 I d\vec{L} \times \vec{r}}{4\pi r^3}. \qquad (7.9)$$

The field element dB_x rotates as it progresses around the loop and, due to symmetry, gives a net zero field along this axis and correspondingly for dB_z too. Consequently, we shall further consider only the field component in the y direction. Developing Eq. (7.9), one obtains:

$$dB_y = \frac{\mu_0}{4\pi} \frac{I}{r^3} \left(d\vec{L} \times \vec{r}\right)_y = \frac{\mu_0}{4\pi} \frac{I}{r^2} dL \sin\theta. \qquad (7.10)$$

Integrating over all the current trajectory, one obtains the well-known result for the magnetic field in the center of the current loop, where $\sin\theta = 1$ and r becomes R:

$$B_y = \frac{\mu_0 I}{4\pi R^2} \oint dL = \frac{\mu_0 I}{2R}, \qquad (7.11)$$

and elsewhere on an arbitrary point on the axis passing through the loop center

$$dB_y = \frac{\mu_0 I dL}{4\pi} \frac{R}{\left(y^2 + R^2\right)^{3/2}}, \qquad (7.12)$$

where $\sin\theta = \dfrac{R}{\sqrt{y^2 + R^2}}$.

The field at this point is always in the z direction along the center line. When integrating along the whole current path, all the terms are constant except for dL, which when summed up, gives the circumference of the circle. The magnetic field components are:

$$B_x = 0$$

$$B_y = \frac{\mu_0}{2} \frac{R^2 I}{\left(y^2 + R^2\right)^{3/2}} \cdot \quad (7.13)$$

$$B_z = 0$$

It is useful to also have the most general case of the magnetic field induced by a circular loop at an arbitrary point in space P(x,y,z), as shown in Fig. 7.6. The magnetic field components are now expressed in terms of cylindrical coordinates and elliptic integrals of first and second order, as shown below. These formulae are useful for calculating magnetic field distribution, especially for the class of surface coils that have in great majority, but not only, a circular geometry. They are also used for calculating the magnetic field components given by all circular components belonging to a given probe.

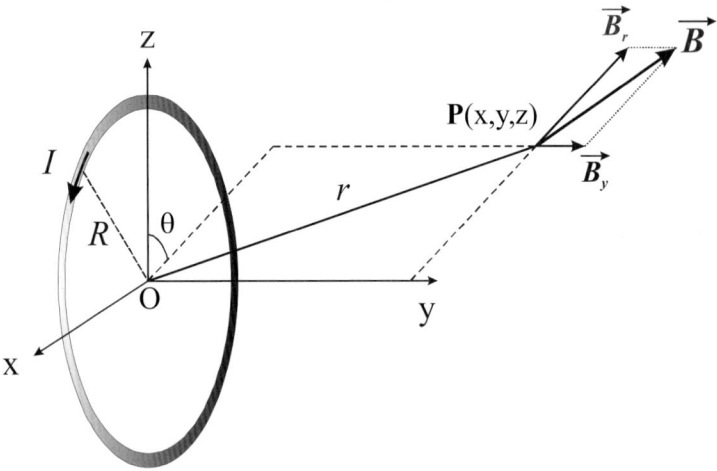

Fig. 7.6 Magnetic field components generated by a current loop at an arbitrary point in space P(x,y,z).

$$B_r = \frac{\mu_0 I}{2\pi} \frac{y}{r\left[(R+r)^2 + y^2\right]^{1/2}} \left[\frac{R^2 + r^2 + y^2}{(R-r)^2 + y^2} E(k) - K(k)\right]$$

$$B_\theta = 0 \qquad (7.14)$$

$$B_y = \frac{\mu_0 I}{2\pi} \frac{1}{\left[(R+r)^2 + y^2\right]^{1/2}} \left[\frac{R^2 - r^2 - y^2}{(R-r)^2 + y^2} E(k) + K(k)\right]$$

where E and K are the complete elliptic integrals of second and first kind, respectively.

$$E(k) = \int_0^{\pi/2} \left(1 - k^2 \sin^2 \phi\right)^{1/2} d\phi$$
$$K(k) = \int_0^{\pi/2} \left(1 + k^2 \sin^2 \phi\right)^{-1/2} d\phi \qquad (7.15)$$

defined on the k variable

$$k^2 = \frac{4Rr}{(R+r)^2 + y^2}. \qquad (7.16)$$

The elliptic integrals may be developed in series and thus Eq. (7.14) may be evaluated at every point in space in the volume of interest.

$$E(k) = \frac{\pi}{2}\left\{1 + \sum_{i=1}^{\infty}\left[\frac{\prod_{t=1}^{i}(2t-1)^2}{\prod_{t=1}^{i}(2t)^2} \frac{k^{2i}}{2i-1}\right]\right\}$$

(7.17)

$$K(k) = \frac{\pi}{2}\left\{1 - \sum_{i=1}^{\infty}\left[\frac{\prod_{t=1}^{i}(2t-1)^2}{\prod_{t=1}^{i}(2t)^2} k^{2i}\right]\right\}$$

An efficient numerical method (King's method [Fan *et al.*, 1987]) will be described in the appendix to evaluate these integrals accurately.

7.2 Effective Field for NMR Experiments

The aim of the NMR probe design is to provide the radio frequency field distribution of the designed probe, calculated prior to experimental trials.

In practice, every conductive element of the probe is producing magnetic field components over the three principal directions, x, y, z. These magnetic field components are varying in time, and have variable amplitudes and phases, due to different current distributions over various conductive geometries. Considering an arbitrary conductive segment, k, the generated magnetic field is given by:

$$\Delta\vec{B}_k = \vec{b}^k_{geometric} I_k \exp i(\omega t + \varphi_k).$$

(7.18)

Generally, for calculation purposes, the terms in Eq. (7.18) may be separated, evidencing the current contributions and the geometrical factor imposed by specific probe design.

Due to the complex form of the currents, the corresponding field components may be expressed as $\Delta\vec{B}^k_u = \vec{b}^k_{geometrical} I^k_u$ for the real part and, similarly, $\Delta\vec{B}^k_v = \vec{b}^k_{geometrical} I^k_v$, for the imaginary one.

Summing up all magnetic field contributions, over all current paths corresponding to the whole probe, may become rather complex if calculated in the laboratory frame. In this respect, one has to add all oscillating vectors describing the magnetic field, generated by all conductive elements in the probe:

$$\vec{B}_u = \left(\sum_k \vec{b}^k_{geometric} I^k_u \right) \cos \omega t$$
$$\vec{B}_v = \left(\sum_k \vec{b}^k_{geometric} I^k_v \right) \sin \omega t \qquad (7.19)$$

At this point, it is useful to stress that Eq. (7.19) contains information about current dephasing. The spatial dephasing is due to geometrical projections of the conductive elements on the three principal axes while the current dephasing in time is due to the electrical configuration of the probe. In this representation (laboratory frame) it is difficult to separately analyse the time dephasing from the geometrical one. A solution to obtain a separation of variables is to transfer all equations describing the field distribution to the rotating frame, as proposed by Hoult [Hoult, 2000]. Firstly, in the rotating frame one gets rid of the time varying components, keeping only the amplitudes and phases as variable parameters. The generated magnetic field $\tilde{B}_1$ now behaves as a static magnetic field whose components are easy to calculate. Moreover, with only the x and y components[2] being involved in getting an NMR signal, in both rotating or laboratory frames, it is much simpler to consider them via increasing and decreasing vectors: − representing the clockwise rotation at an angular frequency ω and + representing the rotation counterclockwise.

Generally, the components of the $\tilde{B}_1$ field in the rotating frame are given by:

$$\tilde{B}^\pm_{1x} = B_{1x} \cos(\pm\omega t) + B_{1y} \sin(\pm\omega t)$$
$$\tilde{B}^\pm_{1y} = -B_{1x} \sin(\pm\omega t) + B_{1y} \cos(\pm\omega t) \qquad (7.20)$$

---

[2] Assuming B_0 is aligned in the z direction.

where B_{1x} and B_{1y} are the magnetic field components in the laboratory frame and the ± signs describe the sense of field component rotation.

Finally, by neglecting the fast time varying terms in $2\omega t$, we get the field components in the rotating frame expressed only in terms of amplitude and phase components in the laboratory frame. Expressed in the terms used above for the field components in the laboratory frame, one obtains:

$$\begin{aligned}\tilde{B}_{1x}^{+} &= \frac{1}{2}\left(B_u^x - B_v^y\right) \\ \tilde{B}_{1y}^{+} &= \frac{1}{2}\left(B_v^x + B_u^y\right)\end{aligned} \quad (7.21)$$

The field components in the negatively rotating frame are obtained easily, replacing the angle by $-\omega t$. The components for the negatively rotating frame $\tilde{B}_1^-$ are given by:

$$\begin{aligned}\tilde{B}_{1x}^{-} &= \frac{1}{2}\left(B_u^x + B_v^y\right) \\ \tilde{B}_{1y}^{-} &= \frac{1}{2}\left(-B_v^x + B_u^y\right)\end{aligned} \quad (7.22)$$

There are three cases to discuss upon designing probes: linearly, circularly, and elliptically polarized probes. At this stage, it will be interesting to determine the rotating field components for each of these designs.

In the case of linearly polarized probes, there are no imaginary components in the laboratory frame, hence $B_v^x = B_v^y = 0$, or in other words, all currents for all conductive segments have the same phase in time. Under these conditions, the rotating field components are given by:

$$\begin{aligned}\tilde{B}_{1x}^{+} &= \frac{1}{2}B_u^x \\ \tilde{B}_{1y}^{+} &= \frac{1}{2}B_u^y\end{aligned} \quad (7.23)$$

and similarly

$$\tilde{B}_{1x}^- = \frac{1}{2} B_u^x$$
$$\tilde{B}_{1y}^- = \frac{1}{2} B_u^y$$
(7.24)

As expected, $\tilde{B}_1^+ = \tilde{B}_1^-$, which corresponds to the well-known decomposition of a linearly polarized field into two counter rotating fields.

In the case of circularly polarized probes, the dephasing in time and space is $\pi/2$ and the amplitudes for the real and imaginary components are equal. Under these conditions, the relationships between the field components are $B_u^x = -B_v^y \quad B_v^x = B_u^y$, giving:

$$\tilde{B}_{1x}^+ = B_u^x \quad \tilde{B}_{1x}^- = 0$$
$$\tilde{B}_{1y}^+ = B_u^y \quad \tilde{B}_{1y}^- = 0$$
(7.25)

Of course, for these designs, the symmetry of the problem also allows $B_u^x = B_v^y \quad B_v^x = -B_u^y$ for the negative sense of rotation, giving:

$$\tilde{B}_{1x}^+ = 0 \quad \tilde{B}_{1x}^- = B_u^x$$
$$\tilde{B}_{1y}^+ = 0 \quad \tilde{B}_{1y}^- = B_u^y$$
(7.26)

The most common case is when the field is elliptically polarized. In this case the time and spatial dephasing is again $\pi/2$, but the field amplitudes are no longer equal and thus an elegant expression cannot be derived.

These formulae were used for calculation of the field maps generated by various probe designs, presented later in the book. An example of a calculated field map is shown in Fig. 7.7 for a given probe design, including a resonator (a birdcage in this case) driven by a coupling loop. The probe is expected to induce a linearly polarized field but characteristic field distortions close to the coupling loop may be seen in all field plots. The plots (a) and (b) shown in Fig. 7.7 are calculated in the rotating frame using the field components for $\tilde{B}_1^+$ and $\tilde{B}_1^-$, respectively, while the plot in (c) is calculated using the field components in the laboratory frame. The different positions of the field distortion for the two rotating components are to be noted. They are in evidence since the spatial dephasing is better resolved in the rotating frame.

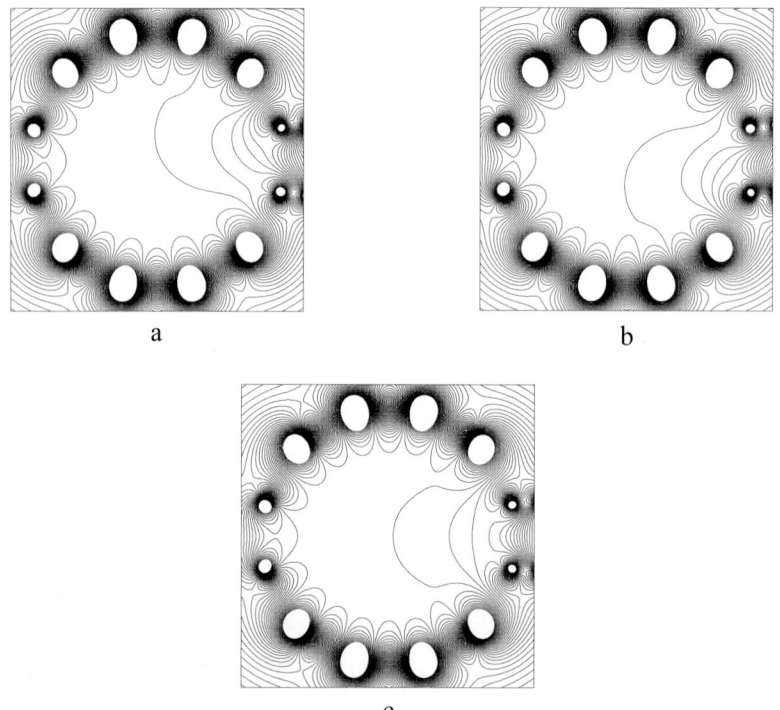

Fig. 7.7 Maps of the magnetic field components generated by a probe in the (a) positive and (b) negative rotating frame as well as (c) in the laboratory frame. Note the asymmetric distribution of the rotating frame components. A similar effect has already been evidenced by Hoult and Tomanek [Hoult and Tomanek, 2002.

7.3 Estimation of the Current Distribution

7.3.1 *Limits and usefulness of the thin wire approximation*

Up to here, we were able to decompose the probe into simple geometrical straight segments and/or loops whose contribution to the total magnetic field is easy to calculate. More importantly, we were able to identify which components of the magnetic field are effective for an NMR experiment. As far as the wavelength of the driving radio frequency is large compared to the linear dimension of the probe coil, the effect of the induced far field (retardation effects) is negligible. In this limit, the induced magnetic field is calculated directly by the Biot–Savart law.[3] Hence the magnetic field estimation requires integration over an already determined current distribution.

It is known that the thin wire approximation gives good estimations when the dimensions of the conductive sheets are small compared to the distance at which the field is estimated. Better results can be obtained when taking into account the so-called "geometric mean distance" [Grover, 2004] that in fact considers the proximity effect (Chapter 3). This effect disturbs the symmetrical current distribution (see below), obtained for an isolated conductive straight wire. The proximity effect is leading, in some cases, to an important displacement of the equivalent thin wire position from the center of the current distribution of interest.

However, there are a number of cases where the conductive parts of the probe are close and of similar dimensions to the sample volume itself. This is especially the case for the "loop gap" resonator that is essentially a winding of a copper sheet that completely encloses the sample volume.

Calculating the magnetic field inside the volume is still possible using the Biot–Savart law but the current density must be estimated first. Another similar situation exists when the probe is enclosed in a

---

[3] The approximation can still be used when the dimensions become a fraction or more of the wavelength. In this case [Crozier *et al.*, 1997], the sum of the magnetic field components should be done taking into account the phase of each current element.

conductive shield in which a certain current density is induced, perturbing the magnetic field distribution of the whole probe. Finally, many other probe designs make use of important conductive sheets where the current density is influenced by the current circulating in the vicinity (proximity effects).

For all these cases, the current distribution must be known prior to applying the Biot–Savart law, taking into account the precautions already discussed, especially regarding the relative probe dimensions and wavelengths in play.

Solving this problem is done nowadays using sophisticated (and expensive) programs that can provide numerical solutions for the complete Maxwell equations, in the presence of the sample.

Fortunately, more simple solutions are offered, especially in cases when the probe has simple symmetry properties.

For example, many probes creating a "transverse" magnetic field can be approximated as infinitely long conductive sheets aligned parallel to the main axis of the whole assembly. This is the case for birdcage, saddle-shaped, slotted-cylinder type coils, among others. The current distributions over all conductive parts can be sampled as parallel current filaments. The magnetic field components created by each filament can be estimated simply from the thin wire approximation formulae, then added together, taking into account the corresponding amplitudes and phases carried by each current. In this case, the contribution of an eventual axial shield can similarly be taken into account.

Other simple geometries that can be treated by the same approach correspond to the "loop gap" and the flat circular coil geometry. In these cases, the current distribution will be sampled as thin wire loops carrying a given current. Again, the thin wire approximation may be used to estimate the magnetic field created by each current loop separately. The total field will be obtained by adding all loop contributions, taking into account the amplitude and phase carried by each current.

Finally, the determination of the current distribution for a specific design also allows calculation of the power loss in the coil, another important issue for all NMR experiments.

A simple method will be described here to estimate the current distributions for simple geometries, such as the single flat strip or two close strips, in order to put in evidence the proximity effects, the cylinder case, and the large conductive plane case. Although very simplified, the method gives accurate estimations for all these simple cases. It can be easily applicable to many probe geometries providing, in particular, a comprehensive description of the field distributions created by different types of coils. In particular, all the examples considered here are common for the large family of axial resonators that are generating the exciting magnetic field parallel to the z axis. The same concepts will also be used, in following chapters, to describe the field distributions created by current loops (loop gaps, flat circular coil).

7.3.2 Current distribution in the isolated flat strip

To provide a comparative basis for the method described later, the current distribution problem of an infinitely long flat strip (Fig. 7.8) will firstly be solved by Maxwell's equations subjected to boundary conditions. The reasoning essentially follows the principles introduced by Carlson [Carlson, 1986].

It should be remarked here that the key assumption of the formalism is that the field at any point near the conductor is solely determined by the local properties of the current distribution.

In this case, the current distribution on the strip should be independent of the location of the strip ends provided they are sufficiently far away (case of a long strip). This consideration allows reconciling of the assumption of an infinitely long strip and the approximation of no retardation effects.

The conductor is assumed perfect, thus the tangential components of the electric field vanish, as well as the normal components of the magnetic field:

$$\vec{E} \times \hat{n} = 0 \\ \vec{B} \cdot \hat{n} = 0 \quad , \tag{7.27}$$

where $\hat{n}$ is the normalized vector perpendicular to the conductive plane.

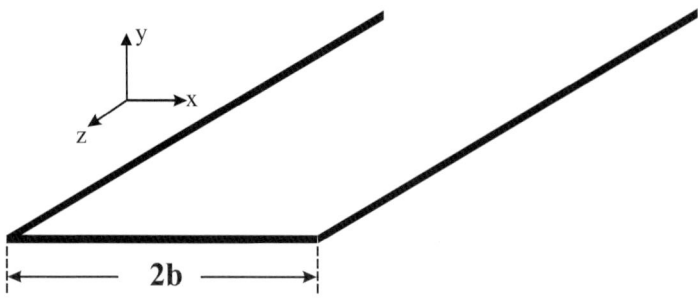

Fig. 7.8 Geometry of the infinitely long flat strip.

Now, it is assumed that a surface current flows on the top and bottom of the strip in the z direction (there are no transverse components due to current conservation law). At this point, it should be remarked that, due to the symmetry of the problem, the current density function $j(x,y,z)$ is even respective to the x axis. Furthermore j is assumed to be independent of the y (thin strip) and the z (long wavelength) axes. The current density may thus be described by a number of parallel filaments, each carrying a given current $j(x)dx$. The choice of the filament number, and of its distribution,[4] is just a matter of accuracy required to correctly sample the current density.

The magnetic field at any point $P(x',0,z')$ on the surface of the conductor (Fig. 7.9), due to a current placed between x and dx, is given, using Eq. (7.3), by:

$$d\vec{B}(x') = \frac{\mu_0}{2\pi} \frac{j(x)dx}{(x-x')} \hat{n} \ . \qquad (7.28)$$

---

[4] For simplicity, the distribution or filaments will be assumed uniform, without lacking in generality.

The total magnetic field at any point P is perpendicular to the strip and is given by:

$$B(x') = \frac{\mu_0}{2\pi}\left[-\int_{x=-b}^{x'-\varepsilon} \frac{j(x)}{|x-x'|}dx + \int_{x'+\varepsilon}^{x=b} \frac{j(x)}{|x-x'|}dx\right]. \quad (7.29)$$

The only approximation used by this formalism is to consider the physical strip as a set of infinitely long conductors (filaments), parallel to the z axis. The integration domain is split in order to avoid discontinuity around x'.

At the center of the conductor ($x' = 0$), the normal component of the magnetic field obviously vanishes on the surface, for any symmetric current distribution $j(x)$.

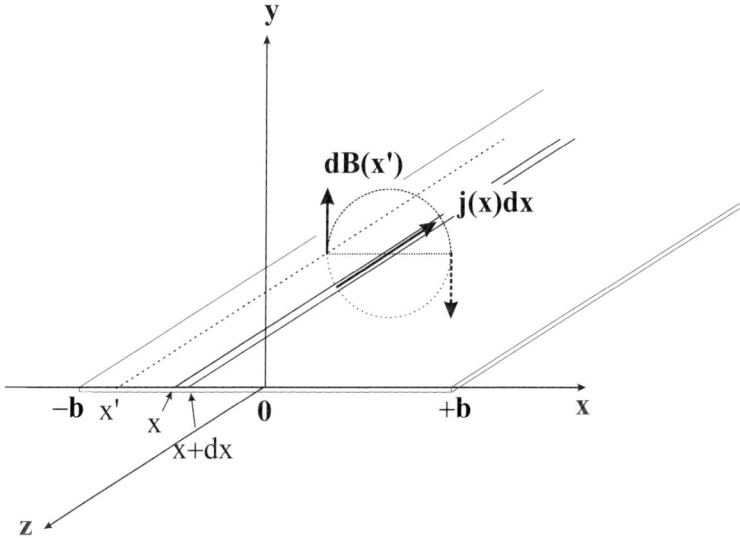

Fig. 7.9 Magnetic field perpendicular to the conductive plane and created by a current filament $j(x)dx$.

Assuming a uniform current density (equal current in each filament) and moving away from the center, the resulting current imbalance results in a net normal component of magnetic field **B** [Fig. 7.10(a)]. As a consequence, the current distribution on the flat strip cannot be uniform. When approaching the strip edges, the magnetic field created by all the filaments on one side along x should be compensated by very few located on the opposed side [Fig. 7.10(b)]. Clearly, the current should increase when approaching the strip edges.

Quantitatively, the current density $j(x)$ can in principle be obtained by solving the equation expressing that the magnetic field component perpendicular to the conductor should be zero at any point [second condition of Eq. (7.27)]

$$B(x') = 0 \quad \forall x': -b \leq x' \leq b, \quad (7.30)$$

where $B(x')$ is given by Eq. (7.29).

Integrating Eq. (7.29) over the conductor surface and solving Eq. (7.30) in order to obtain an analytic expression for the current density $j(x)$ is not easily feasible.

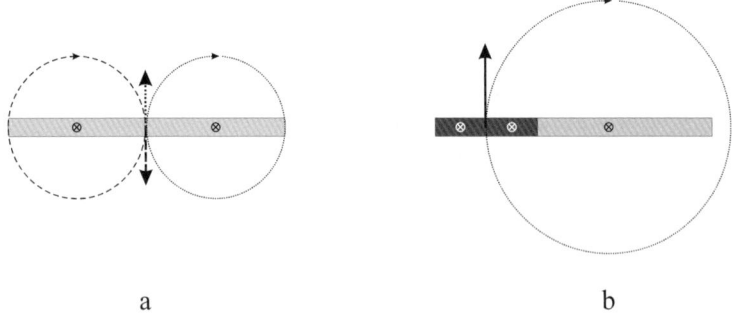

a b

Fig. 7.10 The magnetic field created on the surface of a plane, assuming a uniform current distribution. The current is assumed to flow perpendicularly to the figure, from front to back. In (a), the fields created at the center by equal currents on both sides are cancelling out. In (b), the contributions from the dark grey marked currents, symmetrically placed respective to the field sampling position, cancel together. A total nonzero magnetic field is created by the remaining light grey marked current density.

The problem can be solved, however, as a standard boundary value electrostatics problem [Carlson, 1986]. Near the surface of the conductors, $\vec{E}$ is perpendicular to the surface and proportional to the surface charge.

The line integral of $\vec{B}$ along a loop just outside the conductor is:

$$\oint \vec{B} \cdot \vec{dl} = \oint \left(\hat{z} \times \vec{E}_t \right) \cdot \vec{dl} , \qquad (7.31)$$

where the quantity $\left(\hat{z} \times \vec{E}_t \right)$ is parallel to $\vec{dl}$ near the surface and the charge on the conductor is proportional to the current flowing along the strip. In these conditions the only source of field is the time-varying current density. The electric field evaluated at the surface of the conductor gives the surface current, expressed as:

$$j(x) = \frac{I}{2\pi} \frac{1}{\sqrt{b^2 - x^2}} \qquad (7.32)$$

for $-b < x < b$, where I represents the magnitude of the total current.

A plot of this current distribution is shown in Fig. 7.11. The results reveal, as already expected, a net increase of the current distribution near both edges of the flat strip while it tends to a uniform distribution near the center.

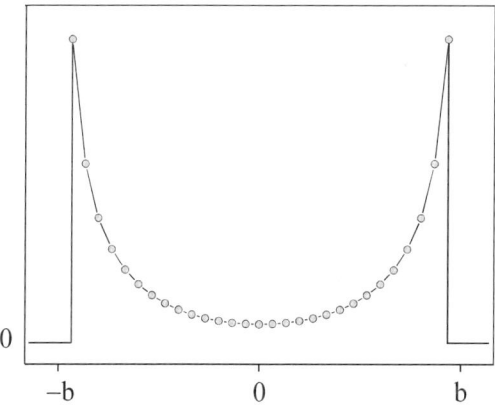

Fig. 7.11 Current distribution over the flat strip surface [as given by Eq. (7.32)], showing an important increase of current density close to the edges.

Although an analytical solution can be obtained quite easily for the simple isolated flat strip, the problem will rapidly become cumbersome for more complicated conductor designs.

In fact Eq. (7.30) can be written as a system of linear equations that is easily resolved in order to get a numerical solution.

Let us describe the current density by a set of discrete longitudinal current filaments I_k located at a certain position x_k (Fig. 7.12).

Then, let us define a set of positions x_j along the x axis, perpendicular to the current direction, where the magnetic field created by all currents is given by:

$$\vec{B}_j = \frac{\mu_0}{2\pi}\left(\sum_{k(x_j>x_k)} I_k \frac{1}{|x_j - x_k|} - \sum_{k(x_j<x_k)} I_k \frac{1}{|x_j - x_k|}\right)\hat{n}. \quad (7.33)$$

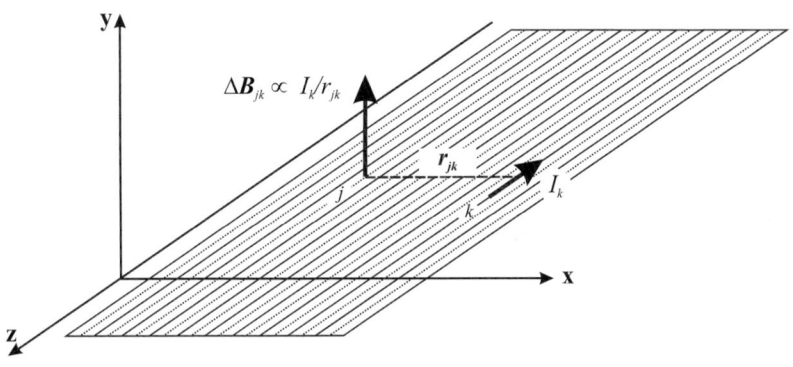

Fig. 7.12 Sampling of the strip current density by parallel linear segments of current I_k (full lines). The test points where the magnetic field is calculated are located on dotted lines placed in-between the current filaments. The field contribution of filament k at position j (ΔB_{jk}) is proportional to the current I_k and inversely proportional to the distance between the two lines k and j.

Equation (7.33) can be rewritten in a more general formulation that will later allow the formalism to be extended to more complicated conductive structures than the simple flat strip:

$$B_j^\perp = \sum_k a_{jk} I_k . \qquad (7.34)$$

In Eq. (7.34), a_{jk} is a geometrical factor that depends only on the relative positions of the current filament k contributing to the perpendicular component of the magnetic field at position j. Thus a_{jk} can be calculated for any conductor configuration using the simple Biot–Savart law. I_k is the unknown current that flows in the filament k. All I_k form a current basis set describing the current density. Writing now that the magnetic field $B_j^\perp$ is equal to zero for all j,

$$B_j^\perp = 0 \quad \forall j = 1, n \qquad (7.35)$$

and adding the condition that the sum of all currents I_k is equal to the total current flowing in the strip, Eq. (7.36),

$$\sum_{k=1}^{m} I_k = I_{port} \qquad (7.36)$$

one obtains a set of linear equations that should be solved for $\{I_k\}_{k=1,m}$

$$\begin{pmatrix} a_{11} & \cdots & a_{1m} \\ \vdots & & \vdots \\ a_{n1} & \cdots & a_{nm} \\ 1 & \cdots & 1 \end{pmatrix} \begin{pmatrix} I_1 \\ \vdots \\ \vdots \\ I_m \end{pmatrix} = \begin{pmatrix} 0 \\ \vdots \\ 0 \\ I_{port} \end{pmatrix}. \qquad (7.37)$$

I_{port} is the total current that is issued in the corresponding "port" by an external source. In the case of the single isolated flat strip, there is only one port giving rise to the matrix form shown in Eq. (7.37).

Solving the linear system of equations above requires some additional constraints in the choice of parameters that have not been discussed until now, namely the number (m) of current filaments and the number of positions (n) where the component of the magnetic field perpendicular to the conductor is evaluated.

In the case of the flat strip, a simple way to sample the current density is to consider m filaments uniformly distributed over x. The positions for the field calculations are naturally chosen in-between the current filaments in order to avoid singularities in Eq. (7.33). Furthermore, with the magnetic field amplitude being independent of z, it is sufficient to consider n positions (testing points) along a line parallel to x to completely determine the problem. In this way, the matrix of dimension $(n + 1) \times m$ is squared ($n = m - 1$), in principle allowing the equation to be solved by inverting the matrix in Eq. (7.37).

The matrix being nearly singular, a stable and robust computational technique is required. Solving the linear system by the Singular Value Decomposition (SVD) method [Press et al., 1992, Chapter 2] appeared to be very efficient. This approach also allows some freedom in the choice of the **A** matrix dimensions that could no longer be square.

Considering again the flat strip case, the matrix elements are given by:

$$a_{jk} = \frac{-1}{x_j - x_k} \quad j < m, \ k \leq m$$
$$a_{mk} = 1$$
(7.38)

where the factor $\mu_0/2\pi$ has been dropped out.

The linear system of equations to be solved can be written as:

$$\mathbf{A}I = b,$$
(7.39)

where I is the vector of unknown currents $\{I_k\}_{k=1,m}$ and b the right hand side vector in Eq. (7.37)

$$b_j = \begin{cases} 0 & j < m \\ I_{port} & j = m \end{cases}.$$
(7.40)

The solution is given by [Press et al., 1992]:

$$\mathbf{I} = \left[\mathbf{V} \begin{pmatrix} 1/w_1 & \cdots & 0 \\ \vdots & \ddots & \vdots \\ 0 & \cdots & 1/w_m \end{pmatrix} \mathbf{U}^\perp \mathbf{b} \right], \tag{7.41}$$

where the matrices $\mathbf{V}$, $\mathbf{U}$ ($\mathbf{U}^\perp$ is the transposition of $\mathbf{U}$) and the central diagonal matrix $\{1/w_j\}$ are obtained from the SVD of $\mathbf{A}$

$$\mathbf{A} = \mathbf{U} \begin{pmatrix} w_1 & \cdots & 0 \\ \vdots & \ddots & \vdots \\ 0 & \cdots & w_m \end{pmatrix} \mathbf{V}^T. \tag{7.42}$$

w_j are the singular values and $\mathbf{U}$ and $\mathbf{V}$ are orthogonal matrices.

Apart from the robustness of the SVD that can handle singular matrices, rectangular matrices can be handled as well. Thus, the matrix $\mathbf{A}$ does not need to be squared. If it is squared, it can be shown that its inverse is given by:

$$\mathbf{A}^{-1} = \mathbf{V} \left[diag(1/w_j) \right] \mathbf{U}^\perp, \tag{7.43}$$

hence Eq. (7.41). This formalism may be generalized even for rectangular matrices. It should be mentioned here that some singular values, w_i, could be equal, or very nearly equal, to zero. In those cases, the inverse $1/w_j$ tends to infinity. This problem is simply bypassed by replacing $1/w_j$ by zero! The reason for this unusual procedure is that the nonzero singular values in fact represent the rank of the matrix. In other words, the zero (or nearly zero) singular values correspond to the part of the matrix that does not contain any significant information. The corresponding contribution could thus be eliminated (zeroed) in the final solution (see Press *et al.*, 1992 for a complete discussion).

The current density calculated by this method for the flat strip is compared in Fig. 7.13 with those obtained from the exact formula Eq. (7.20). The agreement is good and could be even better using a larger number of sampling points of the conductive surface.

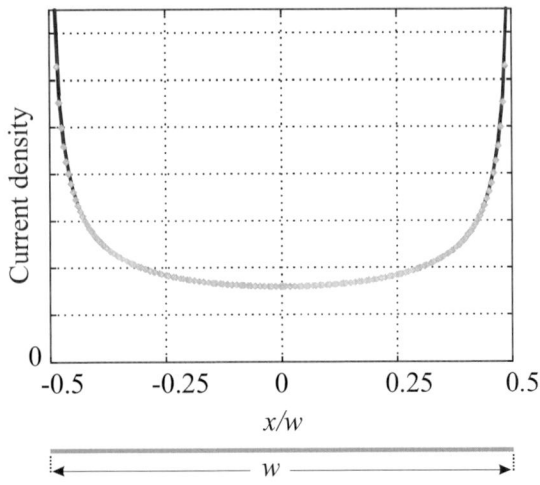

Fig. 7.13 Current distribution for a flat strip, calculated by applying the exact formula given in Eq. (7.32) (continuous line) and by SVD (scatter).

7.3.3 Proximity effects

7.3.3.1 Two coplanar strips

In order to illustrate the effect of proximal current density we consider firstly the case of a planar strip in which a slit is made, separating the whole conductor into two planar strips located close together. As with the single strip case, the current density is sampled by m filaments and the perpendicular component of the magnetic field is calculated on $m-1$ test points. The matrix elements a_{jk} are still given by Eq. (7.37) for $j < m$. The remaining row(s) of **A** depends on how the currents are supplied in the two conductors (one or two ports). In the one-port case [Fig. 7.14(a)], the strips are assumed to be driven by the same current source I_{port}. The linear system of equations has the same form as in Eq. (7.37), and the matrix **A** is rectangular ($m \times m-1$). The result is shown in Fig. 7.15. The total current splits equally (as expected) between the two conductive sheets.

Magnetic Field Amplitude Estimation 453

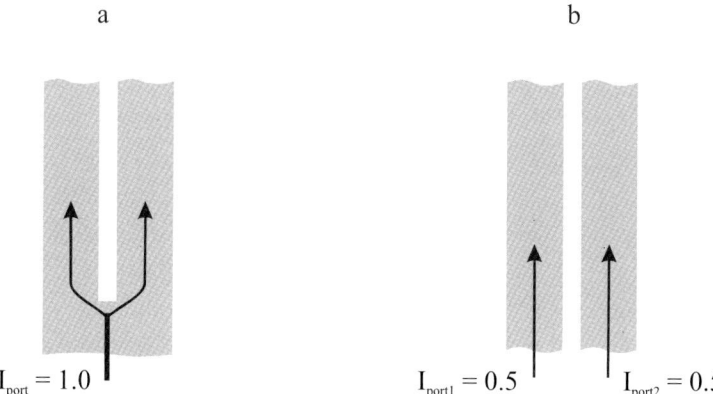

Fig. 7.14 Two possibilities to drive a strip split into two identical parts. In (a) the total current is applied to a single port. It splits equally (by symmetry) into the two strips. In (b) two current sources drive each strip separately.

Another possibility to drive the two strips is to have two current sources [two-port case, Fig. 7.14(b)]. Let I_{port1} and I_{port2} be the corresponding currents. The linear system of equations takes the following form:

$$\begin{pmatrix} & & & & & \\ & & a_{jk} & & & \\ & & & & & \\ 1 & \cdots & 1 & 0 & \cdots & 0 \\ 0 & \cdots & 0 & 1 & \cdots & 1 \end{pmatrix} \begin{pmatrix} I_1 \\ \vdots \\ \vdots \\ I_m \end{pmatrix} = \begin{pmatrix} 0 \\ \vdots \\ 0 \\ I_{port1} \\ I_{port2} \end{pmatrix}, \quad (7.44)$$

where the matrix **A** is square in this example.

In the present case, (two parallel strips driven by the same voltage source) both currents I_{port1} and I_{port2} are equal to $I_{port}/2$. The results are thus identical to those obtained for the one-port case.

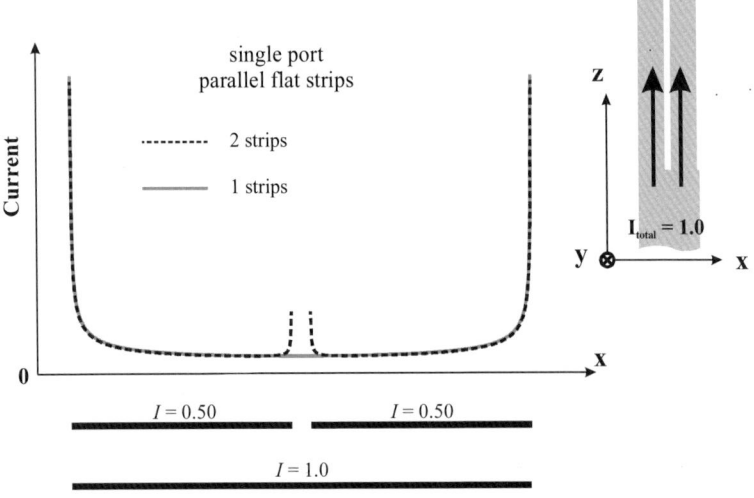

Fig. 7.15 Current distribution calculated by SVD for two neighboring flat strips, showing the proximity effect between the strips (dotted line). The full grey line represents the current density on a single strip carrying the same total current as the two separate strips.

The current density on the x axis coincides with that already calculated for the single strip, except in the proximity of the gap. At near edges, the current density increases, depending on the gap dimensions in an obvious manner. As the gap increases, the current density of the proximal edges will also increase, to reach the distribution expected for an isolated single strip. In contrast, the corresponding singularity will disappear as the slit dimension will decrease down to the continuous conductor strip limit.

The proximity effects become more important when increasing the geometrical complexity of the conductive parts, inducing important changes in the corresponding field maps. The changes induced in the current distribution by close conductive parts of the probe also have important consequences for local resistive changes and thus in the Q factor.

7.3.3.2 Three or more strips

When more than two strips are involved, the correct way to obtain the current distribution is to use a multi-port approach.

In other words, N conductors are sampled by a given number m of filament currents and the component of magnetic field perpendicular to the conducting surface is calculated at n given test points (usually in-between the filaments) in order to build the coefficients a_{jk} of the matrix **A**. The current density is given by the solution of the following linear system of equations:

$$\sum_k a_{jk} I_k = 0 \quad j = 1, n, \tag{7.45}$$

$$\sum_{k \in strip\_l} I_k = I_{port\_l} \quad l = 1, N, \tag{7.46}$$

or, explicitly,

$$\begin{pmatrix} & & & a_{jk} & & & \\ 1 & \cdots & 1 & 0 & \cdots & \cdots & \cdots & \cdots & \cdots & 0 \\ \vdots & \vdots & \vdots & \vdots & \vdots & \vdots & \vdots & \vdots & \vdots & \vdots \\ 0 & \cdots & \cdots & \cdots & 1 & \cdots & 1 & 0 & \cdots & \cdots & 0 \\ \vdots & \vdots & \vdots & \vdots & \vdots & \vdots & \vdots & \vdots & \vdots & \vdots \\ 0 & \cdots & \cdots & \cdots & \cdots & \cdots & \cdots & 1 & \cdots & 1 \end{pmatrix} \begin{pmatrix} I_1 \\ \vdots \\ \vdots \\ I_m \end{pmatrix} = \begin{pmatrix} 0 \\ \vdots \\ 0 \\ I_{port\_1} \\ \vdots \\ I_{port\_l} \\ \vdots \\ I_{port\_N} \end{pmatrix}. \tag{7.47}$$

The current I_{port_l} is the total current that flows on conductor l, driven by a given source. It is worth mentioning here that even if the strips are all connected to the same voltage source, the current sharing is not straightforwardly predicted, except for the two strips case, as described in the previous section. In general, one must look for a circuit analysis method, taking into account the strip impedance[5] and the voltage source induced through the mutual inductances between each conductive part. Examples will be provided with a number of probe circuits in the following chapters.

7.3.3.3 Wire in proximity of a conductive plane

We now consider the case of a conductor placed in close proximity to a large ground conductive plane that does not carry any current (i.e., is not the return path for the current carried by the strip wire). This is a two-port configuration, the wire carrying a current $I_{port\_1} = I$ and a large strip (ground plane) carrying a total current $I_{port\_2} = 0$. The corresponding linear system of equations is [similar to Eq. (7.44)] given by:

$$\begin{pmatrix} & & & & \\ & a_{jk} & & & \\ & & & & \\ 1 & \cdots & 1 & 0 & \cdots & 0 \\ 0 & \cdots & 0 & 1 & \cdots & 1 \end{pmatrix} \begin{pmatrix} I_1 \\ \vdots \\ \vdots \\ I_m \end{pmatrix} = \begin{pmatrix} 0 \\ \vdots \\ 0 \\ I \\ 0 \end{pmatrix}. \qquad (7.48)$$

Being mainly interested in the current density on the large conductive plane, we assume the thin strip to be defined by a small number of current filaments (only one at the limit if the current distribution is of no interest). On the contrary, the sampling of the plane is made out of a large number of segments (>1000). Furthermore we calculate the magnetic field distribution of this configuration and compare it to the

---

[5] At high frequencies, the impedance is generally dominated by the inductance and thus proportional to the frequency. The voltage induced by neighboring currents is also proportional to the frequency. As a result, the shared current distribution will not depend on the frequency above the limit where the real part impedance dominates (below some hundreds of kHz or even at DC with supraconductive materials).

field that will be obtained from the well-known "image" method. The width of the conductive plane is made much larger (100 times) than the distance of the wire to the plane. If not, the field distribution would exhibit "distortions" resulting from the field created by the large current density at the plane edges.

The current distribution on the ground plane (Fig. 7.16), satisfying the zeroed integral condition, is showing a strong reversal of currents in the proximity of the wire.

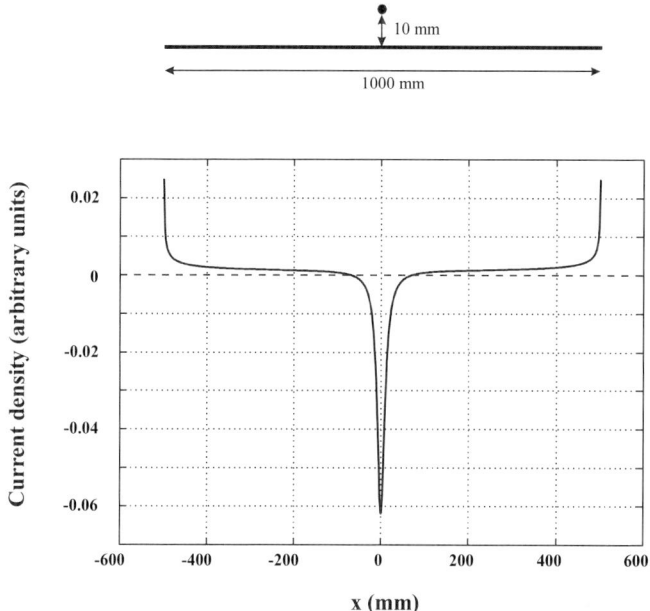

Fig. 7.16 Current distribution on a ground plane, induced by a neighboring current filament.

This effect is similar to that for a virtual identical conductor, placed in a symmetric position on the other side of the ground plane and bearing opposed currents (image concept).

This is demonstrated by the field map obtained from the above current density compared to that calculated from the image method (Fig. 7.17).

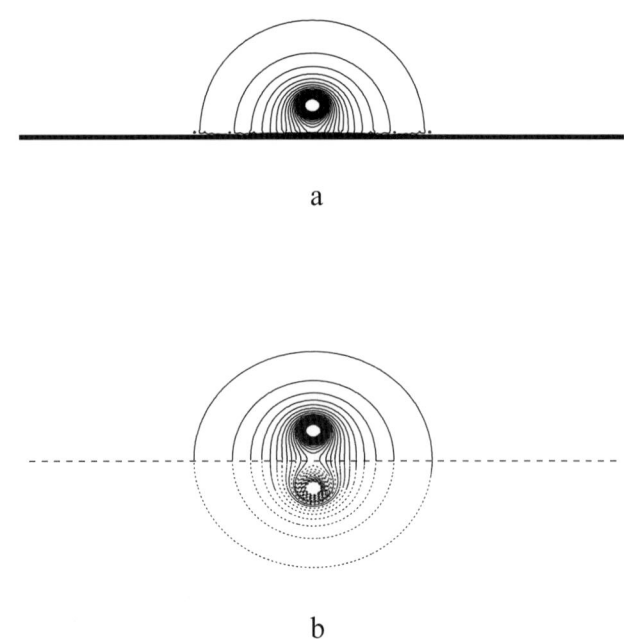

Fig. 7.17 (a) The field map calculations obtained from the current distribution shown in Fig. 7.16 and (b) using the virtual image concept. Both methods give identical results.

Image theory permits the removal of the ground plane, placing a virtual image filament, symmetrically opposed, on the other side of the ground plane, as shown in Fig. 7.18.

For many probe designs a current filament is located in the vicinity of a conductive plane. Consequently, the image theory is frequently invoked when calculating the magnetic field generated inside shielded probes, for instance. In this situation the shield is replaced by the reciprocal current filament for which the contribution is calculated in the region of interest.

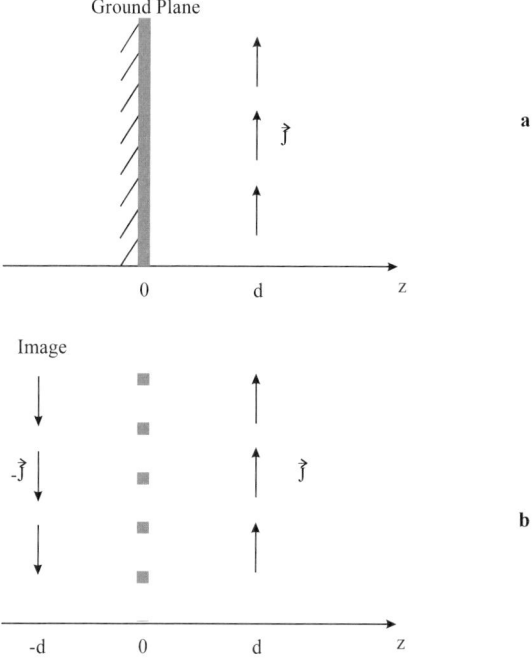

Fig. 7.18 Image theory for a current filament near a large ground plane. The ground plane in (a) is replaced by a symmetrical opposed current filament in (b).

In addition, the image method allows an easy estimation of the inductance of part of the conductor inside the shield in order to analyze the frequency-dependent properties of the corresponding probe circuit.

7.3.4 Current density in round wires

The current distribution on tubular conductors can be obtained using the same calculation approach as for the flat strips.

Assuming an isolated round wire, the current density is uniformly distributed around the tubular conductor [Fig. 7.19(a)], as expected. This result is also obtained by resolving Eq. (7.35). The geometrical factors a_{jk} do not show a simple form as for the flat strip but can still be calculated (see Resources for Readers, page X).

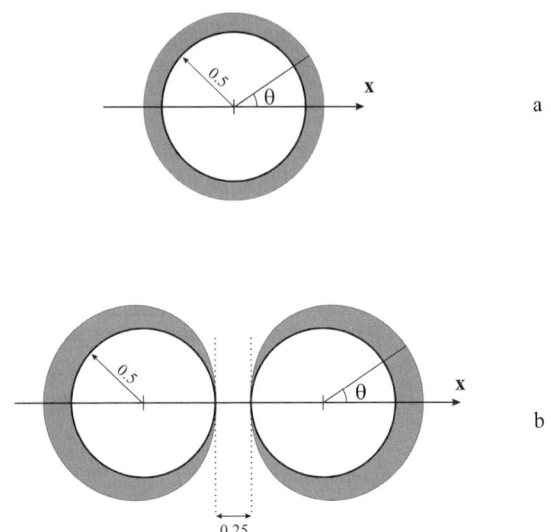

Fig. 7.19 (a) Current distribution on a round wire and (b) on two neighboring round wires, showing the proximity effect in between (the current is flowing in the same direction). The current density around the wire surface is represented by the width of the grey region.

At this point it should be mentioned that the calculated current density is a surface density representing how the current is distributed around the cylindrical coordinate θ. Because one assumes a perfect conductor, the current does not penetrate inside the wire. In case of realistic finite conductivity, one should add the current density in the radial dimension that describes the "skin depth effect." Both proximity and skin depth effects are, however, independent from each other. In particular, the surface density does not depend on frequency (at high frequency) while the penetration of current inside the conductor depends on both the frequency and conductivity of the material.

When approaching two identical tubular conductors [Fig. 7.19(b)] the current distribution diminishes strongly between them, while slightly increasing on the rest of the surface. In fact, the effect is similar to that giving rise to an increased distribution of currents when approaching the edges of a flat strip. This effect has important consequences on the resistance of the wire, depending not only on the skin effect but also on the proximity effect. In particular the effect dictates the well-known rules that optimize the Q factor of solenoid coils.

7.3.5 *Concluding remarks*

The current distribution on the conductor surface and the proximity effects have been well-known since the pioneering works of Butterworth and others that were developed during the first decades of the 20th century [Terman, 1943, pp. 36]. The method presented here allows these results to be obtained in a simple and inexpensive way. Although limited to particular geometries it can, anyway, be applied to the current estimation and field distributions in many probe structures. This can be done essentially due to the fact that the field at any point near the conductor is solely determined by the local properties of the current distribution.

On the other hand, the method has a pedagogical benefice. It clearly shows the origin of the current distribution on the conductor surface. In particular, it is worth noting that the frequency does not play any role in the surface current density. At first sight this can be curious and contrary to immediate expectations. In fact, the current "preference" to flow on the edge of the conductor is due to the "proximity effect" that tends to push away the neighboring current density, as was exemplified for some cases in the previous section. This effect is not to be merged with the so-called skin effect which describes the ability of the oscillating current to penetrate inside a conductor. Needless to say, here we assume a perfect conductor. In this case, the current does not penetrate the conductor, regardless of its frequency. The current flows on the surface of the conductor but the density is not uniform, except for a circular isolated wire. For other cases, the surface density is given by the numerical computation described here and is always frequency-independent.

Let us assume now that the conductor is not perfect and has a finite conductibility. The skin depth is finite, although very small at frequencies encountered in NMR (Chapter 2). On the other hand, the current density profile will depend on frequency only at the very low frequency range where the resistance dominates the reactance impedance. As demonstrated from accurate calculations using full electromagnetic simulation methods, the profile will reach a frequency-independent state for relatively low frequencies of less than 1 MHz (assuming the copper conductibility properties). As a result, the "high frequency" profile is very similar to the one obtained by the simple method outlined here.

7.4 Survey of Modern Electromagnetic Simulation Methods

Technological and scientific advances have allowed a different approach to the calculation of different generated magnetic fields through a better definition of the electromagnetic properties of the investigated systems. Complex problems may be modeled to a large extent via differential equations and appropriate boundary conditions. This approach works very well for rather simple geometrical designs. For any real-life problems, the development of advanced numerical methods allows, on one hand, a more precise definition of local conditions and on the other, allows the manipulation of huge matrices or very complex differential equations. In this way, almost any problem may be defined and solved in small surface or volume steps. The final result is obtained by putting together all intermediate solutions.

One such method, the *Finite Elements Method* (*FEM*), consists in principle of defining a solution that satisfies the partial differential equation on average over a finite domain (element). In this kind of method the problem domain is subdivided into cells (elements) and an approximate solution is sought, which satisfies suitable matching conditions across the common boundaries of adjacent elements. Every element is then connected to the neighboring elements and finally the whole domain (field) is analyzed by propagating the calculated values from one element to another via connecting points (nodes). In this way the *FEM* is developed as a very precise technique for defining a problem since its definition is taking place in a restricted domain, in a single element. The finite difference method is an alternative approach to derive difference equations for physical systems. It is based on the direct conversion of the governing differential equations for a physical system by the substitution of difference operators. These operators act over small divisions of space ($\Delta x, \Delta y, \Delta z$) or time ($\Delta t$). There are several ways to make the conversion but one has to keep in mind that the computation solutions approach the exact results in the limit that $\Delta x, \Delta y, \Delta z, \Delta t \to 0$.

For example, the *Finite Difference Time Domain* method (*FDTD*) [Taflove and Hagness, 2000] is based on the discretization of Maxwell's curl equations directly in time and spatial domains:

$$\nabla \times \vec{E} = -\mu_0 \frac{\partial \vec{H}}{\partial t}$$
$$\nabla \times \vec{H} = \varepsilon \frac{\partial \vec{E}}{\partial t} + \sigma \vec{E} + \vec{J}$$
(7.49)

These equations are solved in a volume V, divided into unit cells (Yee cells), as shown in Fig. 7.20.

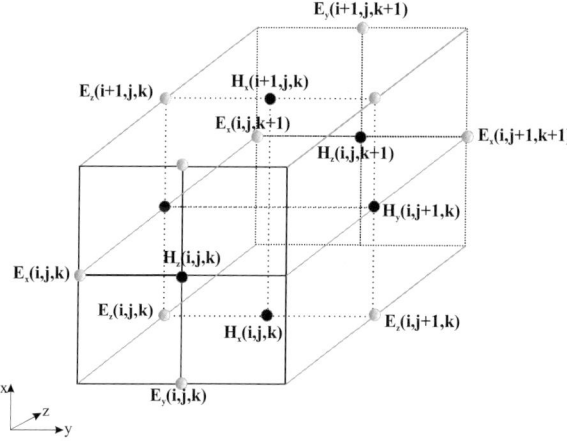

Fig. 7.20 A few electromagnetic field components are represented for the whole volume V, divided into Yee cells.

Each Yee cell is characterized by three electric and three magnetic field components, distinguishable via (i,j,k) unit volume indexing and also by the corresponding time and space discretization, Δt, Δx, Δy, Δz. Thus, even for field components having the same (i,j,k) label, they differ through time and space location. The electric field components are assigned at the center of each edge, while the magnetic field components are assigned at the center of each face of the cells describing the volume V. In this way, there is a $\Delta t/2$ time difference between E and H field components in the cells. The purpose of this exact discrimination in space and time is, given the initial conditions, to calculate the propagation of the electromagnetic field at any point and at any time, inside the volume V, via medium parameters, ε, μ, σ.

The basic ideas of Yee's formalism are:

The Yee algorithm solves for both electric and magnetic fields in time and space, using the coupled Maxwell curl equations rather than solving separately for electric and magnetic fields via wave equations. Using both E and H equations at the same time, the solution appears to be more robust than using either alone.

Yee's algorithm centers the E and H components in three-dimensional space, so that every E component is surrounded by four circulating H components and *vice versa* (Fig. 7.20). This provides a simple picture of three-dimensional space, filled by an interlinked array of Faraday and Ampere laws.

The Yee algorithm centers the H and E components in time in a so-called leapfrog arrangement. All the E computations in this modeled space are completed and stored in the computer's memory for a particular time point using previously stored H data. All the H computations in the space are completed and stored using the E data just computed. The cycle begins again until time-stepping is concluded. In this way a numerical wave mode is propagating through the whole mesh, which will thus be defined in terms of electromagnetic field parameters (Fig. 7.21).

Since all radio frequency coils are open structures, the generated field radiates unlimited in the space which is, obviously, impossible to simulate, hence the method is strongly subject to the proper boundary conditions surrounding the initially defined volume V.

For high frequency circuits, the lattice resolution is usually too fine for that needed for solving the spectral wavelength propagation through the circuit. As a result, it may be necessary to run *FDTD* simulations for tens of thousands of time-steps in order to fully simulate the impulse responses needed for calculating impedances, S parameters, and resonant frequencies of all the structures in the electronic circuits.

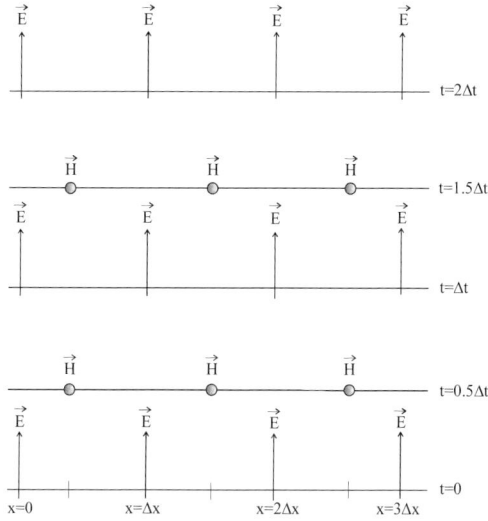

Fig. 7.21 Space–time chart of the Yee algorithm for one-dimensional wave propagation.

The principal drawback of the method is determined by the fact that the grid fits the round shapes poorly, usually requiring a stair-stepping approximation. To achieve a certain accuracy, small grid spacing and short time-step size have to be used, making the method cumbersome and time-consuming. The strategy used to overcome this situation is to extrapolate the electromagnetic field time waveform beyond the actual *FDTD* time-step, thus allowing a good estimation of the circuit parameters accompanied by a considerable reduction of the calculation time. This is the main idea of the *Transmission Line Matrix* (*TLM*) method based on the network theory, where voltage and currents are the parameters to calculate. The *FDTD* computed E and H fields are processed in a transmission line system to obtain the line's characteristic impedance, load, inductance, and capacitance. From these values the equivalent series lumped elements are obtained and thus all the characteristics of the electronic circuit. The current propagation is simulated via uncoupled transmission lines. In this case, the medium parameters ε, μ, σ are modeled by open and short circuited stubs.

One of the most powerful techniques for the computation of electromagnetic fields in the frequency domain is represented by the *Method of Moments* (*MoM*) [Harrington, 1993; Chen *et al.*, 1999]. The technique is based on solving the Maxwell equations by reducing them to a system of linear equations whose solutions provide a numerical solution to the problem. Since all *MoM* techniques need to first establish a set of trial solutions, this method is prone to a choice of initial values. The difference between the trial and the true solutions, the residuals, is minimized in order to get the best fit of the trial functions whose parameters are giving the values for ε, μ, σ in the calculated volume. Depending on the form of the integral equations, *MoM* technique can be applied to the electromagnetic field propagation in conductors only, in homogeneous dielectrics only, or in very specific conductor–dielectric configurations. It is not very effective when applied to arbitrary configurations with complex geometries or structures.

Another approach belonging to the class of methods that simulate the generated magnetic field by solving the Maxwell equations is the *Method of Lines* (*MoL*). This is a frequency-domain, full-wave method where the calculation space is sampled in only two directions. Another simplification, characteristic to *MoL* is that the vector wave equation is solved analytically in one dimension and extrapolated numerically in the other two. These approximations allow a better description of the probe design and a large variety of coil configurations to be simulated very fast. But, due to its geometrical limitations (simulations only of cylindrical geometries), *MoL* cannot replace other full-wave methods such as *FEM* or *FDTD*.

Although computational power is always increasing, enabling the fast manipulation of large data matrices and more and more detailed meshes applied to the surfaces, the trend in numerical simulation techniques is toward using some hybrid forms of approximate analytical and numerical methods.

Chapter 8

Homogeneous Resonators

Ideally, a homogeneous magnetic field will be created, either by a uniform distribution of loop currents on a sphere[1] [Fig. 8.1(a)] or by a cosine dependent distribution of line currents, flowing parallel to a cylinder axis [Fig. 8.1(b)] [Smythe, 1950]. The magnetic field generated by the resonator is always perpendicular to the current direction (Fig. 8.1).

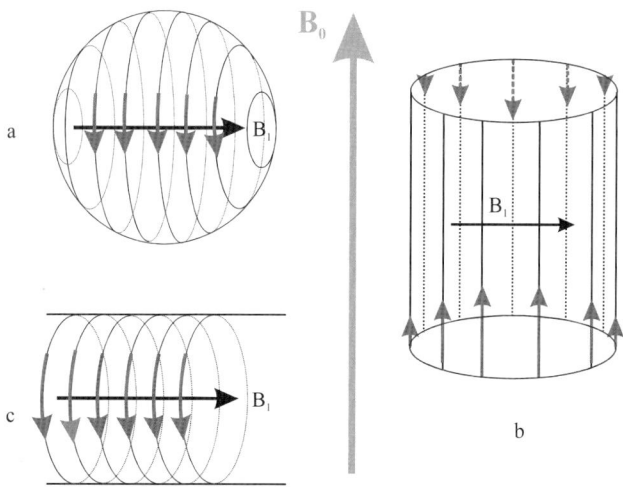

Fig. 8.1 Ideal current distributions that create a homogeneous magnetic field inside the coil volume. Note that for the cylinder cases, the magnetic field is heterogeneous near the edges.

---

[1] It corresponds to a constant current density per unit length, thus to a surface current density that follows a cosine distribution.

Spherical geometry has well-known properties that are widely used in building air core low field resistive magnets. It creates a magnetic field parallel to the symmetry axis of the current density. On the other hand, the cylinder geometry creates a magnetic field perpendicular to the line currents. It is used, for example, in the design of the so-called "birdcage" resonators [Hayes, 1987] or in the Varian "millipede®" coil design [Wong, 2001]. Another case to be considered, similar to the spherical current distribution, is the long solenoid [Fig. 8.1(c)]. It also provides a uniform magnetic field, parallel to the resonator axis. In all of these cases, the current is assumed to be constant in time, requiring very small coil dimensions compared to the wavelength. Obviously, none of these ideal conditions can be realized in practice, but a number of probe designs have been described in order to approach them as closely as possible.

According to these general considerations, two main classes of designs may be proposed, depending on the orientation of the magnetic field respective to the current's direction [Link, 1992]. A particular design will be chosen in order to satisfy the experimental constraints, depending specifically on the sample accessibility to the resonator. Another point that should be taken into account is the fact that the main B_0 field homogeneity also depends on the geometry and the orientation of the sample respective to its direction.

Roughly speaking, the "axial resonators"[2] that approach the spherical current distribution (or the solenoid-like configuration) provide a very good sensitivity but have the disadvantage of poor access, especially for superconducting magnets. The B_0 magnetic field homogeneity may also be a problem due to the field distortion created by a cylindrical sample axially perpendicular to B_0. On the other hand, the "transverse resonators," generally characterized by a lower sensitivity compared to the previous configuration, instead have very good accessibility.

---

[2] The current distribution on the homogeneous resonators is generally axially symmetric and determined by the resonator geometry. The corresponding resonators will be named according to the direction of the generated magnetic field respective to the current symmetry axis.

Furthermore, the cylindrical sample orientation, having its main axis parallel to B_0, is favorable for obtaining good magnetic field homogeneity.

Having chosen between the axial or transverse type of resonator, the design will now depend on the ratio of the coil dimensions to the wavelength. For example, the solenoid could be used at relatively low frequencies, while the "loop gap" resonator will be preferred at higher frequencies. On the other hand, the well-known saddle-shaped coil, which is mainly used as a transverse resonator inside superconducting magnets, will be replaced by the slotted tube resonator at very high frequencies.

Some of the resonator designs that will be encountered in NMR work are described in this chapter according to the orientation of the generated magnetic field, going progressively from low to high frequency designs. For most cases, a practical example will be given in detail for frequencies around 200 MHz.

8.1 Axial Resonators

This class of resonators corresponds to those shown in Figs. 8.1(a) and 8.1(c). The generated magnetic field is parallel to the resonator axis.

8.1.1 *Magnetic field amplitude*

In order to have a basis for the probe evaluation it is convenient to know the magnetic field at the probe center. The field can be estimated in other regions of the resonator volume, knowing the characteristic field distribution of a given design.

The magnetic field created at the center of a uniform distribution of current circulating on parallel circles on the surface of a conducting sphere can be obtained after integrating Eq. (7.10) where r is constant and equal to the sphere radius R (Fig. 8.2).

Hence, a given loop current i of radius a and centered on the x axis at a position defined by θ ($a = R \sin\theta$) gives an elementary field expressed as:

$$dB_x = \frac{\mu_0 i}{4\pi R^2} \sin\theta \oint dL = \frac{\mu_0 i}{2R^2} a \sin\theta = \frac{\mu_0 i}{2R} \sin^2\theta. \quad (8.1)$$

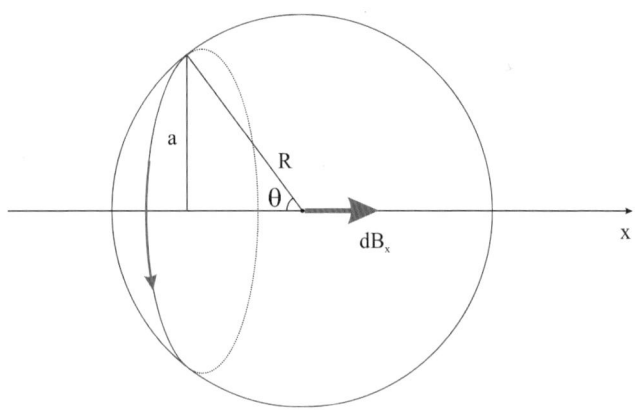

Fig. 8.2 Integration of the Biot–Savart law for the spherical current distribution.

The magnetic field created by the whole spherical current distribution is given, summing up the contribution of all the loop currents, as[3]

$$B_x = \frac{\mu_0 i}{2R} \int_{-R}^{R} \sin^2\theta \, dx = \frac{\mu_0 i}{2} \int_0^{\pi} \sin^3\theta \, d\theta, \quad (8.2)$$

hence

$$B_x = \frac{2\mu_0 i}{3}, \quad (8.3)$$

where i is the current density per unit of length along the sphere diameter. It will be of interest to express the magnetic field in terms of the total current that flows in the resonator. Because the current density is expected to be uniform, the total current I is given by:

$$I = 2Ri, \quad (8.4)$$

---

[3] Note that θ depends on x as $x = R\cos\theta$.

hence

$$B_x = \frac{2}{3}\frac{\mu_0 I}{d},\qquad (8.5)$$

where d is the diameter of the sphere. When the dimensions are in mm, and the current in A, the magnetic field, in gauss, is given by:

$$B_x = \frac{8\pi}{3}\frac{I}{d} = 2B_1^+.\qquad (8.6)$$

Note that this is the amplitude of the magnetic field in the laboratory frame. In the case of a probe design producing a linearly polarized field, the value should be divided by a factor of two in order to estimate the optimum pulse length.

If the circular currents are uniformly distributed on a long (infinite) cylinder, the magnetic field is also uniform everywhere inside the resonator volume. Its magnitude can either be calculated after integration of Eq. (7.10) for an infinite set of circular current, but more efficiently by using the Ampere Law. Hence (Fig. 8.3):

$$lB_x = \mu_0 il,\qquad (8.7)$$

where l is an arbitrary length of the cylinder and i is the current density per unit length.

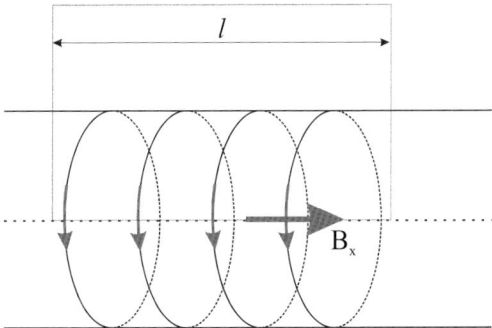

Fig. 8.3 Geometry of the cylindrical, uniform, current distribution.

Thus, the product il represents the total current enclosed by the integration pathway of the Ampere Law. The magnetic field created inside the ideal cylindrical current distribution is given by:

$$B_x = \mu_0 i \,. \tag{8.8}$$

Obviously this equation also applies for the solenoid case replacing the current density i by nI, where n is the number of turns per unit length and I is the current driving the solenoid coil.

It is interesting to note that the magnetic field created by the cylindrical current distribution [Eq. (8.8)] is greater than that created by the spherical distribution [Eq. (8.3)] by a factor of 3/2, assuming the same current density per unit length.

To express the magnetic field of the cylindrical configuration as a function of the *total* current supplied to the probe structure is not as straightforward because the total current is infinite! A first approximation is to assume a "long" cylinder of length l, which is expected to produce, near its center, a uniform magnetic field of the same amplitude as the ideal case. In this case, the total current is given by:

$$I = li \,, \tag{8.9}$$

and the corresponding magnetic field is

$$B_x \approx \frac{\mu_0 I}{l} \,. \tag{8.10}$$

An exact mathematical formulation will be deduced later. The interesting result here is that the ratio of B_1 field (at the center of the coil) per total current is inversely proportional to the dimensions of the probe coil. This is a general result which also applies for a flat circular coil where the amplitude of the magnetic field is inversely proportional to the coil diameter [Eq. (7.11)]. Hence, when designing a given probe, one should keep in mind that, as the dimensions increase, the required current (hence the driving power) should be increased accordingly to obtain the desired rate of rotation of the magnetization. Reciprocally, the NMR sensitivity is expected to decrease when the probe dimensions increase.

However, to compare between various designs, it should be reiterated that the coil resistance, eventually including the sample loss contributions, should be taken into account. Hence,

$$S/N \propto \frac{B_1}{I}\frac{1}{\sqrt{r}} = \frac{B_1}{\sqrt{P}}, \qquad (8.11)$$

where P is the source power, I is the total current supplied to the probe (assumed to be perfectly matched), and r is the coil resistance.

It appears, therefore, that the ability of the coil structure to produce a given magnetic field amplitude per unit current is an important factor for the signal intensity, but the noise contribution, provided by the square root of the coil resistance, must be taken into account for an evaluation of the overall sensitivity of the probe.

8.1.2 Approximations of the spherical uniform current density

8.1.2.1 Helmholtz coil

The simplest approximation to the desired uniform current distribution on a sphere is provided by the so-called Helmholtz pair that is formed by two parallel circular coils driven by in-phase current (Fig. 8.4).

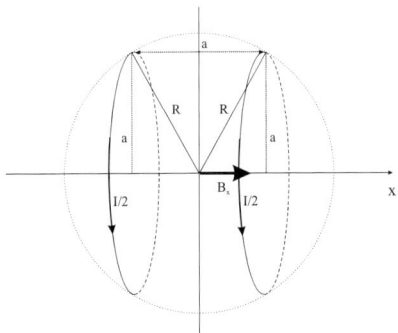

Fig. 8.4 Geometry of the Helmholtz coil configuration. The static magnetic field is assumed to be perpendicular to the figure plane.

This open configuration has the advantage of very good accessibility for the sample. The best homogeneity, hence the best approximation for the ideal current density, is given when the distance between the two coils is equal to the coil radius a. In these conditions, the magnetic field, at the center, is oriented along the coil axis. Its amplitude in the laboratory frame is given by:

$$B_x = \frac{4}{5}\frac{\mu_0 I}{d} = 2B_1^+, \qquad (8.12)$$

where d is the diameter of the sphere enclosing the Helmholtz coil and I is the total current flowing on the sphere surface (i.e., each coil is driven by $I/2$). d is related to the radius a of the Helmholtz coils by (Fig. 8.4):

$$d = 2R = a\sqrt{5}. \qquad (8.13)$$

The magnetic field amplitude is slightly higher than that provided by the ideal resonator, but at the expense of the volume where the field is homogeneous. The spatial distribution of the magnetic field obtained with the Helmholtz configuration is shown in Fig. 8.5.

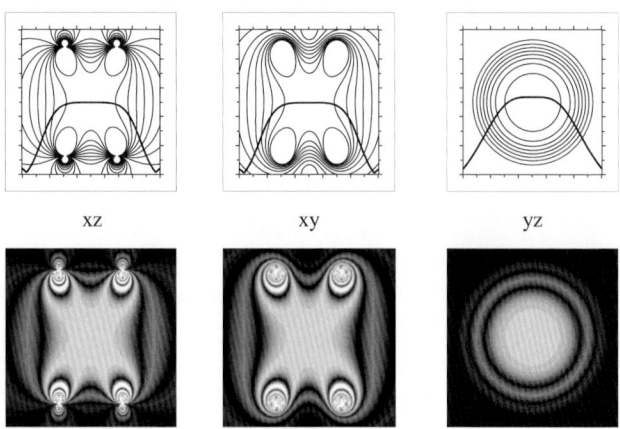

xz   xy   yz

Fig. 8.5 Mapping of the magnetic field created by the Helmholtz coils (x is along the resonator axis, $z \parallel B_0$). Upper: contour plots for B_1^+. Lower: simulation of the B_1^+ mapping experiment using an initial 360° rotation [Crozier et al., 1995].

8.1.2.2 Four coil configuration

This configuration (Fig. 8.6) is a better approximation of the spherical current density than the Helmholtz coil, being used for building air core low field resistive magnets [Garrett, 1951; Franzen, 1962; Mansfield and Morris, 1982], but it has also been proposed as an almost perfect NMR probe design [Hoult et al., 1990a].

The geometrical alignment of this configuration is known to be very critical with respect to the field homogeneity obtained. However, this reputation comes from the fact that the requirement is severe with the B_0 field homogeneity (better than 10^{-5}). The B_1 field homogeneity that is generally required in NMR probe design is less critical (±5%).

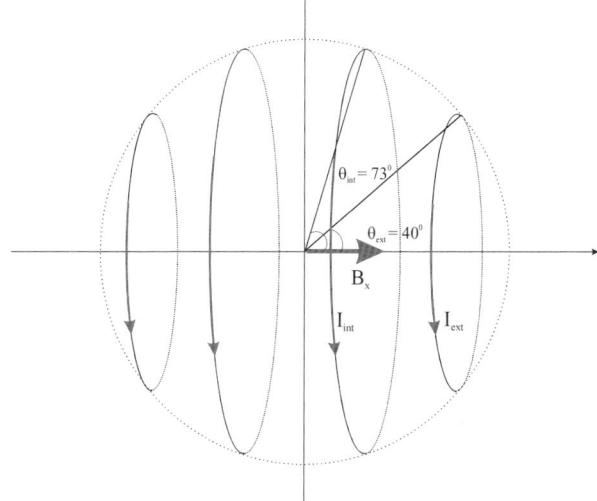

Fig. 8.6 Geometry of the four coil configuration. The static magnetic field is assumed to be perpendicular to the figure plane.

Hoult and Deslauriers expanded the magnetic field in a harmonic series of Legendre polynomials and demonstrated that a homogeneous field distribution can be obtained up to the eighth order when the dimensions and currents follow the given relationships:

$$\cos^2\theta_{ext} = \frac{1}{3}\left(1+\frac{2}{\sqrt{7}}\right) \quad \cos^2\theta_{int} = \frac{1}{3}\left(1-\frac{2}{\sqrt{7}}\right)$$
$$\frac{I_{ext}}{I_{int}} = -\frac{\cos^4\theta_{int} + 10\cos^2\theta_{int} - 3}{\cos^4\theta_{ext} + 10\cos^2\theta_{ext} - 3}$$
(8.14)

where I_{ext} and I_{int} are the currents in the external and internal loops, respectively. θ_{ext} and θ_{int} are the semi-angles subtending the two loops, expressed in radian.

The spatial distribution of the magnetic field obtained with the four coil configuration is shown in Fig. 8.7, in the xy, xz, and yz planes.

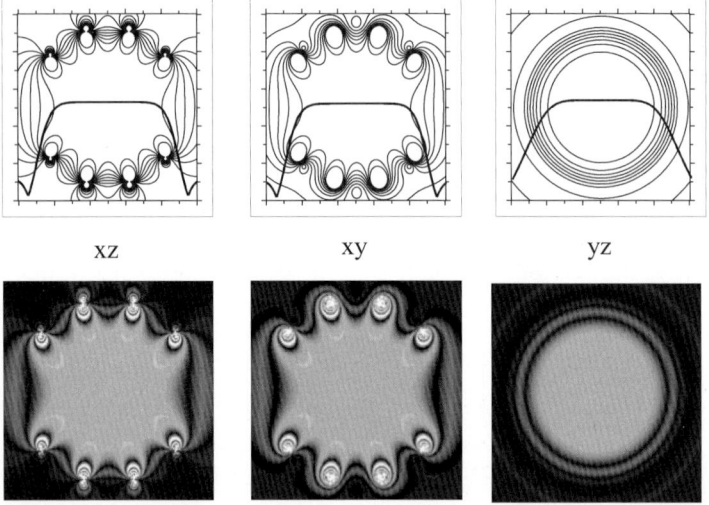

Fig. 8.7 Mapping of the magnetic field created by the optimum four coil configuration. Upper: contour plots for B_1^+. Lower: simulation of the B_1 mapping experiment using an initial 360° rotation [Crozier et al., 1995].

The magnetic field at the center of such a symmetric configuration can be estimated by adding together the magnetic fields created by each coil [Eq. (7.13)]. Let I be the total current flowing on the sphere surface:

$$I = 2(I_{ext} + I_{int}),$$
(8.15)

then

$$B_1 = \frac{\mu_0 I}{d}\left[1 - 2\left(\frac{I_{int}}{I}\cos^2\theta_{int} + \frac{I_{ext}}{I}\cos^2\theta_{ext}\right)\right], \quad (8.16)$$

where D is the diameter of the sphere that encloses the four coils. When the dimensions and currents verify the relations given by Eq. (8.14), the magnetic field is given by:

$$B_x = 0.714\frac{\mu_0 I}{d} = 2B_1^+. \quad (8.17)$$

The characteristics of the magnetic field produced by the axial configurations is summarized in Fig. 8.8 as a histogram showing both the homogeneity scale and amplitudes [Li et al., 1994].

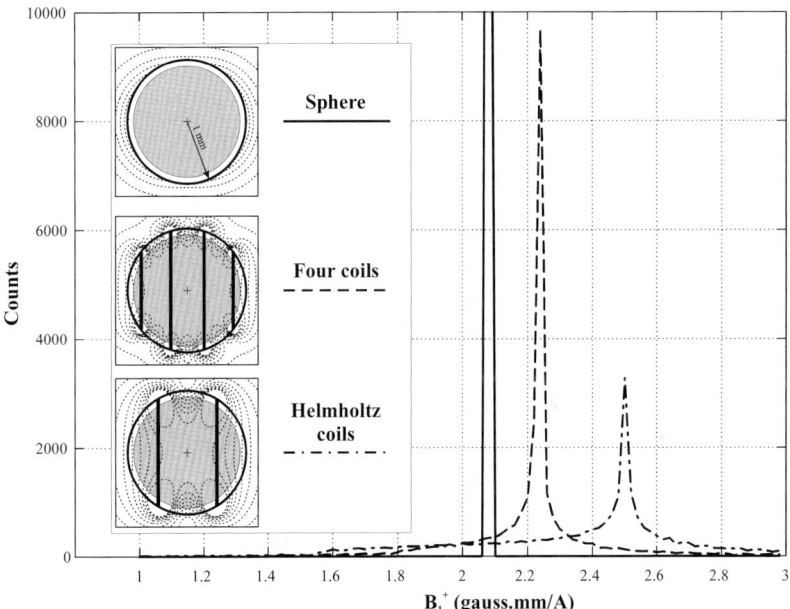

Fig. 8.8 Histograms of the B_1^+ magnetic field distribution in the xy plane (perpendicular to B_0) calculated in an ROI of 90% of the coil diameter (grey region). The best homogeneity (delta function) is given by the spherical current distribution, at the expense of smaller magnetic field amplitude. The largest field amplitude is given by the Helmholtz coils but is associated with the lowest homogeneity. The horizontal scale is expressed in gauss per A and assumes a 1 mm coil diameter.

The corresponding values of the magnetic field at the center (the 90° pulse for a specific design) are given in Table 8.1, in gauss, for a resonator of radius 1 mm and driven with a total current of 1 A.

Table 8.1 Magnetic field B_1^+ amplitude in the coil center. The corresponding 90° pulse length for the proton is calculated for a coil having a 1 mm diameter and a current of 1 A.

	Sphere	Four coils	Helmholtz coils
B_1^+ (gauss mm A^{-1})	$(4/3)\pi$	1.42π	$(8/5)\pi$
Pw$_{90}$ ^{1}H	14 µs	13 µs	13 µs

8.1.2.3 Guidelines for a practical design of Helmholtz probes and four coil probes

Building multiple coil resonators requires taking care of the geometry and current ratio in the coils. Although more or less complex circuits could be designed to distribute power in order to obtain the required ratio, it could be better to let the currents share themselves "naturally" among the coils.

The current sharing for the Helmholtz coil resonators is quite easy to obtain. The resonator is constituted of two identical planar coils, magnetically coupled and each resonating at the same frequency. Hence (Chapter 6) two resonant modes will be obtained, one corresponding to a gradient mode (upper frequency) and one corresponding to the homogeneous mode (lower frequency). Choosing the lower frequency mode assures that the predicted RF homogeneity is obtained, provided the two coil circuits are identical. The latter condition will also provide the required equality of the amplitude and phase of the currents in both coils. Because the distance between the two coils is imposed by the coil dimensions, the mutual inductance is given; hence the resonant frequency of the whole resonator must be adjusted by equally tuning each coil. This equilibration may be quite cumbersome to maintain, especially when loading the probe with a large variety of samples. Thus the whole resonator may be preferably coupled to the spectrometer using a third coupling loop. In this case, the tuning and matching procedures could be made independent, the latter being easily realized by displacing the coupling loop or better, using a tuned coupling loop.

A similar scheme should be employed in the design of the four coil probes. In this case, another complication arises due to the required current ratio that should be taken into account. As the geometry of the coils has been set, the current ratio can be controlled through the loop's own resonant frequency. Let X_p be the impedance at the desired frequency ω_0 of one given loop resonator $L_p C_p$:

$$X_p = j\left(L_p \omega_0 - 1/C_p \omega_0\right). \tag{8.18}$$

To obtain the required current ratio, the impedance X_p should be adjusted such that [Hoult and Deslauriers, 1990a]:

$$X_p = -j\omega_0 \sum_{q \neq p} M_{pq} \frac{i_q}{i_p}, \tag{8.19}$$

where p and q stand for the coil number and M_{pq} is the mutual inductance between coil p and q. The self and mutual inductances can be calculated with a good accuracy (see Chapter 10), hence the resonant frequency of each coil (in fact, the two inner and outer coils) can be estimated. Each resonator can then be built separately. When the four resonators are put together, four resonant modes will be observed. Among these, only the lower frequency mode gives the homogeneous RF field. Accurate alignment of the resonator requires a lot of work on the RF workbench but, if done carefully, will prevent a need to tune it again in the spectrometer.

Finally, the question arises of how many turns should define each coil design? First of all, the NMR sensitivity [Eq. (8.11)] is independent of the number of turns for a given design, in first approximation. This is a general property that will be considered in more detail for the solenoid (next paragraph) and for the surface coil (Chapter 9). Secondly, and probably more importantly, the NMR probe should produce high magnetic fields and low electric fields. This implies that the coil should have a low impedance (high current and low voltage); hence the lowest number of turns compatible with the design is to be preferred (the inductance is proportional to the square of the number of turns). Using a one-turn coil is therefore preferred in this case. Finally, the coil must be segmented (Section 4.2) if necessary to minimize the antenna effect.

This class of resonators would work at relatively low frequencies, of the order of 100 MHz. Due to their opened structures, cylindrical samples, having their axis parallel to the main magnetic field, can be easily accommodated.

8.1.3 *Solenoid types*

These coil types are cylindrically-shaped and produce a magnetic field parallel to the cylinder axis. They fit cylindrical samples having the axis perpendicular to the main static field. In these cases, the B_0 homogeneity may be a problem.

8.1.3.1 *The solenoid coil*

The solenoid is probably the coil design that allows the best sensitivity for NMR. However, it has some disadvantages, especially respective to the B_1 direction. The resonator being perpendicular to the main magnetic field axis may lead to both B_0 homogeneity and sample accessibility problems in the modern superconductive axial magnets. Furthermore, the resonator inductance increases rapidly with its dimensions and becomes self-resonant for frequencies well below those used nowadays. Therefore, it is preferably employed with small samples and/or at low frequencies. Particularly, the microcoil design that is fast developing allows NMR experiments to be performed on a very small quantity of soluble chemicals or macromolecules, [Olson *et al.*, 1995].

A limit to the solenoid coil usage can be roughly estimated from the general rule of thumb stating that the coil impedance should not be greater than 100 to 150 Ω. This statement puts the coil self-resonance (that arises for an impedance around 600 Ω for an optimized solenoid design) well above the working frequency and limits the electric field generated by the probe. Eventually the solenoid coil could be segmented allowing larger dimensions for ordinarily accepted designs.

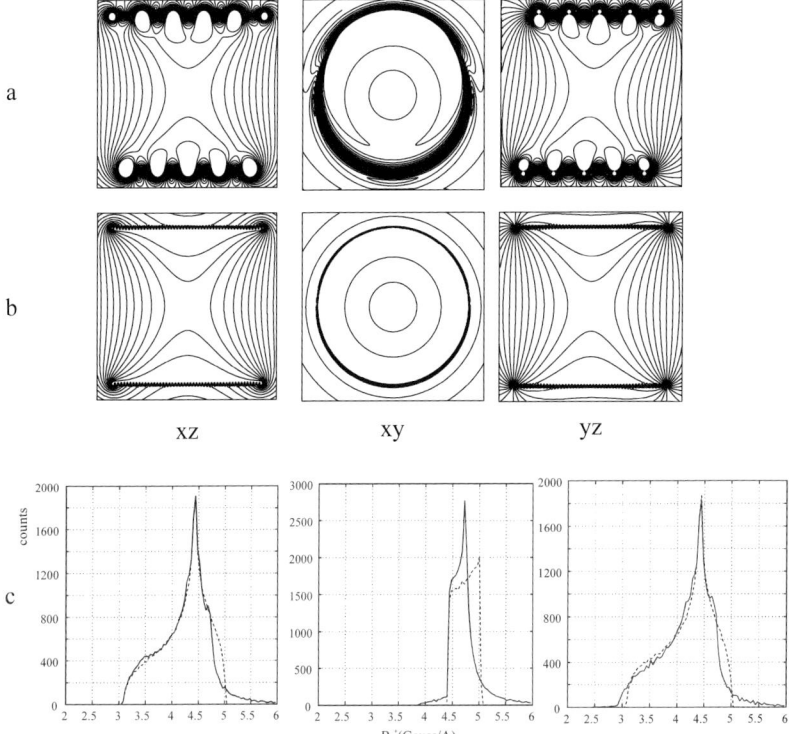

Fig. 8.9 B_1^+ field maps in the three orthogonal planes of a typical solenoid coil having a geometry optimized for the best Q factor. z is the direction of the static magnetic field and x is the solenoid axis. (a) 5 turn solenoid. (b) 50 turn solenoid. This solenoid mimics the ideal cylindrical, uniform, current distribution. (c) Magnetic field histograms calculated in the ROI corresponding to 90% of the solenoid volume. The continuous and dashed lines are for the solenoid having 5 and 50 turns, respectively.

As already mentioned, the magnetic field produced inside a solenoid coil is uniform when it is infinitely long. As the length of the coil decreases, the magnetic field distribution tends toward the highly heterogeneous distribution produced by the surface coil (Chapter 9). For given dimensions (length to diameter ratio), the homogeneous volume can be increased by winding the coil tightly and with a nonuniform stepping. Typically the windings are pushed closer together at the extremities than at the center. Unfortunately, this winding type is not suitable for a high Q design.

Another source of magnetic field heterogeneity is due to the helical winding with a finite step. As far as the turn separation is kept small respective to the coil diameter, the field distortion resulting from this effect is generally acceptable. This is no longer the case for a realistic design to be used in NMR experiments. This is shown in Fig. 8.9 where the field maps for a 5 turn (a realistic case) and a 50 turn (unrealistic case[4]) solenoid coil having the same optimum geometry (ratio of the coil length to its diameter equals unity) are compared.

The on-axis magnetic field amplitude created by the solenoid coil can be calculated from the integration of the Biot–Savart law on a helical winding as follows (Fig. 8.10).

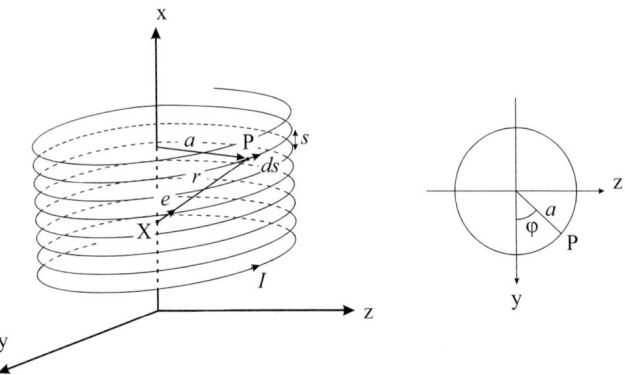

Fig. 8.10 Coordinate system used in the integration of the Biot–Savart law for the helical current of a solenoid.

Let a and l be the radius and length of the solenoid, n the number of turns, and s the winding step. The magnetic field produced at a given point X $(x, 0, 0)$ on the solenoid axis is given by:

---

[4] For this design, it is probable that the self-resonance appears at a much lower frequency than the expected NMR frequencies and that the coil resistance will be high due to the proximity effect. These two factors will contribute to significantly reducing the Q factor of this coil.

Homogeneous Resonators 483

$$B_x = \frac{\mu_0 I}{4\pi} \int_{solenoid} \frac{(d\vec{s} \times \vec{e})_x}{r^2}, \qquad (8.20)$$

where ds is an element of segment along the coil wire at a given point P and e is a unit vector along the segment, of length r, joining P and X. The coordinates (x_P, y_P, z_P) of P as a function of the angle φ (Fig. 8.10) are given by:

$$x_P = s\varphi/2\pi + x \quad y_P = a\cos\varphi \quad z_P = a\sin\varphi. \qquad (8.21)$$

Taking the derivatives with respect to φ, one obtains the components of ds as:

$$dx = s/2\pi \, d\varphi \quad dy = -a\sin\varphi \, d\varphi \quad dz = a\cos\varphi \, d\varphi. \qquad (8.22)$$

The components of e are given by:

$$e_x = \frac{(\varphi s/2\pi + x)}{r} \quad e_y = \frac{a\cos\varphi}{r} \quad e_z = \frac{a\sin\varphi}{r}. \qquad (8.23)$$

Integration on the solenoid wire is done in the range $\varphi = [-n\pi, +n\pi]$. Then:

$$B_x = \frac{\mu_0 I a^2}{4\pi} \int_{-n\pi}^{+n\pi} \frac{d\varphi}{\left[a^2 + (x + \varphi s/2\pi)^2\right]^{3/2}}. \qquad (8.24)$$

Finally:[5]

$$B_x = \frac{\mu_0 n I}{2l} \left[\frac{x + l/2}{\sqrt{a^2 + (x + l/2)^2}} - \frac{x - l/2}{\sqrt{a^2 + (x - l/2)^2}} \right]. \qquad (8.25)$$

The magnetic field at center ($x = 0$) is given by:

---

[5] A similar equation has been previously obtained by Idziak and Haeberlen [Idziak and Haeberlen, 1982].

$$B_x = \frac{\mu_0 n I}{l_D} = 2B_1^+, \qquad (8.26)$$

where l_D is the length of the solenoid diagonal. If the coil is much longer than its diameter, l_D is close to l [long solenoid case, Eq. (8.10)]. On the contrary, if the coil is very short, then l_D is equal to the coil diameter [simple loop case, Eq. (7.11)]. The magnetic field at the center is, as already expected, inversely proportional to the coil "dimensions."

The magnetic field being proportional to the number of turns, one could be tempted to increase n^6 in order to obtain the largest field for a given current. Obviously, this reasoning is incorrect because the wire resistance will increase with the number of turns and thus the noise contribution increases accordingly [Eq. (8.1)].[7] Let P be the source power, assumed to be entirely dissipated in the probe coil (perfect matching):

$$P = r I_{eff}^2 = r I^2 / 2, \qquad (8.27)$$

where I_{eff} is the "effective" current and I its amplitude (the last equality of Eq. (8.27) is valid only if the current is purely sinusoidal in time). Expressing r as a function of the coil impedance and quality factor, one obtains:

$$I = \sqrt{\frac{2PQ}{L\omega}}, \qquad (8.28)$$

where L is the coil inductance. The inductance is proportional to the square of the number of turns and to the geometry of the coil. Using Eq. (8.26), one obtains:

---

[6] Independent of any frequency values.

[7] The coil resistance is proportional to the square of the number of turns (Chapter 2) and the ratio B_1/I is proportional to n. Hence, the sensitivity, which is the ratio of B_1/I to the square root of r, is independent of n.

$$\frac{B_1}{\sqrt{P}} = F_{geo}\sqrt{\frac{Q}{\omega}}, \qquad (8.29)$$

where F_{geo} is a factor that depends only on the dimensions of the coil. When the frequency of interest is well below the self-resonance of the coil, Q depends only on the volume of the resonator, on the square root of frequency, and is independent of the number of turns (Chapter 10) for a given geometry. Hence, the sensitivity is independent of the number of turns. This is an important result that allows the designer to concentrate on the geometry of the coil and on the choice of the design. With the dimensions of the sample generally being imposed, the final choice for the design of the solenoid resonator depends solely on its impedance at the expected working frequency. As a result, a multiple turn solenoid will be well adapted for experiments on "small" samples ("microcoils") while a single turn coil ("loop gap") will be better for "large" samples. The distinction between "large" and "small" samples depends on the working frequency and, to a lesser extent, the coil shape. For example, a capillary sample (providing a very high sensitivity with a minimal quantity of product) requires a long solenoid of very small diameter.[8] A larger sample (spherical or tubular) would require a solenoid designed with a high Q, hence having its length nearly equal to the diameter. As the diameter and length (volume) of the solenoid or the frequency increase, the coil impedance increases. In this case, the number of turns should be reduced to unity leading to the so-called loop gap resonator.

8.1.3.2 *The loop gap*

The loop gap resonator (Fig. 8.11) was initially designed for very low field Electron Paramagnetic Resonance (EPR), working in the range of several hundred megahertz. It is basically designed as a cylinder having one or multiple slits (the gaps) along the generator. The gaps play the role of a tuning capacitor [Froncisz and Hyde, 1982]. A very similar

---

[8] Extended discussion on the design of (solenoid) microcoils can be found in the literature, for example, Webb, 1997; Seeber *et al.*, 2001; Minard and Wind, 2001ab; Webb, 2012; Webb, 2013.

design, called a "stripline resonator" [Decorps and Fric, 1969, 1972], was proposed earlier. In this design, the tuning slit is replaced by a piece of transmission line acting as a tuning capacitor. The resonant frequency range of the loop gap covers all the NMR frequencies. Hence, they are perfectly suited for high frequency NMR probe designs [Hardy and Whitehead, 1981; Najim and Grivet, 1992; Piasecki et al., 1998; Fan et al., 1987; Kneeland et al., 1986; Hall et al., 1985; Grist and Hyde, 1985]. At the lower NMR frequencies, the tuning slit (or the stripline) can be replaced by lumped capacitors [Hornak et al., 1986].

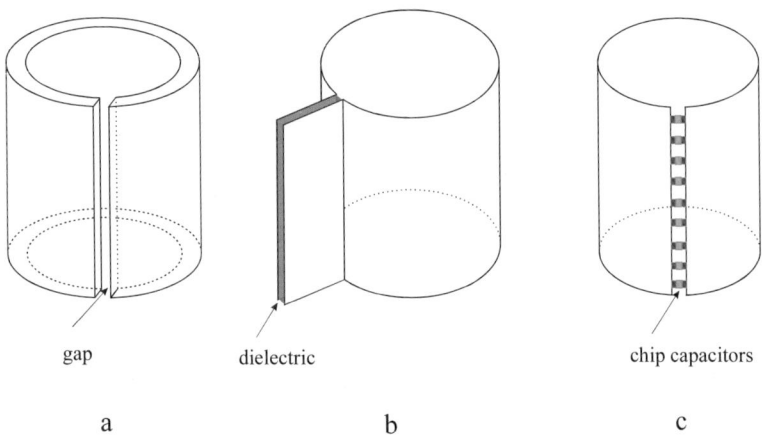

gap   dielectric   chip capacitors

a   b   c

Fig. 8.11 Three basic loop gap resonators designed for decreasing frequencies, from left to right. The higher frequency cylindrical cavity (a) is tuned by the capacitor formed by the gap. The two other configurations are designed for lower frequency/dimension ratios. At intermediate frequency/dimension ratio (b), the loop inductance is tuned by a capacitor formed by the continuation of the copper foil and a sheet of dielectric. In (c), the inductance is tuned by a parallel combination of ship capacitors soldered directly to the copper foil and distributed all along the gap.

Other similar designs (Fig. 8.12), such as the concentric loop gap [Koskinen and Metz, 1992] or the "bracelet" coil [Kan et al., 1994] have also been proposed, considerably extending the usefulness of the loop gap resonator concept. These probes are built on the transmission line principle, wound up on a cylindrical shape. The whole resonator is specially designed in order to favor the common mode currents, thus allowing the creation of a magnetic field outside the transmission line while confining the electric field inside the line dielectric. Other designs, such as the bridged loop gap resonator [Weis *et al.*, 1997] and the solenoid scroll coil [Stringer *et al.*, 2005], fit these properties, that constitute the characteristics of an efficient NMR probe (the magnetic field should be generated in the sample volume and the electric field should be confined outside).

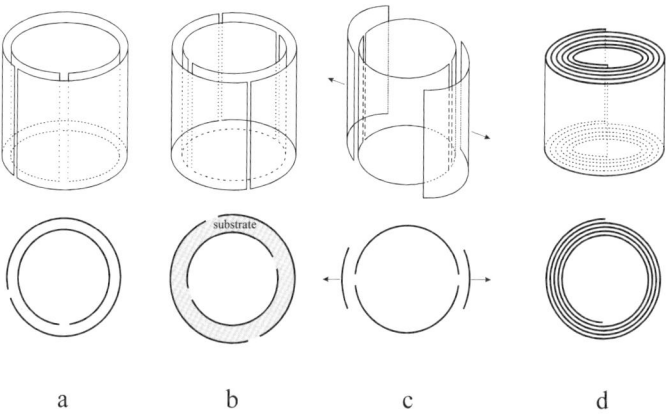

a   b   c   d

Fig. 8.12 Other designs minimizing the electric field in the sample are similar to the loop gap resonator. In (a) the resonator is made of two coaxial loop gaps that are rotated respective to one another, allowing a wide tuning range [Koskinen and Metz, 1992]. A similar design (b), having more gaps, is based on transmission line concepts [Kan *et al.*, 1994]. (c) The bridged loop gap resonator [Pfenninger *et al.*, 1988; Forrer *et al.*, 1990; Weis *et al.*, 1997] is also intended to confine the electric field outside the sample while preserving a good magnetic field homogeneity inside the resonator. The "scroll coil" [Stringer *et al.*, 2005] in (d) is similar to a self-resonating solenoid tuned by the distributed capacitance formed by the copper foils and dielectric sheet that are wound on the cylindrical shape. It is a version of the Swiss-Roll design used in metamaterials (Chapter 2).

It should be stressed at this point that the current density on the cylinder is not uniform as would be the case for a tightly wound solenoid coil [Fig. 8.9(b)] or for the ideal cylinder case. As suggested in Section 7.3, the current density is expected to increase considerably at the edges of the copper foil, similar to the copper flat strip case. The calculated current density for a given loop gap is shown in Fig. 8.13.

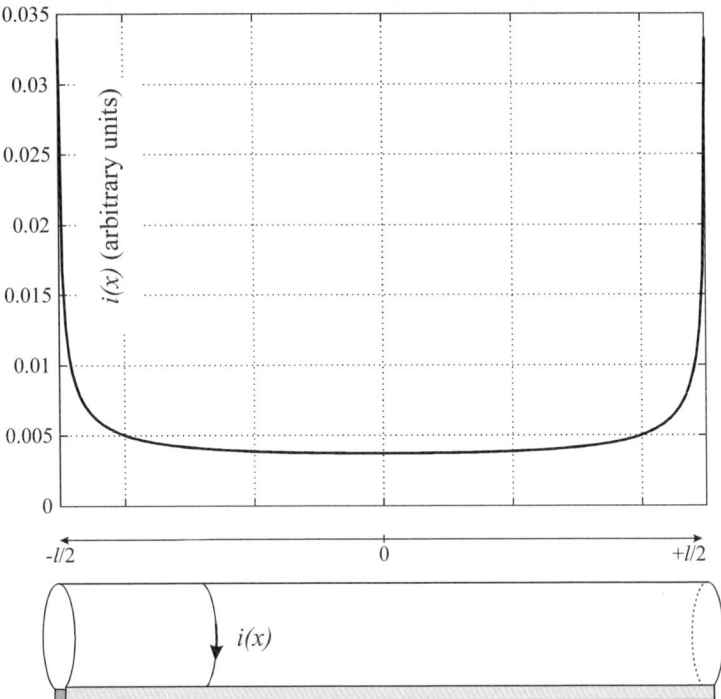

Fig. 8.13 Circular current density along the axis of the loop gap or "hollow cylinder" [Labiche *et al.*, 1999]. The density is calculated here as explained in Chapter 7 and in [Mispelter and Lupu, 2008].

The current distribution is frequency independent, at least when the coil reactive impedance dominates the resistance (high Q). The current density is roughly uniform near the center coil where one expects a uniform magnetic field distribution. Near the edges of the finite length cylinder the field amplitude decreases strongly. This is partially rectified by the large current distribution increase in the same regions. Hence, the magnetic field homogeneity inside the loop gap resonator is expected to improve compared to the solenoid. Figure 8.14 shows the magnetic field distribution and homogeneity calculated for a realistic solenoid (5 turns and $l = d$) and for a loop gap of the same dimensions.

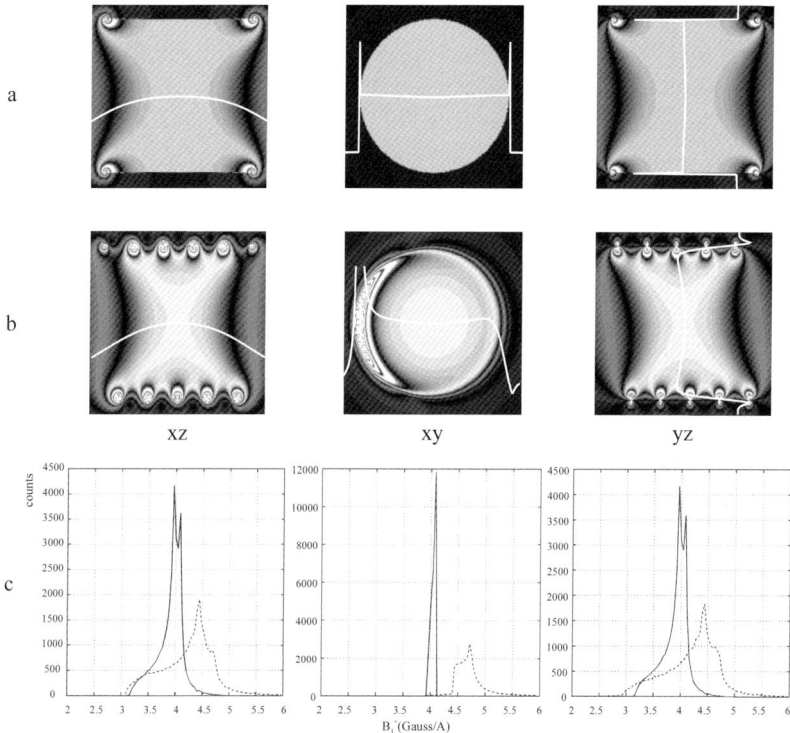

Fig. 8.14 B_1^+ mapping (using an initial 360° rotation) in the three orthogonal planes for two solenoid coils having the same dimensions (equal length and diameter). z is the direction of the static magnetic field and x is the solenoid axis. (a) A loop gap. (b) A 5 turn solenoid coil. (c) Magnetic field histograms calculated in the ROI corresponding to 90% of the coil volume. Note that the loop gap histograms (continuous lines) indicate a better homogeneity than for the solenoid coil (dashed line) at the expense of the field amplitude.

Clearly, the loop gap exhibits a better field homogeneity but at the price of a slight decrease of the B_1 amplitude for the same current density. A definitive comparison of sensitivity would require an estimation of the Q factor. It is expected to be roughly equal for the solenoid and loop gap of the same dimension (Chapter 10). However, to the second order, the loop gap may have a better quality factor, especially when loaded, due to the reduction of dielectric losses (lower electric field amplitude). This will be especially true at high frequency and for relatively large sample dimensions. On the contrary, for very small samples, the impedance of the loop gap becomes so small that it no longer dominates the resistance of the tuning capacitor and of the soldered connections. In these cases, the Q factor of the loop gap configuration drops dramatically and the solenoid has to be preferred.

8.1.4 Practical designs of solenoid type coils

Two solenoid coils will firstly be described. One is a small "microcoil," working at 200 MHz, designed to record static magnetic field maps on a 4.7 tesla horizontal magnet. The second is a solenoid coil designed to perform high sensitivity ^{31}P experiments on a Wide-Bore 9.4 tesla vertical magnet ($f = 162$ MHz). This coil allows the recording of spectra on muscle tissues of the leg of a living mouse in only one pulse. In this way, the metabolite kinetic could be sampled in real time with a sampling rate of about 10 s (the delay required to let the ^{31}P magnetization recover thermodynamic equilibrium).

On the other hand a double tuned ($^{1}H/^{31}P$) loop gap resonator working at 200/81 MHz on a 4.7 tesla horizontal magnet will be also described. It was designed to investigate perfused organs (the sample volume being of the order of 150 cm^3).

8.1.4.1 A microcoil for static magnetic field mapping

The coil is made out of a 5.5 turn solenoid wound on a small glass tube (2.3 mm outer diameter) filled with a small quantity of water. The winding step is adjusted so that the length of the solenoid coil is 2.0 mm, using 0.3 mm enameled copper wire. The inductance calculated using Eq. (2.84) is 57.5 nH. Taking into account the connections, the total inductance is estimated to be 60 nH. A symmetric capacitive coupling [Fig. 4.34(b)] is used to match the coil to the spectrometer 50 Ω.

The required capacitor values are calculated from Eq. (4.12), assuming a Q value of around 200,[9] as:

$$C_{tune} = 9.7\,pF$$
$$C_{match} = 1.3\,pF \quad . \qquad (8.30)$$

The tuning capacitor is made out of the parallel combination of a fixed chip capacitor of 5.1 pF and an adjustable small sapphire dielectric capacitor of 0–4.5 pF. Taking account of the parasitic capacitors the variable capacitor allows the coil to be tuned at the operating frequency. The matching capacitor was split into a fixed 2.0 pF chip and an adjustable sapphire dielectric 0–2.5 pF, according to the circuit shown in Fig. 4.34(b). The Q factor of the completed probe loaded with $CuSO_4$ doped water was 240. Shielding the probe was not required. The whole probe (Fig. 8.15) may be built in a couple of hours.

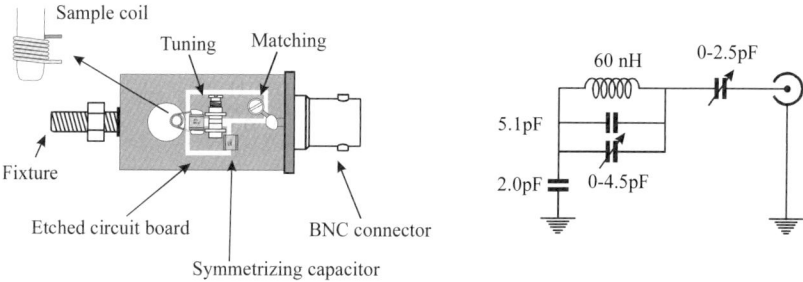

Fig. 8.15 Layout and electrical circuit of a probe used for static magnetic field sampling. The coil probe is a small solenoid enclosing a drop of pure water contained in a glass capillary. The probe is designed for 200 MHz. The chip capacitors are ceramic ATC (case B) and the adjustable are Sprague–Goodman UHF capacitors.

8.1.4.2 *A 0.4 ml high sensitivity phosphorus coil (162 MHz)*

The second solenoid probe was of a larger volume in order to fit a mouse leg. It was wound with enameled copper wire (1.0 mm diameter) on a

---

[9] The Q factor value expected here is what is usually observed for most common probe designs and is therefore a starting point allowing all component values to be calculated.

plastic cylinder, terminated by a thin plastic sheet, which protects the coil from direct contact with the animal (Fig. 8.16). This simple precaution avoided a considerable Q factor collapse when loading the probe. It should be noted that the electric field was quite intense for this design due to its rather high impedance.

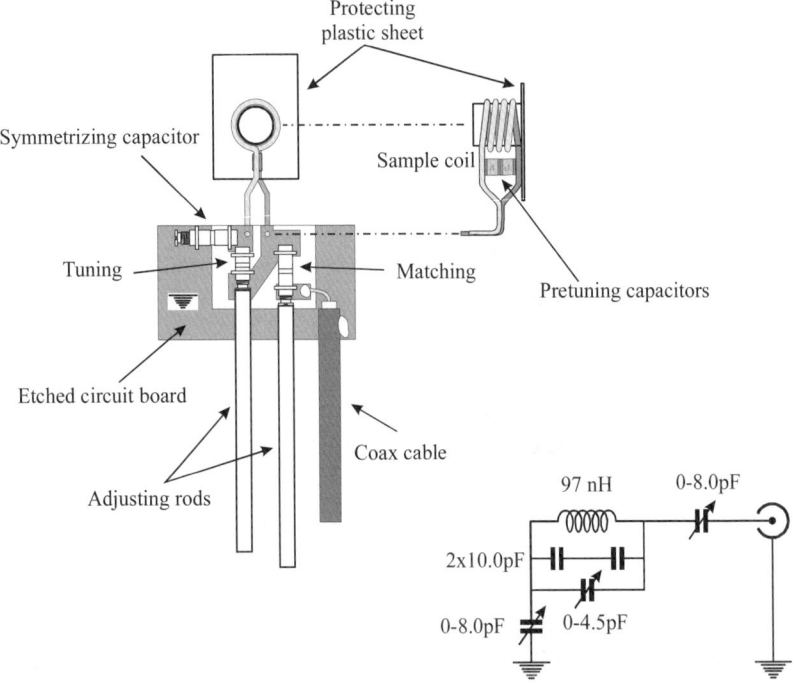

Fig. 8.16 Layout and electrical circuit of a solenoid coil probe designed for 162 MHz (^{31}P spectroscopy in a 9.4 tesla static magnetic field). The circuit board is mounted vertically in an alumina tube that fits in the bore (89 mm) of the vertical magnet. Chip and adjustable capacitors are from ATC (case B) and from Sprague–Goodman (UHF series), respectively.

As a result, the coil was pre-tuned by two small 3 mm chip capacitors connected in series in order to increase the breaking voltage. The voltage that appears on the coil terminals during the high power RF pulse may indeed be large enough to induce sparking or, sometimes, to burn the capacitor. A flashing probe emits a recognizable sound during RF pulsing, but the phenomenon can be better put in evidence by sampling of the nuclear magnetization rotation, using incremented pulse lengths [Keifer, 1999].

The solenoid was made out of 4 turns of 1 mm enameled copper wire wound on a form of 8 mm outer diameter and having a total length of 8 mm. Accordingly, the inductance is 97 nH and its impedance at 162 MHz is about 100 Ω. The tuning/matching circuit was still the equilibrated circuit of Fig. 4.34b. From Eq. (4.12), the following component values, again assuming a Q value of 200, are obtained:

$$C_{tune} = 8.98\,pF$$
$$C_{match} = 1.0\,pF \quad . \quad (8.31)$$

The tuning and matching components are again made using a combination of fixed and adjustable capacitors (Fig. 8.16). The probe head was mounted on an aluminum tube that fitted in the wide bore vertical magnet. Because the diameter of the shield was large respective to the coil dimensions, no significant influence on the electrical properties of the probe could be observed.

8.1.4.3 *A 150 ml double tuned ($^1H/^{31}P$) coil operating at 4.7 T*

This probe design is a little bit more elaborate due to its size and to the double tuning. The required operating frequencies were 200 MHz for proton imaging and 81 MHz for recording the ^{31}P spectra in order to sample the cellular metabolism. The sensitivity at this frequency is particularly critical due to the relatively low concentration of the nuclear species to be observed.

As the sample is a container that holds a suspended living organ and the magnet is horizontal, a vertical solenoid-like coil producing an axial RF field perpendicular to the main static field would obviously be the best choice. However due to the required size (diameter of up to 60 mm) and the frequency (200 and 81 MHz) a loop gap should be chosen.[10]

---

[10] A solenoid of the size needed here would have an inductance greater than 1 µH, corresponding to an impedance greater than 1200 Ω at 200 MHz. Furthermore, the

A copper foil of 50 mm width is wound on a cylinder of 64 mm outer diameter. The inductance is 51.5 nH, resulting in an impedance of 64.7 Ω and 26.2 Ω, at 200 and 81 MHz, respectively. These impedance values fit perfectly in the acceptable range expected for an NMR coil design. Due to the shape and low impedance of the inductor, a particular attention should be paid to the connections between the coil and the matching network. In particular, it is desirable that the coil is pre-tuned, thus becoming a resonator, with a capacitor that maintains the electrical continuity along the slit. As already mentioned, at microwave frequencies, the gap constitutes the resonating capacitor by itself, the cylinder wall being sufficiently thick [Fig. 8.11(a)].

At NMR frequencies, the coil is made out of a thin copper foil which is extended on a dielectric sheet in order to make the required capacitance [Fig. 8.11(b)]. In this case, the capacitance should be considered as a piece of transmission line [Fan et al., 1987; Najim and Grivet, 1991]. Another method is to solder a parallel arrangement of chip capacitors all along the slit [Fig. 8.11(c)]. The best way is to fit, as closely as possible, the current distribution filaments with the corresponding capacitor value instead of an equal sampling along the slit. However, soldering a capacitor at one point near the center, or eventually in an unsymmetrical manner, should be avoided. Obviously to optimize the sensitivity at the lowest frequency, the coil will be tuned optimally to 81 MHz. In order to provide a second resonance at 200 MHz, while accepting some degradation of the sensitivity at this frequency, the solution is to incorporate a trap circuit [Fig. 6.23(b)] having a small inductance value compared to that for the main coil. The sole difficulty is to maintain the continuity of currents along the copper coil foil and to minimize the connection resistance due to soldering. Hence the whole resonator is built using a single copper sheet, as shown schematically in Fig. 8.17.

The small impedance trap inductor was constituted of a small diameter loop gap located at the opposed side of the phosphorous tuning

---

required length of wire would approach one wavelength! Clearly this solution is not viable, unless a segmented solenoid, split into 10 to 20 parts, is used.

capacitor. Its inductance value, free of the tuning metal rod, is 7.1 nH (a cylinder of 20.6 mm diameter and 50 mm long).

The trap is tuned to ω_{off} = 184 MHz, Eq. (6.37), with a brass piston moving inside the trap inductance. The insertion of a conductive material inside the coil reduces the volume in which the magnetic energy is stored, resulting in a decrease of the total inductance.

Fine tuning the coil to 81 MHz is realized by the addition of an adjustable capacitor soldered at both edges of the resonator where the current density is higher.

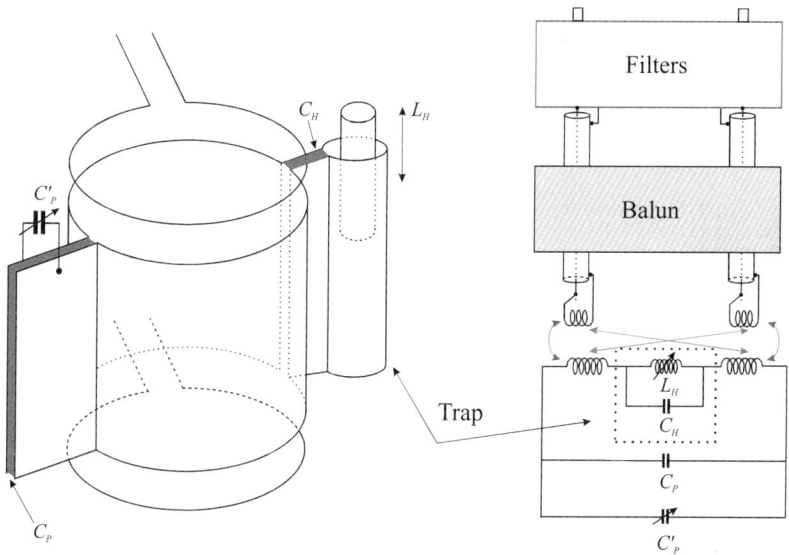

Fig. 8.17 Schematic drawing of a double tuned loop gap resonator. The resonator is inductively coupled to the spectrometer using two coupling loops. Each loop is connected through a balun and a filter. The balun is a broadband cable trap formed by winding the coax cable connecting the loop to the input filter port. Each filter is a stop band filter, permitting a good isolation of both ports and maintaining a 50 ohms impedance transfer at the operating frequency.

Although this version had good electrical characteristics at both frequencies and was easy to tune and match (an adjustable magnetic coupling was used and associated with appropriate narrow stop band filters, Fig. 8.18), the probe was temperature-sensitive, leading to tuned frequency instabilities.

A second version has been built (Fig. 8.19), based on a similar principle except that the coil was made of copper adhesive tap firmly glued on a Delrin® support and the capacitors were replaced by high voltage ceramic models. The trap was divided into three circuits (each trap inductance was 18.6 nH tuned by a microwave chip capacitor), two of which were adjustable.

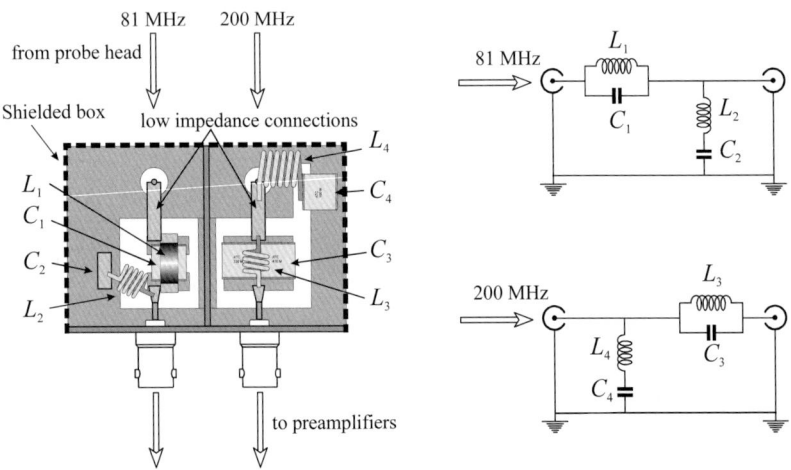

Fig. 8.18 Layout and electrical circuit drawings of the high power stop band filters used with the double tuned loop gap resonators shown in Figs. 8.17 and 8.19. The component values are: L_1 = 13.5 nH (1 turn "hairpin" coil, thin copper foil 5 × 40 mm); L_2 = 135 nH (6 turns, internal diameter 5.5 mm, length 7 mm, 1.0 mm wire); L_3 = 48 nH (3 turns, internal diameter 5.5 mm, length 3.8 mm, 1.0 mm wire); L_4 = 193 nH (6 turns, internal diameter 8 mm, length 9.7 mm, 1.0 mm wire); C_1 = 47 pF, C_2 = 4.7 pF, C_3 = 80 pF (33 + 47pF), C_4 = 20 pF (ATC chips case E).

The coupling was also changed to be more flexible. A fixed position tuned coupling loop (Section 4.1.3.4) was used to match the resonator at 81 MHz and a classical capacitive coupling was realized at 200 MHz. The matching capacitor was directly connected on the trap circuit. Connections to the filters were done through a balun formed by winding coax cables, as in the previous design of Fig. 8.17. In this way, the probe connections (the coupling loop and the 200 MHz matching capacitors) are isolated from each other and balanced respective to the common ground. The realization of this probe is simple, no critical adjustments are required. The S-parameters of the probe are shown in Fig. 8.20.

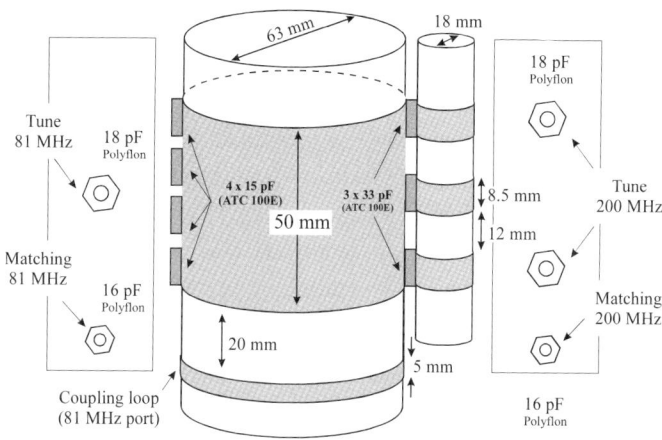

Fig. 8.19 Drawing of the double tuned (81/200 MHz) loop gap resonator. The electrical circuit is as in Fig. 8.17, except that the 200 MHz trap is tuned by two variable capacitors (18 pF). The tuning and matching capacitors are mounted on two etched printed circuit boards located at diametrically opposed positions respective to the central resonator. The connections to the appropriate ATC 100E chip capacitors are made using low impedance copper foil sheets. The lower loop is connected to the corresponding 81 MHz matching capacitor (16 pF) in series with the coax cable.

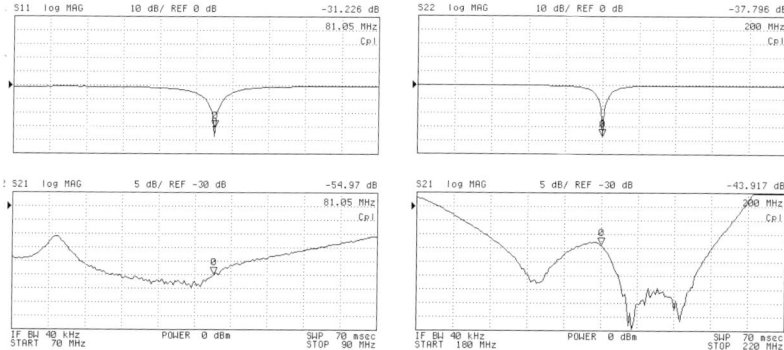

Fig. 8.20 The electrical characteristics of the completed double tuned, loaded, loop gap probe, including the stop band filters. Upper: S_{11} and S_{22}, represent the reflection coefficient at the ^{31}P and ^{1}H ports, respectively. Lower: $S_{21}(= S_{12})$ represents the isolation between ports.

8.2 Transverse Resonators

The transverse resonators are cylindrically shaped and produce a magnetic field perpendicular to their main axis. They are well suited for working with any size axial magnets, either horizontal or vertical. The sample accessibility is convenient. The NMR sensitivity of these probe designs is, however, generally worse than that obtained with the previously described type of resonator. The largest difference is probably given by the saddle coil which has, in theory, a sensitivity of only one third compared to the solenoid, [Hoult and Richards, 1976]. In practice the difference may be smaller depending on the design and loading of the resonator.

8.2.1 *Magnetic field amplitude*

As outlined in the introduction, the magnetic field is created by a current distribution that flows along the cylinder generator [Fig. 8.1(b)] and has, ideally, a "cosine" dependence in amplitude.

Again, this ideal current distribution, producing a perfectly homogeneous magnetic field inside the resonator, can only be approximated in practice, to levels depending on the different types of resonators, described below.

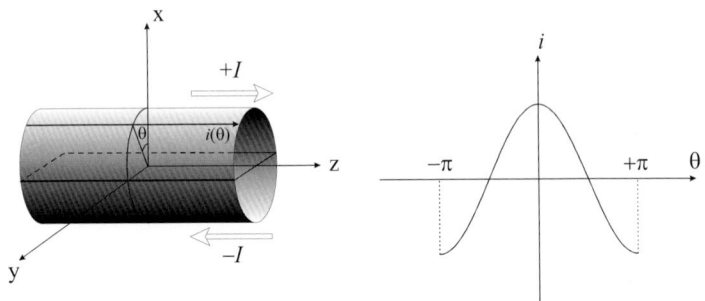

Fig. 8.21 Ideal cylindrical linear current distribution that creates a uniform RF magnetic field perpendicular to the static magnetic field B_0. B_0 is assumed to be parallel to the cylinder axis (Oz). The current sheet is constituted of two half cylinder shells with a "cosine" dependency on the radial angle θ. Each shell is driven by a total current I, of opposite sense.

A common feature of all of these resonators is that the current distribution on the half shells of the cylinder is symmetrical respective to the *xz* plane bisecting each shell, and anti-symmetric respective to the *xy* plane separating both shells (Fig. 8.21). Let $\pm I$ be the amplitude of the total current circulating in the cylinder half shell:

$$|I| = \int_{-\pi/2}^{\pi/2} i(\theta) d\theta. \tag{8.32}$$

Owing to the symmetrical properties, the field created at the center of the current distribution is oriented along the *y* axis, perpendicular to the symmetry plane of the current distribution. The components B_x cancel out due to the symmetry of currents respective to the *xz* plane. Adding all the contributions from each pair of current filaments opposed in phase, the magnetic field at the center of a finite length cylinder is obtained as [see Eq. (7.6)]:

$$B_y = \frac{2\mu_0}{\pi} \frac{l}{d\sqrt{l^2+d^2}} \int_{-\pi/2}^{\pi/2} i(\theta) \cos\theta d\theta, \tag{8.33}$$

where *l* and *d* are, respectively, the length and diameter of the cylinder.

The crudest approximation of the cosine current distribution is the delta function:

$$i = \begin{matrix} 0 & \text{if } \theta \neq 0, \pi \\ \pm 1 & \text{otherwise} \end{matrix}. \tag{8.34}$$

This case corresponds to two diametrically opposed wires. The amplitude of the magnetic field at the center is the largest among all the configurations:

$$B_y = \frac{2\mu_0 I}{\pi} \frac{l}{d\sqrt{l^2+d^2}} \tag{8.35}$$

but the homogeneity is obviously the worst.

For the ideal cosine current distribution,

$$i = I_0 \cos\theta \tag{8.36}$$

the total current I is given by:

$$I = I_0 \int_{-\pi/2}^{\pi/2} \cos\theta \, d\theta = 2I_0 \tag{8.37}$$

and the magnetic field after integration of Eq. (8.33) at center is

$$B_y = \frac{\mu_0 I}{2} \frac{l}{d\sqrt{l^2 + d^2}}. \tag{8.38}$$

The magnetic field amplitude is the smallest among all configurations, and the homogeneity is the best. To obtain a uniform field inside the resonator, the cosine distribution is required but the cylinder should also be infinitely long (at least about three times its diameter). In the ideal case, the magnetic field inside an infinitely long cylinder is equal to:

$$B_y = \frac{\mu_0 I}{2d} = 2B_1^+. \tag{8.39}$$

Compared to the spherical distribution of the same dimension (diameter) the created magnetic field is 33% lower in amplitude and even lower than that of the solenoid. On the other hand, the current pathways are much longer in this configuration compared with the spherical current distribution or with the solenoid-like coil. In the latter case, the current circulates on circular paths of the coil circumference length at most. On the contrary, for the present resonator, the current pathway is longitudinal and twice the resonator length.

Owing to the fact that it must be long to get a homogeneous RF field, this is not in favor of sensitivity, which requires a high B_1/I ratio and a small resistance. Finally, the return paths of current from one sheet to the other add supplementary resistance that depends on the coil diameter. All in all, the sensitivity for this type of resonator is worse than that obtained with the previous axial coils.

The estimation of the magnetic field amplitude given by Eq. (8.38) does not take into account the contribution from the current return paths that connect the two half cylindrical shells. The corresponding circular currents will create a magnetic field that adds to the field created by the linear currents inside the resonator improving the B_1/I ratio, if the resonator is short enough. If it is long, the contribution from the return path currents will become negligible. This will be considered in detail, separately for each particular case, in the following paragraphs.

8.2.2 The saddle-shaped coil

8.2.2.1 The optimum geometry and RF magnetic field

The saddle coil is a crude approximation of the ideal current density but it is widely used as RF and gradient coils. Two identical current wires are located on each half shell. The longitudinal wires are connected in series and driven by the same source of current $I/2$.[11] Hence the connections are made as shown in Fig. 8.22.

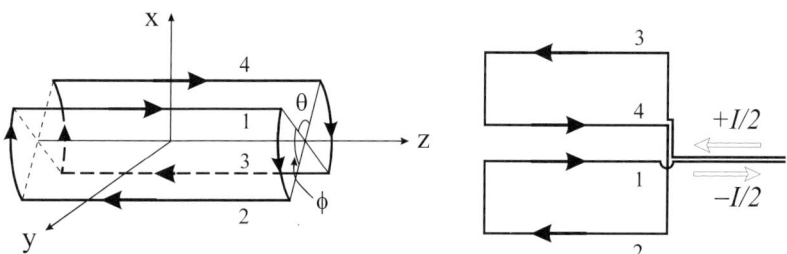

Fig. 8.22 Geometry of the saddle coil (left) and current paths (right). The magnetic field B_1 created by this current configuration is aligned along the y axis.

---

[11] The *total* current I that flows in each half cylinder is equal to the sum of current flowing in the two wires 1 and 3 or 2 and 4 (Fig. 8.22), hence twice the current driving the coil.

The magnetic field created by the longitudinal wires is given by:

$$B_y = \frac{2\mu_0 I}{\pi d} \frac{l}{\sqrt{l^2 + d^2}} \cos\left(\frac{\theta}{2}\right), \qquad (8.40)$$

where θ is the angle subtended by two wires of same current (Fig. 8.22).

The best RF field homogeneity inside the coil is obtained when θ is equal to 60°. This can be demonstrated by expanding the magnetic field equations at all points as a harmonic series of spatial coordinates. The magnetic field created by the optimum saddle coil, at the center, is equal to:

$$B_y = \frac{\sqrt{3}\mu_0 I}{\pi d} \frac{l}{\sqrt{l^2 + d^2}}. \qquad (8.41)$$

The magnetic field amplitude is found in-between the two extreme cases considered until now and represented by the two-wire case [delta function in Eq. (8.34)] and the cosine distribution [Eq. (8.36)]. The magnetic field homogeneity is comprised between the two cases too.

It should be noted at this point that the amplitude and homogeneity of the RF field also depend on the resonator length. When the resonator is infinitely long, the amplitude is at a maximum and the RF field distribution is identical in all the planes perpendicular to the cylinder axis. In other words, the field is obviously constant along any line parallel to the axis. In practice, the resonator has a finite length that reduces the amplitude at the center (and consequently inside the resonator volume of interest) and degrades the RF field homogeneity along the cylinder axis. In fact, the return path current of the end arcs of the resonator has an influence on both the RF field amplitude and homogeneity inside the resonator. Taking into account the contribution from the return path currents, the magnetic field amplitude at the center of the saddle coil is increased by a factor that obviously depends on the ratio d/l (diameter to the length) of the saddle coil. The integration of the Biot–Savart law on a circular segment subtended by an angle ϕ gives the magnetic field created, on axis, at a distance z_0 from the ring center as[12] (Fig. 8.23):

$$B_x = \frac{\mu_0 (I/2)}{2\pi} \frac{az_0}{\left(a^2 + z_0^2\right)^{3/2}} \cos(\phi/2) \qquad (8.42)$$

$$B_y = \frac{\mu_0 (I/2)}{2\pi} \frac{az_0}{\left(a^2 + z_0^2\right)^{3/2}} \sin(\phi/2) \qquad (8.43)$$

$$B_z = \frac{\mu_0 (I/2)}{4\pi} a^2 \phi \qquad (8.44)$$

Note that the sin(ϕ/2) factor is equal to the cos(θ/2) factor of Eq. (8.40) where θ is the angle subtended by the arc joining two equal current wires (Fig. 8.22).

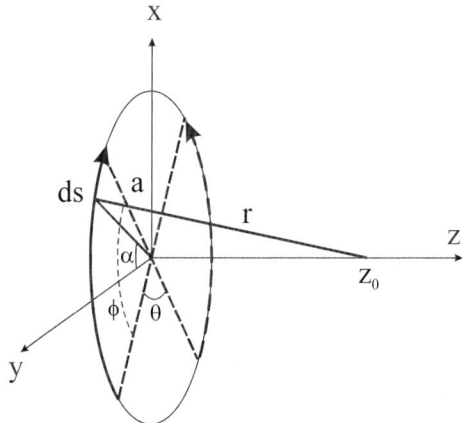

Fig. 8.23 Geometry of the circular current return path of the saddle coil. θ is the angle subtended by the two equal current wire positions and ϕ is the angle subtended by the return path connections.

---

[12] Remember that the *total* current carried by each half cylinder is twice the current flowing in each wire.

The contributions from the four return paths will add for the B_y field component but cancel out for both B_x and B_z. Calculating all the contributions, the total field created by the completed saddle coil, at its center, is obtained as:[13]

$$B_y = \frac{2\mu_0 I}{\pi} \frac{l}{d\sqrt{l^2+d^2}} \left[1 + \frac{d^2}{d^2+l^2}\right] \cos(\theta/2) = 2B_1^+ . \qquad (8.45)$$

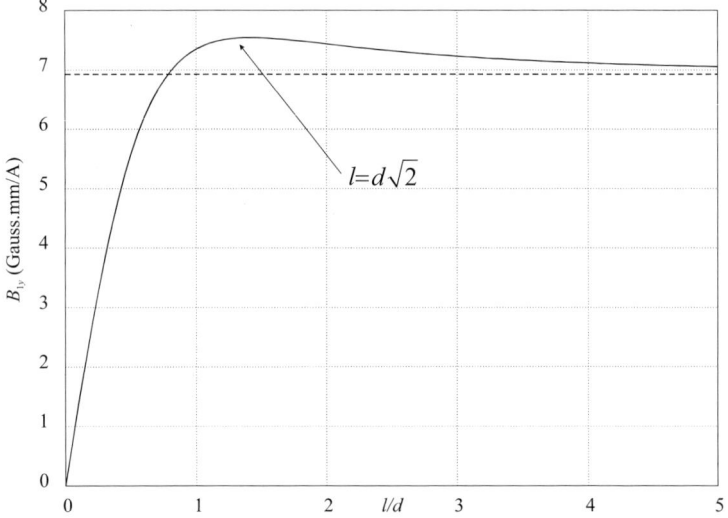

Fig. 8.24 Amplitude of the laboratory frame magnetic field created at the center of a $\theta = 60°$ ($\phi = 120°$) saddle coil, taking into account the circular return current paths (Eq. 8.45), as a function of the length to diameter ratio. The dashed line represents the plateau value for an infinitely long coil.

Due to the return currents, the amplitude of the magnetic field at the center of the coil is at a maximum when its length is equal to 1.41 times its diameter (Fig. 8.24). On the other hand, the optimum homogeneity over the coil volume is obtained for a larger ratio l/d of 1.661 [Mansfield and Morris, 1982, p. 266].

---

[13] The equation is similar to that already obtained by Hoult [Hoult and Richards, 1976] and Harman [Harman, 1988].

These two values are easily reconcilable, as the maximum of the function $B_1(l/d)$ is flat for a length to diameter ratio on the optimum range.

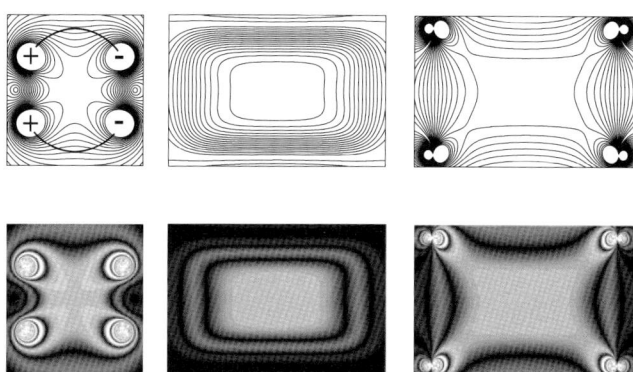

Fig. 8.25 The magnetic field created by the ideal saddle coil, having l/d = 1.661. Upper: contour plot of B_1^+. Lower: simulation of B_1 mapping experiment (after an initial 360° rotation). From left to right, the field is mapped in the xy, xz, yz plane. z is the direction of the static field. The phase of the currents flowing in the straight wires is represented in the drawing upper left.

The magnetic field maps obtained with such an optimum configuration are shown in Fig. 8.25.

8.2.2.2 *A practical design*

The saddle-shaped coil design is particularly suited for experiments on nuclei having low gyromagnetic ratios and when an axial sample access is required. This could be the case for vertical supraconducting magnets, as used in high resolution NMR. As an example, a ^{39}K probe is described here, designed for a magnet of 9.4 tesla. The corresponding Larmor frequency is 18.7 MHz for the most abundant isotope (^{39}K). The probe diameter is 10 mm, a standard dimension for liquid NMR experiments. Due to the small probe size and low frequency, a multi-turn coil (three turns) has been chosen in order to get "reasonable" impedance at 18.7

MHz (64.6 Ω, 551 nH[14]). The coil windings should fit in a small space and the wire diameter should not be too small in order to minimize the coil resistance. The sample size imposes that the coil probe should be a little bit longer than the optimum giving the best field homogeneity. The final coil design of the coil is represented in Fig. 8.26. It was made out of copper clad glued on a quartz tube (outer diameter of 11.4 mm) and shaped using conventional printed circuit board etching technology. The final insert (coil wound on the quartz tubing) was then fitted in a commercial probe replacing the initial high frequency saddle coil. The probe performance (90° pulse width and field homogeneity) was similar to that of another commercial probe of the same diameter.

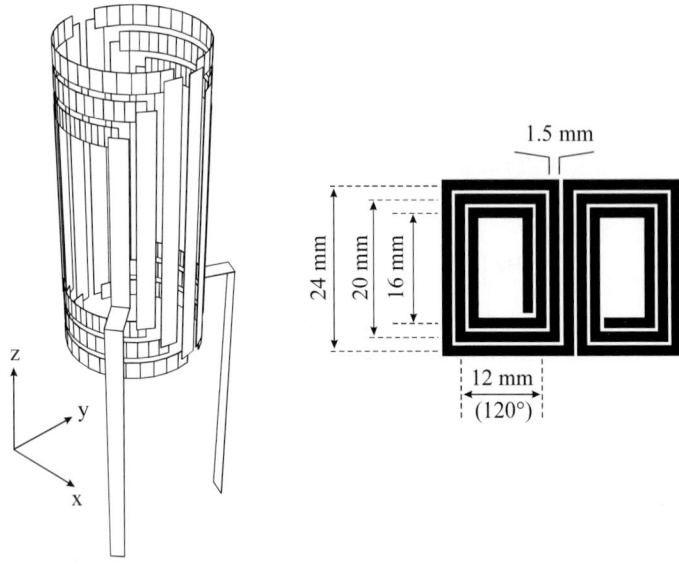

Fig. 8.26 Drawing of the ^{39}K coil (left) made with copper tape (right) wound on a 11.4 mm outer diameter quartz tube. The drawing to the left has been made using the draw utility from the FastHenry package [Kamon et al., 1994] from which the coil inductance was calculated.

---

[14] Estimation of the inductance of the coil is not as straightforward as for a simple solenoid or single coil. Relevant methods have been outlined in Chapter 3. In the present case, the inductance was estimated using the FastHenry package [Kamon et al., 1994].

The magnetic field maps calculated for the coil are shown in Fig. 8.27 and the B_1 homogeneity is evaluated in Fig. 8.28 for a sample volume that fitted the total coil space. Despite the compromised geometry of the resonator, the RF field homogeneity is quite acceptable. It is comparable to the ideal saddle coil field distribution (Fig. 8.25), using the same leveling. Finally, the sensitivity proved to be quite satisfactory.

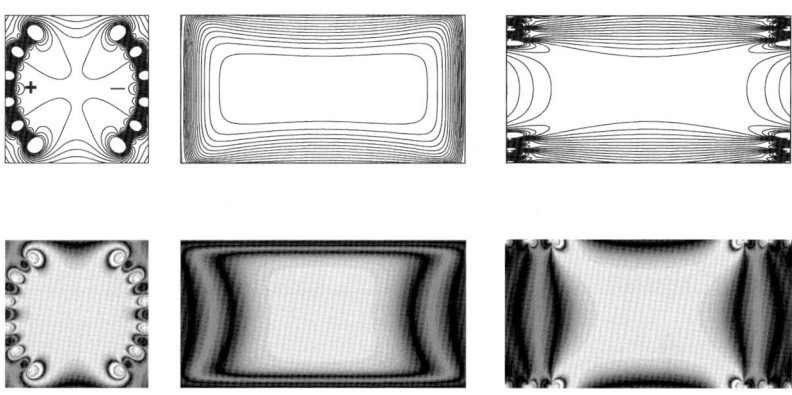

Fig. 8.27 The 18.7 MHz RF magnetic field created by the 3 turn saddle coil of Fig. 8.26. Upper: contour plot of B_1^+. Lower: simulation of the B_1 mapping experiment (using an initial 360° rotation). From left to right, the field is mapped in the *xy*, *xz*, *yz* plane. *z* is the direction of the static field.

In spite of its intrinsic lower sensitivity compared to the solenoid configuration, the saddle coil is of frequent use in axial magnets. It is indeed well adapted for low frequency and/or for small samples. As a result, it is still proposed for high resolution proton experiments well over 500 MHz. For these applications, the coil diameter is usually small (5 to 10 mm), sometimes up to 20 mm. The limit is again determined by the wire length compared to the wavelength and by the coil impedance at the working frequency.

For very low gyromagnetic nuclei experiments (for example ^{39}K), the number of turns[15] can be greater than unity. For other nuclei (proton, phosphorous, and eventually carbon) the number of turns should not be greater than unity and the sample size should not exceed 10 mm in diameter when the static magnetic field is greater than about 7 tesla (300 MHz for the proton). For higher sample size and frequency, the coil should be replaced by other designs more suited to high frequencies (next paragraphs).

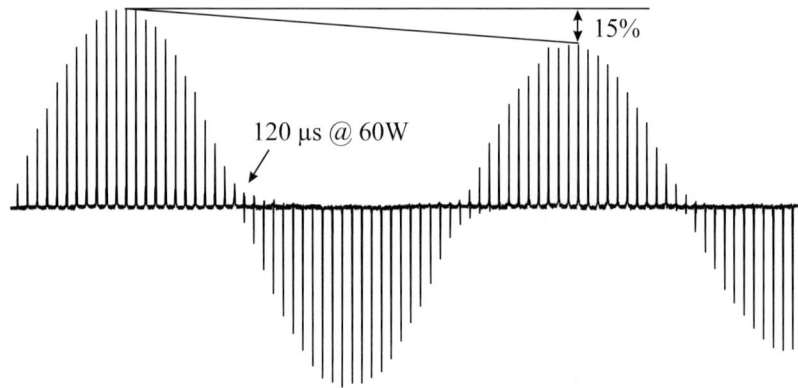

Fig. 8.28 Rotation of the ^{39}K nuclei magnetization obtained experimentally at 18.7 MHz. The sample is a highly conducting solution of KCl in water (1 M). The pulse length increases from left to right in steps of 5 µs and the transmitter power is 60 W. A 15% decrease in signal amplitude observed after a 450° rotation is due to the RF heterogeneity within the sample volume (that fills in the whole probe coil volume). It corresponds to satisfactory performances. A π rotation is obtained with a 120 µs pulse length corresponding to $B_1^+ = 21$ gauss. This is in good agreement with the estimated magnetic field amplitude at the center of 22.7 gauss. The estimation is done using the program "field" (see Resources for Readers, page X) assuming a peak current of 12.3 A (the input power is 60 W, the measured probe Q factor is 76, the coil inductance is 511 nH, hence the resistance is 0.79 Ω). The field amplitude at the center estimated from Eq. (8.45), using a mean geometry ($d = 11.4$ mm, $l = 20$ mm) and a *total* current I of 3 turns × 24.6 A on each cylinder half shell is $B_1^+ = 24.3$ gauss.

---

[15] As for the solenoid coil, it can be demonstrated that the performance of the coil is independent of the number of turns. Rather, it will be chosen according to the impedance of the coil at the working frequency.

8.2.3 UHF saddle coil-like resonators

At frequencies above 100 MHz, the wire length and coil impedance may become a problem. For example, a saddle coil that fits a 10 mm sample has a diameter of about 12 mm and a length of 20 mm. The total wire length of a 1 turn coil, including connections, is about 170 mm; that is greater than $\lambda/6$ at 300 MHz. Clearly the length is above the acceptable limit of $\lambda/10$. The design of proton decoupling coils that are used in combination in a 10 mm X-nuclei-observing coil in a high resolution spectrometer is more critical. In this case, the decoupling coil has a diameter of about 18 mm and a length of 30 mm. The length of wire approaches $\lambda/4$ at 300 MHz, which is not acceptable.

On the other hand, as the coil impedance increases, the electric field created in the sample increases too, leading to important dielectric losses and consequently important heating of the sample. In order to avoid such problems, some designs have been proposed in which the current is distributed over the surface of curved copper foil instead of flowing on a long path of thin wire.

The corresponding resonators [Fig. 8.29(a)] are based on the earlier designs of the slotted cylinder [Kan et al., 1973] and the similar Slotted Tube Resonator, (STR), [Schneider and Dullenkopf, 1977].

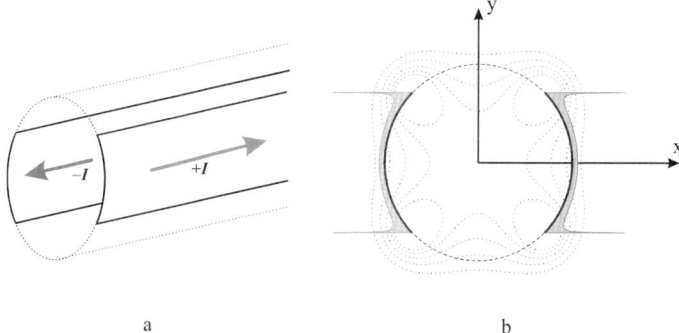

Fig. 8.29 (a) The slotted tube is made out of two cylindrical sheets driven by currents of equal amplitude and opposed phase. (b) The greyed surface represents the current density (absolute intensity) on the conductive sheets.

The slotted tube is a shielded, symmetrical transmission stripline made out of two arched conductors. The well-known "Alderman–Grant" coil [Alderman and Grant, 1979] is based on this concept. These resonators have been demonstrated to be efficient at high field NMR [Leroy-Willig *et al.*, 1985; Cross *et al.*, 1985] and even in the microwave range for EPR [Koptioug *et al.*, 1997], having the advantage of being simple to build.

Similar to the flat strip case presented in Chapter 7, the current on the curved strip circulates predominantly on the borders of the copper foil [Fig. 8.29(b)]. Hence, the current distribution is still not the ideal cosine distribution but more resembles that of a saddle coil.[16] The corresponding coils are therefore considered as Ultra High Frequency (UHF) versions of the saddle coil. The important difference, however, is that the length of the current pathway is shorter, hence more suitable for higher frequencies. The length is now roughly equal to the coil length, allowing the coil dimensions to be increased up to about 100 mm long for a 300 MHz design, for example.

An attempt to approach the ideal cosine dependence more closely has been proposed in a design where equal currents are tentatively forced in wires (or strips) disposed nonuniformly around the cylinder [Bolinger *et al.*, 1988]. In fact, the magnetic field distribution obtained for this design resembles that of a slotted tube because the current sharing between each wire is not uniform. It depends on the ratio of the mutual and self inductances within the wire network.[17]

For all the designs presented now, the optimum angle subtending the curved copper foil is different from the angle subtending the "in-phase current paths" of the saddle coil (60°). This is due to the particular current distribution on the conductive surfaces (not uniform).

All these designs have both the simplicity and the ease of tuning and matching in common. Compared with the saddle coil, they also considerably improve the usable frequency range up to several hundreds of megahertz.

---

[16] The magnetic field distribution too.

[17] This design has been described with more detail in the first edition.

8.2.3.1 *The Alderman–Grant coil; a version of the slotted cylinder*

This resonator was designed [Alderman and Grant, 1979] to reduce heating in conductive samples due to proton decoupling in classical ^{13}C observing experiments at high magnetic fields.

The Alderman–Grant Resonator (AGR) is basically formed by two H shaped copper foils wound on a cylinder of dielectric material (Fig. 8.30).

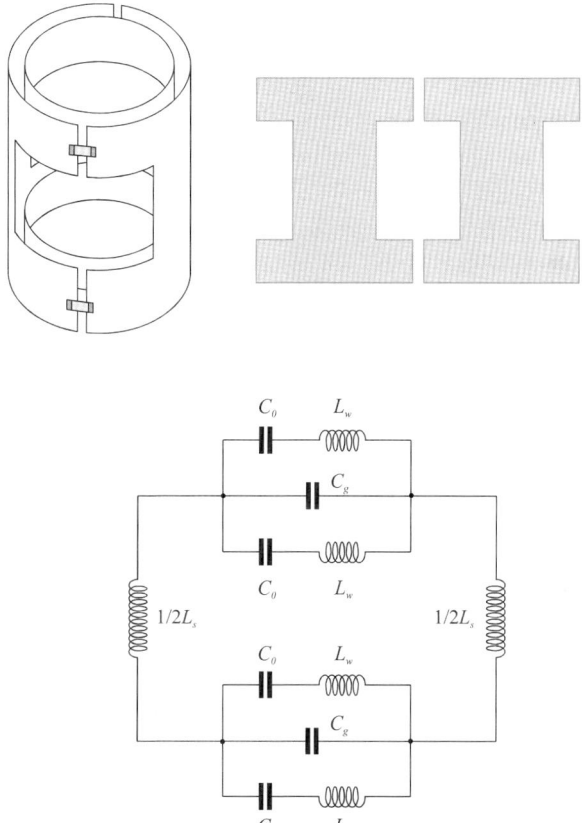

Fig. 8.30 The Alderman–Grant resonator (upper) and its equivalent circuit (lower). L_s is the slotted tube inductance. L_w is the inductance of one half end ring (one branch of the H). C_g is the distributed capacitance between the end and guard rings.

The resulting inductance is tuned by capacitors linking the two H. Two guard rings (electrically continuous rings) are added to protect the sample from the high electric field regions, near the tuning capacitors. The distributed capacitance between the guard rings and the branches of the H contributes to the total tuning capacitance.

The magnetic field inside the resonator is essentially created by the current distribution that flows along the vertical cylindrical copper sheets. Carlson [Carlson, 1988] has calculated that the optimum angle Ω subtended by the "vertical" copper sheets should be equal to 94°. This is slightly higher than the usual value of 80°, as deduced for the STR by Schneider from electrostatic methods [Schneider and Dullenkopf, 1977]. Solving the two-dimensional Laplace equation for a discrete array, Leroy-Willig [Leroy-Willig et al., 1985] found that the best homogeneity for the equivalent slotted cylinder is obtained for an angle of 85°, for the shield diameter three times the slotted tube diameter.

The influence of the shielding will be discussed later. Remembering that the optimum angle is 60° for the saddle coil, the difference is due to the current density spreading over the copper sheet of the slotted tube resonator while it is concentrated in the wires of the saddle coil.

The magnetic field created by the slotted tube of the AGR is calculated, for different angles Ω, in the median xy plane using the current density estimation method outlined in Chapter 7. It is valid for a slotted tube of infinite length. The optimum homogeneity is obtained for $\Omega = 90°$ (Fig. 8.31). In fact, this angle appears to not be too critical and a value comprising between 85° and 90° will provide an optimum homogeneity.

The current circulating in the wings also contributes to the generated magnetic field, in a similar manner to the return current paths of the saddle coil, already discussed in previous paragraphs. The corresponding contribution can be estimated assuming that the current flows in thin wires located approximately in the middle of the wing copper sheets and subtending an angle of $180° - \Omega$ (Fig. 8.32).

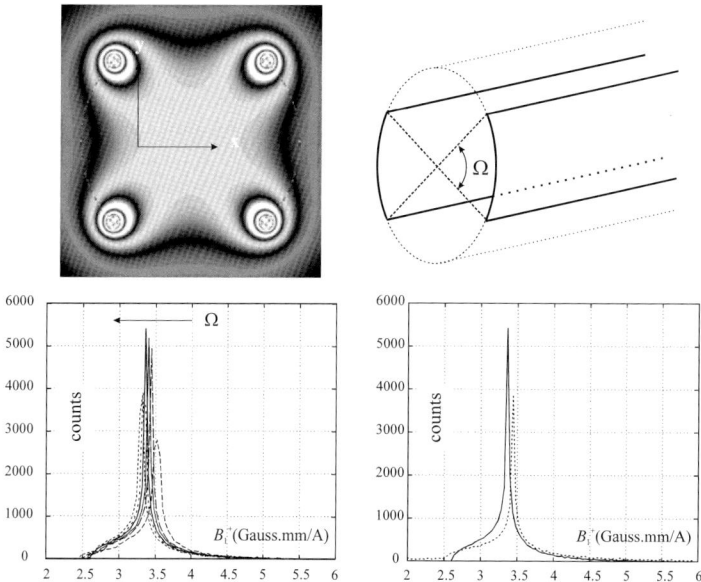

Fig. 8.31 Magnetic field map (upper left) created by the slotted tube (upper right). Lower: histogram of the magnetic field in the transverse plane calculated in a ROI equal to 90% of the cylinder diameter, for different Ω values, from 80° to 95°. The best homogeneity is obtained for $\Omega = 90°$ (full line). It is compared (lower right) to that obtained for the best saddle coil (dotted line).

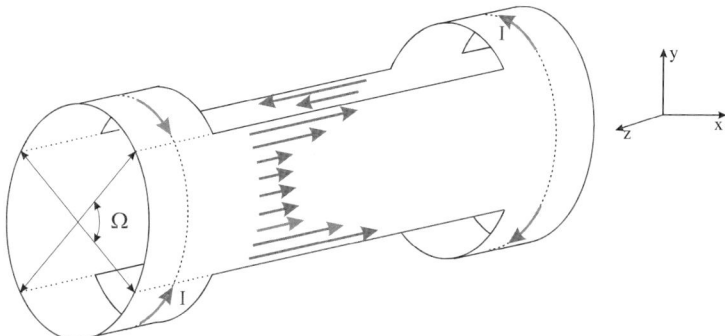

Fig. 8.32 The geometry of currents in the Alderman–Grant resonator, used for magnetic field estimation [Eq. (8.46)].

As a result, the magnetic field at the center of the AGR (without the shield) can be estimated from the following semi-empirical formula:

$$B_{1y} = 6.772 \frac{I}{d} \frac{l}{\sqrt{l^2 + d^2}} \left[1 + 0.835 \frac{d^2}{d^2 + l^2} \right], \qquad (8.46)$$

where l is the length (mm) of the resonator (Fig. 8.32), d its diameter (mm) and the angle Ω is assumed to be equal to 90°. The magnetic field is expressed in gauss and the *total* current I in A. It is interesting to note that the magnetic field amplitude at the center is almost equal to that created by the saddle coil, the field created by the AGR being only slightly lower. The field value is 6.772 gauss mm A^{-1} for an infinitely long slotted tube compared to 6.928 for the saddle coil [Eq. (8.45)]. Also, the correction factor resulting from the branches of the H is slightly lower than the contribution of the return currents in the saddle coil. As expected, the field homogeneity is somewhat better for the slotted tube design (Fig. 8.31).

The comparison is made for a *total* current of 1 A in each half cylinder sheet. This corresponds to a driving current of 0.5 A for the saddle coil. Concerning the power needed for creating an RF magnetic field, it should be remarked that the current travels twice the length of the saddle coil in each direction, while only once in the case of the slotted tube resonator. All in all, the efficiency of the slotted tube is expected to be of the same order of magnitude as that provided by a well-designed saddle coil. In fact, the choice between the two categories of designs will depend on the frequency and on the coil size. At low frequency, the saddle coil will be a good choice, especially for small samples, in a similar manner as the microsolenoid is better than a loop gap coil for very small sample tubes. At higher frequencies, the choice of the slotted tube is obviously favorable, especially for large samples.

The dimensions of the AGR are to be chosen in a similar way as already discussed for the saddle coil. In particular, the length to diameter ratio should not be too small. The magnetic field amplitude decreases quickly when l/d is lower than unity. Above this limit, the length of the coil should be optimized to obtain a compromise between good magnetic field homogeneity inside the required sample volume and the coil resistance that is proportional to *l*. A value for *l*/*d* comprising between 1 and 1.5 appears to be a good compromise. The upper limit of the coil length is provided by the wavelength. This corresponds to a coil with a maximum diameter of the order of 100–150 mm at 200 MHz.

8.2.3.2 Coupling the Alderman–Grant resonator to the spectrometer; a practical design

In their original work, Alderman and Grant proposed to drive the resonator between a "hot point" located on one of the wings and the guard ring, which is connected to the ground [Fig. 8.33(a)].

However, if the impedance provided by the distributed capacitance between the wing and the guard ring is too small (or if the guard ring is absent), an alternative coupling scheme is necessary [Fig. 8.33(b)]. Obviously, the resonator can also be fed through a coupling loop, providing a very simple way of equilibrating the resonant structure [Fig. 8.33(c)].

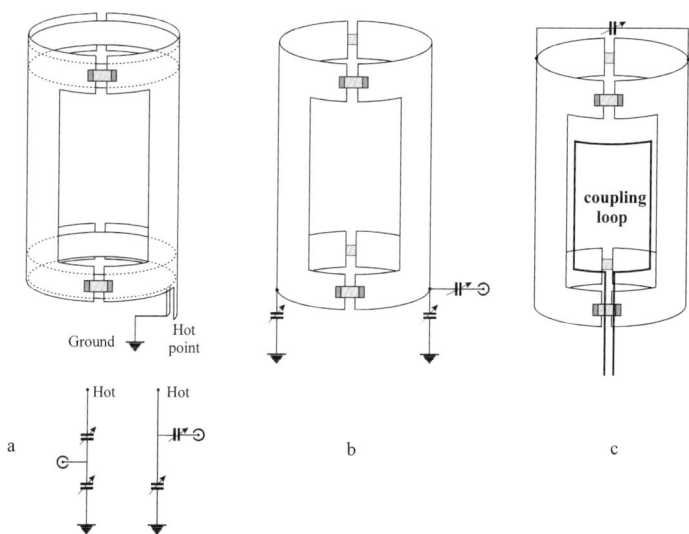

Fig. 8.33 Different ways to connect the slotted cylinder to the spectrometer. Left: the method proposed by Alderman and Grant. Middle and right: equilibrated configurations. The variable capacitor (right) can be made with a moveable guard ring or with another ring isolated from the resonator end ring by a foil of dielectric.

516 NMR Probeheads for Biophysical and Biomedical Experiments

The practical design which will be described here was used for many years for small animal imaging at 4.7 tesla. It proved to be efficient and very cheap. No expensive capacitors were used. It requires only copper and Teflon sheets and a few lengths of wire to make the coupling loop. Eventually a tuning capacitor can be added in series with the coupling loop to avoid the complicated mechanical parts required for the matching loop displacement. Finally, fine tuning of the resonator can be done by a copper ring, isolated from the wings of the AGR resonator by a dielectric sheet (Teflon). The tuning ring is moved along the cylinder, in order to adjust the resonant frequency of the AGR. The completed design is represented in Fig. 8.34.

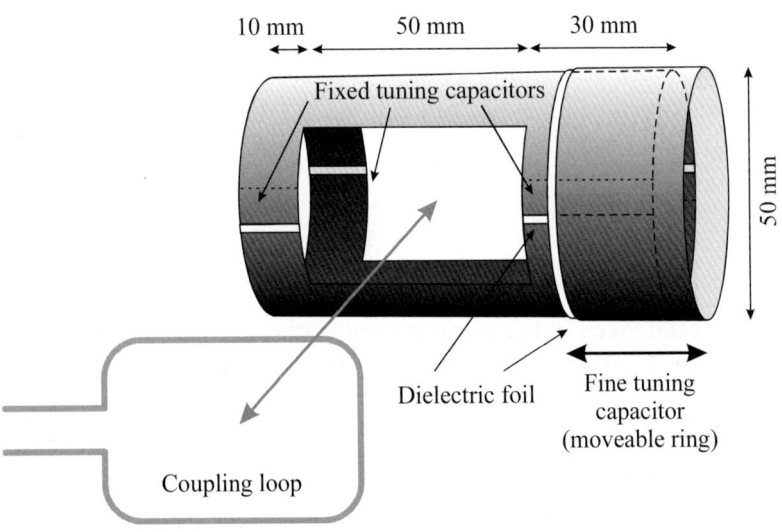

Fig. 8.34 A cheap, but fully functional, slotted tube resonator (200 MHz) that does not employ expensive capacitors. Matching is achieved by moving the coupling loop above the coil window. The magnetic field is aligned along an axis perpendicular to the window.

Its performance was comparable to a birdcage resonator, although the B_1 homogeneity improved for the latter, as expected.

The AGR was not shielded. Consequently, it detuned slightly when inserted inside the magnet. The frequency shift can be compensated anyway by moving the tuning ring such that no further matching is required. No extra noise was detected during receiving, suggesting that the balancing provided by the inductive coupling was efficient.

The capacitance value C_0 required to tune the AGR at 200 MHz can be estimated from the analysis of the equivalent circuit in Fig. 8.30 [Nijhof, 1990]. This circuit is an approximate representation of the resonator, hence the calculated capacitance values represent only a starting base for subsequent empirical adjustments. The main difficulty is probably in estimating the contribution of the equivalent capacitance C_g resulting from the guard rings.

Assuming, firstly, that the guard rings are absent, the whole structure is equivalent to an inductance $L_w + L_s$. In this case, the tuning capacitance C_0 is given by:

$$(L_w + L_s)C_0\omega^2 = 1, \tag{8.47}$$

where L_s is the inductance of the slotted tube. Some formulae have already been given in the literature, but they proved to be unsatisfactory. In contrast, FastHenry [Kamon et al., 1994] appeared to be a convenient approach to estimate the slotted tube inductance. The results of inductance values thus calculated are shown in Fig. 8.35. Except for very long resonators, the inductance depends strongly on the l/d ratio, but is always directly proportional to the length. Because the NMR probe will mainly be built with a ratio comprising between 2 and 3, the extended view of the dependence of L as a function of l/d is given in Fig. 8.35 (lower) for an optimum Ω angle of 90°.

L_w, in Eq. (8.47), represents the effective inductance of the copper sheet that links the two halves of the slotted tube. Neglecting the mutual inductance between the wing segments at the opposed sides of the AGR, L_w can be approximated roughly as one half the inductance of a complete ring of the same diameter and width.

Hence, an approximate formula for L_w is given by:

$$L_w = \frac{1}{8}\mu_0\pi \frac{d^2}{w+0.45d}. \tag{8.48}$$

Considering the dimensions of the design in Fig. 8.34, the equivalent total inductance is estimated to be 61.8 nH. The slotted tube contributes to only 23.8 nH of the total inductance. Hence the required tuning capacitance should be of 10 pF to tune the resonator to 200 MHz, in absence of the guard rings.

On the contrary, if the guard ring capacitance dominates, the contribution of L_w to the total inductance can be neglected[18] and the resonance equation becomes:

$$L_s\left(C_0 + \frac{1}{2}C_g\right)\omega^2 = 1. \tag{8.49}$$

The capacitance C_g can be estimated approximately as [Nijhof, 1990]:

$$C_g = \frac{1}{4}\pi\varepsilon_0\varepsilon_r \frac{wd}{e}. \tag{8.50}$$

For the design shown in Fig. 8.34, most of the tuning capacitance is concentrated in C_g (estimated to be about of 40 pF), thus requiring only small additional capacitance (of the order of 6.6 pF). The capacitance C_0 is formed, in part, by overlapping the end rings (Fig. 8.34).

Finally, a moveable copper ring, separated from the wing by a thin foil of Teflon® allowed the resonance frequency to be finely adjusted.

---

[18] In fact, the wings and the guard rings form a transmission line of low characteristic impedance, if C_g dominates. If the line is short compared to the wavelength, the line behaves essentially as a capacitor with a value close to the distributed capacitance C_g.

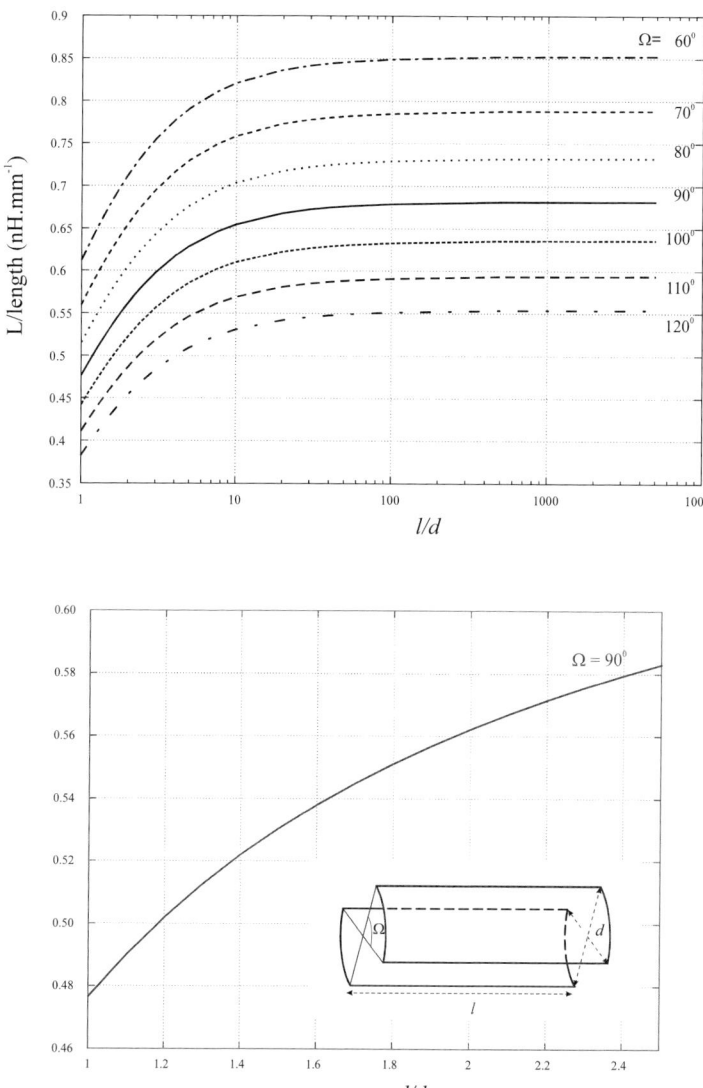

Fig. 8.35 Unshielded slotted tube inductance per unit length (nH/mm) as a function of the length to diameter ratio and for different values of Ω. Lower: an extension of the graph for optimum designs.

The original AGR or slotted cylinder structure can be modified in different ways (Fig. 8.36). In particular, the "mortarboard" implementation [Leroy-Willig *et al*., 1985], terminated by a conductive plane [Fig. 8.36(a)] increases the effective resonator length and improves the B_1 field homogeneity in the longitudinal plane.

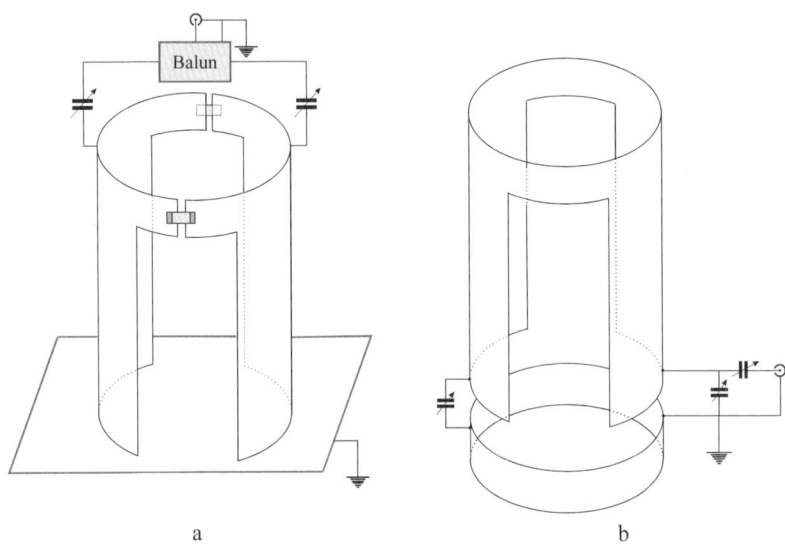

Fig. 8.36 (a) The mortarboard version of the slotted cylinder. (b) A simple effective implementation of the slotted tube uses only two symmetrical tuning capacitors.

8.2.3.3 *Shielding the UHF saddle coil-like resonators*

Shielding a saddle coil-like probe (or any axial resonator) represents a general requirement of probe isolation from the electromagnetic interactions with the surroundings. Beside the electromagnetic isolation, the shield has the important role of stabilizing the tuning and matching positions when moving the coil assembly inside the magnet. On the other hand, the shield presence has some consequences on the resonator electrical properties. In particular, the inductance of the coil is reduced, changing more or less the tuning and matching positions of the resonator.

Furthermore, the RF magnetic field tends to be concentrated in the space between the coil and the shield, diminishing the useful field in the sample volume. Consequently, the coil B_1/I ratio is reduced and the overall sensitivity of the probe is degraded.

In order to estimate the magnetic field decrease in the presence of the shield, one should estimate the current distribution on both the slotted tube and the shield conductive surface. This can be done using the method outlined in Chapter 7.

Let us consider the case of a slotted cylinder (diameter d) and shielded by a conductive cylinder of diameter d_s. The corresponding three cylindrical sheets (the shield and two halves of the slotted tube) are decomposed in a discrete set of linear current filaments I_k, perpendicular to the figure plane (Fig. 8.37).

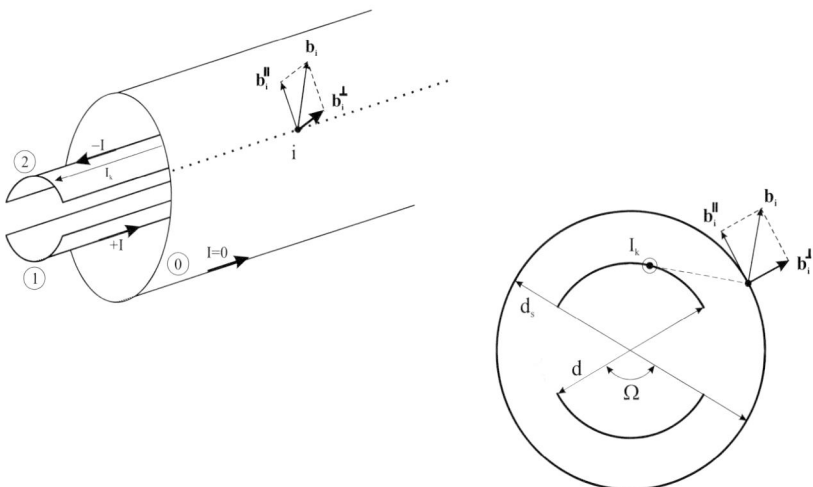

Fig. 8.37 Geometry of the shielded slotted cylinder. The conductive sheets of the slotted tube are connected at ports 1 and 2 to current sources. The sources are supposed to deliver $+I$ and $-I$, respectively. The shield is driven by a zero current source. A filament of current I_k creates a magnetic field $\mathbf{b}_{ik}$ at a given point i on the shield conductive sheet. $\mathbf{b}_{ik}$ is decomposed into a parallel and perpendicular component (upper right). The total perpendicular component should be equal to zero [Eq. (8.51)].

A current source of $\pm I$ drives the slotted tube. The shield is connected to a current source that delivers no current at all. Hence, the total current in the shield is equal to zero. The sets of relevant equations are (refer to Chapter 7, Section 7.3):

$$\left\{\sum_{k\in\{filaments\}} a_{ik}I_k = 0\right\}_{i\in\{test\ points\}}, \qquad (8.51)$$

$$\left\{\sum_{k\in\{filaments\}} I_k\delta_{jk} = Iport_j\right\}_{j\in\{ports\}}, \qquad (8.52)$$

where δ_{jk} is the Kronecker symbol (equal to 1 or 0), describing the condition that the filament k does or does not belong to the conductive sheet connected to port j.

The dimension of the set of equations is given by the total number of current filaments describing the current density on all conductive sheets is the dimension of the set {*test points*} in Eq. (8.51). The number of independent ports (three in the present case) to which the conductive sheets are connected is the dimension of the set {*ports*} in Eq. (8.52). The coefficients a_{ik} represent the magnetic field amplitude perpendicular to the copper foil at a given test point i, created by the filament of current I_k (Fig. 8.37).

Figure 8.38 shows the magnetic field maps obtained for different ratios of the shield to resonator diameter ($\Delta = d_s/d$). As Δ decreases, an important decrease in the generated magnetic field intensity at the coil center is noticed. From (a)–(f), there is an obvious decrease of magnetic field amplitude inside the probe, accompanied by an important increase of the magnetic field amplitude in-between the resonator and the shield. This aspect should be taken into account very carefully when designing a probe, since a very close shield may be capable of chasing the RF field out from inside the resonator (sample volume) to regions uninteresting from the NMR point of view.

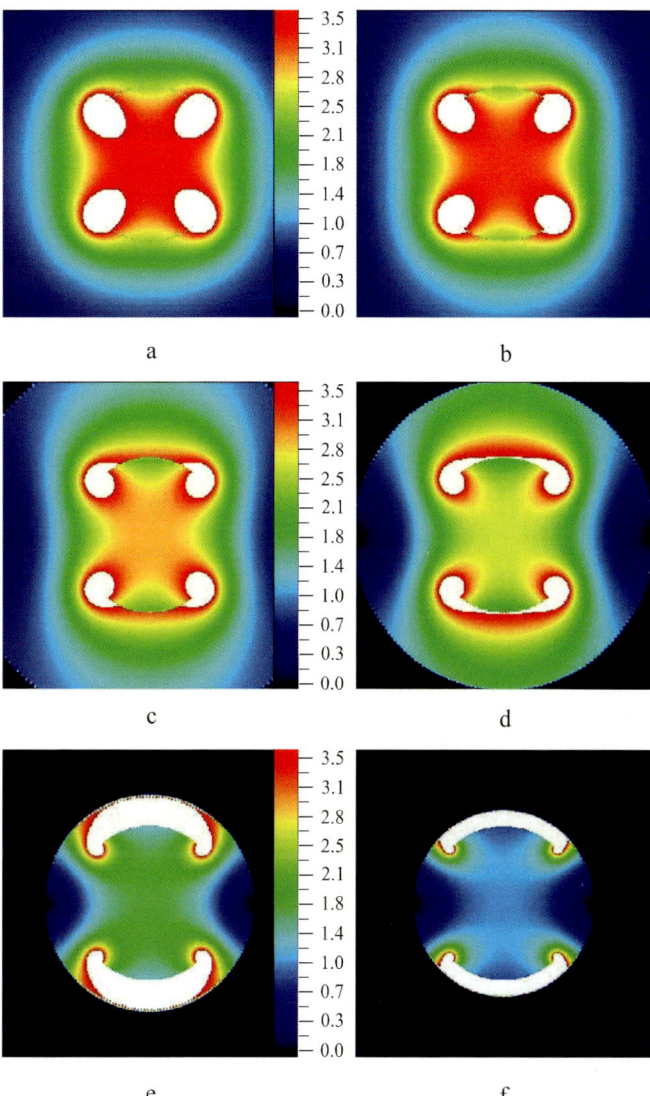

Fig. 8.38. RF magnetic field map (B_1^+) created by a shielded slotted tube, driven by a constant current source of 1 A, as a function of the shield to slotted tube diameter ratio (a – 10, f – 1.2). Note the decrease of the magnetic field at the center and the increase in amplitude in-between the shield and the resonator.

Figure 8.39 shows the corresponding histogram and field amplitude at the center. The field amplitude is given for a total current of unity (+/− 1 A in each shell of the slotted tube), a constant resonator diameter of 1 mm, and assuming a very long coil ($l/d > 5$). Δ is varied from 1.2 to 10.

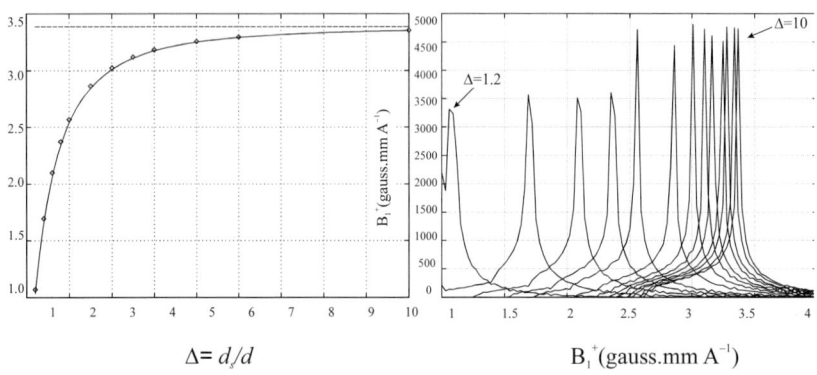

Fig. 8.39 Left: magnetic field amplitude at the center of a shielded slotted tube (of diameter 1 mm and driven by 1 A) as a function of Δ. Right: corresponding histograms of the magnetic field distribution in a Region Of Interest (ROI) of 90% of the slotted tube diameter.

As the shield diameter decreases, the magnetic field energy concentrates between the shield and the conductive sheet of the slotted tube,[19] at the expense of the field amplitude at the center which decreases accordingly.

This result can be qualitatively understood using the images concept (Fig. 8.40). Each filament of current, surrounded by a cylindrical conductive surface, creates a mirrored image constituted by a filament of current circulating in the opposed direction and located outside the shield cylinder. The magnetic field created near the shielded coil center by the filament inside the shield is therefore diminished by the field created by its image. In contrast, it is reinforced in the space between the resonator and the shield.

---

[19] This is a general effect that can, for example, also be observed for shielded axial solenoids, such as the loop gap resonator [Hardy and Whitehead, 1981].

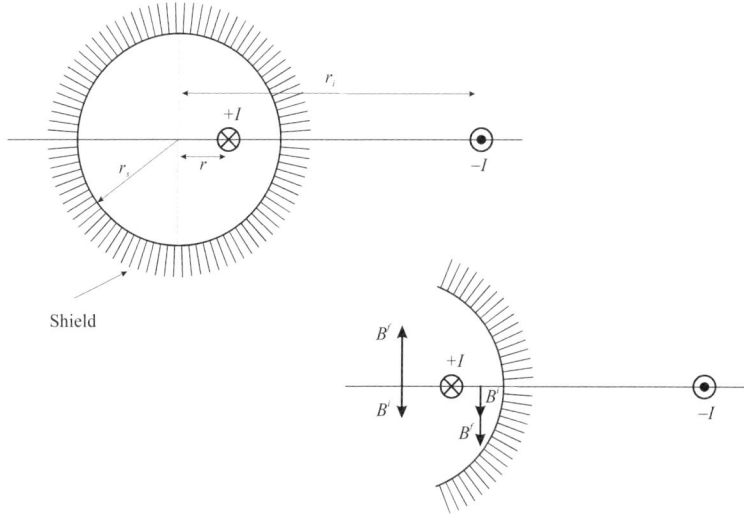

Fig. 8.40 The magnetic field created by a filament of current (perpendicular to the figure plane) placed in a cylindrical shield. The magnetic field is the superposition of field created by the filament itself (B^f) and its image (B^i). The two components subtract at the center of the shield and add in-between the filament and the shield (lower).

Quantitatively, the magnetic field created in the center by a given current filament can be easily estimated knowing the position of its image.

For a filament located at a distance r from the center (corresponding to the resonator radius), the radial position r_i of the image is given by:

$$r_i = \frac{r_s^2}{r}, \tag{8.53}$$

where r_s is the shield radius. The RF magnetic field created, at the center, by one particular current filament of the resonator and its image, is given by [using Eq. (7.5)]:

$$B_y = \frac{\mu_0 I}{2\pi}\left(\frac{1}{r} - \frac{1}{r_i}\right) = \frac{\mu_0 I}{\pi d}\left(1 - \left(\frac{d}{d_s}\right)^2\right), \qquad (8.54)$$

where d and d_s are the diameters of the resonator and of the shield, respectively.

Integrating all the current filaments of a resonator enclosed in a coaxial conductive cylinder, the magnetic field at the center is obtained as:

$$B_y^{shielded} = B_y^{unshielded}\left(1 - \left(\frac{d}{d_s}\right)^2\right). \qquad (8.55)$$

Equation (8.55) fits well the field values shown in Fig. 8.39 for the shielded slotted tube, using the numerical method described in Chapter 7. This result is still valid for any shielded axial resonator [Jin, 1999; Collins et al., 1997] such as, for example, the birdcage resonators described in the following paragraphs.

The RF field homogeneity is almost invariant down to a diameter ratio of 2.0 (Fig. 8.39). Below this value, the best field homogeneity is generally obtained for a slightly higher Ω angle (100° instead of 90° for the slotted tube), but the main effect remains a dramatic decrease of the field amplitude when the shield approaches the resonator surface.

In these conditions, one would expect that the sensitivity will decrease dramatically too. However, this effect depends on all coil resistance contributions in the presence of the shield. The "pure" resistance of the slotted tube is essentially unaffected by the presence of the shield, although the current distribution is slightly modified. More importantly, the radiation resistance of the slotted tube will considerably decrease.

This effect would be particularly beneficial at high frequencies. Furthermore, the electric field is also occupying the space between the shield and the surface of the resonator, reducing the electric losses in the sample accordingly. As a result, the sensitivity (B_1 per unit current divided by the square root of the resistance) will not decrease as rapidly as expected from the B_1/I decrease alone. The prediction of the actual sensitivity is not straightforward. However, the presence of the shield is sometimes beneficial and sometimes even required!

8.2.4 *The birdcage resonator*

The slotted tube resonator allowed operation at higher frequencies than the classical saddle coil, while the magnetic field homogeneity remained similar. This is due to the non-ideal cosine current distribution on the cylinder surface.

The concept of the "birdcage resonator" provides an improvement toward higher frequencies and closer approximation to the cosine distribution. It was introduced by Hayes in 1985 [Hayes *et al.*, 1985] and appeared as a great improvement in NMR probe designs [Hayes, 2009]. The price to be paid is an increased complexity in tuning the resonator and a greater dependence of the RF magnetic field homogeneity on the coil design. Another difficulty will be encountered in multiple tuning the resonator. However, it can be naturally driven in the quadrature mode which provides another important advantage of this design.

Basically, the birdcage is a loop network of identical filter cells. A typical birdcage design is given in Fig. 8.41.

When excited by a current source, "waves" are propagated along the network. For some particular frequencies, the waves combine constructively to create stationary states, corresponding to resonant modes of the ladder network. From the spatial point of view, each cell contains a straight conductor that contributes to the magnetic field inside the sample volume. The conductors are arranged parallel to the cylinder generator, equally spaced. For one particular stationary state of the network, the ideal cosine current distribution is obtained. Hence, the magnetic field distribution created by the birdcage is expected to reach the ideal homogeneity. Obviously, this depends on the accuracy in sampling the cosine distribution, hence on the number of leg, or rung conductors.

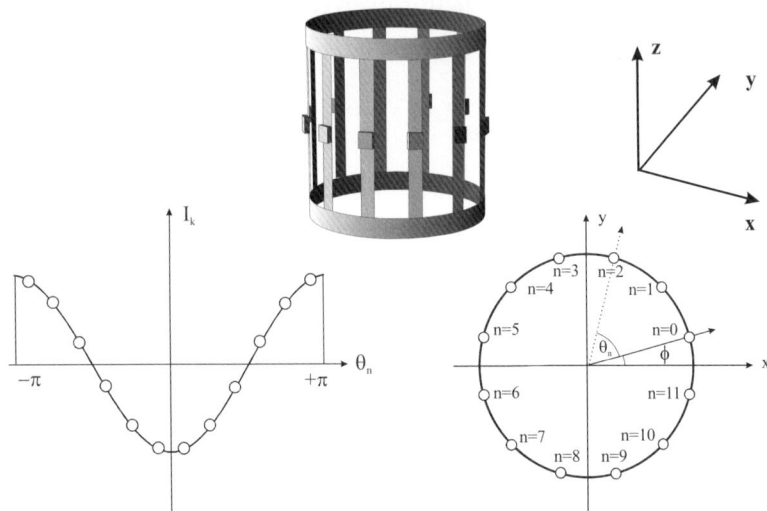

Fig. 8.41 Spatial representation of an LP 12 leg birdcage (upper). The current distribution in the vertical legs (lower left) is a cosine function of the azimuthal angle (lower right).

In contrast to the previous design, the birdcage resonator has a plurality of resonant modes. Among all these modes, only one, the so-called "$k = 1$" mode, approaches the cosine distribution creating a homogeneous magnetic field in the coil volume. This is also the sole mode that gives a nonzero magnetic in the center of the coil. This property helps in the identification of this mode among all the resonant modes of the birdcage.

In the following, the magnetic properties of the birdcage design are described, assuming that the ideal current distribution is shared among all its rungs. Then, the electrical circuits that are able to produce this current distribution are considered in detail.

8.2.4.1 *RF field map of the k = 1 mode*

Consider a $2N$ rung ideal birdcage. It is constituted of $2N$ current filaments, equally arranged on a cylinder of diameter d and length l. The current amplitude of each filament is expected to follow the cosine relationship:

$$I_n = I(2N)\cos(\theta_n + \phi),$$
$$n = 0, 2N-1$$
(8.56)

where θ_n is the relative angle position of leg n (Fig. 8.41). Because the legs are arranged equally around the cylinder, θ_n is given by:

$$\theta_n = n\pi/N.$$
(8.57)

The spatial phase factor ϕ in Eq. (8.56) is defined by the driving mode of the birdcage and determines the direction (polarization) of the magnetic field created in the plane perpendicular to the coil symmetry axis. The axis system can be chosen so that the magnetic field component created at the center is aligned with y. This geometry defines ϕ as:

$$\phi = \pi/2N.$$
(8.58)

The current amplitude factor $I(2N)$ is related to the total current driving the probe. As is the case for all probes of the axial type, a total current of $\pm I$ is expected to flow in each half cylindrical sheet (Fig. 8.21),

$$\sum_{n=0}^{2N-1} |I_n| = 2I.$$
(8.59)

Hence $I(2N)$ is given by [from Eqs. (8.56) to (8.58)]:

$$I(2N) \sum_{n=0}^{2N-1} \left|\cos\left[\frac{\pi}{2N}(2n+1)\right]\right| = 2I.$$
(8.60)

In these conditions, the magnetic field at the center is oriented along y and its amplitude is given by integration of Eq. (8.33). Using the symmetry properties of the design, one gets:

$$B_y = \frac{2\mu_0 I}{\pi d} \frac{l}{\sqrt{l^2 + d^2}} \varsigma$$
(8.61)

with

$$\zeta = \frac{\sum_{n=0}^{N/2-1} \cos^2\left[\frac{\pi}{2N}(2n+1)\right]}{\sum_{n=0}^{N/2-1} \cos\left[\frac{\pi}{2N}(2n+1)\right]}. \quad (8.62)$$

The field amplitude at the center of the birdcage depends slightly on the number of legs, through the factor ζ (Eq. 8.62). This factor is to be compared to the $\cos(\theta/2)$ of Eq. (8.40), which is equal to 0.866 for an optimized saddle coil ($\theta = 60°$). The corresponding value for the AGR [Eq. (8.46)] is 0.8465. For the birdcage coil, the corresponding factor ranges from $\sqrt{2}/2$ (0.707) for a four leg birdcage to $\pi/4$ (0.785) for an infinite number of legs (Fig. 8.42).

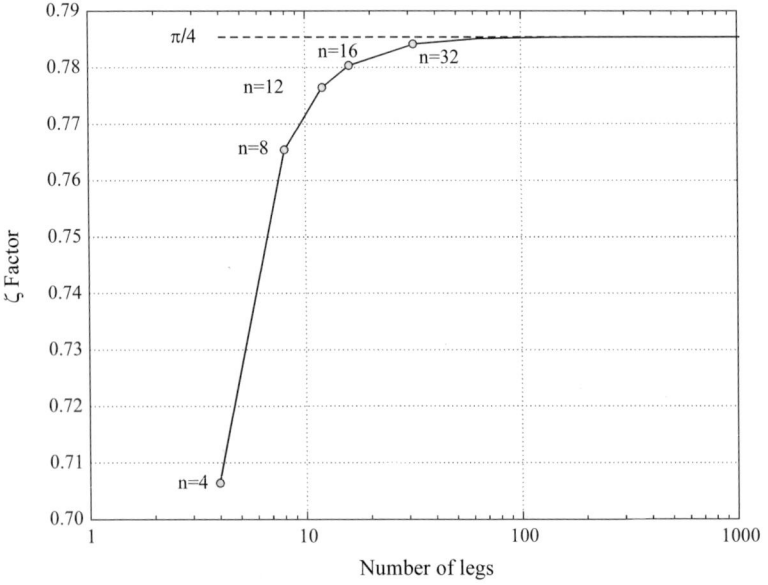

Fig. 8.42 The factor ζ of Eq. (8.69) as a function of the birdcage number of legs.

The field amplitude given by the birdcage with a high number of legs is, as expected, equal to the value already predicted for a continuous cosine distribution of current [Eq. (8.38)].

Homogeneous Resonators 531

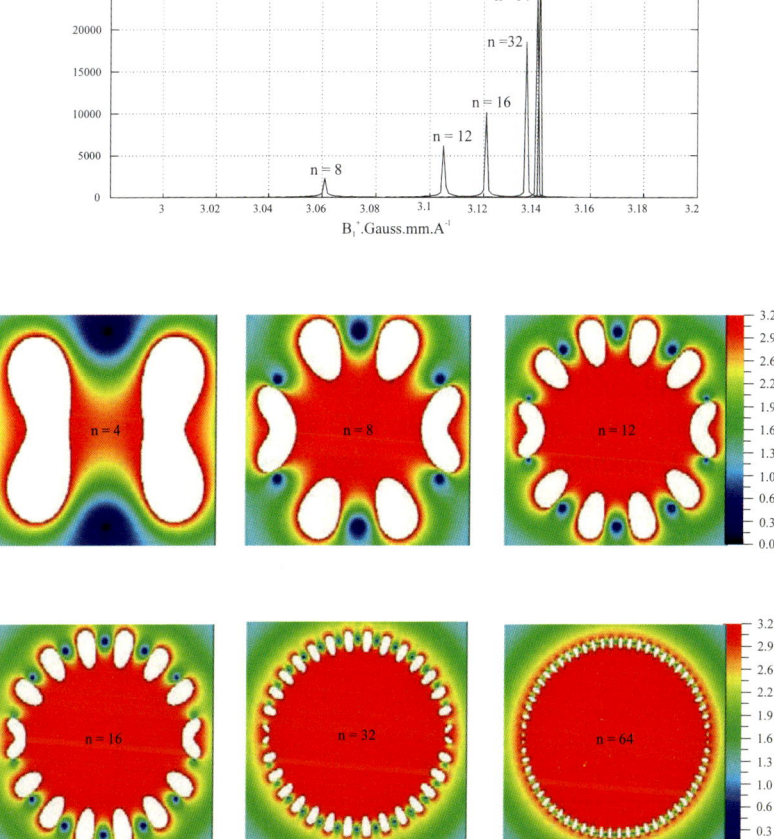

Fig. 8.43 RF field amplitude and homogeneity of the birdcage as a function of leg number. Upper: histograms of the magnetic field created by an ideal birdcage, calculated in a ROI of 90% of the coil diameter, as a function of the number of legs ($n = 4$ to 128). The histogram for $n = 4$ is too flat to be visible. Lower: corresponding magnetic field maps for $n = 4$ to 64. The magnetic field map for $n = 128$ looks similar to the $n = 64$ case. The magnetic field amplitudes are given in gauss, assuming a coil of 1 mm in diameter, infinite length, and driven by a total current of 1 A.

The magnetic field homogeneity increases with the number of rungs (Fig. 8.43). It is therefore advantageous to design the coil with a maximum number of rungs. In practice, a birdcage with 12 rungs is easy to build and adjust. It provides a better homogeneity than a slotted tube, but at the expense of a 10% lower magnetic field for the same input current. A 64 to 128 leg birdcage is ideally the best compromise that is still accessible to the well equipped electronic lab.

These considerations apply to the ideal birdcage, meaning a coil in which the current and position of the legs are perfect. In practice, it is difficult to maintain the symmetry while tuning and matching the coil under the conditions of a particular sample loading.

This will be considered later from a practical point of view. At this point we will solely examine the effect of the precision of currents on the field homogeneity. As an example, a 12 and a 128 leg birdcage will be considered, assuming they have a perfect geometry but an error in current distribution of 2% and 10%. The results are shown in Fig. 8.44.

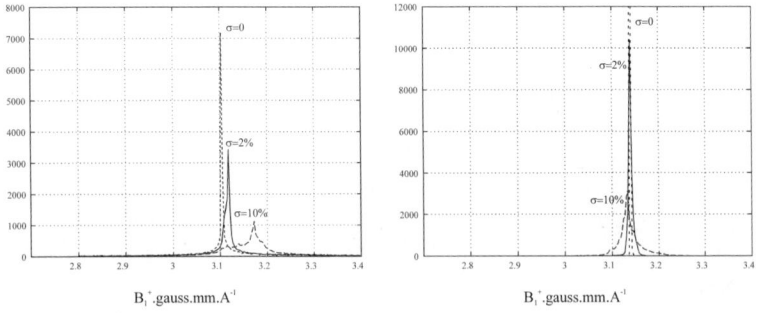

Fig. 8.44 Effects of current amplitude errors on the RF field homogeneity. Histograms of the magnetic field (ROI 90% of coil diameter) created by a birdcage of 12 (left) and 128 (right) legs. Both are driven by ideal ($\sigma = 0$) and perturbed current distributions ($\sigma = 2\%$ and 10%).

It appears that the 128 leg birdcage is less sensitive to an error in current than the 12 leg birdcage. As the number of legs increases the averaging of the current errors is clearly beneficial. In any case, the precision in the current distribution should be better than 10%.

The analysis above did not take account of the return current path that contributes to the magnetic field inside the resonator as was the case with the saddle coil and the AGR. From Eq. (8.43), each arc of return current I_n (of angle $\pi - 2\theta_n$, Fig. 8.45), contributes to the component y of the field as:

$$B_y^{arc} = \frac{\mu_0 I_{2N} \cos(\theta_n)}{\pi d} \frac{l}{\sqrt{l^2 + d^2}} \sin\left(\frac{\pi - 2\theta_n}{2}\right) \frac{d^2}{l^2 + d^2}. \qquad (8.63)$$

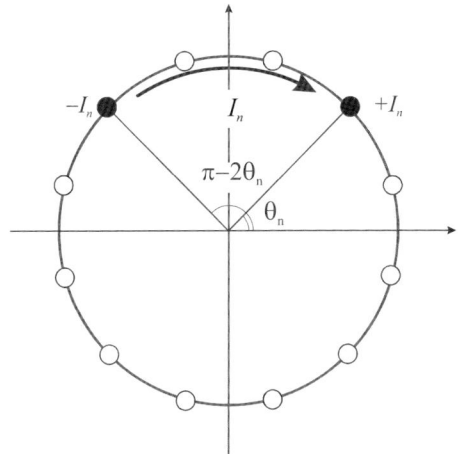

Fig. 8.45 Geometry of arc of return currents for field calculations in the birdcage.

Summing up the contributions of the four sets of $N/2$ arcs of return current, the B_x and B_z components cancel out and the B_y component adds to that given by Eq. (8.61) to give the final field amplitude at the center of the birdcage as:

$$B_y = \frac{2\mu_0 I}{\pi d} \frac{l}{\sqrt{l^2 + d^2}} \left(1 + \frac{d^2}{l^2 + d^2}\right) \zeta = 2B_1^+. \qquad (8.64)$$

The factor ζ is given in Eq. (8.62). The functional dependence of the magnetic field value at the center is similar to that already obtained for the saddle coil.[20] In particular, the magnetic field amplitude at the center is at a maximum when the length to diameter ratio is equal to $\sqrt{2}$ (Fig. 8.23).

8.2.4.2 Network analysis of the birdcage circuit

The electrical circuit representation of the birdcage design is a circular ladder network made of elementary filter cells. These cells can be described by a two-port L network, as shown in Fig. 8.46).

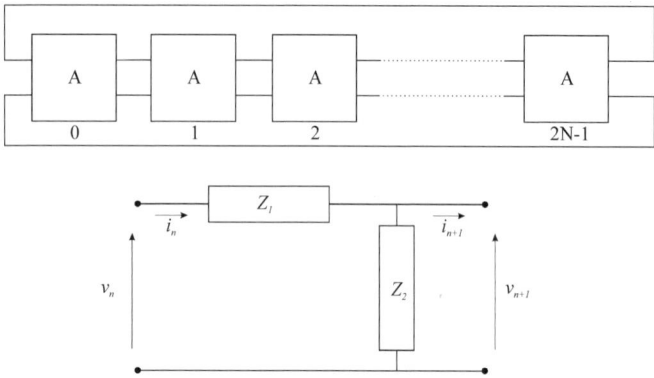

Fig. 8.46 Upper: schematic representation of a $2N$ leg birdcage as a circular ladder network made out of identical filter cells. Lower: details of each filter section.

The correspondence between the lumped element network and the physical birdcage is shown, for some particular examples, in Fig. 8.47.

Note that the simple L two-port network description assumes that the coil has a symmetry plane, virtually connected to the ground. In the case of an asymmetric design (expected or not!) a complete description of the birdcage as a complex network of lumped RLC elements is still possible. It can be analyzed using a linear circuit analysis software that solves the problem numerically. Alternatively, elegant analytic descriptions of the birdcage resonant modes, including explicit formulation of all the mutual interactions between the elementary cells can be found in the literature [Leifer, 1997ab; Tropp, 1997; Pascone et al., 1991.

---

[20] The factor ζ is equivalent to the $\cos(\theta/2)$ factor of Eq. (8.45).

These approaches still assume a perfect symmetry of the birdcage. In case of moderate deviations from symmetry, a perturbation theory can be applied [Tropp, 1991].

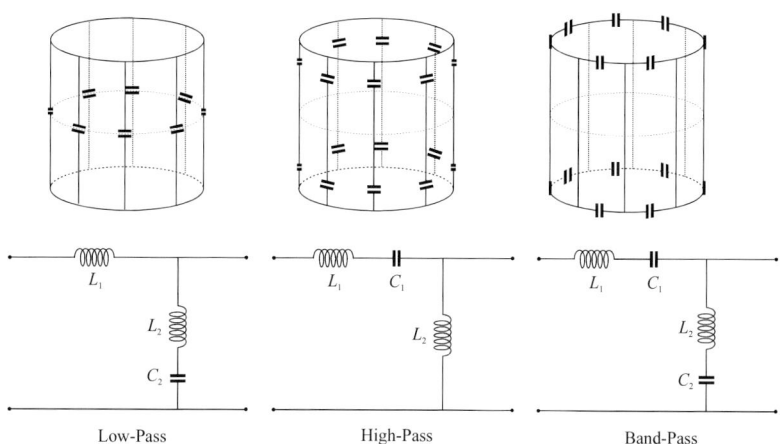

Fig. 8.47 The spatial representation (upper) and elementary cell circuits (lower) corresponding to the so-called low-pass, high-pass and hybrid birdcage [Pimmel and Briguet, 1992] types.

Here, it will be given a simpler representation in order to deduce some general properties of the birdcage circuits. The mutual interactions between cell elements (inductors) can be taken into account using the effective inductance concept [Chin et al., 2002].

With each network cell being connected in series, as shown in Fig. 8.46, the ABCD chain matrix formalism (Chapter 3) is particularly suited for analyzing the whole network. Let **A** be the chain matrix of one elementary cell that relates the input (v_1, i_1) and output (v_2, i_2) vectors as:

$$\begin{pmatrix} v_1 \\ i_1 \end{pmatrix} = \mathbf{A} \begin{pmatrix} v_2 \\ i_2 \end{pmatrix}. \qquad (8.65)$$

The matrix **A** depends on the frequency ω, as do the impedances Z_1 and Z_2.

$$\mathbf{A}(\omega) = \begin{bmatrix} 1 + \dfrac{Z_1}{Z_2} & Z_1 \\ \dfrac{1}{Z_2} & 1 \end{bmatrix} \qquad (8.66)$$

The input and output (v, i) vectors are related by the following relationship:

$$\begin{pmatrix} v_0 \\ i_0 \end{pmatrix} = \mathbf{A}^{2N}(\omega) \begin{pmatrix} v_{2N-1} \\ i_{2N-1} \end{pmatrix}, \qquad (8.67)$$

where $2N$ is the total number of cells (or rungs). Due to the fact that the input and output are connected together, the total chain matrix of the network must verify the following equation:

$$\mathbf{A}^{2N}(\omega) = \mathbf{I}, \qquad (8.68)$$

where $\mathbf{I}$ represents the identity matrix. This is the resonant equation of the birdcage.

Equation (8.68) can be solved [for example, Pimmel, 1990] by calculating the eigenvalues of the matrix $\mathbf{A}$. Let $[\lambda]$ be the diagonal matrix of the eigenvalues of $\mathbf{A}$:

$$\mathbf{A} = \mathbf{U}^{-1}[\lambda]\mathbf{U}. \qquad (8.69)$$

Equation (8.68) can then be replaced by:

$$[\lambda]^{2N} = \mathbf{I}. \qquad (8.70)$$

Hence, the two eigenvalues of $\mathbf{A}$ must verify:

$$\lambda_1^{2N} = \lambda_2^{2N} = 1. \qquad (8.71)$$

Equation (8.71) is valid for:

$$\begin{aligned} \lambda_{1,2} &= e^{j\varphi} \\ \varphi &= k\pi/N \quad k \in \mathbb{N} \end{aligned}, \qquad (8.72)$$

where φ is the transfer phase shift function of the elementary cell voltage and current. From Eq. (8.72), there are $2N$ distinct solutions. However, as shown below, some of the solutions are frequency degenerated (degenerate mode of resonance). The mode corresponding to $k = 1$ is the mode of interest from the NMR point of view. The phase shift changes at the slowest rate, from 0 to 2π, along the network. It corresponds to the desired homogeneous RF field. For the higher order modes, the currents change more rapidly in-phase, corresponding to RF field gradient modes. Finally, the mode $k = 0$ corresponds to a constant phase along the network.

The eigenvalues of the matrix $\mathbf{A}$ are easily obtained from:

$$\det(\mathbf{A} - [\lambda]) = 0 \tag{8.73}$$

or

$$\lambda^2 - \lambda\left(2 + \frac{Z_1}{Z_2}\right) + 1 = 0. \tag{8.74}$$

Replacing the solutions of Eq. (8.72) into Eq. (8.74), one obtains the relationship that relates Z_1 and Z_2 for each mode k:

$$\frac{Z_1}{Z_2} = -4\sin^2\left(\frac{k\pi}{2N}\right). \tag{8.75}$$

Hence, the resonant frequencies for each mode of the birdcage can be solved from the known dependences of Z_1 and Z_2 as a function of frequency. Reciprocally, the required Z_1 and Z_2 (hence the probe components) can be calculated from knowledge of the desired mode and frequency (next paragraph).

The first mode, $k = 0$, corresponds to either Z_2 infinity or Z_1 equal to zero. Only the latter case is possible for the low-pass (LP), high-pass (HP) or band-pass (BP) birdcage (Fig. 8.47. The condition $Z_1 = 0$ is realised at DC (zero frequency) for the LP birdcage, and corresponds to the resonance of the end ring in the case of the HP or BP birdcage. This mode is useless because there is no current in Z_2 (in the legs).

Each other mode, corresponding to $0 < k < N$ is identical to mode $2N - k$, according to:

$$\frac{Z_1}{Z_2}(\omega_k) = \frac{Z_1}{Z_2}(\omega_{2N-k}). \qquad (8.76)$$

Finally, mode $k = N$ is distinct and non-degenerate. There is therefore a total of $N + 1$ resonant modes described by this formalism.[21]

The resonant frequencies corresponding to these modes depend on the particular type of circuit. Assuming the simplest, basic, LP and HP ladder network (Fig. 8.48), the resonant frequencies are respectively given by:

$$\omega_k^2 = \frac{4\sin^2(k\pi/2N)}{LC} \qquad (8.77)$$

and

$$\omega_k^2 = \frac{1}{4LC\sin^2(k\pi/2N)}. \qquad (8.78)$$

Although the circuits of Fig. 8.48 are very crude approximations of the birdcage resonators, they correctly describe the general behavior of the different birdcage designs. For a quantitative analysis (next paragraph) the impedances Z_1 and Z_2 must include the effective inductances of each conductive element (legs and end ring segments).

---

[21] As will be shown later, the HP design will show another resonant mode, corresponding to the resonance of the two coupled end ring circuits. This resonance corresponds to the "Helmholtz" mode in which the currents in both end rings are in-phase. It cannot obviously be described by the present formalism in which only one end ring is assumed. The resonant mode ($k = 0$) of the end ring is, however, contained in the present description. For a symmetric birdcage, having two end rings, this mode corresponds to out-of-phase end ring currents, by symmetry to the virtual ground plane. The corresponding resonance corresponds to a "Maxwell" mode.

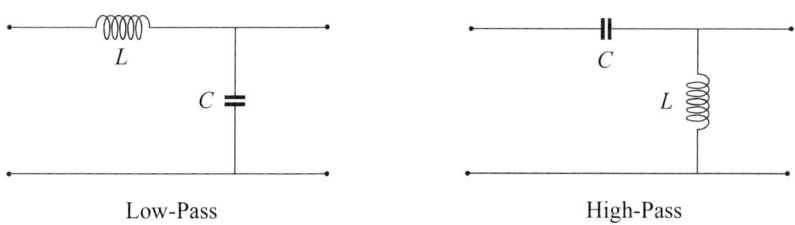

Fig. 8.48 Simplified elementary cell of the ladder network describing the two basic birdcage types.

The general behavior of each mode, already outlined, can be retrieved from Eqs. (8.77) and (8.78). In particular, the $k = 0$ mode corresponds to the lowest ($\omega_0 = 0$) frequency for the LP birdcage while it corresponds to the highest frequency ($\omega_0 = \infty$) mode for a HP.

The next mode $k = 1$ is the mode of interest for the homogeneous probe design (Fig. 8.49, scheme for an eight leg frequency resonant modes). It appears at the lowest nonzero frequency for the LP birdcage and at the highest finite resonance for the HP birdcage. The other modes appear in increasing frequency order for the LP birdcage and decreasing frequency order for the HP birdcage.

It should be mentioned at this point that the HP birdcage presents another resonant mode, close to the $k = 0$ mode [Leifer, 1997a]. This mode corresponds to a so-called Helmholtz resonance of the two coupled end rings in which the currents are in-phase (Co-Rotational mode, CR). The pure $k = 0$ mode is a Maxwell resonant mode in which the end ring currents are out-of-phase (Anti-Rotational mode, AR). These two modes generally appear at the high frequency edge of the resonant spectrum of the birdcage, well above the frequency corresponding to the "NMR useful" $k = 1$ mode. From the well-known properties of coupled resonators (Chapter 6), one can predict that the AR mode will be resonant at a higher frequency than the CR mode.

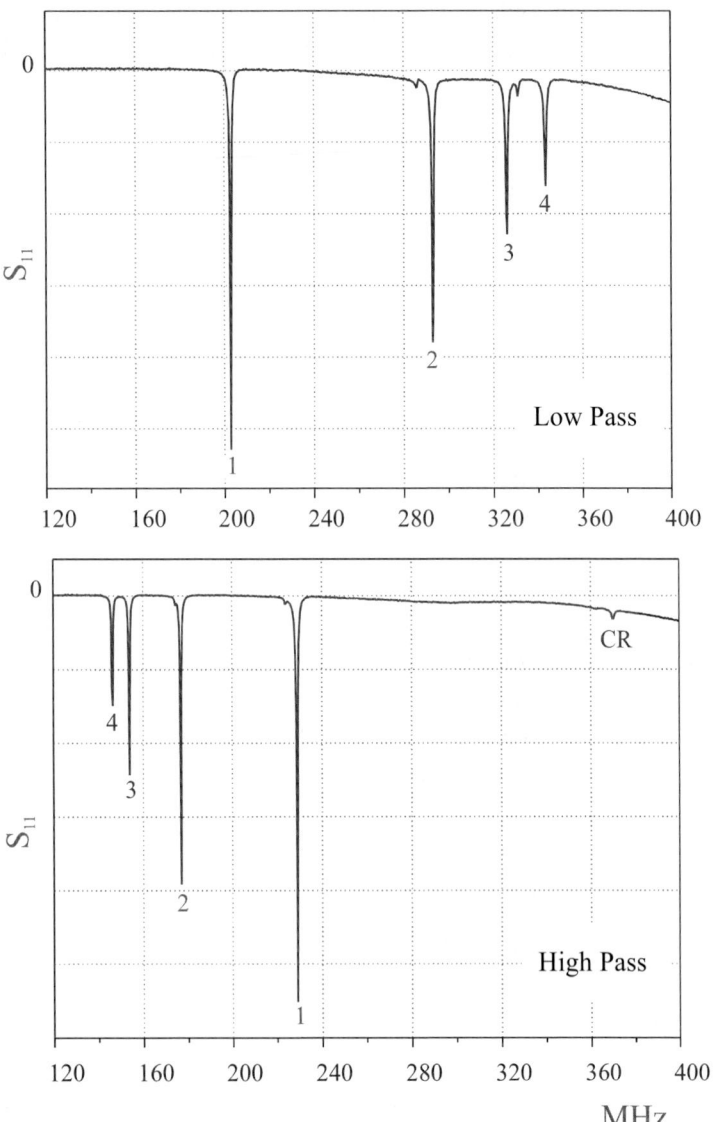

Fig. 8.49 Resonant modes of a low-pass and high-pass eight leg birdcage, obtained using a small coupling loop connected to a network analyzer (see Chapter 10).

Homogeneous Resonators 541

The leg currents, which are responsible for the RF field created by the birdcage, can be generally written as [Hayes *et al.*, 1985; Leifer, 1997a; Tropp, 1989]:

$$I_n = I_0 \cos(\pi k n / N + \phi), \qquad (8.79)$$

where n and k ($1 \leq k \leq N$) are the leg number and frequency mode, respectively.

Note the close similarity of this equation with Eq. (8.56), for the $k = 1$ homogeneous mode. Equation (8.86) is the nth sampling point of a periodic function of frequency $k/2N$. The uniform spatial sampling is realized by the $2N$ legs. This is illustrated in Fig. (8.50) for the four distinct resonant modes ($k = 1, 4$) of an eight leg birdcage, with the corresponding resulting field map, assuming $\phi = 0$.

ϕ is a spatial phase factor, identical to the phase factor in Eq. (8.56), which depends on the way the birdcage resonator is excited. Changing this phase obviously has the effect of rotating the orientation of the RF magnetic field.

This property of the birdcage easily allows driving the resonator in order to produce a circularly polarized magnetic field. The circular polarization can be done by exciting the birdcage at spatial positions 90° apart, using two oscillating sources having a 90° phase difference in time (quadrature excitation).

8.2.4.3 *Estimation of the current in a birdcage coil*

To evaluate the probe, the amplitude of the magnetic field B_1 could be estimated from measured quantities such as the transmitter average power P and the Q factor of the probe (Chapter 10). Then, the calculated value could be compared with that obtained from the 90° pulse duration in an NMR experiment or measured with a calibrated magnetic probe.

The magnetic field amplitude is given by Eq. (8.64). The current amplitude I will now be evaluated from the known RF power P dissipated in the coil and from the Q factor of the probe.

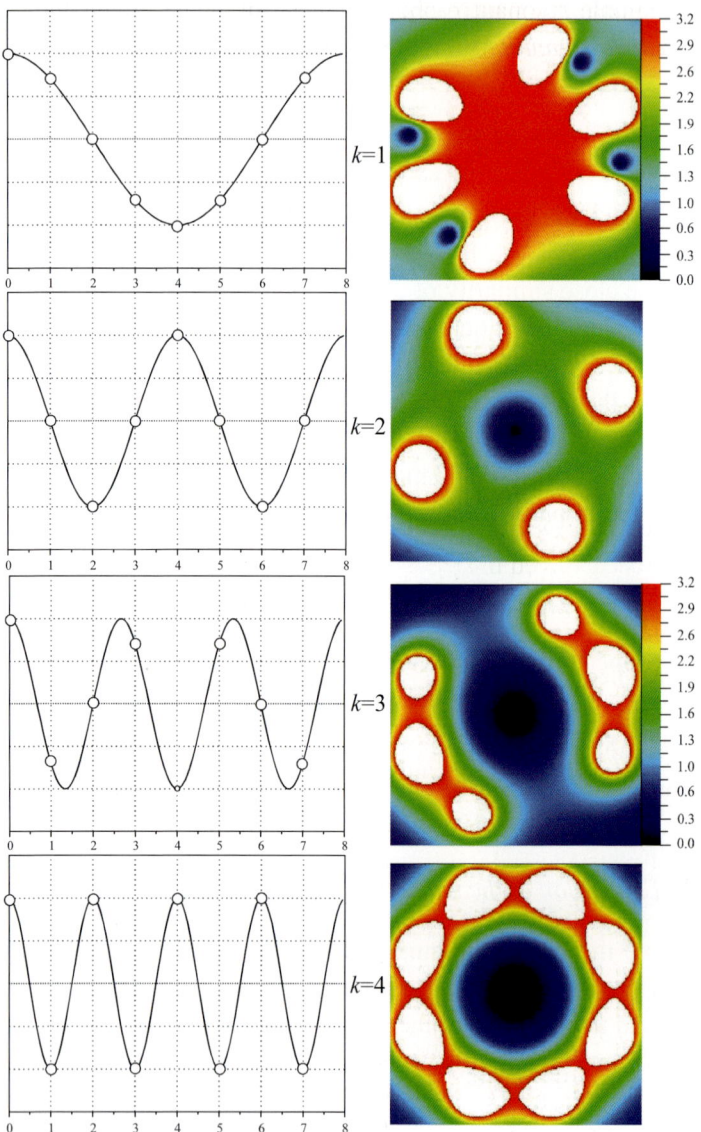

Fig. 8.50 Currents in an eight leg birdcage for the four distinct resonant modes $k = 1$ to 4 (left). Legs are numbered from 0 to 7. The excitation source is assumed to be connected to leg 0 ($\phi = 0$). The corresponding B_1^+ field maps are shown to the right. Note that for $k > 1$ a gradient is obtained, resulting in B_1^+ (at the center) = 0.

For a simple resonant probe coil that can be described by an RLC circuit, the *peak amplitude* current[22] is given by:

$$2P = rI^2, \qquad (8.80)$$

where the coil resistance r can be estimated from the measured Q value and from the known value for either the coil or the tuning capacitor impedance:

$$r = \frac{L\omega}{Q} = \frac{1}{QC\omega}. \qquad (8.81)$$

L can be calculated or measured, but C is generally known from the coil construction process.

The above reasoning applies to a single tuned, single coil probe. The problem seems *a priori* more complicated for a birdcage resonator. In fact, it has been solved in an elegant manner by Tropp [Tropp, 2002]. We followed a similar idea, but within a different formalism in order to get the peak amplitude I required in Eq. (8.64) to estimate the magnetic field created by the birdcage. The main idea is deriving equations for the birdcage similar to Eqs. (8.80) and (8.81).

Firstly, the dissipated power in a birdcage is expressed as a function of the current amplitude $I(2N)$ and of the legs and end ring segment resistances (respectively R_L and R_E). This allows the definition of an equivalent resistance, R_{eq}, representing the losses of the birdcage, in a similar manner as in Eq. (8.98):

$$P_R = R_{eq} I_{rms}^2 (2N), \qquad (8.82)$$

where $I_{rms}(2N)$ is the amplitude factor [see Eq. (8.56)] of the cosine dependence of the leg I_{rms} currents.[23] If the current is a pure sinusoidal function in time, the peak amplitude is given by:

---

[22] For a sinusoidal time dependent current, the peak amplitude is √2 times the RMS amplitude. Note that I here is the amplitude factor $I(2N)$ in Eq. (8.56).

[23] *RMS* means Root Mean Square. The power dissipated in a pure resistance is proportional to the square of this quantity.

$$I(2N) = I_{rms}(2N)\sqrt{2}. \tag{8.83}$$

Secondly, R_{eq} is related to the measured Q factor and the tuning capacitance C known from the birdcage design. Specifically, the dissipated power in the legs is given by:

$$P_R^{legs} = \sum_n R_L I_n^2, \tag{8.84}$$

where I_n is the RMS current that dissipates energy in the resistance R_L[24] of leg n. From Eq. (8.56) (the driving of the birdcage is assumed to be symmetric such that the phase ϕ is equal to $\pi/2N$):

$$\begin{aligned}P_R^{legs} &= R_L I_{rms}^2(2N) \sum_{n=0}^{2N-1} \cos^2\left[\frac{\pi}{2N}(2n+1)\right] \\ &= NR_L I_{rms}^2(2N)\end{aligned} \tag{8.85}$$

The dissipated power in the end ring can be similarly estimated from the evaluation of the current flowing in each segment of the end rings (Fig. 8.51).

Assuming a symmetric driving of the birdcage, the current in segment 1 (that joins legs 0 and 1) is equal to the current in leg 0. The current in segment 2, joining legs 1 and 2, is equal to the sum of the current in leg 0 and 1, and so forth. The maximum current is thus generated in segment $N/2$ (linking leg $N/2 - 1$ and $N/2$) and is equal to the sum of currents in all preceding legs. Finally, the current decreases in the next segments (linking leg $N/2$ and $N - 1$) until reaching the segment number $N - 1$ where the current is equal to the current flowing in segment 1. No current flows in the next segment (N[25]) as well as in the segment (0) preceding segment 1.

Obviously, the four sets of end ring segments have the same current distribution, except for a π phase factor, irrelevant for the Joule effect.

---

[24] All legs are assumed to have the same resistance.
[25] At this point, it should be recalled that N is half the total number of legs of the birdcage.

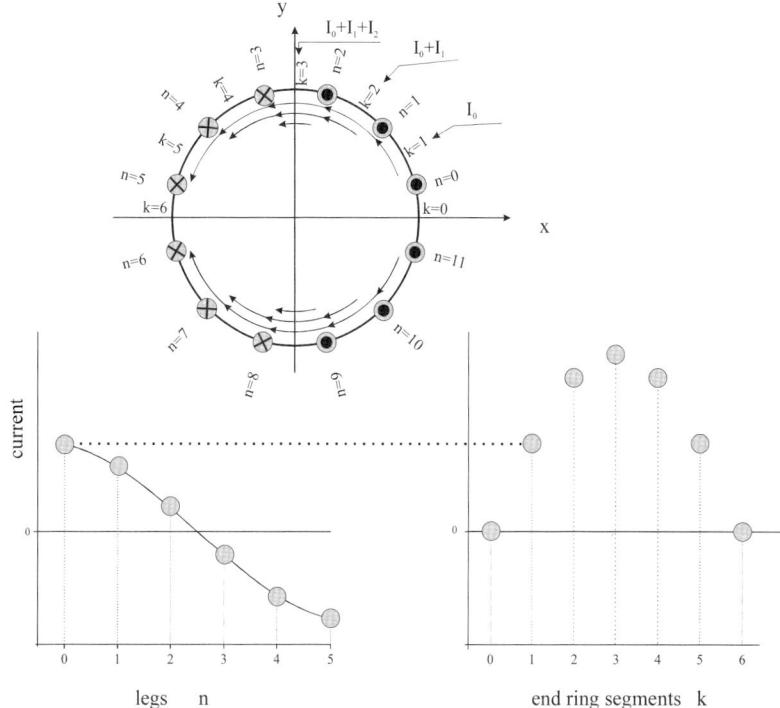

Fig. 8.51 Leg and end ring segment currents for a 12 leg birdcage. It is assumed that excitation is done such that the symmetry plane is aligned with the y axis.

Formally, the total power dissipated in all end ring segments characterized by the common resistance R_E is given by:

$$P_R^{er\_segments} = 4R_E I_{rms}^2 (2N) \sum_{k=1}^{N-1} i_k^2, \qquad (8.86)$$

where i_k is the normalized current in end ring segment k. The corresponding current is given by:

$$1 \leq k \leq \frac{N}{2}, \quad i_k = \sum_{n=0}^{k-1} \cos\left[\frac{\pi}{2N}(2n+1)\right]$$
$$\frac{N}{2} < k \leq N-1, \quad i_k = i_{N-k} \tag{8.87}$$

As a result:

$$4 \sum_{k=1}^{N-1} i_k^2 = \frac{N}{2\sin^2(\pi/2N)}. \tag{8.88}$$

The total power dissipated in the birdcage is equal to:

$$P_R = P_R^{legs} + P_R^{er\_segments}$$
$$= I_{rms}^2 (2N) \left[N \left(R_L + \frac{R_E}{2\sin^2(\pi/2N)} \right) \right], \tag{8.89}$$

where the term in brackets represents the equivalent resistance, R_{eq}, of the birdcage describing its losses.[26] The equation is valid for both the LP and HP birdcage designs.

Let us write R_{eq} as:

$$R_{eq} = aR_L + bR_E, \tag{8.90}$$

where a and b are the sum of squared normalized currents in the leg and end ring segments, respectively. The equivalent resistance depends only on the geometry of the resonator, assumed to be perfectly symmetric and subjected to the cosine current distribution. Hence it does not depend on the particular birdcage design (either HP, LP, or BP).

Following a similar reasoning as Tropp [Tropp, 2002], the Q factor of the birdcage resonator is obtained, by definition, as:

$$Q = \frac{\omega E}{P_R}, \tag{8.91}$$

---

[26] Note that R_{eq} does have not the same physical meaning as R_{net} introduced by Tropp [Tropp, 2002].

where E is the stored energy in the reactive components. Let C be the tuning capacitance for either the LP or HP configuration. Hence:

$$Q = \frac{\sum i_C^2}{R_{eq} C \omega} \qquad (8.92)$$

assuming all capacitances are equal. The numerator is the sum of the squared normalized currents that flow in the capacitors, i.e., in either the legs or the end ring segments of the LP or HP configuration, respectively. The corresponding sum is therefore equal to a or b, depending on the particular circuit that tunes the resonator.

The Q factor for the LP configuration is therefore equal to:

$$Q_{LP} = \frac{a}{R_{eq} C_{LP} \omega}, \qquad (8.93)$$

while for the HP birdcage

$$Q_{HP} = \frac{b}{R_{eq} C_{HP} \omega}. \qquad (8.94)$$

For a perfectly symmetrical birdcage, the term a (equal to half the number of legs) and b are given by Eq. (8.95):

$$a = N = N_{rungs}/2$$
$$b = \frac{N}{2\sin^2(\pi/2N)}. \qquad (8.95)$$

At this point, it should be pointed out that the Q value for a given birdcage does not depend on the particular configuration, whether it is LP or HP, as would be expected from physical arguments. This can be easily demonstrated from the above equations [Eqs. (8.93) and (8.94)] and Eqs. (8.100) and (8.104) which give the capacitance values for tuning a given birdcage in the LP or HP configuration. As a result, the capacitance that tunes the LP circuit is smaller by the factor $2\sin^2(\pi/2N)$ than the capacitance value that tunes a HP configuration:

$$C_{LP} = 2\sin^2\left(\pi/2N\right)C_{HP}.\qquad(8.96)$$

This is exactly the ratio a/b that describes the relative contribution of the legs and end rings to the total losses of the birdcage.

8.2.4.4 Estimation of the tuning capacitance

Tuning the homogeneous mode of the birdcage resonator to the required frequency needs an accurate estimation of the capacitor(s) value. The inductance of each leg and circular segment that constitutes the end rings is easy to estimate using formulae or numerical methods. However, the obtained values do not represent the values for L_1 and L_2. In other words, the mutual inductances between all segments[27] should be taken into account. Two main approaches can be used to deal with the effect of the mutual.

The first one consists of using the birdcage theory [Pascone et al., 1991 Harpen, 1993; Tropp, 1997; Leifer, 1997a], taking into account the complete matrix impedance including all nonzero mutual inductance between all the cells. This approach has the advantage of providing a method that allows the description of any real case. This includes the treatment of all asymmetry in the design, either as a perturbation theory or in solving numerically the complete system of linear equations, as formed by the complete (self and mutual) inductance matrix. A number of simulation programs [Jin, 1999; Giovannetti et al., 2002] have been developed for calculating the birdcage resonant modes with good accuracy, knowing the geometry of the conductive sheets. Eventually, numerical simulations, solving the full Maxwell equations, permit a complete evaluation of the birdcage in presence of both sample and shield.

The second approach [Chin et al., 2002] is probably closer to the practical point of view of the probe designer. In this case, one is interested in estimating the capacitance values that are needed to tune the resonator in one particular mode (generally the homogeneous mode) at a

---

[27] The mutual inductance between the legs and the circular ring segments is zero because they are perpendicular to each other.

given frequency. This provides a (very) good starting point to build and finally adjust a loaded probe.

The key of this approach is to assume a perfect symmetry of the probe and a current distribution on the legs that fits the theoretical periodic distribution, depending on the desired mode. The inductances L_1 and L_2 of the elementary cells are equated to an *effective* inductance that takes into account the self inductance of the isolated pieces of conductor (partial inductance) as well as the mutual inductances. The latter contribution depends on the mutual M_{ij} weighted by the ratio of the currents I_i and I_j on the conductors (see Eq. (8.97) below). Thus, the effective inductance depends on the chosen mode. The required capacitance can then be easily calculated from analytic formulae deduced from the simple theory. Another advantage of this approach is the possibility of dealing with the shield effect by the image theory. Accordingly, each image acts as a new conductive segment that has a mutual with all other segments, reducing the effective inductance of the legs. The position of the images is described by Eq. (8.53). This efficient approach is employed in the Birdcage Builder software from Penn State University [Chin *et al.*, 2002] that has been proved to give valuable results.

Quantitatively, the self inductance of each leg is given by formulae such as Eqs. (A.1) and (A.5), depending on the shape of the conductor (either round wire or flat strip). The mutual inductances between legs, and eventually between the images, are given by Eq. (8.A.14). The self and mutual inductances of the end ring segments are a little bit more difficult to evaluate. One approach is to approximate the ring segment that connects the legs as straight segments. In this case, formulae can be used for the self inductance [Eqs. (A.1), (A.5)] and for the mutual inductance [Grover, 2004]. Another approach is to use FastHenry, already cited in this chapter, or numerical evaluation of the relevant integrals (Chapter 10) as proposed by Giovannetti [Giovannetti *et al.*, 2002]. Our preferred method is to evaluate the self inductance of a given arc of end ring segment by a combination of formulae and numerical evaluation of the Neumann equation (Chapter 10). The latter is also used to evaluate the mutual inductances between all segments (straight or curved).

Whatever the particular method used to evaluate the self and mutual inductance, the effective inductances are given by:

$$L_i^{eff} = L_i + \sum_{\substack{j=1 \\ j \neq i}}^{j=n} M_{ij}\left(I_j/I_i\right), \qquad (8.97)$$

where i is a given conductor of partial inductance L_i, M_{ij} is the mutual inductance between the conductor i and all the n other conductors. I_i and I_j are the current flowing in the conductors i and j, respectively.

Knowing the effective inductance values L_{ER}^{eff} and L_{Leg}^{eff}, eventually taking account of a cylindrical shield, the capacitance values can be obtained from the expressions of the impedance values Z_1 and Z_2.

Neglecting the resistances, for an LP birdcage one obtains:

$$Z_1 = jL_{ER}^{eff}\omega_k \qquad (8.98)$$

and

$$Z_2 = j\left(\frac{L_{Leg}^{eff} C\omega_k^2 - 1}{2C\omega_k}\right). \qquad (8.99)$$

Using Eq. (8.75), C is obtained from:

$$C_{LP} = \frac{\alpha}{\omega_s^2\left(L_{ER}^{eff} + \alpha L_{Leg}^{eff}\right)} \qquad (8.100)$$

where

$$\alpha = 2\sin^2\left(\frac{k\pi}{2N}\right). \qquad (8.101)$$

ω_s is the desired resonance frequency and k is the desired resonant mode.[28] Generally one wants the homogeneous mode, $k = 1$.

For a HP birdcage:

---

[28] Remember that L^{eff} depends on k.

$$Z_1 = j\left(\frac{L_{ER}^{eff} C\omega_k^2 - 1}{C\omega_k}\right) \quad (8.102)$$

and

$$Z_2 = j L_{Leg}^{eff} \omega_k / 2, \quad (8.103)$$

and C is given by:

$$C_{HP} = \frac{1}{\omega_s^2 \left(L_{ER}^{eff} + \alpha L_{Leg}^{eff}\right)}. \quad (8.104)$$

At this point, it is interesting to note that, for a given coil geometry, the tuning capacitance value for the LP birdcage is α times the value for the HP configuration. Because α is a small value, decreasing as the number of legs increases, a HP configuration may generally be preferred. However, when the number of legs is moderate (≤ 12), the dimensions are small, and/or the frequency is low (200 MHz), the LP configuration will be a better choice. As a rule, the configuration should be chosen such that the impedance of the capacitor is comprised between about 20 to 200 Ω. Exceptions can be encountered, however, in which a birdcage design with a large number of legs can be tuned by the distributed capacitances between the end of the legs and the end rings (that act as guard rings).

For a BP birdcage, one of the capacitance values should firstly be chosen, either in the leg or the end-ring branch. The effective "inductance" of the corresponding branch is reduced by the impedance of the capacitance, the new effective inductance being given by:

$$L^{eff} = L - 1/C\omega_s^2. \quad (8.105)$$

The capacitance value for the other branch is finally calculated using the appropriate equations [Eq. (8.77) or (8.78)]. This configuration allows an increase in the number of possible geometries that can be tuned using reasonable capacitance values.

8.2.5 Practical use of the birdcage

8.2.5.1 Tuning the birdcage resonator

The main issue of tuning and matching the birdcage resonator is getting the ideal cosine distribution of currents among its rungs which provides the RF field homogeneity. This can be obtained if the symmetry of the birdcage is satisfied.

The accuracy of the calculated capacitance values depends on the precision with which the effective inductances have been estimated. The method already described provides a sufficiently good starting value for the capacitance, which should be, in any case, adjusted on the final design. Indeed, whatever the accuracy of the inductance calculations, imprecision in the geometry and in the component values are to be expected. Perturbations due to the probe coupling to the spectrometer or to the presence of the sample make adjusting the tuning of the birdcage in its final environment necessary. The simplest method, frequently used, is to adjust two diametrically opposed capacitors (Fig. 8.52).

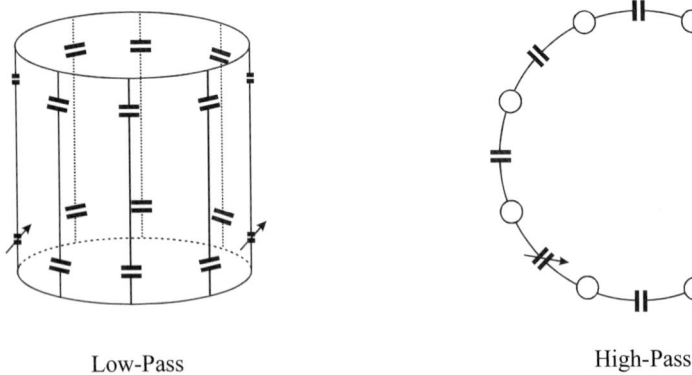

Low-Pass · High-Pass

Fig. 8.52 A simple method to tune a low-pass or high-pass linearly polarized birdcage. For the high-pass case, only one end ring is shown. The legs are perpendicular to the figure plane.

Another capacitive solution, which will be described in detail in the following paragraph, consists of connecting two diametrically opposed points of the resonator circuit by a tuning capacitor (Fig. 8.53).

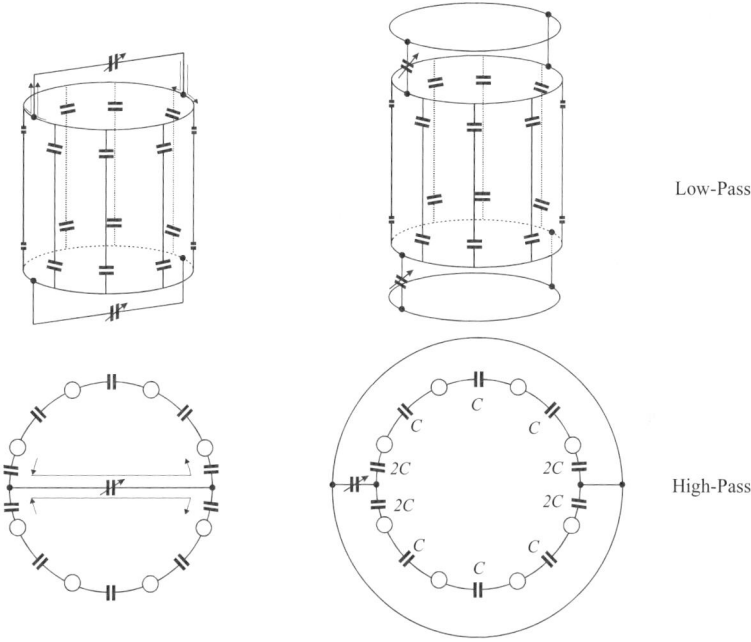

Fig. 8.53 Tuning a linearly polarized low-pass (upper) or high-pass (lower) birdcage by connecting a variable capacitor across diametrically opposed points of an end ring. To preserve the symmetry plane of the birdcage design, the polarization should be perpendicular to the diameter joining the two points. In this case, the current flowing in the tuning capacitor splits equally, as shown by arrows, maintaining a null total current in the corresponding end ring segments, as required for mode $k = 1$. To preserve the longitudinal RF field homogeneity, a tuning capacitor should be connected across both end rings, as shown in the upper right.

Appropriately choosing the location of these points,[29] the initial symmetry plane characteristic of the birdcage design is maintained over a large range of the tuning capacitance value. The RF field homogeneity in the transverse plane is preserved over a broad frequency range. To also

---

[29] This depends on the driving of the birdcage coil.

preserve the field homogeneity in the longitudinal plane, the same tuning circuit should be connected to the opposed end ring (Fig. 8.53). In practice, the connection of the diametrically opposed points is made by adding a second circular loop, parallel to the end ring (Fig. 8.53). A similar solution was proposed earlier, adding a tuned loop in parallel to the end rings of a HP birdcage [Barberi and Rutt, 1993]. The tuned loop is coupled to the corresponding end ring by an appropriate reactive network that allowed a wide tuning range (i.e., from proton to fluorine resonance).

"Inductive" solutions have been proposed too. These methods consist of changing the effective inductance of the legs. This can be done using a moveable conductive cylinder, axial to the birdcage resonator [Dardzinski *et al.*, 1998]. The cylinder (Fig. 8.54) acts as a partial shield, providing the required inductance modification but at the expense of the B_1 distribution change in the longitudinal plane.

The tuning range obtained for a shield diameter of 1.5 times larger than that of the birdcage was 20 MHz for a 400 MHz resonance frequency. A shield to coil diameter ratio of 1.3 provides a larger tuning range, up to 40 MHz, but with the consequence of B_1 field homogeneity and amplitude degradation [see Eq. (8.55)].

Fig. 8.54 A moveable shield changes all the leg inductance values, hence the resonant frequency, while keeping the cylindrical symmetry.

Another solution has been proposed in which each leg is replaced by a resonant rectangular loop [Wen *et al.*, 1994; Fakri *et al.*, 1996]. The birdcage-like structure is obtained by arranging identical loops around a cylinder (Fig. 8.55). The propagation along the ladder network is realized by the inductive coupling between each loop, without physical connections. Tuning the resonant modes of the structure is achieved by modification of the mutual inductance, hence by simultaneously rotating the rectangular loops around their long axis.

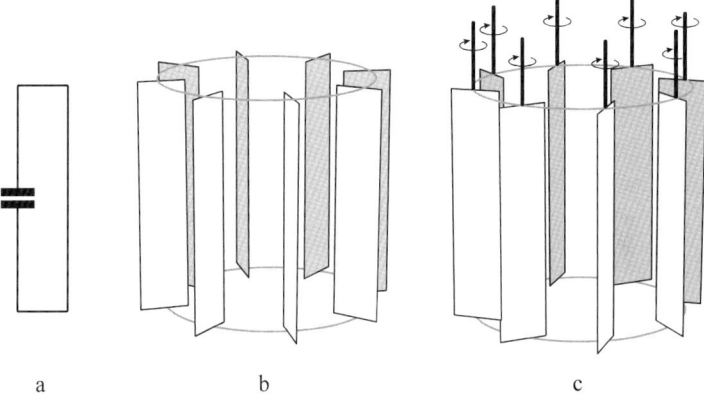

a b c

Fig. 8.55 The "free element" birdcage design [Wen *et al.*, 1994]. Each "leg" is a resonator (a) that couples through inductive interaction to the other ones (b). The birdcage resonant modes are obtained through this coupling. The rotation of each resonator plane (c) changes the mutual inductance, hence the resonance frequency. If all resonators are rotated simultaneously, the cylindrical symmetry is kept, hence the magnetic field homogeneity is preserved. This design functions in a similar manner to the TEM resonators described below.

The resonant frequency of the homogeneous mode is given by:

$$\omega = \omega_0 \frac{1}{\sqrt{1 + \sum_{n \neq 0} \frac{M_n}{L} \cos(\pi n / N)}}, \quad (8.106)$$

where ω_0 is the resonant frequency of the isolated rectangular loop and L its inductance. M_n is the mutual inductance between the rectangular loop

n and an arbitrary one, used as reference ($n = 0$). This expression is obviously valid in the case of perfect cylindrical symmetry only.

8.2.5.2 Asymmetry effects

In spite of all care taken in practical realization of the birdcage, the imprecisions in capacitance values and in the resonator geometry inevitably lead to asymmetry. In these cases, resonant spectra become more complicated than expected (Fig. 8.56).

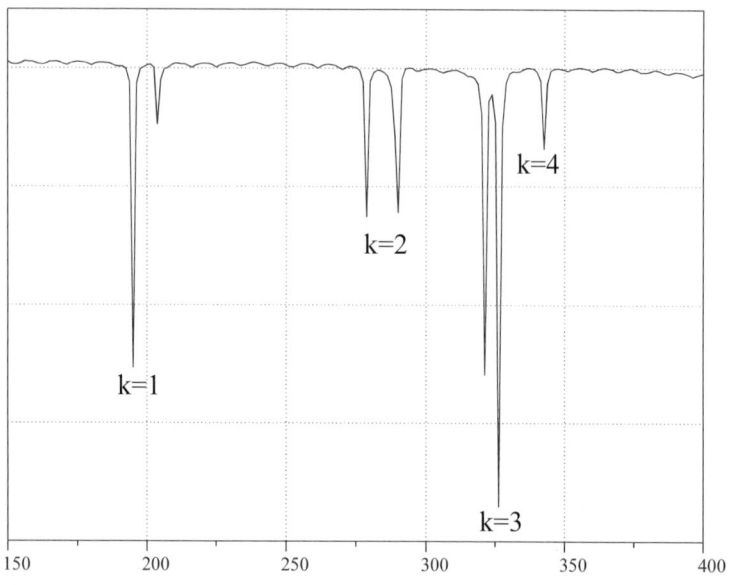

Fig. 8.56 Resonant spectrum of an eight leg birdcage that has been perturbed on purpose.

Joseph *et al.* [Joseph, *et al.*, 1989] and Tropp [Tropp, 1989; 1991] have analyzed the effect of component error on the resonant spectrum of both the HP and LP birdcage, using a theoretical approach based on the perturbation theory. The general result is the splitting of the degenerate resonant modes. The split modes producing a particular field polarization are excited depending on the driving geometry. This geometry dependence results in a disconcerting change of the resonant spectrum when moving a coupling loop around the coil.

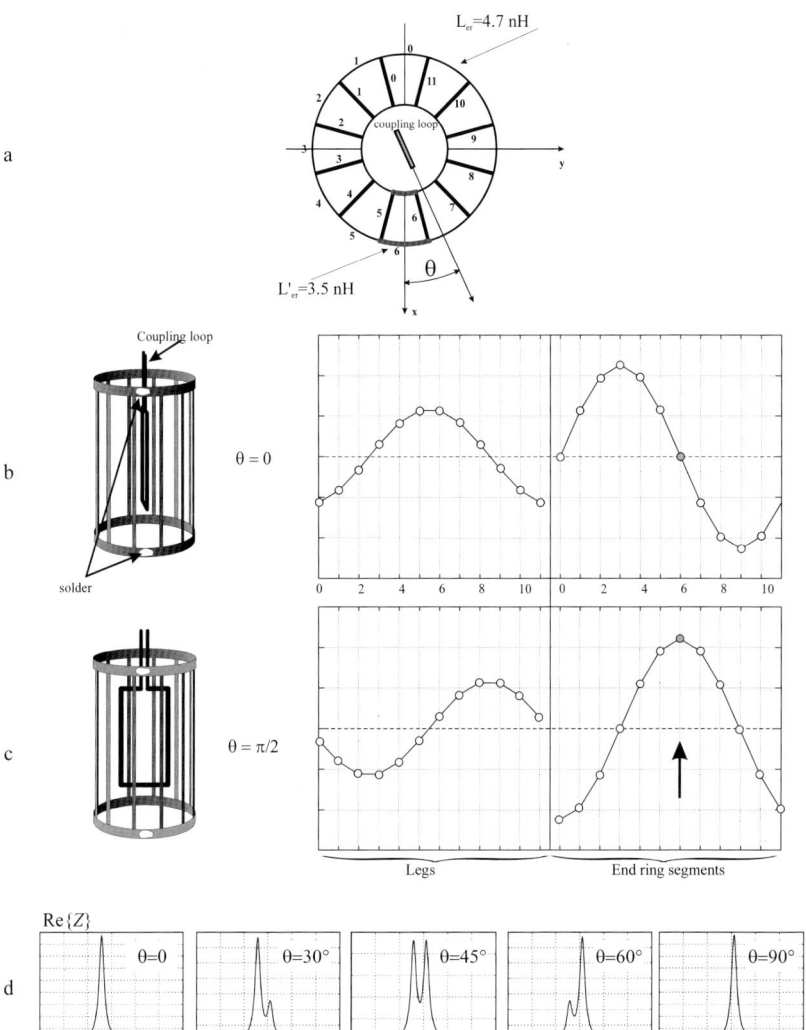

Fig. 8.57 Effect of a local perturbation on the resonant spectrum (near mode $k = 1$) of a 12 leg birdcage (a) as a function of a coupling loop orientation (b,c). Pure modes are excited if the coupling loop is oriented respective to the perturbed segment as indicated (b,c). Notice the current in the perturbed end ring segment indicated by an arrow. Impedance changes relative to the coupling loop orientations are shown in d.

In order to illustrate this effect, we have chosen a 12 leg birdcage, carefully built with copper tape glued on a cylinder. The resonance spectrum is recorded on an RF workbench using an inductive coupling loop placed in the coil center (Fig. 8.57 and connected to a network analyzer.

The impedance that appears on the loop terminals is the resonator impedance multiplied by a factor that depends on the coupling strength (Chapter 4, Fig. 4.20). We are particularly interested in the $k = 1$ mode, appearing near 200 MHz for the shown design. If the resonator were perfectly symmetric, the resonant frequency would not change during rotation of the loop. In fact, two distinct resonant modes are to be observed, depending on the rotation angle of the loop (Fig. 8.57). A "pure" mode is obtained when the coupling loop is oriented at particular angles, 90° apart. For intermediate positions, two modes are simultaneously excited and visible on the resonant spectra.

This effect is attributed to a local inductance change of the end ring segments due to the soldering metal that closes the end ring circuit. In this place, the thickness of the segment increases, resulting in a lower inductance. The resonant spectrum obtained in these conditions can be easily simulated, by linear circuit analysis, using the values quoted in Fig. 8.57(a). In this case, the two modes were separated by about 1 MHz, but, in practice, much larger splitting may be observed.

The effect can be well understood qualitatively. When the coupling loop is oriented at $\theta = 0$, the perturbed inductance of the end ring segment has no effect on the resonance because, due to symmetry, there is no current in the corresponding branch [Fig. 8.57(b)]. The resonant frequency mode is that of the perfect birdcage, thus producing a homogeneous magnetic field. When the coupling loop is rotated by 90°, the excited $k = 1$ mode implies a large current in the perturbed end ring segment. Because its inductance is lower than it should be, the resonant frequency slightly increases, as shown in Fig. 8.57(c). In this situation, one expects a slight distortion of the magnetic field distribution. When the position of the loop is placed in-between, both resonant modes are excited and appear on the resonant spectrum [Fig. 8.57(d)]. A similar effect was already described while perturbing one capacitor of an LP birdcage [Tropp, 1989].

Coping with the birdcage dissymmetry means finding a coupling geometry that favors one pure resonant "$k = 1$" mode at the desired frequency. When the dissymmetry is well controlled, eventually intentionally introduced to create either a given field polarization or double tuning design (see below), the split modes still exhibit the cosine current distribution, at least to the first order of the perturbation [Joseph et al., 1989; Tropp, 1989].

At this point it should be mentioned that the dissymmetry described here as an example can be easily diverted by cutting the corresponding end ring segments. In this case, the polarization of the homogeneous $k = 1$ mode is forced with B_1 aligned along the axis perpendicular to the diameter joining the two lacking ring segments (along y in Fig. 8.58). The coupling geometry must then be chosen accordingly.

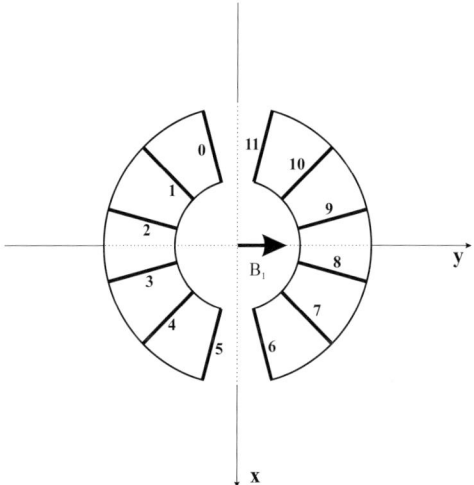

Fig. 8.58 Field polarization obtained when cutting diametrically opposed end ring segments.

Finally, the splitting of the resonant $k = 1$ mode may be corrected by two adjustable capacitors placed at an azimuth angle differing by $\pi/4$ [Tropp, 1991]. The corrective capacitance values should however remain within 7 to 8% of the nominal birdcage capacitance value. This correction, eliminating any specific field polarization, is very useful for adjusting the isolation between the quadrature ports of a circularly polarized birdcage (Fig. 8.59).

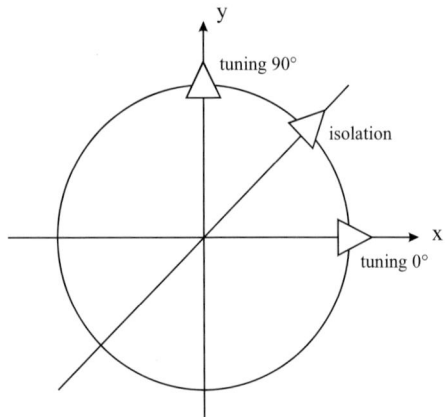

Fig. 8.59 Positions of tuning and correction capacitors in a quadrature birdcage design. The "isolation" capacitor corrects for any parasitic linear polarization of the magnetic field introduced by an occurring dissymmetry.

8.2.5.3 *Interfacing the birdcage to the spectrometer*

The coupling of the birdcage resonator to the spectrometer through an impedance matching network must fulfill at least two conditions. Firstly it should introduce the least possible dissymmetry. Secondly it should excite the desired field polarization.

The simplest method is probably the inductive coupling. This can be realized using a rectangular coupling loop conveniently oriented against the resonator legs (Fig. 8.60). If the birdcage has been designed with a perfect cylindrical symmetry, the position of the coupling loop can be chosen arbitrarily and will determine the final magnetic field polarization. On the contrary, if the resonator design is not symmetric (for example in intentionally cutting end ring segments), as the field polarization is already defined, the coupling loop position must be defined accordingly. In all these cases, the matching procedure consists of adjusting the coupling by a mechanical displacement of the loop, or preferably, using a tuned coupling loop. The inductive coupling has the advantage of providing an automatic electrical equilibration of the probe resonator and being easy to build. On the other hand, it may create some distortion of the B_1^+ field distribution and may sometimes be encumbering, hence a (balanced) capacitive coupling has also frequently been chosen.

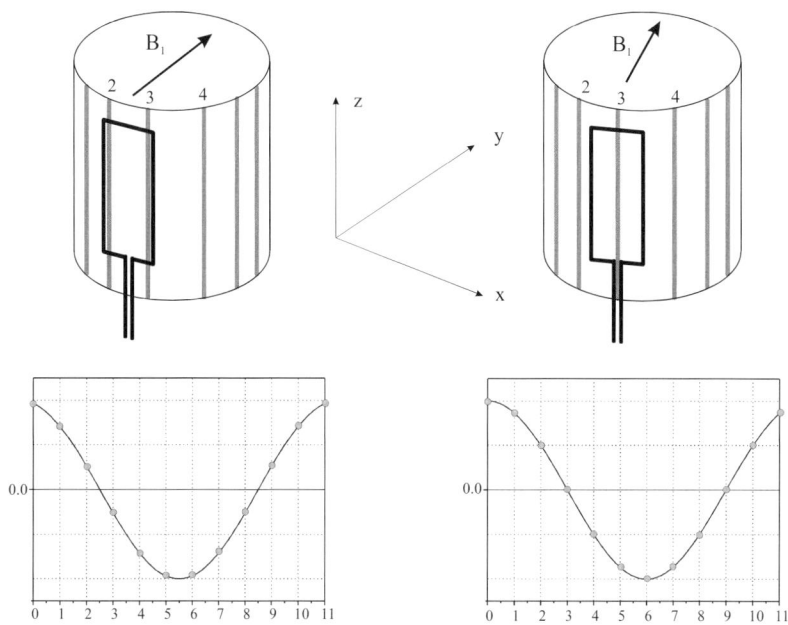

Fig. 8.60 Current distribution on the birdcage legs and magnetic field polarization as a function of the matching coupling loop position. Left: legs 2 and 3 are excited with equal currents of opposed phase. Right: legs 2 and 4 are excited with equal currents of opposed phase, hence leg 3 has a null current.

Capacitive coupling can be realized in a variety of ways, depending mainly on the type of birdcage (HP or LP). In this case, it is almost impossible to keep the original perfect cosine current distribution among the legs, but it can be approached very closely. The capacitive coupling may use either lumped capacitors or take advantage of the distributed capacitances existing between the legs and the shield cylinder. Some examples are given in Fig. 8.61.

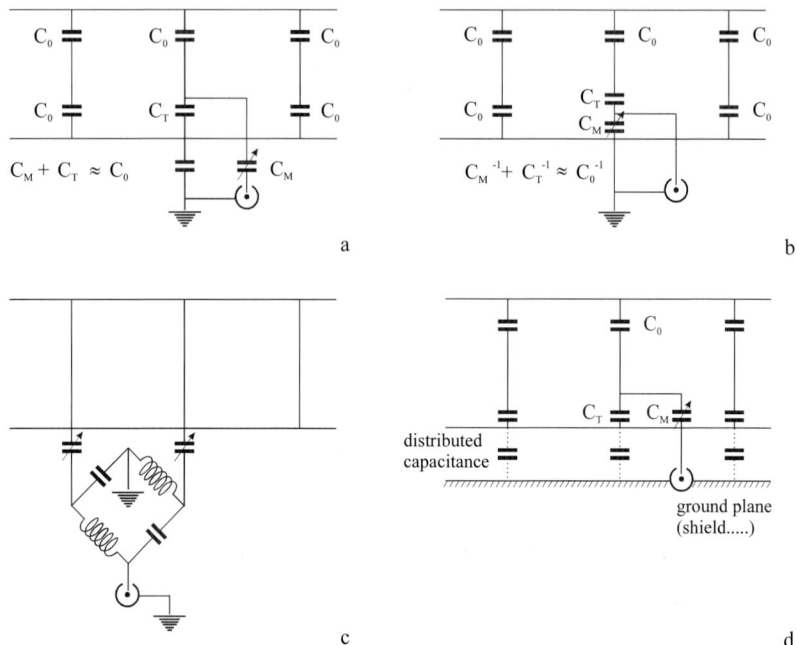

Fig. 8.61 A few capacitive coupling schemes of the birdcage. Upper: one leg of a low-pass birdcage is driven by a classical capacitive matching network [(a) Fig. 4.8 and (b) Fig. 4.5 of Chapter 4]. A symmetric coupling is shown in (c) [Barberi *et al.*, 2000] using an LC balun. It applies for all types of birdcage design. The scheme shown in (d) benefits from the distributed capacitance existing between the birdcage conductors and the shield (ground plane). These capacitors may be replaced by discrete elements.

8.2.5.4 *A practical design*

A simple and easily reproducible LP birdcage design, producing a linearly polarized RF field, is described here as an example [Lupu *et al.* 2004]. The resonator has 12 legs, providing a reasonably good homogeneity. It is designed for small animal imaging, the useful sample volume being about 50 cm^3. The resonator is tuned to 200 MHz, but prototypes of the same dimensions have been successfully built for a 7 tesla imager (300 MHz), still the LP type. For higher frequencies, a HP design is preferable.

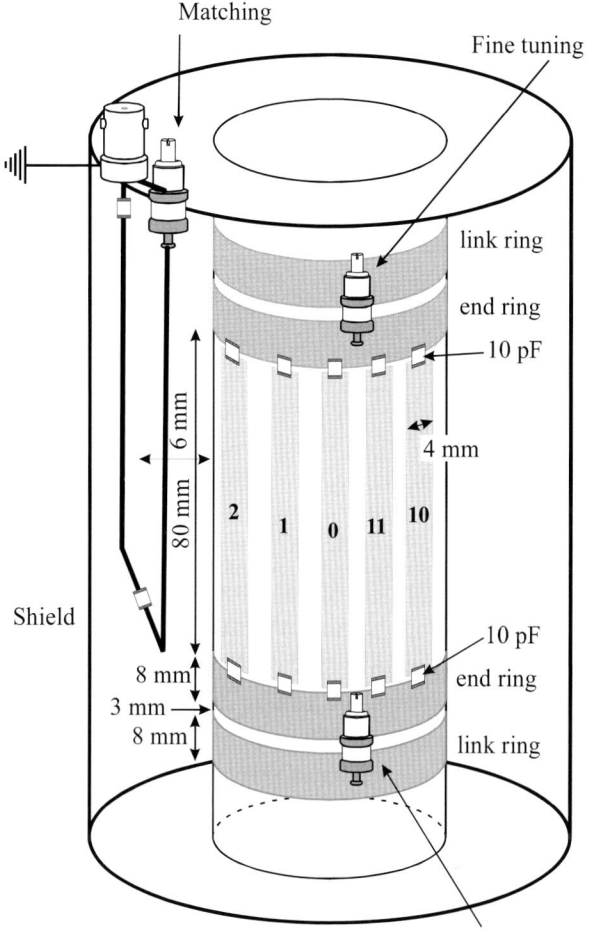

Fig. 8.62 Design of the 12 leg low-pass birdcage, tuned at 200 MHz. The coil diameter is 52 mm and the shield diameter is 80 mm. The rectangular coupling loop (100 × 15 mm) is positioned right above legs 2 and 3 (legs are numbered from 0 to 11). It is segmented, as shown, with 10 pF capacitors. The rotor of the matching capacitor (adjustable 25 pF) is connected to the ground. The tuning and field compensation capacitors are adjustable 25 pF. They are balanced by 26.7 pF (22 + 4.7 pF) ship capacitors soldered diametrically on the rear side.

The coupling of the birdcage to the spectrometer is done by a tuned inductive loop. The fine tuning adjustment is achieved using the method already outlined in Fig. 8.53. It allows the original cosine current distribution to be kept within the birdcage legs for a reasonably large tuning range. From the point of view of the experimenter, the coil is easy to use, having only two capacitors to adjust (one for the matching and one for the tuning). The adjustment accommodates a large range of sample loading conditions.

The final shielded birdcage resonator has a similar $B_1/\sqrt{P}$ ratio to the (unshielded) slotted tube resonator already described in Fig. 8.34. The two resonators have the same diameter, but the homogeneous field volume is larger for the birdcage design, particularly in the longitudinal direction (the birdcage is longer than the slotted tube).

The coil design and dimensions are shown in Fig. 8.62. The resonator has a 52 mm diameter and it is enclosed in an 80 mm diameter shield cylinder. Using Eq. (8.55), the magnetic field amplitude at the center per unit current is expected to be reduced by a factor of 0.58 compared to an unshielded coil. Assuming that the effective length of the coil is equal to the distance joining the median line of the end rings (we neglect for the moment the effect of the link rings, Fig. 8.62), the rotating frame component of the magnetic field is given by [from Eq. (8.64)]:[30]

$$B_1^+ = 0.0377 \cdot I \text{ gauss}, \quad (8.107)$$

where I (in A) is the *peak amplitude* of the *total* current flowing in the half cylinder sheet. For a 12 leg birdcage, I is related to the current amplitude of the cosine function [Eq. (8.60)] by:

$$I(12) = 0.259 \cdot I. \quad (8.108)$$

The tuning capacitance at 200 MHz is estimated to be 5.11 pF, using, for example, the Birdcage Builder software [Chin *et al.*, 2002]. Assuming a Q factor of 340 (as measured on the empty probe), R_{eq} is equal to 2.75 Ω [Eq. (8.93)]. With a 10 W transmitter power, the peak

---

[30] ζ is equal to 0.776 for a 12 leg birdcage.

amplitude $I(2N)$ of the current [Eqs. (8.82) and (8.83)] is estimated to be equal to 2.7 A, in conditions of perfect impedance matching. The total peak current I in one half cylinder sheet is therefore equal to 10.4 A [Eq. (8.108)]. Finally, the magnetic field at the center (and in all the sample volume, the magnetic field being homogeneous) is $B_1^+ = 0.39$ gauss [Eq. (8.107)]. It corresponds to a proton precession frequency in the rotating frame of 1671 Hz. In other words, the expected 90° pulse width that will be obtained with that probe is 149 µs, for 10 W.

Experimentally, the measured Q factor of the final probe, loaded with a conductive water solution of sodium chloride decreases to 150, mostly due to the magnetic losses. Hence, the 90° pulse width, which is proportional to the square root of Q, is expected to be 225 µs for a 10 W input power. We measured 200 µs on this sample. Part of the difference may be due to errors in estimating the power delivered by the transmitter. Another source of errors might be an overestimation of the shield effect on the contribution to the magnetic field generated by the end ring currents. Indeed, the reduction factor of Eq. (8.55) was calculated considering the leg currents alone. It should be smaller for the end ring contribution. The agreement is nevertheless very satisfactory owing to the simplicity of the calculations. In any case, the tools developed here should help in estimating the final behavior of the probe. This is a prerequisite estimation that allows its evaluation and especially the transmitter power requirements.

Fine tuning the birdcage is done using the method outlined in Section 8.2.5.1. In this method, the tuning capacitor connects two diametrically opposed points located on the end ring segments where the total current is zero when the homogeneous field $k = 1$ mode is excited. Hence, the correct functioning of the tuning procedure requires that the leg current phase ϕ is settled up conveniently. This is done through an appropriate inductive coupling, using a rectangular loop placed symmetrically above legs 2 and 3 (or legs 8 and 9) where there is a change of phase in the current (change of sign). Alternatively, the resonator may be excited through capacitive coupling.

To understand the tuning method proposed here and its limitation, it is interesting to have a look at the current values in the different conductive parts of the birdcage. Rather than attempting an analytical calculation of the new circuit, it is simpler to do a numerical analysis on the network of Fig. 8.63, using the linear circuit simulator described in Chapter 3.

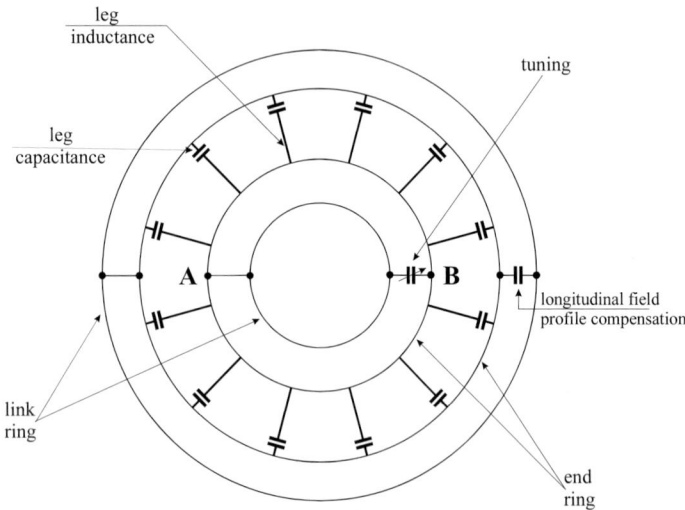

Fig. 8.63 Schematic circuit of a 12 leg tunable low-pass birdcage used in the simulation. The leg and end ring segment inductance are equal to 46 and 3.77 nH. The inductance of half the link ring is 31.8 nH. The end ring segments connecting the tuning and compensating capacitors are split into two equal inductances of 1.39 nH. The leg capacitance value is 5.22 pF for the on resonance (200 MHz) and 4.3 pF for off resonance (220 MHz) birdcage. The required tuning capacitance value to force the 220 MHz birdcage to 200 MHz is equal to 27.45 pF for the non-compensated configuration. When the field profile compensation is activated, the required tuning and compensating capacitance values are equally set to 21.96 pF. The resistance values of the legs and end rings are chosen such that the resonator Q value is 340. For the purpose of the simulation, the capacitor diametrically opposed to both the tuning and compensating capacitors, existing in some designs, is replaced by a short circuit.

The complete inductance matrix (including self and mutual) was introduced, assuming the geometry described in Fig. 8.62.

Starting with a birdcage initially tuned to 220 MHz, its resonance is shifted down to 200 MHz by a single capacitor connected across one of the end rings. Fig. 8.64 shows the corresponding current in the legs and in the end rings. They are compared to those calculated for an ideal birdcage of the same geometry but already tuned to 200 MHz. Both resonators are assumed to be impedance matched by a tuned coupling loop and driven by a constant power supply (10 W).

As a result, the leg currents of the "forced" birdcage decrease slightly, but the correct cosine dependence is maintained.

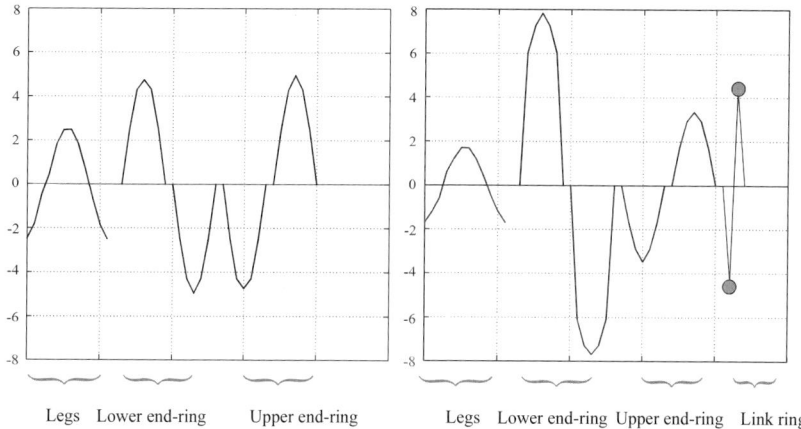

Fig. 8.64 Currents (A) in the legs and end rings for the ideal (left) and forced (right) birdcage. Note the higher current in the end ring which is connected to the tuning capacitor.

The most striking effect is a large current increase in the end ring (Fig. 8.64) where the tuning capacitor is connected, leading to a distortion of the magnetic field distribution along the resonator axis (Fig. 8.65). This can be easily corrected by symmetrically connecting an identical capacitor on the opposed end ring (Fig. 8.66). In practice, the equilibrating capacitance value is not too critical. It can be adjusted only once, before the routine use of the resonator.

This tuning method easily allows a 10% tuning range without gross perturbation of the magnetic characteristics of the probe. This range is sufficient for shifting the resonator frequency from proton to fluorine, for example. In theory, a larger frequency shift can still be obtained, but at the price of a large reduction of the magnetic field generated by the legs.[31]

---

[31] A moderate deviation of the current ratio from the ideal cosine dependence will also be observed.

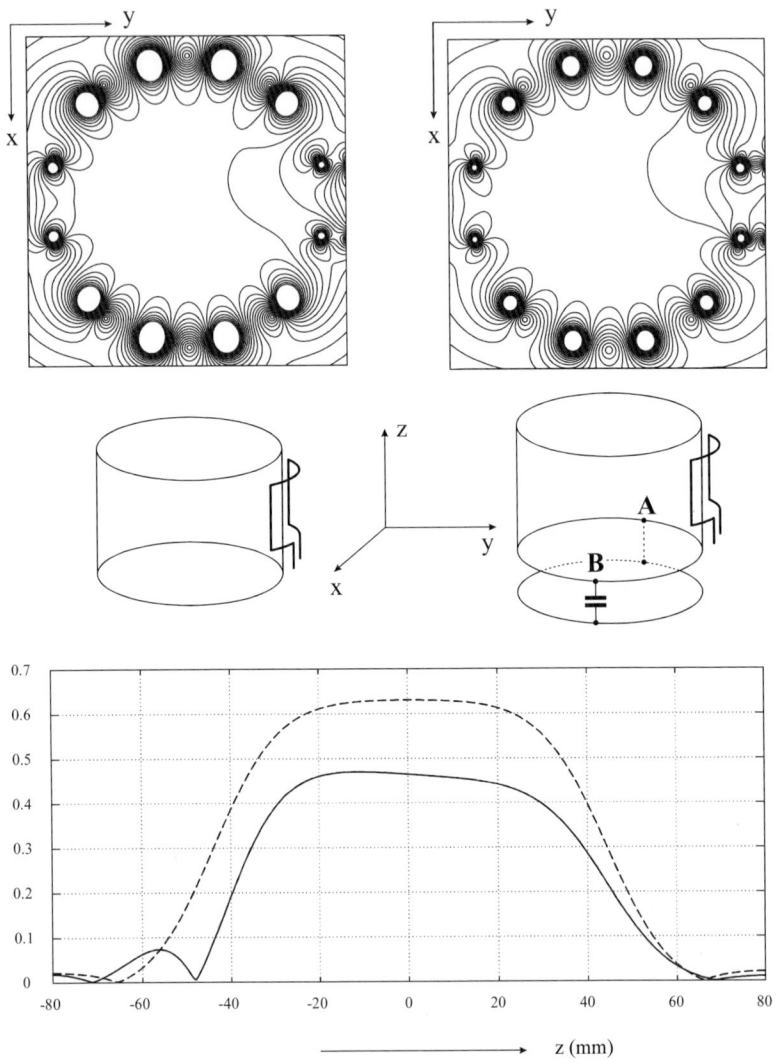

Fig. 8.65 B_1^+ field map (upper) in the transverse plane for the ideal (right) and forced (left) birdcage. The field distortion observed in the map (at the right) is due to the coupling loop. Although the field amplitude at the center is lower for the forced birdcage, the homogeneity in the transverse plane looks very similar. The decrease of the field amplitude along the longitudinal axis for the forced birdcage (full line, lower) is due to the imbalance of the end ring currents.

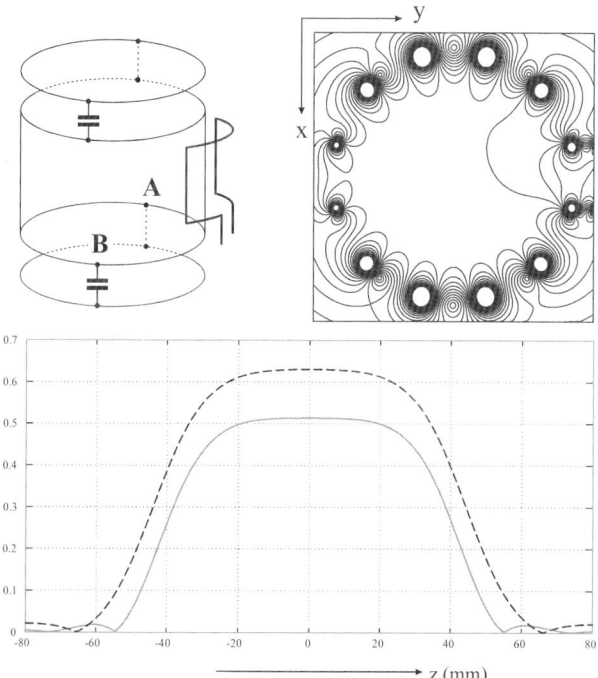

Fig. 8.66 Correction of the field profile along the resonator axis (lower) by addition of a second link ring and capacitor to equilibrate the end ring currents of the birdcage.

This is due to the fact that the CR (and, eventually, the AR) resonance mode of the end rings is favored, as the tuning capacitor is increased, while the current across the legs is diminishing. Indeed, as the frequency lowers, the capacitor impedance of the legs increases to the point that they finally behave as insulators supporting the end ring!

Until now, we have considered only the case of a tuning capacitor that forces the birdcage to a lower frequency than the one for which it was initially designed. In fact, the resonator frequency can be shifted up or down, depending on the nature of the impedance connected to points A and B. In contrast to a capacitive impedance that has the effect of lowering the frequency, an inductive impedance has the effect of increasing the resonant frequency. For both cases, the frequency shift increases as the tuning impedance decreases. In the present design, the reactive impedance between points A and B is due to the inductance of the link ring in series with the tuning capacitance (Fig. 8.67). Therefore, the tuning reactance may be either inductive or capacitive depending on the relative impedance values of both components. As a result, the tuning capability of the resonator may be changed according to the design of the link ring. Furthermore, each link ring contributes to the magnetic field distribution along the cylinder axis in addition to the end ring contribution and should be taken into account at the optimization stage of the probe design.

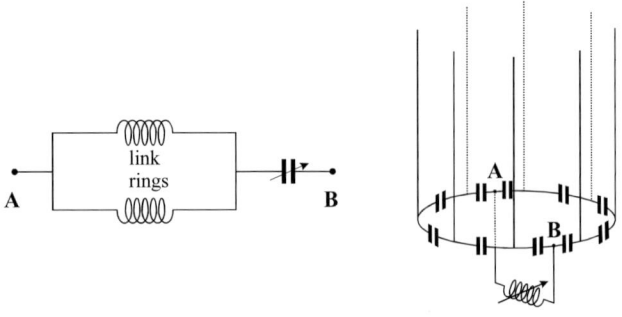

Fig. 8.67 The tuning reactive impedance that links point A and B on the end ring birdcage (left). It can be used as a variable inductor as required for a HP birdcage (right).

This tuning method, applied to the LP resonator, can be similarly adapted to a HP birdcage design. In this case, it is preferable to tune the $k = 1$ mode of the initial birdcage at a lower frequency than desired. The shift to a higher frequency will be obtained using a variable inductive reactance. This can still be realized with a variable capacitor in series with the fixed inductance link rings (Fig. 8.67).

8.2.6 *Practical design of a quadrature birdcage*

The $k = 1$ mode of an ideal birdcage resonator is intrinsically degenerate. The corresponding magnetic component of the RF field is the combination of two orthogonal linearly polarized magnetic fields oscillating in quadrature. This leads to the well-known "circular polarization" which is a property of the birdcage resonator. This permits the design of a highly efficient probe for NMR.

To build a quadrature probe implies firstly that the birdcage has the required symmetry, keeping the degenerate characteristic of the $k = 1$ mode. Secondly, the two components of the magnetic RF field must be properly excited, at 90°.

The following birdcage coil design is a small probe, specifically adapted for small animal imaging at a moderately high field (4.7 T). The dimensions of the probe resonator are given in Fig. 8.68. The small size of the probe permits a classical quasi-static approach for its design. Similar probes can be built without difficulties for a higher field, up to 7 T (300 MHz).

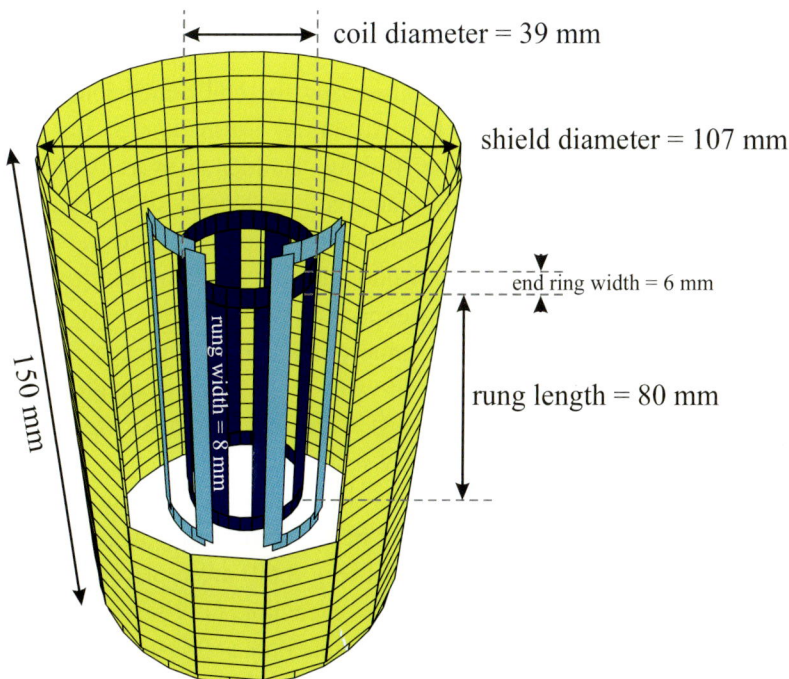

Fig. 8.68 Dimensions of the shielded birdcage coil. The coupling loops (105 × 20 mm) are made of coper strips (width = 6 mm). The distance between the coupling loop and the birdcage is 6 mm.

A design with eight rungs was chosen as it represents a good compromise between sensitivity, RF magnetic field homogeneity, and ease of construction. The resonator will be inductively coupled providing a simple and efficient balancing of the coil. The coupling loops could be fitted easily in the available space between the coil and the shield.

The birdcage was shielded [Hayes and Each, 1997; Rzedzian and Martin, 1993; Alecci and Jezzard, 2002; Spence and Wright, 2003; Weyers and Liu, 2006] to be conveniently decoupled from the gradient coils and isolated from RF interferences. The shield was split along the cylinder suppressing the eddy currents induced by the gradient coils. RF continuity was maintained with capacitors linking the shield strips.

8.2.6.1 *Designing and adjusting the birdcage resonator for quadrature operation*

A HP design has been chosen considering the coil dimensions and the operating frequency (200 MHz). An initial estimation of the tuning capacitances is obtained with the Birdcage Builder software [Chin *et al.*, 2002] giving the coil and shield diameters, the lengths of the rungs, and the diameter or width of the conductors. The calculated capacitance proves to be a valuable starting value, generally leading to a resonance frequency slightly higher than the expected one.

As a result, the capacitance value predicted by Birdcage Builder is 25.5 pF. It is close to the calculated value of 25.25 pF as predicted using the program NMRP (Chapter 3). The capacitance used in the experimental design was 26.7 pF (22 + 4.7 pF).

Before connecting any matching circuit, it is recommended that the coil is tuned such that the two modes become effectively degenerate. The capacitors on three consecutive rings (Fig 8.69) are replaced by a fixed 22 pF in parallel with a variable capacitor (1–10 pF, Voltronics Corporation NMAT10HVE). The two capacitors at 90° tune the so-called I and Q modes and the capacitor in the middle (at 45°) adjusts the isolation between the resonant modes. These variable capacitors are initially set midway.

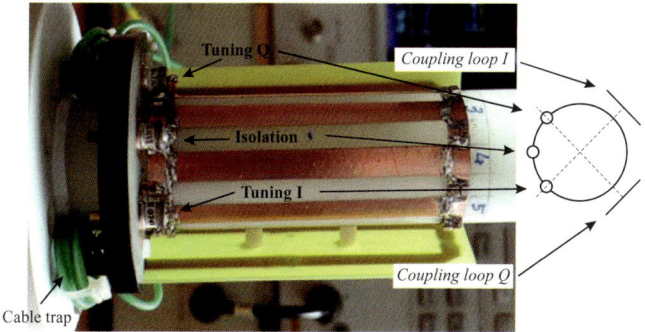

Fig 8.69 Localization of the tuning and isolation capacitors relative to the coupling loops.

The adjustment is straightforward using the setup shown in Fig 8.70. A small shielded loop (Chapter 10) is positioned at the center of the coil with its axis perpendicular to the birdcage symmetry axis. The $k = 1$ resonant mode is the only one producing a nonzero RF magnetic field in the center of the coil. In this way, the test loop is sensitive only to this very mode. The plane of the test loop is oriented approximately parallel to the mesh containing the Q (or I) tuning capacitor (Fig. 8.78). In this way, the loop detects the magnetic field component oriented perpendicular to the mesh which is expected to be a pure Q (or I) mode. The test loop is connected to port 1 of a network analyzer (S_{11}) or any similar device able to detect energy absorption.

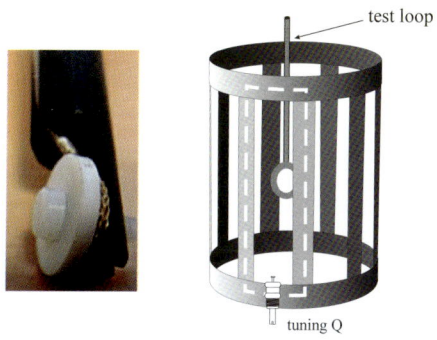

Fig. 8.70 Experimental setup for detecting and adjustment of the $k = 1$ mode of the birdcage.

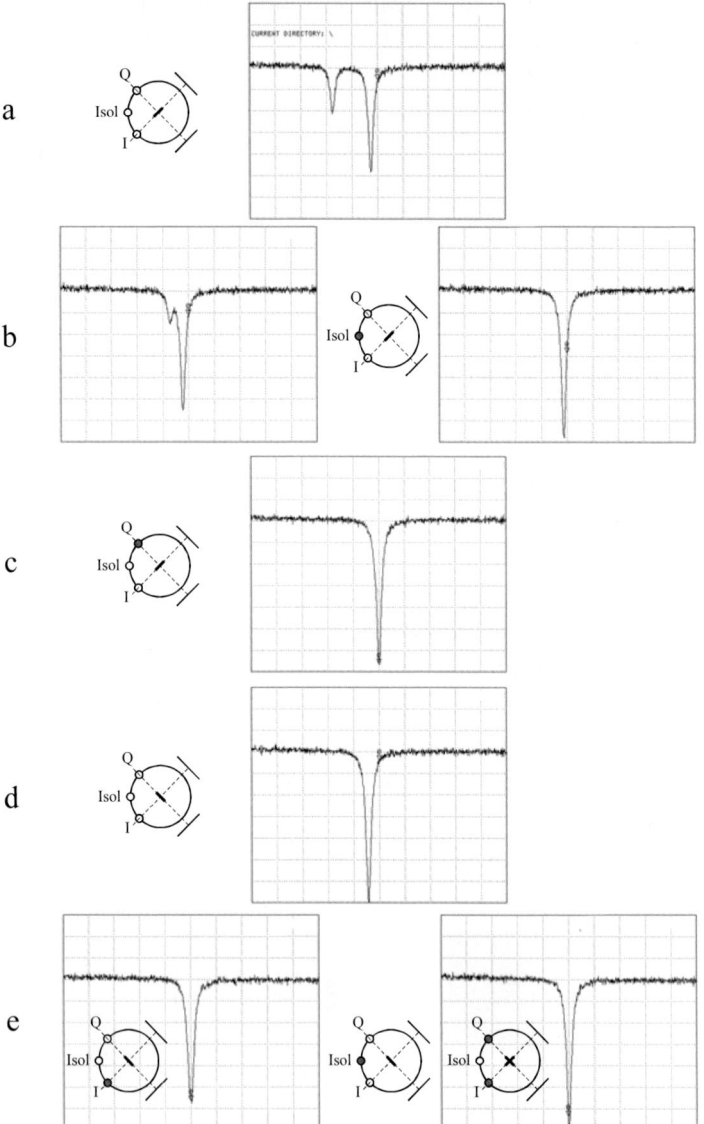

Fig. 8.71 Reflection coefficient (S_{11}) seen by the small test coil during the adjustment of the quadrature birdcage. The capacitor in use is outlined in grey. The orientation of the test coil is figured by the black rectangle in the coil center.

Fig. 8.71(a) shows a typical scattering spectrum that would be observed before any adjustment. Two peaks are visible in a similar manner as already described in Section 8.2.5.2, indicating that the two $k = 1$ modes resonate at different frequencies. At this point, an attempt to adjust both modes to the same frequency results in a displacement of the two resonances.

For the chosen direction, it is a simple matter to obtain a pure Q mode (and consequently a pure I mode in the orthogonal direction) by firstly adjusting the "isolation" capacitor. The "parasitic" resonance disappears progressively leaving only one peak in the S_{11} spectrum [Fig. 8.71(b)]. At this time, it can be verified that adjusting the I capacitor has only a small effect on the S_{11} spectrum. It is now possible to precisely adjust the Q mode to the target operating frequency [Fig. 8.71(c)].

Rotating the pickup coil by 90°, one should observe only one peak, generally at a different frequency [Fig. 8.71(d)], corresponding to the I mode. Its resonance is now adjusted to the target frequency, using the I capacitor.

At this point, the rotation of the test loop around the birdcage axis must leave the S_{11} spectrum almost invariant, indicating that the RF field is circularly polarized. However, one could observe a small variation of the peak amplitude. This may be corrected by a slight readjustment of the isolation capacitor. Generally, this results in a dipper peak in the S_{11} spectrum [Fig. 8.71(e)]. The last adjustment of the isolation capacitor will be made on the final setup.

8.2.6.2 *Interfacing the resonator to the console*

The previous adjustment permits perfectly orthogonal resonant modes to be obtained and a completely independent adjustment of the I and Q resonant modes. The matching network can now be connected to the resonator. After connection, especially using a capacitive coupling, the corresponding I and Q capacitors must be readjusted to recover the desired operating frequency. The final adjustment of the isolation capacitor will be done after properly tuning and matching the two ports of the birdcage. The corresponding scattering parameters are shown in Fig. 8.72.

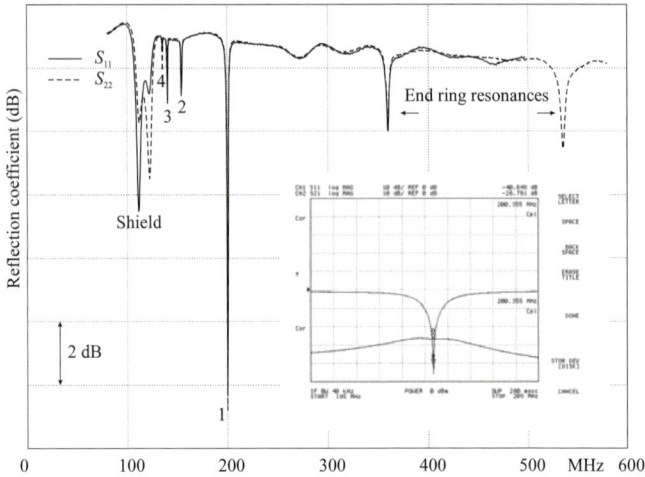

Fig. 8.72 Typical S-parameters of a birdcage coil showing all resonant modes. The insert shows an extended view of S_{11} and S_{12} close to the operating resonant mode (200 MHz, $k = 1$).

The isolation between ports is around 25 dB. This is a typical figure that could be obtained only after insertion of a broadband cable trap (choke inductance). This trap is realized by winding four turns of the coaxial cable (visible in Fig. 8.69). Without the cable trap, the isolation between ports decreased to about 15 dB for this design.

It is worth mentioning here that a "ground" connection to the cable braid, just before the cable pops out of the birdcage, allows elimination of most of the issues related to parasitic currents that may enter the probe resonator (including its shield).

Fig. 8.73 Interface circuits of the birdcage. The cable braid connections are shown by arrows.

This "ground point" should be tightly connected to the shield (Fig. 8.73).Finally, a quadrature hybrid should be associated with the coil. A device like the one already described in Chapter 5 (Fig. 5.12) was used with this probe.

8.2.6.3 Design of the shield

The shield [Fig. 8.74(a)] is constituted of eight thin (35 µm) strips aligned along the cylinder axis [Alecci and Jezzard, 2002]. This arrangement minimizes the eddy currents produced by the pulsed gradients. To maintain the continuity of the shield at RF, a number of capacitors (15 × 8) are soldered along the slits. This shield constituted a resonator, having its own resonant modes [Fig. 8.74(b)]. Using 1 nF capacitance, these modes can be set outside the resonant spectrum of the birdcage.

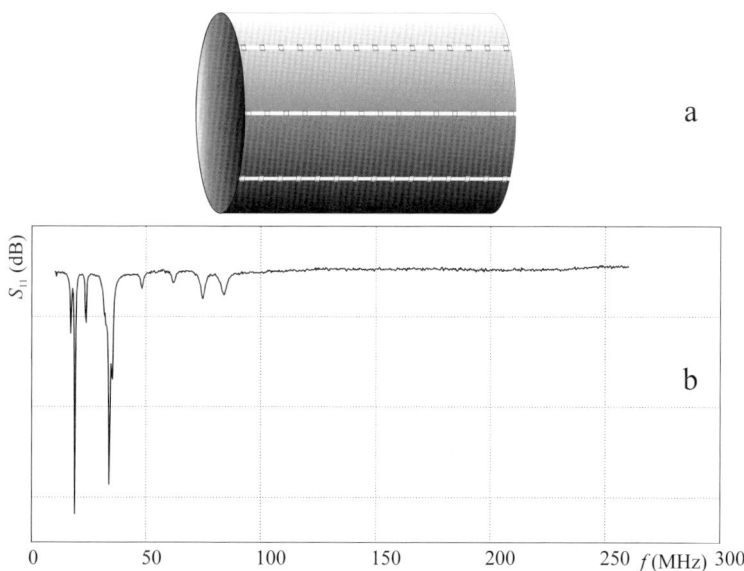

Fig. 8.74 (a) The split shield is transparent for the gradient pulses and opaque for RF. The shield continuity at RF is given by a set of 15 × 1 nF capacitors (case 1210) soldered along the slits. (b) The resonance spectrum of the shield, measured between 10 and 300 MHz with a pickup coil connected to port 1 of a network analyzer (S_{11}).

8.2.6.4 Evaluation of the probe

The port assignment of the 90° hybrid can be easily determined by recording an image (or a signal) while exchanging the TX and RX connections to the console or exchanging the I and Q connections to the probe. Obviously, a brighter image will be obtained with the correct assignment. It should be noted at this point that this depends on the static magnetic field orientation, on the probe orientation within the magnet, and on the nucleus under investigation.

The RF magnetic field B_1^+ for this probe was estimated, from quasi-static simulations, to be about three times larger (Fig. 8.75) than the linearly driven 12 rung birdcage described before. There are at least three factors that explain this difference: a smaller coil diameter (× 1.33), a higher shield to probe diameter ratio (× 1.5) and the quadrature driving (× 1.41). The improvement of the amplitude of the magnetic field per unit current created by a 12-rung birdcage as compared to an eight-rung is negligible (× 1.01, from Eq. 8.62).

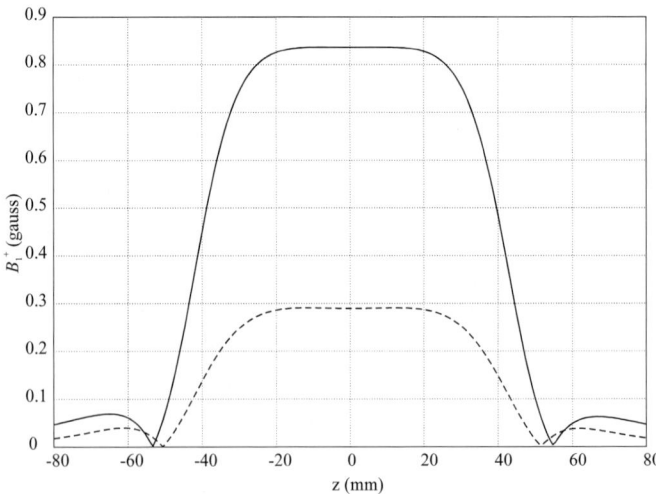

Fig. 8.75 Amplitude of the B_1^+ field along the z axis of the quadrature birdcage (full line) compared to the linear coil (Fig. 8.70). The magnetic field amplitude is calculated for the shielded coils, assuming a 10 W transmitter power and $Q = 185$. The probes are tuned and matched at 200 MHz.

The B_1^+ field amplitude is roughly inversely proportional to the diameter of the coil. Hence a gain of 1.33 is in favor of the new design (52/39). The field amplitude reduction factor due to the shield [Eq. (8.55)] is expected to be 0.58 for the previous design having $\Delta = 1.54$ (80/52). The reduction factor is only 1/0.87 for the new design ($\Delta = d/d_S = 2.74$), resulting in another gain in field amplitude of 1.5. Finally, the gain expected from the quadrature driving is 1.41. The total gain is therefore expected to be equal to 2.81.

The measured power gain for this design compared to the 12-rung birdcage was 9.1 dB (scaled to the same Q factor). This corresponds to an improvement of the amplitude of B_1^+ (and of the sensitivity) equal to 2.85 ($10^{(9.1/20)}$), which is very close to the predicted value.

8.2.7 Double tuning the birdcage resonator

Multiple tuning the birdcage resonator should be done while preserving, at least, the cosine current distribution around the coil cylinder. Coupling between the resonant modes at the different frequencies should also be taken into account.

A number of working solutions have been proposed. They may be distinguished into three different classes. The first one involves a single birdcage subjected to the perturbations of the tuning capacitors allowing the appearance of multiple "NMR useful" $k = 1$ resonant modes. This can be done using either a direct replacement of the capacitors by a multiple pole reactance (Chapter 6) or a controlled perturbation of the cylindrical symmetry by changing a few capacitance values. These designs have the advantage of keeping the same filling factor for all resonant frequencies, although the magnetic field distribution may differ to a certain extent. The second and third classes involve a birdcage design having either four end rings or using two imbricate resonators, respectively. For both designs, the filling factor differs for each resonant frequency mode. The choice between these methods depends on a number of parameters including the available space, and the filling factor and field homogeneity requirements. It also depends on the range covered by the resonant frequency values.

8.2.7.1 Pole insertion methods

One method, of the first class, is conceptually obvious (Fig. 8.76) but difficult to realize practically. It can be used essentially when the resonant frequency ratio is high (for example, $^1H/^{31}P$ or $^1H/^{23}Na$). The magnetic field homogeneity is expected to be identical for both resonant modes. The method consists of replacing the tuning capacitor by a reactive network like the one already described in Chapter 6, either in the legs for an LP or in the end rings for a HP birdcage. Generally, the series combination of a capacitor with a trap circuit[32] [Fig. 6.23(b)] is preferred in order to preserve the low frequency sensitivity. In this case, the trap should be tuned near the highest frequency (Chapter 6).

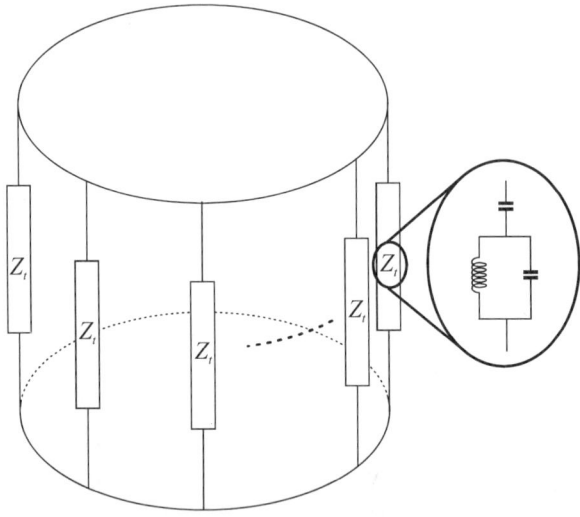

Fig. 8.76 An obvious method, but difficult to realize practically, for double tuning a birdcage resonator. It consists of replacing all tuning capacitors by a reactive impedance Z_t that includes a trap, as described in Chapter 6.

---

[32] A related method has also been proposed [Rath, 1990b]. The basic cell of the birdcage ladder network is either a BP or a stop band filter rather than an LP or HP. One elegant realization of this design is the four end ring dual birdcage described next.

The main difficulty in applying this method is that the equivalent capacitor that tunes the coil to ω_{HF} is critically dependent on ω_{off}.[33] When using a single coil, this is not a problem as far as only one trap has to be adjusted to tune the probe to the right frequency. In the case of a birdcage, all the traps should be tuned very precisely to the same frequency (ω_{off}). In addition they must be shielded from each other and from the birdcage itself. In spite of these difficulties, working designs have been described in the literature [for example, Isaac et al., 1990, Shen et al., 1997], but others recognize that the trap design is far from being encouraging [Matson et al., 1999], at least when applied to the birdcage concept.

8.2.7.2 Alternate rung method

The double tuned $^1H/^{31}P$ quadrature LP birdcage described by Matson et al. [Matson et al., 1999] is an alternative to the previous design, eliminating the difficulties of adjusting the trap circuit. Indeed, when the traps are used as current blocking devices (Fig. 8.77) their adjustment is far less critical than when they are used as tuning elements.

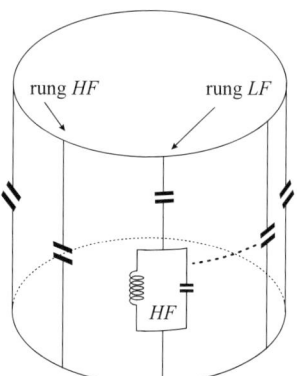

Fig. 8.77 Double tuned birdcage suited to a high frequency ratio. The trap isolates the high frequency (*HF*) legs from being perturbed by the low frequency ones (*LF*).

---

[33] The impedance of the trap circuit varies greatly with the frequency near its resonance (ω_{off}), hence for ω_{HF}.

In this design, the double resonance is obtained by alternating the tuned frequency of the legs [Vaughan et al., 1994; Pang et al., 2012]. A blocking trap introduced in the low frequency (*LF*) legs isolates the corresponding legs from any current at high frequency (*HF*). If the ω_{HF}/ω_{LF} frequency ratio is high, the resonator can be described as two imbricate independent LP birdcages. Each birdcage produces a practically ideal[34] RF field, but with half the total number of rungs.

When the frequencies are close, as is the case for example with a dual ^{1}H/^{19}F probe, the method outlined in Fig. 8.78 should be preferred.

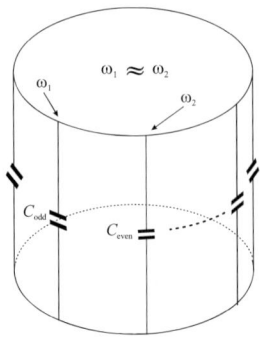

Fig. 8.78 Double tuned birdcage best suited to a low frequency ratio.

The perturbations of the birdcage introduced by the alternate rungs split the degenerate $k = 1$ mode into two orthogonal resonant modes. If the difference in frequency is small enough, the ideal current cosine distribution is still maintained for both modes, the RF field distribution being close to optimum homogeneity at both frequencies.

When the ω_{HF}/ω_{LF} ratio is intermediate, the two resonant modes, called the optical mode (at *HF*) and acoustic mode (at *LF*) [Amari et al., 1997] create distinguishable B_1 field distribution at *HF* and *LF*. The B_1 field distribution of the ideal birdcage is obtained at *LF* only, except if the birdcage is limited to four rungs [Lanz et al., 2001].

---

[34] At *LF*, the impedance of the *HF* rungs is high but finite. A small current may flow in the corresponding legs leading to some distortion of the B_1 map, unless the ratio ω_{HF}/ω_{LF} is high.

In this case, the acoustic and optical resonant modes are orthogonal, thus creating identical field distributions at both frequencies. However, the price to be paid is a poor RF field homogeneity as each mode corresponds to a two rung birdcage coil.

8.2.7.3 *Four ring double resonant birdcage*

This class of design uses a four ring birdcage resonator [Murphy-Boesh et al., 1994] as shown in Fig. 8.79.[35]

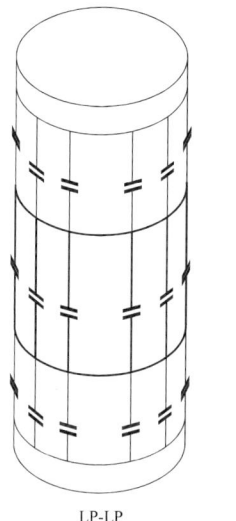
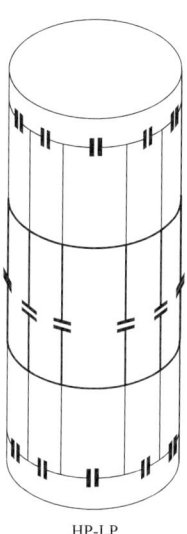

LP-LP   HP-LP

Fig. 8.79 Two possible four ring double tuned birdcage designs.

---

[35] An intermediate solution, including the four end rings, was previously reported [Derby et al., 1990]. In this design, an inner structure was a conventional eight leg LP birdcage tuned to ^{31}P. Two outer end rings were added, tuned, and connected through a capacitor to diametrically opposed legs of the inner birdcage. The design resembles the tuning method described in the previous section. Because the outer structure is tuned to *HF*, the tuning capacitors have a high impedance which does not significantly disturb the inner birdcage design. The outer structure can be viewed as a *HF* Alderman–Grant coil that shares a pair of legs with the *LF* birdcage. In this design the inner birdcage can be driven in quadrature, in contrast to the *HF* coil which produces a linearly polarized field.

These coils can be driven in quadrature to produce a circularly polarized magnetic field at one or both frequencies if desired. Generally, the outer birdcage structure functions at *HF* and the shorter one, involving the inner end rings, resonates at *LF*. It results that the sample volume "seen" by the coil depends on the frequency mode. However, this is not always a problem as far as the high frequency mode (usually at the proton frequency) is generally used for image localization and/or decoupling purposes.

The outer *HF* birdcage can either be HP or LP, while the inner *LF* birdcage is generally of LP type. The HP configuration will preferably be used when the frequency-diameter product *fd* [Doty et al., 2007] is large (>15 MHz m). For smaller *fd* an LP type may be advantageously used. The resulting four ring birdcage will therefore be described as LP–LP or HP–LP depending on the outer structure configuration. At this point, it is worth mentioning that the four ring configuration LP–HP electrically resembles the so-called "band-stop" birdcage already described by Rath [Rath, 1990b].

As the outer structure is tuned to *HF*, the corresponding capacitors have a high impedance at *LF*. Hence, the inner structure functions as an isolated *LF* birdcage, generating a homogeneous RF field. This is particularly true for the LP–LP configuration, for which the high frequency LP capacitance values are much smaller than for a HP configuration. At *HF*, the structure can be viewed as two short birdcage resonators (the outer structures) (over)coupled together through the leg impedances and, eventually, by the mutual induction between the outer and inner structures. In these conditions, one expects two distinct resonant modes resulting (Chapter 6) from the counter-rotating and co-rotating $k = 1$ modes of the two outer structures. The currents of the corresponding modes develop voltages across the common inner end rings (Fig. 8.80). These voltages are either in-phase or out-of-phase among the legs, for the counter-rotating or co-rotating mode. Hence there is no leg current when the counter-rotating mode is excited. In contrast, when the co-rotating mode is excited, currents are induced in the legs producing the required magnetic field. If the outer structures are symmetric, and if the legs are identical, the $k = 1$ mode of the resonant outer structures will respect the cosine law. It results that the magnetic field produced by this design will have the ideal homogeneity in the transverse plane, at both frequencies. However, the field distributions in the longitudinal plane differ at *HF* and *LF*.

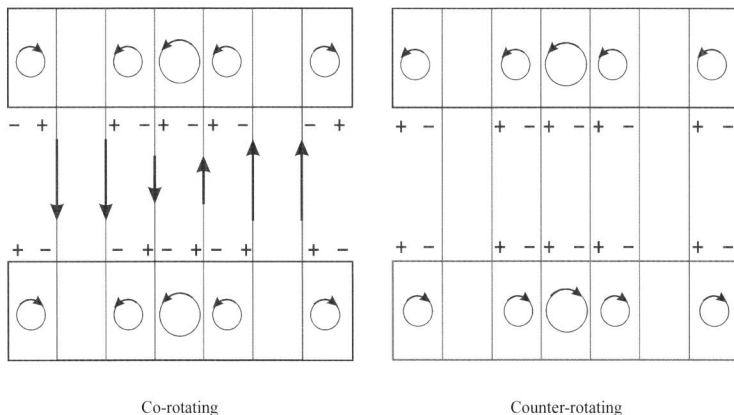

Fig. 8.80 High frequency resonant modes of the four ring birdcage design. A current in the legs (producing the useful NMR mode) appears only when the two outer structures resonate in a co-rotating mode. In this case, the corresponding currents develop opposed phase voltages across the legs (left).

8.2.7.4 *Crossed-coil resonators*

These dual tuned birdcage designs are constituted of concentric resonators [Fitzsimmons *et al*., 1993], as represented in Fig. 8.81. Accordingly, the filling factor is different among the resonant frequencies. The coil could be driven in quadrature at both frequencies. The design forms a transformer device that is operating in opposite phase at high frequency, resulting in a corresponding loss of sensitivity.[36] At low frequency, both resonators operate in-phase, providing the optimum sensitivity. For most applications where the *HF* mode is used for imaging purposes only, this design is valuable. The coupling between the two birdcages may still be minimized by rotating the inner coil respective to the outer one. Also, the outer coil is made longer than the inner one,

---

[36] In other words, the *LF* birdcage screens the RF field produced by the *HF* coil.

minimizing the coupling between the pairs of end rings. According to the dimensions and frequencies of both resonators, one coil is generally HP and the other is LP.

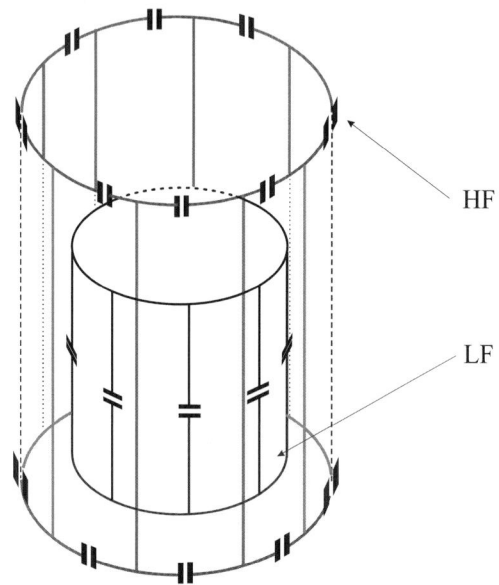

Fig. 8.81 Imbricate birdcage design. Coupling between the two resonators screens the RF magnetic field created by the high frequency birdcage.

Another version of this dual tuned resonator has been proposed recently; the so-called rung-pair birdcage [Habara et al., 2007]. Each rung of the design is made of parallel strips and capacitors constituting a dual tuned resonator. As with the imbricate *HF* and *LF* classical birdcages, the efficiency is much better at *LF* than at *HF*. The probe can also be driven in quadrature at both frequencies.

An improvement of the design concerning the *HF* sensitivity has recently been described as the cross-cage resonator [Lanz et al., 2000; Weisser et al., 2001]. In this design the resonant $k = 1$ mode of the lower frequency birdcage is suppressed in one direction by cutting the zero current end ring segments (Fig. 8.82). Hence, this direction can be used for exciting the *HF* birdcage. The excitation of the *LF* coil, at 90°, produces a linearly polarized $k = 1$ mode that does not induce a significant current in the *HF* coil due to its high impedance at *LF*. The sensitivity is then recovered for both coils, but they cannot be driven in quadrature. This design is especially well adapted for microimaging resonators.

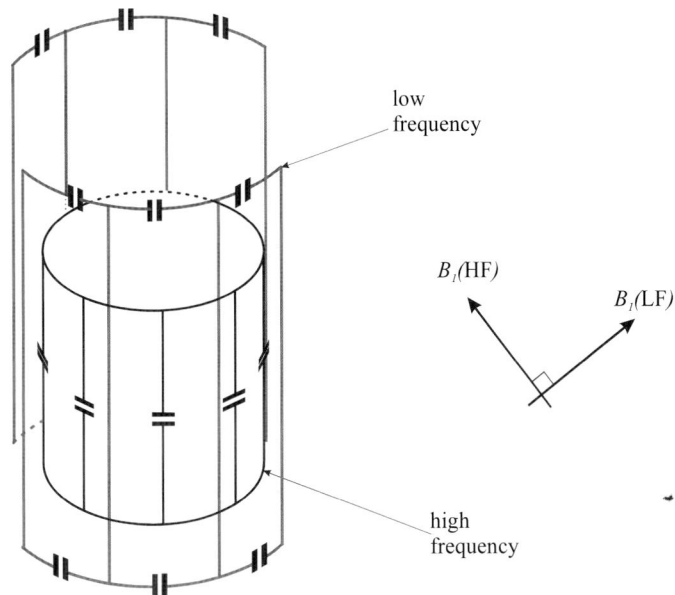

Fig. 8.82 The cross-cage resonator design [Lanz *et al.*, 2000; Weisser *et al.*, 2001]. The magnetic field polarization is forced at low frequency by construction, allowing an efficient decoupling between both resonators.

8.2.7.5 *A practical design*

A double tuned (^{23}Na/^{1}H) birdcage coil [Mispelter and Lupu, 2007] operating at 53 MHz and 200 MHz is briefly described as an example of the four ring design (Section 8.2.7.3).

One of the drawbacks of this design is that it creates a high B_1 field in the sample region close to the outer structure of the birdcage. This is due to the high current flowing, at the proton frequency, in the outer loops (Fig. 8.80). As these loops are not indispensable in creating the magnetic field inside the coil volume, they are folded up on a plane perpendicular to the coil axis, in a wheel-like configuration (Fig. 8.83).

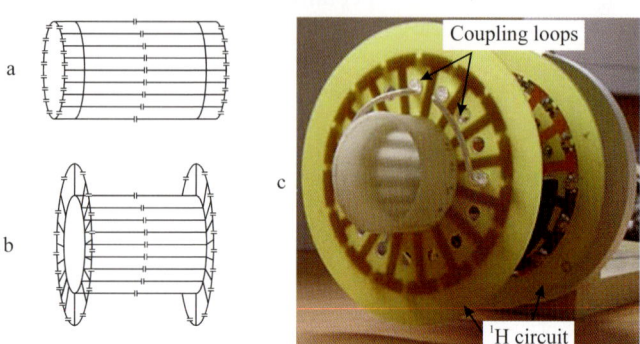

Fig. 8.83 (a) The classical four ring configuration. (b) The wheel-like configuration. (c) Practical realization.

The magnetic field profile along the axis of the classical configuration is shown in Fig. 8.84(a) at both frequencies. The B_1^+ field profile at proton frequency (*HF*) extends greatly over the field profile created at the *LF*. With the wheel-like configuration, the B_1^+ field at *LF* is almost unperturbed [Fig. 8.84(b)], while the field at *HF* is significantly attenuated in the regions where the *LF* B_1^+ becomes negligibly small.

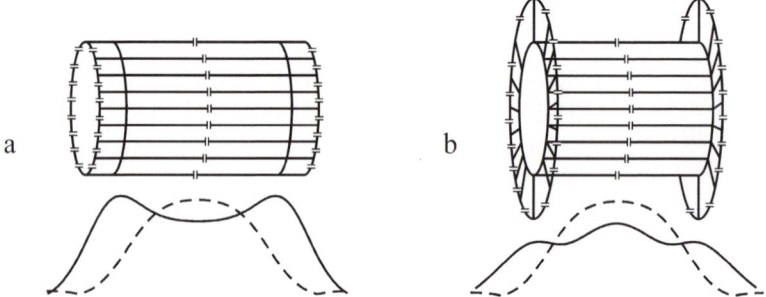

Fig. 8.84 B_1^+ profile along the coil axis at 200 MHz (full line) and 53 MHz (dashed line). (a) The four ring configuration. (b) The wheel-like configuration.

The probe performance had been simulated and compared to the performances expected for the classical four ring configuration and for reference probes having the same dimensions (Fig. 8.85). Confidence in the simulation is supported by the good agreement with the measured 90° pulse width duration (Fig. 8.85) as well as with the measured S_{11} and S_{22} spectra (Fig. 8.86).

The proton field amplitude, at the center of the coil, is considerably attenuated for the double tuned configurations, as expected. But, for the wheel configuration, it is significantly lowered. As far as the proton coil is used for localization, the sensitivity is not an issue.

The sodium field amplitude is fortunately almost independent of the design, providing optimum sensitivity for the low-gamma nucleus.

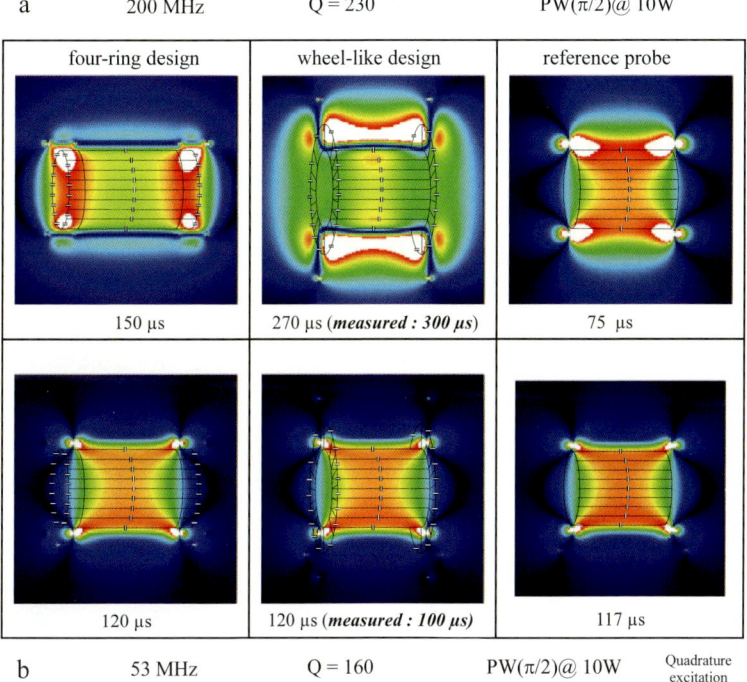

Fig. 8.85 Simulation of the B_1^+ distribution and amplitude (90° pulse duration at the center) in the zx or zy planes for the four ring and wheel-like designs compared with single tune probes of the same dimensions. (a) 200 MHz, linear excitation, $Q = 230$, transmitter power 10 W. (b) 53 MHz, quadrature excitation, $Q = 160$, transmitter power 10 W.

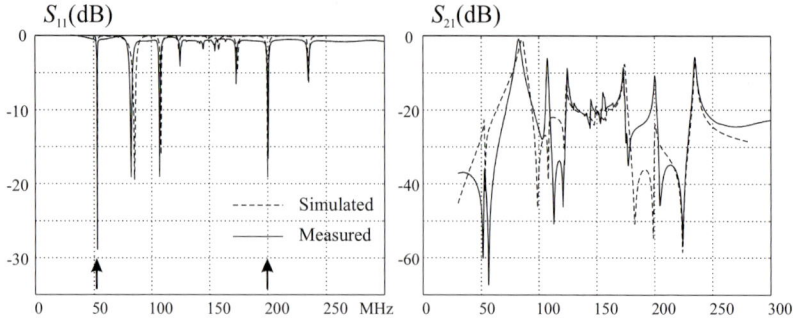

Fig. 8.86 Simulated (dashed line) and measured (full line) S-parameters for the wheel-like, double tuned shielded birdcage. The arrows indicate the operating frequencies (^{23}Na - 53 MHz, ^{1}H - 200 MHz). The simulation was done using the NMRP software described in Chapter 3 (see also the Resources for Readers, page X).

An inductive coupling with quadrature driving, similar to the proton quadrature birdcage coil already described, has been chosen for the sodium channel. The proton frequency mode has been excited by one of the coupling loops, leading to a linear polarization and lower sensitivity, but still sufficient for tissue localization purposes. Details on the probe construction are given in Fig. 8.87.

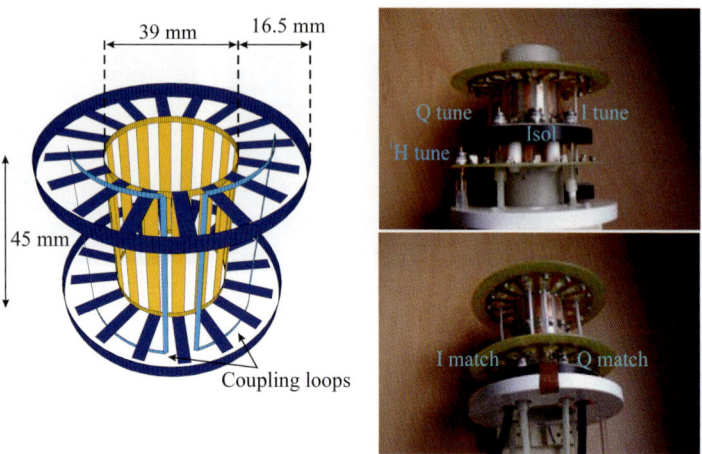

Fig. 8.87 Geometry of the double tuned, proton–sodium birdcage coil.

Identification of the $k = 1$ mode for both frequencies is easily obtained with a small pickup coil, centered in the coil where a nonzero magnetic field is created. The assignment has been further confirmed by simulation (Fig. 8.88).

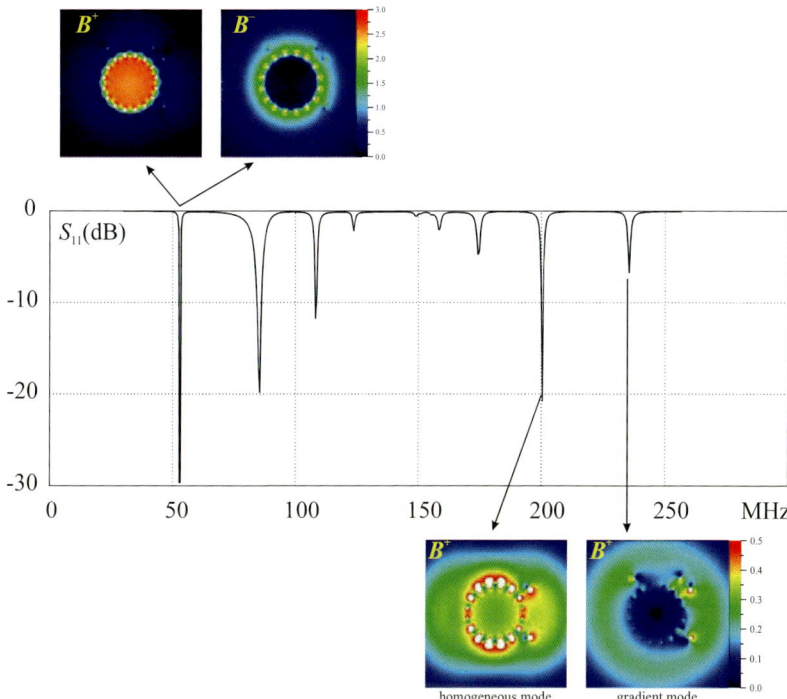

Fig. 8.88 Simulation of the B_1^+ field maps in the xy plane (orthogonal to the coil axis) when the probe is tuned and matched to 50 Ω at the indicated frequencies. The RF field at 53 MHz is circularly polarized (upper). The probe has $Q = 160$ and it is excited via a 90° hybrid (7.07 W in each port of the quadrature coil). The high frequency modes at 200 and 230 MHz correspond to a homogeneous and gradient RF field distribution, respectively. The probe has $Q = 230$ at 200 MHz and is excited with 10 W.

Finally, this design has the advantage that it considerably decreases the unnecessary power deposition of RF energy at the proton frequency near the extremity of the probe. Furthermore, it minimizes the magnetic coupling of resonant modes which simplifies the frequency spectrum of the resonator, hence its adjustment.

8.3 Transmission lines resonator

8.3.1 *TEM resonators*

It is well-known that on increasing the frequency, the quality of an NMR probe may degrade rapidly, depending on its size. The radiation resistance of the coil, the associated electric field, and the losses in the sample are increasing. Also, the current amplitude and phase are no longer constant along the coil, inducing important RF field heterogeneities. A few possible designs that attempt to increase the usable frequency range of NMR probes have already been described in previous sections. The Ultra High Frequency (UHF) versions of the saddle coil have less inductance and smaller current pathways, affording a significant improvement of the useful frequency range. On the other hand, the birdcage design provides a further improvement toward the ideal current distribution around the cylinder creating a homogeneous RF field. Finally, shielding the resonator reduces the radiating resistance, and the capacitive shortening of the conductive elements of the coil (segmentation) reduces the electric field and phase shift effects.

Despite all these improvements, the size of imaging probes still remains limited.

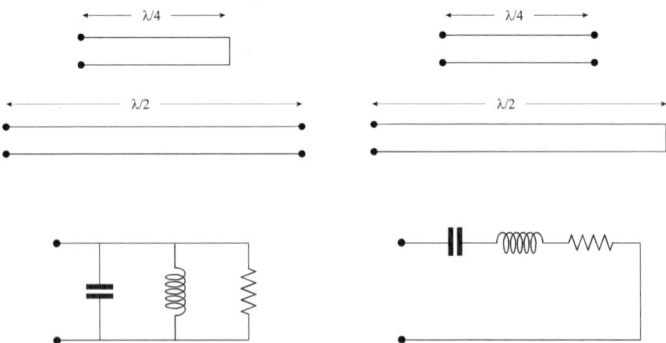

Fig. 8.89 Resonators built on quarter- or half-wave transmission lines and their equivalent lumped element circuits.

The problem is especially critical for whole body (even for human head) imaging probes that should function at continuously increasing magnetic fields. The development of small animal microimaging at very high fields (above 10 T) led to the search for new solutions. This research field rapidly evolved during the last decade.

Most of the UHF designs are based on transmission line resonators instead of lumped element LC resonant circuits, a common practice in microwave engineering. For example, a half- or quarter-wave line, either short circuited or open ended, forms a good quality resonator, equivalent either to a parallel or series LC resonator (Fig. 8.89).

These resonators can be cylindrical coaxial lines or cavities or striplines producing a transverse electromagnetic field (Fig. 8.90).

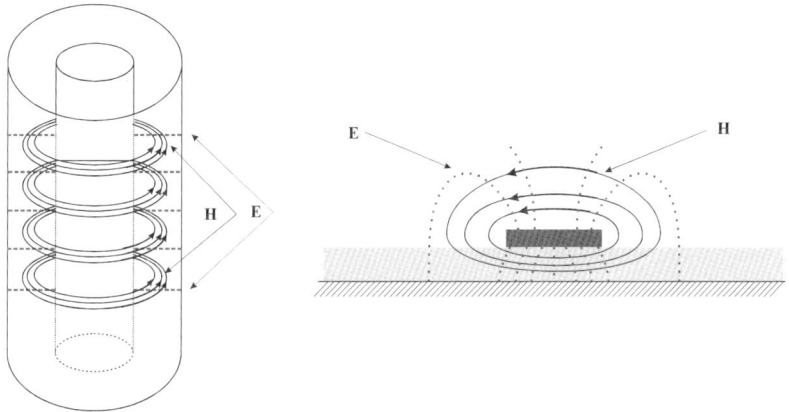

Fig. 8.90 Electromagnetic field created by a coaxial cavity (left) or a stripline (right) resonator. These resonators can be used for both NMR excitation or receiving.

They have two major characteristics in common: the electromagnetic field is confined in the space between the two conductors and the current is not constant along the resonator axis. To be useful for NMR excitation and reception, the resonator should be modified. Firstly, a hole is made in the coaxial resonator (Fig. 8.91, [Kan et al., 1973]) in such a way that a transverse magnetic field is created as required.

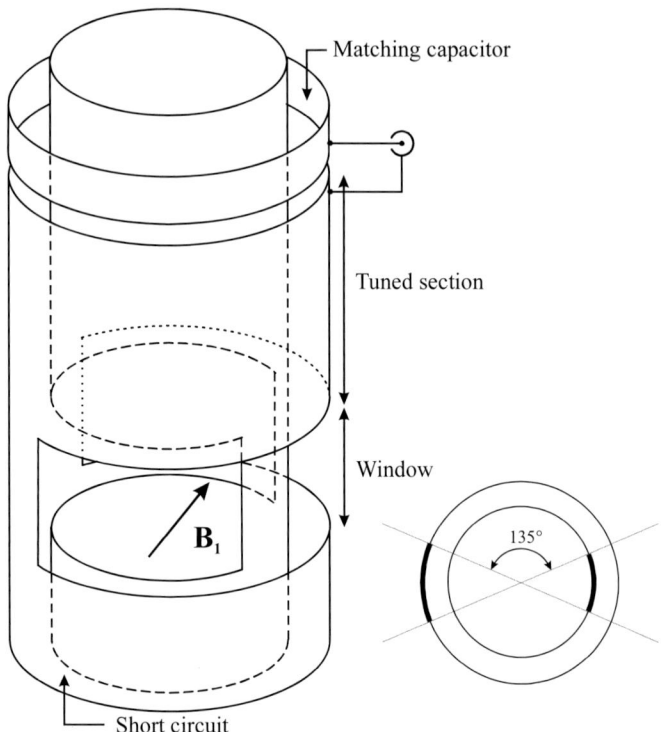

Fig. 8.91 Early design of an NMR probe based on transmission line concepts [Kan *et al.*, 1973]. The window aperture angle should be 135° (right) and its height should be roughly equal to the mean diameter of the holder tube in order to optimize the sensitivity (ηQ product). The length between the window and the short circuit end can take any convenient value without altering the resonant frequency. The matching capacitor is constituted by the upper band to which the cable is connected and the inner cylinder. Originally, the probe was designed for a high resolution, 5 mm sample tube, spectrometer working at 240 MHz. In these conditions, the tuned section was of 50 mm length and the window height was of 6 mm (the drawing is not to scale).

Another possibility is to enlarge the space comprised between the central and outer conductors to make room for the sample.

However, the length of the slot should be short enough respective to the wavelength and correctly positioned in such a way that the current amplitude is roughly constant and maximum (Fig. 8.92).

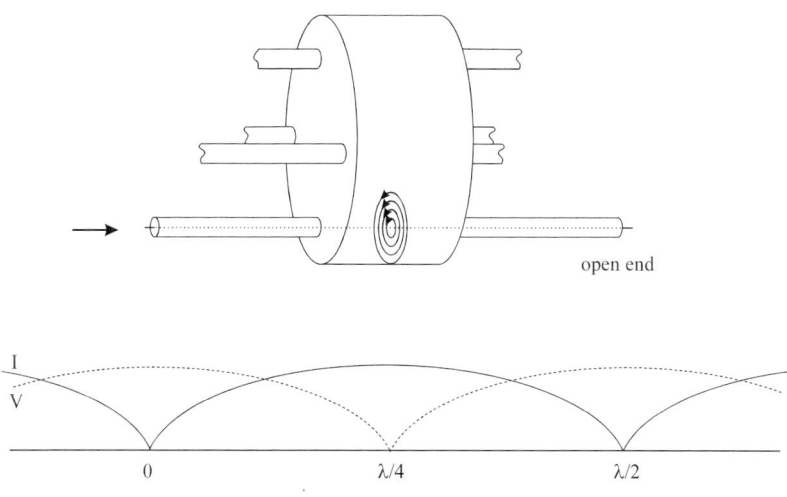

Fig. 8.92 In the design by van Waals and Bergman [van Vaals and Bergman, 1990], the RF field is created in the central section of an open ended half-wave coaxial resonator, when the current is maximum. The space between the central and outer conductor is enlarged to make room for the sample volume. Several lines can be added to enhance the magnetic field characteristics in the region of interest. The current phase in each additional line is controlled by insertion of transmission lines of appropriate length.

A major drawback of this design is that the whole resonant structure extends well out of the useful sample region, making the probe quite encumbering. A solution is to shorten the line resonator by capacitors connected at one or both ends. This led to the Alderman–Grant [Alderman and Grant, 1979] and slotted cylinder [Leroy-Willig et al., 1985] resonators, as described in preceding sections, and finally to the so-called TEM resonator (Fig. 8.93) [Krause, 1985; Dürr and Rauch, 1991; Vaughan et al., 1994; Bogdanov and Ludwig, 2003] which is the basis of the most recent designs.

All the homogeneous TEM resonator designs make use of the birdcage concept in order to naturally get the cosine current distribution among the cylinder. This configuration is obtained with equally arranged resonators on the cylinder, like the legs of the birdcage design.

These are, for example, the previously mentioned resonant rectangular loops of the "free element" birdcage [Wen et al., 1994; Fakri et al., 1996] (Fig. 8.94).

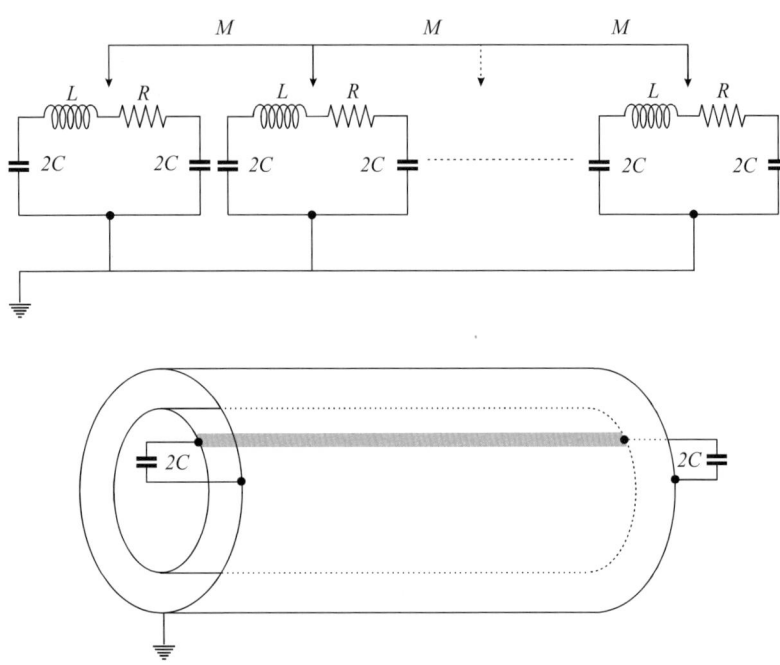

Fig. 8.93 Schematic representation of a TEM resonator [Tropp, 2002]. An elementary cell (lower) is constituted of a resonator formed from lumped and/or distributed inductances and capacitances (some examples are given in Fig. 8.83). The equivalent circuit of each resonator can be simply schematized as a parallel resonant RLC circuit, as shown (upper). Inductive coupling (possibly capacitive, but not preferred in an NMR probe!) between each cell permits the propagation of "waves" along the network in a very similar manner as into the ladder network of a classical birdcage.

For higher frequencies, these resonators are made out of tubular [Krause, 1985; Dürr and Rauch, 1991] or microstrip line [Bridges, 1988; Bogdanov and Ludwig, 2003] conductors connected to the grounded shield by tuning capacitors at both ends.

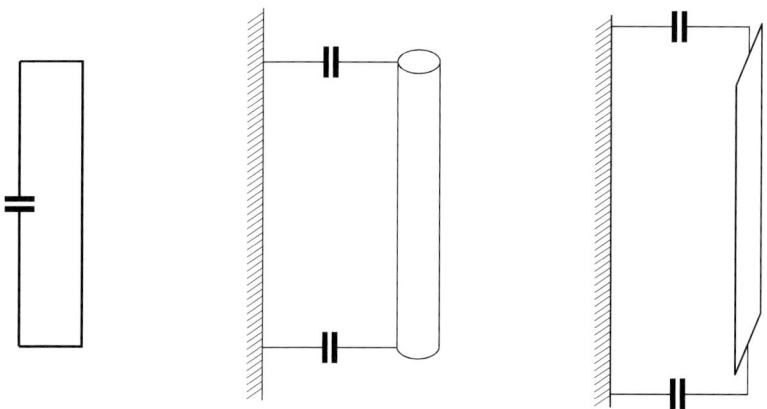

Fig. 8.94 Some resonator designs used in the TEM coil. The rectangular tuned loop of the "free element" birdcage (left) and two capacitively shorted transmission lines. The "central" conductor of the line can be either a round wire (middle) or a flat strip (right). The outer conductor, connected to the ground, is constituted by the cylindrical wall of the TEM resonator shield.

In all cases, the structure has resonant modes very similar to the birdcage design, due to the coupling between the constitutive elementary resonators. Hence, the practice of the classical birdcage design [Tropp, 2002] applies in most cases to the TEM resonators.

8.3.2 *Split transmission line resonators*

The previous designs still use lumped capacitors and eventually distributed capacitance between the shield and the conductor to form the elementary resonator. Another possibility is to use coaxial lines that have a slit in the central conductor [Röschmann, 1988] as shown in the design of Vaughan *et al.* [Vaughan *et al.*, 1994]. The central conductor is directly connected to the shield at both ends, forming the part of a "slotted" re-entrant cavity resonator (Fig. 8.95).

As previously presented, the coupling between the elementary resonant coaxial lines is required in order to obtain the resonant modes of the birdcage.

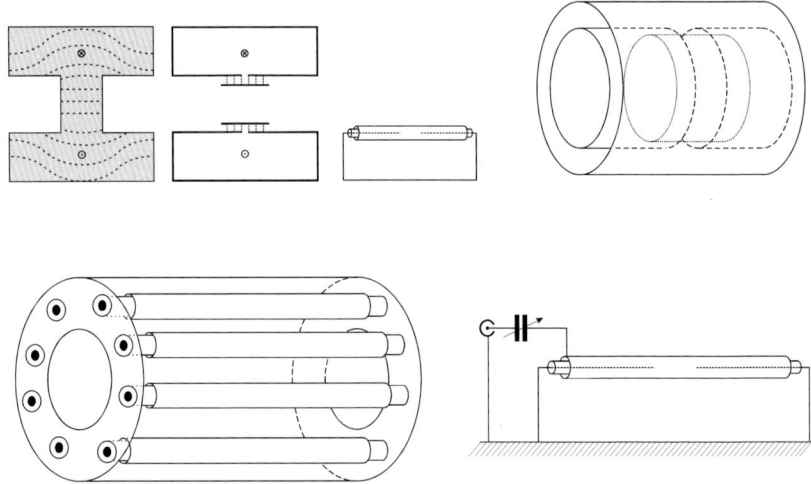

Fig. 8.95 Scheme showing how a re-entrant cavity resonator [Terman, 1955, pp. 160] can be transformed into an NMR resonator. The dashed lines in the cavity (upper left) represent the electric field, perpendicular to the magnetic flux lines in the TEM resonant mode. The central region is replaced by the capacitance that exists between a conductive cylinder and the inner conductor of the re-entrant cavity (upper middle). The corresponding cavity is shown (upper right). It is not usable as an NMR volume coil. Hence, the inner cylinder of the cavity is split into N elementary resonators constituted by a piece of split transmission line having its central conductor connected to the external conductor of the cavity (ground). Implementation of the TEM resonator and driving scheme is shown in the lower figures [Vaughan *et al.*, 1994].

Hence, the coaxial line must create a magnetic field outside its outer conductor. The "normal" mode of operation of a transmission line is the so-called differential mode in which the currents on the inner and outer conductors have opposed phases and equal amplitudes, thus not producing any magnetic field outside. Creating a disruption in one of the conductors disturbs the differential mode and favors the so-called common mode in which the currents flow in-phase. In this case, a magnetic field is created in a substantial space, outside the transmission line, as is required[37] for a homogeneous volume resonator.

In the design by Vaughan et al., the outer conductor is continuous. The central conductor is cut and can be moved axially in order to adjust the slit width, providing a way to tune the TEM resonator.

Alternatively, the slit can be made on the outer shield of the coaxial lines (Fig. 8.96) as, for example, in the magnetic loop antenna [*ARRL Handbook*, 1989], in inductive coupling loops for NMR probes, and loop resonators used as surface coils (Chapter 9).

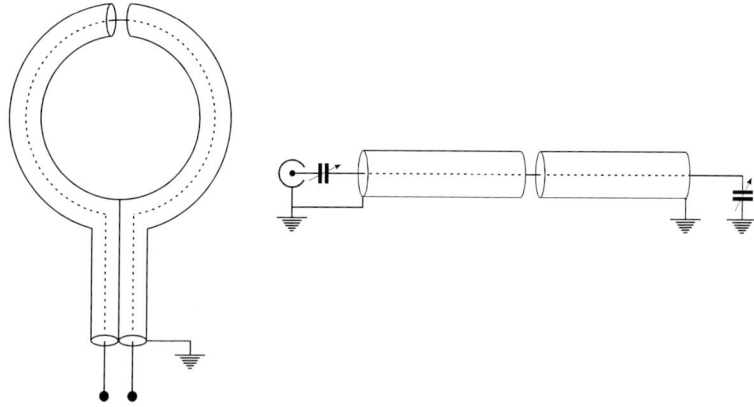

Fig. 8.96 A coaxial line having a slit in the outer braid favors the common mode currents, producing a magnetic field outside. The line can have the form of a loop (left) being used as a "magnetic" antenna, an inductive matching device, or even an NMR probe coil. The linear configuration (right) can be used as an element of a TEM volume resonator.

The microstrip transmission line resonator [Pozar, 1998] is an interesting alternative to the coaxial line, being more compact and allowing more room for the sample space. Such resonators have already been used in NMR [Gonord et al., 1988; 1994; Haziza, 1997; Zhang et

---

[37] In the next chapter, the split conductor transmission line resonators which are able to create magnetic field in a limited volume outside and close to the strip, required for surface examination, will be presented.

al., 2001], provided again that they are split. Some examples will be presented later, in the next chapter.

A volume probe based on microstrip transmission lines (MTL, [Zhang *et al.*, 2003]) is also able to produce an RF magnetic field.

The required resonant mode is still obtained from the coupling between the striplines arranged around a cylinder (Fig. 8.97). The strip conductors are continuous, but the "shield" is cut in the middle line, forming the required slit. One half of the tube is connected to the ground and the other one is floating.

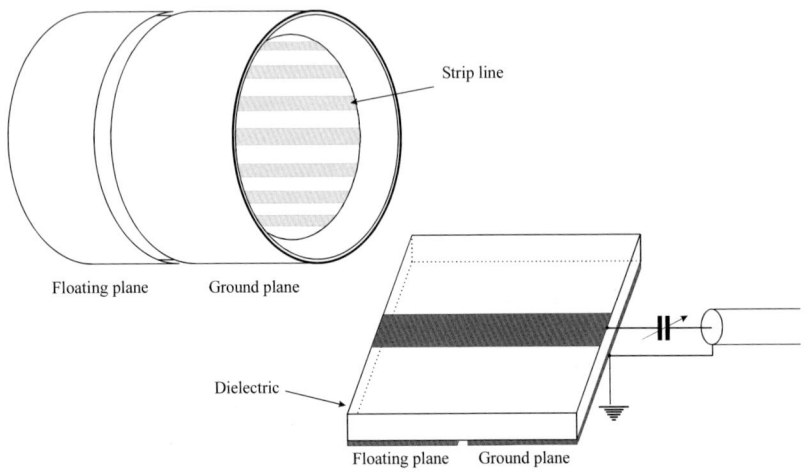

Fig. 8.97 A TEM volume resonator (left) built on microstrip transmission lines [Zhang *et al.*, 2003]. In this case, a slit is made in the external conductor. One half is connected to the ground and the other one is left floating. A flat version of a similar resonator (right) can be used to produce a magnetic field locally. The resonator can be driven as shown and finely tuned by terminating the stripline with adjustable capacitors.

All these probes, based on transmission lines, have been proved to be efficient as human head coils, functioning in quadrature at 4 tesla (170 MHz).

Chapter 9

Heterogeneous Resonators

9.1 The Basic Surface Coil

The surface coil was probably first introduced in 1951 by Suryan and later, in 1959, by Singer for NMR flow measurements [Suryan, 1951; Singer, 1959]. In 1974, when Hoult demonstrated the advantages of NMR for *ex vivo* observation of phosphorus metabolites [Hoult *et al.*, 1974], it was clear that the study of living systems by the noninvasive NMR spectroscopy had great potential. However, a localization tool was required for spatial selection within an intact living organism. A simple and efficient method had been introduced in 1980 by Ackerman *et al.* who designed the so-called "surface coil" to map the metabolism of living tissues *in vivo* in a real time and non-destructive manner, [Ackerman *et al.*, 1980; Bosh and Ackerman, 1992]. Since that time, the surface coil probe has been continuously improved in its design, usage, and success.

Basically, it consists of a simple loop of wire, creating an inhomogeneous magnetic field (Fig. 9.1), that decreases quickly when going away from the coil plane (hence its name). This is therefore a simple means to localize a Region Of Interest (ROI) within a large sample (animals or humans). This method is currently used, especially for *in vivo* laboratory experiments and in many clinical protocols. It has also been extended to an examination of the surface materials using an integrated imaging system which incorporates a small permanent magnet coaxial with a surface coil (MOUSE®, [Eidmann *et al.*, 1996]).

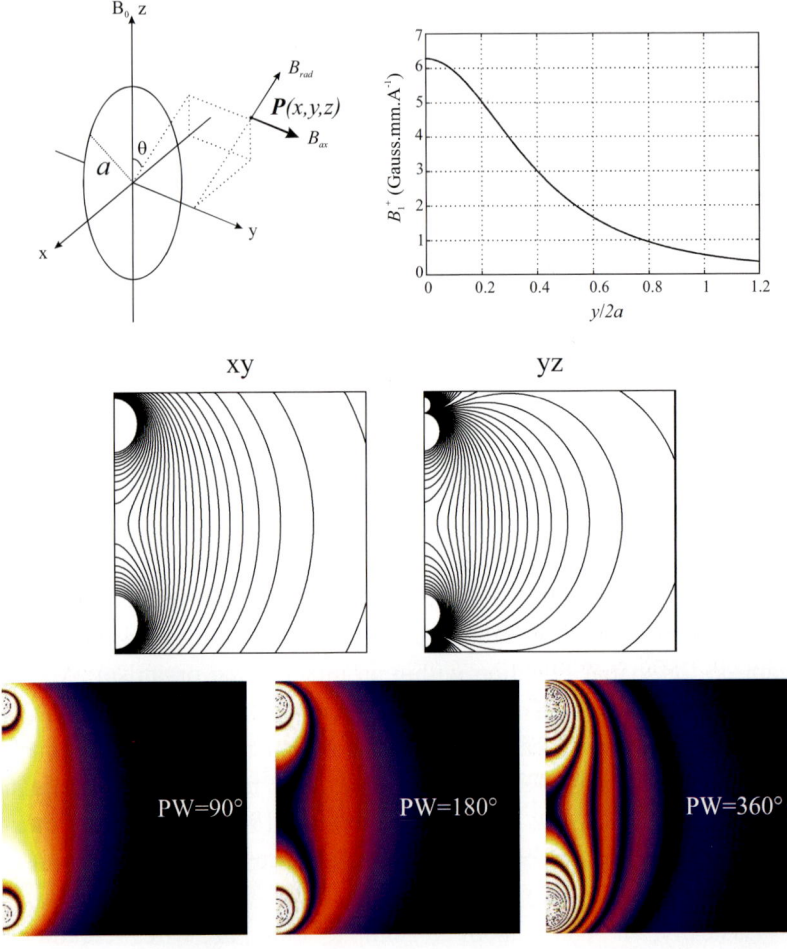

Fig. 9.1 Geometry of the basic surface coil (upper left). Oz is the direction of the main magnetic field B_0. The NMR efficient magnetic component of the RF field (B_1^+) is shown on the y axis (upper right), as a function of the position with respect to the coil diameter. The field amplitude is expressed in gauss (1 G = 0.1 mT) for a coil of 1 mm in diameter, carrying a current of 1 A. The magnetic field amplitude in the xy and yz planes is shown as contours (middle). The signal amplitude that will be obtained from small voxels centered in the xy plane after a one pulse experiment is shown (lower) for different settings of the excitation pulse duration. It corresponds to a 90°, 180°, or 360° rotation at the coil center. The optimum is obtained for a pulse angle of roughly 180° at the coil center.

Heterogeneous Resonators

The filling factor of a surface coil being high and the probe being "small," its sensitivity is very high when one is interested in a small sample volume compared to that of a whole body resonator.

The surface coil is frequently used as a transmitting/receiving probe, but also as a receive-only coil [Crowley *et al.*, 1985; Beck and Blackband, 2001]. The advantage of this last usage is that the receiving sensitivity is high while the excitation, provided by a whole body resonator, is homogeneous. Many pulse sequences are indeed sensitive to the B_1 amplitude, requiring a homogeneous excitation in all the volumes of interest that could not be provided by the simple loop coil.

9.1.1 *The simple loop magnetic field distribution*

9.1.1.1 *"Ideal" case*

Consider a single loop of thin wire at low frequency (quasi-static case). As we have already shown in Chapter 7, the current loop gives two magnetic field components, in radial and axial directions, expressed via complete elliptic integrals of the first and second order[1] (K and E, respectively) as given by Smythe [Smythe, 1950].

The radial component (refer to Fig. 9.1) is given by:

$$B_{rad} = \frac{\mu_0 I}{2\pi} \frac{y}{\left(x^2+z^2\right)\left[\left(a+\left(x^2+z^2\right)^{1/2}\right)^2 + y^2\right]^{1/2}} \times \left[-K(k) + \frac{a^2+x^2+y^2+z^2}{\left[a-\left(x^2+z^2\right)^{1/2}\right]^2 + y^2} E(k) \right] \quad (9.1)$$

and the axial component is:

---

[1] An efficient numerical method for evaluating the complete elliptic integrals of the first and second kind is given in the appendix.

$$B_{ax} = \frac{\mu_0 I}{2\pi} \frac{1}{\left[\left(a+\left(x^2+z^2\right)^{1/2}\right)^2+y^2\right]^{1/2}} \times$$

$$\left[K(k) + \frac{a^2 - x^2 - y^2 - z^2}{\left[a-\left(x^2+z^2\right)^{1/2}\right]^2+y^2} E(k)\right], \quad (9.2)$$

where

$$k^2 = \frac{4a\left(x^2+z^2\right)^{1/2}}{\left[\left(a+\left(x^2+z^2\right)^{1/2}\right)^2+y^2\right]}. \quad (9.3)$$

Assuming that the static magnetic field is aligned along the z axis, the relevant RF field components for an NMR experiment, B_x and B_y, in the laboratory frame, are given by [Fig. 9.1(upper left)]:

$$\begin{aligned} B_x &= B_{rad}\sin\theta \\ B_y &= B_{ax} \end{aligned} \quad (9.4)$$

The corresponding magnetic field is linearly polarized and can be decomposed into two rotating components having equal amplitude. Along the coil axis it takes a simple expression already given in Chapter 7 [Smythe, 1950]:

$$B_y = \frac{\mu_0 I}{2} \frac{a^2}{\left(a^2+y^2\right)^{3/2}} = 2B_1^+. \quad (9.5)$$

The magnetic field amplitude decreases strongly moving away from the loop (Fig. 9.1, upper) providing its specific localization properties. Outside the coil axis, the distribution of the positive rotating frame component B_1^+ is shown in Fig. 9.1 (middle), in the xz and yz planes perpendicular to the loop plane. As a result, the B_1^+ values decrease by a factor of two between distances varying from $a/2$ to a. Many authors have suggested that the B_1 region defined by $\{a/2 \leq y, z \leq a\}$ is the sensitive region of a surface coil. This region is obtained when the surface coil is used as a transmit/receive probe. In this case, it is recommended that the pulse length is set for a 180° rotation angle at the coil center. Using a longer pulse induces a rapidly oscillating signal while a shorter one favors the undesired high flux regions (Fig. 9.1, lower). At this point it is noteworthy that the RF field gradient can be profitably used as a localization mean. This is done, for example, using the so-called "depth pulse" sequence [Bendall and Gordon, 1983].[2] This pulse sequence allows control of the observed volume that may be chosen in any plane parallel to the coil, but at the price of the sensitivity when moving away from the coil. Indeed, the sensitivity is still proportional to the B_1 amplitude in the volume of interest.

Very close to the loop, the RF magnetic field is very high, but an important gradient is produced too. If the sample is placed too close to the coil it has two unpleasant consequences. The first is an unnecessary magnetic loss that arises due to the conduction properties of the sample. The second is that the signals received from these so-called "high flux" regions vary in an uncontrolled manner, precluding any quantitative analysis. In general, a small gap (of the order of one tenth the coil diameter) is kept between the sample and the coil. The parasitic capacitance between the sample and the loop is reduced in favor of a better Signal-to-Noise Ratio (SNR) and reduction of undesired electrical effects.

Hence, the dimensions of the coil are fully determined by the experimental project. These are the first characteristics that will subsequently determine the shape, coupling method, and feasibility of the surface coil resonator.

---

[2] This pulse sequence is based on an EXORCYCLE [Bodenhausen et al., 1977] phase cycling of a spin echo sequence. The signal is selectively obtained from a region where the pulse angle is close to 180°. For a given transmitter power, the depth is varied by modifying the length of the pulses of the $\theta-2\theta$ sequence.

9.1.1.2 *Effect of inductive coupling*

Matching the loop through an inductive coupling is an attractive solution due to its simplicity and efficiency respective to the electrical equilibration. It raises the question of the influence of the coupling loop current on the magnetic field distribution in the sample volume. The question has already been addressed by Hoult Hoult and Tomanek, 2002]. They showed that, in certain circumstances, a dissymmetry of the rotating frame magnetic field components results from the coupling loop. Such an effect has also been demonstrated occasionally in Chapters 7 and 8 for inductively coupled cylindrical resonators. In the case of an inductively coupled surface coil, the effect seems hardly observable except when the coupling coil is almost coplanar with the observing coil (the example used by Hoult and Tomanek). The reason for the different behavior may be that efficient coupling of a closed resonator with an external loop requires that the magnetic flux of the loop enter significantly inside the resonator, disturbing the magnetic field in the sample volume. In contrast, the flux coupling with an open structure like a surface coil resonator is less demanding with respect to the magnetic interaction within the volume sample, hence creating lesser distortions. To quantify this effect, the relative amplitude of the RF field created by both coils should be evaluated. This will be done considering some practical examples.

Let us consider a surface coil of 24 mm inner diameter made out of a round wire of 2.5 mm diameter (L_S = 40.63 nH), tuned to 200 MHz and assuming two different Q values. It will be matched to 50 Ω through a parallel, coaxial, coupling loop. The optimum condition of inductive coupling is probably when the primary coupling loop is tuned to the working frequency [Decorps *et al.*, 1985b]. In this case (Section 4.1.3.2), the matching mutual impedance $M\omega_S$ is given by [Eq. (4.46)]:

$$M\omega_s = \sqrt{rZ_0} \qquad (9.6)$$

The required mutual inductance can be realized using different geometries of the primary loop. Two coupling geometries will be considered, involving either a small loop almost coplanar with the surface coil or a larger loop (same diameter as the surface coil) but pulled back from the surface coil, in order to preserve the same mutual inductance value. Having defined the coupling loop geometry and the M value [Eq. (9.6)], the required distance to the surface coil can be estimated using the "m2d" utility (included in the NMRP software utilities). The geometry is shown in Fig. 9.2 and some simulated parameters are presented in Table 9.1.

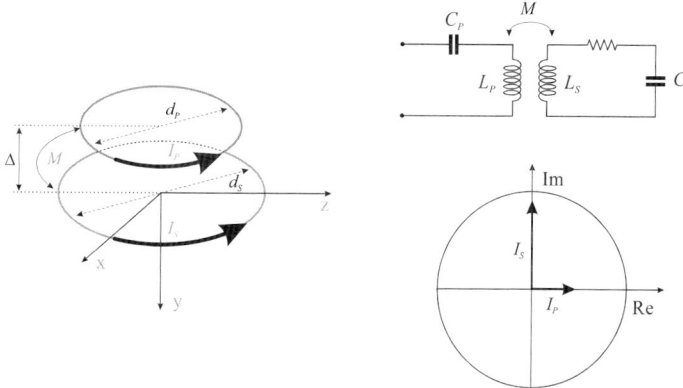

Fig. 9.2 Inductively coupled surface coil. Left: geometry. Right: Equivalent electric circuit and current representation in the complex plane.

The current in the coupling loop is in general much smaller than the current in the probe coil itself

$$\frac{I_{coupling\,loop}}{I_{surface\,coil}} = \sqrt{\frac{r}{Z_0}}, \qquad (9.7)$$

where r is the surface coil resistance[3] (of the order of 1 Ω) and Z_0 is the characteristic impedance of the spectrometer (50 Ω). This suggests that the effect of the coupling loop would be small, especially in the sample volume which is closer to the surface coil than to the coupling loop. In contrast a larger distortion would be observed near the coupling coil, where there is no sample at all.

---

[3] The resistance of the coupling loop, usually very small compared to Z_0, is neglected.

Table 9.1 Specific parameters of the coupling loops shown in Fig. 9.2.

Q	I_P (A)	I_s (A)	M (nH)	d_P (mm)	Δ (mm)
30	0.2	1.084	7.34	13.31	0.0
30	0.2	1.084	7.34	26.5	12.22
100	0.2	1.98	4.02	10.09	0.0
100	0.2	1.98	4.02	26.5	17.96

At this point it should be mentioned that the phase of the currents in the primary and coupling loops differ by 90°. The total magnetic field created by the two-coil system must therefore be calculated using the formalism in Section 7.2. From the corresponding equations, it can be seen that the on-axis magnetic field is linearly polarized ($B_1^+ = B_1^-$) whatever the coupling conditions. The relevant magnetic field amplitude is given by:

$$2B_1^+ = 2B_1^- = \sqrt{\left(B_v^y\right)^2 + \left(B_u^y\right)^2}, \qquad (9.8)$$

where B_u is the magnetic field created by the in-phase current of the coupling loop and B_v is the field created by the 90° out-of-phase current in the surface (secondary) coil. However, out of the coil axis, the magnetic field created by the loops has nonzero components along both x and y. As a consequence, the two rotating components B_1^+ and B_1^- are no longer equal [see Eqs. (7.21) and (7.22)], leading to the dissymmetry already mentioned.

To quantify this effect, the currents in each loop must now be evaluated. Assuming that the probe is perfectly impedance matched to 50 Ω, the peak amplitude of the current in the coupling loop is independent of Q and given by (using the phasor notation):

$$I_P = \sqrt{2P/Z_0} \angle 0°, \qquad (9.9)$$

where P is the transmitter power. The current in the secondary loop (the surface coil) depends on Q and is given by [from Eq. (9.7)]:

$$I_S = \sqrt{\frac{2PQ}{L_S \omega_s}} \angle 90° \qquad (9.10)$$

The corresponding magnetic field profiles and maps are shown in Figs. 9.3 and 9.4, respectively.

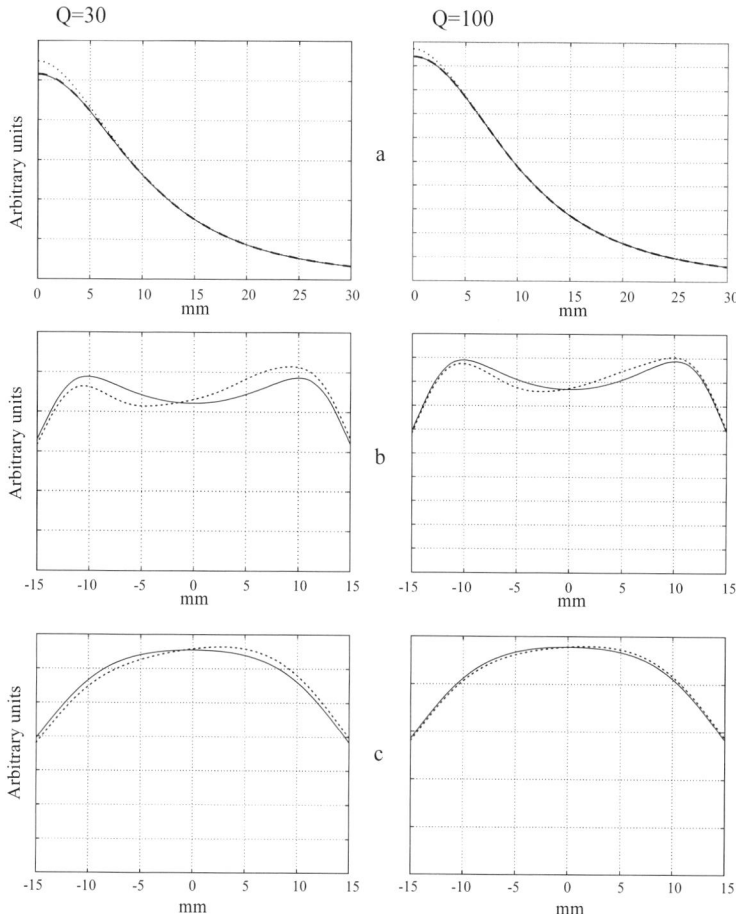

Fig. 9.3 B_1^+ field profiles of the surface coil along the y axis (a) and on an axis parallel to Ox, located at $y = 0.25d$ (b) and $y = 0.5d$ (c), for two different values of Q. Full lines: reference coil (capacitively matched). Dotted line: inductively matched by a small concentric coil. Dashed line: inductively matched by a large coaxial coil. In this case, the coupling coil is pushing back from the surface coil to maintain the required coupling. The field profile is practically identical to that of the reference coil. It is only represented in the upper diagram. The dimensions and current values are given in Fig. 9.2 and Table 9.1.

As a result, the magnetic field amplitude is only slightly changed along the coil y axis when an inductive matching solution is used instead of a capacitive coupling network.[4] The asymmetry in the *xy* plane mentioned is practically insignificant in the sample space except in the extreme case when the coupling loop is quasi coplanar with the surface coil. As already mentioned by Hoult and Tomanek, pulling away a larger loop (to keep the required mutual inductance value) completely eliminates the corresponding field distortion (Fig. 9.4).

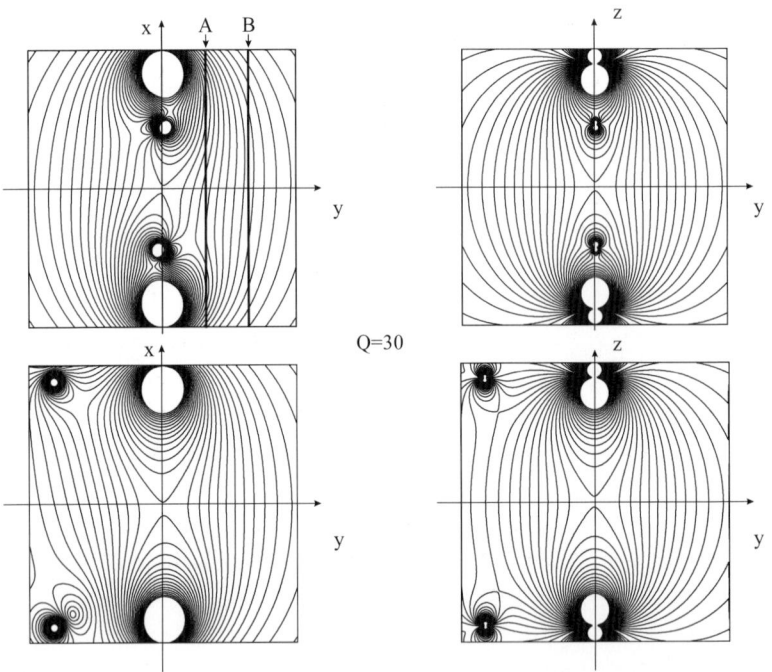

Fig. 9.4 B_1^+ field maps in the *xy* and *yz* plane for the inductively coupled surface coil (worst case, $Q = 30$). A and B show the position of the slices corresponding to the profiles in Fig. 9.3(b) and 9.3(c), respectively. Note that the field map appears distorted, essentially in the region located between the two coils.

---

[4] In this case, the magnetic field is created only by the surface coil current, given by Eq. (9.10).

9.1.2 *Practical design guidelines*

9.1.2.1 *How many turns?*

As already discussed for the solenoid coil, it can be shown that the sensitivity ($B_1/I\sqrt{r}$) is not dependent on the number of turns of the surface coil windings.

Again, the choice should be made regarding the coil impedance at the working frequency. If the impedance is too low, the resistances at the soldering points of the capacitors may unnecessarily dominate the probe losses. If the coil impedance is high, the electric field may be high, increasing the dielectric losses in the sample. On the other hand, at the working frequency the coil may become self-resonant, resulting in a decrease of the Q factor. Thus, the same rules as already quoted for the homogeneous resonator apply here in the case of the single loop.

In practice, the coil impedance should range in-between 20 Ω to 200 Ω.[5] If a one-turn coil of the required dimensions is outside these limits it is still possible either to increase the number of turns (small coil) or to segment the coil conductor (large coil). When the coil approaches its self-resonance the simple wire (round or flat) should be abandoned. Some solutions for very high frequencies are proposed in the next paragraphs.

Finally, it is worth mentioning here that the design of superconducting RF coils, using zero resistance conductors, needs some revision of the above mentioned usual practice (Section 9.1.4).

9.1.2.2 *Spiral windings*

When the coil is made using several turns (coils at low frequency or small coils at high frequency), the question of how to distribute the turns (circular or spiral winding) arises.

---

[5] The lower limit corresponds to a tuning capacitor having a reasonably high Q. The higher limit corresponds to a coil well below its self-resonance (see Section 9.1.3.1).

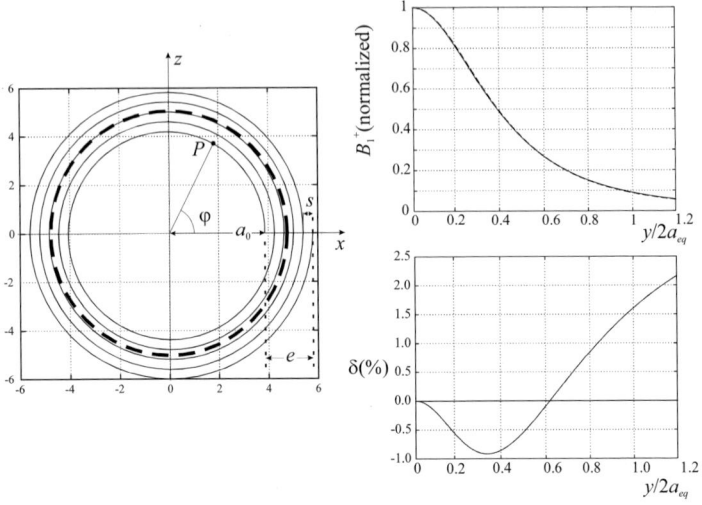

Fig. 9.5 The spiral coil geometry and its on-axis field profile (full line) compared to the field produced by the equivalent circular coil. The equivalent coil is the circular loop that produces the same field at the center when carrying the same current. Its geometry is shown as dashed thick lines to the left. The difference between the field profiles is represented in % (lower right).

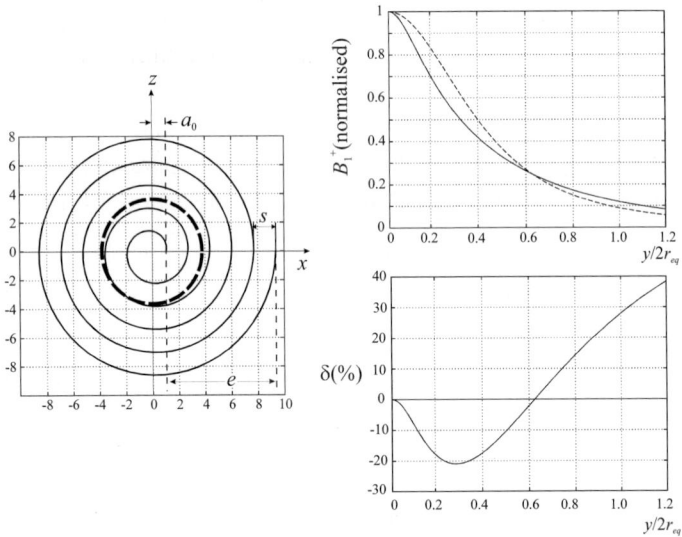

Fig. 9.6 The same as in Fig. 9.5, except that the spiral coil has a larger thickness e.

It can be demonstrated that the on-axis magnetic field created by planar spiral windings is approximately[6] the same as that created by a circular coil having the same number of turns. The diameter of the equivalent circular coil is equal to the mean diameter of the spiral winding (Fig. 9.5). However, this approximation fails when the width of the winding becomes comparable to, or larger than, the coil mean radius. In this case, deviations up to 40% can be observed (Fig. 9.6).

Let a_0 be the inner radius of the winding, s the winding step, and N the number of turns. The width of the coil is equal to:

$$e = Ns. \qquad (9.11)$$

The spiral winding is characterized by the following equations describing the coordinates (x, z) of each point spanning the conducting wire (assumed thin) (Fig. 9.5):

$$x = (a_0 + \varphi s/2\pi)\cos\varphi, \qquad (9.12)$$

$$z = -(a_0 + \varphi s/2\pi)\sin\varphi, \qquad (9.13)$$

where φ is varied from 0 to $2\pi N$. The magnetic field component along the coil axis at a distance y is given by integrating the Biot–Savart equation [Eq. (7.1)]:

$$B_y = \frac{\mu_0 NI}{4\pi} \int_0^{2\pi N} \frac{(a_0 + \varphi s/2\pi)^2}{\left[y^2 + (a_0 + \varphi s/2\pi)^2\right]^{3/2}} d\varphi. \qquad (9.14)$$

The final result obtained after the obvious change of variable

$$\varphi \to u = \sqrt{a_0 + \varphi s/2\pi} \qquad (9.15)$$

is

---

[6] Within 3%.

$$B_y = \frac{\mu_0 NI}{2e} \left\{ \frac{a_0}{\sqrt{y^2 + a_0^2}} - \frac{a_0 + e}{\sqrt{y^2 + (a_0 + e)^2}} + \ln\left[\frac{a_0 + e + \sqrt{y^2 + (a_0 + e)^2}}{a_0 + \sqrt{y^2 + a_0^2}}\right] \right\}. \qquad (9.16)$$

The circular coil that produces the same magnetic field amplitude for the same current NI at center ($y = 0$) has the radius:

$$a_{eq} = \frac{e}{\ln(1 + (e/a_0))}. \qquad (9.17)$$

If e is of the order or smaller than a_0, a_{eq} can be approximated by the mean diameter ($a_0 + e/2$) within less than 4% error. When e is twice the inner diameter of the spiral, the difference reaches 10%.

Assuming a wide spiral winding as in Fig. 9.6, its magnetic field profile, map, and sensitivity in the xy plane are compared with those of the equivalent circular cylindrical coil (Figs. 9.6 and 9.7).

The on-axis magnetic field of the spiral coil is generally lower than for the equivalent circular coil, excepting the points far away from the coil plane where it increases significantly (Fig. 9.6). However, this happens in regions where the sensitivity is intrinsically low. On the other hand, the sensitive volume of the spiral coil is not very different from that obtained for the equivalent circular coil (Fig. 9.7) if there is no specific advantage for one or the other configuration.

The choice will result in practical constraints depending specifically on a given experimental setup. A number of subtle parameters, such as the total length of wire, the proximity effect, or the thickness of the coil,[7] should be evaluated in each case. One should also take care of capacitive

---

[7] A cylindrical coil will in general be thicker than a flat spiral coil with the same number of turns. Hence, the accessibility of the sensitive volume would be favored by the spiral configuration. On the other hand, the total length of wire will be shorter for the cylindrical winding compared to the equivalent spiral one, but the proximity effects increasing the resistance, could be larger.

coupling with the sample that will be larger for the spiral coil and which may increase the noise contribution from the dielectric losses. As a result, the spiral coil is an advantageous design for microcoils used for microimaging applications at high fields [Eroglu *et al.*, 2003; Gimi *et al.*, 2003] or as implanted, inductively coupled, coils [Jow, 2007; Ginefri *et al.*, 2012].

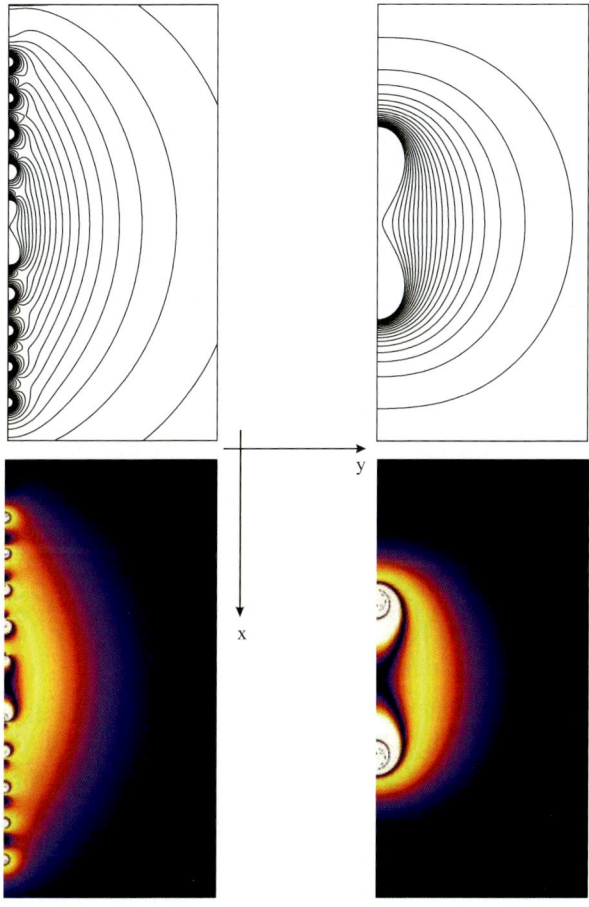

Fig. 9.7 B_1^+ field map (upper) and signal intensity (lower) for a large spiral coil (right) and its equivalent loop (left). The signal intensity is calculated assuming a pulse length such that the flip angle at the center of the coils is 180°.

9.1.2.3 Non-circular winding shapes

Until now, the coil has been assumed to be circularly wound, but in practice the coil is frequently shaped around the ROI in the sample in order to optimize the filling factor. Some simple deformations will be considered here: the elliptical, rectangular, or polygonal windings (Fig. 9.8).

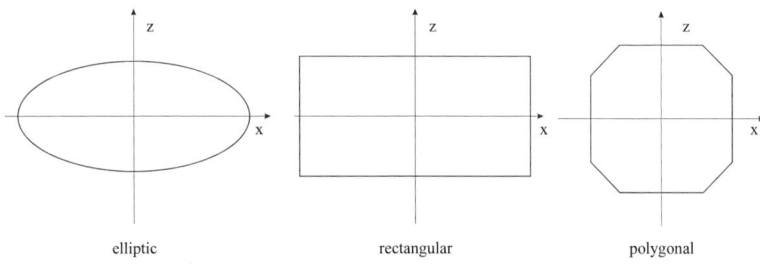

Fig. 9.8 Simple, non-circular shapes of the surface coil.

When compared to the basic circular coil, these shapes could be characterized by at least three parameters regarding the sensitivity (apart from optimization regarding the working frequency):
- the length of wire that determines the resistance;
- the shape of the sensitive volume;
- the penetration depth.

While the polygonal shape is obviously similar to the circular coil, the rectangular or elliptical shape would be preferred for elongated samples. In this case, the penetration depth is less than for the circular coil, but it can still be controlled in shaping the planar coil into a curved form adapted to the sample.

Quantitatively, the magnetic field created, on axis, by a planar, rectangular loop of current I can be obtained by summing the contribution of each side of the rectangle.

Using Eq. (7.6), one obtains:

$$B_y = 2B_1^+ = \frac{2\mu_0 I}{\pi} \frac{ab}{\sqrt{a^2+b^2+4y^2}} \times \left(\frac{1}{a^2+4y^2} + \frac{1}{b^2+4y^2}\right), \quad (9.18)$$

where a, b are the rectangle sides and y is the distance on the axis perpendicular to the rectangle plane where the field is calculated.

In order to compare the efficiency of field penetration of this shape with that of a circular coil, let us define the equivalent circular coil that produces the same field at the center when carrying the same current. For a given rectangular geometry, the diameter of the equivalent circular coil is given by:

$$d_{eq} = \pi/2 \left(\frac{1}{a^2} + \frac{1}{b^2}\right)^{-1/2}. \quad (9.19)$$

In the case of a square coil ($a = b$):

$$d_{eq} = (\pi/4) diag, \quad (9.20)$$

where *diag* is the diagonal of the square. It may be noted at this point that the surface of the equivalent circle differs from that of the square by only 3%, while the wire length is higher by about 14% for the square.

The profile of the magnetic field along axis y is shown in Fig. 9.9 and compared with the reference profile of the equivalent circular loop.

As a result, there is no significant difference between the square and the circular loops. In contrast, the magnetic field created by a rectangular loop ($b/a = 5$) does not penetrate as deeply as that produced by its equivalent circular coil. The rectangular shape (or equivalently the elliptical shape) may be preferred whenever an elongated region is to be observed, but at the price of the penetration depth.

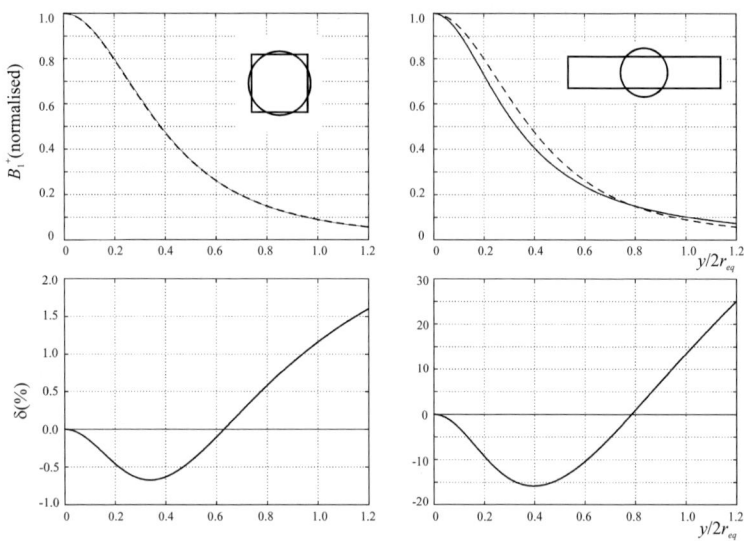

Fig. 9.9 On-axis field profile for a square and a rectangular coil compared to that created by their equivalent circular loop. The equivalent coil is the circular loop that produces the same field at the center when carrying the same current. Note the different scale in $\delta(\%)$ for the square loop (left) and rectangular loop (right).

9.1.2.4 *Wiring shape*

Having defined the dimensions, the winding shaping, and the number of turns from experimental constraints, the shape of the wire conductor remains to be chosen.

Usually, a simple round wire will be used in most routine work. At usual NMR frequencies, of the order or higher than several tens of megahertz, the skin depth is much smaller than the wire diameter. Hence it can be either solid or tubular without any consequences. Because the coil inductance decreases as the wire dimensions increase, one is tempted to prefer large diameter wires, but its size may be inconsistent with the required coil diameter. For example, it is illusory to try winding a 10 mm diameter coil with a 2.5 mm wire. This wire is also inadequate to wind a 300 mm surface coil. In the former case, a 1 mm wire would be preferred while in the latter case, an 8 mm diameter plumbing tube will be best adapted. Instead of a round wire, a thin flat conductor also allows a large, low inductance coil to be built.

If the dimensions of the round or rectangular wire are much smaller than the coil diameter, the magnetic field created in the sample volume can still be described using the thin wire approximation,[8] using an equivalent mean geometry.

However, if the wire dimensions increase respective to the coil diameter, the nonuniform current distribution on the conductor must be taken into account for an estimation of the magnetic field intensity. This is particularly true when the coil is formed using thin copper foil.

When the coil is made out of a thin cylinder, the current distribution is that of a short Loop Gap (LG) resonator [Fig. 9.10(a)]. The magnetic field profile is similar to the one created by a thin wire coil but located far away from the sample, in the middle plane of the LG [Fig. 9.10(b)]. When compared with the "equivalent" loop that would create the same magnetic field in the center of the plane, close to the sample Fig. (9.10b), it is obvious that the field penetration of the LG is much smaller.

On the other hand, if the coil is a flat ring, the current circulates preferably on the inner edge of the copper foil, due to the proximity effect [Fig. 9.11(a)]. The magnetic field profile is still similar to that created by a circular thin wire coil having a diameter smaller than the mean diameter of the flat annular ring.

The inductance of the LG design is generally smaller than a similar loop of the same diameter (realized with a wire of "reasonable" gauge). The inductance of the flat ring is, on the contrary, similar to the equivalent loop that produces the same magnetic field at the center.

An earlier comparison of the intrinsic sensitivity of surface coils shaped with flat copper foil or round wire [Balaban *et al.*, 1986] has shown that the flat ring is not to be favored, due to large capacitive coupling with the sample. Electrically shielded structures, such as those described later (Section 9.1.3.3), would prevent this effect.

Finally, a two-wire conductor with a crossover like in the litz coil element (Fig. 2.19) can replace, with some advantages, a flat annular conductor. In this case, the inner and outer currents are equal.

---

[8] It is assumed that the sample is far from the high flux regions.

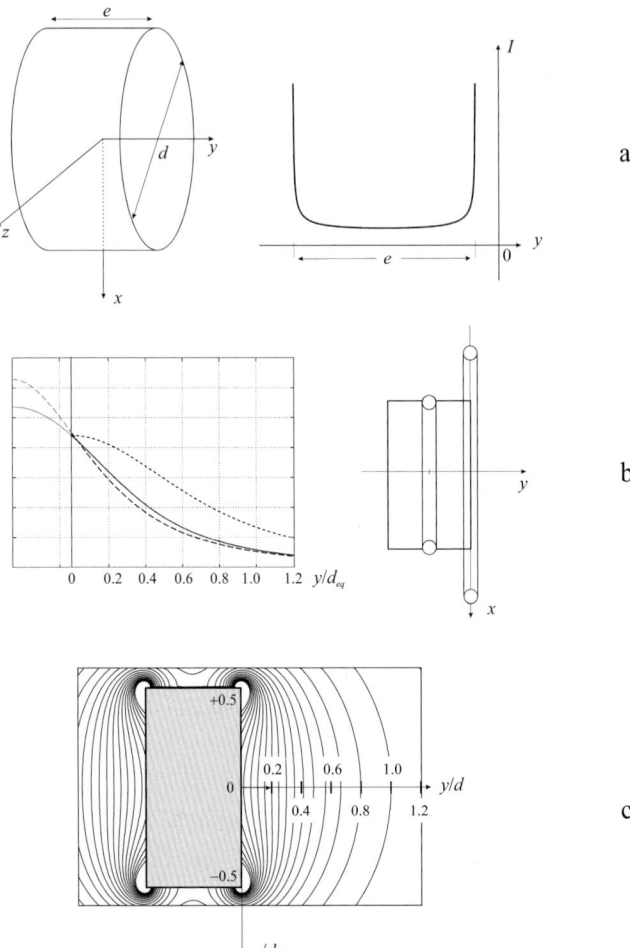

Fig. 9.10 (a) The short loop gap geometry (left) used as a surface coil and the current distribution on the copper foil (right). (b) Relative positions and dimensions of the loop gap foil, of its thin circular coil approximation, and of the equivalent coil. The equivalent coil is located at the left edge of the loop gap, close to the sample region. Its dimensions are such that it produces the same field as the loop gap, at the same position at the center. Both coils are assumed to carry the same total current. To the left is shown the on-axis field profile, inside the sample region, calculated for the loop gap (full line), for the thin wire coil approximation of the LG (dashed lines) and for the equivalent coil (dotted line). (c) The magnetic field map, in the xy plane, of the LG as it appears outside the resonator volume, within the sample region.

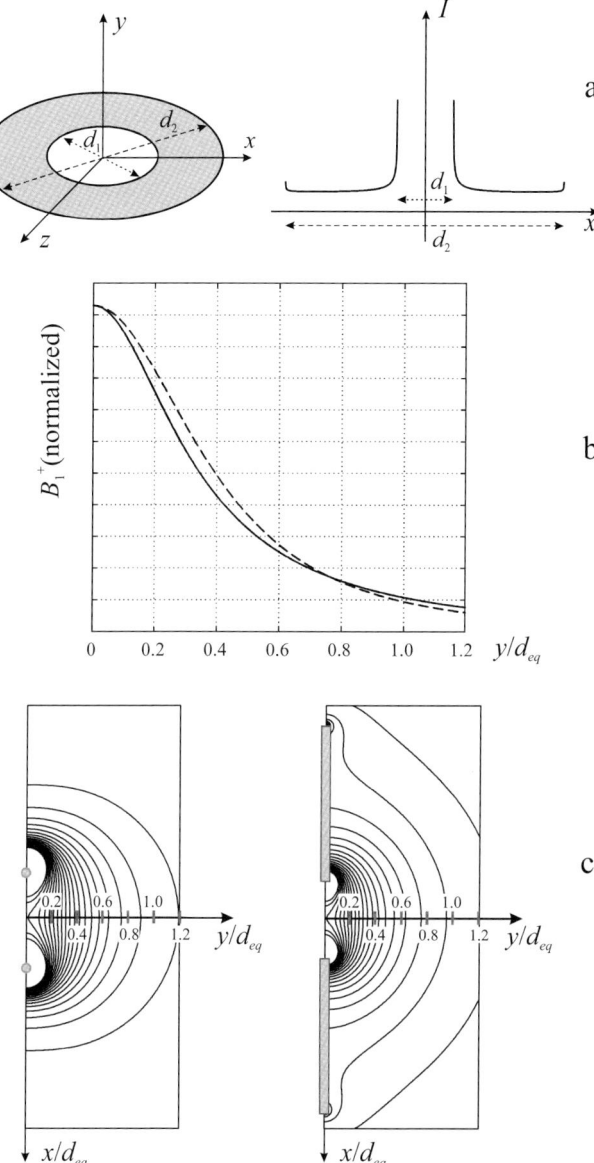

Fig. 9.11 Same as in Fig. 9.10, for a flat annular coil. In this case, the equivalent coil is concentric with the ring coil. It is the circular loop that produces the same field at the center when carrying the same total current.

The above considerations essentially apply to a one-turn coil having an impedance within the quoted range of 20 to 200 Ω. If the impedance of the one-turn coil is too low, it requires more turns that can be shaped either cylindrically or as a flat spiral, as already discussed. In these cases, the round wire as well as the flat copper foil can still be used, depending on practical considerations. However, at very low frequencies (below 10 MHz), the use of the so-called litz wire[9] instead of a solid copper wire of the same diameter is advantageous [Croon *et al.* 1999; Grafendorfer *et al.*, 2005; Dominguez-Viqueira *et al.*, 2010].

9.1.2.5 Opened resonators

The surface coil is basically an opened resonator. Hence, any opened homogeneous resonators can be used as a "surface coil." In particular, the single surface coil can be viewed as a split Helmholtz coil. The four coils of Section 8.1.2.2 can be split to form two coil resonators similar to those described later, in order to improve the depth at which observation can be done.

Similarly, all the cylindrical resonators already described in the previous chapter can be used to form a half resonator, open coil, allowing observations to be made on any volume located more or less in depth, in any sample. The literature is full of a variety of shapes. The number of possibilities is limited only by the experimentalist imagination. Only a very limited number of designs will be quickly reviewed here (Fig. 9.12).

As a first example, consider the single wire. It is able to create a magnetic field and receive a signal from excited nuclei in its close proximity. A combination of parallel wires assembled in loops may be used as a planar surface coil (zig-zag coil [Nakada *et al.*, 1987]).

A half saddle coil is nothing other than a curved rectangular coil that can be shaped to fit a given sample volume. A particular mention should be made of the half birdcage resonator. It has interesting properties regarding the usable frequency range and depth of sensitive volume.

---

[9] Litz wires, also termed « Litzendraht conductors » [Terman, 1943, pp. 37] are described in Chapter 2 (Section 2.4.1.3).

All these shapes have in common that they create a heterogeneous magnetic field distribution that can be profitably used as a localization mean and eventually as an imaging tool (imaging in RF field gradients [Hoult, 1979; Friedrich and Freeman, 1988; Canet, 1997; Baril *et al.*, 2000]).

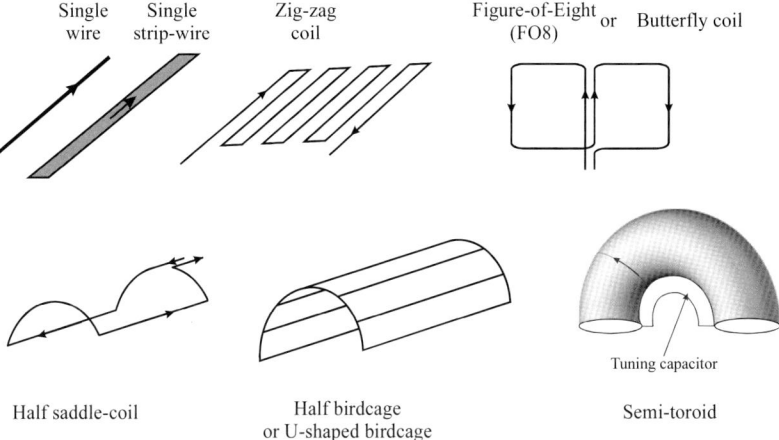

Fig. 9.12 Some shapes that may be used as an open resonator (the single wire or the flat strip, the zig-zag coil [Nakada *et al.*, 1987], the Fo8 [Bottomley *et al.*, 1989; Alfonsetti *et al.*, 2005], the half saddle coil, the half birdcage or U-shaped birdcage coil [Gasson *et al.*, 1995; Hudson *et al.*, 2000] and the semi-toroid [Assink *et al.*, 1986; Grist *et al.*, 1986]).

9.1.3 *Surface coils for ultra high frequency*

9.1.3.1 *Segmented loop*

As already extensively discussed, the impedance of the coil should not be "too large." A limit can be established as a function of the coil dimensions respective to the wavelength. The impedance of a circular coil of mean diameter d_m wound with a round wire of diameter ϕ can be expressed as (Eq. 2.97):

$$Z = L\omega = 0.4\pi^2 d_m \nu K , \qquad (9.21)$$

where Z is in Ω, the dimensions in mm and ν in GHz. K is a number that depends on the ratio of d_m and ϕ:

$$K = \ln\frac{8d_m}{\phi} - 2. \qquad (9.22)$$

For "reasonable" practical designs, K ranges from about 2.5 (corresponding, for example, to a small coil of $d_m = 11$ mm made with a wire of $\phi = 1$ mm) to a maximum of 3.7 (corresponding to a large coil with a ratio of d_m/ϕ of 40; for example, a surface coil of 10 cm diameter and a wire of 2.5 mm, or a 30 cm diameter coil made of plumbing tube of $\phi = 8$ mm). The impedance as a function of the diameter d_m to wavelength λ ratio is [from Eq. (9.21)]:

$$Z = 120\pi^2 K \frac{d_m}{\lambda}. \qquad (9.23)$$

Assuming a limiting value of d_m/λ of 1/20, the maximum impedance that is acceptable for being well below the self-resonance limit is therefore comprised between 150 to 220 Ω (depending on K). This is obviously an order of magnitude, justifying the already quoted limiting range of 20 to 200 Ω. It is also to be noted that an impedance of 200 Ω corresponds to a capacitance of 5.3 pF at only 200 MHz. This provides an idea of the order of capacitance magnitude acceptable for tuning an NMR coil.

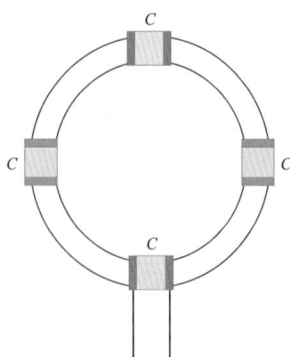

Fig. 9.13 Segmentation of the loop by capacitors along its length reduces its electrical length, hence allowing a large loop to be used at high frequency.

As a rule, the wire dimension of the coil[10] should be kept to a minimum of $\lambda/20$. If this is not the case, a routine method consists of just segmenting the coil. All the consecutive segments are connected together in series through capacitors (Fig. 9.13).

If L is the total, unsegmented inductance, the required capacitance is therefore equal to:

$$C = n/L\omega^2, \qquad (9.24)$$

where n is the number of segments. If the number of segments is unnecessarily increased, the resistance of the probe may increase due to the capacitors and the soldering connections.

9.1.3.2 *The crossover coil*

Reducing the inductance of a circular coil may be done by using copper foil instead of round wire but at the expense of an increase of the dielectric loss or a reduction of the magnetic field penetration. An improvement in designing such a coil, known as the "crossover coil" (Fig. 9.14), has been proposed by Nagel *et al.* [Nagel *et al.*, 1990]. The main advantage of the design is that it is explicitly balanced with respect to the ground. Furthermore, the electric fields are effectively shielded from the tissues, minimizing the dielectric losses and the high flux regions that do not enter the sample. Finally, the coil can be built using a thin stripline glued on a flexible support allowing the probe to be shaped as close as needed around a given sample.

The center point of the coil is effectively grounded, providing a "true" balancing. The capacitive matching network must therefore be symmetric and a balun should be inserted to connect the symmetric probe to the asymmetric coaxial cable.

---

[10] This is valid for a "classical" coil. The microstrip line resonator (Section 9.1.3.4) has, in contrast, a conductor length of $\lambda/2$! Due to the resulting nonuniform current distribution along the coil conductor, one expects (and indeed observes [Zhang *et al.*, 2001b]) a distortion of the magnetic field distribution.

The crossover coil can be viewed as a curved strip transmission line split in the middle, where the conductors are crossed together. Hence one upper strip is connected to the bottom ground connection of the opposed line. The ground strips close to the sample are made larger in order to effectively shield it from electric fields. The magnetic field in the sample is created by the common mode currents that flow in opposite directions in the two halves of the stripline (Fig. 9.14). The magnetic field distribution in such a structure should therefore be similar to that of the simple loop.

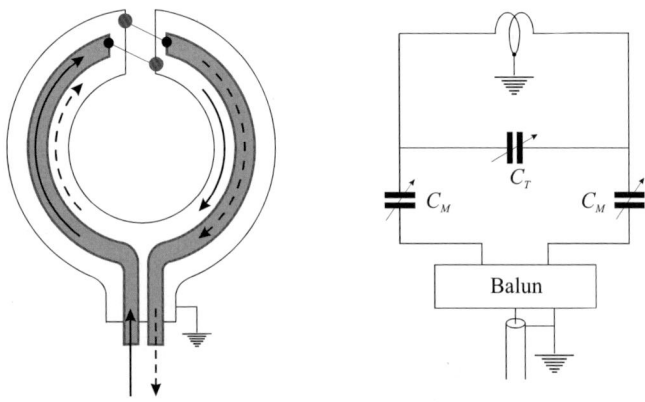

Fig. 9.14 A schematic view of the crossover coil (left) and its equivalent circuit (right). It is made of two copper striplines separated by a dielectric material. The strip on the down face is connected to the ground and is closest to the sample. The upper strip forms a piece of transmission line with the opposed grounded strip. A slit is made in the transmission line path. At this place, connections are realized as shown, favouring the common mode currents (currents circulate in the same direction in the two parallel conductors of the transmission line pieces).

9.1.3.3 *Split ring resonator (common mode)*

The crossover design is to be compared with the split coaxial loop [Harpen, 1994; Stensgaard, 1996], which is also similar to the "Twin-Horseshoe" Resonator (THR), [Gonord and Kan, 1994]. In this case, the balancing is "implicit." The latter is a particular case of the Parallel-Plate Split Conductor Resonator (PPR), [Gonord et al., 1988]. These resonators are shown in Fig. 9.15.

All these designs share the advantage that the electric field is confined in the line dielectric and the slit in the transmission line is designed in order to unshield the magnetic field. Furthermore, they are expected to have dimensions much smaller than a wavelength (in vacuum), a condition similar to that which applies to any "classical" wound coil. The main difference from a wired circular coil is that the resonance is given by the distributed inductance and capacitance rather than by lumped elements.

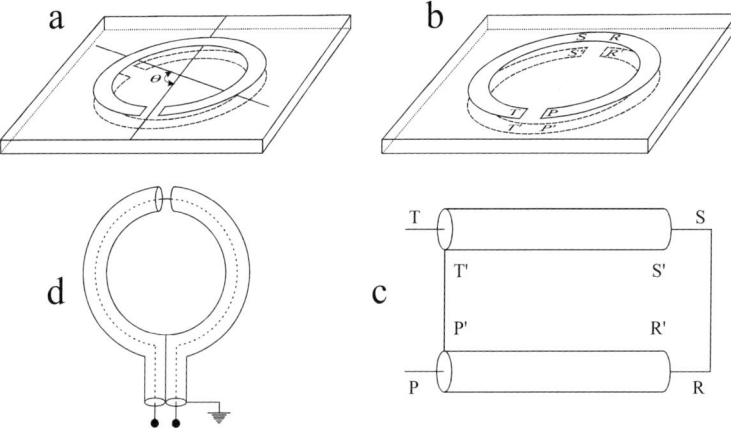

Fig. 9.15 (a) The split ring resonator and (b) its "twin-horseshoe" implementation. (c) The equivalent representation (using coaxial lines) of the resonator showing the similarity with (d) the split coaxial circular coil.

As already suggested, the common mode currents are favored in these resonators in order to get a magnetic field in the sample space, outside the line dielectric. In the special case of the symmetric THR where the slits are diametrically opposed [Fig. 9.15(b)], the resonant frequency f_0 ($\omega_0/2\pi$) of the common mode is given by [Ginefri et al., 2007]:

$$\frac{L_t \omega_0}{4Z_0} \tan\left(\frac{\beta l_t}{4}\right) = 1, \qquad (9.25)$$

where β is the propagation constant of the line ($2\pi/\lambda$) at the resonance frequency, l_t is the circumference of the resonator. Z_0 is the characteristic impedance of the line, which depends on the permittivity and thickness of the substrate and on the width of the strip (Chapter 2). These characteristics do not depend on the line shaping which may be linear, rectangular, or circular. L_t is the "total inductance" of the circuit. For example, in the case of a one-turn THR, L_t is given by the series inductance of the two rings, taking account of the mutual between the rings ($L_t = 2(L + M)$).

The THR can be tuned at any frequency lower than its self-resonance, given by Eq. (9.25) by adding two capacitors C_1 and C_2 across the slits (TP and SR in Fig. 9.15). The corresponding resonant frequency is given by an equation already derived by Gonord and Kan [Gonord and Kan, 1994] as:

$$\begin{aligned}&[1-\Lambda(y_1+y_2)/2]\cos(\beta l_t/2)\\&-[\Lambda/4-\Lambda y_1 y_2+y_1+y_2]\sin(\beta l_t/2)\\&+1=0\end{aligned} \qquad (9.26)$$

where

$$\begin{aligned}\Lambda &= L_t \omega / Z_0\\ y_1 &= C_1 \omega Z_0\\ y_2 &= C_2 \omega Z_0\end{aligned} \qquad (9.27)$$

This kind of resonator has a wide range of applications, with the main advantage that the electric field is confined in the dielectric, not in the sample. They are also easy to build and are not cumbersome.

9.1.3.4 *Microstrip coils (differential mode)*

The slit in the split transmission line design is introduced in order that a substantial magnetic field exists outside the inner space of the line.[11] In fact, a non-slit transmission line can still be used as an NMR probe component taking advantage of the good quality factor of half-wave resonators. A substantial magnetic field outside a microstrip transmission

---

[11] The line functions in the common mode.

line can indeed be obtained when the thickness of the dielectric material is sufficient [Zhang *et al.*, 2001].

When the distance between the strip conductor and the ground plane is of the order of the width of the strip, the lines of forces extend outside producing the required magnetic field in the sample volume [Fig. 9.16(a)].

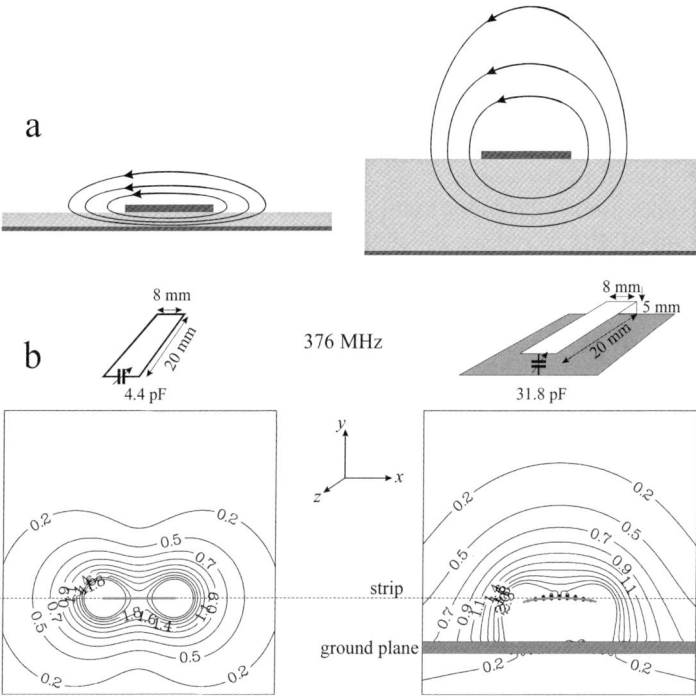

Fig. 9.16 (a) The magnetic field created by a microstrip line, functioning in the differential mode and having much larger width than the substrate thickness, is concentrated close to the line (left). When the dielectric is thicker, the magnetic field extends away from the strip (right). (b) This property allows the design of surface coils based on microstrip lines. The example here is a strip coil tuned and matched to 376 MHz (^{19}F, 9.4 T). The field plots (B_1^+ field amplitude in gauss, assuming an excitation power of 10 W and $Q = 100$) show that the magnetic field of the strip probe extends farther than that created by a rectangular coil of similar dimensions. Furthermore, the tuning capacitor is about eight times greater, indicating a large reduction of the electric field in the space around. Note that the strip is slightly curved to move the sample away from the high amplitude RF field existing at the strip edges.

The resonant frequency is simply given by the condition that the total length of the line is equal to $\lambda/2$.

$$f = c/2l\sqrt{\varepsilon_{eff}}, \qquad (9.28)$$

where c is the speed of light in a vacuum and ε_{eff} is the effective permittivity of the substrate. Note that ε_{eff} differs from the dielectric constant ε_r of the material due to the fact that the electric field lines are partly in air and in the substrate. As a result, ε_{eff} is close to ε_r when the strip width is much larger than the substrate thickness, which is definitely not the case for a usable NMR microstrip probe. Expressions for ε_{eff} are given in the Appendix.

In contrast with the other mentioned resonators, the dimensions are of the order of the wavelength, hence the current amplitude and phase varies along the coil. This has the consequence that the magnetic field distribution is asymmetric in directions parallel to the coil plane. Some correction of this effect can be obtained with a two-turn microstrip winding [Fig. 9.16(b)], but obviously at the price of the reduction of the frequency × dimensions product [Doty, 1999].

This type of resonator is best adapted for a surface coil design that is intended to limit the observation of volume close to the probe. Furthermore it allows the considerable improvement of the accessible upper limit of usable frequencies. It also has the advantage that the probe does not need to be shielded as would be the case with a large classical coil.

9.1.4 Superconducting surface coils

In NMR applied to biological samples, the main sources of noise acting on the probe coil are related to the intrinsic coil resistivity (coil noise) and sample conductivity.

In an attempt to get an increased SNR the easiest way is to accumulate the NMR signal for a sufficient period of time, which is often a technique employed for spectroscopic experiments or *ex vivo* MRI ones. Reducing the time for an *in vivo* MRI experiment represents a normal constraint nowadays, to which an experimentalist should adhere.

Under these conditions one should think about reducing the noise, instead of trying to increase the signal. Since the noise sources are directly related to the intrinsic resistance of the probe and to the conductive sample losses, two solutions appeared to solve the overall noise problem: reducing the probe temperature (thus the thermal noise) and the probe size.

One of the first implementations of cryogenic probes was presented by Styles [Styles *et al.*, 1984], reporting a theoretical improvement of the SNR by a factor of 22 for a liquid He-cooled NMR high resolution probe. This first experiment at liquid He temperature was performed with a normal size, Cu-made probe. Superconductive materials provide a much lower coil noise than the normal conductors, at the same temperature. The magnetic field penetrates deeply inside the material, forming an array of quantum flux vortices. Normal free electrons occupy the vortices giving rise to normal electrical conduction mixed with the Cooper pair, superconductive one. The theory predicts an RF surface resistance proportional to ω^2, the same dependence as the sample losses. This is why the very large Q factors are obtained with cryogenic probes for low frequency precessing nuclei at high B_0 fields. NMR surface coils of different frequencies and sizes were described in the literature with unloaded Q factors up to 500,000, as reported by Black, Odoj and others [Black *et al.*, 1993; Odoj *et al.*, 1998].

On the other hand, reduced sample dimensions can greatly increase the RF sensitivity if the dominant noise comes from the sample. The noise induced in very small size coils varies more rapidly than the NMR signal at close distances from the sample. An extreme limit of such very small size coils is represented by the so-called pinpoint probe developed by Serfaty [Serfaty *et al.*, 1994]. This probe is considered as having truly infinitesimal dimensions compared to sample or placement distances. Calculating the behavior of pinpoint probes, for optimum performance, one gets results only applicable for such cases. For example, Serfaty got the best SNR when the pinpoint probe was placed away from the sample (at a distance s), as a function of the penetration depth, (d), $s = d/5$.

Combining the small size probe design (loops of a few mm diameter) with low temperature conditions, the cryogenic MRI probes attempt high resolution performances, arriving now at a Field Of View (FOV) of less than 4 mm, as reported by Darrasse and Ginefri [Darrasse and Ginefri, 2003].

Diminishing the coil dimension induces a decrease of its inductance value. In order to keep up the frequency of interest, one has to increase the capacitance value correspondingly, increasing the losses. The solution chosen by Darrasse and Ginefri is to replace the classical coil by monolithic, self-resonant probes. They are constituted by two microcircuits separated by a dielectric surface. At current MRI frequencies, High Temperature Superconductor (HTS) thin films offer a surface resistance of a few orders of magnitude lower than copper surfaces of the same size, offering new possible developments for the monolithic probes. High quality films are grown on a single crystal of a low-loss dielectric, such as lanthanum aluminate or sapphire. Each element of the circuit is constituted by an inductive loop and two circular concentric strips constituting the corresponding capacitance. In this way, high resolution images (up to 10 μ) were obtained.

9.1.5 *Interfacing the coil to the spectrometer*

Interfacing the probe to the spectrometer has already been discussed in Chapter 4. In the case of the surface coil, some particular points need further development.

Because the coil is generally intended to be used with living systems, particular attention should be paid to the coil equilibration. This can be done in a convenient way through an inductive coupling, owing to the fact that the magnetic field perturbation is rather limited (Section 9.1.1.2). However this solution is not always possible for limited space reasons or when the surface coil is used as a receive-only probe placed in a homogeneous resonator. In these cases, a capacitive coupling scheme is more convenient.

When the coil is implanted in tissues, the tuning and matching capacitors are no longer accessible. Hence, a remote impedance matching control is required. The original scheme, [Fig. 9.17(a)], proposed by Murphy-Boesh and Koretsky, [Murphy-Boesh and Koretsky, 1983] has been proved to be efficient.

However, care should be taken of the signals (due to common mode currents) that may be received from the imperfect transmission line[12] which relates the implanted coil to the remote matching network.

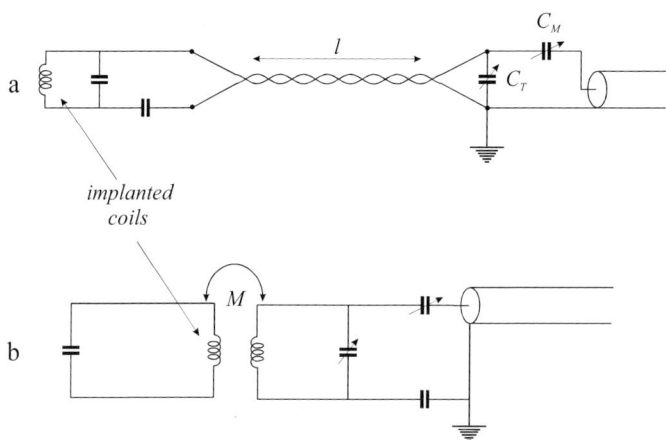

Fig. 9.17 Two solutions for relating an *in vivo* implanted coil to the external electronic circuits. (a) In the original version proposed by Murphy-Boesh and Korestsky [1983], the high resistance of the probe resonator is lowered to a value close to the system characteristic impedance by a "pre-matching" capacitor and connected through a piece of symmetric transmission line to an external circuit that finalizes the impedance matching. The series matching capacitors are positioned in order to take care of the electric balancing. (b) In the design of Schnall *et al.* [Schnall *et al.*, 1986a] the probe resonator is connected to the external remote control circuit through an inductive coupling.

Another interesting solution has been presented by Schnall *et al.* [Schnall *et al.*, 1986a] where the implanted resonator is inductively coupled to a remote resonator that is externally impedance matched to the spectrometer [Fig. 9.17(b)]. In this case, undesired signals may arise from the tissues comprised between the two coils. These can be minimized by choosing the "RF gradient mode" of the coupled resonator (corresponding to coil currents circulating out-of-phase, appearing at high frequency). Elimination of parasitic signals can be subsequently improved with "crushing gradients" of the static magnetic field.

---

[12] This could be a twisted pair of isolated wires or a coaxial cable of very small diameter (less than 1 mm).

9.2 Extending the Observed Volume (Multi-Ring Coils)

The SNR of localized spectroscopy in a given volume, or of an MR imaging experiment, relies on the spatial uniformity of the magnetic field in the volume of interest. On the other hand, the single surface coil exhibits a large RF field gradient and a depth-limited sensible volume. This property was at the origin of the surface coil design. The need for increasing the observation depth while retaining the enormous gain in sensitivity of the surface coil compared to a whole body resonator, led to new solutions for improving the simple loop properties [Adriany and Gruetter, 1997]. To this end, it has been proposed to combine the magnetic field created by two or more inhomogeneous coils [Styles *et al.*, 1985; Bendall *et al.*, 1986b; Volotovskyy *et al.*, 2003]. The characteristics of these coils have been examined in detail in a paper by King *et al.* [King *et al.*, 1999].

Generally, when the magnetic field is profiled in this way, it results in a decrease of the magnetic field amplitude. But an in-depth larger homogeneous volume is gained.

9.2.1 *Coaxial rings*

9.2.1.1 *RF field profiling*

The principle of profiling the magnetic field distribution of a single inhomogeneous surface coil is to compensate for its inherent RF gradient, in a given volume. The simplest method is to use two coaxial, surface coils [Fig. 9.18(a)], driven by currents of predetermined amplitude and phase. The magnetic field created on-axis by the two-coil system can be expressed as:

$$B_y(\psi) = B_{coil1}(0) \times \left\{ \left[1 + (2\psi)^2\right]^{-3/2} + \alpha/\beta \left[1 + \left(2(\psi - \delta)/\beta\right)^2\right]^{-3/2} \right\}, \quad (9.29)$$

where ψ is the on-axis position, normalized respective to the diameter of coil 1 (d_1). The characteristics of the field profiling coil 2 (dimensions, positions, and currents) are normalized respective to the corresponding characteristics of coil 1. Specifically, α and β are the normalized current and diameter of coil 2; δ is the displacement of the plane containing coil 2 respective to coil 1; n_1 and n_2 are the number of turns of coil 1 and 2:

$$\alpha = n_2 I_2 / n_1 I_1$$
$$\beta = d_2 / d_1$$
$$\psi = y / d_1 \quad . \quad (9.30)$$
$$\delta = \frac{y_{02} - y_{01}}{d_1}$$

As an example, Fig. 9.18(b) shows the on-axis profile of a double concentric coil system for a particular diameter ratio and different current ratio. In this way, the RF field can be made homogeneous in a predetermined volume.

The two-coil system can be used in different configurations. In their original paper, Styles *et al.* [Styles *et al.*, 1985] used two concentric coils, the larger coil as a transmitter coil and the smaller one as a receiver. The current induced in the smaller one (the receiver coil) during the transmitting period profiles the magnetic field allowing a homogeneous nuclear spin excitation in a large volume, close to the receiver coil. The signal collected by the smaller receiver coil benefits from its good SNR characteristics. When combined with depth pulses, their localization properties are improved compared with those obtained with a single coil system [Bendall, 1983].

Another possibility is to use both coils as a transmit/receive probe [Tomanek *et al.*, 1997]. In this case, the two-coil system operates midway between the volume coil and the surface coil while retaining the advantages of both [King *et al.*, 1999].

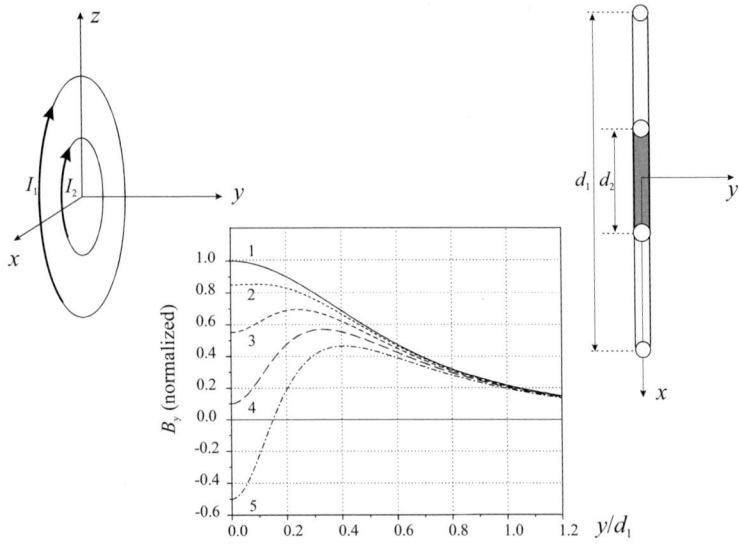

Fig. 9.18 On-axis profiling of the magnetic field created by a two concentric coil system, as a function of the current ratio [$\alpha = I_2/I_1$, Eq. (9.30)]. The dimension (y) is normalized with respect to coil 1. The magnetic field amplitude is normalized with respect to the field created at the coil center when $I_2 = 0$. In decreasing order from the center coil, the ratios are $\alpha_1 = 0.0$; $\alpha_2 = -0.05$; $\alpha_1 = -0.15$; $\alpha_1 = -0.3$; $\alpha_1 = -0.5$. The coil diameter ratio $\beta = d_2/d_1$ is equal to 1/3.

The design of Tomanek *et al.* and King *et al.* includes two coaxial coils (Fig. 9.19), driven by opposed phase currents. The on-axis field profile is similar to the one presented in Fig. 9.18. The current ratio is controlled by the mutual and the self inductances of the two coils, as well as by the resonant frequencies of the isolated circuits. The two-coil system is impedance matched through an inductive coupling loop (the magnetic field contribution of the coupling coil is neglected).

Another design [Hernandez *et al.*, 2003] combines two concentric coils, one rectangular and one circular, which are coupled inductively and, via added capacitors, capacitively too. The whole system was used as a receive-only probe placed inside the whole body transmitter resonator. The use of a complex coupling network between the two coils may be an efficient means to better control the currents and phases all around the circuit in order to meet the localization and sensitivity requirements.

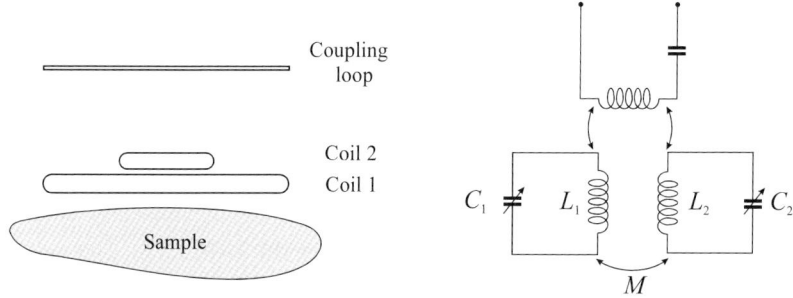

Fig. 9.19 Schematic geometry of the two coaxial coil magnetic field profiling system of Tomanek [Tomanek *et al.*, 1997]. The equivalent electrical circuit is shown to the right.

Whatever the configuration used, the optimization of the coil system geometry and of the current ratio can be made from estimation of the field distribution predicted by simple theory such as the Biot–Savart law. Once the magnetic field distribution has been optimized, the current ratio can be obtained from adjustment of the electrical parameters (inductances, mutual, tuned frequencies for each coil etc.) of the multiple coil circuit. This can easily be done using a simple linear circuit analysis tool.

9.2.1.2 *X-observed, proton decoupled system*

When a small surface coil is used for X-nuclei (e.g., ^{13}C) and a larger concentric coil is used for proton decoupling, the impedance presented by the observing coil is so small at the proton frequency that the induced current creates a magnetic field that tends to cancel the decoupling field in the sensitive volume. One cannot use an electrical decoupling of both coils to solve the problem because the ^{1}H RF field is required during the receiving period. In fact, it has been shown that such a configuration can benefit from a homogeneous decoupling RF field[13] in a region that coincides with the optimum sensible volume of the X-nuclei coil [Tiffon et al., 1986]. In this case, the small X coil plays the role of the RF decoupling field profiling coil.

The optimization of the two-coil system consists, as for any multiple coil system, of determining the dimensions and the current ratio into the two coils. The optimization has been proved to be dependent on only two parameters. The first one is the diameter ratio of the two coils that controls the relative magnetic field created by both coils (for a given current) and the mutual inductance between the two coils. The second parameter is the inductance, hence the wire diameter, of the X-nuclei coil. The inductance of the proton coil is free, having practically no influence on the magnetic field distribution at the X-nuclei frequency.

Quantitatively, the on-axis field profile of the coil system is given by Eq. (9.29). The current ratio may be expressed as:

$$\frac{I_2}{I_1} = -\frac{M}{L_2}\left[\frac{x^2}{x^2-1}\right], \qquad (9.31)$$

where x is the ratio of the resonant frequencies of the coupled resonators ($x = v_1/v_2$). I_2 is the current induced in the field profiling coil of inductance L_2 at frequency v_1.

Hence, at the carbon frequency ($x = 1/4$), the factor in brackets in Eq. (9.31) is small (1/15). Consequently, the current induced in the proton coil can be neglected. In contrast, at the proton frequency ($x = 4$) the corresponding term is close to unity (16/15) and the current induced in the smallest carbon coil efficiently changes the magnetic field distribution created by the proton coil. M does not depend on the coil wire dimensions, but the carbon coil inductance L_2 depends critically on its wire diameter. At this point, it should be noted that the field profiling contribution does not depend on the *number of turns for both coils*! (M is proportional to $n_1 n_2$, L_2 to n_2^2, hence I_2/I_1 is proportional to n_1/n_2. It results that α, in Eq. (9.30), is independent of n_1/n_2).

---

[13] Allowing phase modulated pulse sequences like WALTZ [Shaka et al., 1983] to be efficient.

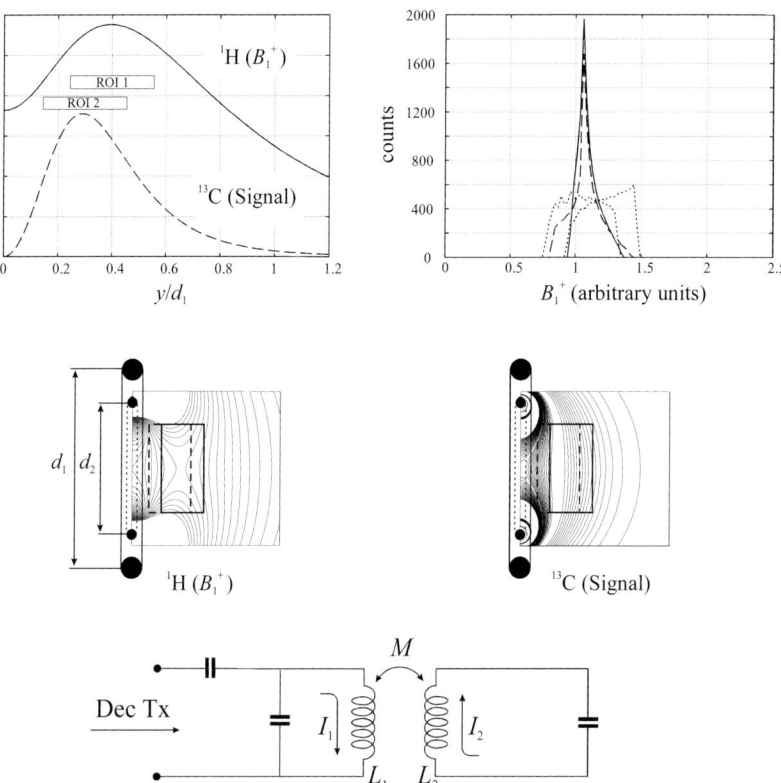

Fig. 9.20 An optimized two-coil {^{1}H}–^{13}C system. The larger coil is the decoupling proton coil. The smaller is the transmit/receive coil at the ^{13}C frequency. The on-axis (y) RF decoupling field profile (full line) is shown upper left (y is normalized with respect to the larger coil diameter). The dotted line represents the ^{13}C signal intensity that would be obtained assuming an excitation pulse angle of 180° at the coil center. ROI1 corresponds to the region where the proton field is homogeneous and ROI2 where the ^{13}C signal is maximum using the 180° pulse excitation at the center. The two regions almost coincide for this design. The histograms of the decoupling RF magnetic field are shown upper right. The homogeneity of the two-coil systems in ROI1 (full line) and ROI2 (dashed line) is similar. It is however much improved, in the same ROIs, compared to the one that would be obtained with a single proton coil of the same geometry (histograms in dotted lines). The proton magnetic field map in the xy plane is shown lower left and the carbon sensitivity map is shown lower right, together with the location of the ROIs considered here.

As a result, the optimum configuration is given by the following inequalities [Mispelter *et al.*, 1989]

$$\phi/d_C < 0.1$$
$$1.4 \leq d_H/d_C \leq 2.0 \qquad (9.32)$$

where ϕ is the diameter of the inner coil wire (assuming round wire), d_C its radius and d_H the radius of the external proton decoupling coil (β of Eq. (9.30) is d_C/d_H). The corresponding magnetic field distribution for an optimized design is shown in Fig. 9.20.

9.2.2 *Array coils*

As shown before, the surface probes are local sensors, receiving the NMR signal from a definite region, sometimes smaller than the ROI. For specific imaging purposes, like spine imaging or other extended organs, a complex system was developed: the NMR phased array [Roemer, 1990]. Generally, it consists of an assembly of small loop coils,[14] each acquiring the NMR signal from a small region but all covering surfaces usually associated with volume imaging coils.

Each coil is associated with an independent preamplifier and receiver chain. The outputs of the receiver chains are digitized, phase shifted and combined in an optimum manner, dependent on the point in space that originated the signal.

Generally, a two-dimensional array of surface coils may be arranged on a straight plane or on curved surfaces when they surround the whole sample volume.

The use of array coils is subject to solving two major problems: inductive coupling between the coils and signal processing for obtaining the overall accurate image.

Throughout this book, we made use of the inductive coupling arising from proximal conductive parts in which a current is flowing. For array

---

[14] The original array coil is probably the Mansfield "petal coil" [Mansfield, 1988; 1992]. A variant of this coil was recently described and proved to provide a better SNR compared to a single-loop surface coil [Hidalgo *et al.*, 2001; Rodriguez *et al.* 2005].

coils the inductive coupling is something to be avoided, being a major cause of a low SNR and low sensitivity of such probes. First of all, if identical resonant loops are placed close to each other, the mutual inductance causes resonant frequency splitting, as shown in Fig. 9.21.

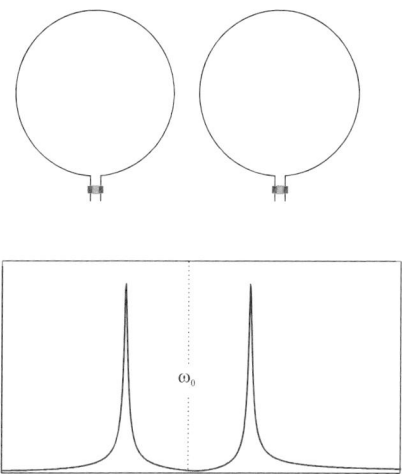

Fig. 9.21 Due to the mutual induction between the two coplanar coils they cannot be tuned independently to the same resonant frequency. The whole resonator has two resonant frequency modes for which both coils share their noise contribution, hence behaving as a single, larger coil.

The splitting results, first of all, in a loss of sensitivity at the resonant frequency. Moreover, signal and noise are also transferred from one coil to another via the mutual inductance, thus creating more problems than the array coil is intended to solve.

A well-known way to force the inductive coupling of close coils to zero is their relative positioning. For planar surface coils, the technique of overlapping is used to eliminate the mutual inductance of adjacent coils. Figure 9.22 shows the mutual dependence on the distance between two planar coils for two specific geometries: circular loop and square loop. As expected, the mutual inductance depends on the coil geometry, thus the optimum distance for overlapping that forces the mutual to zero, is different for the two presented examples, even for identical linear dimensions (the circle diameter and square length are considered equal).

Placing two or more coils in such a way as to minimize their mutual inductance, makes the signal acquisition possible both simultaneously and independently.

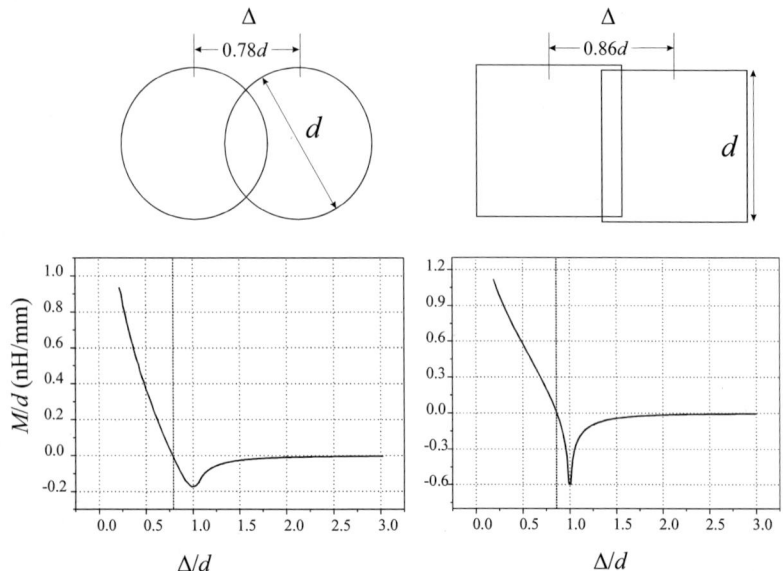

Fig. 9.22 Mutual inductance as a function of the center separation between two circular (left) and square (right) loops. In both cases, an optimum distance (when the loops overlap) exists, for which the mutual is equal to zero.

Overlapping the adjacent coils does not solve the mutual induction problem entirely since a small but significant coupling still exists between farther coils. Finally, this interaction is reduced to negligible levels by connecting the coils to low input preamplifiers. If the input impedance of the preamplifiers is very small, there is no net current flowing in the surface coils, the NMR signal being fed into the receiving chain mainly via the electric field. The absence of the current in the coils forces the remaining magnetic coupling between the coils to zero. The whole system is now behaving like multiple independent coils, thus keeping the benefit of very large field-of-view probes without the loss of sensitivity induced by the mutual. This technique is presented in detail by Roemer *et al.* in a paper dedicated to array coils and data processing [Roemer *et al.*, 1990].

9.3 The Surface Coil as a Receive-Only Probe

Another possibility to improve the surface coil properties (increased depth of observation and better B_1 homogeneity in the volume of interest) while retaining its advantage of an inherent high sensitivity is to use it as a receive-only probe. In this case, the spins are efficiently excited in a large volume by a homogeneous volume resonator, and the magnetization created in a given smaller volume is detected by the small surface coil positioned inside the resonator. It should be mentioned at this point that the combination of a receiving heterogeneous surface coil with a homogeneous excitation resonator has some difficulties in certain circumstances when the signal collected by the surface coil is given by the superposition of signals of varying phases, depending on their spatial localization [Haase, 1985]. This is the case in non-localized spectroscopy experiments and can be explained using the principle of reciprocity. For imaging (or localized spectroscopy) experiments, this effect will be of minor importance provided the voxel size is sufficiently small that the phase does not vary too much inside (the surface coil RF field is homogeneous in a small voxel).

The coupling between the two coils also has deleterious consequences for the B_1 field distribution and coil noise. During transmission, the current induced in the surface coil creates a magnetic field that adds or subtracts from the primary exciting RF field created by the resonator, hence producing a heterogeneous excitation in the vicinity of the surface coil (Fig. 9.23).

During receiving, part of the electromagnetic energy received by the surface coil will be dissipated in the resistance of the large resonator. Hence, the interaction of the resonator and the surface coil must be canceled out in this type of experiment.

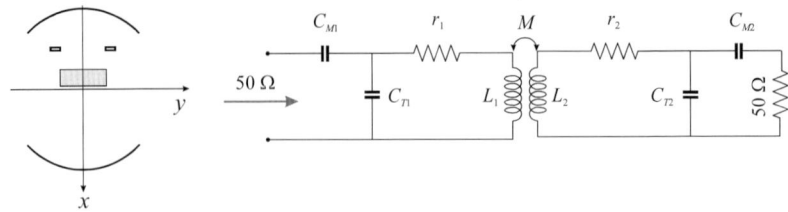

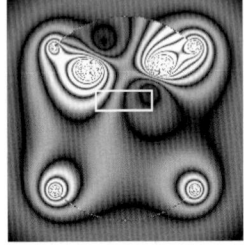 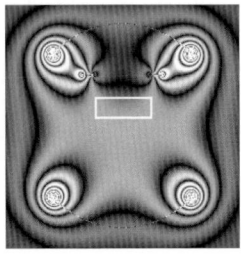 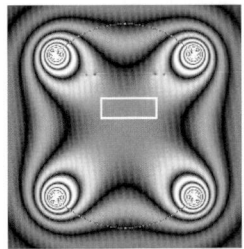

Fig. 9.23 Simulation of the magnetic field map (360° + 90° – imaging) that would be obtained with a homogeneous resonator (slotted cylinder) and a rectangular loop inside. The geometry is shown upper left and the equivalent circuit upper right. The surface coil is assumed "geometrically decoupled," i.e., its plane is parallel to the resonator field direction (Oy). It is clear that the mutual inductance is different from 0 (it would be zero if the rectangular coil were at the center of the slotted cylinder). Three cases have been considered. Firstly, the rectangular coil has the same length as the slotted cylinder. In this case, $k = 0.01$, the B_1^+ field distortion is very high (down left). If its length is reduced to 0.1 times the resonator length, k decreases to 0.001. There is still some distortion visible, even with so small a mutual inductance. Finally, the same rectangular loop is opened (i.e., $I_2 = 0$); the field distortion is practically undetectable, except very close to the coil copper foil, due to eddy currents. The simulation has been done using the utilities provided with the book (current calculations, magnetic field maps). The self and mutual inductance have been calculated using FastHenry [Kamon et al., 1994].

9.3.1 *Passive decoupling*

Passive decoupling of a linearly polarized coil and resonator can, in principle, be achieved geometrically. However, in practice it is difficult to perfectly decouple the two coils simply by adjusting their mutual orientation, except when their dimensions are very different. In addition, the sample itself can provide an efficient coupling pathway. Finally, capacitive coupling again couples both coils. Hence, in most cases, a geometric decoupling should be accompanied by another effective means that cancels out or, at least, minimizes the induced currents in each coil. This is done using switching diodes that will shift the resonant frequency of one resonator away with respect to the other during appropriate periods of the NMR sequences. In a more efficient design, the current in the resonator is canceled out at the working frequency, improving the isolation to a high level.

The principle of passive decoupling, using diodes, is based on the non-linearity of its current/voltage characteristics (Fig. 9.24).

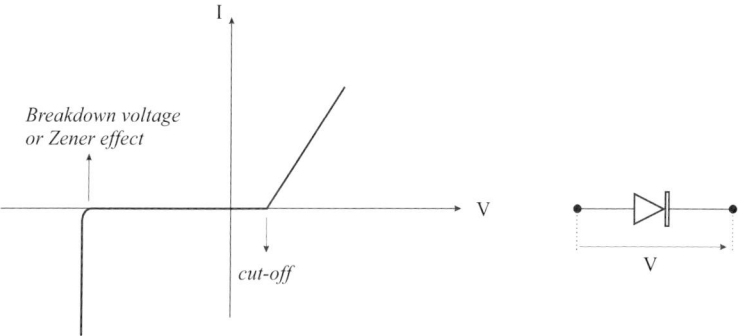

Fig. 9.24 Typical voltage/current characteristic of a semiconductor diode.

Typically, if the voltage across the diode is larger than a certain limit (for example, 0.6 V for silicon diodes), it presents a low resistance (typically of the order of 1 Ω). Below this limit, the resistance increases to several kΩ (the slope of the current/voltage characteristic is low). Finally, if the diode is zero or reverse-biased, it presents a very high resistance in parallel to a small capacitance. Hence, it acts as a switching device depending on the applied peak amplitude voltage.

During the transmitting period, the diode is biased to a high RF voltage and presents a low impedance. During the receiving period it presents a high impedance because the very small nuclear magnetization induction voltage is much lower than the cut-off voltage.

In a passive decoupling scheme, no external source is added to bias the diode resulting in a shape distortion (rectification) of the RF current [Fig. 9.25(a)]. Hence a crossed-diode circuit will be used instead of a single diode [Fig. 9.25(b)].

From a practical point of view, the diode operating at high frequency must have the smallest possible resistance in the conductive state and the smallest possible leakage capacitance in the opened state. Classical silicon diodes, such as the well-known 1N4118[15] (or 1N914), can be used for this application.

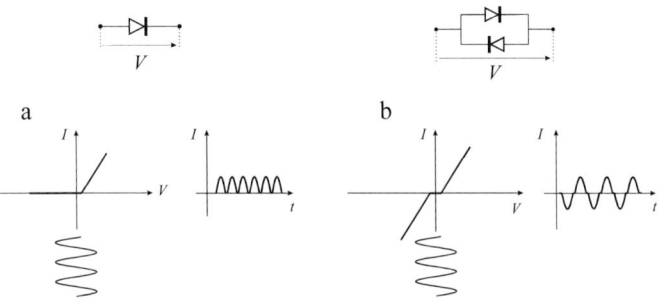

Fig. 9.25 When a sinusoidal voltage is applied to a single diode, the current is rectified (a), resulting in a large distortion of the RF current (and a large power loss at the frequency of interest). When crossed-diodes are used instead there is less distortion, depending on the ratio of the RF peak to the cut-off voltage. The ratio is very high (100 or more) during transmitting in an NMR experiment.

Better performances are nowadays obtained with the so-called PIN diode (see Chapter 2). This device operates as a variable resistance at RF and microwave frequencies, and can handle the high power that is sometimes delivered by NMR transmitters. Some manufacturers have even developed a particular line of nonmagnetic products, especially devoted to medical MRI. The properties of these diodes are essentially similar to those of a classical diode, but they are much improved respective to the decoupling needs. In particular, when the diode is forward-biased, its resistance is as low as 0.1 Ω provided the bias current is of the order of at least 100 mA, a condition easily reached during the transmission pulses. In contrast, the resistance increases to several kΩ

---

[15] These diodes are "small signal." The 1N4148 can handle a much higher current.

when driven by 1 μA.[16] Furthermore, when zero-biased, the capacitance is of the order of 1 pF and even less when reverse-biased.

A typical circuit for passive decoupling of the receiver coil is shown in Fig. 9.26.

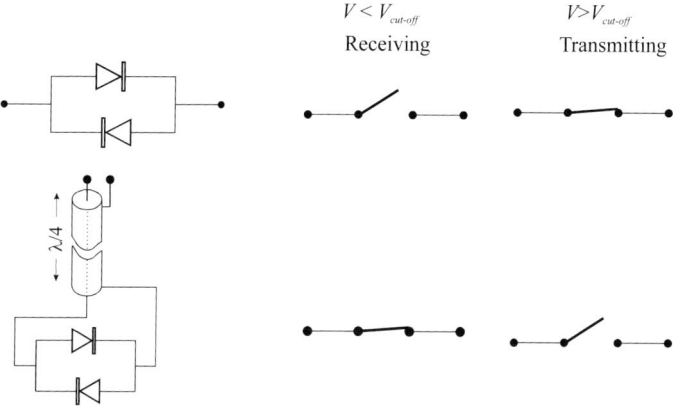

Fig. 9.26 Two basic circuits in which the crossed-diodes are used in a passive decoupling scheme. A quarter-wave transmission line allows the switch positions to be reversed.

It uses a quarter-wave transmission line terminated by crossed-diodes [Bendall *et al.*, 1986a]. During the transmission pulse, the induced voltage across the circuit is sufficient to induce a current in the diodes. A short at the transmission line end where the diodes are connected is thus realized, so a high impedance appears at the opposed end of the line. The circuit is inserted in the surface coil, opening the pathway for induced currents. When the diodes are non-conducting (receiving period) the impedance presented by the switching circuit is low (a small inductance), closing the receiver coil.

Due to the quarter-wave line, the switch functions correctly for a discrete set of frequencies; in fact on a relatively limited band around the resonant frequencies of the transmission line. Hence, the quality of

---

[16] The resistance of a PIN diode can be well controlled over four decades by the forward current. The PIN diodes have applications not only for switching but also in controlled, very wide band, attenuators.

decoupling depends on the transmission line adjustment. Finally, a variable capacitor can be soldered in parallel to the crossed-diodes (Fig. 9.27).

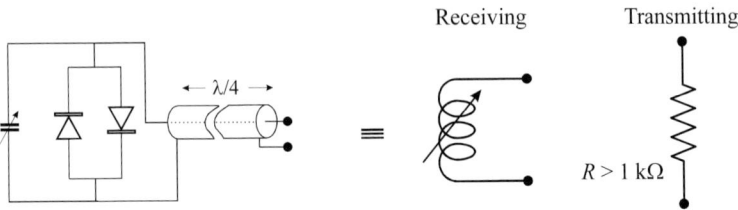

Fig. 9.27 A passive decoupling circuit of the receiver coil, using crossed-diodes (left), that also allows a remote control of the coil tuning. The equivalent circuit of the switch during receiving and transmitting is shown to the right.

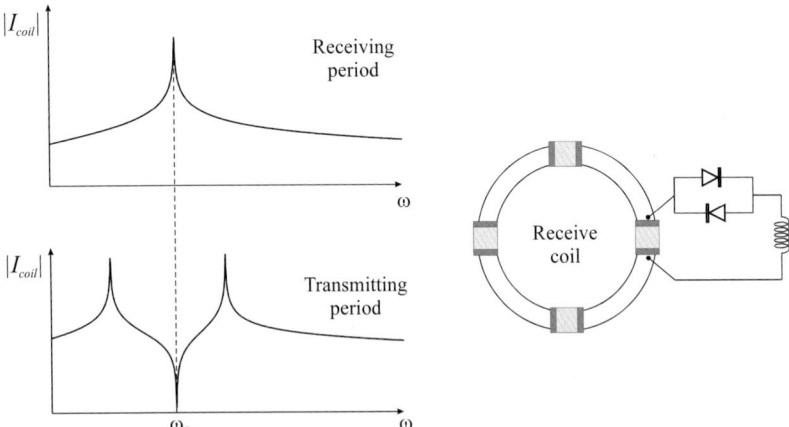

Fig. 9.28 The pole insertion method for passive decoupling of the receiver coil [Edelstein et al., 1986]. During receiving the diodes are opened, the inductor of the decoupling circuit is disconnected. During transmitting, the diodes conduct. The inductor forms a blocking resonant circuit with one of the tuning capacitors. Insertion of a resonant circuit into the coil resonator leads to two well-separated resonant frequencies (as already seen in Chapter 6). In addition, there is a dip in the current at the operating frequency.

This provides a means to remotely tune the coil during the receiving period, while this capacitor has practically no influence during the transmitter pulses (it is short circuited by the conducting diodes).

Another clever circuit has been proposed by Edelstein [Edelstein et al., 1986; Hyde et al., 1990]. An inductance is connected by the switching diodes in parallel to one of the surface coil tuning capacitors (Fig. 9.28), constituting a parallel resonant circuit (pole insertion). If the pole is tuned to the working frequency, the resulting circuit becomes an efficient blocking current device. This circuit also constitutes the basis of an efficient active decoupling device.

The decoupling of the transmitter coil can be done by simply inserting two crossed-diodes in the resonator circuit [Edelstein et al., 1986]. The diodes are fired by the transmit pulses but are opened during receiving. However, these diodes should absorb a very high current during the pulse.

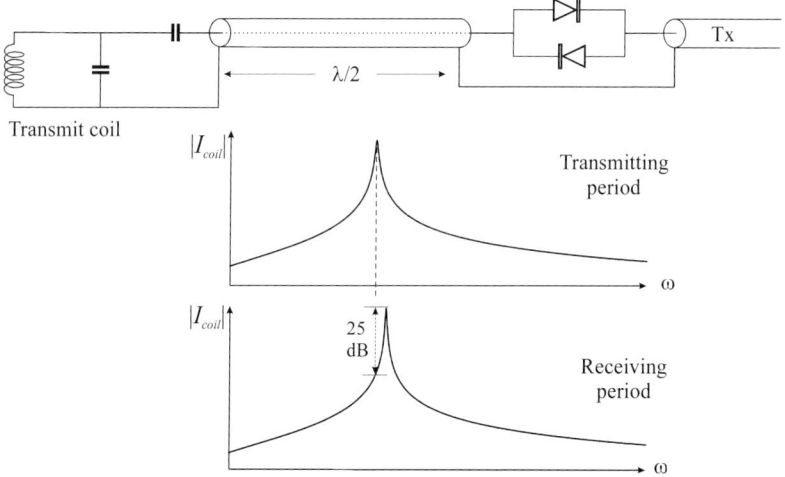

Fig. 9.29 Passive decoupling of the transmitter coil. The diode switch is inserted in the matched connection, where the current is much smaller than in the resonator. During the transmit pulse, the diodes connect the transmitter to the coil. During receiving, the opened diodes present a high impedance that is reflected at the input of the $\lambda/2$ line. The matching capacitor being out of the circuit, the resonant frequency of the transmitter coil increases slightly. This results in a moderate isolation.

In the circuit of Fig. 9.29 the crossed-diodes are inserted in the matching circuit [Haase, 1985; Bendall *et al.*, 1986ab], where the current is much less. During the transmit pulses the resonator is driven as required, while it is detuned during the receiving period.

The circuits described above are not balanced, as is generally required. This problem has been addressed by Picard *et al.* [Picard *et al.*, 1995ab]. The authors propose a passive decoupling scheme that pays particular attention to the electrical balancing of the probe, using twin coaxial lines (Fig. 9.30), and a balanced tuning/matching scheme [Chen and Hoult, 1989].

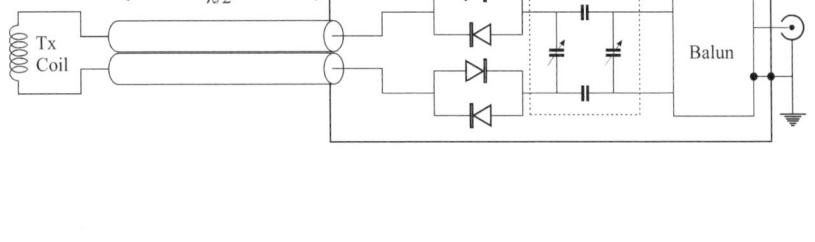

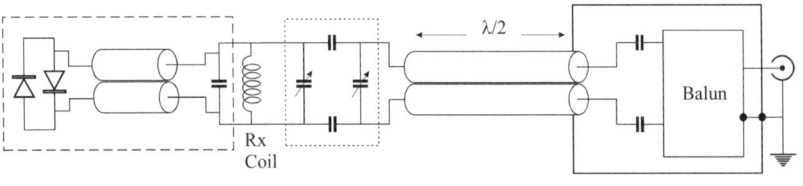

Fig. 9.30 An elaborate dual transmitter/receiver surface coil circuit that includes a passive decoupling of both coils [Picard *et al.*, 1995ab]. Passive decoupling of the transmitter coil is done by opening the coil circuit. Decoupling of the receiving coil is done by the pole insertion method.

Another balanced scheme is proposed in Fig. 9.31. It is very simple and efficient and provides a means to remotely fine tune the probe inside the magnet.

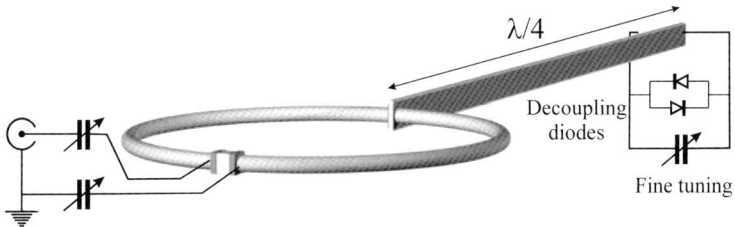

Fig. 9.31 Another electrically balanced decoupling circuit. The receiver surface coil is opened at the opposed side of the tuning capacitor. A quarter-wave stripline is inserted with two crossed-diodes and a remote tuning capacitor. The quarter-wave line, made using a piece of standard printed circuit board on FR4 quality epoxy, is relatively short (about 15 cm at 200 MHz) due to the high permittivity of the dielectric. At higher frequency, a substrate of better quality would be preferable.

9.3.2 Active decoupling

Active decoupling still makes use of the diode properties[17] as described above, but the method becomes particularly efficient when using a PIN diode. In this case, the diodes are biased by an external circuit [Edelstein *et al.*, 1986] that delivers current pulses, synchronized with the pulse sequence. Most of the actual commercial spectrometer and imager machines include the required circuitry.

The connections of the switching diodes to the diode driver device require a choke inductance (Fig. 9.32). This inductance presents a high impedance to the RF currents (Chapter 2) at the operating frequency.

A simple circuit is presented in Fig. 9.33 [Li and Sotak, 1991]. The circuit uses two PIN diodes of opposed polarity, one in the transmitter circuit, the other in the receiver circuit. In this configuration, the diodes being inserted directly into the resonant circuit may induce excessive losses, especially with small coils of intrinsically very low resistance.

---

[17] In previous designs [Bendall *et al.*, 1984] a reed relay is used in place of the crossed-diodes. However these devices are difficult to use in a strong magnetic field.

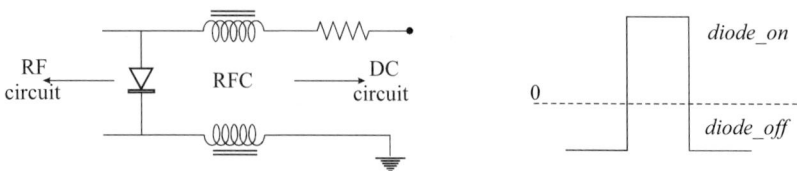

Fig. 9.32 The basic active decoupling circuit uses a PIN diode, biased by an external power supply. The RF circuit in which the diode is inserted is isolated from the current supply by the RF choke (RFC) inductance.

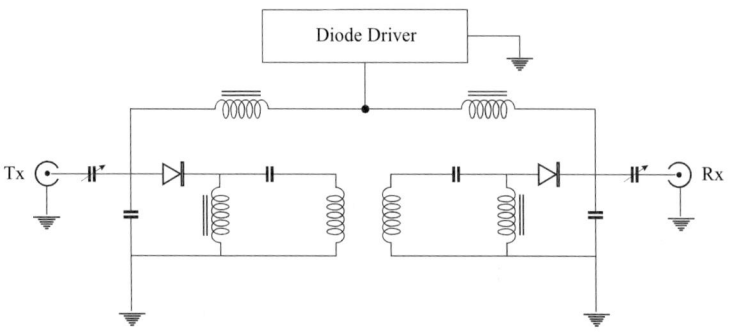

Fig. 9.33 A simple active decoupling scheme [Li and Sotak, 1991].

Another scheme is proposed in Fig. 9.34 which makes use of the "pole insertion" technique already described in the previous section. In this case, the symmetrizing capacitor of the matching network of the receiver coil is part of the pole insertion circuit that functions during the transmit period.

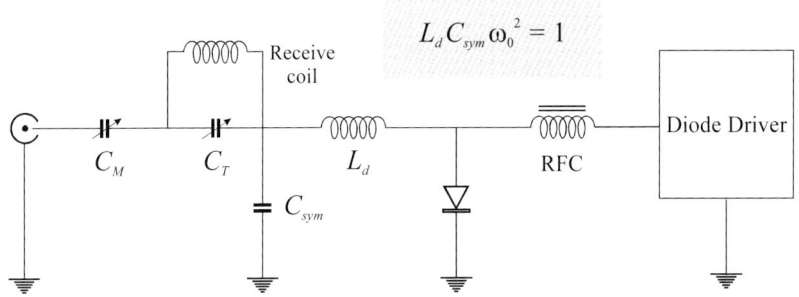

Fig. 9.34 An active decoupling circuit of a receiver surface coil that makes use of the pole insertion technique. The coil is assumed to be impedance matched by a classical capacitive network, as shown. The symmetrizing capacitor is used together with L_d to form a blocking resonant circuit when the diode is activated.

The advantage of such a scheme is that the conducting diode enters the receiver probe only during the transmit period. During receiving, the diode is opened, isolating the coil from additional noisy resistance and preserving the electrical balancing of the circuit. The typical resonant spectrum of such a circuit is shown in Fig. 9.35, where the diode is either conducting or opened. In the former case, two well-separated resonant modes are clearly observed.

The decoupling efficiency can be evaluated from the scattering parameter[18] S_{21} (or S_{12}) either in the final resonators or simply using a small coupling loop approached parallel to the surface coil [Fig 9.35(c)]. In the active decoupling schemes, care should be paid to the resonance frequency of the RFC tuned by the reverse capacitance of the diode. It has sometimes been observed that this circuit may couple with the surface coil through L_d with undesired effects if their resonant frequencies are close enough.

The transmitter coil can simply be decoupled using a passive scheme, while an active circuit may be used to improve the decoupling during receiving. In this case, the active detuning circuit should be inserted at a

---

[18] Or simply by looking at the transmission characteristics between the surface coil and a coupling loop.

point where the peak voltages are not too large, precluding that the diodes are fired during the pulses.

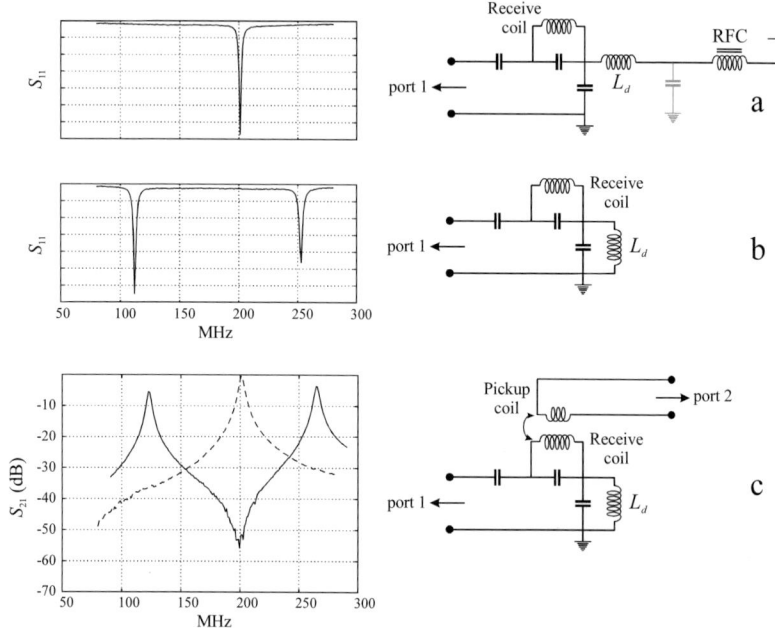

Fig. 9.35 The evaluation of the circuit in Fig. 9.34 using a (scalar or vector) network analyzer. During the receiving period, the diode is reverse-biased (a). It behaves as a small capacitance that can be neglected, provided it is not resonating with the RFC. The S_{11}-parameter is typical for a probe circuit tuned and matched at the desired frequency. When the diode is switched on (b), the inductor L_d comes in parallel with C_{sym}, resulting in a splitting of the resonant spectrum. In order to evaluate the efficiency of the decoupling, one may use a small coupling coil connected to port 2 of the network analyzer (transmission measurement). In this case, the S_{21} transmission scattering parameter allows the decoupling circuit to be evaluated and precisely adjusted (c, full line, the diode is on; dotted line, the diode is opened).

An example of a versatile transmission/receiver circuit, representative of the state-of-the-art, has been already described by Barberi *et al.* [Barberi *et al.*, 2000].

Chapter 10

Probe Evaluation and Debugging

The first step in construction of a probe design is a simulation of the circuit using the specific software available. Some have been described succinctly in Chapters 3 and 7. This step allows estimation of the component values required for tuning and matching the probe. The behavior of the probe at Radio Frequency (RF) could also be evaluated, providing information about the RF field distribution associated with each resonant mode, coupling between channels, etc... For a simple project, however, this is not necessarily required, but in every case the simulation will be of great help to guide the building process.

In a second stage, characterization of the probe on the RF workbench will be performed during each step of the construction. Again, a simple simulation could help to understand the observations.

In a third step, before the probe is delivered to the NMR experimentalist, one may evaluate its behavior in the presence of the sample loading (frequency shift, losses, tuning and matching capabilities, etc.).

Finally, the probe performance on the NMR machine will be compared with that expected in order to validate the probe design.

The information presented in this chapter is designed to help the reader find, in one place, the information that may be needed to complete the construction of a designed probe.

The very last step, especially important for coils that are to be used in a clinical MRI environment, is an evaluation of its safety (SAR). This requires elaborate software and detailed models of the human body [Nyenhuis, 2012; Shrivastava and Vaughan, 2012]. However this, in itself unique, topic will not be addressed in this book.

10.1 Instrumentation

From an electrical point of view, the probe, in its simplest form, behaves as a one-port network. The impedance presented at the input should be matched to a real 50 Ω value, at a particular frequency. Hence we need some method to visualize and eventually measure its impedance. This measurement requires a few instruments that are normally found in the RF engineer lab, but not necessarily in the NMR environment. Some of the instrumentation and measurement methods available for the adjustment and evaluation of a given probe are presented here.

Characterization of transmission line properties is also briefly described in this section. This could be useful, especially when designing circuits like quadrature hybrids, cable traps, $\lambda/4$ or $\lambda/2$ stubs. Also, the evaluation of the quality of a cable is important when it is intended to be used in a remote matching circuit.

Finally, the evaluation of the noise factor (F) of a preamplifier or of an entire receiver channel will be considered in the last section.

10.1.1 *The pick-up coil*

The first instrument which may be found on the probe designer workbench is a small "pick-up coil" (Fig. 10.1). It is very useful at every stage of the probe construction.

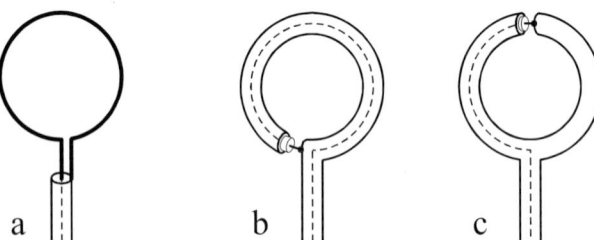

Fig. 10.1 Pick-up coils. (a) A simple small loop of wire soldered to a coaxial cable. (b) A shielded pick-up coil [Ott, 2009, pp. 696] which should be insensitive to the electric field component. The version shown here is a simplified (working) version of the one already proposed by Hoult [1978]. (c) A shielded and balanced pick-up coil [Carobbi and Millanta, 2004].

The pick-up coil is sensitive to the magnetic component of the RF field created by the probe. It must therefore be electrically screened [Fig. 10.1(b)]. To avoid any difficulties and errors arising from common mode currents on the service line,[1] it should be balanced [Fig. 10.1(c)] [Carobbi et al., 2000; Carobbi and Millanta, 2004].

The probe (or any resonator) resonant frequencies can be sampled by a peak-up coil loosely coupled to the probe coil and related to a frequency sweep generator (Fig. 10.2).

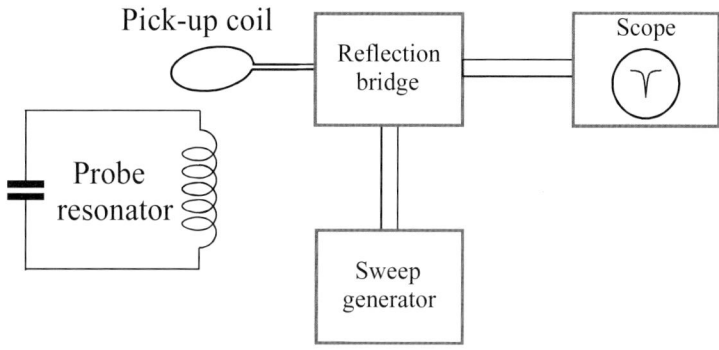

Fig. 10.2 Determination of the resonant frequencies of a given probe resonator using a pick-up coil and relatively cheap instruments.

The pick-up coil is also an invaluable tool for sampling the magnetic field amplitude all around a probe circuit. For example, the efficient resonant mode of a homogeneous probe can be quickly identified.

10.1.2 Impedance bridge

The impedance bridge is the obvious tool for measuring the input port impedance of the probe resonator. For the NMR frequencies (lower than one GHz), resistive impedance bridges (Fig. 10.3) are available. The 50 Ω reference is generally incorporated in the device. The voltage that results from the bridge imbalance, due to the difference between the

---

[1] The service line is the portion of transmitter line connecting the loop to the measuring instrument.

probe impedance and the reference value (50 Ω), is externally available on a connector through a broadband balun or sometimes through an incorporated crystal detector.

An impedance bridge, or any other device used for RF impedance evaluation, is characterized by its "directivity," directly related to the quality of the null obtained when impedances are matched together.

More precisely, the directivity represents the dynamic range available in evaluating the unknown impedance. It can be evaluated by the difference in the response of the device when the test port is connected to a short (or an open) and to a 50 Ω load. The directivity ranges between 30 dB for a fairly good to more than 45 dB for an excellent device.

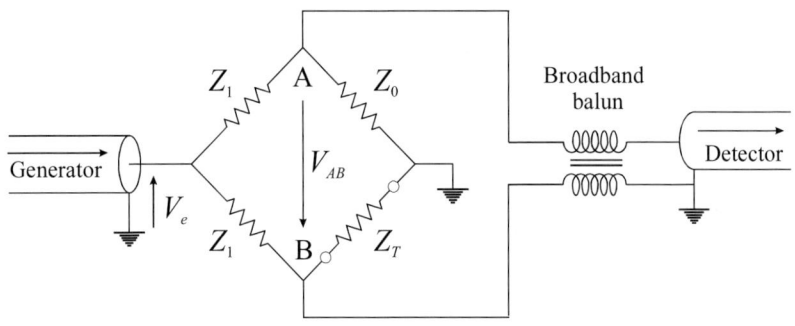

Fig. 10.3 Scheme of a typical impedance bridge. Resistive bridges are available up to 1 GHz. The balun is the most critical part.

Quantitatively, the output to input voltage ratio of a perfect resistive bridge is given by:

$$\frac{V_{AB}}{V_e} = \frac{Z_1(Z_0 - Z_T)}{(Z_1 + Z_0)(Z_1 + Z_T)}, \qquad (10.1)$$

where Z_T is the impedance to be measured, Z_0 is the reference impedance and Z_1 is the opposed branch impedance of the bridge.

Usually the response of the bridge is calibrated using a test set comprising a reference (50 Ω), a short and an open load. When the reference load is connected to the test port ($Z_T = Z_0$), the voltage ratio is ideally equal to 0. When the short ($Z_T = 0$) or the open ($Z_T = \infty$) loads are connected, the voltage responses are, respectively, given by:

$$\frac{V_{AB}}{V_e}(short) = \frac{Z_0}{Z_1 + Z_0} \qquad (10.2)$$

and

$$\frac{V_{AB}}{V_e}(open) = -\frac{Z_1}{Z_1 + Z_0}. \qquad (10.3)$$

When measuring impedances at high frequency, it is of interest to read directly the so-called reflection coefficient, which is related to the unknown impedance Z_T by:

$$\Gamma = \frac{Z_T - Z_0}{Z_T + Z_0}. \qquad (10.4)$$

It can be seen from the above equations that if $Z_1 = Z_0$, Γ is directly given by the imbalance voltage ratio:

$$\Gamma = 2\frac{V_{AB}}{V_e}. \qquad (10.5)$$

The absolute values for the voltage ratio obtained when the bridge is terminated by a short or an open load ($\Gamma = 1$) are equal:

$$\frac{V_{AB}}{V_e}(short) = -\frac{V_{AB}}{V_e}(open) = \frac{1}{2}. \qquad (10.6)$$

The above equations are exact only if the detector impedance is infinite. This is generally not the case, especially at high frequencies. If the impedance across points A and B is Z_D, the voltage ratio becomes:

$$2\frac{V_{AB}}{V_e} = \Gamma \left(\frac{2Z_D}{2Z_D + Z_0 + \frac{2Z_T Z_0}{Z_0 + Z_T}} \right). \qquad (10.7)$$

It is no longer exactly proportional to Γ. In practice, if using a matched detector to Z_0, the correction factor ranges from 2/3 for a short (or open) load to 1/2 when the load approaches Z_0. Using a 4 : 1 transformer balun to connect the matched detector, the corresponding deviation from the ideal response of the bridge is sensibly decreased.

10.1.3 Power divider, hybrid, and directional coupler

Apart from the "resistive bridge," other devices may be used to evaluate the reflection coefficient, which is a measure of the relative difference between the measured and the reference impedance.

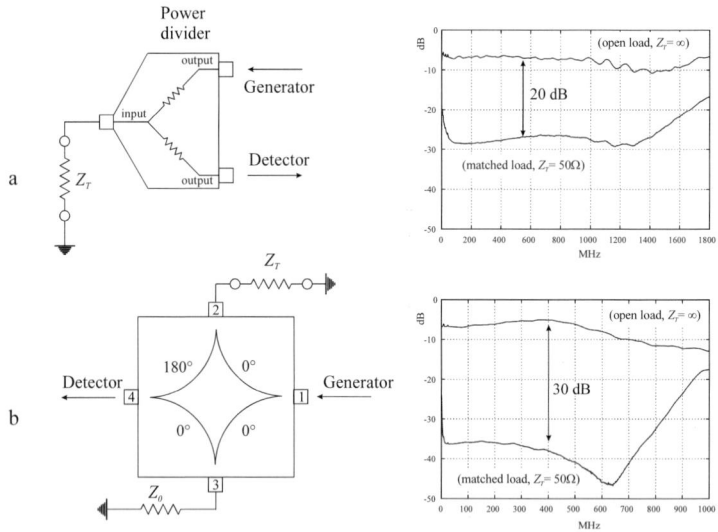

Fig. 10.4 Reflection bridges that can be used for evaluating an unknown impedance Z_T. (a) The two "output" ports of a power divider are isolated from each other only if the "input" port is matched. (b) A 180° hybrid is one version of a power splitter. The detector to source power ratios are shown to the right for matched and unmatched (open or short) situations.

A cheap device is the power splitter/combiner. This device is designed to share a given source signal between two (or more) loads. The output ports are isolated from each other by generally 10 to 20 dB, only if the input port is terminated by its characteristic impedance [Fig.

10.4(a)]. If this is not the case the output ports are not isolated at all. As an impedance bridge, the power divider has a moderate directivity. In this configuration, the generator and the detector are connected to the output ports, while the unknown load is connected to the input port [Fig. 10.4(a)]. The detected voltage has the lowest value when the load impedance matches the characteristic impedance value of 50 Ω.

A 180° hybrid (Chapter 5) is a particular case of a power splitter/combiner and is also equivalent to an impedance bridge. Its characteristic impedance is defined by the impedance value of the load connected to the isolated port. A practical broadband 180° hybrid can be made with a directivity of 30 dB over more than 500 MHz [Fig. 10.4(b)], using two broadband ferrite transformers (Fig. 10.5).

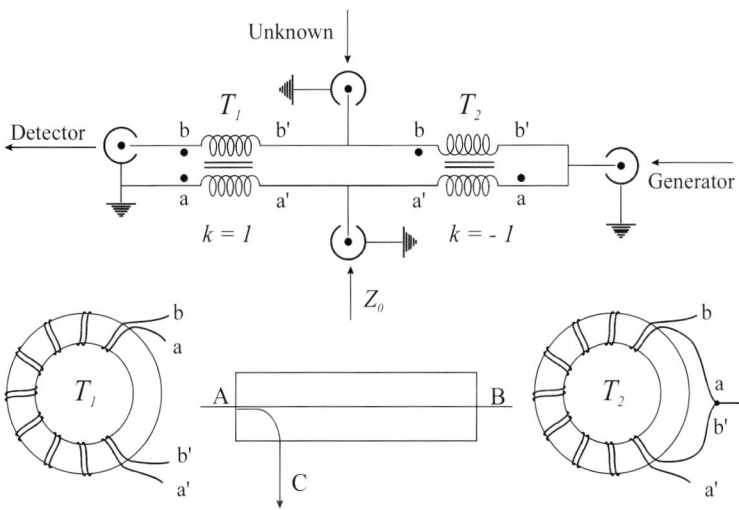

Fig. 10.5 A typical broadband (10 to 500 MHz) 180° hybrid. The two transformers are built using a two wire line wound on a ferrite core and connected as shown. A simple version, using Amidon FT23-64 ferrite cores (five turns of twisted bifilar pairs) functions easily up to 500 MHz.

The transmission loss of some of these devices is "only" of the order of 3–6 dB, from the source to the detector, hence they are particularly suited to being connected to insensitive detectors and/or to sources with limited power output.

A better, generally more expensive, device is the precision directional coupler[2] (Fig. 10.6). Its response is close to the true reflection coefficient, even if the detector impedance is low, thus providing a better precision, especially for the Q measurements. Furthermore its directivity is generally better than a general purpose splitter (40–45 dB over a broad frequency band).

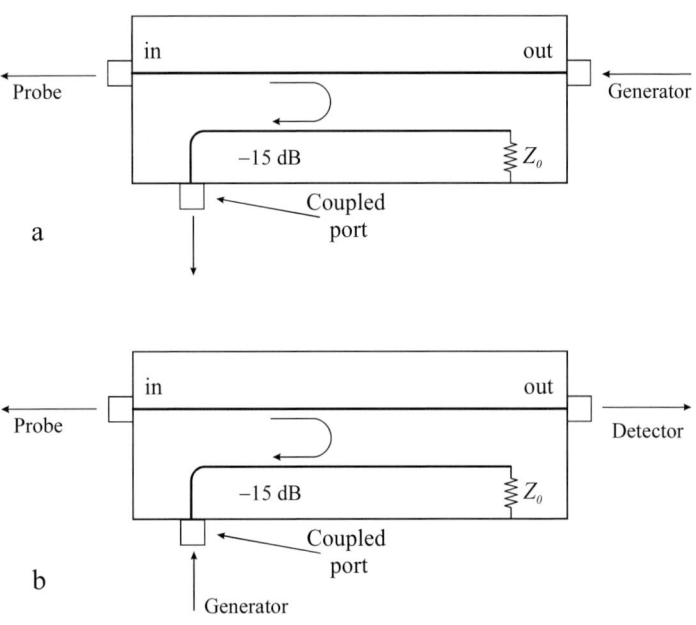

Fig. 10.6 Two ways of connecting a directional coupler for evaluating the probe impedance by measurement of the reflection coefficient.

10.1.4 Sweep generator, crystal detector, and spectrum analyzer

Probe impedance measurements with a bridge or an equivalent device require an RF generator and a detector. The generator is preferably a

---

[2] Some inexpensive directional couplers have limited precision. They are used, for example, to sample a small fraction of energy from a transmitter line in order to drive a frequency counter or a power meter.

sweeping oscillator that allows the direct display of the frequency response on an oscilloscope. Depending on the investment available, a crystal or a spectrum analyzer can be used as a detector of the reflected voltage.

The response of a typical crystal detector to a given input power has already been presented in Chapter 2 (Fig. 2.60). Here is just a reminder of its main characteristics. The sensitivity is limited to an input power of about −30 dBm when connected to a standard oscilloscope. The response to a given excitation is either a square law (for an input power of around −17 dBm) or linear (for an input power of 0 dBm, or 1 mW, in 50 Ω). The crystal detector can be associated with a logarithmic amplifier. The dynamic range of such a detector generally improves to about a maximum value of 60 dB. Coaxial crystal detectors are readily available[3] with a flat response over a very broadband frequency range, from DC to several GHz.

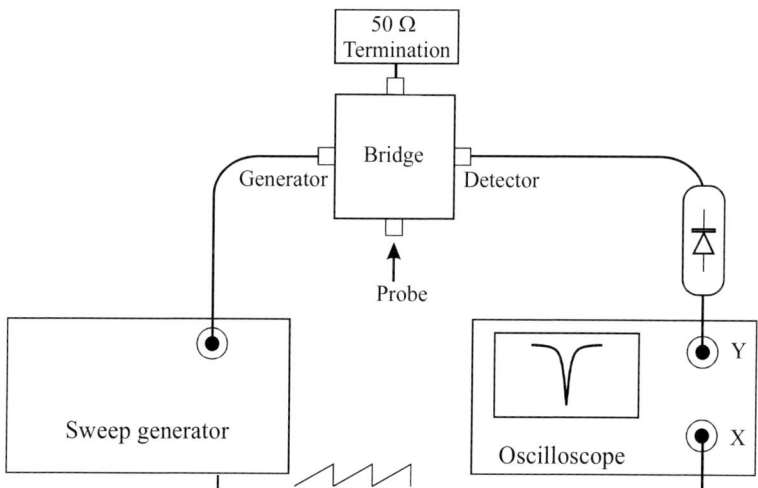

Fig. 10.7 A typical experimental setup for evaluating the probe input reflection coefficient as a function of frequency, using a crystal detector, a sweep generator, and an oscilloscope. The bridge may be replaced by any of the devices already presented (power splitter, 180° hybrid, directional coupler, etc.).

---

[3] The popular HP8470B (Agilent, Keysight Technologies) operates from DC to 18 GHz.

Typically, the sweep generators deliver 0 to 10 dBm max. Hence, the measurements with a crystal detector are, most of the time, in-between the pure quadratic and linear ranges.[4] The Q measurements using the "–3dB method" need a preliminary calibration of the detector response.

A spectrum analyzer has a much better sensitivity and dynamic range than a single crystal detector. It usually displays the value of the input voltage on a logarithmic scale (dB). The minimum detectable signal is determined by the "noise floor" of the spectrum analyzer (about –120 to –130 dBm, depending on the receiver bandwidth).

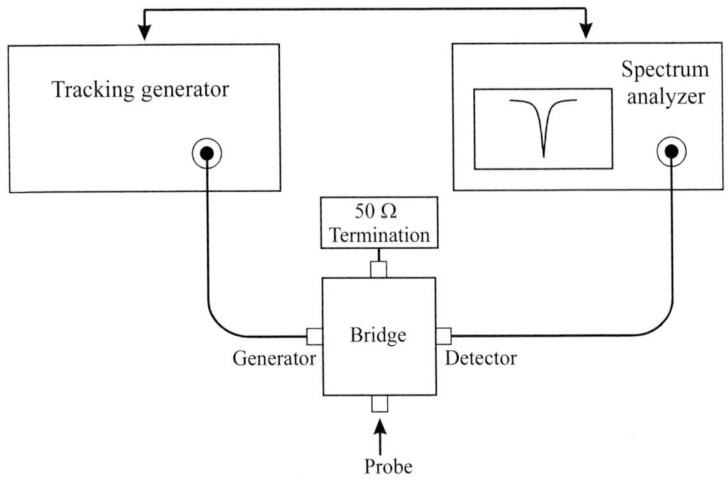

Fig. 10.8 A typical experimental setup for evaluating the probe input reflection coefficient as a function of frequency, using a spectrum analyzer as a detector and a tracking generator. See also the Fig. 10.7 caption.

The dynamic range is typically of the order of 80 dB, but can be better than 100 dB. In contrast to the crystal detector, the spectrum analyzer should be tuned to the frequency at which the measurement is done. The tuning frequency of the analyzer can be spanned automatically over any desired range specified by the user, but the generator frequency

---

[4] The output power of the bridge is between –10 dB and –50 dB below the source power.

must be swept accordingly. One solution is to use a broadband noise source that simultaneously delivers all the frequencies over a very broad range. In this case, the frequency selection is made by the spectrum analyzer.

A better solution is to use the "tracking" generator which is available as an option with most of the spectrum analyzers. In this case, the frequency of the source is automatically synchronized with the receiver frequency of the analyzer (Fig. 10.8).

10.1.5 *Scalar and vector network analyzer*

The previous experimental setup configurations allow the estimation of one network parameter (either the reflection or the transmission coefficients) after a cumbersome measurement of the incident and reflected wave amplitudes separately.

A network analyzer provides a direct reading of the reflection coefficient and/or the transmission parameters. This measuring device incorporates the tracking generator as well as, at least, two receivers: one measuring the reflected wave and the other the incident wave (Fig. 10.9). If a third receiver module is added, both the reflection and transmission parameters of a two-port network can be simultaneously obtained. In this case, the association with a so-called "S-parameter test set" allows the simultaneous measurement of all four scattering matrix coefficients of a two-port network.

When the receivers are sensitive only to the magnitude of the input signal, as it is the case with the so-called scalar network analyzer (SNA), only the magnitude of the S-matrix coefficients are obtained.

The detectors (or receivers) are either a crystal detector associated with a logarithmic DC amplifier, or the super-heterodyne receivers of a spectrum analyzer. Setups such as those shown in Figs. 10.7 and 10.8 are typically SNAs.

It is, however, sometimes desirable to know the phase component of the probe input impedance in order to compensate its reactive component efficiently. In this case, a vector analyzer is required. It incorporates phase sensitive receivers. Such an analyzer represents a must for the probe evaluation.

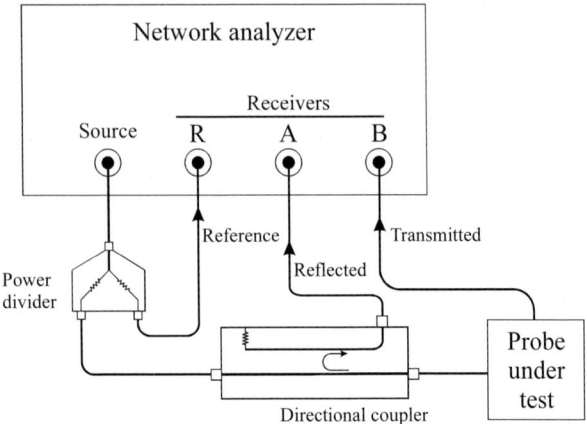

Fig. 10.9 Typical experimental setup for measuring the probe input impedance (input reflection coefficient or S_{11}-parameter) using a network analyzer. The "R" and "A" receivers measure the reference and reflected waves, respectively. Transmission properties (S_{21}, port isolation, etc.) of a multiple port probe can also be simultaneously measured using a third receiver "B."

10.1.5.1 Transmission/reflection and S-parameter test sets

Strictly speaking, the network analyzer is the electronic part including the sources,[5] the receivers, and the signal processing units. The processing units perform the calculation of the ratios A/R and B/R, and display the results on a linear or log scale as a function of frequency. On the Vector Network Analyzer (VNA), the display can also be presented as a polar plot (real part versus imaginary part of the complex valued ratios).

The measuring bridges or directional couplers are provided as separate parts, either the so-called "reflection–transmission test unit"[6] or the "S-parameter test set."[7] These parts are nowadays frequently integrated in the instrument.

---

[5] The pioneering instruments such as the HP8410 did not include the source.

[6] Such as the HP8743A, operating in the microwave range.

[7] Such as the HP8746B or the long-lived 85046A(B) (HP, Agilent, Keysight Technologies), operating in the RF range.

10.1.5.2 Calibration

The first, obvious, calibration procedure on a VNA is the compensation of the phase difference between the reference and the measuring channels. In the older analog instruments the compensation was done with a "line stretcher" incorporated in the reference signal pathway. Modern instruments (including digital signal processing) perform this compensation numerically. It corresponds to the so-called "port extension." At this point, it should be noticed that the line length compensation in the reflection measuring mode (S_{11}) is twice the length in the transmission measuring mode (S_{21}). In the S_{11} mode, the reflected wave travels the extended path twice. In the S_{21} mode, the transmitted wave travels the extension only once.

The RF sources, receivers, and directional couplers of a network analyzer are the origins of measurement errors. These errors can be efficiently compensated, considerably improving the measured accuracy.

Measurements with a network analyzer and their error correction are the subject of books [Dunsmore, 2012] and applications notes provided by the manufacturers.[8] Only the popular "Open-Short-Load" (OSL) one-port error calibration [http://www.vnahelp.com/tip20.html] will be briefly presented here.

In this approach, the network analyzer can be viewed as a perfect analyzer connected to the Device Under Test (DUT) through a linear network representing the imperfections of the network analyzer (Fig. 10.10).

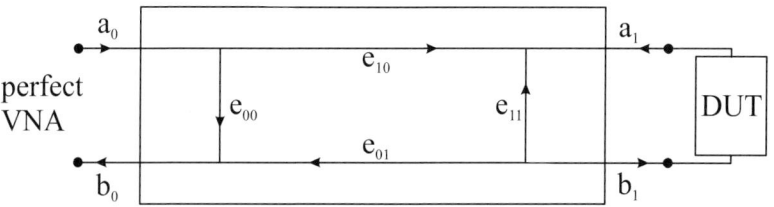

Fig. 10.10 *S*-parameters of the circuit modeling the network analyzer errors.

---

[8] See, for example: http://www.vnahelp.com/appnotes.html.

This representation assumes that the network analyzer is linear. A good VNA receiver exhibits a dynamic range better than 110 dB.

The S-parameters of the circuit representing the errors are usually identified as e_{00}, e_{01}, e_{10} and e_{11} (see Figs. 2.10 and 10.10). These parameters represent the following errors:
- e_{00} is the directivity error,
- e_{11} is the source match error,
- e_{10} and e_{01} are the tracking errors.

The forward and backward wave amplitude and phases are represented by the complex values a_0, a_1 and b_0, b_1 respectively. From Eq. (2.17):

$$\begin{pmatrix} b_0 \\ b_1 \end{pmatrix} = \begin{pmatrix} e_{00} & e_{01} \\ e_{10} & e_{11} \end{pmatrix} \begin{pmatrix} a_0 \\ a_1 \end{pmatrix}. \quad (10.8)$$

The scattering parameters of the circuit representing the errors are completely described by three parameters A, B, and C, defined by:

$$\begin{aligned} A &= e_{10}e_{01} - e_{00}e_{11} \\ B &= e_{00} \\ C &= -e_{11} \end{aligned} \quad (10.9)$$

Using the error model described above, it can be easily demonstrated that the measured reflection coefficient $\Gamma_{measured}$ of a DUT is related to its "true" reflection coefficient Γ_{DUT} by:

$$\Gamma_{measured} = \frac{A\Gamma_{DUT} + B}{C\Gamma_{DUT} + 1}. \quad (10.10)$$

This equation can be reverted to get the corrected reflection coefficient of the DUT as:

$$\Gamma_{DUT} = \frac{\Gamma_{measured} - B}{A - C\Gamma_{measured}}. \quad (10.11)$$

The three coefficient A, B, C should now be determined. This is done using three "standards" with known characteristics (reflection coefficients). Let Γ_i be the (known) reflection coefficients of the standard i and $\Gamma_{i,measured}$ the reflection coefficient measured with the VNA to be calibrated. The constants A, B, and C can be obtained by solving the following set of three linear equations:

$$A\Gamma_i + B - C\Gamma_i\Gamma_{i,measured} = \Gamma_{i,measured}$$
$$i = 1, 2, 3$$
(10.12)

The usual practice is to use an open ($\Gamma = 1$), a short ($\Gamma = -1$) and a load ($\Gamma = 0$) as standards. The equations determining the constants A, B, and C become:

$$A + B - C\Gamma_O = \Gamma_O$$
$$-A + B + C\Gamma_S = \Gamma_S$$
$$B = \Gamma_L$$
(10.13)

where Γ_O, Γ_S, and Γ_L are the reflection coefficients measured with the open, short, and load standards, respectively.

Eq. (10.13) assumes perfect standards. Some expensive calibration sets are provided with accurate Γ values, improving the calibration.

Typically, the directivity error of an old S-parameter test set was around -50 dB or better. Recent S-parameter test sets exhibit a substantially greater error. Anyway, the OSL calibration allows improvement of the measuring accuracy.

In practice, the OSL calibration procedure is useful when designing a probe. The directivity error is probably the least important, but correction of the other errors is essential for evaluating the Q factor of a resonant circuit. Furthermore, the procedure automatically compensates the phase difference between the reference and measuring channels. This allows precise setting of the reference plane. The reactive component at the input port of a given probe is also determined, facilitating the matching process.

Finally, a simple two-port calibration (thru) is sufficient in almost all cases as this allows, with good accuracy, evaluation of the isolation (S_{21}) between the two ports of a quadrature or a double tuned probe.

10.1.6 Other impedance measuring instruments

Some other instruments are available. They are not specifically designed for the RF laboratory. They are generally lighter, cheaper and may serve the fieldworker.

Antenna analyzers are cheap instruments incorporating a bridge and a crystal detector. The bridge is manually adjustable, indicating the impedance directly – either the magnitude, or sometimes the real and imaginary components. An RF generator is sometimes included. These instruments are generally limited to 200 MHz, except the most expensive ones that can be compared to handheld VNA.

Morris Instruments Inc.[9] proposes probe tuner devices that are specifically designed for operating in the magnetic field environment of an NMR instrument (MRI). These light handheld devices are battery operated. They integrate an RF synthesized sweeper, a bridge, and an LCD display. They can operate as a one-port tuner (reflection) or as a two-port transmission measuring instrument.

Finally, an LCR meter is a useful instrument in the lab. The cheaper instruments operate at low frequency (less than 1 MHz). They are not suited to measuring the probe impedance but are very useful to sort out capacitors, estimate the capacitance of a transmission line, estimate an inductance, etc...

10.2 Characterization of a Transmission Line

Probe designs make use of transmission lines, either commercial with sometimes ill-defined characteristics or possibly homemade with nonstandard impedance.

The construction of hybrids (Chapter 5) also requires precise cutting of a line to a given fraction of a wavelength.

In this paragraph, some procedures to characterize a line (electrical length, characteristic impedance, loss parameters) are described.

Firstly, the wavelength in the line (velocity coefficient) will be measured, secondly, the characteristic impedance will be determined,

---

[9] http://morrisinstruments.com/.

and thirdly, the *R*, *G*, *L*, *C* parameters (Chapter 2) of the line will be estimated.

10.2.1 *Velocity coefficient*

The velocity coefficient *k* is defined by the ratio of propagation speed in the line to the velocity of light in the vacuum. Eq. (2.52) is repeated here for completeness.

$$k = \frac{v_p}{c} \tag{10.14}$$

For any line, the propagation speed in the line is slower than in a vacuum due to the dielectric. For standard coaxial cable, *k* ranges between 0.6 and 0.7. Fig. 10.11 shows a simple setup to estimate this coefficient.

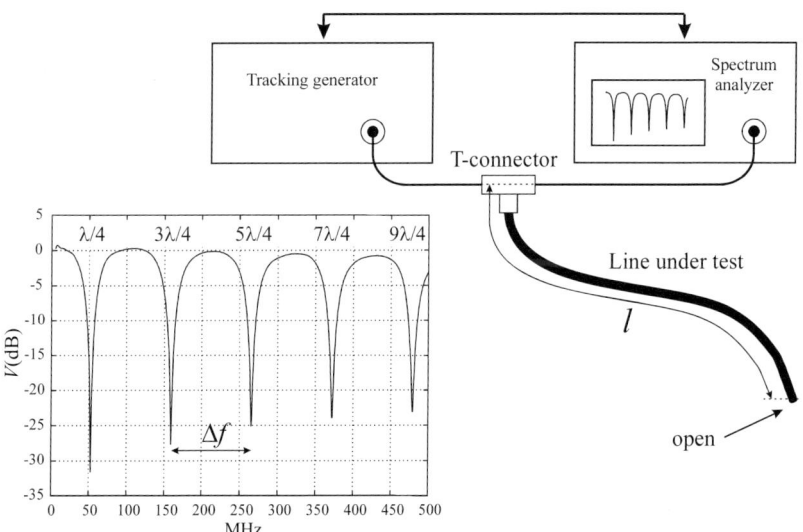

Fig. 10.11 Experimental setup to determine the electrical length of a given transmission line. The figure shows the response of a 94 cm line of RG58 coax. The dips are separated by 107 MHz, hence *k* is equal to 0.67. Using this cable, a λ/4 line measures 25 cm at 200 MHz.

The line (open ended) is inserted in parallel to the connection (of any length) between a sweep generator and a detector. When the length of the unknown line is an odd multiple of $\lambda/4$, the impedance at the connection point is close to zero, producing a dip in the frequency response displayed on the detector.

Similar measurements can be done using a network analyzer (S_{21}), the T-connector being inserted between ports 1 and 2 of an SNA (or a VNA). No accurate calibrations are required.

In the example shown in Fig. 10.11 the cable is not terminated.[10] In a case where the cable is terminated by a short, the minimum values will appear at a multiple of $\lambda/2$.

The coefficient k is obtained from the frequency separation Δf between the dips as:

$$k = \frac{2\Delta f}{c} l, \qquad (10.15)$$

where l is the physical length of the line and c is the speed of light.[11]

A slight difference in the measured value of k can be found when the line is opened or shorted due to a small difference of the physical line length. This difference can be due to fringing effects at the end of the non-terminated cable and to an extra length of conductor in the case of the shorted line. Using a long cable of the order of several meters minimizes these errors.

10.2.2 Evaluation of the characteristic impedance

The determination of the velocity coefficient, as described above, is independent of both the characteristic impedance of the measuring system and the cable. Another independent measurement of the capacitance of the cable allows calculation of its characteristic impedance.

---

[10] Equivalent to an open.
[11] $c = 2.99792458 \ 10^8$ m s^{-1} (old determination, before 2012!)

The characteristic impedance of a low-loss transmission line is given by the square root of the ratio of the inductance and the capacitance per unit length [Eq. (2.49)].

Both quantities can be measured with an LCR meter, but at low frequency, the inductance increases.[12] In contrast, the capacitance is generally a constant value from a few kHz up to several GHz.

Let C_0 be the capacitance per unit length of a cable, measured with an LCR meter at low frequency. A capacitance of several hundreds of pF is easily measured with confidence at 100 kHz.[13]

The high frequency characteristic impedance of the cable can be determined from this measurement and from the previous measurement of k, using the following equation:

$$Z_c = \frac{1}{ckC_0}. \qquad (10.16)$$

This equation can be obtained by combining Eqs. (2.49) and (2.51) and using Eq. (10.14). c is the speed of light in a vacuum and C_0 is the capacitance per unit length of the cable.

As an example, Table 10.1 below shows some examples of measurements done on common coaxial cables.

Table 10.1 Measurements on some common 50 Ω coaxial cables.

Cable	l(m)	C_l(pF)	L_l(μH) at 1 kHz	Z_c(Ω) at 1 kHz	C_0(pF m^{-1})	k	Z_c(Ω)
RG174	4.527	435	1.38	56.42	96.09	0.694	50.03
RG58	7.003	698	2.33	57.73	99.67	0.661	50.60
RG223	10.097	1042	3.26	55.91	103.2	0.661	48.90
RG214	6.00	604.5	2.08	58.60	100.75	0.655	50.54

---

[12] At high frequency, the current flows on the surface of the conductor due to the skin effect. The inductance becomes a constant above about 10 MHz when the skin depth becomes negligible compared to the thickness of the conductors.

[13] The capacitance of a 5 m length of a standard 50 Ω cable is of the order of 500 pF. The corresponding impedance at 100 kHz is 3.2 kΩ. It is easily measured with an LCR meter. Such a capacitance can also be measured with great accuracy using a good impedance bridge, even operating at 1 kHz.

10.2.3 Estimation of the loss parameters

To complete the characterization of the cable, the resistance R and conductance G remain to be determined (refer to the model shown in Fig. 2.22).

The relevant equations have already been discussed in detail in Chapter 2. They involve the attenuation constant which can be written, assuming a low-loss transmission line as:

$$\alpha = \frac{1}{2}\left(\frac{R}{Z_c} + GZ_c\right) = a\sqrt{f} + bf, \qquad (10.17)$$

where a and b are constant values that are generally given by the cable manufacturer. They can be easily determined connecting a known length of cable between port 1 and port 2 of a network analyzer. The plot of S_{21} versus frequency can be fitted to Eq. (10.17) which proves to be valid, at least in the RF range. An example has already been shown in Fig. 2.24. It will just be recalled here that this approach is valid for a low-loss cable, assuming that the resistance varies as the square root of frequency and that the conductance is proportional to the frequency.

10.3 Noise Figure Measurement

The Noise Figure (NF)[14] evaluation of the receiver channel(s) of the NMR machine could be an important step, especially when a degradation of the sensitivity is suspected, while the probe appears well constructed.

An evaluation of the expected magnetic field amplitude will be presented later in this chapter. The estimated value can be compared with measurements on the NMR machine. Provided the transmitter power is known, comparable values of the B_1 field indicate that the probe is functioning as expected. From the Principle of Reciprocity, the sensitivity is that expected with the given probe design. In fact, this assumes that the NF of the receiver channel, principally the very first

---

[14] The noise figure (NF, in dB) is related to the noise factor (F, dimensionless number) by NF(dB) = $10\log_{10}(F)$ [Eq. (2.142)].

components of the chain (the cable, the filter(s), and the preamplifier) is good.[15]

The Y-factor method presented here is a well-established method described in an application note AN57-1 by Agilent[16] (now Keysight Technologies and previously Hewlett-Packard Company). It has several advantages, the measurement being independent of the gain and bandwidth of the receiver under test. The test should however be performed in a region of linear response of the device.

In this method a switchable noise source is connected to the input of the device. The output noise power is measured when the source is switched off (N_1) and switched on (N_2). The ratio Y:

$$Y = \frac{N_2}{N_1} \qquad (10.18)$$

is related to the noise factor F by a very simple relationship:[17]

$$F = \frac{ENR}{Y-1}.$$
$$NF_{dB} = 10\log_{10}(F) \qquad (10.19)$$

ENR is a dimensionless ratio, the "excess noise ratio," characterizing the noise delivered by the source when it is powered on or off.

Common noise sources are constituted of a special diode generating noise when reverse biased into avalanche breakdown. An attenuator is added in order to establish the source impedance. The commercial sources are usually calibrated (they should be) as a function of frequency. The ENR is frequently given in dB, from which the dimensionless ratio of Eq. (10.19) can be obtained using:

$$ENR = 10^{\frac{ENR_{dB}}{10}}. \qquad (10.20)$$

---

[15] Say, lower than 2 dB.

[16] http://cp.literature.agilent.com/litweb/pdf/5952-8255E.pdf.

[17] Note that the measurement is independent of the gain of the device under test.

Typical commercial noise sources have an *ENR* of 6, 15, or 30 dB. The choice depends on the *F* to be measured. A noise source with an *ENR* of 15 dB is a good compromise for most of the measurements to be done on an NMR machine. Note that the *ENR* of such a noise source corresponds to an equivalent noise temperature difference of the order of 10000 K. It is given by:

$$ENR_{dB} = 10\log_{10}\left(\frac{T_{hot} - T_{cold}}{T_0}\right), \qquad (10.21)$$

where T_{hot} and T_{cold} are the equivalent noise temperature corresponding to the source on and off, respectively. T_0 is the reference temperature defined as the "room" temperature (290 K). When an attenuator is included in the noise source, T_{cold} is about 290 K.

To estimate the *NF* of an NMR receiver chain, a calibrated noise source (of known *ENR*) and a convenient DC power supply[18] are sufficient. Connect the source to the probe cable including, possibly, the filters. Power the noise source to the on position. Record a time domain signal using a receiver gain and bandwidth sufficient to fill about one half the dynamic range of the Analog to Digital Converter, taking care not to overload the receiver. This will give the N_2 noise level. Power off the source while leaving it connected. Record a new time domain signal with the same acquisition condition (Fig. 10.12). This will give the N_1 noise level.

Evaluation of the noise power level is straightforward from the digitized signals. The *Y* ratio is directly given by the ratio of the square[19] of the standard deviation of the recorder signals obtained with the source on and off. Applying Eq. (10.19) gives the *F* (an ENR_{dB} of 15 dB corresponds to *ENR* = 31.62).

---

[18] The noise source generally requires a source of 21 V to be switched in the on position.

[19] The power level is proportional to the square of the amplitude.

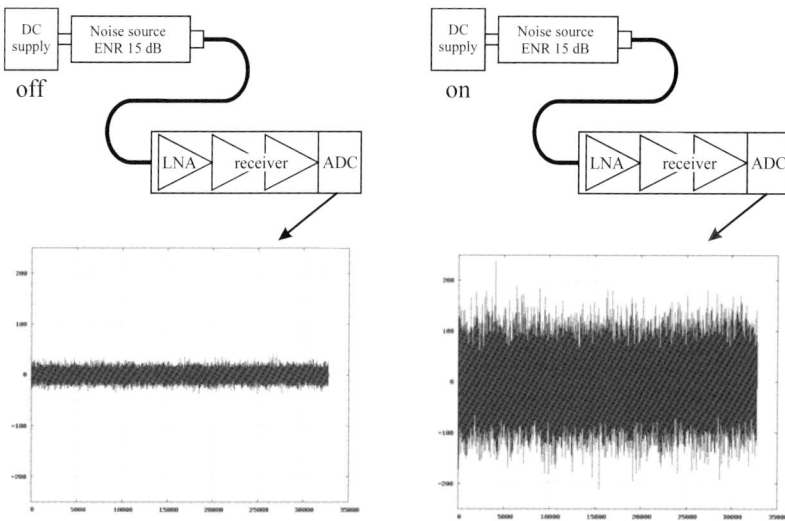

Fig. 10.12 A typical recording of noise at the output of the receiver chain of an NMR machine connected to a noise source switched off (left) and on (right). The signal displayed here is the real part of the complex valued acquisition. The imaginary component would give similar results. The receiver was tuned to 400 MHz. The bandwidth was 6 kHz. 32k complex values were recorded in 1.3 s. The Y factor was estimated from the square of the ratio of the standard deviations ($\sigma = 9.2$ and $\sigma = 50.6$) calculated on the recorded signals ($Y = 20.3$). F is obtained from Eq. (10.19) as 1.56, corresponding to an NF of 1.9 dB.

On the RF workbench, the F of a preamplifier can be measured comfortably using a Noise Figure Analyzer in addition to the calibrated noise source. This instrument is capable of simultaneously measuring the gain and the NF of the device under test. It calculates the Y factor internally and displays the corresponding NF. Automatic switching of the source is done using a pulsed DC supply integrated in the instrument. The noise level at the output of the device is measured with a superheterodyne receiver. The latter does not require extreme performance in terms of intrinsic NF. A receiver with an NF of around 6 dB is indeed able to determine the NF of an ultra low noise preamplifier ($NF > 0.4$ dB).

10.4 Evaluating the Probe on the RF Workbench

10.4.1 *Matching the probe input impedance to 50 Ω*

The first step in adjusting a new probe is to tune and match it at the operating frequency (ω_S). Any instrument, as described above, giving even an indication of the "reflected voltage" or of the Standing Wave Ratio (swr) will allow the probe to be conveniently adjusted [Hoult, 1978; Daugaard *et al.*, 1981ab].

The output of the detector is frequency swept. For a perfect match, a dip appears at the operating frequency, Fig. 10.13(a). A polar plot recorded around the resonant frequency is shown in Fig. 10.13(b).

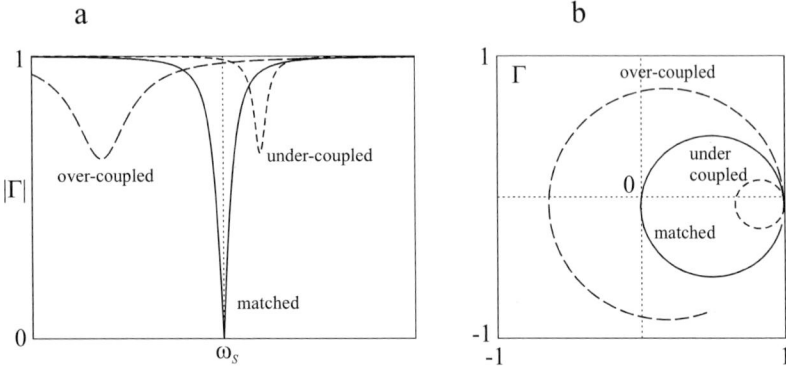

Fig. 10.13 Typical reflection coefficients obtained when the probe is matched, over coupled, and under coupled (capacitive matching case assumed). (a) The magnitude of the reflection coefficient ($|\Gamma|$) is represented on a linear scale as a function of frequency. (b) Representation of the complex valued Γ on a polar plot.

If the probe is over coupled, the reflected voltage increases. Its minimum generally appears at a lower frequency than that of the perfect match and, correspondingly, the frequency response broadens. This corresponds to a large circle on the polar plot that encompasses the origin ($\Gamma = 0$). On the contrary, if the probe is under coupled, the minimum of the swr voltage (vswr) appears at a higher frequency, and the frequency response sharpens [Fig. 10.13(a)]. On the polar plot, the $\Gamma = 0$ lies outside the circle with a smaller diameter [Fig. 10.13(b)].

On the RF workbench, the matching procedure is done as shown in Fig. 10.14. When the adjustment is done on the spectrometer, at a constant frequency, the procedure is different. Firstly a minimum vswr is quickly obtained by adjusting the "matching" capacitor. Secondly, the tuning capacitor is varied in order to move the resonance across the desired frequency, slightly above or below, depending on the previous situation. The procedure is repeated until the correct match is obtained at the desired frequency. This can be obtained after a small number of cycles. Because the matching varies only slightly with the frequency, the ultimate step consists of adjusting the resonant frequency.

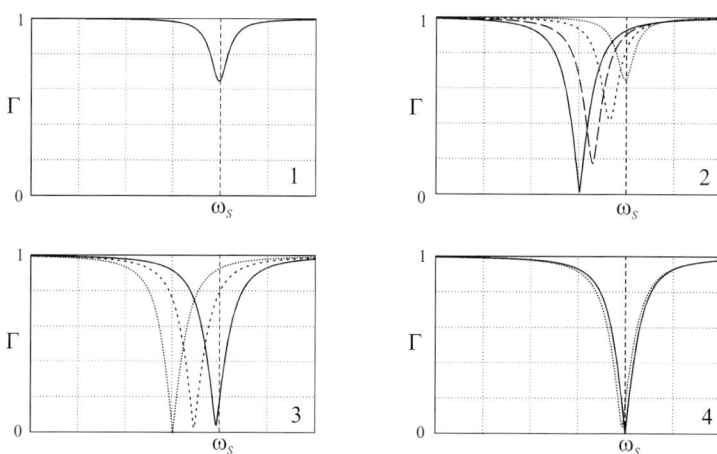

Fig. 10.14 Impedance matching ($\Gamma = 0$) procedure using the experimental setup of Figs. 10.7 to 10.9. The resonance is shifted to ω_0 (1). The matching capacitor is adjusted to the best match (2). The frequency has moved to a lower frequency than ω_0, it is readjusted close to ω_0 (3). The final adjustment of the matching is accompanied by a slight frequency shift very close to ω_0 (4).

Sometimes, a satisfactory matching solution cannot be found. This situation may appear for complex probe circuits, when the input port impedance is not that expected from the simple equivalent circuit described in Chapter 4.

For example, a split transmission line used as a coupling loop in a matching network may present a capacitive impedance at high frequency instead of an inductive one as could be expected.

Another example is when searching a convenient feed point for a complex resonator. A simple magnitude detector or even an SNA cannot solve the problem *a priori*. In these cases a vector impedance meter, or an analyzer, is useful. Knowing the complex impedance at the desired frequency, the matching components can be calculated either manually or with the help of a Smith Chart. The fine adjustments can finally be made while looking at the vswr response, using any convenient instruments (from the vector network analyzer equipped with an S-parameter test set to the simple combination of a sweeper and a crystal detector).

10.4.2 Evaluation of the Q factor

At this point, the definition of the "Q factor" should be specified as some confusion is frequently encountered in the literature (related to NMR or to electronic devices). In the following, "Q" means the "intrinsic" Q factor of the coil measured in defined conditions (empty or sample loaded). The intrinsic Q factor is defined in Eq. (2.5), which is repeated here:

$$Q = \frac{|X|}{r}, \tag{10.22}$$

where $|X|$ is the absolute value of the reactive impedance of the "coil" and r is the equivalent resistance representing the losses.

When measured from the reflection coefficient of a matched probe, the "Q factor," sometimes named "Q_{loaded}," is equal to half the above Q value. This Q factor, defining the bandwidth of the probe circuit, is preferably named $Q_{matched}$.

$$Q_{matched} = \frac{Q}{2} \tag{10.23}$$

The coupling of the probe with the sample may be characterized by the ratio of the Q factors when the probe is empty or sample loaded (filling factor). In this case, the Q factor is named as Q_u (probe unloaded or empty) and Q_l (probe loaded with a sample), respectively.

The Q factor measurement for the empty or loaded probes allows a preliminary evaluation of the probe sensitivity in various situations. As already seen, the sensitivity is approximately related to $\sqrt{\eta Q}$, where η is the filling factor. When combined with an evaluation of the frequency shifts observed after loading, it allows an optimization of the probe equilibration circuit and/or of Faraday screening of the sample respective to the electric field. Furthermore, the Q factor value allows the estimation of the magnetic field created by the probe. It represents valuable information, to be predicted in order to evaluate the correct functioning of the probe in its final environment.

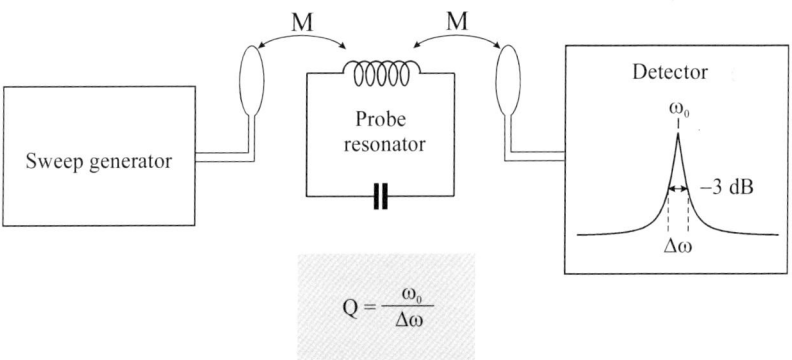

Fig. 10.15 Measurement of the probe coil resonator Q factor using two small loosely coupled coils. The width at −3 dB of the response curve gives the Q value.

The Q factor can be evaluated by loosely coupling the probe resonator to two small pick-up coils [Hoult, 1978]. One is connected to a sweep generator while observing the voltage induced in the other, connected to a (very) sensitive detector (Fig. 10.15). The Q factor is obtained from the frequency difference that exists between the −3 dB points of the circuit response. This is probably the most precise method to evaluate the Q factor, but the more delicate to set up.

Another method that can be applied when the probe coil is not directly accessible to the pick-up coil has been proposed by Jiang [Jiang, 2000]. After carefully matching the probe (previous paragraph), the Q factor is deduced from the resonant frequency and the tuning and matching capacitance values (inverting the equations given in Table 4.4). However, the method cannot be applied to any resonator structures, such as, for example, to a birdcage or a transmission line resonator. In fact it applies only to the basic capacitive matched coil.

More frequently, the Q value is evaluated directly on the RF workbench from the frequency response of $\Gamma(\omega)$ at the –3 dB level (Fig. 10.16), when the probe has been matched to the characteristic impedance Z_0. In this case, the intrinsic Q value of the probe resonator is twice the Q measured at the –3 dB level [Eq. (10.23)].

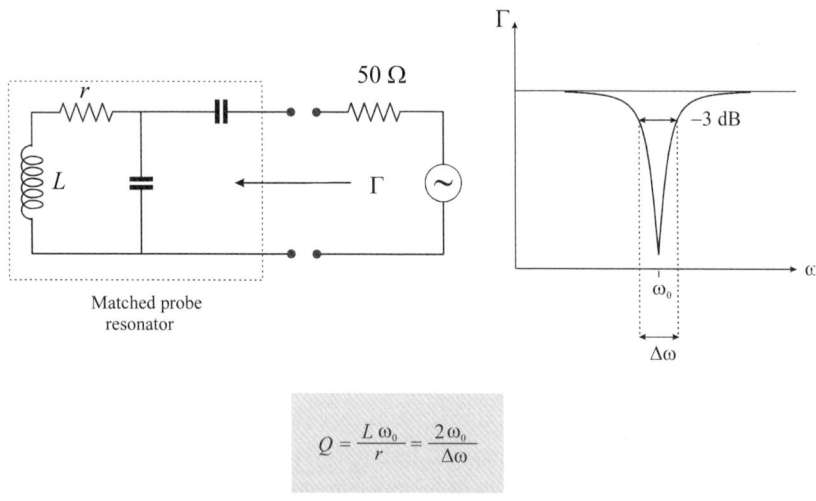

Fig. 10.16 Q measurement performed on a matched probe. The corresponding value is obtained from the frequency dependence of the reflection coefficient.

In practice, the –3 dB level may be difficult to obtain, especially on a logarithmic scale, if the response of the whole circuit is not ideal [Fig. 10.17(a)]. A crystal detector provides a larger dynamic around the –3 dB value than a logarithmic detector [Fig. 10.17(b)], but in practice, the maximum power level that is delivered by standard sweeper generators (around 10 dBm[20]) is such that the detector is neither in the linear nor in the quadratic region (Fig. 2.60). Hence, in contrast to a frequent expectation, the −3 dB level is not at half the maximum level of the response curve as it should be if the response were quadratic.

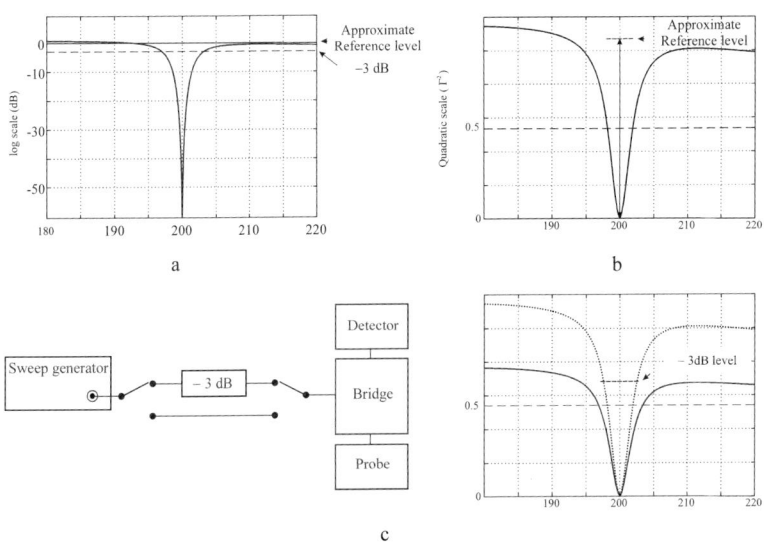

Fig. 10.17 Estimation of the −3 dB level for Q measurements. (a) The −3 dB level is hardly detectable on an uncorrected reflection coefficient. A crystal detector provides a greater dynamic around the −3 dB level. (b) The corresponding level is expected to be at half the reference level if the detector is purely quadratic (low incident level). (c) Because the detector practically never has a purely quadratic response, the −3 dB level must be calibrated independently using the experimental setup.

The −3 dB level can be accurately determined anyway by temporarily inserting an attenuator in the source line [Fig. 10.17(c)]. In this way, the Q factor can still be evaluated with good confidence.

Using an S-parameter set and an expensive network analyzer (either scalar or vector), the reference level can be accurately defined by the OSL calibration procedure (Section 10.1.5.2). Hence, the −3 dB points of

---

[20] 0 dBm = 1 mW.

the S_{11}-parameter are well-defined and the Q value may be deduced with reasonable precision.

10.4.3 $B_1/\sqrt{P}$ evaluation methods

Attention has already been paid to the fact that in certain circumstances (multiple-tuned probe for example) the measured Q factor at the probe ports does not necessarily represent the sensitivity that would finally be obtained. Therefore an accurate evaluation of the sensitivity requires an estimation of the magnetic field directly produced by the coil probe. The ultimate test is obviously a measure of the 90° pulse length which provides a direct measurement of the B_1 field amplitude. This parameter could be evaluated initially on the RF workbench without the need for an NMR machine. A comparison between the expected B_1 amplitude[21] and the measured one is a very valuable evaluation for tracking an eventual malfunction of the realized probe.

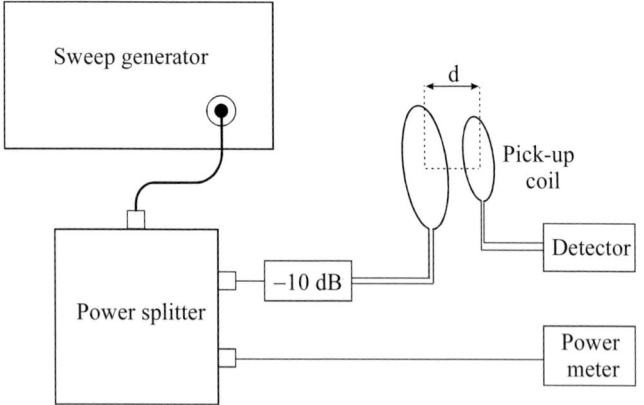

Fig. 10.18 Calibration of the effective area of a pick-up coil. The driving source coil is expected to have a much larger diameter than the pick-up coil.

The simplest, but very efficient, method is to use a pick-up, or sense, coil. It may be calibrated using a well-defined experimental setup, for

---

[21] Calculated from the formulae already given and from the current deduced from the measurement of the Q factor and the known power delivered by the test generator.

example, a simple circular loop of accurately known geometry and driven by a known source current [Chen *et al.*, 1986] (Fig. 10.18). This "broadband" method allows the calibration of the pick-up coil over a wide range of frequencies, much lower than the self-resonance of the reference loop.

Another possibility is to tune the reference circular coil to a frequency close to the desired one. The induced voltage can be predicted accurately from the geometry of the experimental setup and from the electrical characteristics of the reference coil resonator (Q factor, its inductance, and the RF power supplied to the circuit).Other methods that allow a more or less direct evaluation of the $B_1/\sqrt{P}$ value have been proposed in the literature. Among these, one clever method [Darrasse and Kassab, 1993] uses two identical small coils that are geometrically decoupled in free space by a slight overlapping (as in array coils, Section 9.2.2). One of the coils is connected to a generator and the other one to a detector (Fig. 10.19). In free space, the detected voltage is small due to the decoupling between the two coils. When placed close (or inside) a probe resonator, a signal is observed in the receiver channel that directly provides the equivalent signal that would be induced by a precessing magnetization at the same position.

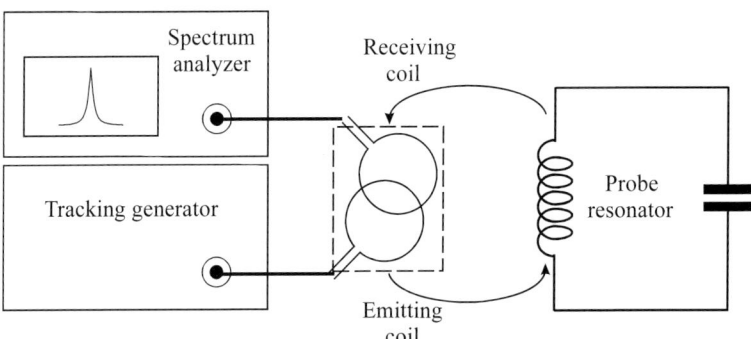

Fig. 10.19 Schematic representation of the two-coil method for probe efficiency evaluation. The two sampling coils are expected to be much smaller than the probe coil resonator. The coupling between them is adjusted far from the probe resonator in order to minimize the received signal. When approached to the probe resonator, an induced voltage appears on the receiver coil, from which one obtains $B_1/\sqrt{P}$.

For the evaluation of small coil resonators, the two-coil system is difficult to build. In this case, a single loop-probe is preferable [Ginefri et al., 1999].

The new method (Fig. 10.20) is based on the measurement of the reflection coefficients of the small loop-probe in free space (Γ_0) and when it is approached to the resonator under test (Γ_c), matched, and loaded by Z_0. The ratio $B_1/\sqrt{P}$ is directly given from the difference in the reflection coefficients ($\Gamma_{compensated} = \Gamma_c - \Gamma_0$) as [Ginefri et al., 1999]:

$$\frac{B_1}{\sqrt{P}} = \frac{(R+Z_0)}{A\omega_0}\sqrt{\frac{2\Gamma_{compensated}}{2Z_0 - (R+Z_0)\Gamma_{compensated}}}, \quad (10.24)$$

where R is the damping resistance of the loop-probe (Fig. 10.17) and A its area (it is assumed that the sampling coil is sufficiently small that the RF magnetic field is homogeneous all over its surface).

Finally, the Q factor can be obtained from the frequency variation of $\Gamma_{compensated}$ (at –3 dB). As shown by other methods in which the probe is loaded by its matched impedance, the Q value obtained should be doubled to get the intrinsic probe coil quality factor [Eq. (10.23)].

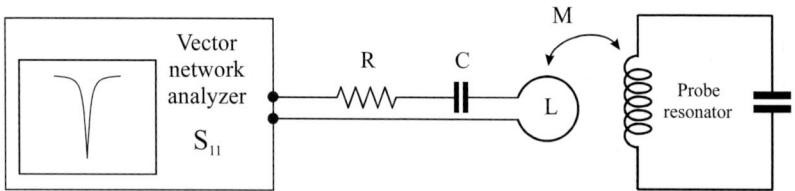

Fig. 10.20 Experimental setup for probe efficiency evaluation, using the single loop method. The method is based on the measurement of the reflection coefficient change that occurs when approaching the single loop to the probe resonator.

10.5 Evaluating the Probe on the NMR Instrument

The ultimate test of a probe is obviously when connected to the NMR spectrometer. The final tests that should be done in order to characterize the probe are regarding its sensitivity, the magnetic field distribution in the sample space, and the maximum transmitter power that the probe can handle. The magnetic field map may be evaluated either by imaging a given sample, as occasionally described in the previous two chapters, or by recording a spectrum line under conditions of varying the pulse angle (see for example Fig. 8.28). The latter experiment allows measurement of the pulse length required to rotate the nuclear magnetization by 90°, hence it is a direct measure of the amplitude of the RF magnetic field in the rotating frame. This measurement provides a direct evaluation of the sensitivity of the built probe [Hoult and Richards, 1976] that could be compared with other probes.

The experiment can also be done by varying the transmitter power, eventually showing the limits that the probe can accept without any damage. Indeed, if the power, hence the voltage across the capacitors, is too high, the probe will be arcing during pulses, leading to a random phase and amplitude variation of the spectrum recorded in these conditions. An extensive discussion of the characteristics obtained within these experiments was recently given in a paper that the reader is encouraged to read [Keifer, 1999]. We will only discuss the comparison of the measured and expected B_1 field amplitude here.

As already discussed, for each type of probe design, the amplitude of the RF field created by a given current depends mainly on the probe dimensions. Hence, a comparison with probes of similar dimensions (essentially the diameter of the probe) provides an initial evaluation of the built device. On the other hand, an absolute evaluation of its performance compared to the expected one is easily done by simple calculations requiring only a few formulae and knowledge of the probe dimensions and its Q factor. A large discrepancy between the expected and measured values should raise the question of the origin of this disagreement to the designer. At this point it should be noted that, owing to the many uncertainties (in the transmitter power and approximation in the theory, etc.), a 30 to 50% error may be acceptable. An experimental comparison between different probes should, however, be more accurate.

	B_1^+ (gauss) $I = \sqrt{2PQ\omega C_{eq}}$	V_{peak}	C_{eq}
 Helmoltz coils	$\dfrac{8\pi I}{5d}$	$\dfrac{I}{2C\omega}$	$2C$
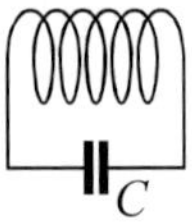 Solenoid (or loop gap)	$\dfrac{2\pi n I}{\sqrt{d^2+l^2}}$	$\dfrac{I}{C\omega}$	C
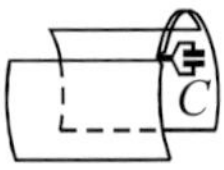 Saddle coil	$\dfrac{3.46 n I}{\sqrt{d^2+l^2}} \dfrac{l}{d}\left(1+\dfrac{d^2}{d^2+l^2}\right)$	$\dfrac{I}{2C\omega}$	$4C$
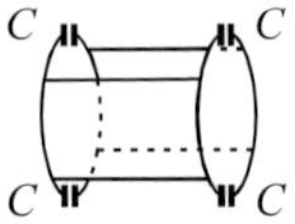 AGR	$\dfrac{3.39 I}{\sqrt{d^2+l^2}} \dfrac{l}{d}\left(1+0.835\dfrac{d^2}{d^2+l^2}\right)$	$\dfrac{I}{2C\omega}$	C
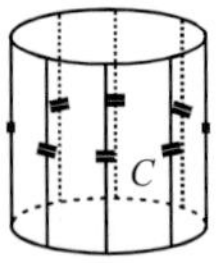 Birdcage (low-pass)	$\dfrac{4\zeta I}{\sqrt{d^2+l^2}} \dfrac{l}{d}\left(1+\dfrac{d^2}{d^2+l^2}\right)$	$\dfrac{I}{\beta C\omega}$	$\dfrac{\beta^2 C}{N}$
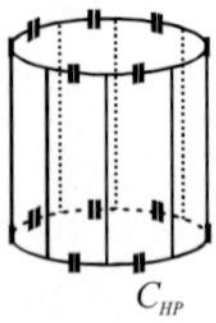 Birdcage (high-pass)	Same as above replace C by αC_{HP}		
 Circular surface coil	$\dfrac{2\pi n I}{d}$	$\dfrac{I}{C\omega}$	C

Fig. 10.21 Estimation of the rotating frame magnetic field at the center of various probe geometries. The length l and diameter d are in mm. n is the number of turns. N is half the number of legs in the birdcage case. V_{peak} is the peak voltage across the capacitor for a given transmitter power P. The total current I is obtained from the value for C_{eq} given in the last column. C is the capacitance value used in the design.

Fig. 10.21 summarizes the equations that allow estimation of the expected magnetic field amplitude for a given probe (assuming known dimensions, the Q factor when loaded with a sample, and the incident power P delivered by the transmitter). This evaluation is done for a given product PQ (the RF B_1 field amplitude is proportional to the square root of PQ). The first column gives the B_1 RF field amplitude in the rotating frame in gauss, when the probe dimensions are in mm and the current I in A. I is obtained from the value of C_{eq} given in the third column. The second column provides the voltage expected across the tuning capacitor, also a function of $\sqrt{PQ}$. The formulae for the birdcage design involves parameters [$\beta = I/I(2N)$, Eq. (8.60) and ζ in Eq. (8.62)] that depend on the number of legs. Table 10.2 gives numerical values for α, β, and ζ as a function of the total number of legs, $2N$.

Table 10.2 Coefficients in Fig. 10.21 for the birdcage design.

Number of legs	8	12	16	24	32
$\beta = I/I(2N)$	2.613	3.863	5.126	7.661	10.202
ζ	0.765	0.776	0.780	0.783	0.784

Appendix A

Physical Constants and Useful Formulae

A.1 Physical Constants

Table A.1 Physical constants in SI units.

Constant	Abbreviation	Value	Unit
Speed of light[1]	c	2.99792458×10^8	m s^{-1}
Vacuum permittivity	ε_0	8.85419×10^{-12}	F m^{-1}
Vacuum permeability	μ_0	$4\pi \times 10^{-7}$	H m^{-1}
Planck's constant	h	6.62607×10^{-34}	J s
Boltzmann's constant	k	1.38065×10^{-23}	J K^{-1}
Electron charge	e	1.60218×10^{-19}	C
Nuclear magneton	μ_N	5.050785×10^{-27}	J T^{-1}
Nuclear magneton	μ_N/h	7.62259	MHz T^{-1}
Proton gyromagnetic ratio	$\gamma_H/2\pi$	42.57748	MHz T^{-1}

Note: (1) $c = 1/\sqrt{\varepsilon_0 \mu_0}$

A.2 Self and Mutual Inductance

Some formulae have been included and evaluated in Chapter 2 for specific inductance models. These formulae and others will be repeated here for convenience.

In addition, numerical methods combining formulae and linear network analysis have been presented in Chapter 3. This approach will be further discussed here with the example of the FastHenry software [Kamon et al., 1994].

Integration methods such as those presented and evaluated by Jin [Jin, 1999, pp. 58] and Giovanetti [Giovanetti et al., 2002] will be briefly described together with the Neumann formula for calculating mutual inductances.

Finally, an efficient numerical method [Fan et al., 1987] will be presented for calculating the complete elliptic integrals of the first and second kind that are involved in many magnetism-related equations.

A.2.1 Inductance formulae

All inductance values are in nH and the dimensions in mm (see Fig. A.1 for definitions of dimensions). All formulae assume that the dimensions are much smaller than a wavelength (see Section 2.4.4).

Straight length l of round wire (Fig. A.1(a), high frequency formula):

$$L = 0.2l\left[\ln(4l/d) - 1.0\right]. \tag{A.1}$$

Flat strip in air [Fig. A.1(b)], accurate formula [Ruelhi, 1972; Wu et al., 1992; FastHenry, Kamon et al., 1994]:

$$L = 0.8l \left\{ \begin{aligned} & \frac{1}{4}\left[\frac{1}{w_n}\sinh^{-1}\left(\frac{w_n}{\alpha_t}\right) + \frac{1}{t_n}\sinh^{-1}\left(\frac{t_n}{\alpha_w}\right) + \sinh^{-1}\left(\frac{1}{r_n}\right)\right] \\ & + \frac{1}{24}\left[\begin{aligned} & \frac{t_n^2}{w_n}\sinh^{-1}\left(\frac{w_n}{t_n\alpha_t(r_n+\alpha_r)}\right) + \frac{w_n^2}{t_n}\sinh^{-1}\left(\frac{t_n}{w_n\alpha_w(r_n+\alpha_r)}\right) \\ & + \frac{t_n^2}{w_n^2}\sinh^{-1}\left(\frac{w_n^2}{t_n r_n(\alpha_t+\alpha_r)}\right) + \frac{w_n^2}{t_n^2}\sinh^{-1}\left(\frac{t_n^2}{w_n r_n(\alpha_w+\alpha_r)}\right) \\ & + \frac{1}{w_n t_n^2}\sinh^{-1}\left(\frac{w_n t_n^2}{\alpha_t(\alpha_w+\alpha_r)}\right) + \frac{1}{t_n w_n^2}\sinh^{-1}\left(\frac{t_n w_n^2}{\alpha_w(\alpha_t+\alpha_r)}\right) \end{aligned}\right] \\ & - \frac{1}{6}\left[\frac{1}{w_n t_n}\tan^{-1}\left(\frac{w_n t_n}{\alpha_r}\right) + \frac{t_n}{w_n}\tan^{-1}\left(\frac{w_n}{t_n\alpha_r}\right) + \frac{w_n}{t_n}\tan^{-1}\left(\frac{t_n}{w_n\alpha_r}\right)\right] \\ & - \frac{1}{60}\left[\begin{aligned} & \frac{(\alpha_r+r_n+t_n+\alpha_t)t_n^2}{(\alpha_r+r_n)(r_n+t_n)(t_n+\alpha_t)(\alpha_t+\alpha_r)} \\ & + \frac{(\alpha_r+r_n+w_n+\alpha_w)w_n^2}{(\alpha_r+r_n)(r_n+w_n)(w_n+\alpha_w)(\alpha_w+\alpha_r)} \\ & + \frac{(\alpha_r+\alpha_w+1+\alpha_t)}{(\alpha_r+\alpha_w)(\alpha_w+1)(1+\alpha_t)(\alpha_t+\alpha_r)} \end{aligned}\right] \\ & - \frac{1}{20}\left[\frac{1}{r_n+\alpha_r} + \frac{1}{\alpha_w+\alpha_r} + \frac{1}{\alpha_t+\alpha_r}\right] \end{aligned} \right\}. \tag{A.2}$$

In Eq. (A.2), w_n and t_n are the normalized width and thickness, respectively:

$$w_n = w/l$$
$$t_n = t/l \quad , \quad (A.3)$$

the parameters r_n, α_w, α_t, and α_r being defined as:

$$\begin{aligned} r_n &= \sqrt{w_n^2 + t_n^2} \\ \alpha_w &= \sqrt{w_n^2 + 1} \\ \alpha_t &= \sqrt{t_n^2 + 1} \\ \alpha_r &= \sqrt{r_n^2 + 1} \end{aligned} \quad (A.4)$$

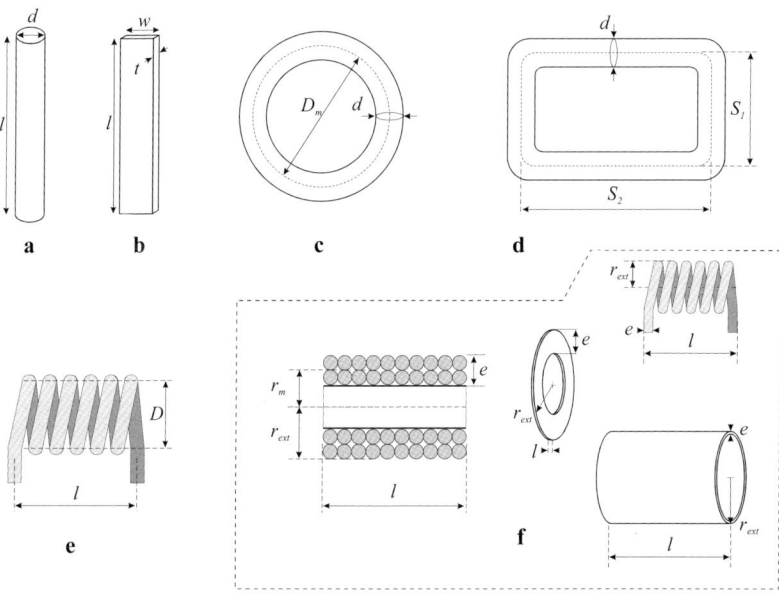

Fig. A.1 Geometries used for the self inductance formulae.

Flat strip in air [Fig. A.1(b)], approximate formula:

$$L = 0.2l\left[\ln\left(2l/(w+t)\right) + 0.50\right]. \tag{A.5}$$

Circular loop of round wire [Fig. A.1(c)]:

$$k = \frac{D_m}{D_m - d/2}\sqrt{1 - \frac{d}{D_m}}$$

$$L = (\pi/5) D_m \sqrt{1 - \frac{d}{D_m}} \left[\left(\frac{2}{k} - k\right) K(k) - \frac{2}{k} E(k)\right] \tag{A.6}$$

Circular loop of round wire [Fig. A.1(c)], approximate formula:

$$L = (\pi/5) D_m \left[\ln(8D_m/d) - 2\right]. \tag{A.7}$$

Square loop of round wire (Fig. A.1(d), $S_1 = S_2 = S$):

$$L = 0.8S\left[\ln(2S/d) + d/2S - 0.774\right]. \tag{A.8}$$

Rectangular loop of round wire [Fig. A.1(d)], empiric formula:

$$x = \frac{1 + S_2/S_1}{2\sqrt{S_2/S_1}}$$

$$L = 0.8x\left[1 - 0.1845(x-1)\right] \times \tag{A.9}$$

$$\sqrt{S_1 S_2}\left[\ln\left(2\sqrt{S_1 S_2}/d\right) + d/2\sqrt{S_1 S_2} - 0.774\right]$$

x is the ratio of the rectangular loop perimeter to the square of same area.

Rectangular loop of round wire [Fig. A.1(d)], accurate formula [Grover, 2004, pp. 60]:

$$L = 0.4 \begin{bmatrix} S_1 \ln\left(\dfrac{4S_1}{d}\right) + S_2 \ln\left(\dfrac{4S_2}{d}\right) + 2\sqrt{S_1^2 + S_2^2} \\ -S_1 \sinh^{-1}\left(\dfrac{S_1}{S_2}\right) - S_2 \sinh^{-1}\left(\dfrac{S_2}{S_1}\right) - \dfrac{7}{4}(S_1 + S_2) \end{bmatrix}. \quad \text{(A.10)}$$

Single layer solenoid [Lorentz formula, Fig. A.1(e)]:

$$k = D\big/\sqrt{D^2 + l^2}$$
$$L = (4\pi/30) n^2 \sqrt{D^2 + l^2} \times \qquad \text{(A.11)}$$
$$\left[K(k) - E(k) + \left(E(k) - k\right) D^2 / l^2 \right]$$

Cylindrical coil inductance [Fig. A.1(f)]:

$$L = 0.4\pi^2 n^2 \dfrac{r_m^2}{l + e + r_{ext}} AB$$
$$A = \dfrac{10l + 12e + 2r_{ext}}{10l + 10e + 1.4r_{ext}} \qquad \text{(A.12)}$$
$$B = 0.5 \log_{10}\left[100 + \dfrac{14 r_{ext}}{2l + 3e} \right]$$

A.2.2 *Mutual inductance formulae*

Mutual inductance formulae are given here for two simple geometries. The equations are valid if the wire dimensions are much smaller than the distance between the interacting elements. The formulae give accurate results if the distance is at least the thickness of the wire. More accurate formula, can be found in the Grover book if required.

Mutual inductance between two coaxial loops [Fig. A.2(a)], Maxwell formula:

$$k = 2\sqrt{\frac{d_1 d_2}{4\Delta^2 + (d_1 + d_2)^2}}$$

$$M = 0.2\pi\sqrt{d_1 d_2}\left[\left(\frac{2}{k} - k\right)K(k) - 2E(k)/k\right]$$

(A.13)

Mutual inductance between two parallel wires side by side [Fig. A.2(b)]:

$$M = 0.2l\left[\ln(2l/\Delta) - 1 + \Delta/l\right]. \quad (A.14)$$

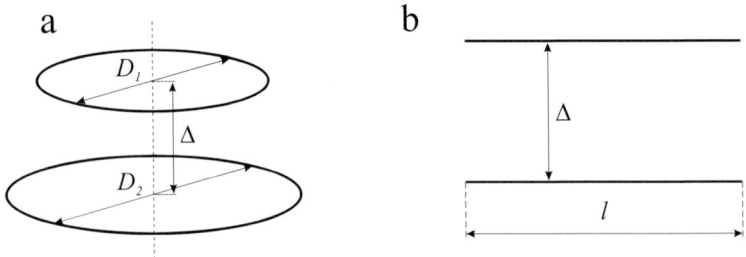

Fig. A.2 Geometries used for the mutual inductance formulae.

The mutual inductance between two straight wires of any relative position has been given by Grover [Grover, 2004]. It is quite complicated and will not be extensively written here. This formula is used in Birdcage Builder [Chin *et al.*, 2002] and is programmed in the FastHenry software [Kamon *et al.*, 1994] and in the NMRP program (accompanying CD, "buildprobe" utilities). The interested reader may refer to the corresponding sources.

A.2.3 *Combined numerical and formulae methods*

In complex situations, involving geometries that do not fit the application range of the above formulae, the mutual and self inductance values can still be obtained from a decomposition of the conductive structure into elementary components, essentially flat strips or straight elements of round wires.

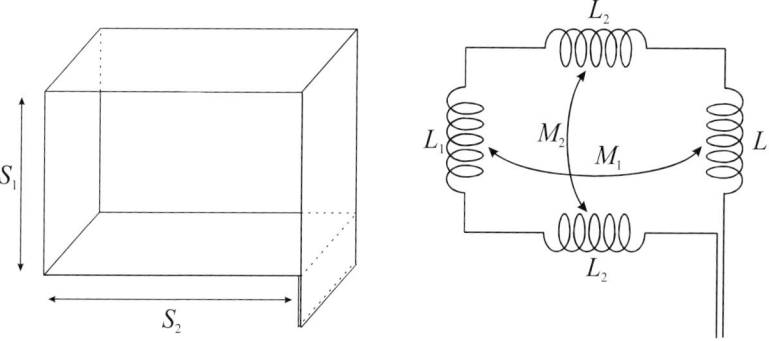

Fig. A.3 Decomposition of a rectangular inductance into elementary shapes of known self and mutual inductance (partial inductance).

The self inductance for each element is given from the above formulae and the mutual inductances involved in the structure can be estimated by the thin wire approximation. For example, the inductance of a rectangle of flat strips can easily be obtained from a formula like (Fig. A.3):

$$L = 2(L_1 - M_1) + 2(L_2 - M_2), \qquad (A.15)$$

where L_1 and L_2 are the self inductances of the side conductor of the rectangle [Eq. (A.2) or Eq. (A.5)] and M_1 and M_2 are the mutual inductances between the opposed sides S_1 and S_2 [Eq. (A.14)], respectively.

More complex structures can be treated in a similar way, but if the structure becomes too complicated, a numerical method is more efficient. The program FastHenry [Kamon *et al.*, 1994], free to use, was designed to calculate the inductance of the connections in microelectronic circuits. It proved to be well adapted for a large number of problems related to NMR probe design. The structure to be calculated is decomposed in conducting planar sheets connected together accordingly to the 3-D geometry of the probe by nodes. A simple example is given is Fig. A.4 for the slotted tube.

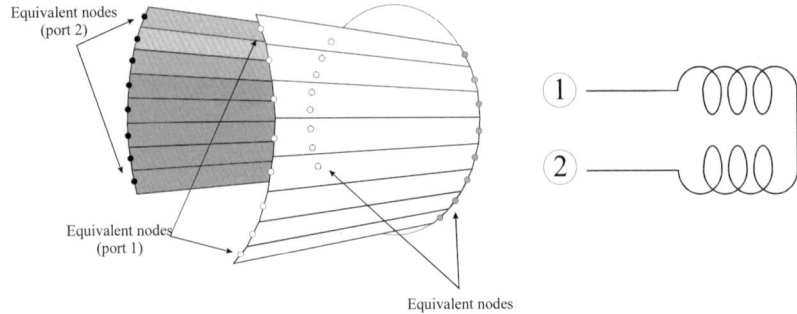

Fig. A.4 An example of the decomposition of a slotted tube into strips that are used by FastHenry [Kamon, M., et al., 1994] for inductance calculation of the equivalent circuit (left). The potential at each strip end is defined by a node. Nodes are connected together by an "equiv." declaration (represented here as white and black circles). The inductance is calculated between ports 1 and 2.

The program provides the self and mutual inductances (the complex impedance matrix) that exist between the external nodes, i.e., the input port of the coil structure. FastHenry decomposes the flat conductor strips in a number of filaments forming a complex network circuit. The impedance matrix of the network is calculated using Grover formulae. The corresponding huge matrix is solved by an original and efficient numerical method. The program "simprobe," part of the NMRP, can do a similar job (Section 3.2.1.4), although the resolution of the linear network equations is less efficient than in FastHenry. The advantage is that it is integrated in the (quasi-static) simulation suite for NMR probes.

A.2.4 Integration methods

The self and mutual inductance can also be evaluated from numerical integration over a number of current filaments spanning the conductors. The inductance of a volume conductor carrying a total current I is given by [Jin, 1999, p. 58]:

$$L = \frac{\mu_0}{4\pi I^2} \iiint_{V'} \iiint_{V} \frac{J(\vec{r}) \cdot J(\vec{r'})}{|\vec{r} - \vec{r'}|} dv dv', \qquad (A.16)$$

where $J(\vec{r})$ the current density at the point defined by the vector $\vec{r}$. At high frequency, the skin effect is such that the volume integral may be replaced by a two-dimensional integral over the conductor surface.

Similarly, the mutual inductance between two volume conductors carrying currents I_1 and I_2 is written as:

$$M_{12} = M_{21} = \frac{\mu_0}{4\pi I_1 I_2} \iiint_{V_1} \iiint_{V_2} \frac{J_1(\vec{r_1}) \cdot J_2(\vec{r_2})}{|\vec{r_1} - \vec{r_2}|} dv_2 dv_1. \quad (A.17)$$

These equations have been used, for example, in the birdcage simulator written by Giovannetti [Giovanetti et al., 2002].

In many cases, the calculation of the mutual inductance can be greatly simplified using the Neumann formula. When separated by a distance, at least of the order of their thickness, the mutual inductance between two conductors can be assimilated to that of two thin wires. The mutual inductance between the two wires is given by:

$$M_{12} = M_{21} = \frac{\mu_0}{4\pi} \oint \oint \frac{\overrightarrow{dl_1} \cdot \overrightarrow{dl_2}}{\Delta}, \quad (A.18)$$

where $\overrightarrow{dl_1}$ and $\overrightarrow{dl_2}$ are two elementary filaments separated by the distance Δ (Fig. A.5).

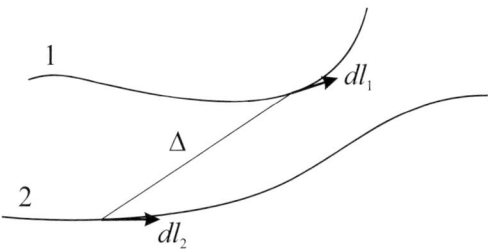

Fig. A.5 Geometry of the Neumann integral (Eq. (A.16)) for mutual inductance calculation.

Note that if $\vec{dl_1}$ and $\vec{dl_2}$ are perpendicular the integrand is zero. This formula proved to give accurate results when compared to more sophisticated calculations.

A.2.5 *Evaluation of the complete elliptic integrals*

Complete elliptic integrals of the first and second kind are involved in the calculation of the magnetic field of a loop, in the Lorentz formula for the inductance of the solenoid, or in the calculation of the mutual inductance between two coaxial loops. Tabulated values are not practical for use in computer programs, but a number of algorithms have been designed to perform the estimation from a series development.

An efficient, stable, and precise algorithm has been published by Fan as a complement to their paper [Fan *et al.*, 1987]. The algorithm can be easily implemented as a simple computer program. The algorithm is derived from the King method. It consists of a rapidly converging series development.

Starting with initial values a_i and b_i, given by:

$$a_i = 1.0$$
$$b_i = \sqrt{1-k^2},$$
(A.19)

the complete elliptic integral of the first and second kind ($K(k) = F(k,\pi/2)$ and $E(k) = E(k,\pi/2)$) are calculated as follows.

Let c_i and p_2 two variables initialized, respectively to k and 1. The loop, Eq. (A.20), is repeated until the value of c_i becomes as small as desired. The limitation relies only on the precision of the computer. In practice, it has been found that with a "double" precision floating variable the algorithm converges in any circumstances when the limiting value for c_i is set to 10^{-7}:

$$\begin{aligned}
&sum = 0 \quad \{ \\
&sum = sum + p_2 c_i^2 \\
&a_{i+1} = (a_i + b_i)/2 \\
&b_{i+1} = \sqrt{a_i b_i} \\
&c_{i+1} = (a_i - b_i)/2 \\
&p_2 = 2p_2 \\
&\}
\end{aligned} \qquad (A.20)$$

After convergence, the elliptic integrals are given by:

$$\begin{aligned} K(k) &= \pi/2a_i \\ E(k) &= K(k)(1 - sum/2) \end{aligned} \qquad (A.21)$$

A.3 Capacitance formulae

A useful source of capacitance formulae is the literature concerning microstrip lines [Gupta *et al.*, 1996]. Capacitance or inductance values can also be obtained using some simple algebra related to the characteristic impedance of transmission lines (Section A.4). The book by Charles S. Walker gives a large set of formulae related to various forms of transmission lines [Walker, 1990].

In this section, only very basic formulae will be given evaluating the capacitance value for some common configurations. All capacitance values are given in pF and the dimensions are in mm.

Two parallel strips of length *l*, width *w*, and separated by *e* [Fig. A.6(a)]. The thickness of the conductor is assumed to be negligible with respect to *w* and *e*

$$C = 8.854 \cdot 10^{-3} \varepsilon_r \frac{wl}{e}. \qquad (A.22)$$

The formula does not take account of the fringing effect and assumes a homogeneous dielectric medium (relative susceptibility ε_r) surrounding the two strips.

Two parallel round wires of length l, diameter d, and separated by e [Fig. A.6(b)]:

$$C = 8.854 \cdot 10^{-3} \varepsilon_r \frac{\pi l}{\ln\left[\frac{e}{d} + \sqrt{\left(\frac{e}{d}\right)^2 - 1}\right]}. \quad (A.23)$$

The wires are assumed to be immersed in a homogeneous medium with a relative permittivity ε_r.

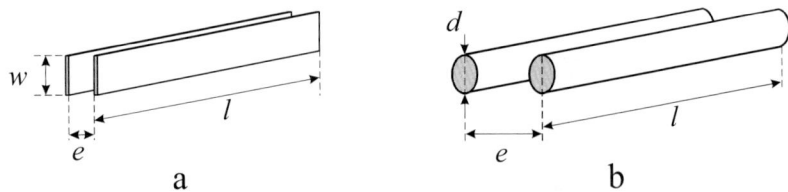

Fig. A.6 Geometry of the two-conductor configurations relevant to Eqs. (A.22) and (A.23).

Two coplanar strips of length l, width w, and separated by s (distance between inner sides), from [Gupta et al., 1996, pp. 400].

$$k = \frac{s}{s + 2w}$$

$$C = 8.854 \cdot 10^{-3} \frac{(\varepsilon_r + 1)}{2} \frac{2lK\left(\sqrt{1-k^2}\right)}{K(k)} \quad (A.24)$$

$K(u)$ is the complete elliptic integral of the first kind. The plane of the strips separates the space into two halves. One half is a dielectric substrate of relative permittivity ε_r and the other one is air (Fig. A.7).

This configuration introduces the concept of effective dielectric constant. This is particularly important regarding the phase (or propagation) velocity on such a line. The phase velocity is given by:

$$v_p = \frac{c}{\sqrt{\varepsilon_{reff}}}, \quad (A.25)$$

where ε_{reff} is the relative effective permittivity given by the ratio of the total capacitance C to the capacitance of the corresponding line with all the dielectric replaced by air.

In the present case, ε_{reff} is given by:

$$\varepsilon_{reff} = \frac{\varepsilon_r + 1}{2}. \tag{A.26}$$

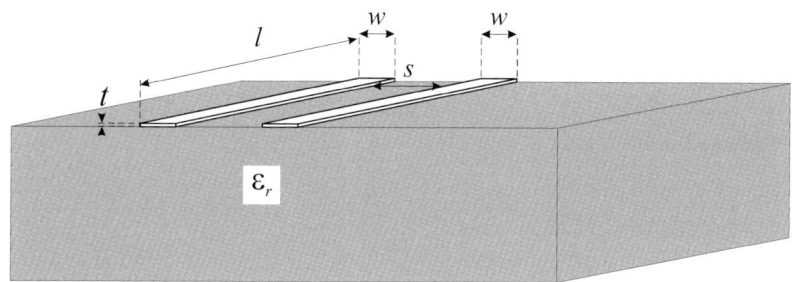

Fig. A.7 Geometry of the coplanar strips configuration.

Two coplanar strips of length l, width w, and separated by s, approximate formula from [Walker, 1990, pp. 51]:

$$C = 8.854 \cdot 10^{-3} \varepsilon_{reff} \frac{\pi l}{\ln\left[\pi\left(\frac{s}{w+t}\right)+1\right]}. \tag{A.27}$$

t is the thickness of the rectangular conductors. ε_{reff} is the relative effective permittivity as defined above. It depends on the thickness of the dielectric substrate.

Two coaxial cylindrical conductors of length l and diameters d and D ($D > d$):

$$C = 8.854 \cdot 10^{-3} \varepsilon_r \frac{2\pi l}{\ln\left(\frac{D}{d}\right)}. \tag{A.28}$$

A.4 Transmission line formulae

In this section, the characteristic impedance of transmission lines built on round and strip conductors is summarized. The formulae are based on Wheeler's papers and on the books by Edward C. Jordan [Jordan, 1989], Gupta *et al.* [Gupta *et al.*, 1996], Charles S. Walker [Walker, 1990] and David M. Pozar [Pozar, 1998].

A.4.1 *Relationship between characteristic constants*

Probably the most important relationship relates the characteristic impedance of the line (Z_c), the capacitance ($C_0 = C/l$) per unit length, and the phase velocity (v_p):

$$Z_c = \frac{1}{v_p C_0}. \tag{A.29}$$

This equation is valid for a low-loss line and can be derived from the two fundamental equations for line [Eq. (2.49) and Eq. (2.51)]:

$$Z_c = \sqrt{\frac{L_0}{C_0}}$$
$$v_p = \frac{1}{\sqrt{L_0 C_0}} \tag{A.30}$$

C_0 and v_p can be easily measured as discussed in Chapter 10, providing a mean to estimate Z_c with good accuracy.

Alternatively, C_0 may be computed accurately from quasi-static approximation, based on conformal mapping [Gupta *et al.*, 1996] or summing up the contribution of elementary cells [Walker, 1990].

The characteristic impedance is finally calculated from:

$$Z_c = \frac{1}{c\sqrt{\varepsilon_{reff}} C_{0a}}, \tag{A.31}$$

where C_{0a} is the capacitance per unit length of the line with the dielectric medium replaced by air. The equation above can be deduced from Eq. (A.23) and the definition of ε_{reff}:

$$\varepsilon_{reff} = \frac{C_0}{C_{0a}}. \qquad (A.32)$$

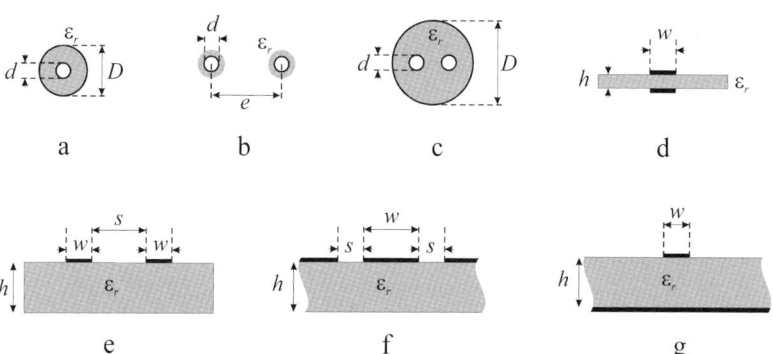

Fig. A.8 Geometry of transmission lines.

A.4.2 Characteristic impedance

In the following formulae, ε_{reff} is the effective relative permittivity as already defined. It is comprised between one (air line) and the relative permittivity e_r of a homogenous dielectric medium encompassing the conductors.

Coaxial line [Fig. A.8(a)]:

$$Z_c = \frac{60}{\sqrt{\varepsilon_r}} \ln\left(\frac{D}{d}\right). \qquad (A.33)$$

Two round wire line [Fig. A.8(b)]:

$$Z_c = \frac{120}{\sqrt{\varepsilon_{reff}}} \cosh^{-1}\left(\frac{e}{d}\right). \qquad (A.34)$$

Balanced shielded line [Fig. A.8(c)]:

$$Z_c = \frac{120}{\sqrt{\varepsilon_r}} \ln\left\{ 2\frac{e}{d}\left[\frac{1-(e/D)^2}{1+(e/D)^2}\right]\right\}. \qquad (A.35)$$

Two strip wire line [Fig. A.8(d)], approximate formulae:

$$\begin{aligned} w < h &\Rightarrow Z_c \approx \frac{120}{\sqrt{\varepsilon_{reff}}} \ln\left(\frac{4h}{w}\right) \\ w \gg h &\Rightarrow Z_c \approx \frac{120\pi}{\sqrt{\varepsilon_{reff}}} \left(\frac{h}{w}\right) \end{aligned}. \qquad (A.36)$$

If the dielectric substrate extends far away from the strip edge, ε_{reff} is equal to $(\varepsilon_r + 1)/2$ in the narrow strip approximation ($w < h$) and to ε_r in the wide strip approximation ($w \gg h$).

Accurate formulae have been derived by Wheeler [Wheeler, 1965]. They are given in Chapter 2 (Eqs. (2.152) and (2.153)).

Coplanar stripline [Fig. A.8(e)]. An approximate formula could be derived from (A.27) as [Walker, 1990, pp.123]:

$$Z_c = \frac{120}{\sqrt{\varepsilon_{reff}}} \ln\left[\pi\left(\frac{s}{w+t}\right) + 1\right]. \qquad (A.37)$$

Coplanar stripline [Fig. A.8(e)]. An accurate formula can be obtained similarly from Eq. (A.24) [Gupta et al., 1996, pp. 400]:

$$k = \frac{s}{s+2w}$$
$$Z_c = \frac{120\pi}{\sqrt{\varepsilon_{reff}}} \frac{K(k)}{K\left(\sqrt{1-k^2}\right)} \qquad (A.38)$$

ε_{reff} ranges from unity (line in air) to $(\varepsilon_r + 1)/2$ if the dielectric substrate is thick.

Coplanar waveguide [Fig. A.8(f)]. This configuration is complementary to the coplanar stripline, exchanging the dimensions s and w. In the following formula, w and s are still assigned to the width of the central strip and to the width of the gap between the strip and the ground plane, respectively, as shown in Fig. A.8(f). If the dielectric thickness is infinite [Gupta et al., 1996, pp. 382] ($h \to \infty$):

$$k = \frac{w}{w+2s}$$

$$Z_c = \frac{30\pi}{\sqrt{\varepsilon_{reff}}} \frac{K\left(\sqrt{1-k^2}\right)}{K(k)}. \tag{A.39}$$

$$\varepsilon_{reff} = \frac{\varepsilon_r + 1}{2}$$

If the thickness of the dielectric is finite, ε_{reff} is generally expressed as [Gupta et al., 1996]:

$$\varepsilon_{reff} = 1 + q(\varepsilon_r - 1), \tag{A.40}$$

where q is a quantity depending on the geometry. In the case of the coplanar waveguide, Gupta et al. give q as:

$$q = \frac{1}{2} \frac{K(k_2)}{K\left(\sqrt{1-k_2^2}\right)} \frac{K\left(\sqrt{1-k_1^2}\right)}{K(k_1)}, \tag{A.41}$$

where

$$\begin{aligned} a &= w/2 \\ b &= (w+2s)/2 \\ k_1 &= a/b \\ k_2 &= \frac{\sinh(\pi a/2h)}{\sinh(\pi b/2h)} \end{aligned} \tag{A.42}$$

Microstrip line [Fig. A.8(g)], approximate formulae depend on the ratio of the width (w) to substrate thickness (h) [Pozar, 1998, pp. 162].

If $w/h \leq 1$:

$$Z_c = \frac{60}{\sqrt{\varepsilon_{reff}}} \ln\left(\frac{8h}{w} + \frac{w}{4h}\right). \quad (A.43)$$

If $w/h > 1$:

$$Z_c = \frac{120\pi}{\sqrt{\varepsilon_{reff}}} \frac{1}{w/h + 1.393 + 0.667 \ln(w/h + 1.444)}. \quad (A.44)$$

ε_{reff} is given by:

$$\varepsilon_{reff} = \frac{\varepsilon_r + 1}{2} + \frac{(\varepsilon_r - 1)}{2\sqrt{1 + 12h/w}}. \quad (A.45)$$

The two equations (A.43) and (A.44) correspond to the common approximation of ratios of complete elliptic integrals.

To conclude, Gupta *et al.* [Gupta *et al.*, 1996] give a complete set of accurate formulae involving these elliptic integrals. Many other configurations are considered in their book to which the interested reader is encouraged to refer.

Bibliography

Ackerman, J.J.H., Grove, T.H., Wong, G.G., Gadian, D.G. and Radda, G.K. (1980). Mapping of Metabolites in Whole Animals by ^{31}P NMR using Surface Coils, *Nature*, **283**, 167–170.

Adriany, G. and Gruetter, R. (1997). A Half-Volume Coil for Efficient Proton Decoupling in Humans at 4 Tesla, *J. Magn. Reson.*, **125**, 178–184.

Advanced Receiver Research, Special Frequency GaAsFET Preamplifiers, web page available at:
http://www.advancedreceiver.com/page12.html [accessed 12/01/2014].

Agilent (2010). Fundamentals of RF and Microwave Noise Figure Measurements (Application Note 57-1) available at:
http://cp.literature.agilent.com/litweb/pdf/5952-8255E.pdf [accessed 12/01/2014].

Agilent (2013) Impedance Measurement Handbook, A guide to measurement technology and techniques, 4th edn. (Application Note 5950-3000) available at: http://cp.literature.agilent.com/litweb/pdf/5950-3000.pdf [accessed 12/01/2014].

Aissaini, S., Guendouz, L. and Canet, D. (2014). CMOS Based Q-Switch for Low-Power Pulsed ^{14}N Quadruple Resonance, *Concepts Magn. Reson.*, **44B**, 12–17.

Alderman, D.W. and Grant, D.M. (1979). An Efficient Decoupler Coil Design which Reduces Heating in Conductive Samples in Superconductive Spectrometers, *J. Magn. Reson.*, **36**, 447–451.

Alecci, M. and Jezzard, P. (2002). Characterization and Reduction of Gradient-Induced Eddy Currents in the RF Shield of a TEM Resonator, *Magn. Reson. Med.*, **48**, 404–407.

Alecci, M., Romanzetti, S., Kaffanke, J., Celik, A., Wegener, H.P. and Shah, N.J. (2006). Practical design of a 4 Tesla double-tuned RF surface coil for interleaved 1H and 23Na MRI of rat brain, *J. Magn. Reson.*, **181**, 203–211.

Alfonsetti, M., Clementi, V., Iotti, S., Placidi, G., Lodi, R., Barbiroli, B., Sotgiu, A. and Alecci, M. (2005). Versatile Coil Design and Positioning of Transverse-Field RF Surface Coils for Clinical 1.5 T MRI Applications, *MAGMA*, **18**, 69–75.

Algarin, J.M., Lopez M.A., Freire, M.J. and Marqués, R. (2011). Signal-to-Noise Ratio Evaluation in Resonant Ring Metamaterial Lenses for MRI applications, *New J. Phys.*, **13**, 115006-1–12.

Allard, M. and Henkelman, R.M. (2006). Using metamaterial Yokes in NMR Measurements, *J. Magn. Reson.*, **182**, 200–207.
Alonzo, G.J., Blackwell, R.H. and Marantz, H.V. (1967). Direct-Reading, Fully Automatic Vector Impedance Meters, *HP Journal*, **Jan 1967**, 12–20.
Amari, S., Ulug, M.A., Bornemann, J., van Zijl, P.C.M. and Barker, P.B. (1997). Multiple Tuning of Birdcage Resonators, *Magn. Reson. Med.*, **37**, 243–251.
Anderson, R.W. and Dennison O.T. (1967). An Advanced New Network Analyzer for Sweep-Measuring Amplitude and Phase from 0.1 to 12.4 GHz, *HP Journal*, **Feb 1967**, 2–10.
Andrew, E.R. and Jurga, K. (1987). NMR Probe with Short Recovery Time, *J. Magn. Reson.*, **73**, 268–276.
Andrews, D. (2006). *Lumped Element Quadrature Hybrids*, Artech House, Inc., Norwood.
ARRL Handbook (1988) Handbook for Radio Communications 65th edn. Wilson, M.J. (ed), American Radio Relay League, Inc. Newington, USA.
ARRL Handbook (2009) Handbook for Radio Communications 86th edn. Wilson, M.J. (ed), American Radio Relay League, Inc. Newington, USA.
Assink, R.A., Fukushima, E., Gibson, A.A.V. Rath, A.R. and Roeder, S.B. (1986). A Nondetuning Surface Coil, the Semitoroid, *J. Magn. Reson.*, **66**, 176–181.
Aussenhofer, S.A. and Webb, A.G. (2012). Design and evaluation of a detunable water-based quadrature HEM_{11} mode dielectric resonator as a new type of volume coil for high field MRI, *Magn. Reson. Med.*, **68**, 1325–1331.
Aussenhofer, S.A. and Webb, A.G. (2014). An Eight-Channel Transmit/Receive Array of Te_{01} Mode High Permittivity Ceramic Resonators for Human Imaging at 7 T, *J. Magn. Reson.*, **243**, 122–129.
Baena, J.D., Jelinek, L., Marqués, R. and Silveirinha, M. (2008). Unified Homogeneization Theory for Magnetoinductive and Electromagnetic Waves in Split-Ring Metamaterials, *Phys. Rev.*, **A78**, 013842-1-5.
Balaban, R.S., Koretsky, A.P. and Katz, L.A. (1986). Loading Characteristics of Surface Coils Constructed from Wire and Foil, *J. Magn. Reson.*, **68**, 556–560.
Barbara, T.M., Martin, J.F. and Wurl, J.G. (1991) Phase Transients in NMR Probe Circuits, *J. Magn. Reson.*, **93**, 497–508.
Barberi, E.A. and Rutt, B.K. (1993). The Variable Frequency Birdcage Resonator, *Proc. Soc. Magn. Reson. Med.*, **43**, 284–289.
Barberi, E.A., Gati, J.S., Rutt, B.K. and Menon, R.S. (2000). A Transmit-Only/Receive-Only (TORO) RF System for High-Field MRI/MRS Applications, *Magn. Reson. Med.*, **43**, 284–289.
Baril, N., Thiaudière, E., Quesson, B., Delalande, C., Canioni, P. and Franconi, J.-M. (2000). Single-Coil Surface Imaging Using a Radiofrequency Field Gradient, *J. Magn. Reson.*, **146**, 223–227.
Barnes, S.H., Oaks, S. and Mann, J.E. (1961) Voltage Sensitive Semiconductor Capacitor, *U.S. Patent*, 2,989,671.

Beck, B., Peterson, D.M., Duensing, G.R. and Fitzsimmons, J.R. (2000). Implications of Cable Shield Currents at 3 Tesla and 4.7 Tesla, *Proc. Int. Soc. Mag. Reson. Med.*, **6**, 641.

Beck, B. and Blackband, S. (2001). Phased Array Imaging on a 4.7T/33 cm Animal Research System, *Rev. Sci. Instrum.*, **72**, 4292–4294

Behr, V.C., Haase A. and Jakob, P.M. (2004). RF Flux Guides for Excitation and Reception in ^{31}P Spectroscopic and Imaging Experiments at 2 Tesla, *Concepts Magn. Reson.*, **23B**, 44–49.

Bendall, M.R. (1983). Portable NMR Sample Localization Method Using Inhomogeneous RF Irradiation Coils, *Chem. Phys. Lett.*, **99**, 310–315.

Bendall, M.R. and Gordon, R.E. (1983). Depth and Refocusing Pulses Designed for Multipulse NMR with Surface Coils, *J. Magn. Reson.*, **53**, 365–385.

Bendall, M.R., McKendry, J.M., Cresshull, I.D. and Ordidge, R.J. (1984). Active Detune Switch for Complete Sensitive-Volume Localization in *In Vivo* Spectroscopy Using Multiple rf Coils and Depth Pulses, *J. Magn. Reson.*, **60**, 473–478.

Bendall, M.R., Connelly, A. and McKendry, J.M. (1986a). Elimination of Coupling between Transmit Coils and Surface-Receive Coils for *In Vivo* NMR, *Magn. Reson. Med.*, **3**, 157–163.

Bendall, M.R., Foxall, D., Nichols, B.G. and Schmidt, J.R. (1986b). Complete Localization of *In Vivo* NMR Spectra Using Two Concentric Surface Coils and RF Methods Only, *J. Magn. Reson.*, **70**, 181–186.

Berkowitz, B.A. and Ackerman, J.H. (1987). Proton Decoupled Fluorine Nuclear Magnetic Resonance Spectroscopy *In Situ*, *Biophys. J.,* **51**, 681–685.

Black, R.D., Early, T.A., Roemer, P.B., Mueller, O.M., Mongro-Campero, A., Turner, L.G. and Johnson, G.A. (1993). A High Temperature Superconducting Receiver for Nuclear Magnetic Resonance Microscopy, *Science*, **259**, 793–795.

Black, R.D., Early, T.A. and Johnson, G.A. (1995). Performance of a High-Temperature Superconducting Resonator for High-Field Imaging, *J. Magn. Reson. A*, **113**, 74–80.

Bloch, F. (1946). Nuclear Induction, *Phys. Rev.*, **70**, 460–474.

Bloch, F., Hansen, W.W. and Packard, M. (1946). Nuclear Induction, *Phys. Rev.*, **69**, 127.

Bode, H.W. (1945) *Network Analysis and Feedback Amplifier Design*, D. Van Nostrand Company, Inc., New York, NY, USA.

Bodenhausen, G., Freeman, R. and Turner, D.L. (1977). Suppression of Artifacts in Two-Dimensional J Spectroscopy, *J. Magn. Reson.*, **27**, 511–514.

Bogdanov, G. and Ludwig, R. (2003). Analysis of High-Field RF Coils Using the Method of Lines, *Concepts Magn. Reson.,* **16B**, 22–37.

Bolinger, L., Prammer, M.G. and Leigh, J.S. (1988). A Multiple-Frequency Coil with a Highly Uniform B_1 Field, *J. Magn. Reson.*, **81**, 162–166.

Bosh, C.S. and Ackerman, J.J.H. (1992). 'Surface Coil Spectroscopy', in Diehl, P., Fluck, E., Günter, H., Kosfeld, R. and Seelig, J. (Eds) *In-Vivo Magnetic Resonance*

Spectroscopy[II]: *Localization and Spectral Editing, NMR Basic Principles and Progress Series*, **vol. 27**, Springer-Verlag, Berlin, pp. 3–44.

Bottomley, P.A., Hardy, C.J., Roemer, P.B. and Mueller, O.M. (1989). Proton-Decoupled, Overhauser-Enhanced, Spatially Localized Carbon-13 Spectroscopy in Humans, *Magn. Reson. Med.*, **12**, 348–363.

Bridges, J.F. (1988). Cavity Resonator with Improved Magnetic Field Uniformity for High Frequency Operation and Reduced Dielectric Heating in NMR Imaging Devices, *U.S. Patent*, 4,751,464.

Bringham, E.O. (1974). *The Fast Fourier Transform and its Applications*, Prentice-Hall, Englewood Cliffs, NJ, USA.

Brunner, D.O., De Zanche, N., Fröhlich, J., Paska J. and Pruessmann, K.P. (2009). Travelling-wave nuclear magnetic resonance, *Nature*, **457**, 994–998.

Buess, M.L., Garroway, A.N. and Miller, J.B. (1991). NQR Detection Using a Meanderline Surface Coil, *J. Magn. Reson.*, **92**, 348–362.

Callendar, M. V. (1947). Q of Solenoid Coils, *Wireless Engineer*, June 1947, 185.

Caloz, C. and Itoh, T. (2005). Electromagnetic Metamaterials, Transmission Line Theory and Microwave Applications, John Wiley and Sons, Inc., Hoboken, NJ, USA.

Canet, D. (1997). Radiofrequency Field Gradient Experiments, *Prog. Nucl. Magn. Reson. Spectrosc.*, **30**, 101–135.

Carlson, J.W. (1986). Currents and Fields of Thin Conductors in rf Saddle Coils, *Magn. Reson. Med.*, **3**, 778–790.

Carlson, J.W. (1988). Radiofrequency Field Propagation in Conductive NMR Samples, *J. Magn. Reson.*, **78**, 563–573.

Carobbi, C.F.M., Millanta, L.M. and Chiosi L. (2000). The High-Frequency Behavior of the Shield in the Magnetic-Field Probes, *IEEE Int. Symp. Electromagn. Compat.*, **1**, 35–40.

Carobbi C.F.M. and Millanta L.M. (2004). Analysis of the Common-Mode Rejection in the Measurement and Generation of Magnetic Fields Using Loop Probes, *IEEE Trans. Instrum. Meas.*, **53**, 514–523.

Chen, C.N., Sank, V.J. and Hoult, D.I. (1983). Quadrature Detection Coils – A Further $\sqrt{2}$ Improvement in Sensitivity, *J. Magn. Reson.*, **54**, 324–327.

Chen, C.N., Sank, V.J., Cohen, S.M., and Hoult, D.I.. (1986). The Field Dependence of NMR Imaging. I. Laboratory Assessment of Signal-to-Noise Ratio and Power Deposition, *Magn. Reson. Med.*, **3**, 722–729.

Chen, C.N. and Hoult, D.I. (1989) *Biomedical Magnetic Resonance Technology*, IOP Publishing Ltd., Philadelphia, PA, USA.

Chen, J-H., Lin, F-H. and Kuan, W-P. (1999). Quantitative Analysis of Magnetic Resonance Radio-Frequency Coils Based on Method of Moment, *IEEE Trans. Magn.*, **35**, 2118–2127.

Chen, X., Grzegorczyk, T.M., Wu, B-I., Pacheco, J. and Kong, J.A. (2004). Robust method to retrieve the constitutive effective parameters of metamaterials, *Phys. Rev.*, **E70**, 016608-1-7.

Chin, C.L., Collins, C.M., Li, S., Dardzinski, B.J. and Smith, M.B. (2002). BirdcageBuilder: Design of Specified-Geometry Birdcage Coils with Desired Current Pattern and Resonant Frequency, *Concepts Magn. Reson.*, **15**, 156–163.

Chingas, G.C. (1983). Overcoupling NMR Probes to Improve Transient Response, *J. Magn. Reson.*, **54**, 153–157.

Collins, C.M., Li, S., Yang, Q.X. and Smith, M.B. (1997). A Method for Accurate Calculation of B_1 Fields in Three Dimensions. Effects of Shield Geometry on Field Strength and Homogeneity in the Birdcage Coil, *J. Magn. Reson.*, **125**, 233–241.

Cook, B. and Lowe, I.J. (1982). A Large-Inductance, High-Frequency, High-Q, Series-Tuned Coil for NMR, *J. Magn. Reson.*, **49**, 346–349.

Corum, K.L. and Corum, J.F. (2001). RF coils, helical resonators and voltage magnification by coherent spatial modes, *5th International Conference on Telecommunications in Modern Satellite, Cable and Broadcasting Service, TELSIKS 2001*, **1**, 339–348.

Croon, J.A., Borsboom, H.M. and Mehlkopf, A.F. (1999). Optimization of Low Frequency Litz-wire RF Coils, *ISMRM*, 740.

Cross, V.R., Hester, R.K. and Waugh, J.S. (1976). Single Coil Probe with Transmission-Line Tuning for Nuclear Magnetic Double Resonance, *Rev. Sci. Instrum.*, **47**, 1486–1488.

Cross, T.A., Müller, S. and Aue, W.P. (1985). Radiofrequency Resonators for High Field Imaging and Double-Resonance Spectroscopy, *J. Magn. Reson.*, **62**, 87–98.

Crowley, M.G., Evelhoch, J.E. and Ackerman, J.J.H. (1985). The Surface-Coil NMR Receiver in the Presence of Homogeneous B_1 Excitation, *J. Magn. Reson.*, **64**, 20–31.

Crozier, S., Luescher, K., Forbes, L.K., and Doddrel, D.M. (1995). Optimized Small-Bore, High_Pass Resonator Designs, *J. Magn. Reson.*, *B*, **109**, 1–11.

Crozier, S., Forbes, L.K., Roffmann, U.W., Luescher, K. and Doddrel, D.M. (1997). A Methodology for Current Density Calculations in High-Frequency RF Resonators, *Concepts Magn. Reson.*, **9**, 195–210.

Dardzinski, B.J., Li, S., Collins, C.M., Williams, G.D. and Smith, M.B. (1998). A Birdcage Coil Tuned by RF Shielding for Application at 9.4 T, *J. Magn. Reson.*, **131**, 32–38.

Darrasse, L. and Kassab, G. (1993). Quick Measurement of NMR-coil sensitivity with a Dual-Loop Probe, *Rev. Sci. Instrum.*, **64**, 1841-1844.

Darrasse, L. and Ginefri, J.C. (2003). Perspectives with Cryogenic RF Probes in Biomedical MRI, *Biochimie*, **85**, 915–937.

Daugaard, P., Ellis, P.D. and Jakobsen, H.J. (1981a). A High-Performance 18 mm Probe System for the Varian XL-100–15 NMR Spectrometer, *J. Magn. Reson.*, **43**, 434–442.

Daugaard, P., Jakobsen, H.J., Garber, A.R. and Ellis, P.D. (1981b). A Simple Method for NMR Probe Tuning and Some Consequences of Improper Probe Tuning, *J. Magn. Reson.*, **44**, 224–227.

Decorps, M. and Fric, C. (1969). Etude Comparative de Divers Types de Volumes Résonnants pour Spectromètres à Résonance Paramagnétique Electronique, en Ondes Métriques, *J. Sci. Instrum.*, **2**, 1036–1040.

Decorps, M. and Fric, C. (1972). Un Spectromètre Basse Fréquence à Haute Sensibilité pour L'étude de la Resonance des Spins Electroniques, *J. Sci. Instrum.*, **5**, 337–342.

Decorps, M., Laval, M., Confort, S. and Chaillout, J.J. (1985a). Signal to Noise and Spatial Localization of NMR Spectra with a Surface Coil and the Saturation Recovery Sequence, *J. Magn. Reson.*, **61**, 418–425.

Decorps, M., Blondet, P., Reutenauer, H., Albrand, J.P. and Remy, C. (1985b). An Inductively Coupled, Series-Tuned NMR Probe, *J. Magn. Reson.*, **65**, 100–109.

Decorps, M. (2011). *Imagerie de Résonance Magnétique, Bases Physiques et Méthodes*, CNRS Editions EDP Sciences, Paris, France.

Deo, N. (1974) Graph Theory with Applications to Engineering and Computer Science, Prentice-Hall of India Learning Private Ltd, New Delhi, India.

Derby, K., Tropp, J. and Hawryszko, C. (1990). Design and Evaluation of a Novel Dual-Tuned Resonator for Spectroscopic Imaging, *J. Magn. Reson.*, **86**, 645–651.

Devasahayam, N., Subramanian, S., Murugesan, R., Cook, J.A., Afeworki, M., Tschudin, R.G., Mitchell, J.B. and Krishna, M.C. (2000). Parallel Coil Resonators for Time-Domain Radiofrequency Electron Paramagnetic Resonance Imaging of Biological Objects, *J. Magn. Reson.*, **142**, 168–176.

Dominguez-Viqueira, W., Carias, M. and Santyr G.E. (2010). Optimization of Multi-turn Litz Wire Radiofrequency Coils for Hyperpolarized Noble Gas Imaging of Rodent Lungs at 73.5mT, *Proc. Int. Soc. Magn. Reson. Med.*, **18**, 1508.

Doty, F.D., Entzminger, G. and Hauck, C.D. (1999). Error-Tolerant RF Litz Coils for NMR/MRI, *J. Magn. Reson.*, **140**, 17–31.

Doty, F.D. (2000). Low-Inductance Transverse Litz Foil Coils, *U.S. Patent*, 6,060,882.

Doty, F.D., Entzminger, G., Kulkarni, J., Pamarthy, K. and Staab, J.P. (2007). Radio Frequency Coil Technology for Small-Animal MRI, *NMR Biomed.*, **20**, 304–325.

Doty, F.D. and Entzminger Jr, G. (2012). 'Litz Coils for High Resolution and Animal Probes, Especially for Double Resonance', in Vaughan, J.T. and Griffiths, J.R. (eds), *RF Coils for MRI*, ch. 21, John Wiley & Sons, Ltd, Chichester, UK, pp. 245–258.

Dunsmore, J.P. (2012) Handbook of Microwave Components Measurements with Advanced VNA techniques, John Wiley & Sons, Ltd, Chichester, UK.

Dürr, W. and Rauch, S. (1991). A Dual Frequency Circularly Polarizing Whole-Body MR Antenna for 69/170 MHz, *Magn. Reson. Med.*, **19**, 446–455.

Edelstein, W.A., Hardy, C.J. and Mueller, O.M. (1986). Electronic Decoupling of Surface-Coil Receivers for NMR Imaging and Spectroscopy, *J. Magn. Reson.*, **67**, 156–161.

Eidmann, G., Savelsberg, R., Blumler, P. and Blumich, B. (1996). The NMR MOUSE, a Mobile Universal Surface Explorer, *J. Magn. Reson., A*, **122**, 104–109.

Ernst, R.R. and Anderson, W.A. (1966). Application of Fourier Transform Spectroscopy to Magnetic Resonance, *Rev. Sci. Instrum.*, **37**, 93–102.

Eroglu, S., Gimi, B., Roman, B. Friedman, G. and Magin, R.L. (2003). NMR Spiral Surface Microcoils: Design, Fabrication and Imaging, *Concepts Magn. Reson.*, **17B**, 195–210.

Fakri, L., Lapray, C. and Briguet, A. (1996). Design and Modeling the Free Element Birdcage Resonator, *ESMRMB*, Prague, pp. 272.

Fan, M., Gonord, P., Kan, S. and Taquin, J. (1987). A UHF Probe for NMR Micro-Imaging Experiments, *Magn. Reson. Med.*, **4**, 591–596.

Fano, R.M. (1948). Theoretical Limitations on the Broadband Matching, *Technical Report N°41*, MIT, Research Laboratory of Electronics, pp. 1–44.

Fish, P.J. (1994). *Electronic Noise and Low Noise Design*, McGraw-Hill, Inc, New York, USA.

Fitzsimmons, J.R., Brooker, H.R. and Beck, B. (1987). A Transformer-Coupled Double-Resonant Probe for NMR Imaging and Spectroscopy, *Magn. Reson. Med.*, **5**, 471–477.

Fitzsimmons, J.R., Beck, B.L., and Brooker, H.R. (1993). Double Resonant Quadrature Birdcage, *Magn. Reson. Med.*, **30**, 107–114.

Forrer, J., Pfenninger, S., Eisenegger, J. and Schweiger, A. (1990). A Pulsed ENDOR Probehead with the Bridged Loop-Gap Resonator: Construction and Performance, *Rev. Sci. Instrum.*, **61**, 3360–3367.

Franzen, W. (1962). Generation of Uniform Magnetic Fields by Means of Air-Core Coils, *Rev. Sci. Instrum.*, **33**, 933–938.

Fratila, R.M., Gomez, M.V., Sykora, S. and Velders, A.H. (2014). Multinuclear Nanoliter One-Dimensional and Two-Dimensional NMR Spectroscopy with a Single Non-Resonant Microcoil, *Nat. Commun.*, **5**, http://dx.doi.org/10.1038/ncomms4025.

Freire, M.J., Marqués, R. and Jelinek, L. (2008). Experimental Demonstration of a $\mu=-1$ Metamaterial Lens for Magnetic Resonance Imaging, *Appl. Phys. Lett.*, **93**, 231108-1–3.

Freire, M.J., Jelinek, L., Marqués, R. and Lapine, M. (2010). On the Application of $\mu r=-1$ Metamaterial Lenses for Magnetic Resonance Imaging, *J. Magn. Reson.*, **203**, 81–90.

Freire, M.J., Lopez, M.A., Meise, F., Algarin, J.M., Jakob, P.M., Bock, M. and Marques, R. (2013). A Broadside-Split-Ring Resonator-Based Coil for MRI at 7 T, *IEEE Trans. Med. Imaging*, **32**, 1081–1084.

Friedrich, J. and Freeman, R. (1988). Spatial Localization Using a Straddle Coil, *J. Magn. Reson.*, **77**, 101–118.

Froncisz, W. and Hyde, J.S. (1982). The Loop-Gap Resonator: A New Microwave Lumped Circuit ESR Sample Structure, *J. Magn. Reson.*, **47**, 515–521.

Gadian, D.G. (1982). Nuclear Magnetic Resonance and its Applications to Living Systems, Oxford University Press, UK.

Galassi, M., Davies, J., Theiler, J., Gough, B., Jungman, G., Alken, P., Booth, M. and Rossi, F. (2009). *GNU Scientific Library. Reference Manual*, 3rd edn., Network Theory Ltd., U.K.

Garrett, M.W. (1951). Axially Symmetric Systems for Generating and Measuring Magnetic Fields. Part I, *J. App. Phys.*, **22**, 1091–1107.

Gasson, J., Summers, I.R., Fry, E.M. and Vennart, W. (1995). Modified Birdcage Coils for Targeted Imaging. *Magn. Reson. Imaging*, **7**, 1003–1012.

Gimi, B., Eroglu, S., Leoni, L., Desai, T.A., Magin, R.L. and Roman, B.B. (2003). NMR Spiral Surface Microcoils: Applications, *Concepts Magn. Reson.*, **18B**, 1–8.

Ginefri, J.C., Durand, E. and Darrasse, L. (1999). Quick Measurement of Nuclear Magnetic Resonance Coil Sensitivity with a Single-Loop Probe, *Rev. Sci. Instrum.* **70**, 4730–4731.

Ginefri J.C., Darrasse, L. and Crozat, P., (2003). High Temperature Superconducting Surface Coil for *In Vivo* Microimaging of the Human Skin, *Magn. Reson. Med.*, **45**, 376–382.

Ginefri, J.C., Poirier-Quinot, M., Girard, O. and Darrasse, L. (2007). Technical Aspects: Development, Manufacture and Installation of a Cryo-Cooled HTS Coil System for High-Resolution *In-Vivo* Imaging of the Mouse At 1.5 T. *Methods*, **43**, 54–67.

Ginefri, J.C., Rubin, A., Tatoulian, M., Woytasik, M., Boumezbeur, F., Djemaï, B., Poirier-Quinot, M., Lethimonnier, F., Darrasse, L. and Dufour-Gergam, E. (2012). Implanted, Inductively-Coupled, Radiofrequency Coils Fabricated on Flexible Polymeric Material: Application to *In Vivo* Rat Brain MRI at 7 T, *J. Magn. Reson.*, **224**, 61–70.

Giovannetti, G., Francesconi, R., Landini, L., Santarelli, M.F. and Positano V. (2002). A Fast and Accurate Simulator Design of Birdcage Coils in MRI, *MAGMA*, **15**, 36–44.

Giovannetti, G., Viti V., Positano, V., Santarelli, M.F., Landini, L. and Benassi, A. (2007). Magnetostatic Simulation for Accurate Design of Low Field MRI Phased-Array Coils, *Concepts Magn. Reson.*, **31B**, 140–146.

Giovannetti, G., Hartwig, V., Landini, L. and Santarelli, M.F. (2010). Low-Field MR Coils: Comparison between Strip and Wire Conductors, *Appl. Magn. Reson.*, **39**, 391–399.

Giovannetti, G., Hartwig, V., Landini, L. and Santarelli, M.F. (2011). Sample-Induced Resistance Estimation in Magnetic Resonance Experiments: Simulation and Comparison of Two Methods, *Appl. Magn. Reson.*, **40**, 351–361.

Glover, G.H., Hayes, C.E., Pelc, N.J., Edelstein, W.A., Mueller, O.M., Hart, H.R., Hardy, C.J., O'Donnell, M. and Barber, W.D. (1985). Comparison of Linear and Circular Polarization for Magnetic Resonance Imaging, *J. Magn. Reson.*, **64**, 255–270.

Gonnella, N.C. and Silverman, R.F. (1989). Design and Construction of a Simple Double-Tuned, Single-Input Surface-Coil, *J. Magn. Reson.*, **85**, 24–34.

Gonord, P., Kan, S. and Leroy-Willig, A. (1988). Parallel-Plate Split-Conductor Surface Coil: Analysis and Design, *Magn. Reson. Med.*, **6**, 353–358.

Gonord, P. and Kan, S. (1994). Twin-Horseshoe Resonator, *Rev. Sci. Instrum.,* **65**, 509–510.

Gonord, P., Kan, S., Leroy-Willig, A. and Wary, C. (1994). Multigap parallel Bracelet Resonator Frequency Determination and Applications, *Rev. Sci. Instrum.,* **11**, 3363–3366.

Gorss, C.G. (1967). Methods of Measuring Impedance, *HP Journal*, **Jan 1967**, 2–11.

Grafendorfer, T., Conolly, S.M., Sullivan, C.R., Macovski, A. and Scott, G. (2005). Can Litz Coils Benefit SNR in Remotely Polarized MRI, *Proc. Int. Soc. Magn. Reson. Med.,* **13**, 923.

Grafendorfer, T., Conolly, S.M., Matter, N.I., Pauly, J. and Scott, G. (2006). Optimized Litz Coil Design for Prepolarized Extremity MRI, *Proc. Int. Soc. Magn. Reson. Med.,* **14**, 2613.

Grandi, G., Kazimierczuk, K.M., Massarini, A. and Reggiani, U. (1999). Stray Capacitances of Single-Layer Solenoid Air-Core Inductors, *IEEE Trans. Ind. Appl.,* **35**, 1162–1168.

Grist, T.M. and Hyde, J.S. (1985). Resonators for *In Vivo* ^{31}P NMR at 1.5 T, *J. Magn. Reson.,* **61**, 571–578.

Grist, T.M., Jesmanowicz, A., Froncisz, W. and Hyde, J.S. (1986). 1.5 T *In Vivo* ^{31}P NMR Spectroscopy of the Human Liver Using a Sectorial Resonator, *Magn. Reson. Med.,* **3**, 135–139.

Grover, F.W. (2004). *Inductance Calculations*, Dover Publication, Inc, Mineola, NY, USA.

Guanella, G. (1944). New Method of Impedance Matching in Radio Frequency Circuits, *Brown–Boveri Rev.,* **31**, 327–329.

Gupta, K.C., Garg, R., Bahl, I. and Bhartia, P. (1996). *Microstrip Lines and Slotlines*, 2nd edn.., Artech House, Inc, Norwood, MA, USA.

Haase, A. (1985). A New Method for Decoupling of Multiple-Coil NMR Probes, *J. Magn. Reson.,* **61**, 130–136.

Haase, J., Curro, N.J. and Slichter, C.P. (1998). Double Resonance Probes for Close Frequencies, *J. Magn. Reson.,* **135**, 273–279.

Haase, A., Odoj, F., von Kienlin, M., Warnking, J., Fidler, F., Weisser, A., Nittka, M., Rommel, E., Lanz, T., Kalusche, B. and Griswold, M. (2000). NMR Probeheads for *In Vivo* Applications, *Concepts Magn. Reson.,* **12**, 361–388.

Habara, H., Ochi, H., Soutome, Y. and Bito, Y. (2007). The "Rung Pair" Birdcage Coil that has the Transmission Line Resonance Mode, *Proc. Int. Soc. Magn. Reson. Med.,* **15**, 3274.

Haines K.N. (2010). *Applications of high dielectric materials in high field magnetic resonance*, Dissertation thesis, Pennsylvania State University, USA.

Haines K., Smith, N.B. and Webb, A.G. (2010). New High Dielectric Constant Materials for Tailoring the B1+ Distribution at High Magnetic Fields, *J. Magn. Reson.,* **203**, 323–327.

Hall, L.D., Marcus, T., Neale, C., Powell, B., Sallos, J. and Talagala, S.L. (1985). A Modified Split-Ring Resonator Probe for NMR Imaging at High Field Strengths, *J. Magn. Reson.*, **62**, 525–528.

Hall, A.S., Alford, N.M., Button, T.W., Gilderdale, D.J., Gehring, K.A. and Young, I.R. (1991). Use of high temperature superconductor in a receiver coil for magnetic resonance imaging, *J. Magn. Reson.*, **20**, 340–343.

Hardy, W.N. and Whitehead, L.A. (1981). Split-Ring Resonator for Use in Magnetic Resonance from 200–2000 MHz. *Rev. Sci. Instrum.*, **52**, 213–216.

Harman, R.R., Buston, P.C., Hall, A.S., Young, I.R. and Bydder, G.M. (1988). Some Observations of the Design of rf Coils for Human Internal Use. *Magn. Reson. Med.*, **6**, 49–62.

Harpen, M.D. (1993). Cylindrical Coils Near Self-Resonance. *Magn. Reson. Med.*, **30**, 489–493.

Harpen, M.D. (1994). The Theory of Shielded Loop Resonators. *Magn. Reson. Med.*, **32**, 785–788.

Harrington, R.F. (1993) *Field Computation by Moment Methods*, IEEE Press, Piscataway, NJ, USA.

Hayes, C.E., Edelstein, W.A., Schenck, J.F., Mueller, O.M. and Eash, M. (1985). An Efficient, Highly Homogeneous Radiofrequency Coil for Whole-Body NMR Imaging at 1.5 T, *J. Magn. Reson.*, **63**, 622–628.

Hayes, C.E. (1987). Radio frequency coil for NMR, *U.S. Patent*, 4,692,705 and 4,694,255.

Hayes, C.E. and Eash, M.G. (1997). Shield for decoupling RF and gradient coils in an NMR apparatus, *U.S. Patent*, 4,642,569.

Hayes, C.E. (2009). The development of the birdcage resonator: a historical perspective. *NMR Biomed.*, **22**, 908–918.

Haziza, N., Bittoun, J. and Kan, S. (1997). Multiturn Split Conductor Transmission-Line Resonator, *Rev. Sci. Instrum.*, **68**, 1995–1997.

Hernandez, R., Rodriguez, A., Salgado, P. and Barrios, F.A. (2003). Concentric Dual Loop RF Coil for Magnetic Resonance Imaging, *Rev. Mex. Fis.*, **49**, 107–114.

Hidalgo, S., Rodriguez, A.O., Rojas, R., Sanchez, J., Reynoso, G. and Barrios, F. (2001). Petal Resonator Surface Coil, *Proc. Int. Soc. Magn. Reson. Med.*, **9**, 1112.

Hill, H.D.W. and Richards, R.E. (1968). Limits and Measurements in Magnetic Resonance, *J. Sci. Instrum.*, **1**, 977–983.

Hong, J-S. and Lancaster, M.J. (1996). Couplings of Microstrip Square Open-Loop Resonators for Cross-Coupled Planar Microwave Filters, *IEEE Trans. Microwave Theory Tech.*, **44**, 2099–2109.

Hornak, J.P., Ceckler, T. and Bryant, R.G. (1986). Phosphorus-31 NMR Spectroscopy Using a Loop-Gap Resonator, *J. Magn. Reson.*, **68**, 319–322.

Hoult, D.I., Busby, S.J., Gadian, D.G., Radda, G.K., Richards, R.E. and Seeley, P.J. (1974). Observation of Tissue Metabolites using ^{31}P Nuclear Magnetic Resonance, *Nature*, **252**, 285–287.

Hoult, D.I. and Richards, R.E. (1976). The Signal-to-Noise Ratio of the Nuclear Magnetic Resonance Experiment, *J. Magn. Reson.*, **24**, 71–85.
Hoult, D.I. (1978). The NMR Receiver: A Description and Analysis of Design, *Prog. Nucl. Magn. Reson. Spectrosc.*, **12**, 41–77.
Hoult, D.I. (1984). Fast Recovery with a Conventional Probe, *J. Magn. Reson.*, **57**, 394–403.
Hoult, D.I., Chen, C-N. and Sank, V.J. (1984). Quadrature Detection in the Laboratory Frame, *Magn. Reson. Med.*, **1**, 339–353.
Hoult, D.I. and Deslauriers, R. (1990a). A High-Sensitivity, High-B_1 Homogeneity Probe for Quantitation of Metabolites, *Magn. Reson. Med.*, **16**, 411–417.
Hoult, D.I. and Deslauriers, R. (1990b). Elimination of Signal Strength Dependency upon Coil Loading – An Aid to Metabolite Quantitation when the Sample Volume Changes, *Magn. Reson. Med.*, **16**, 418–424.
Hoult, D.I. (2000). The principle of reciprocity in Signal Strength Calculations – A Mathematical Guide, *Concepts Magn. Reson.*, **12**, 173–187.
Hoult, D.I. and Phil, D. (2000). Sensitivity and Power Deposition in a High-Field Imaging Experiment, *J. Magn. Reson. Imaging*, **12**, 46–67.
Hoult, D.I. and Tomanek, B. (2002). Use of Mutually Inductive Coupling in Probe Design, *Concepts Magn. Reson.*, **15**, 262–285.
Hu, S., Reimer, J.A. and Bell, T.A. (1998). Single-Input Double-Tuned Circuit for Double Resonance Nuclear Magnetic Resonance Experiments, *Rev. Sci. Instrum.*, **69**, 477–478.
Hudson, A.M.J., Köckenberger, W. and Bowtell, R.W. (2000). Open Access Birdcage Coils for Microscopic Imaging of Plants at 11.7 T, *Magn. Reson. Mater. Phys., Biol. Med.*, **10**, 69–74.
Hyde, J.S., Rilling, W. and Kusumi, A. (1982). Dispersion *Electron* Spin Resonance with Loop-Gap Resonator, *Rev. Sci. Instrum.*, **89**, 485–495.
Hyde, J.S., Froncisz, R.J. and Jesmanowicz, A. (1990). Passive Decoupling of Surface Coil by Pole Insertion, *J. Magn. Reson.*, **89**, 1934–1937.
Idziak, S. and Haeberlen, U. (1982). Design and Construction of a High Homogeneity rf Coil for Solid-State Multiple-Pulse NMR, *J. Magn. Reson.*, **50**, 281–288.
Isaac, G., Schnall, M.D., Lenkinski, R.E. and Vogele, K. (1990). A Design for a Double-Tuned Birdcage Coil for Use in an Integrated MRI/MRS Examination, *J. Magn. Reson.*, **89**, 41–50.
Jiang, J.Y. (2000). A Simple Method for Measuring the Q Value of an NMR Sample Coil, *J. Magn. Reson.*, **142**, 386–388.
Jiang, J.Y., Pugmire, R.J. and Grant, D.M. (1987). An Efficient Double-Tuned ^{13}C/^{1}H Probe Circuit for CP/MAS NMR and its Importance in Linewidths, *J. Magn. Reson.*, **71**, 485–494.
Jin, J.J. (2012). 'Practical Electromagnetic Modeling Methods', in *RF Coils for MRI*, Vaughan, J.T. and Griffiths, J.R. (eds), ch. 27, John Wiley & Sons, Ltd, 339–362.

Jin, J. (1999). Electromagnetic Analysis and Design in Magnetic Resonance Imaging, CRC Press, Boca Raton, FL, USA.

Jordan, E.C. (1989). *Reference Data for Engineers: Radio, Electronics, Computer and Communications*, 7th edn., Howard W. Sams & Comp., Indianapolis, IN, USA.

Joseph, P.M. and Lu, D. (1989). A Technique for Double Resonant Operation of Birdcage Imaging Coils, *IEEE Trans. Med. Imaging*, **8**, 286–294.

Jow, U-M. (2007). Design and Optimization of Printed Spiral Coils for Efficient Transcutaneous Inductive Power Transmission, *IEEE Trans. Biomed. Circuits Syst.*, **1**, 193–202.

Jurga, K., Reynhardt, E.C. and Jurga, S. (1992). NMR Transmit-Receive System with Short Recovery Time and Effective Isolation, *J. Magn. Reson.*, **96**, 302–306.

Kamon, M., Tsuk, M.J. and White, J. (1994). Fasthenry: A Multipole-Accelerated 3-D Inductance Extraction program, *IEEE Trans. Microwave Theory Tech.*, **42**, 1750–1758.

Kan, S., Gonord, P., Salset, C. and Vibet, C. (1973). A Versatile and Inexpensive Electronic System for a High Resolution NMR Spectrometer, *Rev. Sci. Instrum.*, **44**, 1725–1733.

Kan, S. and Courtieu, J. (1980). A Single-Coil Triple Resonance Probe for NMR Experiments, *Rev. Sci. Instrum.*, **51**, 2427–2429.

Kan, S. and Gonord, P. (1992). Q Optimization of RF Inductors for Use in NMR Probes, *Magn. Reson. Med.*, **23**, 372–375.

Kan, S., Jehenson, P. and Leroy–Willig, A. (1992). Single-Input Double-Tuned Foster-Type Circuit, *Magn. Reson. Med.*, **26**, 7–15.

Kan, S., Leroy-Willig, A., Gonord, P., Wary, C., Jehenson, P., Syrota, A. and Sauzade, M. (1994). A Multi-Gap, Parallel-Plate Bracelet Resonator: Theory and Application to ^{1}H and ^{13}C NMR, *Proceedings of SMR 2nd Meeting*, pp. 1126.

Keifer, P.A. (1999). 90° Pulse Width Calibrations: How to Read a Pulse Width Array, *Concepts Magn. Reson.*, **11**, 165–180.

Kendrick, R.D. and Yannoni, C.S. (1987). High-Power ^{1}H-^{19}F Excitation in a Multiple-Resonance Single-Coil Circuit, *J. Magn. Reson.*, **75**, 506–508.

Khennouche, M.S., Gadot, F., Belier, B. and de Lustrac, A. (2012). Different Configurations of Metamaterials Coupled with an RF Coil for MRI Applications, *Appl. Phys.*, **A109**, 1059–1063.

King, S.B., Ryner, L.N., Tomanek, B., Sharp, J., C. and Smith, I.C.P. (1999). MR Spectroscopy Using Multi-Ring Surface Coils, *Magn. Reson. Med.*, **42**, 655–664.

Kneeland, J.B., Jesmanowicz, A., Froncisz, W., Grist, T.M. and Hyde J.S. (1986). High-Resolution MR Imaging Using Loop-Gap Resonators, *Radiology*, **158**, 247–250.

Knight, D.W. (2013). An Introduction to the Art of Solenoid Inductance Calculation with Emphasis on Radio-Frequency Applications, available at:
http://www.g3ynh.info/zdocs/magnetics/Solenoids.pdf [accessed 12/01/2014].

Kodibagkar, V.D. and Conradi, M.S. (2000). Remote Tuning of NMR Probe Circuits, *J. Magn. Reson.*, **144**, 53–57.

Koptioug, A.V., Reijerse, E.J. and Klaassen, A.K. (1997). New Transmission-Line Resonator for Pulsed EPR, *J. Magn. Reson.*, **125**, 369–371.

Koskinen, M.F. and Metz, K.R. (1992). The Concentric Loop-Gap Resonator – A Compact, Broad Tunable Design for NMR Applications, *J. Magn. Reson.*, **98**, 576–588.

Kozlov, M. and Turner, R. (2009). Fast MRI Coil Analysis Based on 3-D Electromagnetic and RF Circuit Co-Simulation, *J. Magn. Reson.*, **200**, 147–152.

Krause, N. (1985). High-Frequency Field System for Nuclear Magnetic Resonance Apparatus, *U.S. Patent*, 4,506,224.

Kubo, A. and Ichikawa, S. (2003). Ultra-Broadband NMR Probe: Numerical and Experimental Study of Transmission Line NMR Probe, *J. Magn. Reson.*, **162**, 284–299.

Kuhns, L.P., Lizac, M.J., Lee, S.H. and Conradi, M.S. (1988). Inductive Coupling and Tuning in NMR Probes; Applications, *J. Magn. Reson.*, **78**, 69–76.

Labiche, A., Kan, S., Leroy-Willig, A. and Wary, C. (1999). Investigation of the Radio Frequency Magnetic Field inside a Hollow Cylinder Used for Nuclear Magnetic Resonance Imaging, *Rev. Sci. Instrum.*, **70**, 2113–2115.

Lanz, T., Weisser, A. and Haase, A. (2000). The Double Tuned $^1H^{23}Na$ Crosscage Resonator for High Field NMR Microscopy, *Proc. 8th Int. Soc. Magn. Reson. Med.*, pp. 1390.

Lanz, T., Weisser, Ruff, J., A. and Haase, A. (2001). Double Tuned ^{23}Na ^1H Nuclear Magnetic Resonance Birdcage for Application on Mice *In Vivo*, *Rev. Sci. Instrum.*, **72**, 2508–2510.

Lee, T.H. (2004). Planar Microwave Engineering. A Practical Guide to Theory, Measurements and Circuits, Cambridge University Press, New York, NY, USA.

Leifer, M.C. (1997a). Resonant Modes of the Birdcage Coil, *J. Magn. Reson.*, **124**, 51–60.

Leifer, M.C. (1997b). Theory of the Quadrature Elliptic Birdcage Coil, *Magn. Reson. Med.*, **38**, 729–732.

Lemdiasov R.A., Obi, A.A. and Ludwig, R. (2011). A Numerical Postprocessing Procedure for Analyzing Radio Frequency MRI Coils, *Concepts Magn. Reson.*, **38A**, 133–147.

Lemdiasov, R.A. and Ludwig, R. (2012). 'Radiofrequency MRI Coil Analysis: A Standard Procedure', in Vaughan, J.T. and Griffiths, J.R. (eds), *RF Coils for MRI*, ch. 26, Wiley & Sons, Ltd, Chichester, UK, pp. 327–338.

Leroy-Willig, A., Darrasse, L., Taquin, J. and Sauzade, M. (1985). The Slotted Cylinder: An Efficient Probe for NMR Imaging, *Magn. Reson. Med.*, **2**, 20–28.

Li, L. and Sotak, C.H. (1991). An Efficient Technique for Decoupling NMR Transmit Coils from Surface-Coil Receivers, *J. Magn. Reson.*, **93**, 207–213.

Li, S., Yang, Q.X. and Smith, M.B. (1994). RF Coil Optimization: Evaluation of B_1 Field Homogeneity Using Field Histograms and Finite Element Calculations, *Magn. Reson. Imaging*, **12**, 1079–1087.

Link, J. (1992). The Design of Resonators Probes with Homogeneous Radiofrequency Fields, *NMR Basic Princ. Prog.*, **26**, 3–31.

Lipworth, G., Ensworth, J., Seetharam, K., Huang, D., Lee, J.S., Schmalenberg, P., Nomura T., Reynolds, M.S., Smith, D.R. and Urshumov Y. (2014). Magnetic metamaterial superlens for increased range wireless power transfer, *Sci. Rep.*, **4**, 3642-1–6.

Lopez, M.A., Freire, M.J., Algarin, J.M., Behr, V.C., Jakob, P.M. and Marqués, R. (2011). Nonlinear split-ring metamaterial slabs for magnetic resonance imaging, *Appl. Phys. Lett.*, **98**, 133508-1–3.

Lowe, I.J. and Tarr, C.E. (1968). A Fast Recovery Probe and Receiver for Pulsed Nuclear Magnetic Resonance Spectroscopy, *Rev. Sci. Instrum.*, **1**, 320–322.

Lowe, I.J. and Engelsberg, M. (1974). A Fast Recovery Pulsed Nuclear Magnetic Resonance Sample Probe Using a Delay Line, *Rev. Sci. Instrum.*, **45**, 631–639.

Lupu, M., Dimicoli, J.L., Volk, A. and Mispelter, J. (2004). An Efficient Design for Birdcage Probes Dedicated to Small-Animal Imaging Experiments, *MAGMA*, **17**, 363–371.

MacLaughlin, D.E. (1989). Coaxial Cable Attenuation in NMR Sample Coil Circuits, *Rev. Sci. Instrum.*, **60**, 3242–3248.

Mager, D., Peter, A., Del Tin, L., Fischer, E., Smith, P.J., Hennig, J. and Korvink, J.G. (2010). An MRI Receiver Coil Produced by Inkjet Printing Directly on to a Flexible Substrate, *IEEE Trans. Med. Imag.*, **29**, 482–487.

Mansfield, P. and Morris, P.G. (1982). *NMR Imaging in Biomedicine*, Academic Press, New York, NY, USA.

Mansfield, P. (1988). The Petal Resonator: A New Approach to Surface Coil Design for NMR Imaging and Spectroscopy, *J. Phys. D*, **21**, 1643–1644.

Mansfield, P. (1992). Surface electrical coil structures, *U.S. Patent*, 5,143,688.

Martin, J.F. and Daly C.P. (1986). Transmission Line Matching of Surface Coils, *Magn. Reson. Med.*, **3**, 346–351.

Matson, G.B., Vermathen, P. and Hill, T.C. (1999). A Practical Double-Tuned $^1H/^{31}P$ Quadrature Birdcage Headcoil for ^{31}P Operation, *Magn. Reson. Med.*, **42**, 173–182.

Matsuki, M. and Matsushima, A. (2012). Improved Numerical Method for Computing Internal Impedance of a Rectangular Conductor and Discussions of its High Frequency Behavior, *Prog. Electromagn. Res. M*, **23**, 139–152.

Matthaei, G., Young, L. and Jones, E.M.T. (1980). *Microwave Filters, Impedance-Matching Networks, and Coupling Structures*, Artech House, Inc., Norwood, MA, USA.

McDermott, R., Trabesinger, A. H., Muck, M., Hahn, E.L., Pines, A., and Clarke, J. (2002). Liquid-State NMR and Scalar Couplings in Microtesla Magnetic Fields, *Science*, **295**, 2247–2249.

McLachlan, L.A. (1980). Lumped Circuit Duplexer for a Pulsed NMR Spectrometer, *J. Magn. Reson.*, **39**, 11–15.

McNichols, R.J., Wright, S.M., Wasser, J.S. and Coté, G.L. (1999). An Inductively Coupled, Doubly Tuned Resonator for in vivo Nuclear Magnetic Resonance Spectroscopy, *Rev. Sci. Instrum.*, **70**, 3454–3456.
Medhurst, R.G. (1947a). H.F. Resistance and Self-Capacitance of Single-Layer Solenoids, *Wireless Engineer*, February 1947, 35–43.
Medhurst, R.G. (1947b). H.F. Resistance and Self-Capacitance of Single-Layer Solenoids (continued), *Wireless Engineer*, March 1947, 80–92.
Medhurst, R.G. (1947c). Q of Solenoid Coils, *Wireless Engineer*, September 1947, 281.
Michels, H. (2014). DISLIN Home Page, avalaible at: http://www.dislin.de/ [accessed 12/01/2014].
Miller, J.B., Suits, B.H., Garroway, A.N. and Hepp, M.A. (2000). Interplay among Recovery Time, Signal, and Noise: Series- and Parallel-Tuned Circuits are not always the Same, *Concepts Magn. Reson.*, **12**, 125–136.
Minard, K.R. and Wind, R.A. (2001a). Solenoidal Microcoil Design. Part I: Optimizing RF Homogeneity and Coil Dimensions, *Concepts Magn. Reson.*, **13**, 128–142.
Minard, K.R. and Wind, R.A. (2001b). Solenoidal Microcoil Design. Part II: Optimizing Winding Parameters for Maximum Signal-to-Noise Performance, *Concepts Magn. Reson.*, **13**, 190–210.
Mispelter, J., Tiffon, B., Quiniou, E. and Lhoste, J.M. (1989). Optimization of ^{13}C-{^{1}H} Double Coplanar Surface Coil Design for the WALTZ-16 Decoupling Sequence, *J. Magn. Reson.*, **82**, 622–628.
Mispelter, J. and Lupu, M. (2007). An Easy-to-Made Double Tuned ^{23}Na / ^{1}H, Inductively Driven, Quadrature Birdcage, for Small Animal MRI at 4.7 Tesla, *Proc. Int. Soc. Magn. Reson. Med.*, **15**, 3272.
Mispelter, J. and Lupu, M. (2008). Homogeneous resonators for magnetic resonance: A review. *C.R. Chim.*, **11**, 340–355.
Mohan, S.S., del Mar Hershenson, M., Boyd, S.P. and Lee, T.H. (1999). Simple Accurate Expressions for Planar Spiral Inductances, *IEEE J. Solid-State Circuits*, **34**, 1419–1424.
Mohorič, A. and Stepišnik, J. (2009). NMR in the Earth Magnetic Field, *Prog. Nucl. Magn. Reson. Spectrosc.*, **54**, 166–182.
Mosig, J., Bahr, A., Boltz, T. and Ladd, M.E. (2009). A Novel Metamaterial Transmit/Receive Coil Element for 7 T MRI – Design and Numerical Results, *Proc. Int. Soc. Mag. Reson. Med.*, **17**, 4743.
Murphree, D., Cahn, S.B., Rahmlow, D and DeMille, D. (2007). An Easily Constructed, Tuning Free, Ultra-Broadband Probe for NMR, *J. Magn. Reson.*, **188**, 160–167.
Murphree, D., DeMille, D. and Cahn, S.B. (2012). Transmission Line Probe for NMR, *U.S. Patent*, 8,164,336 B1.
Murphy-Boesch, J. and Koretsky, A.P. (1983). An *In Vivo* Probe Circuit for Improved Sensitivity, *J. Magn. Reson.*, **54**, 526–532.

Murphy-Boesch, Srinivasan, R., Carvajal, L. and Brown, T.R. (1994). Two Configurations of the Four-Ring Coil for ^{1}H Imaging and ^{1}H-Decoupled ^{31}P Spectroscopy of the Human Head, *J. Magn. Reson.*, **103B**, 103–114.

Nagel, T.L., Stolk, J.A., Aderman, D.W. Schoenborn, R. and Schweizer, M.P. (1990). The Crossover Surface Coil: An Efficient *In Vivo* NMR Detection, *Magn. Reson. Med.*, **13**, 271–278.

Najim, E. and Grivet, J-P. (1991). Efficiency Estimation for Single-Coil, Separate-Input, Double-Tuned NMR Probes, *J. Magn. Reson.*, **93**, 27–33.

Najim, E. and Grivet, J-P. (1992). A Double-Tuned Probe for Metabolic NMR Studies, *Magn. Reson. Med.*, **23**, 367–371.

Nakada, T., Kwee, I.L., Miyazaki, T., Iriguchi, N. and Maki, T. (1987). ^{31}P NMR Spectroscopy of the Stomach by Zig-Zag Coil, *Magn. Reson. Med.*, **5**, 449–455.

Ngspice, (1999). last release (2014) available at: http://ngspice.sourceforge.net/presentation.html [accessed 12/01/2014].

Nijhof, E.J. (1990). Slotted Resonator: Principles and Applications for High-Frequency Imaging and Spectroscopy on Electrically Conducting Samples, *Magn. Reson. Imaging*, **8**, 345–349.

Niknejad, A.M. (2007, reprinted with corrections 2008). *Electromagnetics for High-Speed Analog and Digital Communication Circuits*, Cambridge University Press, New York, NY, USA.

Nyenhuis, J. (2012). 'RF Device Safety and Compatibility', in Vaughan, J.T. and Griffiths, J.R. (eds), *RF Coils for MRI*, ch. 32, John Wiley & Sons, Ltd, Chichester, UK, pp. 409–423.

Odoj, F., Rommel, E., van Kienlin, M. and Haase, A. (1998). A Superconducting Probehead Applicable for Nuclear Magnetic Resonance Microscopy at 7 T, *Rev. Sci. Instrum.*, **69**, 2708–2712.

Olson, D.L., Peck, T.L., Webb, A.G., Magin, R.L. and Sweedler, J.V. (1995). High Resolution Microcoil ^{1}H-NMR for Mass-Limited, Nanoliter-Volume Samples, *Science*, **270**, 1967–1970.

Ott, H.W. (2009) Electromagnetic Compatibility Engineering, John Wiley & Sons, Inc., Hoboken, NJ, USA.

Pang, Y., Xie, Z., Xua, D., Kelley, D.A., Nelsona, S. J., Vigneron, D. B. and Zhang, X. (2012). A Dual-Tuned Quadrature Volume Coil with Mixed $\lambda/2$ and $\lambda/4$ Microstrip Resonators for Multinuclear MRSI at 7 T, *Magn. Reson. Imaging*, **30**, 290–298.

Parisi, S.J. (1989). 180° Lumped Element Hybrid, *IEEE Microwave Th. The. Int. Microwave Symp.*, pp. 1243–1245.

Pascone, R.J., Garcia, B.J., Fitzgerald, T.M., Vullo T., Zipagan, R. and Cahill, P.T. (1991). Generalized Electrical Analysis of Low-Pass and High-Pass Birdcage Resonators, *J. Magn. Reson. Imaging*, **9**, 395–408.

Pascone, R.J., Vullo T., Farrelly, J. and Cahill, P.T. (1992). Explicit Treatment of Mutual Inductance in Eight-Column Birdcage Resonators, *Magn. Reson. Imaging*, **10**, 401–410.

Pendry, J.B., Holden, A.J., Robbins, D.J. and Stewart W.J. (1999). Magnetism from Conductors and Enhanced Nonlinear Phenomena, *IEEE Trans. Microwave Theory Tech.*, **47**, 2075–2084.

Pendry, J.B., (2000). Negative Refraction Makes a Perfect Lens, *Phys. Rev. Lett.*, **85**, 3966–3969.

Peshkovsky, A.S., Forguez J., Cerioni L. and Pusiol, D.J. (2005). RF Probe Recovery Time Reduction with a Novel Active Ringing Suppression Circuit, *J. Magn. Reson.*, **177**, 67–73.

Peterson, D.M., Beck, B.L. and Duensing, G.R. (2002). Reduction of Cable Shield Currents Generated by High Field Body Coils at 3 Tesla and Above, *Proc. Int. Soc. Mag. Reson. Med.*, **10**, 850.

Peterson, D.M., Beck, B.L., Duensing, G.R. and Fitzsimmons, J.R. (2003). Common Mode Signal Rejection Methods for MRI: Reduction of Cable Shield Currents for High Static Magnetic Fields Systems, *Concepts Magn. Reson.*, **19B**, 1–8.

Peterson, D.M. (2012). 'Impedance Matching and Baluns', Vaughan, J.T. and Griffiths, J.R. (eds), *RF Coils for MRI*, ch. 25, John Wiley & Sons, Ltd, Chichester, UK, pp. 315–323.

Pfenninger, S., Forrer, J., Schweiger, A. and Weiland, T. (1988). Bridged Loop-Gap resonator: A Resonant Structure for Pulsed ESR Transparent to High-Frequency Radiation, *Rev. Sci. Instrum.*, **59**, 752–760.

Piasecki, W., Froncisz, W. and Hubbell, W. (1998). A Rectangular Loop-Gap Resonator for EPR Studies of Aqueous Samples, *J. Magn. Reson.*, **134**, 36–43.

Piatek, Z., Baron, B., Szczegielniak, T., Kusiak, D., Pasierbek, A. (2012). Self Inductance of Long Conductor of Rectangular Cross Section, *Przeglad Elektrotechniczny (Electrical Review)*, **88**, 323–236.

Piatek, Z. and Baron, B. (2012). Exact Closed Form Formula for Self Inductance of Conductor of Rectangular Cross Section, *Prog. Electromagn. Res. M*, **26**, 225–236.

Picard, L., Blackledge, M. and Decorps, M. (1995a). Improvements in Electronic Decoupling and Receiver Coils, *J. Magn. Reson. B*, **106**, 110–115.

Picard, L., von Kienlin, M. and Decorps, M. (1995b). An Overcoupled NMR Probe for the Reduction of Radiation Damping, *J. Magn. Reson.*, **117**, 262–266.

Pimmel, P. (1990). Les Antennes en Résonance Magnétique Nucleaire: Fonctionnement et Réalisation – Résonateurs pour l'Imagerie et pour la Spectroscopie *In Vivo*, PhD Thesis, Université Lyon1, France.

Pimmel, P. and Briguet, A. (1992). A Hybrid Bird-Cage Resonator for Sodium Observation at 4.7 T, *Magn. Reson. Med.*, **24**, 158–162.

Poole, C.P. (1967). Electron Spin Resonance. A Comprehensive Treatise on Experimental Techniques, John Wiley & Sons, Inc, New York, NY, USA.

Pozar, D.M. (1998). *Microwave Engineering*, 2nd edn., John Wiley & Sons, New York, NY, USA.

Press, W.H., Vetterling, W.T., Teukolsky, S.A. and Flannery, B.P. (1992). *Numerical Recipes in C. The art of Scientific Computing*, 2nd edn., Cambridge University Press, Cambridge, USA.

Purcell, E.M., Torrey, H.C. and Pound, R.V. (1946). Resonance Absorption by Nuclear Magnetic Moments in a Solid, *Phys. Rev.*, **69**, 37–38.

Qian, C. and Brey, W.W. (2009). Impedance Matching with an Adjustable Segmented Transmission Line, *J. Magn. Reson.*, **199**, 104–110.

Qian, C., Murphy-Boesch, J., Dodd, S. and Koretsky, A. (2012). Sensitivity Enhancement of Remotely Coupled NMR Detectors Using Wirelessly Powered Parametric Amplification, *Mag. Reson. Med.*, **68**, 989–996.

Qian, C., Yu, X., Chen, D-Y., Dodd, S., Bouraoud, N., Pothayee, N., Chen, Y., Beeman, S., Bennett, K., Murphy-Boesch, J. and Koretsky, A. (2013). Wireless Amplified Nuclear MR Detector (WAND) for High-Spatial-Resolution MR Imaging of Internal Organs: Preclinical Demonstration in a Rodent Model, *Radiology*, **268**, 228–236.

Radkovskaya, A., Shamonin, M., Stevens, C.J., Faulkner, G., Edwards, D.J., Shamonina, E. and Solymar L. (2005). Resonant Frequencies of a Combination of Split Rings: Experimental, Analytical and Numerical Study, *Microwave Opt. Technol. Lett.*, **46**, 473–476.

Radu, X., Lapeyronnie, A. and Craeye, C. (2008). Numerical and Experimental Analysis of a Wire Medium Collimator for MRI, *Electromagnetics*, **28**, 531–543.

Radu, X., Garray, D. and Craeye C. (2009). Toward a Wire Medium Endoscope for MRI Imaging, *Metamaterials*, **3**, 90–99.

Raffin, R.A. (1979). *L'Emission et la Réception d'Amateur*, 9th edn, Editions Techniques et Scientifiques Françaises, Paris, France.

Ramo, S., Whinnery, J. R. and Van Duzer, T. (1994) *Fields and Waves in Communication Electronics*, 3rd edn, John Wiley & Sons, Inc., New York, NY, USA.

Rath, A.R. (1990a). Efficient Remote Transmission Line Probe Tuning, *Magn. Reson. Med.*, **13**, 370–377.

Rath, A.R. (1990b). Design and Performance of a Double-Tuned Bird-Cage Coil, *J. Magn. Reson.*, **86**, 488–495.

Reed, J. and Wheeler, G.J. (1956). A Method of Analysis of Symmetrical Four-Port Networks, *IEEE Trans. Microwave Theory Tech.*, **4**(4), 246–252.

Rennings, A., Svejda, J.T., Otto, S., Solbach, K. and Ernil, D. (2013). A MIM/Coaxial Stub-Line CRLH Zeroth-Order Series-Mode Resonator used as an RF Coil Element for 7-Tesla Magnetic Resonance Imaging, *IEEE IMS2013 Proc.*, pp. 1–4.

Rodriguez, A.O., Hidalgo, S.S, Rojas, R. and Barrios F.A. (2005). Experimental Development of a Petal Resonator Surface Coil, *Magn. Reson. Imaging*, **23**, 1027–1033.

Roemer, P.B., Edelstein, W.A., Hayes, C.E., Souza, S.P. and Mueller, O.M. (1990). The NMR Phased Array, *Magn. Reson. Med.*, **16**, 192–225.

Rohde, U.L. and Bucher, T.T.N. (1988). *Communications Receivers. Principles & Design*, McGraw-Hill Book Comp., New York, NY, USA.
Rosa, E.B. (1906). Calculation of the Self-Inductance of Single-Layer Coils. *Bull. Bur. Stand. (U.S.)*, **2**, 161–187.
RSGB Handbook (1982) Radio Communication Handbook 5th edn, Radio Society of Great Britain, Potters Bar, U.K.
Ruheli, A.E. (1972). Inductance Calculations in a Complex Integrated Circuits Environment, *IBM J. Res. Dev.*, **16**, 58–69.
Ruthroff, C.L. (1959). Some Broad Band Transformers, *Proc. IRE*, **47**, 1337–1342.
Rzedzian, R. and Martin, C. (1993). Split Shield for Magnetic Resonance Imaging, *U.S. Patent*, 5,243,286.
Sakellariou, D., Le Goff, G. and Jacquinot, J.-F. (2007). High-Resolution, High-Sensitivity NMR of Nanolitre Anisotropic Samples by Coil Spinning, *Nature*, **447**, 694–697.
Sank, V.J., Chen, C.N. and Hoult, D.I. (1986). A Quadrature Coil for the Adult Human Head, *J. Magn. Reson.*, **69**, 236–242.
Schnall, M.D., Barlow, C., Subramanian, V.H. and Leigh, J.S. (1986a). Wireless Implanted Magnetic Resonance Probes for *In Vivo* NMR, *J. Magn. Reson.*, **68**, 161–167.
Schnall, M.D., Subramanian, V.H. and Leigh, J.S. (1986b). The Application of Overcoupled Tank Circuits to NMR Probe Design, *J. Magn. Reson.*, **67**, 129–134.
Schnall, M.D. (1992). 'Probes Tuned to Multiple Frequencies for In-Vivo NMR', in Diehl, P., Fluck, E., Günter, H., Kosfeld, R. and Seelig, J. (Eds) *In-Vivo Magnetic Resonance Spectroscopy[1]: Probeheads and Radiofrequency Pulses, Spectrum Analysis, NMR Basic Principles and Progress Series*, **vol. 26**, Springer-Verlag, Berlin, pp. 35–63.
Schneider, H.J. and Dullenkopf, P. (1977). Slotted Tube Resonator: A New NMR Probe Head at High Observing Frequencies, *Rev. Sci. Instrum.*, **48**, 68–73.
Scott, E., Stettler, J. and Reimer, J.A. (2012). Utility of a Tuneless Plug and Play Transmission Line Probe, *J. Magn. Reson.*, **221**, 117–119.
Seeber, D.A., Cooper, R.L., Ciobanu, L. and Pennington, C.H. (2001). Design and Testing of High Sensitivity Microreceiver Coil Apparatus for Magnetic Resonance and Imaging, *Rev. Sci. Instrum.*, **72**, 2171–2179.
Seeber, D.A., Jevtic, J. and Menon, A. (2004). Floating Shield Current Suppression Trap, *Concepts Magn. Reson.*, **21B**, 26–31.
Serfaty, S., Darrasse, L. and Kan, S. (1994). The Pinpoint NMR Coil, *Proc. Soc. Magn. Reson.*, **1**, 219.
Serfaty, S., Haziza, N., Darrasse, L. and Kan S. (1997). Multi-Turn Split-Conductor Transmission-Line Resonators, *Magn. Reson. Med.*, **38**, 687–689.
Sevick, J. (1994). Building and Using Baluns and Ununs. Practical Designs for the Experimenter, CQ Communic., Inc., Hicksville, NY, USA.

Sevick, J. (2001). *Transmission Line Transformers*, 4th edn., Noble Publishing Corp., Thomasville, GA, USA.

Shaka, A.J., Keeler, J. and Freeman, R. (1983). Evaluation of a New Broadband Decoupling Sequence, *J. Magn. Reson.*, **53**, 313–340.

Shelby, R.A., Smith, D.R. and Schultz, S. (2001). Experimental Verification of a Negative Index of Refraction, *Science*, **292**, 77–79.

Shelkunoff, S.A. and Friis H.T. (1952) *Antennas. Theory and Practice*. John Wiley & Sons, New York, NY, USA.

Shen, G.X., Boada, F.E. and Thulborn, K.R. (1997). Dual-Frequency, Dual-Quadrature, Birdcage RF Coil Design with Identical B_1 Pattern for Sodium and Proton Imaging of the Human Brain at 1.5 T, *Magn. Reson. Med.*, **38**, 717–725.

Shrivastava, D. and Vaughan, J.T. (2012). 'Radiofrequency Heating Models and Measurements', in Vaughan, J.T. and Griffiths, J.R. (eds), *RF Coils for MRI*, ch. 32, John Wiley & Sons, Ltd, Chichester, UK, pp. 425–436.

Silver, X., Xu Ni, W., Mercer, E.V., Beck, B.L., Bossart, E. L., Inglis, B. and Marecil, T.H. (2001). *In Vivo* ^{1}H Magnetic Resonance Imaging and Spectroscopy of the Rat Spinal Cord Using an Inductively-Coupled Chronically Implanted RF Coil, *Magn. Reson. Med.*, **46**, 1216–1222.

Singer, J.R. (1959). Blood flow rates by nuclear magnetic resonance measurements, *Science*, **130**, 1652–1653.

Smith, D.R., Padilla, W.J., Vier, D.C., Nemat-Nasser, S.C. and Schultz S. (2000). Composite Medium with Simultaneously Negative Permeability and Permittivity, *Phys. Rev. Lett.*, **84**, 4184–4187.

Smith, D.R., Pendry, J.B. and Wiltshire, M.C.K. (2004). Metamaterials and Negative Refractive Index, *Science*, **305**, 788–792.

Smith, D.R., Vier, D.C., Koschny, T. and Soukoulis, C.M. (2005). Electromagnetic parameter retrieval from inhomogeneous metamaterials, *Phys. Rev.*, **E71**, 036617-1–11.

Smythe, W.R. (1950). *Static and Dynamic Electricity*, McGraw-Hill Book Comp., New York, NY, USA.

Spence, D.K. and Wright, S.M. (2003). 2-D Full Wave Solution for the Analysis and Design of Birdcage Coils, *Concepts Magn. Reson.*, **18B**, 15–23.

SPICE, home page available at: http://bwrcs.eecs.berkeley.edu/Classes/IcBook/SPICE/ [accessed 12/01/2014].

Stensgaard, A. (1996). Optimized Design of the Shielded-Loop Resonator, *J. Magn. Reson.*, **122**, 120–125.

Stringer, J.A. and Drobny, G.P. (1998). Methods for the Analysis and Design of a Solid State Nuclear Magnetic Resonance Probe, *Rev. Sci. Instrum.*, **69**, 3384–3391.

Stringer, J.A., Bronnimann, C.E., Mullen, C.G., Zhou, D.H., Stellfox, S.A., Li, Y., Williams, E.H. and Rienstra, C.M. (2005). Reduction of RF-Induced Sample Heating with a Scroll Coil Resonator Structure for Solid-State NMR Probes, *J. Magn. Reson.*, **173**, 40–48.

Stroobandt, S. (2014). Single-Layer Helical Round Wire Coil Inductor Calculator, available at: http://hamwaves.com/antennas/inductance.html [accessed 12/01/2014].

Styles, P., Soffe, N.F., Scott, C.A., Cragg, D.A., Row, F., White, D.J. and White, P.C.J. (1984). A High-Resolution NMR Probe in Which the Coil and Preamplifier Are Cooled with Liquid Helium, *J. Magn. Reson.*, **60**, 397–404.

Styles, P., Smith, M.B., Briggs, R.W. and Radda, G.K. (1985). A Concentric Surface-Coil Probe for the Production of Homogeneous B_1 Fields, *J. Magn. Reson.*, **62**, 397–405.

Suryan, G. (1951). Nuclear Resonance in Flowing Liquids, *Proc. Indian Acad. Sci.*, **A33**, 107–111.

Tabuchi, Y., Negoro, M., Takeda, K. and Kitagawa, M. (2010). Total Compensation of Pulse Transients Inside a Resonator, *J. Magn. Reson.*, **204**, 327–332.

Tadanki, S., Colon, R.D., Moore, J. and Waddell, K.W. (2012). Double Tuning a Single Input Probe for Heteronuclear NMR Spectroscopy at Low Field, *J. Magn. Reson.*, **223**, 64–67.

Taflove, A. and Hagness, S.C. (2000). *Computational Electrodynamics – The Finite-Difference Time-Domain Method*, 2nd edn, Artech House, Norwood, MA, USA.

Takeda, K., Tabuchi, Y., Negoro, M. and Kitagawa, M. (2009). Active Compensation of rf-Pulse Transients, *J. Magn. Reson.*, **197**, 242–244.

Tang, B. and Jelicks, L.A. (2002). Two placements of Cosine Coil Legs, *Proc. Int. Soc. Magn. Reson. Med.*, **10**, pp. 876.

Tang, J.A. and Jerschow, A. (2010). Practical Aspects of Liquid-State NMR with Inductively Coupled Solenoid Coils. *Magn. Reson. Chem.*, **48**, 763–770.

Terman, F.E. (1943). *Radio Engineers' Handbook*, McGraw-Hill Book Comp. Inc. New York, NY, USA.

Terman, F.E. (1955). *Electronic and Radio Engineering*, 4th edn., McGraw-Hill Book Comp. Inc. New York, NY, USA.

Tiffon, B., Mispelter, J. and Lhoste, J-M. (1986). A Carbon-13 *In Vivo* Double Surface Coil NMR Probe with Efficient Low-Power Decoupling at 400 MHz Using the WALTZ-16 Sequence, *J. Magn. Reson.*, **68**, 544–550.

Tomanek, B., Ryner, L., Hoult, D.I., Kozlowski, P. and Daunders, J.K. (1997). Dual Surface Coil with High-B_1 Homogeneity for Deep Organ MR Imaging, *Magn. Reson. Imaging*, **15**, 1199–1204.

Tomanek, B., Hoult, D.I., Chen, X. and Gordon, R. (2000). Probe with Chest Shielding for Improved Breast MRI, *Magn. Reson. Med.*, **43**, 917–920.

Ton That, D.M., Augustine, M.P., Pines, A. and Clarke, J. (1997). Low Magnetic Field Dynamic Nuclear Polarization Using a Single-Coil Two-Channel Probe, *Rev. Sci. Instrum.*, **68**, 1527–1531.

Tropp, J. (1989). The Theory of the Birdcage Resonator, *J. Magn. Reson.*, **82**, 51–62.

Tropp, J. (1991). The Theory of an Arbitrarily Perturbed Birdcage Resonator and a Simple Method for Restoring it to Full Symmetry, *J. Magn. Reson.*, **95**, 235–243.

Tropp, J. (1997). Mutual Inductance in the Birdcage Resonator, *J. Magn. Reson.*, **126**, 9–17.

Tropp J. (2002). Dissipation, Resistance, and Rational Impedance Matching for TEM and Birdcage Resonators, *Magn. Reson. Eng.*, **15**, 177–188.

Van Hecke, P., Decanniere, C. and Vanstapel, F. (1989). Double-Tuned Resonator Designs for NMR Spectroscopy, *J. Magn. Reson.*, **84**, 170–176.

Van Vaals, J.J. and Bergman, A.H. (1990). Novel High-Frequency Resonator for NMR Imaging and Spectroscopy, *J. Magn. Reson.*, **89**, 331–342.

Vaughan, J.T., Hetherington, H.P., Otu, J.O., Pan, J.W. and Pohost, G.M. (1994). High Frequency Volume Coils for Clinical NMR Imaging and Spectroscopy, *Magn. Reson. Med.*, **32**, 206–218.

Veselago, V.G. [1968 (Russian text 1967)]. The Electrodynamics of Substances with Simultaneously Negative Values of ε and μ. *Sov. Phys. Usp.*, **10**, 509–514.

Villa, P., Vaquero, J.J., Chesnick, S. and Ruiz-Cabello, J. (1999). Probe Efficiency Improvement with Remote and Transmission Line Tuning and Matching, *Magn. Reson. Imag.*, **17**, 1083–1086.

VNA Help (2014). Microwave Network Analysis Application Notes, http://www.vnahelp.com/appnotes.html [accessed 12/01/2014].

Volland, N.A., Mareci, T.H., Constantinidis, I. and Simpson, N.E. (2010). Development of an Inductively Coupled MR Coil System for Imaging and Spectroscopic Analysis of an Implantable Bioartificial Construct at 11.1 T, *Magn. Reson. Med.*, **63**, 998–1006.

Volotovskyy, V., Tomanek, B., Corbin, I., Boist, R., Tuor, U. and Peeling, J. (2003). Doubly Tunable Double Ring Surface Coil, *Concepts Magn. Reson.*, **17B**, 11–16.

Walker, C.S. (1990). *Capacitance, Inductance and Crosstalk Analysis*, Artech House, Inc., Norwood, MA, USA.

Walton, J.H. and Conradi, M.S. (1989). Probe Tuning Adjustments-Need They Be in the Probe?, *J. Magn. Reson.*, **81**, 623–627.

Weaver, R. (2012) Numerical Methods for Inductance Calculation, available at: http://electronbunker.ca/CalcMethods.html [accessed 12/01/2014].

Webb, A.G. (1997). Radiofrequency Microcoils in Magnetic Resonance, *Prog. Nucl. Magn. Reson. Spectrosc.*, **31**, 1–42.

Webb, A.G., Collins, C.M., Versluis, M.J., Kan, H.E. and Smith, N.B. (2010). MRI and Localized Proton Spectroscopy in Human Leg Muscle at 7 Tesla Using Longitudinal Traveling Waves, *Magn. Reson. Med.*, **63**, 297–302.

Webb, A.G. (2011). Dielectric materials in Magnetic Resonance, *Concept Magn. Reson.*, **38A**, 148–184.

Webb, A.G., Smith, N.B., Aussenhofer, S. and Kan, H.E. (2011). Use of Tailored Higher Modes of a Birdcage to Design a Simple Double-Tuned Proton/Phosphorus Coil for Human Calf Muscle Studies at 7 T, *Concepts Magn. Reson.*, **39B**, 89–97.

Webb, A.G. (2012). 'Microcoil', Vaughan, J.T. and Griffiths, J.R. (eds), *RF Coils for MRI*, ch. 19, John Wiley & Sons, Ltd, Chichester, UK, pp. 225–232.

Webb, A.G. (2013). Radiofrequency Microcoils for Magnetic Resonance Imaging and Spectroscopy, *J. Magn. Reson.*, **229**, 55–66.
Weekley, A.J., Bruins, P., Sisto, M. and Augustine, M.P. (2003). Using NMR to Study Full Intact Wine Bottles, *J. Magn. Reson.*, **161**, 91–98.
Weis, V., Mittelbach, W., Claus, J., Möbius, K. and Prisner, T. (1997). Probehead with Interchangeable Tunable Bridged Loop-Gap Resonator for Pulsed Zero-Field Optically Detected Magnetic Resonance Experiments on Photoexcited Triplet States, *Rev. Sci. Intrum.*, **68**, 1980–1985.
Weisser, A., Lanz, T. and Haase, A. (2001). Field Analysis in Crossed Birdcage-Coils, *Proc. Int. Soc. Mag. Reson. Med.*, **9**, 1128.
Wen, H., Chesnick, A.S. and Balaban R.S. (1994). The Design and Test of a New Volume Coil for High Field Imaging, *Magn. Reson. Med.*, **32**, 492–498.
Wen, H., Jaffer, F.A., Denison, T.J., Duewell, S., Chesnick, A.S. and Balaban, R.S. (1996). The Evaluation of Dielectric Resonators Containing H_2O and D_2O as RF Coils for High-Field MR Imaging and Spectroscopy, *J. Magn. Reson.*, **110B**, 117–123.
Wetterling, F., Högler, M., Molkenthin, U., Junge, S., Gallagher, L., Macrae, I.M. and Fagan, A.J. (2012). The Design of a Double-Tuned Two-Port Surface Resonator and its Application to *In Vivo* Hydrogen- and Sodium-MRI, *J. Magn. Reson.*, **217**, 10–18.
Weyers, D.J. and Liu, Q. (2006). Shielding Apparatus for Magnetic Resonance Imaging, *U.S. Patent*, 7,102,350.
Wheeler, H.A. (1965). Transmission-Line Properties of Parallel Strips Separated by a Dielectric Sheet, *IEEE Trans. Microwave Theory Tech.*, **13**, 172–185.
White, J.F. (2004) High Frequency Techniques. An Introduction to RF and Microwave Engineering, IEEE Press, John Wiley & Sons, Inc., Hoboken, NJ, USA.
Wiltshire, M.C.K., Pendry, J.B., Young, I.R., Larkman, D.J., Gilderdale, D.J. and Hajnal, J.V. (2001). Microstructured Magnetic Materials for RF Flux Guides in Magnetic Resonance Imaging, *Science*, **291**, 849–851.
Wiltshire M.C.K., Hajnal, J.V., Pendry, J.B., Edwards D.J. and Stevens C.J. (2003a) Metamaterial Endoscope for Magnetic Field Transfer: Near Field Imaging with Magnetic Wires, *Opt. Express*, **11**, 709–715.
Wiltshire, M.C.K., Pendry, J.B., Larkman, D.J., Gilderdale, D.J., Herlihy, D., Young, I.R. and Hajnal, J.V. (2003b). Geometry Preserving Flux Ducting by Magnetic Metamaterials, *Proc. Int. Soc. Magn. Reson. Med.*, **11**, 713.
Wiltshire, M.C.K., Shamonina, E., Solymar L. and Young, I.R. (2004). Development of Metamaterial Components for use in MRI and NMR Systems, *Proc. Int. Soc. Magn. Reson. Med.*, **12**, 1582.
Wiltshire, M.C.K., Pendry, J.B. and Hajnal, J.V. (2006). Sub-Wavelength Imaging at Radio Frequency, *J. Phys.: Condens. Matter*, **18**, L315–L321.
Wiltshire, M.C.K. (2009). 'RF Metamaterials', in Capolino, F. (ed), *Theory and Phenomena of Metamaterials*, ch. 14, CRC Press Boca Raton, FL, USA.

Wong, W.H. (2001). Millipede Coils, U.S. Patent, 628,518.
Wong-Foy, A., Saxena, S., Moulé, A.J., Bitter, H-M.L., Seeley, J.A., McDermott, R. Clarke, J. and Pines, A. (2002). Laser-Polarized ^{129}Xe and MRI at Ultralow Magnetic Fields, J. Magn. Reson., 157, 235–241.
Wu, R-B., Kuo, C-N. and Chang, K.K. (1992). Inductance and Resistance Computations for Three-Dimensional Multiconductor Interconnection Structures, IEEE Trans. Microwave Theory Tech., 40, 263–271.
Xie, Y., Jiang, J. and He, S. (2012). Proposal of Cylindrical Rolled-Up Metamaterial Lenses for Magnetic Resonance Imaging Application and Preliminary Experimental Demonstration, Prog. Electromagn. Res., 124, 151–162.
Yoda, K. and Kurokawa M. (1989). Inductive Coupling to the Slotted-Tube Quadrature Probe, J. Magn. Reson., 81, 284–287.
Zhang, X., Ugurbil, K. and Chen, W. (2001). Microstrip RF Surface Coil Design for Extremely High-Field MRI and Spectroscopy, Magn. Reson. Med., 46, 443–450.
Zhang, X., Ugurbil, K. and Chen, W. (2003). A Microstrip Transmission Line Volume Coil for Human Head MR Imaging at 4 T, J. Magn. Reson., 161, 242–251.

Index

ABCD matrix, 162, 164
active decoupling, 651, 652, 653
adjustable inductor, 103
admittance matrix, 155, 157, 160
admittance, 28,30
Alderman–Grant resonator (AGR), 511, 515, 517, 520
Alembert equation, 61
Ampere law, 472
array coils, 640, 641, 642
attenuation, 65, 70
axial resonator, 468

balancing, 244, 251, 254, 259, 261, 264, 275, 367, 368, 370
balun, 275, 277, 278, 279, 280
Biot–Savart, 427, 449, 482, 613
Birdcage resonator, 349, 527, 532, 535, 537
Birdcage Builder, 196
Bode–Fano limit, 199
broadband matching, 216

cable trap, 277, 281, 282, 283
cables, 58
calibration, 667
capacitance, 118, 701, 702, 703
capacitors, 104, 105, 107
chain matrix, 155, 163, 305
characteristic impedance, 59, 62, 147, 673, 705-707
chip inductor, 102
choke inductor, 98, 99, 100, 101
circularly polarized, 439
coil segmentation, 253
common mode, 276

conductors, 51
coupled resonators, 237, 238
coupling loop, 229, 230, 233
critical coupling, 227, 230
crossover coil, 625
current density, 451, 454, 455, 470
current distribution, 186, 457, 460
current filaments, 448, 449
cylindrical coil, 84

damping circuit, 274
dead time, 273, 275
delay line, 291
dielectric resonators, 152, 153
differential mode, 118, 276
diode, 124, 125, 127
distributed components, 116
double tune, 414, 416

eddy current, 83
effective field, 436
effective permeability, 142
efficiency, 235, 411, 414, 415, 417, 419
electromotive force, 11
elliptic integrals, 435, 436, 603, 700
elliptically polarized, 439
end ring, 545

FastHenry, 178
FDTD, 463, 465
FEM, 462
field polarization, 559
filling factor, 140
flat strip, 451
flux guide, 150, 151

GMD, 190, 191, 193

Helmholtz coil, 473
high-pass filter, 166
hybrid, 299, 301, 303, 309, 313, 315, 317, 319, 323, 324

impedance bridge, 658
impedance matching, 200, 201, 203, 204, 205, 207, 209
impedance matrix, 155, 160
impedance measurement, 30
impedance, 28, 30
inductance, 86, 92, 93, 97, 118
inductive coupling, 228
inductive coupling, 607
inductor, 82
isolation, 346, 348, 367, 387, 576

KCL, 171
KVL, 171, 176, 178

laboratory frame, 438
lenses, 151
linear birdcage, 563
linear network analysis, 155, 169
linearly polarized, 438
Litz wire, 55, 57
LNA, 25, 70, 71, 128, 132, 133, 134, 136, 198, 284, 286, 287, 288
loop gap, 485, 488, 489
losses, 13, 48
low-loss transmission line, 63
low-pass filter, 164

matching circuits, 265, 268
Maxwell's equations, 445
metamaterial, 137, 138
microcoil, 490
microstrip coil, 629
MoM, 466
MSTR, 148
multiple tuning, 361, 416
mutual inductance, 189, 192, 691, 692, 696

network analyzer, 665
noise factor, 70, 128, 130, 675
Noise Figure, 25, 128, 130, 674, 677
null-point, 386

odd–even analysis, 310, 311, 312
Ohm's Law, 175
opened resonator, 623
optimum coupling 232, 233
over coupling, 226, 234

passive decoupling, 645, 646, 647, 649
pick-up coil, 656
PIN diodes, 269
Principle of Reciprocity, 9, 295
proximity effect, 45, 47, 452, 461
proximity factor, 87

Q-spoiling, 271, 272
quadrature birdcage, 571, 588, 590
quadrature hybrid, 295, 309
quadrature, 297
quality factor, 29, 82, 86–89, 104, 366, 547, 680, 681

radiation resistance, 251
radio frequency coil, 13
rat-race, 325, 327, 325
recovery time, 271, 274
rectangular loop, 617
reflexion coefficient, 30, 31, 77
resistor, 81
resistance, 44, 87
resonant modes, 396, 419, 539
resonator, 109, 111, 115, 348
ribbon conductor, 53
rotating frame, 437, 438
round wire, 52
rung, 544, 581

saddle coil, 498, 501, 505, 507
scattering matrix, 157, 167
segmented loop, 624
sensitivity, 16, 246, 247, 424
shield, 521, 522, 577

single loop, 94
skin depth, 45, 47
slotted tube resonator (STR), 509, 510
Smith Chart, 40, 41, 42
SNA, 39
solenoid coil, 85, 472, 480, 490
S-parameters, 36, 38, 180
spiral coil, 96, 612, 615
split ring resonator, 627
SRR, 143, 145, 146
surface coil, 601, 615
Swiss-Rolls, 144, 145
switch, 350, 351, 352, 354, 359, 371, 374, 384, 413, 422
switching diodes, 269

tank circuit, 376, 379, 389, 390
TEM resonator, 592, 595
THR, 146, 147, 148
TLM, 465
T-network, 404, 407
toroidal coil, 91
transmission line, 58, 121, 157, 704
transverse resonator, 468, 498
trap circuit, 371, 387, 416
trap, 123
two-port network, 157, 161

under coupling, 226, 232

varactor, 128
variable capacitor, 108
velocity coefficient, 671
VNA, 36, 40
VSWR, 33

wave propagation, 59

Yee cell, 463
Π-network, 404, 410